Studies in Systems, Decision and Control

Volume 429

The series "Studies in Systems, Decision and Control" (SSDC) covers both new developments and advances, as well as the state of the art, in the various areas of broadly perceived systems, decision making and control–quickly, up to date and with a high quality. The intent is to cover the theory, applications, and perspectives on the state of the art and future developments relevant to systems, decision making, control, complex processes and related areas, as embedded in the fields of engineering, computer science, physics, economics, social and life sciences, as well as the paradigms and methodologies behind them. The series contains monographs, textbooks, lecture notes and edited volumes in systems, decision making and control spanning the areas of Cyber-Physical Systems, Autonomous Systems, Sensor Networks, Control Systems, Energy Systems, Automotive Systems, Biological Systems, Vehicular Networking and Connected Vehicles, Aerospace Systems, Automation, Manufacturing, Smart Grids, Nonlinear Systems, Power Systems, Robotics, Social Systems, Economic Systems and other. Of particular value to both the contributors and the readership are the short publication timeframe and the worldwide distribution and exposure which enable both a wide and rapid dissemination of research output.

Indexed by SCOPUS, DBLP, WTI Frankfurt eG, zbMATH, SCImago.

All books published in the series are submitted for consideration in Web of Science.

Songsak Sriboonchitta · Vladik Kreinovich ·
Woraphon Yamaka
Editors

Credible Asset Allocation, Optimal Transport Methods, and Related Topics

Springer

Editors
Songsak Sriboonchitta
Faculty of Economics
Chiang Mai University
Chiang Mai, Thailand

Vladik Kreinovich
Department of Computer Science
University of Texas
El Paso, TX, USA

Woraphon Yamaka
Faculty of Economics
Center of Excellence in Econometrics
Chiang Mai University
Chiang Mai, Thailand

ISSN 2198-4182 ISSN 2198-4190 (electronic)
Studies in Systems, Decision and Control
ISBN 978-3-030-97272-1 ISBN 978-3-030-97273-8 (eBook)
https://doi.org/10.1007/978-3-030-97273-8

This Springer imprint is published by the registered company Springer Nature Switzerland AG
The registered company address is: Gewerbestrasse 11, 6330 Cham, Switzerland

Preface

Among the main objectives of econometrics are as follows:

- to gauge the quality and efficiency of the current economic situation,
- to predict the future economic situation and
- to provide recommendations on how to make future economic situation as good as possible.

One of the main difficulties in predicting the future economic situation is that this situation depends on many factors beyond our control. As a result, we cannot predict the exact future situation, we can, at best, predict probabilities of different outcomes. To gauge the quality of the future economic situation, we therefore need to measure how close is the corresponding probability distribution to the probability situation describing the ideal outcomes.

In general, a natural measure of closeness between to objects is the amount of efforts that are needed to transform the first object to the second one. For example, a natural way to gauge the closeness between two geographic distributions of goods is by the smallest number of ton-kilometres needed to transform the first distribution into the second one. For goods, it is known how to solve this transportation problem. It turns out that similar techniques can be applied in general economic situations. Because of their origin, these methods are known as optimal transport methods.

How can we improve the economic situation? Our resources, our assets—in particular, our financial resources—are limited. It is therefore important to allocate these resources so as to achieve the best results.

These two directions—asset allocation and optimal transport methods—are the main topics of this book. Several papers also deal with related economic issues. Overall, papers from this volume present a good state-of-the-art slice of what has been done, what can be done, and what should be done in solving the corresponding econometric problems.

We hope that this volume will:

- help practitioners to become better knowledgeable of the state-of-the-art econometric techniques, and
- help researchers to further develop these important research directions.

We want to thank all the authors for their contributions and all anonymous referees for their thorough analysis and helpful comments.

The publication of this volume was partly supported by the Faculty of Economics of the Chiang Mai University, Thailand. Our thanks to the leadership and staff of the Chiang Mai University for providing crucial support. Our special thanks to Prof. Hung T. Nguyen for his valuable advice and constant support.

We would also like to thank Prof. Janusz Kacprzyk (Series Editor) and Dr. Thomas Ditzinger (Senior Editor, Engineering/Applied Sciences) for their support and cooperation in this publication.

Chiang Mai, Thailand
El Paso, USA
Chiang Mai, Thailand
December 2021

Songsak Sriboonchitta
Vladik Kreinovich
Woraphon Yamaka

Contents

Theoretical Results

Why Quantiles Are a Good Description of Volatility in Economics: A Pedagogical Explanation

Sean Aguilar, Vladik Kreinovich, and Uyen Pham

Abstract To make investment decisions, we need to know, for each financial instrument, not only its expected return—but also how the actual return may deviate from its expected value. A numerical measure of such deviations is known as volatility. Originally, volatility was measured by the standard deviation from the expected price, but it turned out that this measure does not always adequately describe our perception of volatility. Empirically, it turned out that quantiles are a more adequate description of volatility. In this paper, we provide an explanation of this empirical phenomenon.

1 Formulation of the Problem

What is volatility and why it is important. One of the main problems in economics is how to best invest money.

Money can be invested in different financial instruments: stocks, bonds, derivatives, etc.

- For some of these instruments—e.g., for some bonds—we know exactly how much interest we will get in the next year.
- For other instruments, e.g., for stocks, we may gain a lot—or we may even lose, if the price of this stock goes down.

In economics, this unpredictability of gain is known as *volatility*.

Comment. Usually, the riskier instruments—i.e., instruments with higher volatility—provide better expected return: otherwise, why would someone invest in a riskier

S. Aguilar · V. Kreinovich (✉)
Department of Computer Science, University of Texas at El Paso, El Paso, TX 79968, USA
e-mail: vladik@utep.edu

S. Aguilar
e-mail: sraguilar4@miners.utep.edu

U. Pham
John Von Neumann Institute, University of Economics and Law, Ho Chi Minh City, Vietnam
e-mail: uyenph@uel.edu.vn

S. Sriboonchitta et al. (eds.), *Credible Asset Allocation, Optimal Transport Methods, and Related Topics*, Studies in Systems, Decision and Control 429,
https://doi.org/10.1007/978-3-030-97273-8_1

instrument if it is possible to invest in a less risky instrument with the same (or even better) expected return on investment.

How can we describe volatility in numerical terms. To make investment decisions, we need to be able to compare the risks of different instruments. For this, we need to be able to describe volatility in numerical terms.

Originally, volatility was described in terms of the standard deviation of the return. However, it turned out that standard deviation does not always adequately describe our intuitive idea of volatility:

- sometimes instruments with higher standard deviations are perceived as less risky, and,
- sometimes, vice versa, instruments with lower standard deviations are perceived as more risky.

It turns out that a better description of intuitive volatility is provided by *quantiles* of the distribution (see, e.g., [4])—i.e., the values x for which the probability $\text{Prob}(r \le x)$ that the return r is smaller than or equal to x is equal to some selected value p:

$$\text{Prob}(r \le x) = p.$$

Here are some example of quantiles:

- for $p = \frac{1}{2}$, we get a median;
- for $p = \frac{1}{4}$ and $p = \frac{3}{4}$, we get quartiles, etc.

Comment. Quantiles are usually defined in terms of the cumulative distribution function (cdf) $F(x) \stackrel{\text{def}}{=} \text{Prob}(X \le x)$, as the corresponding value $F^{-1}(p)$ of the inverse function.

Natural question: why quantiles? A natural question is: why namely quantiles—and not any other characteristics of a probability distribution—provide a description of volatility which is the closest to the intuitive understanding of volatility.

In this paper, we provide an explanation.

2 Our Explanation

Main idea. In our explanation, we use the fact that the effect of money on a person's happiness is unusually nonlinear; see, e.g., [1–3].

This non-linearity is easy to explain:

- if you have no money at all, and you gain a dollar, you are happy, but
- if you already have \$1000 and you gain one more dollar, you will barely notice the difference.

Psychological experiments show that this nonlinear dependence of happiness on money is somewhat different for different people.

How is this related to the intuitive idea of volatility? When people think of volatility, they think in terms of changes in their happiness level—which is related to money by some non-linear transformation. So, what we want is a characteristic that would not change if we consider a different person, with a different non-linear function relating money and happiness level.

In other words, as a volatility characteristic, we want to have a characteristic $c(X)$ of a random variable X (that describes possible gains) to have the following property:

- if the volatility characteristic measured in the money scale is $c(X)$,
- then for any monotonic re-scaling function $f(x)$, the volatility $c(Y)$ of a re-scaled variable $Y = f(X)$ should be equal to the same value $c(X)$ described in the new scale, i.e., we should have

$$c(f(X)) = f(c(X)).$$

This natural invariance requirement leads to quantiles. Let us show that this natural invariance requirement explains the appearance of quantiles.

Indeed, it is known that if we have the cdf $F(x)$, then the variable $Y = F(X)$ is uniformly distributed on the interval [0, 1]; see, e.g., [4]. This fact is used to simulate probability distributions with a known cdf.

Let p denote the value of the desired characteristic $c(U)$ for the case when the random variable is uniformly distributed on the interval [0, 1]. Since $F(X) = U$, we have $c(F(X)) = p$. Thus, the above invariance requirement takes the form

$$c(F(X)) = F(c(X)),$$

i.e., the form $p = F(c(X))$. By applying the inverse function $F^{-1}(z)$ to both sides of this equality, we conclude that

$$c(X) = F^{-1}(p).$$

This is exactly the quantile value that we tried to explain.

Conclusion. We have indeed explained why quantiles are the best description of intuitive notion of volatility.

Acknowledgements This work was supported in part by the National Science Foundation grants 1623190 (A Model of Change for Preparing a New Generation for Professional Practice in Computer Science), and HRD-1834620 and HRD-2034030 (CAHSI Includes). It was also supported by the program of the development of the Scientific-Educational Mathematical Center of Volga Federal District No. 075-02-2020-1478.

The authors are thankful to Hung T. Nguyen for his help and encouragement.

References

1. Kahneman, D. (2011). *Thinking, fast and slow*. Farrar, Straus, and Giroux.
2. Lorkowski, J., & Kreinovich, V. (2015). Granularity helps explain seemingly irrational features of human decision making. In W. Pedrycz & S.-M. Chen (Eds.), *Granular computing and decision-making: Interactive and iterative approaches* (pp. 1–31). Springer Verlag.
3. Lorkowski, J., & Kreinovich, V. (2018). *Bounded rationality in decision making under uncertainty: Towards optimal granularity*. Springer Verlag.
4. Sheskin, D. J. (2011). *Handbook of parametric and non-parametric statistical procedures*. Chapman & Hall/CRC.

An Introduction to Stacking Regression for Economists

Achim Ahrens, Erkal Ersoy, Vsevolod Iakovlev, Haoyang Li, and Mark E. Schaffer

Abstract We present an introduction to "stacked generalization" (Wolpert in Neural Networks 5(2):241–259, 1992 [21]). The increased availability of "Big Data" in economics and the emergence of non-traditional machine learning tools presents new opportunities for applied economics, but it also imposes additional challenges. The full range of supervised learning algorithms offers a rich variety of methods, each of which could be more efficient in addressing a specific problem. Selecting the optimal algorithm and tuning its parameters can be time-consuming due to potential lack of experience in machine learning and relevant economic literature. "Stacking" is a useful tool that can be used to address this: it is an ensemble method for combining multiple supervised machine learners in order to achieve more accurate predictions than could be produced by any of the individual machine learners. Besides providing an introduction to the stacking methodology, we also present a short survey of some of the estimators or "base learners" that can be used with stacking: lasso, ridge, elastic net, support vector machines, gradient boosting, and random forests. Our empirical example of how to use stacking regression uses the study by Fatehkia et al. (PLOS ONE 14(2):1–16, 2019 [6]): predicting crime rates in localities using demographic and socioeconomic data combined with data from Facebook on user interests.

Invited paper for the 15th International Conference of the Thailand Econometric Society, Chiang Mai University, Thailand, 5–7 January 2022. All errors are our own.

A. Ahrens
ETH Zürich, Zürich, Switzerland

E. Ersoy · V. Iakovlev · H. Li · M. E. Schaffer (✉)
Heriot-Watt University, Edinburgh, UK
e-mail: M.E.Schaffer@hw.ac.uk

S. Sriboonchitta et al. (eds.), *Credible Asset Allocation, Optimal Transport Methods, and Related Topics*, Studies in Systems, Decision and Control 429,
https://doi.org/10.1007/978-3-030-97273-8_2

1 Introduction

In the last few decades, the social sciences, including economics, have seen a rise in the availability of Big Data. This type of dataset is characterized by a large number of covariates relative to the sample size. The use of such data presents new opportunities, but it also imposes additional challenges. The full range of supervised learning algorithms offers a rich variety of statistical tools, any one which could be more efficient in addressing a particular problem. Not surprisingly, there is no method that is superior to all the others; some machine learners will perform relatively well in some situations and relatively poorly in others. Selecting the optimal tool and tuning its parameters can be difficult and challenging in practice. In this paper, we present an introduction to "stacked generalization" [21], a strategy that addresses this problem by, in effect, examining the relative performance of different estimators with respect to a given dataset. "Stacked generalization" or "stacking" is an ensemble method for combining multiple machine learners in order to achieve more accurate predictions than could be produced by any of the component estimators. Its flexibility in addressing various problems such as classification, regression and unsupervised learning makes it potentially applicable to a wide range of economic fields.

The structure of the paper is as follows. Stacking relies on cross-validation, and cross-validation is also a commonly used technique to choose tuning parameters of machine learning estimators, so we begin with a brief presentation of this tool. We then present a short survey of the estimators or "base learners" that our stacking regression example will use: lasso, ridge, elastic net, support vector machines, gradient boosting, and random forests. Having set out the main ingredients of stacking, we present the basics of stacking regression. Our empirical example of how to use stacking regression uses the study by Fatehkia et al. [6]: predicting crime rates in localities using demographic and socioeconomic data combined with data from Facebook on user interests. The final section concludes.

2 Cross-Validation

Before formally introducing stacking, we briefly review *cross-validation (CV)*. Stacking can, in fact, be seen as a logical extension of cross-validation.

Training an algorithm and evaluating the predictive performance on the same dataset can result in overfitting, which may yield poor results when a different data set from the same population is used [12]. Testing on new data can deliver an unbiased judgment on the performance of an algorithm [17]. However, most of the time only a limited amount of data are available, which leads to the idea of data splitting. In particular, data are split into the "training sample" that trains the algorithm and the "validation sample" where the performance of the algorithm is evaluated. The validation sample can be regarded as "new data" as long as the data split is random.

One data split delivers a *validation* estimate of the predictive accuracy and averaging over several splits gives a *cross-validation* estimate. If the data are identically distributed and independent between training and validation samples, it can be applied to almost any algorithm in almost any framework [2]. Our main interest here is how CV can be used to find the optimal tuning parameter for a prediction task, but we first briefly discuss how to implement model selection via CV.

Consider an independent dataset D and a family of models $\{m_j\}_{j=1}^J$ with mean squared prediction error (MSPE) as the criterion. For example, each model could correspond to a least squares model with a unique set of predictors, or a lasso model with a specific penalty level. The commonly used V-fold CV randomly split the data into V approximately equal-sized folds. Each time one fold is used as the validation sample and the remainder are used as training sample to fit the models. One MSPE is then obtained for each validation sample. Averaging over the V folds yields the overall MSPE for a specific model m_j. The model with the smallest MSPE is said to have the best predictive performance and will be selected as the final model. This CV procedure is called "CV with averaging" since the estimates of the MSPE are averaged before comparing the models. Alternatively, one can count the number of times each candidate model is preferred under all the splittings and select the one with the highest count; this is known as "CV with voting" [22]. Denoting by n the number of observations in the data D, the special case where $V = n$ is called "leave-one-out" CV procedure [15]. Specifically, each observation is successively left out from the training sample and used for the validation while the remaining $n - 1$ observations construct the training sample.

Cross-validation is computationally intensive due to the need to repeatedly fit the model and assess its performance across different folds and across all the candidate models. Any pre-processing (e.g., standardisation) of the data needs to be done by fold since the training sample should not include any information from the validation sample. Since LOO is computationally more expensive except in the special case of a linear model, V-fold CV is generally used in practice with V ranging from 5 to 10. Zhang [23] also showed that the improvement on performance is small when a V larger than 10 is used. Define the split ratio to be the number of observations in the training sample divided by the number of observations in the validation sample. Then the split ratios for V-fold CV and LOO CV are $V - 1$ and $n - 1$ respectively. Some theoretical studies [16, 22, 23] indicate that in the context of a low-dimensional linear model, CV procedure can only consistently select the correct model if the split ratio tends to zero. However, the conventional split ratios such as 4-to-1 (5-fold CV) and 9-to-1 (10-fold CV) are still popular since a small split ratio dramatically reduces the size of training sample and often leads to much poorer model fitting [13]. Consequently, overfitting becomes a common problem in terms of including irrelevant regressors. In fact, model consistency is not always the objective. If a researcher is especially interested in predictive performance rather than the true model, which is quite common, conventional V-fold CV will be a good choice.

3 Estimators

3.1 *Lasso*

Selecting a subset of predictors and discarding the rest is an attractive approach in many applications. Shrinkage methods such as the *lasso* and its relatives offer an attractive approach [10].

The lasso minimizes the mean squared error subject to a penalty on the absolute size of coefficient estimates:

$$\hat{\beta}_{\text{lasso}} = \underset{\beta}{\operatorname{argmin}} \sum_{i=1}^{n} \left(y_i - \beta_0 - \sum_{j=1}^{p} x_{ij}\beta_j \right)^2 \tag{1}$$

$$\text{subject to} \sum_{j=1}^{p} |\beta_j| \leq t$$

In the constraints expression, t is chosen to minimise an estimate of the mean squared error. Note also that the constant term, β_0, does not appear in the penalty term. This is by design and due to the nature of the optimisation problem: adding a constant to y_i would not shift the predictions by the same amount. Hence, it is common practice to reparametrise Eq. 1 such that each x_{ij} is replaced by $x_{ij} - \bar{x}_j$, and β_0 is estimated as $\bar{y} = \frac{1}{n}\sum_1^n y_i$. Rewriting the lasso problem in the familiar Lagrangian form yields

$$\hat{\boldsymbol{\beta}}_{\text{lasso}}(\lambda) = \underset{\beta}{\operatorname{argmin}} \left\{ \sum_{i=1}^{n} \left(y_i - \boldsymbol{x}_i'\boldsymbol{\beta} \right)^2 + \lambda \sum_{j=1}^{p} \psi_j |\beta_j| \right\} \tag{2}$$

The tuning parameter λ controls the overall penalty level and ψ_j are predictor-specific penalty loadings. These loadings have a key role in the pre-standardisation of the data. Because lasso is not invariant to linear transformations, scaling matters. If the predictors have unequal variance, the data can be pre-standardised such that $\frac{1}{n}\sum_i x_{ij}^2 = 1$ and set $\psi_j = 1$ for $j = 1, \ldots, p$. Alternatively, we can set the penalty loadings to $\hat{\psi}_j = (\frac{1}{n}\sum_i x_{ij}^2)^{-1/2}$. The two methods yield identical results that may vary in practice only due to numerical computation reasons. Further, there is a direct link between the parameters λ in Eq. 2 and t in Eq. 1.

The ℓ_1 lasso penalty makes the solutions nonlinear in y_i—the lasso solution is a quadratic programming problem—and there is no closed form solution. However, it is possible to obtain the trajectory of coefficient estimates as a function of λ with the same computational cost as popular alternatives (e.g., ridge regression as outlined in Sect. 3.2).

Tibshirani [18] motivates the lasso with two major advantages over OLS. First, from a subset selection perspective, the lasso serves as a continuous model selection technique. Due to the nature of the ℓ_1-penalty, the lasso discards some predictors from the model by setting their coefficient estimates to zero. Second, the lasso can outperform least squares in terms of prediction accuracy due to the bias-variance trade-off. To see the link between the lasso and least squares estimates, consider different values of t in the constraint term of Eq. 1. On one extreme, as the penalty goes to zero the lasso estimates approach those of OLS. On the other end of the spectrum, choosing a sufficiently small penalty leads to all coefficients being exactly zero. The intermediate case is one reason for the lasso's popularity: in the lasso solution some coefficients are set exactly to zero. In effect, the lasso combines both model selection and regularisation. A final consideration is the advantage lasso's ℓ_1 penalty offers vis-á -vis the ℓ_0 penalty as employed by Akaike and Bayes information criteria. The lasso solution is computationally attractive due to its convex form. This is in contrast with model selection based on AIC or BIC which requires each potential sub-model to be estimated and quickly becomes computationally taxing unless the model is extremely simple.

3.2 *Ridge Regression*

Ridge regression [11, 19] is a commonly used shrinkage method that imposes a penalty on the size of regression coefficients. It resembles the lasso but has some key differences. The ridge estimate is defined by the expression given in Eq. 1 with the following constraint:

$$\text{subject to} \sum_{j=1}^{p} \beta_j^2 \leq t$$

In other words, the ridge regression replaces the ℓ_1-penalty of the lasso with a ℓ_2-penalty. Like the lasso, the above expression can be reparametrised as follows.

$$\hat{\beta}_{ridge} = \underset{\beta}{\operatorname{argmin}} \left\{ \sum_{i=1}^{n} (y_i - \boldsymbol{x}_i' \beta)^2 + \lambda \sum_{j=1}^{p} \psi_j^2 \beta_j^2 \right\} \tag{3}$$

The interpretation and choice of the penalty loadings, ψ_j, outlined in Sect. 3.1 apply here as well, as does the need to account for uneven variance either through pre-estimation standardisation or by appropriately choosing the penalty loadings ψ_j.

A key difference is that unlike estimators relying on ℓ_1-penalization, the ridge does not perform variable selection. It also does not rely on the assumption of sparsity, which makes the ridge attractive when the assumption of sparsity does not seem plausible.

Unlike the case of the lasso, where there exists no closed form solution, the ridge solution can be expressed as

$$\hat{\boldsymbol{\beta}}_{\text{ridge}} = (\boldsymbol{X}'\boldsymbol{X} + \lambda\boldsymbol{\Psi}'\boldsymbol{\Psi})^{-1}\boldsymbol{X}'\boldsymbol{y}.$$

Here, $\boldsymbol{X}$ is the $n \times p$ matrix of predictors with typical element x_{ij}, $\boldsymbol{y}$ is the response vector, and $\boldsymbol{\Psi} = \text{diag}(\psi_1, \dots, \psi_p)$ is the diagonal matrix of penalty loadings. The ridge regularises the regressor matrix by adding positive constants to the diagonal of $\boldsymbol{X}'\boldsymbol{X}$, and the ridge solution is thus generally well-defined as long as all the ψ_j and λ are sufficiently large even if $\boldsymbol{X}'\boldsymbol{X}$ is rank-deficient.

3.3 Elastic Net

A neat relationship between the lasso and ridge regressions is that they can be viewed as Bayes estimates with different priors. Using the following notation,

$$\tilde{\boldsymbol{\beta}} = \underset{\beta}{\text{argmin}}\left\{\sum_{i=1}^{n}(y_i - \boldsymbol{x}_i'\boldsymbol{\beta})^2 + \lambda\sum_{j=1}^{p}\psi_j|\beta_j|^q\right\} \tag{4}$$

$q = 1$ corresponds to the lasso, and $q = 2$ to ridge regression. The lasso (i.e., $q = 1$) has the advantage of being the smallest q with a convex constraint region. This is attractive because non-convex regions introduce an added challenge to the optimization problem. On the other hand, the ridge is the posterior mean, but the lasso is not—lasso is derived as the posterior mode [10]. It is, therefore, attractive to combine the strengths of lasso and ridge regressions by choosing $q \in (1, 2)$.

The *elastic net* of Zou and Hastie [24] does exactly that by applying an ℓ_q penalty, which is a mix of ℓ_1 (lasso-type) and ℓ_2 (ridge-type) penalization as shown in the expression below.

$$\hat{\boldsymbol{\beta}}_{\text{elastic}} = \text{argmin}\frac{1}{n}\sum_{i=1}^{n}(y_i - \boldsymbol{x}_i'\boldsymbol{\beta})^2 + \frac{\lambda}{n}\left[\alpha\sum_{j=1}^{p}\psi_j|\beta_j| + (1-\alpha)\sum_{j=1}^{p}\psi_j^2\beta_j^2\right] \tag{5}$$

The additional parameter α determines the relative to contribution of ℓ_1 versus ℓ_2 penalisation. In the presence of groups of correlated regressors, the lasso typically selects only one variable from each group, whereas the ridge tends to produce similar coefficient estimates for groups of correlated variables. On the other hand, the ridge does not yield sparse solutions impeding model interpretation. The elastic net is able to produce sparse solutions for some α greater than zero, and retains or drops correlated variables jointly. As might be expected, the behaviour of the elastic net with respect to producing sparse solutions is intermediate to that of the lasso and ridge estimators: it will set some coefficients to exactly zero but not as many as the lasso.

3.4 Support Vector Machine

The linear version of *Support Vector Machine (SVM)* [3] solves a classification problem by producing a decision boundary $f(\mathbf{x})$ based on the training set of labeled data $\mathbf{Z} = \{\mathbf{z}_1, \dots, \mathbf{z}_n\}$, where $\mathbf{z}_i = (y_i, \mathbf{x}_i)$, $y_i \in \{-1, 1\}$ is a categorical dependent variable and $\mathbf{x}_i$ is a vector of k covariates.[1] Each covariate vector $\mathbf{x}_i$ is viewed by a linear SVM as a point on a $k-1$ dimensional *hyperplane* defined by $\{\mathbf{x} : f(\mathbf{x}) = \mathbf{x}^T\boldsymbol{\beta} + \beta_0 = 0\}$. The hyperplane is a geometric representation of the decision boundary produced by the estimator using two parameters: a k-dimensional unit vector, $\boldsymbol{\beta}$, and a constant, $\beta_0 \in \mathbb{R}$. The result of linear SVM's training is a classification rule, $G(\mathbf{x})$, which is given by the side of the decision boundary that an unclassified observation lands on, i.e., $G(\mathbf{x}) = \text{sign}[f(\mathbf{x})] = \text{sign}[\mathbf{x}^T\boldsymbol{\beta} + \beta_0]$.

To make the decision rule efficient, linear SVM maximises the *margin*, $M = \frac{1}{||\boldsymbol{\beta}||}$, between the hyperplane and the closest point on each side of the decision boundary by solving an optimisation problem subject to $y_i(\mathbf{x}_i^T\boldsymbol{\beta} + \beta_0) \geq M\ \forall i$. This constraint can be broken down as

$$\mathbf{x}_i^T\boldsymbol{\beta} + \beta_0 \begin{cases} \geq 1 & \text{if } y_i = 1 \\ \leq -1 & \text{if } y_i = -1, \end{cases} \tag{6}$$

where we implicitly assume that no observations land on the boundary itself.

After linear SVM is trained and validated,[2] the decision rule with optimal parameters $\boldsymbol{\beta}$ and β_0 is applied on a test set of unlabeled data, i.e. where y_i is unknown. Unfortunately, unless we obtain a perfect decision rule, the test set is likely to contain data points that land on the wrong side of the decision boundary. We can incorporate this in the optimisation model by defining a variable ξ_i that takes the value of the distance to the margin if i is misclassified and the value of zero if i is classified correctly. We can then modify the constraint above as

$$y_i(\mathbf{x}_i^T\boldsymbol{\beta} + \beta_0) \geq M(1 - \xi_i)\ \forall i,\ \xi_i \geq 0 \text{ and } \sum_{i=1}^{n} \xi_i \leq C, \tag{7}$$

where C is the *cost parameter*. Finally, one may find it more intuitive to write the optimisation problem in terms of $||\boldsymbol{\beta}||$ rather than in terms of M because setting $\mathbf{x}_i^T\boldsymbol{\beta} + \beta_0$ equal to ± 1 in (6) yields two parallel hyperplanes, the distance between which is given by $\frac{2}{||\boldsymbol{\beta}||}$. Since a lower $||\boldsymbol{\beta}||$ means a higher margin, the problem becomes a minimisation. We can now write the optimisation problem in a computationally convenient manner:

[1] The exposition here as well as in Sect. 3.6 follows in part Hastie et al. [10].

[2] Validation refers to an out-of-sample performance check of the estimator by applying it on labelled data that is not part of the training set.

$$\min_{\boldsymbol{\beta},\beta_0} \frac{1}{2}||\boldsymbol{\beta}||^2 + C\sum_{i=1}^{n} \xi_i \tag{8}$$

$$\text{s.t. } y_i(\mathbf{x}_i^T\boldsymbol{\beta} + \beta_0) \geq 1 - \xi_i \ \forall i, \ \xi_i \geq 0 \text{ and } \sum_{i=1}^{n} \xi_i \leq C. \tag{9}$$

The Lagrange dual function for the problem above is given by

$$\mathcal{L} = \sum_{i=1}^{n} \alpha_i - \frac{1}{2}\sum_{i=1}^{n}\sum_{j=1}^{n} \alpha_i\alpha_j y_i y_j \mathbf{x}_i^T\mathbf{x}_j \tag{10}$$

where α are Lagrange multipliers. It can be shown that solving the function above with a quadratic programming algorithm yields $\widehat{\boldsymbol{\beta}} = \sum_{i=1}^{n} \widehat{\alpha}_i y_i \mathbf{x}_i$. Note that $\alpha_i \neq 0$ only if $y_i(\mathbf{x}_i^T\boldsymbol{\beta} + \beta_0) = 1 - \xi_i$. For such i, the solution is given by $\widehat{\boldsymbol{\beta}}_i = \sum_{i=1}^{n} \widehat{\alpha}_i$ and i is called a *support vector*.

Using the same principle, linear SVM can also predict continuous y_i by treating $f(\mathbf{x}) = \mathbf{x}^T\boldsymbol{\beta} + \beta_0$ as a *regression* model rather than a classifier. In this case we switch the objective function in (8) for

$$\min_{\boldsymbol{\beta},\beta_0} \sum_{i=1}^{n} V(y_i - f(\mathbf{x}_i)) + \frac{1}{2}\frac{1}{C}||\boldsymbol{\beta}||^2 \tag{11}$$

where $V(.)$ is some error measurement function.

SVM can also account for *non-linearity* by increasing dimensionality of the covariates' space using *kernel functions*. One way of transforming the covariates' space to one of a higher dimension would be to use some basis function $h(.)$ that transforms $\mathbf{x}_i$ and $\mathbf{x}_j$ into $h(\mathbf{x}_i) = (h_1(\mathbf{x}_i), h_2(\mathbf{x}_i), \ldots, h_M(\mathbf{x}_i))$ and $h(\mathbf{x}_j) = (h_1(\mathbf{x}_j), h_2(\mathbf{x}_j), \ldots, h_M(\mathbf{x}_j))$, respectively, and then substitute it for $\mathbf{x}_i^T\mathbf{x}_j$ in the Lagrange function above. Solving the updated version of (10) would now involve computing the dot product of $h(\mathbf{x}_i)$ and $h(\mathbf{x}_j)$, which is costly and unnecessary. To solve the new Lagrange function, we only need the inner product, which is provided by a kernel function $K(\mathbf{x}_i, \mathbf{x}_j) = \langle h(\mathbf{x}_i), h(\mathbf{x}_j)\rangle$, a specific choice of which should be based on the structure of the data. Once the kernel is defined, the solution for $f(\mathbf{x})$ can be written as

$$f(\mathbf{x}) = h(\mathbf{x})^T\boldsymbol{\beta} + \beta_0 \tag{12}$$

$$= \sum_{i=1}^{n} \alpha_i y_i \langle h(\mathbf{x}_i), h(\mathbf{x}_j)\rangle + \beta_0 \tag{13}$$

$$= \sum_{i=1}^{n} \alpha_i y_i K(\mathbf{x}_i, \mathbf{x}_j) + \beta_0. \tag{14}$$

3.5 Random Forest

The *random forest* [5] estimator can be thought of as a modified application of bagging on *tree-based models*. Although the latter are powerful tools on their own, trees often suffer from high estimation variance. There exist various tree-based algorithms, the most famous of which are probably *C4.5* that specialises on classification problems and a more universal *Classification and Regression Tree (CART)*. The general principle shared by different tree-based algorithms is to predict the dependent variable by splitting the training set into the subsets based on some threshold values of a selected covariate with every split aiming to maximise the fit of the model.

For each tree $T_b \in \{T_b\}_1^B$ of B such trees, random forest draws with replacement from the training set a bootstrap sample, $\mathbf{Z}^b = \{\mathbf{z}_1^b, \ldots, \mathbf{z}_n^b\}$ of size n, where $\mathbf{z}_i^b = (y_i^b, \mathbf{x}_i^b)$. The algorithm then splits the bootstrap sample into two nodes using a threshold value for some covariate x_i^b that produces the minimum sum of squared errors, and keeps on splitting until a node with the minimum amount of observations is achieved. The minimum size of a node is usually set to 5 observations for a regression and 1 observation for a classification. However, this parameter can be tuned in order to avoid overfitting. The algorithm is then repeated for the rest of B trees, after which the predictions are combined by averaging in case of regression or majority vote in case of classification, i.e.

$$f^B(\mathbf{x}) = \begin{cases} \frac{1}{B}\sum_{b=1}^{B} T_b(\mathbf{x}) & \text{if regression} \\ \text{mode}\{C_b(\mathbf{x})\}_1^B & \text{if classification.} \end{cases} \tag{15}$$

So far we have just described bagging applied to tree-based estimators. One issue with this comes from the fact that the same stronger predictors are likely to be selected for the splits in the majority of the trees. This may lead to high correlation between the trees, which does not contribute to the accuracy of the model. The problem is addressed by selecting a random subset of m out of k covariates to be considered for a split, where m is another parameter alongside B and the minimum node size to be tuned with the common choices being $m = \frac{k}{3}$ for a regression and $m = \sqrt{k}$ for a classification. Furthermore, note that the prediction error for observation i is measured by comparing its true value of y_i to the combined prediction $f^B(\mathbf{x})$ such that $\mathbf{z}_i^b \notin \mathbf{Z}^b\ \forall b$. This feature of random forest is known as *Out-of-Bag (OOB) error estimation*, which approaches n-fold cross-validation error estimation as $B \to \infty$.

3.6 Gradient Boosting

Gradient Boosting [7] is another ensemble algorithm that makes use of the tree-based models. In contrast to random forest focusing on the variance reduction, gradient boosting aims to minimise the bias by starting from an underfit model and improving the residuals with every new tree $m \in 1, \ldots, M$ until it reaches its final form,

$$f(\mathbf{x}) = f_0(\mathbf{x}) + \alpha \sum_{m=1}^{M} f_m(\mathbf{x}), \tag{16}$$

where $\alpha \in (0, 1)$ is the learning rate. This hyperparameter controls the effect of each of the M trees on the final prediction and can be tuned in order to avoid overfitting. This algorithm often produces more accurate estimates than random forest, but it requires more time to train.

To approach a regression problem, gradient boosting starts by computing the first term of (16):

$$f_0(\mathbf{x}) = \arg\min_{\gamma} \sum_{i=1}^{N} L(y_i, \gamma), \tag{17}$$

where $L(y_i, \gamma)$ is a *loss function* and γ are the predictions, each of which corresponds to one of J terminal nodes of tree m given some predictive rule $f(\mathbf{x}) = \gamma_j$. Although there are multiple loss functions available, *squared error loss*, defined as $L(y_i, f(x_i)) = \frac{1}{2}[y_i - f(x_i)]^2$, is convenient for differentiation and, thus, is a popular choice. The algorithm then proceeds to the consequent terms 1 to M. This part involves calculating *pseudo residuals* given by the negative derivative of the loss function with respect to the predictions evaluated at the predictions of the previous tree, i.e.

$$r_{i,m} = -\left[\frac{\partial L(y_i, f(x_i))}{\partial f(x_i)}\right]_{f_m(x_i)=f_{m-1}(x_i)} \quad \forall i, \tag{18}$$

which are then used as respective targets for each of J terminal nodes to fit the tree. Finally, it computes

$$f_m(\mathbf{x}) = f_{m-1}(\mathbf{x}) + \sum_{j=1}^{J} \gamma_{j,m} I(\mathbf{x} \in R_{j,m}), \tag{19}$$

$$\text{where } \gamma_{j,m} = \arg\min_{\gamma} \sum_{x_i \in R_{j,m}} L(y_i, f_{m-1}(\mathbf{x}_i) + \gamma) \; \forall j, \tag{20}$$

and $\sum_{j=1}^{J} \gamma_{j,m} I(\mathbf{x} \in R_{j,m}) = T_m(\mathbf{x})$ is a formal expression for tree m with J terminal nods denoted by $R_{j,m}$. The only difference between the loss function minimised by each of the predictions $\gamma_{j,m}$ from the one in (16) is that it takes into account the prediction of the previous tree. Additionally, gradient boosting can also be applied for a classification problem. The procedure is similar to the one described above with each of $f_m(\mathbf{x})$ being computed K times, where K is the number of classes, and negative log-likelihood being used as the loss function.

4 Stacking

Stacked generalization [21] or simply "stacking" is one of the ensemble learning methods where multiple base learners ("level 0" models) are trained to solve the same optimisation problem and combined to improve the performance. Stacking works well for different problems such as classification, regression and unsupervised learning in practice. In particular, the outputs of the base learners are provided as inputs of a high level model ("level 1" model), which is then trained to gain an additional performance. The main idea is that combining predictions from multiple base learners with optimal weights can often outperform any single learner. There are many estimators (or algorithms) available to decide the base learner: linear regression, lasso, ridge regression, elastic net, support vector machines, gradient boosting, random forest, etc. Intuitively, all the base learners try to predict the same target but different features (predictors) might be used by different algorithms. The final model can combine the noisy predictions into a better prediction. However, it is inefficient and sometimes infeasible to employ all the estimators or algorithms. If extra base learners do not add new features, they may not help improve predictive accuracy and can even hurt by increasing the probability for overfitting. Instead of using trivial methods, such as simple model averaging that assigns equal weights to each base learner and majority voting that considers each base learner as a vote, stacking trains a meta-model that estimates optimal weights to aggregate the base learners. Since restricting the weights to be non-negative and to sum to one can gain better performances [8], models with poor predictive accuracy will usually be given small or zero weights.

Training the base learners and meta-model with the same dataset would also result in overfitting, so cross-validation is used to guarantee the validation sample to be "clean". Stacking can in fact be seen as an extended version of cross-validation [21]. Specifically, we start with a data set $D = (Y_i, X_i)_{i=1}^n$ ("level 0" learning sample) where Y_i is the response variable and X_i is a vector of predictors. First randomly split all the observations into V equal-sized folds $D_1, \dots, D_V$. Define D_v and $D^{(-v)} = D - D_v$ to be the validation and training samples for the vth fold of the V-fold cross-validation. Consider that we have a set of base learners $M_k^{(-v)}, k = 1, \dots, K$ induced by K estimators (or algorithms), which are also called "level 0" generalisers, for each training sample $D^{(-v)}$. To complete the training process, the final base learners $M_k, k = 1, \dots, K$ are derived using all the observations in D.

The best base learner for each algorithm is decided by

$$\hat{f}_k(X_i) = \text{argmin}_{f_k(X_i)} \sum_{i=1}^{N} [Y_i - \tilde{f}_k(X_i)]^2, \; for \; k = 1, \dots, K, \tag{21}$$

where $\hat{f}_k(X_i)$ denotes the prediction of model M_k on X_i. Now we have K predictions for each response variable Y_i, which are used as inputs of the new "level 1" learning sample

$$D_{CV} = (Y_i, \hat{f}_1(X_i), \ldots, \hat{f}_K(X_i)).$$

Traditionally the level 1 sample will be used to select one base learner with the best performance, i.e. $\hat{f}_k(X_i)$ that minimises $\sum(Y_i - \hat{f}_k(X_i))^2$. The stacking idea is to use some algorithms (or "level 1" generaliser) to derive the meta-model $\tilde{M}$ for Y_i as a function of the K predictive values. Consider a linear regression where the stacking estimate of the weights $\hat{w}_s$ is simply obtained by minimizing the sum of squared residuals of the linear regression of Y_i on $\hat{f}_1(X_i), \ldots, \hat{f}_K(X_i)$. In particular,

$$\hat{w}^s = \text{argmin}_w \sum_{i=1}^{N} \left[Y_i - \sum_{k=1}^{K} w_k \hat{f}_k(X_i) \right]^2. \tag{22}$$

The final predictor of the meta-model for Y_i is $\sum_k \hat{w}_k^s \hat{f}_k(X_i)$. All the elements in the vector $\hat{w}^s$ are constrained to be non-negative and to sum to one, which is important to ensure that the predictive accuracy of stacking regression would be better than that of the single best base learner [20]. Suppose there is no restraint on the weights and $f_k(X_i)$ are highly correlated, and the w_k are obtained using least squares. Then it might happen that the final predictor $\sum_k w_k^s f_k(X_i)$ will stay away from the range $[min_k\ f_k(X_i), max_k\ f_k(X_i)]$ and generalization to the population can be poor [4].

Note that in a special case where we restrict the weight vector w^s to have one unit element and the rest zero, it simply becomes a V-fold cross-validation where the base learner with the best predictive accuracy will be selected as the final model. Stacking is more general in the sense that we can use any algorithm other than the linear regression to combine the models in (22). The weights can also depend on the input location X [8].

Variants of stacking can be designed to better adapt to some specific problems. Multi-level stacking is one kind of such extension. For simplicity, a 3-levels stacking is considered. In the first level, some base learners are trained as described above. In the second level, L instead of one meta-model are derived according to L different algorithms. In the last level, one final meta-model take as inputs the predictions made by the L meta-models in the second level. However, this can be time consuming when cross-validation is used. It is better to bear in mind that the best form and algorithms are largely decided by the specific problem we try to solve.

It is interesting to compare stacking with other ensemble learning methods, e.g., bagging and boosting. Most often the base learners do not perform well on their own either due to a high bias or because the variance is too large to be robust. Intuitively, the ensemble methods intend to combine such weak models together into one strong ensemble model to reduce bias and/or variance. This brings up the question of how to combine the base learners. We note here in passing that it is difficult to argue for the superiority in general of any particular ensemble method. For instance, Graczyk et al. [9] show that even though stacking tends to produce more precise predictions, its performance is less stable. They also suggest that different ensemble techniques tend to be more efficient in reducing the prediction error for different estimators.

Bagging usually considers homogeneous base learners and combines them according to some deterministic averaging process. The idea is that training a model depends on the sample used. A different model would have been obtained when observing another dataset. Therefore, multiple bootstrap samples can be created such that each sample can be regarded as an independent dataset drawn from the true population. Then a base learner is fitted for each of the samples and the results are averaged by combing these models in a certain way, producing an ensemble model with less variance than its components.

For example, simple model averaging can be used for a regression problem, while simple majority voting can be used for a classification problem. To ensure that bootstrap samples can be regarded as i.i.d. samples, the size of the original dataset should be large enough to represent the population and should also be large enough compared to the size of bootstrap sample to guarantee that samples are nearly independent.

Boosting fits models iteratively so that the training of a model at a given step depends on the models fitted at the previous steps, generating an ensemble model with lower bias than the base learners. In particular, the observations that are poorly predicted by the previous models in the sequence will be given more importance at the current step. Intuitively, each new model focuses on predicting the most difficult observations. A strong model with lower bias can be obtained at the end of the process. Generally it is difficult to optimally aggregate the base learners during a sequential process in one shot. Instead, the ensemble model can be built as a weighted sum of the base learners by adaptive boosting. More especially, adaptive boosting adds the base learner one by one and selects the best weight for the new base learner at each iteration, which is the best possible improvement over the previous ensemble model. Thus, it may lead to a sub-optimal solution.

Unlike bagging and boosting, stacking usually considers heterogeneous base learners decided by different estimators or algorithms. In addition, stacking trains a meta-model to combine the base learners while bagging and boosting use some deterministic algorithms. Stacking thus provides us with a new way to evaluate the performance of a new estimator or algorithm by looking at how much the corresponding base learner contributes to the meta-model. We can say that the new base learner performs well among all the base learners if it is given relatively a large weight.

5 An Empirical Example: Predicting Crime Rates

We use for our examples the data and basic modelling approach presented in a paper by Fatehkia et al. [6] on prediction of crime rates in US urban neighbourhoods using Facebook and demographic and socioeconomic data.

The traditional approach in quantitative criminology to predicting crime rates in localities is to rely on demographic and socioeconomic characteristics, and possibly supplemented by some addition measures or characteristics of interest (e.g., local laws, elections, or news events). Fatehkia et al. [6] explore whether data available from Facebook on the interests of local Facebook users can usefully improve the predictions for the rates of assault, robbery and burglary in a locality.

Fatehkia et al. [6] use data gathered from three different sources at the US ZIP code level for nine US cities:

1. crime rates calculated using data from each city's data portal;
2. demographic and socioeconomic characteristics of localities deriving from the US Census;
3. data from the Facebook Advertising API.

The nine US cities were Baltimore, Boston, Chicago, Dallas, Los Angeles, New York, Philadelphia, San Francisco and Washington, and were selected from the 25 largest US cities based on data availability and compatibility with US ZIP codes. The dataset comprises data from 432 ZIP code areas.

The outcome variable to be predicted is the crime rate, measured as crime events per 100,000 residents in 2018. We vary from Fatehkia et al. [6] in that we use log crime rates rather than raw crime rates, because the latter are noticeably skewed; the log transformation is in any case commonly used in quantitative criminology. Twelve demographic and socioeconomic measures derive from US Census data relating to 2011–15 five-year estimates sourced from the American Community survey. These data describe the age, income, racial and educational characteristics of the adult population for each locality. Facebook data were collected over the period October–November 2017.

Facebook makes it possible for advertisers to target users based on various characteristics: age, location, gender, interests, etc. This information provided by Facebook can also be used for research purposes, and this is what Fatehkia et al. [6] do: they collected aggregated data by ZIP code on the characteristics and behaviour of Facebook users. These data relate to patterns of media consumption (e.g., type of music) and measures of relationships (e.g., dating patterns).

Fatehkia et al. [6] collected data on the number of users with each interest, in total and segmented by gender and age group (18–24, 18–34, 25+), for each locality. Note that these measures overlapped. These were then combined with demographic data on the total number of users by locality to obtain a measure "fraction of users with interest X", defined as the number of users with that interest as a share of the total number of users in that demographic category. "Interests" were things like categories of games (browser games, shooter games, sports games, etc.), types of music (blues, classical, hip-hop, etc.), types of movies (comedy, science fiction, etc.) and interest in online dating services. In total 554 such interest variables were available.

We implement the model using the *scikit-learn* Python StackingRegressor package[3] with the Stata frontend *pystacked* by Ahrens et al. [1]. We employ eight base learners,[4] all from the *scikit-learn* project:

[3] See Pedregosa et al. [14] for background on the scikit-learn project, and https://scikit-learn.org/stable/modules/generated/sklearn.ensemble.StackingRegressor.html for details about the StackingRegressor package.

[4] A neural net base learner is also available, but we omit this in part because it can require substantial tuning. We did not do such tuning for the other base learners.

1. Lasso with the tuning parameter selected using information criterion (default is AIC).
2. Lasso with the tuning parameter selected by CV.
3. Ridge with the tuning parameter selected by CV.
4. Elastic net with the tuning parameter selected by CV (L2 weight set to 0.95).
5. Random forest.
6. Gradient boosting.
7. SVM.
8. Linear SVM.

We estimate the model using 350 randomly selected observations, and the remaining 76 observation are used as the holdout sample for examining out-of-sample performance. We present two versions of the model: in the first, only levels of predictors are used, and in the second we use levels and squares of predictors.

The stacking weights and in-sample and out-of-sample (holdout sample) MSEs for the linear and quadratic models are presented below, for each of the log crime rate outcomes. The mix of base learners selected via stacking varies by model and crime measure. Gradient boosting frequently attracts a high weight, along with the cross-validated elastic net. The cross-validated lasso in effect replaces the cross-validated elastic net in the robbery models, and support vector machines also enters the robbery models.

As expected, the out-of-sample performance of the stacking regression compares favourably to that of the separate base learners. Of course, because the holdout sample is a random sample (but fixed across estimations), the estimated models and out-of-sample performance would differ for a different holdout sample.

The graphs below show the performance of all eight of the base learners and the stacking regressor in terms of actual versus predicted outcomes for in-sample and out-of-sample. The good performance of some base learners and poor performance of others is clearly visible. Not surprisingly, the base learners that are assigned heavy weights by the stacking algorithm tend also to perform well out-of-sample. The graphs also illustrate the convenience of using stacking regression, in the sense that we did not need to devote any effort to tuning the base learner models and used instead the default settings (with a few exceptions chosen ex ante, e.g., the L1 weight in the elastic net) (Figs. 1, 2, 3, 4, 5, 6, 7, 8, 9, 10, 11, 12; Tables 1, 2, 3).

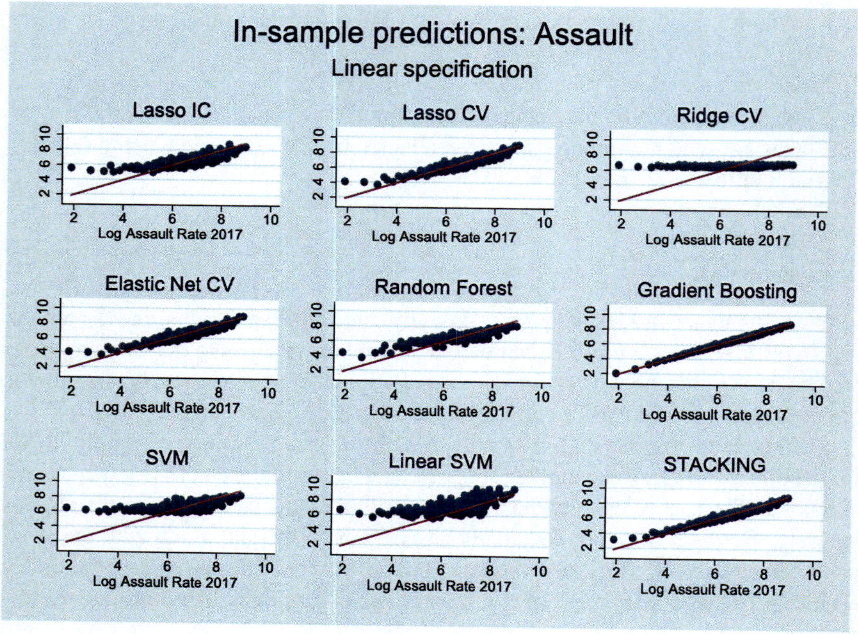

Fig. 1 Assault rate, in-sample, linear specification

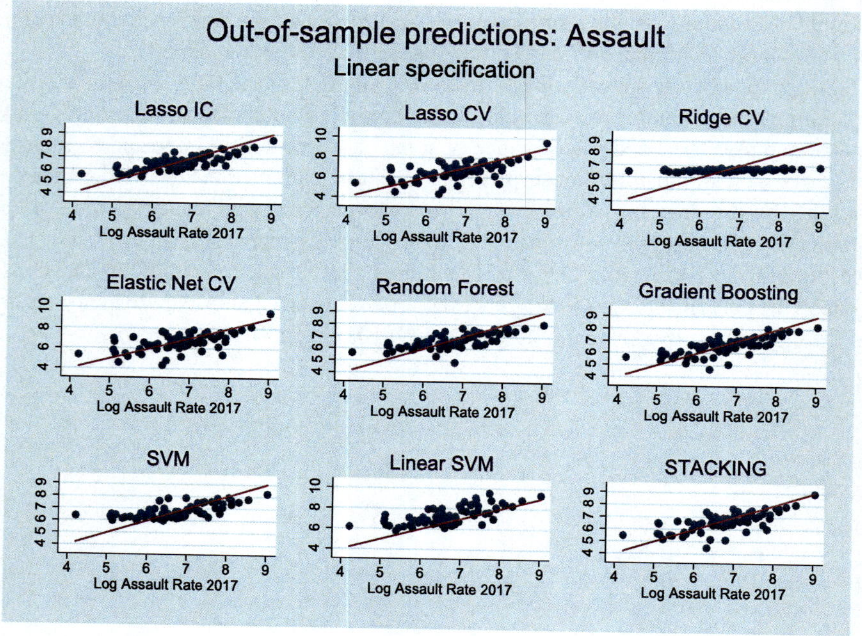

Fig. 2 Assault rate, out-of-sample, linear specification

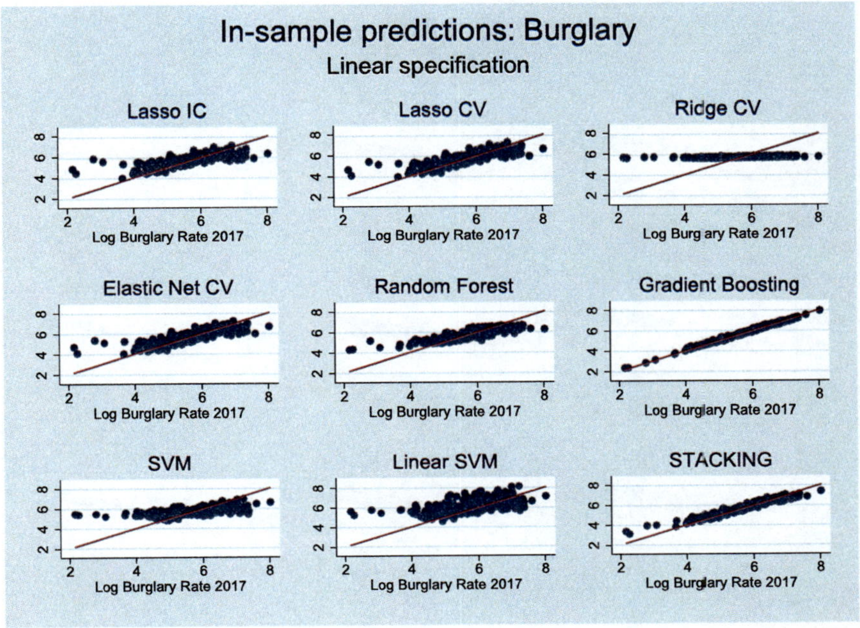

Fig. 3 Burglary rate, in-sample, linear specification

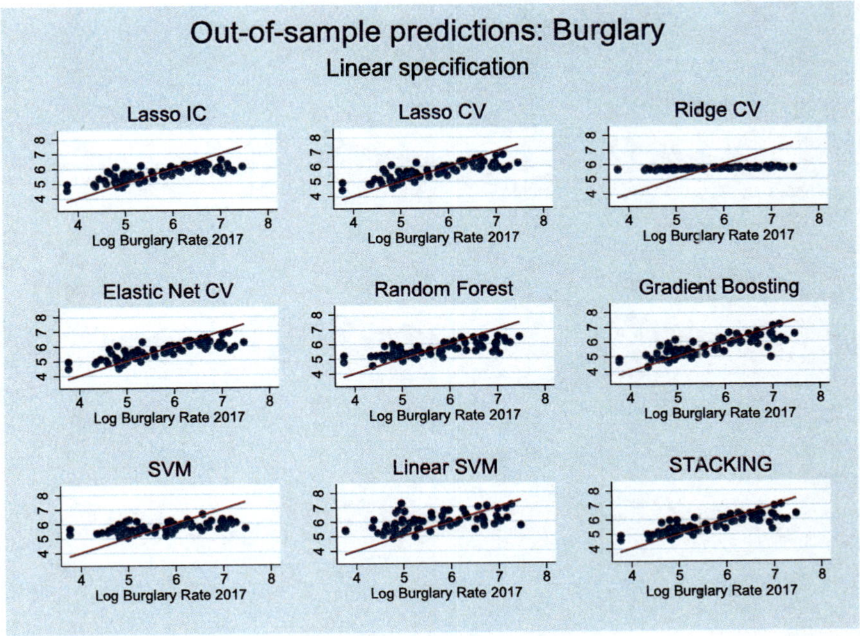

Fig. 4 Burglary rate, out-of-sample, linear specification

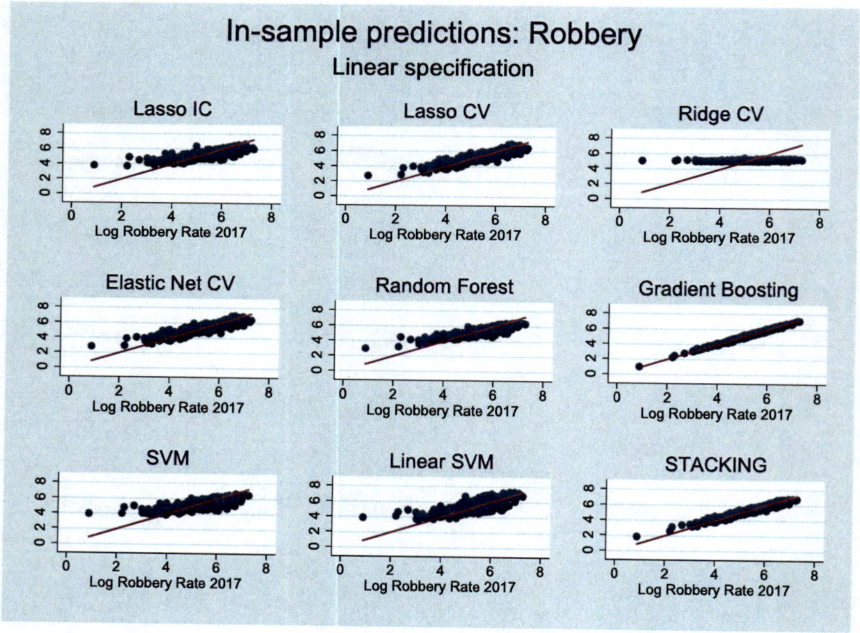

Fig. 5 Robbery rate, in-sample, linear specification

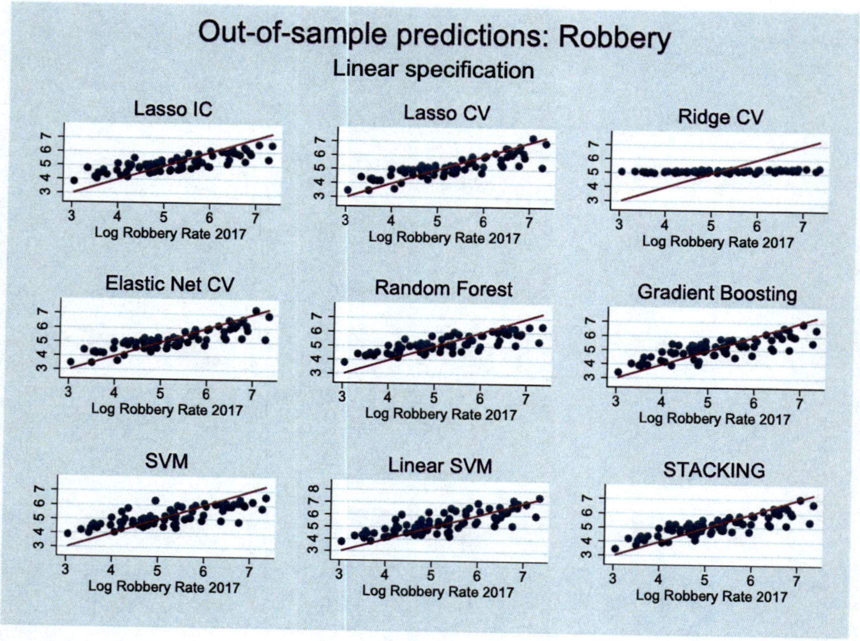

Fig. 6 Robbery rate, out-of-sample, linear specification

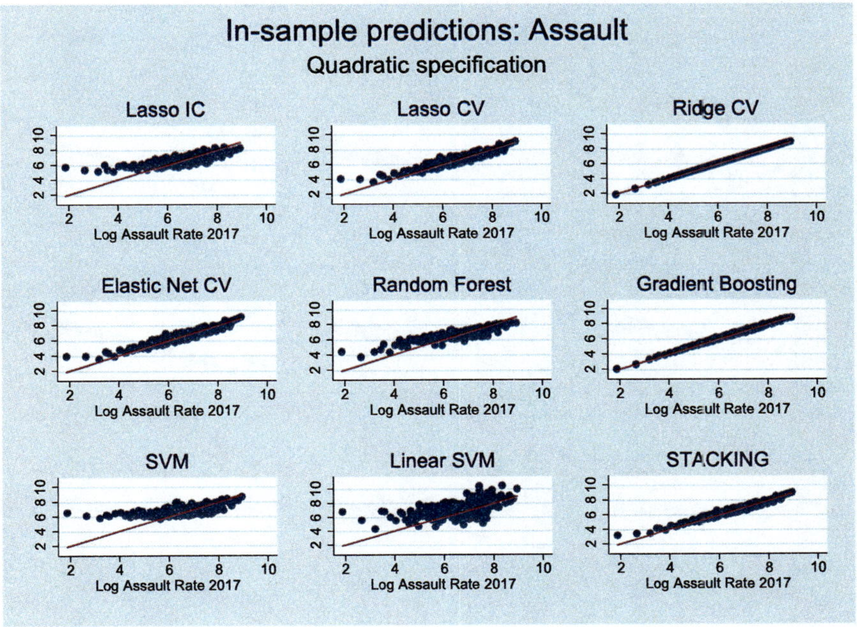

Fig. 7 Assault rate, in-sample, quadratic specification

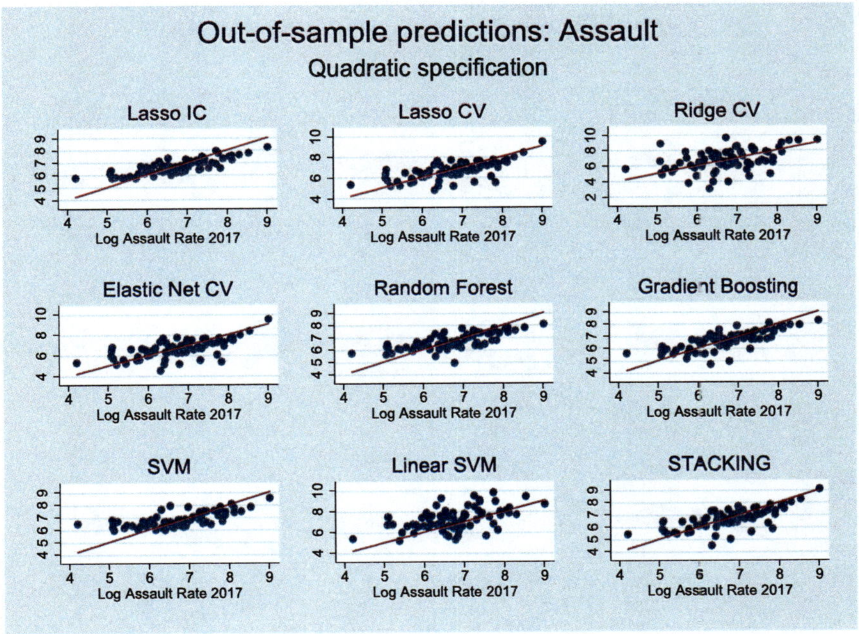

Fig. 8 Assault rate, out-of-sample, quadratic specification

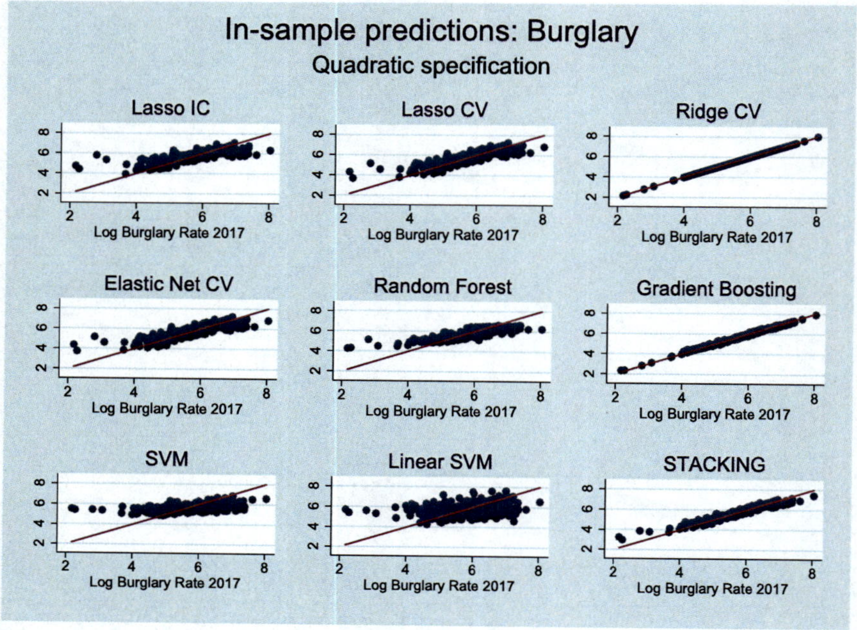

Fig. 9 Burglary rate, in-sample, quadratic specification

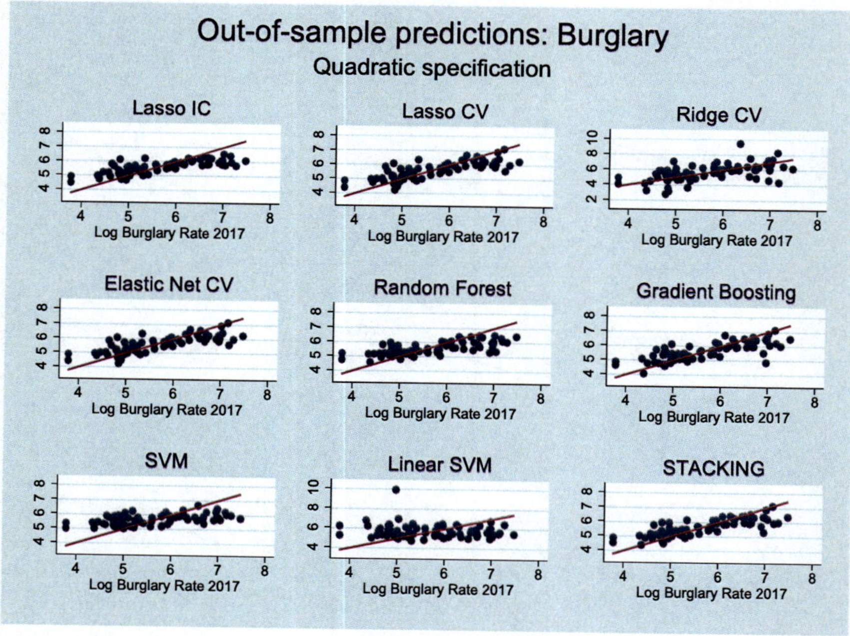

Fig. 10 Burglary rate, out-of-sample, quadratic specification

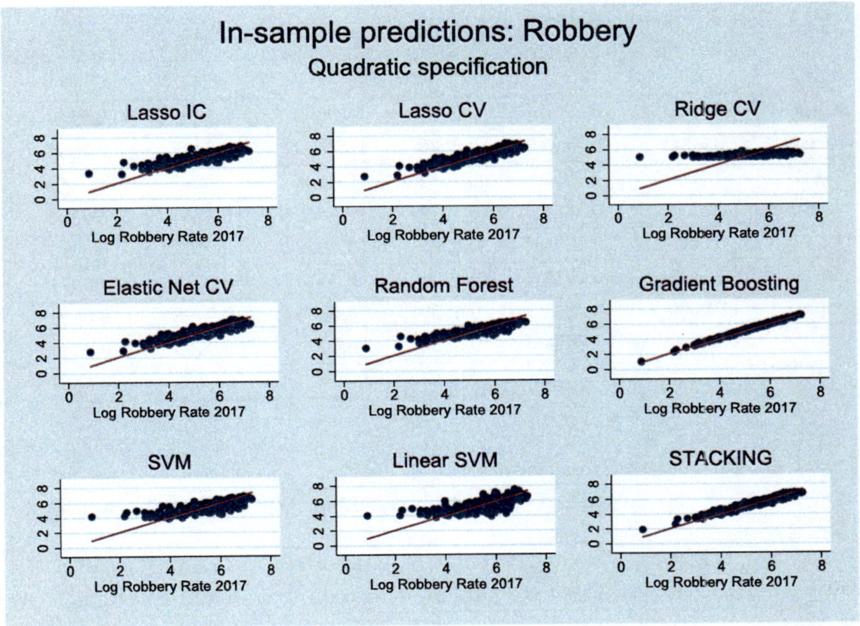

Fig. 11 Robbery rate, in-sample, quadratic specification

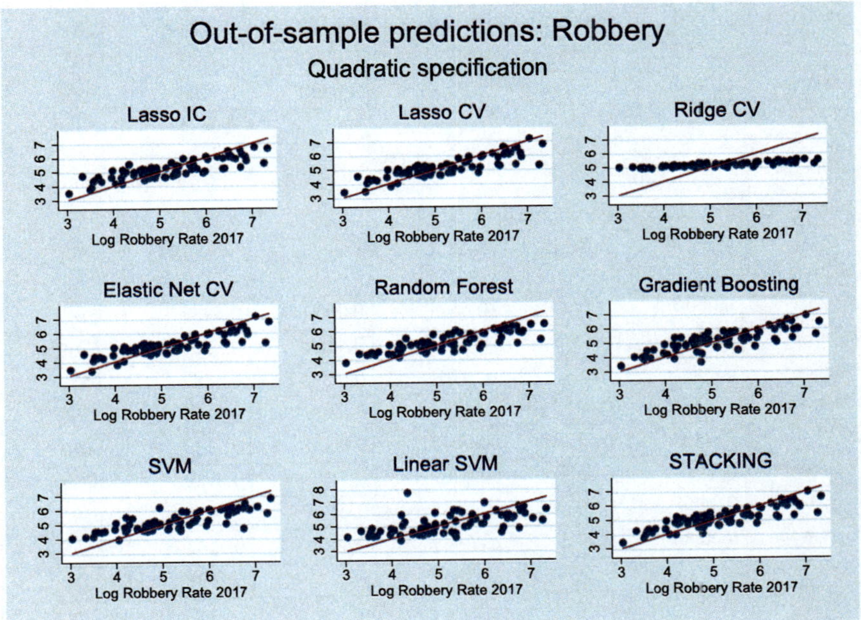

Fig. 12 Robbery rate, out-of-sample, quadratic specification

Table 1 Assault, linear and quadratic models

	Weights	MSE IS	MSE OOS	Weights Q	MSE IS Q	MSE OOS Q
lassoic	0.0000	0.5900	0.5409	0.0000	0.6399	0.5568
lassocv	0.0000	0.3519	0.7580	0.0000	0.3448	0.7055
ridgecv	0.0000	0.9254	0.8235	0.0604	0.0021	1.2646
elasticcv	0.5556	0.3485	0.7633	0.6048	0.3234	0.7340
rf	0.0000	0.5604	0.6297	0.0000	0.5592	0.6317
gradboost	0.4444	0.0914	0.5905	0.3348	0.0914	0.5971
svm	0.0000	0.7543	0.6569	0.0000	0.7496	0.6689
linsvm	0.0000	0.7614	0.6834	0.0000	0.9699	0.9904
STACKING	–	0.2222	0.6412	–	0.2172	0.6643

Table 2 Burglary, linear and quadratic models

	Weights	MSE IS	MSE OOS	Weights Q	MSE IS Q	MSE OOS Q
lassoic	0.0000	0.5916	0.6217	0.0000	0.5725	0.6122
lassocv	0.0000	0.5144	0.5724	0.0000	0.4558	0.5641
ridgecv	0.0000	0.9105	0.8702	0.0000	0.0018	0.9630
elasticcv	0.4345	0.5066	0.5676	0.4520	0.4589	0.5637
rf	0.0000	0.5017	0.6576	0.0000	0.5016	0.6560
gradboost	0.5655	0.0856	0.5731	0.5480	0.0856	0.5869
svm	0.0000	0.7581	0.7791	0.0000	0.7831	0.7527
linsvm	0.0000	0.7468	0.7997	0.0000	0.9293	1.1756
STACKING	–	0.2544	0.5490	–	0.2411	0.5500

Table 3 Robbery, linear and quadratic models

	Weights	MSE IS	MSE OOS	Weights Q	MSE IS Q	MSE OOS Q
lassoic	0.0000	0.5940	0.6521	0.0000	0.5437	0.6283
lassocv	0.4118	0.4337	0.5739	0.4364	0.4527	0.5762
ridgecv	0.0000	0.8805	0.9701	0.0000	0.8482	0.9278
elasticcv	0.0000	0.4452	0.5742	0.0000	0.4556	0.5798
rf	0.0000	0.5139	0.6613	0.0000	0.5139	0.6617
gradboost	0.5393	0.0989	0.6209	0.5172	0.0989	0.6221
svm	0.0488	0.6033	0.6685	0.0464	0.6059	0.6596
linsvm	0.0000	0.6019	0.6449	0.0000	0.7141	0.8856
STACKING	–	0.2474	0.5834	–	0.2638	0.5828

References

1. Ahrens, A., Hansen, C. B., & Schaffer, M. E. (2021). pystacked: Stata program for stacking regression. https://statalasso.github.io/docs/pystacked/
2. Arlot, S., & Celisse, A. (2010). A survey of cross-validation procedures for model selection. *Statistics Surveys, 4*, 40–79.
3. Boser, B. E., Guyon, I. M., & Vapnik, V. N. (1992). A training algorithm for optimal margin classifiers. In *Proceedings of the Fifth Annual Workshop on Computational Learning Theory* (pp. 144–152).
4. Breiman, L. (1996). Stacked regressions. *Machine Learning, 24*(1), 49–64.
5. Breiman, L. (2001). Random forests. *Machine Learning, 45*(1), 5–32.
6. Fatehkia, M., O'Brien, D., & Weber, I. (2019). Correlated impulses: Using Facebook interests to improve predictions of crime rates in urban areas. *PLOS ONE, 14*(2), 1–16. https://doi.org/10.1371/journal.pone.0211350
7. Friedman, J. H. (2002). Stochastic gradient boosting. *Computational Statistics & Data Analysis, 38*(4), 367–378.
8. Friedman, J. H. (2017). *The elements of statistical learning: Data mining, inference, and prediction*. Springer Open.
9. Graczyk, M., Lasota, T., Trawiński, B., & Trawiński, K. (2010). Comparison of bagging, boosting and stacking ensembles applied to real estate appraisal. In *Asian Conference on Intelligent Information and Database Systems* (pp. 340–350). Springer.
10. Hastie, T., Tibshirani, R., & Friedman, J. H. (2009). *The elements of statistical learning: Data mining, inference, and prediction*. Springer Series in Statistics. Springer. ISBN: 9780387848846. https://books.google.co.uk/books?id=eBSgoAEACAAJ
11. Hoerl, A. E., & Kennard, R. W. (1970). Ridge regression: Applications to nonorthogonal problems. *Technometrics, 12*(1), 69–82. https://doi.org/10.1080/00401706.1970.10488635
12. Larson, S. C. (1931). The shrinkage of the coefficient of multiple correlation. *Journal of Educational Psychology, 22*(1), 45.
13. Lei, J. (2020). Cross-validation with confidence. *Journal of the American Statistical Association, 115*(532), 1978–1997.
14. Pedregosa, F., Varoquaux, G., Gramfort, A., Michel, V., Thirion, B., Grisel, O., Blondel, M., Prettenhofer, P., Weiss, R., Dubourg, V., & Vanderplas, J. (2011). Scikit-learn: Machine learning in Python. *Journal of Machine Learning Research, 12*, 2825–2830.
15. Picard, R. R., & Cook, R. D. (1984). Cross-validation of regression models. *Journal of the American Statistical Association, 79*(387), 575–583.
16. Shao, J. (1993). Linear model selection by cross-validation. *Journal of the American statistical Association, 88*(422), 486–494.
17. Stone, M. (1974). Cross-validatory choice and assessment of statistical predictions. *Journal of the Royal Statistical Society: Series B (Methodological), 36*(2), 111–133.
18. Tibshirani, R. (1996). Regression shrinkage and selection via the Lasso. *Journal of the Royal Statistical Society. Series B (Methodological), 58*(1), 267–288. ISSN: 00359246. https://doi.org/10.2307/2346178
19. Tikhonov, A. N. (1963). On the solution of ill-posed problems and the method of regularization. *Doklady Akademii Nauk, 151*(3), 501–504.
20. Ting, K. M., & Witten, I. H. (1999). Issues in stacked generalization. *Journal of Artificial Intelligence Research, 10*, 271–289.
21. Wolpert, D. H. (1992). Stacked generalization. *Neural Networks, 5*(2), 241–259.
22. Yang, Y. (2007). Consistency of cross validation for comparing regression procedures. *The Annals of Statistics, 35*(6), 2450–2473.
23. Zhang, P. (1993). Model selection via multifold cross validation. *The Annals of Statistics*, 299–313.
24. Zou, H., & Hastie, T. (2005). Regularization and variable selection via the elastic net. *Journal of the Royal Statistical Society. Series B: Statistical Methodology, 67*(2), 301–320. ISSN: 13697412. https://doi.org/10.1111/j.1467-9868.2005.00503.x

Economics of Reciprocity and Temptation

Laxman Bokati, Olga Kosheleva, Vladik Kreinovich, and Nguyen Ngoc Thach

Abstract Behavioral economics has shown that in many situations, people's behavior differs from what is predicted by simple traditional utility-maximization economic models. It is therefore desirable to be able to accurately describe people's actual behavior. In some cases, the difference from the traditional models is caused by bounded rationality—our limited ability to process information and to come up with a truly optimal solutions. In such cases, predicting people's behavior is difficult. In other cases, however, people actually optimize—but the actual expression for utility is more complicated than in the traditional models. In such case, it is, in principle, possible to predict people's behavior. In this paper, we show that two phenomena—reciprocity and temptation—can be explained by optimizing a complex utility expression. We hope that this explanation will eventually lead to accurate prediction of these phenomena.

1 Formulation of the Problem

Behavioral economics: a brief reminder. Traditional economic models assumed that people thoroughly analyze all their options and make optimal decisions based on this analysis.

L. Bokati · O. Kosheleva · V. Kreinovich (✉)
University of Texas at El Paso, El Paso, TX 79968, USA
e-mail: vladik@utep.edu

L. Bokati
e-mail: lbokati@miners.utep.edu

O. Kosheleva
e-mail: olgak@utep.edu

N. N. Thach
Institute for Research Science and Banking Technology, Banking University HCMC,
39 Ham Nghi District 1, Ho Chi Minh City, Vietnam
e-mail: thachnn@buh.edu.vn

S. Sriboonchitta et al. (eds.), *Credible Asset Allocation, Optimal Transport Methods, and Related Topics*, Studies in Systems, Decision and Control 429,
https://doi.org/10.1007/978-3-030-97273-8_3

In many decision-making situations, this assumption works reasonably well and leads to a reasonably accurate description of an overall economic process. However, many research results performed in the last decades—some of them of Nobel Prize quality—have shown that in many practical situations, the actual people's behavior differs from the assumed ideal one. The analysis of such behavior and its economic consequences is known as *behavioral economics*.

Challenges for behavioral economics. Behavioral economics provides convincing and impressive examples of people's non-optimal behavior. However, in many cases, it does not provide us with quantitative models that would predict actual people's behavior. Coming up with such models is an important challenge for behavioral economics.

One way to come with such predictions is to better understand why people's behavior differs from the predictions of traditional economic models—so that, hopefully, this understanding will lead us to the desired predictions.

Why people's behavior differs from the traditional economic predictions: two main reasons. There are two main reasons why people's behavior differs from the traditional economic models.

The first reason is that when making decisions, people often have limited ability (and limited time) to make a decision. As a result, they sometimes make a sub-optimal decision. In such situations, it is, in general, not easy to come up with the adequate model of people's behavior—this requires a deep knowledge of how exactly we process limited information in our brains.

However, there is another reason why people's behavior is often different from what the traditional economic models would prescribe. Namely, many traditional economic models—models that assume that each person wants to maximize his/her gain (usually formalized as utility)—provide an oversimplified understanding of how people gauge the gains from different possible actions. In such situations, in principle, we can come up with quantitative models of human behavior—for this, we need to provide more adequate, more accurate models of human utility.

Such situations are the "low-hanging fruits" of this research area, topics in which there is the biggest hope of reaching quantitative descriptions of human behavior.

What we do in this paper. In this paper, we provide two examples of such phenomena, examples that correspond to (seemingly unrelated) phenomena of reciprocity and temptation. It turns out that while, from the economic viewpoint, these are two different behaviors, they can be explained by using similar ideas and similar techniques.

2 How to Make Traditional Models More Adequate: Empathy and Discounting

Utility in the traditional economic models. In the traditional economic models, it is usually assumed that a decision maker maximizes his/her gain (numerically expressed as utility u), and this utility value describe the effect of this decision on this person at this particular moment of time; see, e.g., [7, 15, 17, 23, 27].

Need to go beyond traditional models. In these models, person's decisions are not affected by gains (utilities) of others and/or by gains of the same person at future moments of time. To some extent this is true, but one can easily find examples where gains of others (and/or future gains of the same person) do affect our behavior.

Maybe a proverbial greedy capitalist would gladly earn an extra million by making his workers work more and thus, get less utility, but in general, hardly anyone would prefer, e.g., \$101 to \$100 if this increase is accompanied by someone's severe suffering. Some people spend all their money like there is no tomorrow and retire in poverty, but most people do limit somewhat their current expenses to save for retirement. It is all a matter of degree. Some people are not empathetic enough, some do not save enough—but to some degree, practically everyone is empathetic and practically everyone saves (at least something).

How to describe dependence on other's utilities. Let $u_i^{(0)}$ be approximate utilities that come only from this person's consumption. How can we describe the actual utilities u_i that take into account other people's feelings—i.e., in precise terms, other people's utilities?

A natural way is to add, to $u_i^{(0)}$, terms proportional to other people's utilities, i.e., to consider expressions of the type

$$u_i = u_i^{(0)} + \sum_{j \neq i} \alpha_{ij} \cdot u_j, \tag{1}$$

where each coefficient α_{ij} describes how the utility of the ith person depends on the utility of the jth person; see, e.g., [1–4, 9, 11, 23, 28, 29, 34].

This phenomenon is known by a polite term *empathy*, since for positive values α_{ij}, this formula describes how people feel better if others around them are happier. However, from the purely mathematical viewpoint, it is also possible to have negative values α_{ij}, when someone's happiness makes the other person unhappy. This is not just a mathematical example, such things like jealousy and hatred are, unfortunately, quite real.

How to describe dependence on utility in different moments of time. In the traditional economic models, we assume that a person's utility at moment t is determined only by his/her consumption at this moment of time. In reality, in addition to this approximate utility u_t, the person also takes into account future utilities u_{t+1}, u_{t+2}, ..., and past utilities u_{t-1}, u_{t-2}, etc., with appropriate coefficients:

$$u = u_t + \sum_{j>0} q_j \cdot u_{t+j} + \sum_{j<0} q_j \cdot u_{t+j}. \tag{2}$$

This phenomenon is known as *discounting*, since a person usually considers future experiences as less valuable than the present ones: e.g., people will pay less that a dollar for a chance to get a dollar a year from now; see, e.g., [6, 8, 12–14, 18, 19, 26, 35].

What we will do now. Let us show that these two phenomena explain people's behavior corresponding to reciprocity and temptation.

3 Reciprocity: What It Is and How It Can Be Explained

What is reciprocity. Usually, people have reasonably fixed attitude to others: they feel empathy towards members of their family, members of their tribe, usually citizens of their country—and may be consistently negative towards their country's competitors. However, in addition to these consistent feelings, they also have widely fluctuating attitudes towards people with whom they work—or at least with whom they are teamed up in a experiment set up by a behavioral economics researcher.

It turns out that while it is difficult to predict how these attitudes will evolve—even in what direction they will evolve, positive or negative—there is a general phenomenon that people are nice to those who treat them nicely and negative to those who treat them badly. In terms of the coefficients α_{ij} it means that:

- if α_{ji} is positive, then we expect α_{ij} to be positive too, and
- if α_{ji} is negative, then we expect α_{ij} to be negative too;

see, e.g., [25, 30].

This *reciprocity* phenomenon is intuitively clear—this is, after all, a natural human behavior—but how can we explain it in economic terms?

Let us formulate the problem in precise terms. To explain the reciprocity phenomenon, let us consider the simplest case of formula (1), when we have only two people. In this case, the formula (1) for these two people takes the following form:

$$u_1 = u_1^{(0)} + \alpha_{12} \cdot u_2; \tag{3}$$

$$u_2 = u_2^{(0)} + \alpha_{21} \cdot u_1. \tag{4}$$

Since each person tries to maximize his/her utility, a natural question is as follows:

- suppose that Person 1 knows the attitude α_{21} of Person 2 towards him/her;
- what value α_{12} describing his/her attitude should Person 1 select to maximize his/her utility u_1?

Analysis of the problem. If we replace, in the right-hand side of the equality (3), the value u_2 with the right-hand side of the expression (4), we get

$$u_1 = u_1^{(0)} + \alpha_{12} \cdot u_2^{(0)} + \alpha_{12} \cdot \alpha_{21} \cdot u_1.$$

If we move all the terms containing u_1 into the left-hand side, we get

$$u_1 \cdot (1 - \alpha_{12} \cdot \alpha_{21}) = u_1^{(0)} + \alpha_{12} \cdot u_2^{(0)},$$

hence

$$u_1 = \frac{u_1^{(0)} + \alpha_{12} \cdot u_2^{(0)}}{1 - \alpha_{12} \cdot \alpha_{21}}. \tag{5}$$

This expression can take infinite value—i.e., as large a value as possible—if we take the value

$$\alpha_{12} = \frac{1}{\alpha_{21}}, \tag{6}$$

for which the denominator is 0. We can make it positive—and as large as possible—if we take α_{12} close to the inverse $1/\alpha_{21}$, so that the difference $1 - \alpha_{12} \cdot \alpha_{21}$ will not be exactly 0, but be close to 0, with the same sign as the expression $u_1^{(0)} + \alpha_{12} \cdot u_2^{(0)}$.

This explains reciprocity. Indeed, according to the formula (6):

- if α_{21} is positive, then the selected value α_{12} is also positive, and
- if α_{21} is negative, then the selected value α_{12} is also negative.

4 Temptation: What It Is and How It Can Be Explained

What is temptation. A popular book [33] by a Nobelist Richard H. Thaler starts the chapter on temptation (Chap. 2) with a simple example: a group of friends are given a big bowl of nuts before dinner. As they eat more and more nuts, they realize that if they continue, they will have no appetite for the incoming tasty dinner, so they decided to put away the bowl.

All this sounds reasonable, until we start analyzing it from the economic viewpoint. From this viewpoint, the more options we have, the better, so how come the elimination of one of the options made everyone happier?

This is just one example; for other examples and for a general analysis of this phenomenon, see, e.g., [5, 8, 10, 16, 20, 24, 31–33].

What if we take discounting into account. Let us try to resolve this puzzle by taking discounting into account. Let us denote the overall amount of food that a person can eat in the evening by a (e.g., by a grams), the utility for eating one gram of nuts by n, the utility of eating one gram of dinner by d, the discounting coefficient from dinner to now by q_+, and the amount of nuts that we eat now by x. The variable x can take any value from the interval $[0, a]$.

In terms of these notations, when we eat x grams of nuts and $a - x$ grams of actual dinner, then, taking into account discounting, the overall utility now is equal to

$$n \cdot x + q_+ \cdot d \cdot (a - x). \tag{7}$$

According to the usual decision making idea, we want to select the amount x for which this utility is the largest. But the expression (7) is linear in x, so its largest value on the interval $[0, a]$ is attained at one of the endpoints of this interval, i.e., either for $x = 0$ or for $x = a$. In the first case, we do not eat any nuts at all, in the second case, we only eat nuts and do not eat any dinner. This may be mathematically reasonable, but this is *not* how people behave! How can we explain how people actually behave?

Taking into account that at different moments of time, people have different preferences. In the previous text, we assumed that the only way a person takes into account future events is by discounting. This would make sense if the same person at different moment of time has the same preferences. In reality, people's preferences change. To some extent, the same person at different moments of time is a kind of a different person. So, a proper way to take that into account is to realize that when a person makes decision, he or she needs to find a compromise between his/her today's interests and his/her interests at other moments of time.

This situation is similar to situation of joint decisions making, when several people with somewhat different interests try to come up with a group decision—the only difference is that different people can decide not to cooperate at all, while here, "agents" (i.e., the same person at different moments of time) are "joined at the hip"—decisions by one of them affect another one. Thus, to properly describe decision making, we need to view the problem as a problem of group decision making—group decision making by agents representing the same person at different moments of time.

According to decision theory, a group decision of several cooperating agents should be maximizing the product of their utilities. This is known as *Nash's bargaining solution*; see, e.g., [17, 21, 22]. So, in our case, a person making a decision should be maximizing the product of the utilities at different moments of time.

Let us show, on the above example, that this indeed helps us avoid the above un-realistic prediction that we should have $x = 0$ or $x = a$.

How this idea help. Let us consider the simplest case of two moments of time: the original moment of time when we are eating (or not eating) nuts, and the future moment of time when we will be eating dinner. In the original moment of time, the utility is described by the formula (7). Similarly, at the next moment of time, the utility is described by a formula $q_- \cdot n \cdot x + d \cdot (a - x)$, for an appropriate discounting coefficient q_-. Thus, the correct value x is the one that maximizes the product

$$(n \cdot x + q_+ \cdot d \cdot (a - x)) \cdot (q_- \cdot n \cdot x + d \cdot (a - x)).$$

This function is quadratic, and, in contrast to linear functions, the maximum of a quadratic function on an interval is not necessarily attained at one of the interval's endpoints.

Let us illustrate it on a simplified example where computations are easy: $a = 1$, $n = 1$, $d = 2$, and $q_+ = q_- = 0.25$. In this case, we maximize the function

$$(x + 0.5 \cdot (1 - x)) \cdot (0.25 \cdot x + 2 \cdot (1 - x)) = (0.5 \cdot x + 0.5) \cdot (2 - 1.75 \cdot x).$$

Differentiating this expression with respect to x and equating the derivative to 0 leads to

$$0.5 \cdot (2 - 1.75 \cdot x) + (0.5 \cdot x + 0.5) \cdot (-1.75) = 0,$$

i.e., to $0.125 = 1.75 \cdot x$ and

$$x = \frac{0.125}{1.75} = \frac{1/8}{7/4} = \frac{1}{14} \approx 0.07.$$

The values a, n, etc., were kind of random, but the resulting proportion of nuts snack in the food—about 7%—is quite reasonable.

Comment. So why is everyone happy that the temptation was taken away? Because this allowed everyone not to violate their social contract—in this case, a social contract (as described by Nash's bargaining solution) between a person now and the same person in the future.

Acknowledgements This work was supported in part by the National Science Foundation grants 1623190 (A Model of Change for Preparing a New Generation for Professional Practice in Computer Science) and HRD-1242122 (Cyber-ShARE Center of Excellence). It was also supported by the program of the development of the Scientific-Educational Mathematical Center of Volga Federal District No. 075-02-2020-1478.

References

1. Becker, G. S. (1991). *A treatise on the family*. Harvard University Press.
2. Bergstrom, T. (1989). Love and spaghetti, the opportunity cost of virtue. *Journal of Economic Perspectives, 3*, 165–173.
3. Bergstron, T. (1991). *Systems of benevolent utility interdependence* (Technical report). University of Michigan.
4. Bernheim, B. D., & Stark, O. (1988). Altruism within the family reconsidered: Do nice guys finish last? *American Economic Review, 78*(5), 1034–1045.
5. Camerer, C. F. (2007). Neuroeconomics: Using neuroscience to make economic predictions. *Economic Journal, 117*, C26–C42.
6. Critchfield, T. S., & Kollins, S. H. (2001). Temporal discounting: Basic research and the analysis of socially important behavior. *Journal of Applied Behavior Analysis, 34*, 101–122.

7. Fishburn, P. C. (1969). *Utility theory for decision making*. Wiley.
8. Frederick, S., Loewenstein, G., & O'Donoghue, T. (2001). Time discounting and time preference: A critical review. *Journal of Economic Literature, 40,* 222–226.
9. Friedman, D. D. (1986). *Price theory*. South-Western Publishing Company.
10. Gruber, J. (2002). Smoking's 'internalities'. *Regulation, 25*(4), 52–57.
11. Hori, H., & Kanaya, S. (1989). Utility functionals with nonpaternalistic intergerenational altruism. *Journal of Economic Theory, 49,* 241–265.
12. King, G. R., Logue, A. W., & Gleiser, D. (1992). Probability and delay in reinforcement: An examination of Mazur's equivalence rule. *Behavioural Processes, 27,* 125–138.
13. Kirby, K. N. (1997). Bidding on the future: Evidence against normative discounting of delayed rewards. *Journal of Experimental Psychology (General), 126,* 54–70.
14. Konig, C. J., & Kleinmann, M. (2005). Deadline rush: A time management phenomenon and its mathematical description. *The Journal of Psychology, 139*(1), 33–45.
15. Kreinovich, V. (2014). Decision making under interval uncertainty (and beyond). In P. Guo & W. Pedrycz (Eds.), *Human-centric decision-making models for social sciences* (pp. 163–193). Springer Verlag.
16. Laibson, D. (1997). Golden eggs and hyperbolic discounting. *Quarterly Journal of Economics, 112,* 443–477.
17. Luce, R. D., & Raiffa, R. (1989). *Games and decisions: Introduction and critical survey*. Dover.
18. Mazur, J. E. (1987). An adjustment procedure for studying delayed reinforcement. In M. L. Commons, J. E. Mazur, J. A. Nevin, & H. Rachlin (Eds.), *Quantitative analyses of behavior: Vol. 5. The effect of delay and intervening events*. Erlbaum.
19. Mazur, J. E. (1997). Choice, delay, probability, and conditional reinforcement. *Animal Learning Behavior, 25,* 131–147.
20. McClure, S. M., Laibson, D. I., Loewenstein, G., & Cohen, J. D. (2004). Separate neural systems value immediate and delayed monetary rewards. *Science, 306,* 503–507.
21. Nash, J. (1950). The bargaining problem. *Econometrica, 18*(2), 155–162.
22. Nguyen, H. P., Bokati, L., & Kreinovich, V. New (simplified) derivation of Nash's bargaining solution. *Journal of Advanced Computational Intelligence and Intelligent Informatics (JACIII),* to appear.
23. Nguyen, H. T., Kosheleva, O., & Kreinovich, V. (2009). Decision making beyond Arrow's 'impossibility theorem', with the analysis of effects of collusion and mutual attraction. *International Journal of Intelligent Systems, 24*(1), 27–47.
24. O'Donoghue, T., & Rabin, M. (1999). Studying optimal paternalism, illustrated by a model of sin taxes. *American Economic Review, 89*(1), 103–124.
25. Rabin, M. (1993). Incorporating fairness into game theory and economics. *American Economic Review, 83*(5), 1281–1302.
26. Rachlin, H., Raineri, A., & Cross, D. (1991). Subjective probability and delay. *Journal of the Experimental Analysis of Behavior, 55,* 233–244.
27. Raiffa, H. (1997). *Decision analysis*. McGraw-Hill.
28. Rapoport, A. (1956). Some game theoretic aspects of parasitism and symbiosis. *Bulletin of Mathematical Biophysics, 18.*
29. Rapoport, A. (1964). *Strategy and conscience.*
30. Thaler, R. H. (2015). *Misbehaving: The making of behavioural economics*. W. W. Norton.
31. Thaler, R. H., & Johnson, E. J. (1990). Gambling with the house money and trying to break even: The effects of prior outcomes on risky choice. *Management Science, 36,* 643–660.
32. Thaler, R. H., & Shefrin, H. M. (1981). An economic theory of self-control. *Journal of Political Economy, 89,* 392–406.
33. Thaler, R. H., & Sunstein, C. R. (2009). *Nudge: Improving decisions about health, wealth, and happiness*. Penguin Books.
34. Tipler, F. J. (1994). *The physics of immortality*. Doubleday.
35. Zapata, F., Kosheleva, O., Kreinovich, V., & Dumrongpokaphan, T. (2018, March 13–15). Do it today or do it tomorrow: Empirical non-exponential discounting explained by symmetry ideas. In V.-N. Huynh, M. Inuiguchi, D.-H. Tran, & T. Denoeux (Eds.), *Proceedings of the International Symposium on Integrated Uncertainty in Knowledge Modelling and Decision Making IUKM'2018*, Hanoi, Vietnam.

The Most Infamous Coronavirus Forecast

William M. Briggs

Abstract A brief survey of the strategies used to produce coronavirus forecasts. And a detailed look at the Imperial College of London death forecast that was responsible for the start of the panic. It was shockingly unrealistic but presented in a way its flaws were not easily seen. Forecasts improved after that, but by then it was too late.

Keywords Coronavirus · COVID-19 · Forecasts · Prediction · Verification

1 Outline of the Crisis

There are many places in which the beginnings of history of the coronavirus crisis of 2019–2021, and possibly 2022, can be found. The entire history is not yet written, of course, a project that will not be completed until long after this paper sees print, especially since the crisis, defined here as overt governmental and corporate responses and mandates, is still on-going in late 2021.

Our interest here is in the predictions and forecasts made about the pandemic that were used by citizens, the media, and especially government to make crucial decisions. We can only deal with the early and mid-pandemic forecasts, because as of this writing (late September 2021), all the data is not yet in.

Even so, the literature is already voluminous. It is now so large that it is impossible to do a complete survey. A cursory search of Medrxiv, the medical preprint server, revealed 28,499 articles referencing *COVID-19 SARS-CoV-2* alone on 21 September 2021. Google Scholar, with some overlap, listed over 4 million (there is no typo) articles with just the *COVID-19* keyword. Adding in *coronavirus* and other obvious keywords and the count only goes up.

Obviously, it will not be possible even to limit a survey to just predictions made about the pandemic. All we can do is to highlight general trends, and give criticisms and cautions to various models used in making predictions.

W. M. Briggs (✉)
Charlevoix, MI, USA
e-mail: matt@wmbriggs.com

S. Sriboonchitta et al. (eds.), *Credible Asset Allocation, Optimal Transport Methods, and Related Topics*, Studies in Systems, Decision and Control 429,
https://doi.org/10.1007/978-3-030-97273-8_4

There were many attempts at modeling different aspects of the pandemic. There were standard epidemiological models, e.g. [1–5]; frequentist time series modeling, e.g. [6]; Bayesian time series, e.g. [7]; machine-learning techniques, e.g. [8]; so-called deep learning, e.g. [9]. Some models predicted totals, other predicted peak attributed death times [10].

Our interest in mainly in attributed Covid death totals. Since the literature is so large even here, we settle for a smaller task. Our purpose is to discuss in general terms the use of one consequential forecast during the pandemic, especially how this prediction was used and abused by governments, and the accuracy of this predictive model.

Most of the studies making various predictions were completed only after the outbreak was well under way by late March 2020. By that time, many governments had already made decisions regarding their response—which became ossified and immune from updated information, or so it appeared in many cases. This reflects the well known observation that politicians are loathe to be seen changing course, as doing so can be seen as admitting to error, and error is weakness. This ancient truth has enormous consequences for forecasting, especially official forecasting. It can be too easy to commit to a bad model and to then lack the willpower to repudiate it.

At any rate, earlier predictions, especially those that caught the public interest at the panic's beginning, are our main focus. How those early predictions were later modified is of secondary interest.

As said above, a complete survey of forecasts produced, and their verification, cannot now be accomplished, though some small work on the problem has started. If there has to be one term to describe how models performed, it would be over-forecasted. Predictions for things like death and hospitalizations were generally much larger than reality, in some cases criminally larger. This was, we diagnose, mainly a problem of the zealousness of modelers. I have no proof of it, but perhaps the desire to be recognized drove some modeling decisions, and nobody was going to be recognized by issuing the forecast "It won't be that bad, relative to the history of pandemics." The techniques used, too, almost guaranteed over-forecasts. For example, the asymmetric prediction bounds purposely emphasizing higher upper bounds, as in [4].

Roda et al. [11] stated a truism that "predictions using more complex models may not be more reliable compared to using a simpler model." This is, admittedly, common statistical wisdom, which all acknowledge. But which many do not practice.

Wynants and others [12] did an early paper (submitted 31 March 2020) reviewing 145 coronadoom forecasts. They examined these models in a general sense, the data not yet being available to formally verify them. They instead looked into diagnostic measures used as input to the models, such as "age, body temperature, lymphocyte count, and lung imaging features." Their conclusion was this: "This review indicates that proposed models are poorly reported, at high risk of bias, and their reported performance is probably optimistic. Hence, we do not recommend any of these reported prediction models for use in current practice."

Many were used, alas. In this paper we examine what was perhaps the most influential of all coronadoom forecasts.

2 Term Confusion

Before looking at this most infamous forecast, some terms must be understood. There appeared, during the crisis, to be little interest by any official entity in defining basic words, or in using language which was consistent with previous medical literature.

There are, or were before 2020, in medicine three disease fatality measures of interest, now too often (depressingly) conflated: population, infection, and case fatality rates. The population fatality rate (PFR) is, as the name implies, the number in a population who were killed by a disease divided by the total population. The infection fatality rate (IFR) is the number infected who died divided by the number infected. The case fatality rate (CFR) is the number of cases who died divided by the number of cases. The increase in severity of these rates is obvious.

The media and government consistently misnamed cases. Cases, prior to COVID-19 at any rate, were active illnesses of patients seeking or requiring treatment. True cases were patients who had infections: it took more than just an infection to be labeled a "case".

Unfortunately, throughout the global panic, mere positive tests were called "cases"; see [13] for a detailed analysis of this. When the public, including many government officials, heard "case", they must have imagined patients being strapped to ICU beds hooked to ventilators, because these wrongly named "cases" were used as justification for new and continuing restrictions meant as "solutions" to ending the pandemic (which didn't end). There was no appreciation that, for the majority, especially those under 65, the infection was unnoticed or accompanied by only mild illness. The CDC, for example, estimated that 40% of infected persons were entirely asymptomatic; see [14].

3 The Panic Begins

The most influential forecast was by the Imperial College of London, and its COVID-19 modeling effort led by Neil Ferguson. Their forecast caused an immense public and governmental reaction (to say the least). One paper in *JAMA*, discussing Ferguson's comments on this global reaction [15], said "For predictions, 'models are not crystal balls,' as Ferguson noted in a recent overview of the role of modeling." The article went on to correctly state that "Nevertheless, consumers of epidemiological models, including politicians, the public, and the media, often focus on the quantitative predictions of infections and mortality estimates." Experience shows they focus on the *worst* of these predictions as well, either failing to understand the uncertainty inherent in extreme predictions, or not caring.

It's safe to say that simulations, models, and forecasts, terms which I take as synonymous, drove both the panic and government responses. *Nature* reported that "epidemiologists rushed to model the coronavirus pandemic" [16]. The same paper

Fig. 1 IC1

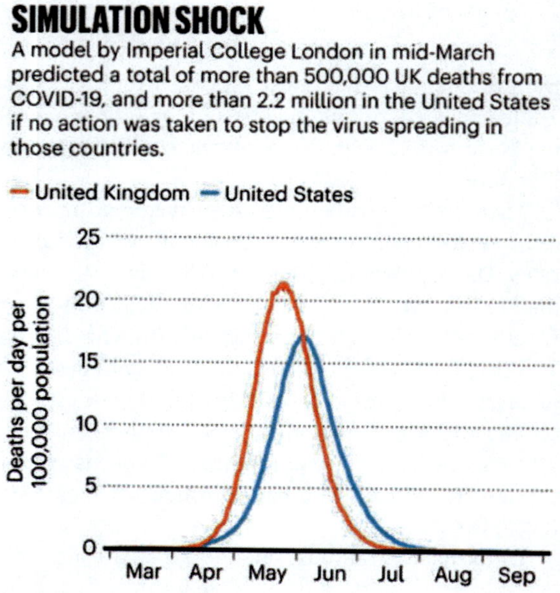

reproduced a figure from a prediction by the COVID-19 response form the UK's Imperial College (IC), a figure that was relied upon heavily both in the UK and USA, as well as elsewhere. It is shown here in Fig. 1.

At first glance, the model does not look unusual or odd, though the accompanying numbers of 500,000 deaths in the UK and 2.2 million in the USA are alarming—and indeed caused public turmoil when the forecast was released. A second, serious gaze reveals the flaws.

The figures are oddly bell-shaped, which is most unusual for epidemic/pandemics. The usual shape is a rapid increase followed by a slow tailing off. This can be seen in the actual data in Fig. 2.

The most massive flaw in the model are the numbers themselves. At the peaks of the forecasted outbreak, the UK was projected to have about 21 per 100,000 deaths *per day*, and the USA about 17 per 100,000—again, *per day*. Again, this might not seem unusual. Until it is recalled the UK had a population in 2020 of about 66.7 million, and the USA around 328.2 million.

The model thus translates into 14,000 peak deaths *per day* in the UK, and 56,000 peak *per day* in the USA! These are, or should have been seen as, astonishing numbers, so astonishing they never should have been believed. For one, they are far worse than historical reports for the Spanish Flu, which has been estimated at killing as many as 50 million worldwide in 1917; see [17] for details.

The IC model projected that in just a few months time coronadoom would be 3.3 times worse than the Spanish flu in the USA, a full century after, when medicine had improved dramatically. I stress all this was to happen in a very brief period of time. As of 27 September 2020, a date keyed to the end date of Ferguson's prediction, the CDC estimated 190,076 deaths attributed to COVID-19 [18]. One year after this, deaths did approach the Spanish flu level, for the USA, but not for many other countries. But this time 18–20-month time frame is much longer than the Ferguson model allowed.

Ferguson's model, extrapolated to the world, thus implied 165 million dead from the coronadoom in by the end of summer 2020. This was an absurd number with no other justification except for the model itself. As of 27 September 2021, a year after Ferguson's model projections ended, the total is about 4.5 million.

These simple calculations, changing the rate per 100,000 to actual totals, appear never to have been done, though if they were they would have cast deep skepticism on the IC model. Instead, the modeled numbers were widely believed, both in the UK and in the USA. The desire to believe was there.

The actual data revealed peak attributed deaths of about 1400 per day in the UK and 5000 in the USA, numbers which occurred in January 2021 (ONS and CDC data), and *not* May of 2020. That represents model errors of over a thousand percent. This may be the worst blown forecast in history.

The counter-argument will be that the lockdowns which were implemented in the UK and USA *caused* the smaller numbers that were seen, and that the (now) counterfactual forecast was indeed accurate. But this argument fails when it is recalled eight states in the USA never locked down, and they had the best deaths per capita, and that other countries in the world never locked down (e.g. Sweden), and they fared at least as well, or better, than countries which had very restrictive policies.

4 SAGE Times

In the UK, the Scientific Advisory Group for Emergencies (SAGE) prepared COVID-19 death forecasts. The media reported a "leak" on 29 August 2020 which said that as many as 85,000 additional people could die in a second wave through the fall of 2020 and winter 2021; see [19]. This was based, as before, on a modeled forecast. Many of the same people involved in this modeling effort were also involved in the first, noted above. Had any lessons been learned from the first blown prediction?

The data is from the UK's Office of National Statistics (ONS) [20]. According to the ONS, through the week of 14 September 2021, just under 145,000 deaths were attributed to COVID-19, at least in part. The time series of daily deaths are in Fig. 2.

The SAGE second forecast was therefore leaked as attributed deaths were rapidly declining in the late summer of 2020. The first forecast was issued as deaths were first rising in early 2020. The forecasted second wave was therefore expected to be 63% larger and deadlier than the first.

The first wave of deaths did occur in roughly the same period of the Ferguson forecast, but it was vastly smaller. The second wave happened in about the same period of time projected in the second model. The actual totals, from the end of September 2020 to 30 April 2021 were 85,614, according to the ONS. The second forecast was therefore right on the money.

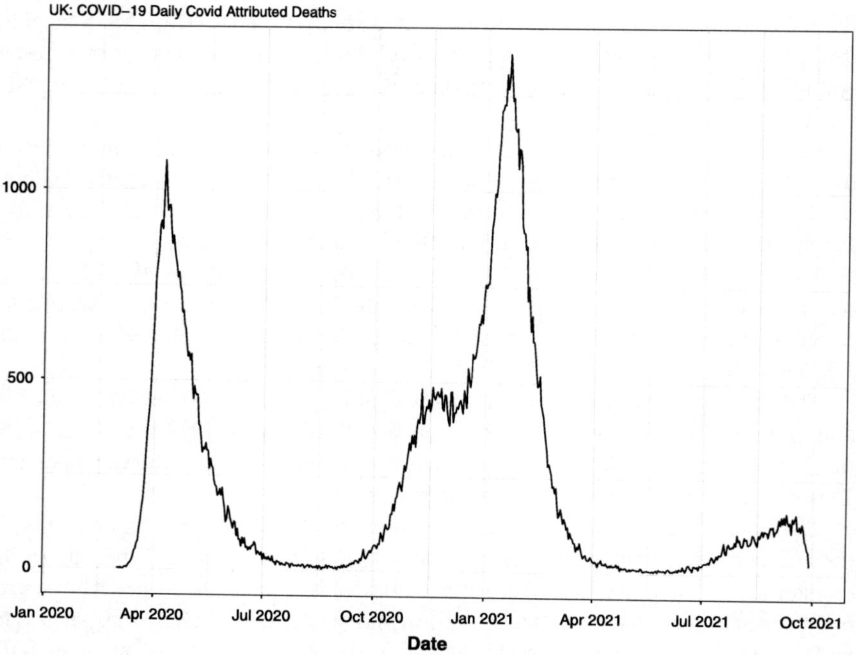

Fig. 2 Daily deaths attributed to COVID-19 in England and Wales according to the ONS from January 2020 to the end of September 2021. The period of Ferguson's first forecast ended in September 2020, a year earlier. The forecast of a second wave of deaths came at the end of August 2020

The abrupt drop at the end of this, and the other series below, is due to late counting. They are not part of the disease signal.

5 Not Cases

As mentioned in the beginning, governments just could not keep straight the difference between a positive test for the virus, and a genuine case. Any positive test, even asymptomatic, was instead taken as a "case". Since genuine cases need medical attention, it was thought more resources would be needed to handle the numbers given than were actually needed.

The UK "cases" are plotted in Fig. 3. These should be directly contrasted with deaths. The two series do not match up in relative size, because "cases" ignores the levels of testing. As mentioned, "cases" are not true cases. They are a function of testing number, test sensitivity and accuracy, and disease severity and prevalence. Testing number is a pure social phenomenon. The greater the panic, or the more

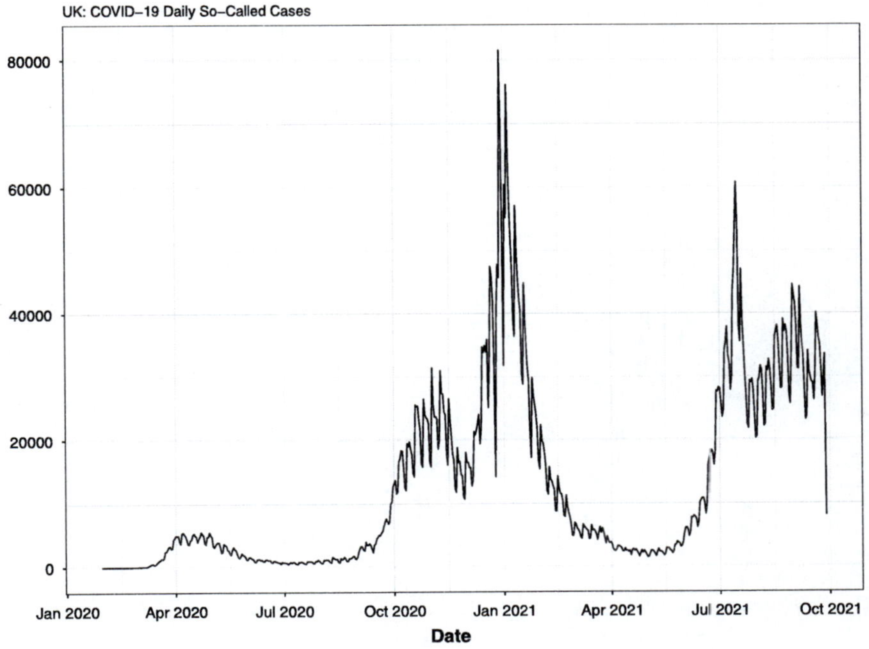

Fig. 3 The number of so-called cases, but in reality the number of coronavirus positive tests, according to the ONS

the panic becomes a typical part of daily life, the more testing. Therefore the more likely marginal, low-grade, and asymptomatic infections are discovered—and touted as "cases".

The first death peak was smaller than the second (really two peaks in one), but not by a great deal. Both the first and second death peaks were much larger than the third peaks, which is only a fraction of the first two.

Almost the opposites of this happened with "cases". The first peak is minuscule. The second (again, two in one) is rather large, but the third is enormous contrasted with the size of the deaths.

This is all seen easily in Fig. 4, which shows the daily deaths per "case". While it's true that the viruses mutate to less virulent forms, it's not likely to have done so this quickly, to the point at which deaths per cases collapses in such short order. The explanation is likely to lie elsewhere, as with testing.

According to the ONS, [21] testing was increasing by leaps and bounds, and with no signs of slowing as of this writing. Here in Fig. 5 are the weekly number of tests through 29 September 2021, the latest data available before this paper went to press. These are, we note, *daily tests*.

Testing was low during the initial outbreak, and so were the so-called cases. These were likely to be closer to true cases, and these people were tested after they became sick. This is why attributed deaths was quite large. Obviously, well people

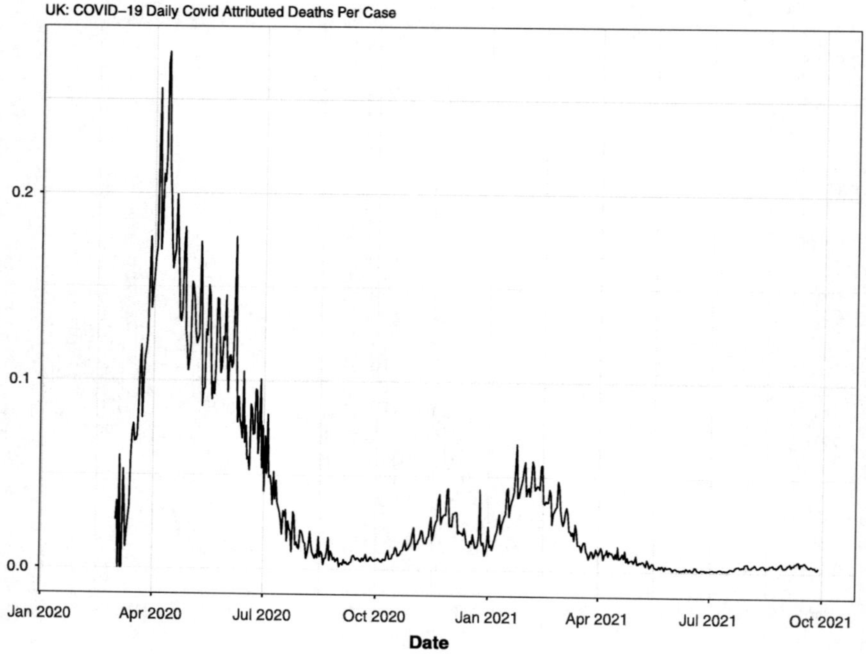

Fig. 4 The number of daily deaths per "case." Given the history of epidemics, the chance of the virus changing its characteristics this quickly is low. The explanation likely lies elsewhere, as with testing

were avoiding hospitals and testing in general. Testing then shot up, and has stayed consistently high ever since, with a new peak seen in the usual January peak in deaths from of all causes.

This means IFRs and CFRs estimates during all parts of the pandemic cannot be well estimated. Not using these official numbers, anyway. There is then some suspicion even the number of deaths said to be caused by this bug were measured incorrectly. This shows the destructive power of panic.

6 Conclusion

As mentioned in the Introduction, the final word on coronavirus predictions is not in, and will not be in for some time. It is already clear, however, that there were many failures in forecasting the pandemic. The Imperial College model, relied upon in the UK and, in part, in the USA and other countries, should never have been believed. It produced, right at the outset, unbelievable numbers, which in turn caused panic. Once embraced, the forecast could not be retreated from without admitting error.

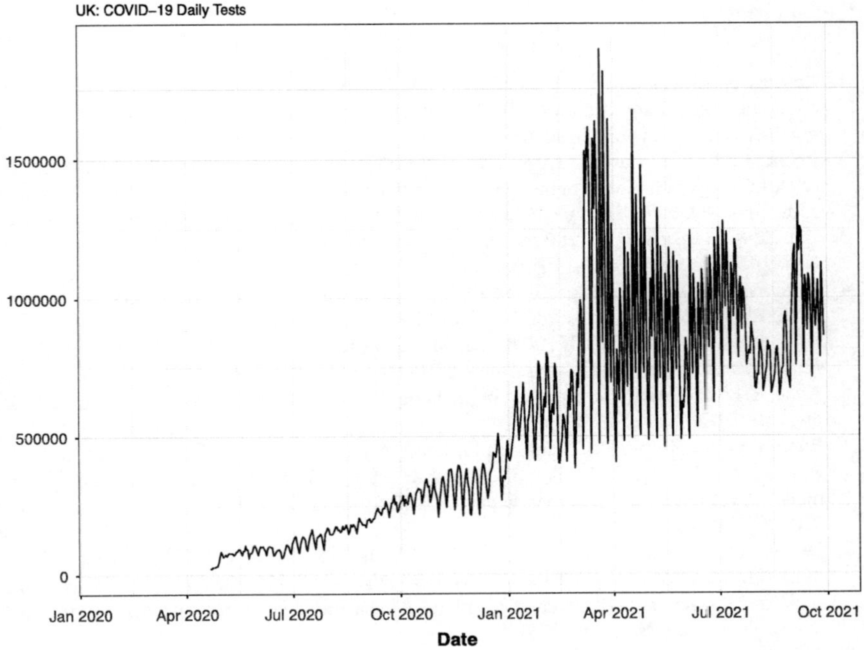

Fig. 5 The number of daily tests across all of the UK, according to the ONS. The date of last reporting in this dataset was 29 September 2021

These screwy forecast numbers, which were as discussed absurd, were never noticed initially, because, for one, an appeal to authority. The research team at IC was in the long habit of producing government forecasts for things like pandemics. As detailed in [13], the record of these predictions was dismal. They have a long habit of predicting consequences far more dire than turn out. That is, they had the same problem every time: over-forecasting.

Government should have noticed this repeated poor performance, and, in effect, sacked IC long before. Yet government appeared to enjoy hearing worst-case scenarios over more plausible ones. Perhaps because these forecasts implied what government wanted to hear. That its "solutions" were needed.

Another reason the poor forecast went unnoticed was the unusual format of the predictions. They were in deaths per 100,000 and not in actual death numbers. The deaths per capita did not seem especially unusual, even though they were thought concerning. When the prediction was converted to simple plain numbers, however, the wild excess is revealed. It is very strange this was never done at the time.

Predictions made after the initial one did better. But there was still a widespread confusion between actual cases (genuine illnesses) and positive tests. This caused in itself a continuation of the panic.

It wasn't only the UK that made these mistakes. Many countries did the same, including the USA.

References

1. Anastassopoulou, C., Russo, L., Tsakris, A., & Siettos, C. (2020). Data-based analysis, modelling and forecasting of the COVID-19 outbreak. *PLOS ONE, 15*(3), e0230405. https://doi.org/10.1371/journal.pone.0230405
2. Roosa, K., Lee, Y., Luo, R., Kirpich, A., Rothenberg, R., Hyman, J., Yan, P., & Chowell, G. (2020). Real-time forecasts of the COVID-19 epidemic in China from February 5th to February 24th, 2020. *Infectious Disease Modelling, 5*, 256. https://doi.org/10.1016/j.idm.2020.02.002
3. Dai, Z., & Locasale, J. W. (2020). Cooperative virus propagation in COVID-19 transmission. *medRxiv*. https://doi.org/10.1101/2020.05.05.20092361
4. Petropoulos, F., & Makridakis, S. (2020). Forecasting the novel coronavirus COVID-19. *PLoS ONE, 15*(3), e0231236. https://doi.org/10.1371/journal.pone.0231236
5. Perc, M., Gorišek Miksić, N., Slavinec, M., & Stožer, A. (2020). Forecasting COVID-19. *Frontiers in Physics, 8*, 127. https://doi.org/10.3389/fphy.2020.00127
6. Anne, R. (2020). ARIMA modelling of predicting COVID-19 infections. *medRxiv*. https://doi.org/10.1101/2020.04.18.20070631
7. Watson, G. L., Xiong, D., Zhang, L., Zoller, J. A., Shamshoian, J., Sundin, P., Bufford, T., Rimoin, A. W., Suchard, M. A., & Ramirez, C. M. (2020). Fusing a Bayesian case velocity model with random forest for predicting COVID-19 in the U.S. *medRxiv*. https://doi.org/10.1101/2020.05.15.20102608
8. Osi, A. A., Dikko, H. G., Abdu, M., Ibrahim, A., Isma'il, L. A., Sarki, H., Muhammad, U., Suleiman, A. A., Sani, S. S., & Ringim, M. Z. (2020). A classification approach for predicting COVID-19 patient's survival outcome with machine learning techniques. *medRxiv*. https://doi.org/10.1101/2020.08.02.20129767
9. Aldhyani, T. H., Alrasheed, M., Alqarn, A. A., Alzahrani, M. Y., & Alahmadi, A. H. (2020). Deep learning and Holt-trend algorithms for predicting COVID-19 pandemic. *medRxiv*. https://doi.org/10.1101/2020.06.03.20121590
10. Tsallis, C., & Tirnakli, U. (2020). Predicting COVID-19 peaks around the world. *Frontiers in Physics, 8*, 217. https://doi.org/10.3389/fphy.2020.00217
11. Roda, W. C., Varughese, M. B., Han, D., & Li, M. Y. (2020). Why is it difficult to accurately predict the COVID-19 epidemic? *Infectious Disease Modelling, 5*, 271. https://doi.org/10.1016/j.idm.2020.03.001
12. Wynants, L., Van Calster, B., Collins, G. S., Riley, R. D., Heinze, G., Schuit, E., Bonten, M. M. J., Dahly, D. L., Damen, J. A. A., Debray, T. P. A., de Jong, V. M. T., De Vos, M., Dhiman, P., Haller, M. C., Harhay, M. O., Henckaerts, L., Heus, P., Kreuzberger, N., Lohmann, A., ... van Smeden, M. (2020). Prediction models for diagnosis and prognosis of covid-19: Systematic review and critical appraisal. *BMJ, 369*. https://doi.org/10.1136/bmj.m1328
13. Axe, D., Briggs, W. M., & Richards, J. (2020). *The price of panic: How the tyranny of experts turned a pandemic into a catastrophe*. Regnery.
14. CDC, Center for Disease Control. (2020). https://www.cdc.gov/coronavirus/2019-ncov/hcp/planning-scenarios.html
15. Jewell, N. P., Lewnard, J. A., & Jewell, B. L. (2020). Predictive mathematical models of the COVID-19 pandemic: Underlying principles and value of projections. *JAMA, 323*(19), 1893. https://doi.org/10.1001/jama.2020.6585
16. Adam, D. (2020). Special report: The simulations driving the world's response to COVID-19. *Nature, 580*(7803), 316. https://doi.org/10.1038/d41586-020-01003-6
17. CDC, Center for Disease Control. (2019). https://www.cdc.gov/flu/pandemic-resources/1918-pandemic-h1n1.html#:~:text=It%20is%20estimated%20that%20about,occurring%20in%20the%20United%20States
18. CDC, Center for Disease Control. (2020). https://www.cdc.gov/nchs/nvss/vsrr/covid19/index.htm
19. Woodcock, A. (2020). *The Independent*. https://www.independent.co.uk/news/uk/politics/coronavirus-winter-second-wave-uk-lockdown-schools-reopen-sage-a9695116.html

20. ONS, Office of National Statistics. (2020). https://www.ons.gov.uk/peoplepopulationandcommunity/birthsdeathsandmarriages/deaths/datasets/weeklyprovisionalfiguresondeathsregisteredinenglandandwales
21. ONS, Office of National Statistics. (2020). https://coronavirus.data.gov.uk/

How to Efficiently Store Intermediate Results in Quantum Computing: Theoretical Explanation of the Current Algorithm

Oscar Galindo, Olga Kosheleva, and Vladik Kreinovich

Abstract In complex time-consuming computations, we rarely have uninterrupted access to a high performance computer: usually, in the process of computation, some interruptions happen, so we need to store intermediate results until computations resume. To decrease the probability of a mistake, it is often necessary to run several identical computations in parallel, in which case several identical intermediate results need to be stored. In particular, for quantum computing, we need to store several independent identical copies of the corresponding qubits—quantum versions of bits. Storing qubit states is not easy, but it is possible to compress the corresponding multi-qubit states: for example, it is possible to store the resulting 3-qubit state by using only two qubits. In principle, there are many different ways to store the state of 3 independent identical qubits by using two qubits. In this paper, we show that the current algorithm for such storage is uniquely determined by the natural symmetry requirements.

1 Formulation of the Problem

Quantum computing is inevitable. While modern computers are several orders of magnitude faster than in the past, for many practical problems, they are still too slow. For example, by using modern high performance computers, in principle, we can compute where a tornado will turn in the next 15 min, but even on the fastest computers, these computations will take several hours, long after the tornado has actually moved—and thus, too late to provide a warning.

O. Galindo · O. Kosheleva · V. Kreinovich (✉)
University of Texas at El Paso, El Paso, TX 79968, USA
e-mail: vladik@utep.edu

O. Galindo
e-mail: ogalindomo@miners.utep.edu

O. Kosheleva
e-mail: olgak@utep.edu

S. Sriboonchitta et al. (eds.), *Credible Asset Allocation, Optimal Transport Methods, and Related Topics*, Studies in Systems, Decision and Control 429,
https://doi.org/10.1007/978-3-030-97273-8_5

One of the main reasons why it is difficult to drastically speed up modern computers is the fact that, according to relativity theory, all speeds are limited by the speed of light. For a usual laptop whose size is about 30 cm, it takes 1 ns for light—the fastest possible process—to pass from one side of the laptop to another. During this time, even the cheapest 4 GHz central processing unit will already perform four different operations. To make computations much faster, we thus need to make all the elements of the computer much smaller. These elements are already comparable to the size of molecules. The only way to make them even smaller is to have elements the size of a few molecules. At such sizes, we need to take into account physical phenomena which are specific for the microworld, i.e., phenomena of quantum physics; see, e.g., [4, 11]. From this viewpoint, quantum computing—i.e., computing by using units that obey laws of quantum physics—is inevitable.

Quantum computing is desirable. At first, quantum effects were viewed by computer engineers as a nuisance: indeed, we want the computer to produce the same desired result every time we ask for the same computation, while in quantum physics, most outcomes are probabilistic—their outcome changes with repetition.

Good news is that computer scientists came up with a way to utilize quantum effects in such a way that we can actually compute several things with guarantee and even faster than by using traditional non-quantum algorithms; see, e.g., [8]. The most widely known quantum algorithm of this type is Shor's algorithm that enables us to factor large integers in polynomial time—and thus, in principle, decode all the messages sent by using the commonly used RSA encryption (since the security of this encryption scheme is based on the fact that the only known non-quantum algorithms for factoring integers would require astronomically large time to factor currently used 200-digit integers). Another well known quantum algorithm is Grover's algorithm for searching for an element in an un-sorted array of n elements: while non-quantum algorithm requires at least n steps (otherwise, it may miss the desired element), Grover's algorithm can find it much faster, in time $\sqrt{n}$.

It is important to mention that most quantum algorithms—including Shor's and Grover's—remain somewhat probabilistic, in the sense that while they produce the correct result with probability close to 1, there is a certain probability of a wrong result. To decrease this probability of the error, a natural idea is to repeat computations several times—either sequentially or, if we want to retain the same computation time, by running several quantum processors in parallel. If the probability that one processor errs is p_0, then the probability that all k parallel quantum processors err is p_0^k, i.e., much smaller.

Need to store intermediate computation results. As we have mentioned earlier, the main motivation for using quantum computing is to solve complex time-consuming problems whose computation requires a lot of time. It is very rare that for such problem, we have a dedicated computer that only solves this problem. Often, when such a time-consuming problem is being solved on a high performance computer, another higher-priority task appears, so the previous computation has to be interrupted, the intermediate computation results have to be temporarily stored—so that the computations can resume when the interrupting task is done.

Thus, for whatever algorithm we use, we need to take this need into account and instruct the computer how to store intermediate results. In particular, this is needed for quantum algorithms.

As we have mentioned in the previous subsection, to decrease the probability of an error, we need to repeat computations in parallel—thus, when an interrupt occurs, we need to store several copies of the same intermediate result.

What exactly do we store. The state of the usual (non-quantum) computer can be described as a sequence of bits, i.e., simple elements that can be only in two different states: 0 and 1. In quantum physics, for every two classical states, we can also have a *superposition* of these states, i.e., the state of the type $a_0|0\rangle + a_1|1\rangle$, where a_0 and a_1 are complex numbers known as *amplitudes* for which $|a_0|^2 + |a_1|^2 = 1$.

For the classical bit, we can measure its state and get 0 or 1. If we apply the same measurement process to the superposition, we get 0 with probability $|a_0|^2$ and 1 with probability $|a_1|^2$. The probabilities of two possible outcomes should add up to 1—this explains the above constraint on the possible pairs (a_0, a_1) of complex values.

The state of several independent particles can be described by using a so-called *tensor product* $\otimes$. Crudely speaking, it means the amplitude of each state of the 2-particle system is equal to the product of the corresponding amplitudes—just like the probability of having two outcomes in two independent events is equal to the product of the corresponding probabilities. In particular, if we have two identical particles in the state $a_0|0\rangle + a_1|1\rangle$, then the state of the corresponding 2-particle system has the form

$$a_0^2|00\rangle + a_0 \cdot a_1|01\rangle + a_1 \cdot a_0|10\rangle + a_1^2|11\rangle,$$

i.e., the state

$$a_0^2|00\rangle + a_0 \cdot a_1 \cdot (|01\rangle + |10\rangle) + a_1^2|11\rangle.$$

Here, the sum $|01\rangle + |10\rangle$ is not a state, since the sum of the squares of the coefficients is equal to 2. We can make it a state if we divide this sum by $\sqrt{2}$. In terms of this state, we get the following expression:

$$a_0^2|00\rangle + \sqrt{2} \cdot a_0 \cdot a_1 \cdot \left(\frac{1}{\sqrt{2}}|01\rangle + \frac{1}{\sqrt{2}}|10\rangle\right) + a_1^2|11\rangle.$$

If we have three identical particles, then we similarly get the state

$$a_0^3|000\rangle + a_0^2 \cdot a_1(|001\rangle + |010\rangle + |100\rangle) +$$

$$a_0 \cdot a_1^2(|011\rangle + |101\rangle + |110\rangle) + a_1^3|111\rangle.$$

To make each of the two sums a state, we can divide it by $\sqrt{3}$. Thus, we get the following expression:

$$a_0^3|000\rangle + \sqrt{3}\cdot a_0^2\cdot a_1\left(\frac{1}{\sqrt{3}}|001\rangle + \frac{1}{\sqrt{3}}|010\rangle + \frac{1}{\sqrt{3}}|100\rangle\right) +$$
$$\sqrt{3}\cdot a_0\cdot a_1^2\left(\frac{1}{\sqrt{3}}|011\rangle + \frac{1}{\sqrt{3}}|101\rangle + \frac{1}{\sqrt{3}}|110\rangle\right) + a_1^3|111\rangle. \quad (1)$$

Good news: we can store this 3-qubit state in 2 bits. The expression (1) is a linear combination of four different states. But every 2-qubit state is also a linear combination of four states, namely $|\hat{0}\hat{0}\rangle$, $|\hat{0}\hat{1}\rangle$ $|\hat{1}\hat{0}\rangle$, and $|\hat{1}\hat{1}\rangle$, where $\hat{0}$ and $\hat{1}$ denote 0 and 1 states of each qubit of the 2-qubit system. Thus, if we, e.g., perform a transformation T_0 that maps:

- the state $|000\rangle$ into

$$T_0(|000\rangle) = |\hat{0}\hat{0}\rangle, \quad (2)$$

- the state $\frac{1}{\sqrt{3}}|001\rangle + \frac{1}{\sqrt{3}}|010\rangle + \frac{1}{\sqrt{3}}|100\rangle$ into

$$T_0\left(\frac{1}{\sqrt{3}}|001\rangle + \frac{1}{\sqrt{3}}|010\rangle + \frac{1}{\sqrt{3}}|100\rangle\right) = |\hat{0}\hat{1}\rangle, \quad (3)$$

- the state $\frac{1}{\sqrt{3}}|011\rangle + \frac{1}{\sqrt{3}}|101\rangle + \frac{1}{\sqrt{3}}|110\rangle$ into

$$T_0\left(\frac{1}{\sqrt{3}}|011\rangle + \frac{1}{\sqrt{3}}|101\rangle + \frac{1}{\sqrt{3}}|110\rangle\right) = |\hat{1}\hat{0}\rangle, \quad (4)$$

and

- the state $|111\rangle$ into

$$T_0(|111\rangle) = |\hat{1}\hat{1}\rangle, \quad (5)$$

then the original 3-qubit state (1) gets transformed—without losing any information—into the following 2-qubit state:

$$a_0^3|\hat{0}\hat{0}\rangle + \sqrt{3}\cdot a_0^2\cdot a_1|\hat{0}\hat{1}\rangle + \sqrt{3}\cdot a_0\cdot a_1^2|\hat{1}\hat{0}\rangle + a_1^3|\hat{1}\hat{1}\rangle. \quad (6)$$

The more qubits we need to store, the more difficult is it, so, from the practical viewpoint, this decrease in number of qubits is a great advantage.

A similar decrease in the number of qubits is possible for any number k of identical qubits; see, e.g., [1–3, 5–7, 9, 10].

Natural question. In the original transformation T_0, we used the basic states $|\hat{0}\hat{0}\rangle$, $|\hat{0}\hat{1}\rangle$ $|\hat{1}\hat{0}\rangle$, and $|\hat{1}\hat{1}\rangle$. In principle, instead of these basic states, we can use any four states

$$a_{i,00}|\hat{0}\hat{0}\rangle + a_{i,01}|\hat{0}\hat{1}\rangle + a_{i,10}|\hat{1}\hat{0}\rangle + a_{i,11}|\hat{1}\hat{1}\rangle, i = 1, \ldots, 4,$$

as long as each of them is a valid state—in the sense that

$$|a_{i,00}|^2 + |a_{i,01}|^2 + |a_{i,10}|^2 + |a_{i,11}|^2 = 1$$

for all i, and that every two different states $i \neq j$ are *orthogonal* in the sense that

$$\sum_{a,b} a_{i,ab} \cdot a^*_{j,ab} = 0,$$

where a^* means a complex conjugate, i.e., $(x + \mathrm{y}i)^* \stackrel{\text{def}}{=} x - y\mathrm{i}$, where $\mathrm{i} \stackrel{\text{def}}{=} \sqrt{-1}$.

So why is the proposed scheme for 3-to-2-qubit compression based on the standard basis and not on any alternative?

What we do in this paper. In this paper, we show that the use of the standard basic can be uniquely determined by natural symmetry requirements.

2 What Are Natural Symmetries?

General (non-quantum) natural symmetries. Let us first consider natural symmetries which are motivated by the problem itself and have nothing to do with quantum physics.

Swapping the objects. A natural symmetry in a system consisting of several similar objects is the possibility to swap these objects.

In the original 3-qubit system, all three qubits are in the same state, so swapping these qubits does not change anything.

On the other hand, in the resulting 2-qubit state, the two qubits are, in general, in different states. Thus, it makes sense to *swap* these qubits. In terms of the corresponding states, this means that:

- we keep the states $|\hat{0}\hat{0}\rangle$ and $|\hat{1}\hat{1}\rangle$ and
- we swap the states $|\hat{0}\hat{1}\rangle$ and $|\hat{1}\hat{0}\rangle$.

In other words, the transformation takes the form

$$N_0(|\hat{0}\hat{0}\rangle) = |\hat{0}\hat{0}\rangle, N_0(|\hat{0}\hat{1}\rangle) = |\hat{1}\hat{0}\rangle, N_0(|\hat{1}\hat{0}\rangle) = |\hat{0}\hat{1}\rangle, N_0(|\hat{1}\hat{1}\rangle) = |\hat{1}\hat{1}\rangle.$$

Swapping the states. Another natural idea is to swap (rename) the states of each object.

In our case, we deal with binary states, i.e., physical systems that can be in two possible states. Which of these two states we identify with 0 and which with 1 is arbitrary.

From this viewpoint, not much should change if we simply swap these two states, i.e., rename 0 as 1 and 1 as 0. In the original 3-qubit system, all three qubits are in the same state, so if we change one, we have to change all the others: $|0\rangle \leftrightarrow |1\rangle$, i.e., we have the transformation n_1 for which:

$$n_1(|0\rangle) = |1\rangle, n_1(|1\rangle) = |0\rangle.$$

In the resulting 2-qubit state, the two qubits are, in general, in two different states. Thus, for the 2-qubit states, we have three options:

- we can swap the states $\hat{0}_1$ and $\hat{1}_1$ of the first qubit:

$$N_1(|\hat{0}_1\rangle) = |\hat{1}_1\rangle, N_1(|\hat{1}_1\rangle) = |\hat{0}_1\rangle, N_1(|\hat{0}_2\rangle) = |\hat{0}_2\rangle, N_1(|\hat{1}_2\rangle) = |\hat{1}_2\rangle;$$

so

$$N_1(|\hat{0}\hat{0}\rangle) = |\hat{1}\hat{0}\rangle, N_1(|\hat{0}\hat{1}\rangle) = |\hat{1}\hat{1}\rangle, N_1(|\hat{1}\hat{0}\rangle) = |\hat{0}\hat{0}\rangle, N_1(|\hat{1}\hat{1}\rangle) = |\hat{0}\hat{1}\rangle;$$

- we can also swap the states $\hat{0}_2$ and $\hat{1}_2$ of the second qubit:

$$N_2(|\hat{0}_1\rangle) = |\hat{0}_1\rangle, N_2(|\hat{1}_1\rangle) = |\hat{1}_1\rangle, N_2(|\hat{0}_2\rangle) = |\hat{1}_2\rangle, N_2(|\hat{1}_2\rangle) = |\hat{0}_2\rangle;$$

so

$$N_2(|\hat{0}\hat{0}\rangle) = |\hat{0}\hat{1}\rangle, N_2(|\hat{0}\hat{1}\rangle) = |\hat{0}\hat{0}\rangle, N_2(|\hat{1}\hat{0}\rangle) = |\hat{1}\hat{1}\rangle, N_2(|\hat{1}\hat{1}\rangle) = |\hat{1}\hat{0}\rangle.$$

We can also have a composition $N = N_1(N_2) = N_2(N_1)$ of these two symmetries, when we swap the states of both qubits:

$$N(|\hat{0}_1\rangle) = |\hat{1}_1\rangle, N(|\hat{1}_1\rangle) = |\hat{0}_1\rangle, N(|\hat{0}_2\rangle) = |\hat{1}_2\rangle, N(|\hat{1}_2\rangle) = |\hat{0}_2\rangle,$$

so

$$N(|\hat{0}\hat{0}\rangle) = |\hat{1}\hat{1}\rangle, N(|\hat{0}\hat{1}\rangle) = |\hat{1}\hat{0}\rangle, N(|\hat{1}\hat{0}\rangle) = |\hat{0}\hat{1}\rangle, N(|\hat{1}\hat{1}\rangle) = |\hat{0}\hat{0}\rangle.$$

Specific quantum symmetries. As we have mentioned, in the quantum case, all we observe are probabilities of different measurement results, and these probabilities are determined only by the absolute values of the amplitudes. Thus, if we multiply each state by a complex number whose absolute value is 1, we will not notice any difference.

In principle, there exist many complex numbers α for which $|\alpha| = 1$. However, all known quantum computing algorithms only use real-valued amplitudes. Because of this, in this paper, we will also restrict ourselves to real-valued amplitudes—and thus, to real-valued factors α. For each numbers, the only two numbers with absolute value 1 are numbers 1 and -1. Multiplying by 1 does not change anything, so the only non-trivial transformations that we should consider are multiplications by -1.

For the 3-qubit states, we have two options:

- we can replace the original 0-state $|0\rangle$ with $-|0\rangle$ and keep the state $|1\rangle$ unchanged:

$$n_2(|0\rangle) = -|0\rangle, n_2(|1\rangle) = |1\rangle;$$

- we can also replace the original 1-state $|1\rangle$ with $-|1\rangle$ and keep the state $|0\rangle$ unchanged:

$$n_3(|0\rangle) = |0\rangle, \quad n_3(|1\rangle) = -|1\rangle.$$

In addition to the transformations n_1, n_2, and n_3, we can also have compositions of these transformations.

For the first qubit of the resulting 2-qubit state, we have two choices:

- we can replace $|\hat{0}_1\rangle$ with $-|\hat{0}_1\rangle$:

$$N_3(|\hat{0}\hat{0}\rangle) = -|\hat{0}\hat{0}\rangle, \quad N_3(|\hat{0}\hat{1}\rangle) = -|\hat{0}\hat{1}\rangle, \quad N_3(|\hat{1}\hat{0}\rangle) = |\hat{1}\hat{0}\rangle, \quad N_3(|\hat{1}\hat{1}\rangle) = |\hat{1}\hat{1}\rangle;$$

- or we can replace $|\hat{1}_1\rangle$ with $-|\hat{1}_1\rangle$:

$$N_4(|\hat{0}\hat{0}\rangle) = |\hat{0}\hat{0}\rangle, \quad N_4(|\hat{0}\hat{1}\rangle) = |\hat{0}\hat{1}\rangle, \quad N_4(|\hat{1}\hat{0}\rangle) = -|\hat{1}\hat{0}\rangle, \quad N_4(|\hat{1}\hat{1}\rangle) = -|\hat{1}\hat{1}\rangle.$$

For the second qubit of the resulting 2-qubit state, we also have two choices:

- we can replace $|\hat{0}_2\rangle$ with $-|\hat{0}_2\rangle$:

$$N_5(|\hat{0}\hat{0}\rangle) = -|\hat{0}\hat{0}\rangle, \quad N_5(|\hat{0}\hat{1}\rangle) = |\hat{0}\hat{1}\rangle, \quad N_5(|\hat{1}\hat{0}\rangle) = -|\hat{1}\hat{0}\rangle, \quad N_5(|\hat{1}\hat{1}\rangle) = |\hat{1}\hat{1}\rangle;$$

- or we can replace $|\hat{1}_2\rangle$ with $-|\hat{1}_2\rangle$:

$$N_6(|\hat{0}\hat{0}\rangle) = |\hat{0}\hat{0}\rangle, \quad N_6(|\hat{0}\hat{1}\rangle) = -|\hat{0}\hat{1}\rangle, \quad N_6(|\hat{1}\hat{0}\rangle) = |\hat{1}\hat{0}\rangle, \quad N_6(|\hat{1}\hat{1}\rangle) = -|\hat{1}\hat{1}\rangle.$$

We can also combine the transformations $N_0 - N_6$.

Natural symmetries: summarizing. Based on the above analysis, there are three natural transformation n_i of the original qubits:

- $n_1(|0\rangle) = |1\rangle$ and $n_1(|1\rangle) = |0\rangle$;
- $n_2(|0\rangle) = -|0\rangle$ and $n_2(|1\rangle) = |1\rangle$;
- $n_3(|0\rangle) = |0\rangle$ and $n_3(|1\rangle) = -|1\rangle$;

and their compositions. For the resulting 2-qubit state, we have transformations N_0 through N_6 and their compositions.

3 Symmetries That Keep the Original Transformation Invariant

Original transformation: reminder. The original transformation T_0 is described by the formulas (2)–(5).

General idea of invariance and how it can be applied here. What does it mean that some dependencies are invariant? For example, the relation $A = s^2$ between the length s of the square's side and its area A is invariant with respect to changing the measuring unit for length (which is equivalent to replacing s with $\lambda \cdot s$), e.g., 2 m = $100 \cdot 2 = 200$ cm. In precise terms, it means that for each such transformation of length, we can find a similar transformation of areas for which the above formula remains true: in this case, this transformation is $A \to \lambda^2 \cdot A$. This notion of invariance is ubiquitous in physics; see, e.g., [4, 11].

Similarly, in our case, invariance would mean that for each of the following four natural transformation n_i of the original qubits, there exists a natural transformation N of the resulting 2-qubit state such that after applying both transformations, we get exactly the same formulas (2)–(5) for the transformation T_0: $N(T_0(n_i)) = T_0$.

What natural symmetry of the 2-qubit state corresponds to swaps. If we first apply the swap n_1 to the original qubits, and then apply the transformation T_0, then we get the following transformation $T_0(n_1)$:

$$T_0(n_1(|000\rangle)) = T_0(|111\rangle) = |\hat{1}\hat{1}\rangle;$$

$$T_0\left(n_1\left(\frac{1}{\sqrt{3}} \cdot (|001\rangle + |010\rangle + |100\rangle)\right)\right) =$$

$$T_0\left(\frac{1}{\sqrt{3}} \cdot (|110\rangle + |101\rangle + |011\rangle)\right) = |\hat{1}\hat{0}\rangle;$$

$$T_0\left(n_1\left(\frac{1}{\sqrt{3}} \cdot (|011\rangle + |101\rangle + |110\rangle)\right)\right) =$$

$$T_0\left(\frac{1}{\sqrt{3}} \cdot (|100\rangle + |010\rangle + |001\rangle)\right) = |\hat{0}\hat{1}\rangle;$$

$$T_0(n_1(|111\rangle)) = T_0(|000\rangle) = |\hat{0}\hat{0}\rangle.$$

One can easily see that to get back the original transformation T_0, it is sufficient to swap 0 and 1 states of both qubits, i.e., consider the transformation $N = N_1(N_2)$ for which

$$N(|00\rangle) = |11\rangle, \quad N(|01\rangle) = |10\rangle, \quad N(|10\rangle) = |01\rangle, \quad N(|11\rangle) = |00\rangle;$$

then, indeed,

$$N_1(N_2(T_0(n_1))) = T_0. \tag{7}$$

What natural symmetry of the 2-qubit space corresponds to changing the sign of the original 0 state. If we first apply the transformation n_2 and then T_0, we will get the following transformation $T_0(n_2)$:

$$T_0(n_2(|000\rangle)) = T_0(-|000\rangle) = -|\hat{0}\hat{0}\rangle;$$

$$T_0\left(n_2\left(\frac{1}{\sqrt{3}}\cdot(|001\rangle + |010\rangle + |100\rangle)\right)\right) =$$

$$T_0\left(\frac{1}{\sqrt{3}}\cdot(|001\rangle + |010\rangle + |100\rangle)\right) = |\hat{0}\hat{1}\rangle;$$

$$T_0\left(n_2\left(\frac{1}{\sqrt{3}}\cdot(|011\rangle + |101\rangle + |110\rangle)\right)\right) =$$

$$T_0\left(-\frac{1}{\sqrt{3}}\cdot(|011\rangle + |101\rangle + |110\rangle)\right) = -|\hat{1}\hat{0}\rangle;$$

$$T_0(n_2(|111\rangle)) = T_0(|111\rangle) = |\hat{1}\hat{1}\rangle.$$

One can see that to get back the original transformation T_0, it is sufficient to replace $|\hat{0}_2\rangle$ with $-|\hat{0}_2\rangle$, i.e., to apply the transformation N_5:

$$N_5(T_0(n_2)) = T_0. \tag{8}$$

What natural symmetry of the 2-qubit space corresponds to changing the sign of the original 1 state. If we first apply the transformation n_3 and then T_0, we will get the following transformation $T_0(n_3)$:

$$T_0(n_3(|000\rangle)) = T_0(|000\rangle) = |\hat{0}\hat{0}\rangle;$$

$$T_0\left(n_3\left(\frac{1}{\sqrt{3}}\cdot(|001\rangle + |010\rangle + |100\rangle)\right)\right) =$$

$$T_0\left(-\frac{1}{\sqrt{3}}\cdot(|001\rangle + |010\rangle + |100\rangle)\right) = -|\hat{0}\hat{1}\rangle;$$

$$T_0\left(n_3\left(\frac{1}{\sqrt{3}}\cdot(|011\rangle+|101\rangle+|110\rangle)\right)\right)=$$

$$T_0\left(\frac{1}{\sqrt{3}}\cdot(|011\rangle+|101\rangle+|110\rangle)\right)=|\hat{1}\hat{0}\rangle;$$

$$T_0(n_3(|111\rangle))=T_0(-|111\rangle=-|\hat{1}\hat{1}\rangle.$$

One can see that to get back the original transformation T_0, it is sufficient to replace $|\hat{1}_2\rangle$ with $-|\hat{1}_2\rangle$, i.e., to apply the transformation N_6:

$$N_6(T_0(n_3))=T_0. \tag{9}$$

4 Main Result: T_0 Is the Only Invariant Transformation

Formulation of our main result. We claim—and we will prove it—that the only real-valued transformation T that is similarly invariant, i.e., for which

$$N_1(N_2(T(n_1)))=T,\ \ N_5(T(n_2))=T,\ \text{ and } N_6(T(n_3))=T, \tag{10}$$

is, in effect, the transformation T_0—modulo rotations of the state of the first qubit of the 2-qubit output and modulo changing signs of some of the resulting four states.

Proof Let is consider a general real-valued transformation:

$$T(|000\rangle)=a_{1,00}|\hat{0}\hat{0}\rangle+a_{1,01}|\hat{0}\hat{1}\rangle+a_{1,10}|\hat{1}\hat{0}\rangle+a_{1,11}|\hat{1}\hat{1}\rangle; \tag{11}$$

$$T\left(\frac{1}{\sqrt{3}}\cdot(|001\rangle+|010\rangle+|100\rangle)\right)=$$
$$a_{2,00}|\hat{0}\hat{0}\rangle+a_{2,01}|\hat{0}\hat{1}\rangle+a_{2,10}|\hat{1}\hat{0}\rangle+a_{2,11}|\hat{1}\hat{1}\rangle; \tag{12}$$

$$T_0\left(\frac{1}{\sqrt{3}}\cdot(|011\rangle+|101\rangle+|110\rangle)\right)=$$
$$a_{3,00}|\hat{0}\hat{0}\rangle+a_{3,01}|\hat{0}\hat{1}\rangle+a_{3,10}|\hat{1}\hat{0}\rangle+a_{3,11}|\hat{1}\hat{1}\rangle; \tag{13}$$

$$T(|111\rangle)=a_{4,00}|\hat{0}\hat{0}\rangle+a_{4,01}|\hat{0}\hat{1}\rangle+a_{4,10}|\hat{1}\hat{0}\rangle+a_{4,11}|\hat{1}\hat{1}\rangle. \tag{14}$$

The condition that $N_5(T(n_2)) = T$ implies, in particular, that

$$N_5\left(T\left(n_2\left(\frac{1}{\sqrt{3}}\cdot(|001\rangle + |010\rangle + |100\rangle)\right)\right)\right) =$$
$$T\left(\frac{1}{\sqrt{3}}\cdot(|001\rangle + |010\rangle + |100\rangle)\right). \qquad (15)$$

The left-hand side of this equality is equal to

$$N_5\left(T\left(n_2\left(\frac{1}{\sqrt{3}}\cdot(|001\rangle + |010\rangle + |100\rangle)\right)\right)\right) =$$

$$N_5\left(T\left(\frac{1}{\sqrt{3}}\cdot(|001\rangle + |010\rangle + |100\rangle)\right)\right) =$$

$$N_5(a_{2,00}|\hat{0}\hat{0}\rangle + a_{2,01}|\hat{0}\hat{1}\rangle + a_{2,10}|\hat{1}\hat{0}\rangle + a_{2,11}|\hat{1}\hat{1}\rangle) =$$

$$-a_{2,00}|\hat{0}\hat{0}\rangle + a_{2,01}|\hat{0}\hat{1}\rangle - a_{2,10}|\hat{1}\hat{0}\rangle + a_{2,11}|\hat{1}\hat{1}\rangle.$$

Thus, due to (12), the equality (15) has the form

$$-a_{2,00}|\hat{0}\hat{0}\rangle + a_{2,01}|\hat{0}\hat{1}\rangle - a_{2,10}|\hat{1}\hat{0}\rangle + a_{2,11}|\hat{1}\hat{1}\rangle =$$

$$a_{2,00}|\hat{0}\hat{0}\rangle + a_{2,01}|\hat{0}\hat{1}\rangle + a_{2,10}|\hat{1}\hat{0}\rangle + a_{2,11}|\hat{1}\hat{1}\rangle.$$

Therefore, $a_{2,00} = a_{2,10} = 0$, and the expression (12) has a simplified form:

$$T\left(\frac{1}{\sqrt{3}}\cdot(|001\rangle + |010\rangle + |100\rangle)\right) =$$

$$a_{2,01}|\hat{0}\hat{1}\rangle + a_{2,11}|\hat{1}\hat{1}\rangle.$$

The right-hand side is a state, so we must have $a_{2,01}^2 + a_{2,11}^2 = 1$, thus there exists an angle α for which $a_{2,01} = \cos(\alpha)$ and $a_{2,11} = \sin(\alpha)$. In terms of this angle, we have

$$T\left(\frac{1}{\sqrt{3}}\cdot(|001\rangle + |010\rangle + |100\rangle)\right) = \cos(\alpha)|\hat{0}\hat{1}\rangle + \sin(\alpha)|\hat{1}\hat{1}\rangle =$$
$$(\cos(\alpha)|\hat{0}\rangle + \sin(\alpha)|\hat{1}\rangle) \otimes |1\rangle. \qquad (16)$$

Similarly, the condition that $N_6(T(n_3)) = T$ implies, in particular, that

$$N_6\left(T\left(n_3\left(\frac{1}{\sqrt{3}}\cdot(|001\rangle+|010\rangle+|100\rangle)\right)\right)\right)=$$
$$T\left(\frac{1}{\sqrt{3}}\cdot(|001\rangle+|010\rangle+|100\rangle)\right). \quad (17)$$

The left-hand side of this equality is equal to

$$N_6\left(T\left(n_3\left(\frac{1}{\sqrt{3}}\cdot(|001\rangle+|010\rangle+|100\rangle)\right)\right)\right)=$$
$$N_6\left(T\left(-\frac{1}{\sqrt{3}}\cdot(|001\rangle+|010\rangle+|100\rangle)\right)\right)=$$
$$N_6(-\cos(\alpha)|\hat{0}\hat{1}\rangle-\sin(\alpha)|\hat{1}\hat{1}\rangle)=$$
$$\cos(\alpha)|\hat{0}\hat{0}\rangle+\sin(\alpha)|\hat{1}\hat{1}\rangle.$$

Thus, due to (16), the equality (17) is always satisfied.
The condition $N_1(N_2(T(n_1)))=T$ then implies that

$$T\left(\frac{1}{\sqrt{3}}\cdot(|011\rangle+|101\rangle+|110\rangle)\right)=$$
$$(\sin(\alpha)|\hat{0}\rangle+\cos(\alpha)|\hat{1}\rangle)\otimes|1\rangle. \quad (18)$$

For the state $|000\rangle$, the condition that $N_5(T(n_2)))=T$ implies that

$$N_5(T(n_2(|000\rangle)))=T(|000\rangle). \quad (19)$$

Here,

$$N_5(T(n_2(|000\rangle)))=N_5(T(-|000\rangle))=$$
$$N_5(-a_{1,00}|\hat{0}\hat{0}\rangle-a_{1,01}|\hat{0}\hat{1}\rangle-a_{1,10}|\hat{1}\hat{0}\rangle-a_{1,11}|\hat{1}\hat{1}\rangle)=$$
$$a_{1,00}|\hat{0}\hat{0}\rangle-a_{1,01}|\hat{0}\hat{1}\rangle+a_{1,10}|\hat{1}\hat{0}\rangle-a_{1,11}|\hat{1}\hat{1}\rangle).$$

Thus, due to (11), the equality (19) takes the form

$$a_{1,00}|\hat{0}\hat{0}\rangle-a_{1,01}|\hat{0}\hat{1}\rangle+a_{1,10}|\hat{1}\hat{0}\rangle-a_{1,11}|\hat{1}\hat{1}\rangle=$$
$$a_{1,00}|\hat{0}\hat{0}\rangle+a_{1,01}|\hat{0}\hat{1}\rangle+a_{1,10}|\hat{1}\hat{0}\rangle+a_{1,11}|\hat{1}\hat{1}\rangle.$$

So, $a_{1,01}=a_{1,11}=0$, and the expression (11) has a simplified form

$$T(|000\rangle) = a_{1,00}|\hat{0}\hat{0}\rangle + a_{1,10}|\hat{1}\hat{0}\rangle.$$

Similar to the above, we can conclude that there exists an angle β for which $\cos(\beta) = a_{1,00}$ and $\sin(\beta) = a_{1,10}$, thus

$$T(|000\rangle) = \cos(\beta)|\hat{0}\hat{0}\rangle + \sin(\beta)|\hat{1}\hat{0}\rangle = (\cos(\beta)|\hat{0}\rangle + \sin(\beta)|\hat{1}\rangle) \otimes |\hat{0}\rangle. \quad (20)$$

The condition $N_1(N_2(T(n_1))) = T$ then implies that

$$T(|111\rangle) = (\sin(\beta)|\hat{0}\rangle + \cos(\beta)|\hat{1}\rangle) \otimes |\hat{1}\rangle. \quad (21)$$

The fact that the states (16) and (21) must be orthogonal means that

$$\cos(\alpha) \cdot \sin(\beta) + \sin(\alpha) \cdot \cos(\beta) = \sin(\alpha + \beta) = 0,$$

so the sum $\alpha + \beta$ is either equal to 0 or to π. If this sum is equal to 0, then $\beta = -\alpha$, $\sin(\beta) = -\sin(\alpha)$, $\cos(\beta) = \cos(\alpha)$, and the formulas (20) and (21) take the form

$$T(|000\rangle) = (\cos(\alpha)|\hat{0}\rangle - \sin(\alpha)|\hat{1}\rangle) \otimes |\hat{0}\rangle;$$

$$T(|111\rangle) = (-\sin(\alpha)|\hat{0}\rangle + \cos(\alpha)|\hat{1}\rangle) \otimes |\hat{1}\rangle.$$

Thus, for the rotated states

$$|\hat{0}'\rangle \stackrel{\text{def}}{=} \cos(\alpha)|\hat{0}\rangle - \sin(\alpha)|\hat{1}\rangle$$

and

$$|\hat{1}'\rangle \stackrel{\text{def}}{=} \cos(\alpha) \cdot |1\rangle + \sin(\alpha)|0\rangle,$$

the formulas (16), (18), (20), and (21) take the desired form

$$T_0(|000\rangle) = |\hat{0}'\hat{0}\rangle,$$

$$T_0\left(\frac{1}{\sqrt{3}}|001\rangle + \frac{1}{\sqrt{3}}|010\rangle + \frac{1}{\sqrt{3}}|100\rangle\right) = |\hat{0}'\hat{1}\rangle,$$

$$T_0\left(\frac{1}{\sqrt{3}}|011\rangle + \frac{1}{\sqrt{3}}|101\rangle + \frac{1}{\sqrt{3}}|110\rangle\right) = |\hat{1}'\hat{0}\rangle,$$

$$T_0(|111\rangle) = |\hat{1}'\hat{1}\rangle.$$

When $\alpha + \beta = -\pi$, we get a similar transformation, but this an additional need to change the sign of the resulting basis states.

The result has been proven.

Acknowledgments This work was supported in part by the US National Science Foundation grants 1623190 (A Model of Change for Preparing a New Generation for Professional Practice in Computer Science) and HRD-1242122 (Cyber-ShARE Center of Excellence). It was also supported by the program of the development of the Scientific-Educational Mathematical Center of Volga Federal District No. 075-02-2020-1478.

References

1. Bacon, D., Chuang, I. L., & Harrow, A. W. (2006). Efficient quantum circuits for Schur and Clebsch-Gordan transforms. *Physical Review Letters, 97*, Paper 170502.
2. Blume-Kohout, R. (2010). Optimal, reliable estimation of quantum states. *New Journal of Physics, 12*, Paper 043034.
3. Eisert, J., Felbinger, T., Papadopoulos, P., Plenio, M. B., & Wilkens, M. (2000). Classical information and distillable entanglement. *Physical Review Letters, 84*, Paper 1611.
4. Feynman, R., Leighton, R., & Sands, M. (2005). *The Feynman lectures on physics*. Addison Wesley.
5. Hayashi, M., & Matsumoto, K. (2002). Quantum universal variable-length source coding. *Physical Review A, 66*, Paper 022311.
6. Keyl, M., & Werner, R. F. (2001). Estimating the spectrum of a density operator. *Physical Review A, 64*, Paper 052311.
7. Marvian, I., & Spekkens, R. W. (2014). A generalization of Schur-Weyl duality with applications in quantum estimation. *Communications in Mathematical Physics, 331*, 431–475.
8. Nielsen, M. A., & Chuang, I. L. (2000). *Quantum computation and quantum information*. Cambridge University Press.
9. Plesch, M., & Bužek, V. (2010). Efficient compression of quantum information. *Physical Review A, 81*, Paper 032317.
10. Rozema, L. A., Mahler, D. H., Hayat, A., Turner, P. S., & Steinberg, A. M. (2014). Quantum data compression of a qubit ensemble. *Physical Review Letters, 113*, Paper 160504.
11. Thorne, K. S., & Blandford, R. D. (2017). *Modern classical physics: Optics, fluids, plasmas, elasticity, relativity, and statistical physics*. Princeton University Press.

Decompositions in Quantum Mechanics—An Overview

John Harding

Abstract There is current interest in using ideas from quantum mechanics in the study of economics. We give an overview of an approach to quantum mechanics rooted not necessarily in Hilbert space, but in the primitive mathematical idea of direct products. This approach includes the standard von Neumann Hilbert space approach. It provides a conceptually simpler understanding of issues from standard quantum mechanics, and offers possibilities beyond the standard Hilbert space formulation. These further possibilities may be of particular interest in consideration of economics where the aim is to exploit quantum principles rather than specific physical situations.

1 Introduction

The standard mathematical treatment of quantum mechanics took its modern form through the work of Dirac [4] and von Neumann [24]. It was von Neumann who fully expressed the theory in terms of Hilbert spaces. A considerable amount of sophisticated mathematics is put into the first steps of the mathematical formulation of the quantum theory of an electron.

In quantum computing, one is content with finite-dimensional Hilbert spaces $\mathbb{C}^n$. But even to consider position and momentum of a single electron, one requires an infinite-dimensional Hilbert space $\mathscr{H}$. This is a complex vector space with inner product $u \cdot v$, that gives a norm $\|u\| = \sqrt{u \cdot u}$, that in turn gives a metric $d(u, v) = \|u - v\|$ under which the vector space is a complete metric space. Observables, such as position and momentum, correspond to self-adjoint operators on $\mathscr{H}$; and states correspond to density operators on $\mathscr{H}$.

A number of substantial mathematical results are the backbone of the theory. The spectral theorem [17, 21] relates a self-adjoint operator to a family of projection operators of $\mathscr{H}$. Gleason's theorem [5, 6, 16, 20] relates density operators to probability

J. Harding (✉)
New Mexico State University, Las Cruces, NM 88003, USA
e-mail: jharding@nmsu.edu

S. Sriboonchitta et al. (eds.), *Credible Asset Allocation, Optimal Transport Methods, and Related Topics*, Studies in Systems, Decision and Control 429,
https://doi.org/10.1007/978-3-030-97273-8_6

measures on the collection $\mathscr{P}(\mathscr{H})$ of all projection operators. Wigner's theorem [23, 25] relates symmetries of the system to automorphisms of $\mathscr{P}(\mathscr{H})$, and Stone's theorem [17, 22, 23] relates the time evolutions of the system to these symmetries.

In this sea of mathematical terminology and results, the question we should not lose is "why?" Why should we attach a Hilbert space to a system, why are self-adjoint operators used for observables, and so forth. The most popular answer to these questions is is because it works; but this is not really an answer at all, it is an instruction to not ask the question.

Building on the idea of a "quantum logic" introduced by Birkhoff and von Neumann [2], Mackey [18, pp. 61–71] gave a simple set of six physically motivated axioms that showed that the collection $\mathscr{Q}$ of Yes/No questions that can be asked of a physical system form a type of structure known as an orthomodular poset (OMP). This is a structure $(\mathscr{Q}, \leq, ', 0, 1)$ comprised of a family of Boolean algebras "glued together", and allows one to reason about events as is done in a Boolean algebra, but allowing for the possibility that certain events might not have a conjunction or disjunction. Mackey's seventh axiom was completely unmotivated, and simply stated that the OMP of questions are the projections $\text{Proj}(\mathscr{H})$ of a Hilbert space $\mathscr{H}$.

In this note, we show how simple properties of direct products can be used to construct an OMP of questions from virtually any type of structure such as a set, or group, or topological space, in place of a Hilbert space. Using the premise of direct products as basic, we give operationally motivated constructions of observables and dynamics. When applied to a Hilbert space, this recreates standard quantum theory. In terms of motivation of the use of direct products—a Yes/No question splits the system into a product of two, one where the event occurs, one where the alternate occurs.

This note is a compendium of results obtained and presented in detail in a series of papers [7–15].

2 Direct Product Decompositions of Hilbert Spaces

We first consider the relationship between projections and products for a Hilbert space. A projection operator P has a closed subspace A as its range, and each closed subspace is the range of a unique projection. A fundamental property of closed subspaces is that each vector $w \in \mathscr{H}$ can be uniquely expressed as $w = u + v$ for some vectors $u \in A$ and $v \in A^{\perp}$, where $A^{\perp}$ is the set of vectors orthogonal to all vectors in A. Thus $\mathscr{H}$ is in bijective correspondence with the set of ordered pairs (u, v) where $u \in A$ and $v \in A^{\perp}$, hence with $A \times A^{\perp}$ (see Fig. 1). This is the idea behind the following result.

Theorem 1 *A closed subspace A of a Hilbert space $\mathscr{H}$ gives a direct product decomposition $\mathscr{H} \simeq A \times A^{\perp}$, and a direct product decomposition $\mathscr{H} \simeq \mathscr{H}_1 \times \mathscr{H}_2$ gives a closed subspace $A = \{(u, 0) : u \in \mathscr{H}_1\}$.*

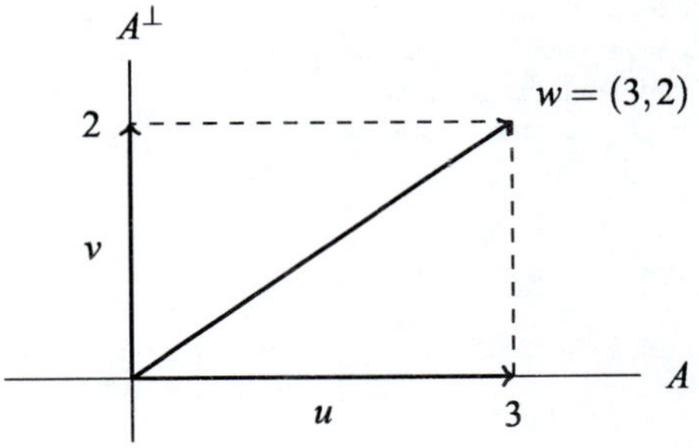

Fig. 1 A closed subspace A inducing a direct product decomposition

We are interested not just in the set of projections of $\mathscr{H}$, but also in its structure as an OMP.

Definition 1 An orthomodular poset (OMP) [18, 20, 23] is a structure $(\mathscr{Q}, \leq, ', 0, 1)$ that consists of a partially ordered set with least element 0, largest element 1, and an order reversing unary operation $' : \mathscr{Q} \to \mathscr{Q}$ of period two called orthocomplementation such that the following hold.

1. if $x \leq y'$, then x, y have a least upper bound that is denoted $x \oplus y$
2. $x \oplus x' = 1$
3. if $x \leq y$ then $x \oplus (x \oplus y')' = y$.

We say that x, y are orthogonal when $x \leq y'$.

Example 1 The closed subspaces of a Hilbert space $\mathscr{H}$ form an orthomodular poset Closed$(\mathscr{H})$. The partial ordering is set containment $\subseteq$, the orthocomplementation A' is given by the orthogonal subspace $A^{\perp}$, the least element 0 is the trivial subspace $\{0\}$, and the largest element 1 is the whole space. When $S \subseteq T^{\perp}$, the least upper bound of S and T is their span $S \oplus T$. It turns out that the closed subspaces are a lattice, that is, that any two closed subspaces have a least upper bound and greatest lower bound, but in physical considerations it is difficult to give meaning to these for arbitrary closed subspaces.

Closed subspaces of $\mathscr{H}$ correspond to direct product decompositions of $\mathscr{H}$. So we can consider the OMP structure in terms of direct product decompositions. The least closed subspace $\{0\}$ and largest closed subspace $\mathscr{H}$ correspond to the direct product decompositions

$$\mathscr{H} \simeq \{0\} \times \mathscr{H} \quad \text{and} \quad \mathscr{H} \simeq \mathscr{H} \times \{0\}$$

The orthocomplement of the decomposition $\mathscr{H} \simeq A \times A^{\perp}$ for A is given by the decomposition for $A^{\perp}$, which is $\mathscr{H} \simeq A^{\perp} \times A^{\perp\perp}$. Since $A^{\perp\perp} = A$, this is the decomposition $\mathscr{H} \simeq A^{\perp} \times A$. Orthocomplements of decompositions are very simple, we just switch the order of the factors.

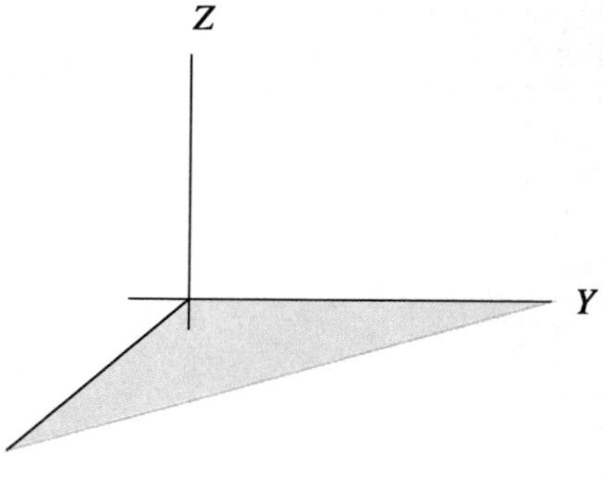

Fig. 2 A ternary decomposition given rise to a comparability between binary decompositions

The situation for the ordering $A \subseteq B$ of closed subspaces is the most interesting of all. Consider the example shown in Fig. 2 of $\mathscr{H}$ being $\mathbb{R}^3$ with X, Y, Z being the three coordinate axes. Let A be the closed subspace X consisting of the x-axis, and let B be the x, y-plane. Then indeed $A \subseteq B$. Note that $\mathscr{H} \simeq X \times Y \times Z$ is a ternary direct product decomposition. Since $A^\perp = Y \times Z$ and $B = X \times Y$, we have

$$\mathscr{H} \simeq A \times A^\perp \quad \text{is} \quad \mathscr{H} \simeq X \times (Y \times Z)$$

$$\mathscr{H} \simeq B \times B^\perp \quad \text{is} \quad \mathscr{H} \simeq (X \times Y) \times Z$$

The partial ordering $\leq$ of binary decompositions is given by all instances of the following that arise from a ternary decomposition $\mathscr{H} \simeq X \times Y \times Z$

$$\mathscr{H} \simeq X \times (Y \times Z) \quad \leq \quad \mathscr{H} \simeq (X \times Y) \times Z$$

That every instance of $A \subseteq B$ is captured by this follows from the third condition in Definition 1, known as the orthomodular law. If $A \subseteq B$, then $B = A \oplus C$ where $C = (A \oplus B^\perp)^\perp$. So $\mathscr{H} = A \times C \times B^\perp$ is the required ternary product decomposition.

3 Direct Product Decompositions in a General Setting

In the previous section, we showed that one can naturally define the structure of an OMP on the binary direct product decompositions $\mathscr{H} \simeq \mathscr{H}_1 \times \mathscr{H}_2$ of a Hilbert space so that the resulting structure is isomorphic to the OMP of closed subspaces Closed($\mathscr{H}$) that plays such a vital role in standard quantum theory. This can be extended to many types of structure. We begin by considering matters for a set X.

Definition 2 A n-ary direct product decomposition of a set X is an indexed family of sets $X_1, \ldots, X_n$ and a bijection $\alpha : X \to X_1 \times \cdots \times X_n$. This n-ary direct product

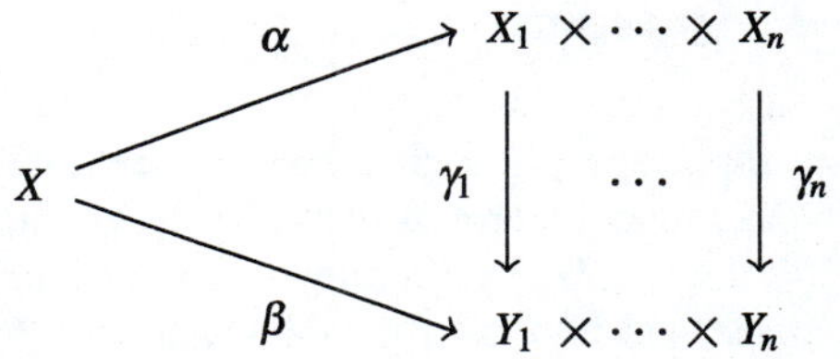

Fig. 3 Equivalence of n-ary decompositions

decomposition is equivalent to another $\beta : Y_1 \times \cdots \times Y_n$ if for $i = 1, \ldots, n$ there are bijections $\gamma_i : X_i \to Y_i$ so that $\beta = (\gamma_1 \times \cdots \times \gamma_n) \circ \alpha$. See Fig. 3.

Example 2 Let $X = \{a, b, c, d\}$ be a 4-element set. Up to equivalence, there is one binary decomposition of $X \simeq \{*\} \times X$ as the product of a 1-element set and a 4-element set X; one binary decomposition $X \simeq X \times \{*\}$ as the produce of a 4-element set and a 1-element set; and 6 binary decompositions $X \simeq \{p, q\} \times \{r, s\}$ as the product of two 2-element sets. The specific elements in the two 2-element sets are of no importance. In each of these six decompositions, each of a, b, c, d is represented as an ordered pair in $\{p, q\} \times \{r, s\}$. The decomposition is specified by stating which element has the same first component as a, and then which has the same second component as a.

Definition 3 Let X be a set. For an n-ary decomposition $\alpha : X \to X_1 \times \cdots \times X_n$, let $[\alpha : X \to X_1 \times \cdots \times X_n]$ be the class of all equivalent decompositions. Let $\mathscr{Q}(X)$ be the set of all equivalence classes of binary decompositions of X, which we call questions.

Definition 4 For a set X, put structure $(\mathscr{Q}(X), \leq, ', 0, 1)$ by setting

1. $0 = [X \simeq \{*\} \times X]$ and $1 = [X \simeq X \times \{*\}]$
2. $[X \simeq X_1 \times X_2]' = [X \simeq X_2 \times X_1]$
3. $[X \simeq X_1 \times (X_2 \times X_3)] \leq [X \simeq (X_1 \times X_2) \times X_3]$ for each $[X \simeq X_1 \times X_2 \times X_3]$

In this definition, we have suppressed the specific bijections, but they are the obvious ones. For $[\alpha : X \to X_1 \times X_2]$, the orthocomplement is $[\alpha' : X \to X_2 \times X_1]$ where $\alpha'(x) = (x_2, x_1)$ if $\alpha(x) = (x_1, x_2)$.

Theorem 2 *For a set X, the structure* $(\mathscr{Q}(X), \leq, ', 0, 1)$ *is an* OMP.

This theorem is established in [8] where it is generalized to other types of structure such as groups, rings, modules, G-sets, topological spaces, uniform spaces, normed groups, topological groups, and so forth. For a group G, an n-ary direct product decomposition consists of an indexed family $G_1, \ldots, G_n$ of groups and a group isomorphism $\alpha : G \to G_1 \times \cdots \times G_n$. The idea for other types of structures are similar, using appropriate notions of products and isomorphisms.

4 Questions

It was Mackey's original argument [18] that showed that the questions of a quantum system should form an OMP. Here we consider matters for the OMP of questions $\mathcal{Q}(X)$ for a set X, but our comments apply equally to the questions of group, topological space, and importantly, to the OMP $\mathcal{Q}(\mathcal{H})$ of questions of a Hilbert space $\mathcal{H}$ that is used so extensively in standard treatments of quantum mechanics.

Definition 5 An n-ary experiment of a physical system is an experiment that has n mutually exclusive and exhaustive outcomes labelled Outcome 1, ..., Outcome n. A question is a binary experiment. We usually use Yes for Outcome 1, and No for Outcome 2 of a question.

Remark 1 An n-ary experiment might consist of n detector bulbs placed in different areas with the guarantee that exactly one bulb will go off. If one wishes to push against this definition and say there will always be the possibility that more than one, or none, of the bulbs will flash, then view an experiment with n bulbs as a 2^n-ary experiment, with one outcome for each possible set of bulbs that can flash. One might also wish to consider experiments with infinitely many outcomes. We don't believe these are physical, and will treat them later as limits of families of finitary experiments.

The fundamental idea of the decompositions approach is that one associates to a physical system a mathematical object such as a set X, and that each n-ary experiment of the system corresponds to an equivalence class of n-ary direct product decompositions $[X \simeq X_1 \times \cdots \times X_n]$.

Definition 6 Let e be an n-ary experiment $[X \simeq X_1 \times \cdots \times X_n]$. Then for a sequence $\sigma = S_1, \ldots, S_k$ of pairwise disjoint sets with union $\{1, \ldots, n\}$ define $\sigma(e)$ to be the k-ary experiment $[X \simeq Y_1 \times \cdots \times Y_k]$ where $Y_j = \prod_{i \in S_j} X_i$.

We say that an experiment f is built from e if $f = \sigma(e)$ for some σ, and let $B(e)$ be the set of all binary experiments, that is, questions, that are built from e. For example, if e is the 4-ary experiment $e = [X \simeq X_1 \times X_2 \times X_3 \times X_4]$, then some of the questions built from e are the following.

$$(\{1,2\}, \{3,4\})e = [X \simeq (X_1 \times X_2) \times (X_3 \times X_4)]$$
$$(\{3\}, \{1,2,4\})e = [X \simeq X_3 \times (X_1 \times X_2 \times X_4)]$$

Note, that the use of the empty set is allowed. Since the product of the empty family is a singleton $\{*\}$, we have $(\emptyset, \{1, 2, 3, 4\})e = [X \simeq \{*\} \times X]$.

Definition 7 A subset B of an OMP P is a Boolean subalgebra if it is closed under orthocomplementation, closed under the join $x \oplus y$ of orthogonal elements, contains the bounds 0, 1, and with the inherited order forms a Boolean algebra under these operations.

If B is a Boolean subalgebra of P, then two elements x, y in B have a join $x \vee y$ and meet $x \wedge y$, and these are their join and meet as taken in B.

Theorem 3 *The finite Boolean subalgebras of the* OMP $\mathcal{Q}(X)$ *are exactly the sets* $B(e)$ *of questions built from some common experiment e. Further, if e has n factors that are not singletons* $\{*\}$*, then* $B(e)$ *has n atoms.*

In quantum mechanics, two questions are compatible if they lie in a Boolean subalgebra of the OMP of questions. Physically, this is interpreted to mean that they can be conducted at the same time. But conducting two binary questions e, f at the same time is the same as conducting one 4-ary experiment g whose outcomes are given by the pairs of outcomes Yes–Yes, Yes–No, No–Yes, and No–No to the questions e, f. This common sense physical reasoning is born in the mathematics.

Proposition 1 *Two questions e, f are compatible, i.e. lie in a Boolean subalgebra, if there is a unique 4-ary experiment* $g = [X \simeq X_1 \times X_2 \times X_3 \times X_4]$ *with*

$$e = (\{1, 2\}, \{3, 4\})g = [X \simeq (X_1 \times X_2) \times (X_3 \times X_4)]$$
$$f = (\{1, 3\}, \{2, 4\})g = [X \simeq (X_1 \times X_3) \times (X_2 \times X_4)]$$

In this case their meet and join are given by

$$e \wedge f = (\{1\}, \{2, 3, 4\})g = [X \simeq X_1 \times (X_2 \times X_3 \times X_4)]$$
$$e \vee f = (\{1, 2, 3\}, \{4\})g = [X \simeq (X_1 \times X_2 \times X_3) \times X_4]$$

Remark 2 There is similarity between Boolean operations on classical questions, which are modeled as subsets of an event space, and Boolean operations on compatible quantum questions viewed as binary direct product decompositions. Instead of taking the union and intersection of sets as in the classical case, we take the union and intersection of indexing sets of the factors of a direct product decomposition.

There is a further property of questions arising from decompositions that points to a physical property of quantum mechanics that is not obvious. Of course, this property applies also to the standard Hilbert space setting, but does not generally hold for questions modeled by an arbitrary OMP. In alternate terminology, the following result [9] shows that the OMP $\mathcal{Q}(X)$ is regular.

Theorem 4 *Let X be a set and S be a set of n questions in* $\mathcal{Q}(X)$*. If any two questions in S can be asked simultaneously, then all of the questions in S can be asked simultaneously. In fact, they can all be built from an experiment with* 2^n *outcomes. This holds also for a group, module, topological space, and so forth.*

5 States

A finite set of pairwise orthogonal elements in an OMP has a join. An OMP is called σ-complete if every countable set of pairwise orthogonal elements has a join [20].

Definition 8 A state on an OMP P is a map $s : P \to [0, 1]$ with

1. $s(0) = 0$ and $s(1) = 1$.
2. If x is orthogonal to y, then $s(x \oplus y) = s(x) + s(y)$.

A state on a σ-OMP is countably additive if the second condition applies to joins of countable pairwise orthogonal families.

Remark 3 Depending on one's perspective, what we call a state is called a finitely additive state, and a state refers to a countably additive state. We shall retain the idea of finite additivity as the primitive notion defining a state. The use of countable additivity is a means to deal with limiting processes of physical situations. As we will see, there are other means.

The idea of a state $s : P \to [0, 1]$ is that when restricted to a Boolean subalgebra B of P it yields a finitely additive probably measure on B. A countably additive state yields a σ-additive probability measure on any Boolean σ-subalgebra of P. The physical interpretation of a state is that for a question e we have

$$s(e) = \text{the probability that } e \text{ yields a Yes answer when in state } s$$

There are some issues [11] with the existence of a good supply of states in the decompositions approach. It seems that one must begin with a structure that has some contact with the real numbers, and to take decompositions that interact well with this structure. While this is an area that requires more investigation, there are several directions that show a wide scope beyond the setting of Hilbert spaces [10].

Definition 9 An η-set is a set X with a map $\eta : X \to [0, \infty)$ and element $0 \in X$ that satisfies $\eta(0) = 0$. A product of a family $(X_1, \eta_1, 0), \ldots, (X_n, \eta_n, 0)$ of η-sets is the set $X = X_1 \times \cdots \times X_n$ with $\eta(x_1, \ldots, x_n) = \sum_{i=1}^{n} \eta_i(x_i)$ and $0 = (0, \ldots, 0)$.

Theorem 5 *For an η-set X, the collection $\mathcal{Q}(X)$ of equivalence classes of it binary direct product decompositions is an* OMP.

The following shows that questions of an η-set have a large supply of states.

Theorem 6 *Let X be an η-set and $x \in X$ with $\eta(x) \neq 0$. There is a finitely additive additive state s_x on $\mathcal{Q}(X)$ such that for a question $e = [X \simeq X_1 \times X_2]$ with $x = (x_1, x_2)$*

$$s_x(e) = \frac{\eta_1(x_1)}{\eta(x)}$$

The standard Hilbert space approach is related to that of η-sets. Given a Hilbert space $\mathscr{H}$, define $\eta(x) = \|x\|^2$. Then questions given by closed subspaces satisfy $\|x\|^2 = \|x_1\|^2 + \|x_2\|^2$, a version of the standard Pythagorean theorem. So $\mathscr{Q}(\mathscr{H})$ is a sub-OMP of that given by considering $\mathscr{H}$ as an η-set. We next consider a situation intermediate to that of η-sets and that of Hilbert spaces.

Definition 10 A norm on a group G is a mapping $\|\cdot\| : G \to [0, \infty)$ with

1. $\|x\| = 0$ iff $x = 0$.
2. $\|x\| = \|-x\|$.
3. $\|x + y\| \le \|x\| + \|y\|$.

We have written the operations of G additively, but G need not be abelian.

A norm on a group induces a metric in the usual way. In fact, there are close connections between normed groups and metric and topological groups [1]. By a complete normed group, we mean one whose metric is complete. A question of a normed group is an equivalence class $[G \simeq G_1 \times G_2]$ that is simultaneously a group decomposition and an η-set decomposition with respect to the map $\eta(g) = \|g\|^2$. For details see [8].

Theorem 7 *For a normed group G, its questions form an* OMP $\mathscr{Q}(G)$ *and for each $g \in G$ with $\|g\| \neq 0$ there is a finitely additive state s_g with*

$$s_g(e) = \frac{\|g_1\|^2}{\|g\|^2}$$

If G is complete, then $\mathscr{Q}(G)$ is σ-complete and the states s_g are σ-additive.

6 Observables

To begin, suppose we have an n-ary experiment $e = [X \simeq X_1 \times \cdots \times X_n]$ for a set X, or some other structure such as a group, or η-set, and so forth. For $i \le n$ we have a question $e_i = [X \simeq X_i \times \prod_{j \neq i} X_j]$. Then for a state $s : \mathscr{Q}(X) \to [0, 1]$ we interpret

$$s(e_i) \;=\; \text{the probability of the } ith \text{ outcome of } e \text{ when in state } s$$

An n-ary experiment is intended to be a purely physical thing, such as detectors placed in various regions with exactly one guaranteed to go off. It is common to introduce a human element into matters as well by associating numerical values to various outcomes of an experiment. This choice of numerical values is a human activity, not a physical one. For instance, deflection through a Stern Gerlach device may have outcomes labelled as $-3/2$, $-1/2$, $1/2$, $3/2$. This may be for what we

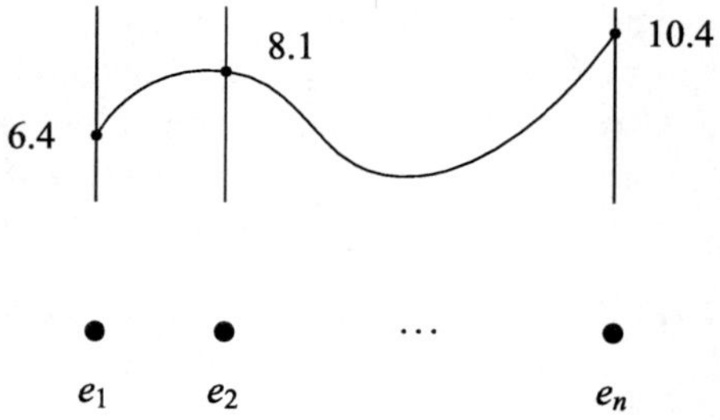

Fig. 4 A finite observable (B, f) where B has atoms $e_1, \ldots, e_n$ and the values of the scaling f on these atoms

consider to be a good reason, but it is still a matter of human choice in giving these numerical values of outcomes. In the following, we separate the physical aspects from the human chosen numerical ones.

Definition 11 A finite Boolean subalgebra of $\mathcal{Q}(X)$ is a finite observable quantity. If A is the set of atoms of a Boolean subalgebra of $\mathcal{Q}(X)$, then a map $f : A \to \mathbb{R}$ is a scaling of this finite observable quantity.

By Theorem 3, there is a correspondence between finite Boolean subalgebras of $\mathcal{Q}(X)$ with n atoms and n-ary experiments $e = [X \simeq X_1 \times \cdots X_n]$ with non-trivial factors. For such B, its atoms correspond to the questions $e_i = [X \simeq X_i \times \prod_{j \neq i} X_j]$. So a scaling of a finite observable quantity amounts to the same thing as an assignment of real numbers to the outcomes of an n-ary experiment.

Definition 12 A finite observable $O = (B, f)$ consists of a finite observable quantity B and a scaling f of B. The expected value of O in state s is $\sum_{i=1}^{n} s(e_i) f(e_i)$ (Fig. 4).

Remark 4 For Hilbert space quantum mechanics, a finite observable $O = (B, f)$ consists of a family of pairwise orthogonal projection operators $P_1, \ldots, P_n$ that sums to unity paired with a family $\lambda_1, \ldots, \lambda_n$ of real numbers. If different numerical values λ_i are given to different events P_i, then this information can be conveyed by the self-adjoint operator $A = \lambda_1 P_1 + \cdots \lambda_n P_n$. The eigenvalues and eigenspaces of A give back the projections and their scaling.

We take the view that finite observable quantities, which correspond to n-ary experiments, are the truly physical entities. We enrich these finite observable quantities with a scaling, but this process of attaching numerical values is a human process. There is a further abstraction of the purely physical to deal with limiting processes.

Definition 13 An observable quantity is a Boolean subalgebra of the OMP $\mathcal{Q}(X)$.

Definition 14 For an observable quantity B, a set $\mathcal{F}$ of questions in B is consistent any finite set of questions in $\mathcal{F}$ can possibly all have a Yes answer at the same time. In other words, $\mathcal{F}$ is consistent if the meet of any finite set of questions in $\mathcal{F}$ is non-zero. An ideal question is a maximally consistent set of questions in B.

The idea of an ideal question is simple, it is a type of limiting process of families of every finer collections of questions. Giving a physical example is challenging, not because of the idea of ideal questions, but because of a general lack of precision when we talk about physical measurements.

Example 3 Suppose we have a classical situation where a particle is located at some point along the real line. Any experiment we conduct might only tell us if the particle is within some given region. Following in this spirit, there should not be difference between testing for the region (2, 3) and the region [2, 3]. A plausible mathematical setup would be a test for each equivalence class, up to sets of measure zero, of regions that are finite unions of possibly infinite intervals with rational endpoints. So our Boolean algebra of questions is the Boolean subalgebra of the algebra of Lebesgue measurable sets modulo measure zero that is generated by the intervals $(-\infty, q)$ for a rational q. For a real number x, the family of questions consisting of equivalences classes of intervals (p, q) containing x is then consistent.

Those familiar with Stone duality for Boolean algebras [3] will recognize that ideal questions of B are the ultrafilters of B. Much is known about this topic. If we let Z be the set of all ideal questions (i.e. ultrafilters) of B, then there is a topology on Z that has as a basis all sets of the form

$$\phi(e) = \{\mathscr{F} : \mathscr{F} \text{is an ideal question and} e \in \mathscr{F}\}$$

With this topology, Z is a compact Hausdorff space, and the clopen subsets of this space are exactly the sets $\phi(e)$ for $e \in B$.

Definition 15 Let B be a Boolean algebra with Stone space Z. The Borel algebra of Z is the σ-algebra generated by its open sets, and the small Borel algebra of Z is the σ-algebra generated by the clopen sets of Z.

Definition 16 Let B be an observable quantity with Stone space Z. A scaling of B is an extended real valued function $f : Z \to \mathbb{R} \cup \{\pm\infty\}$ that is measurable with respect to the Borel algebra of the extended reals and the small Borel algebra of Z. An observable is a pair $O = (B, f)$ consisting of an observable quantity and a scaling (Fig. 5).

Proposition 2 *For a finitely additive state $s : B \to [0, 1]$ on B, there is a unique probability measure μ_s on the small Borel algebra of Z such that for each $e \in B$*

$$\mu_s(\Phi(e)) = s(e)$$

Thus $\mu_s(e)$ gives the probability of a Yes outcome to the question e when in state s.

Proof The collection F of subsets of Z of the form $\phi(e)$ for a question e is a field a sets. Members of F are clopen subsets of a compact space, so have the finite intersection property. So the map $\mu_s : F \to [0, 1]$ given by $\mu_s(\phi(e)) = s(e)$ is a

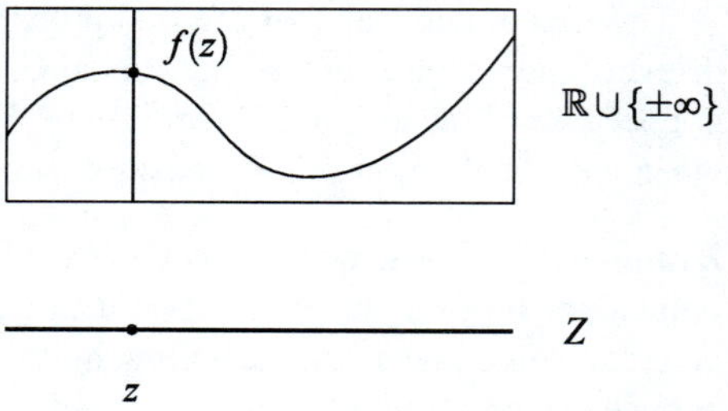

Fig. 5 An observable $O = (B, f)$ where B has Stone space Z and scaling $f : X \to \mathbb{R} \cup \{\pm\infty\}$

probability in the sense of [19, p. 10]. By [19, p. 23], this extends uniquely to a probability measure on the σ-algebra generated by F, which is the small Borel algebra of Z.

Definition 17 Let $O = (B, f)$ be an observable, let s be a finitely additive state on B, and let T be a Borel subset of the extended reals. Then we interpret

$$\mu_s(f^{-1}(T)) = \text{probability a measurement of } O \text{ lies in } T \text{ when in state } s$$

$$\int_Z f(z)\, d\mu_s = \text{expected value of } O \text{ when in state } s$$

Further, there is an obvious functional calculus of observables that are based on the same observable quantity B.

We relate this to the notion of an observable in standard Hilbert space quantum mechanics as a self-adjoint operator. First, we make an observation.

Proposition 3 *Let B be a complete Boolean algebra with Stone space Z and let $f : Z \to \mathbb{R} \cup \{\pm\infty\}$. If f is continuous, then it is measurable with respect to Borel algebra of the extended reals and the small Borel algebra of Z.*

Proof It is enough to show that for any real number λ, that $A = f^{-1}[-\infty, \lambda)$ is in the small Borel algebra of Z. For each n let $A_n = f^{-1}[-\infty, \lambda - 1/n)$. Then $A = \bigcup_n A_n$. Since f is continuous, each A_n is open. Also the closure $\overline{A_n}$ is contained in the closed set $f^{-1}[-\infty, \lambda - 1/n]$ which is contained in A. So $A = \bigcup_n \overline{A_n}$. Since Z is the Stone space of a complete Boolean algebra, the closure of an open set in Z is clopen. So each $\overline{A_n}$ belongs to the small Borel algebra, and hence so does A.

Example 4 To each self-adjoint operator A on a Hilbert space $\mathcal{H}$ there is a complete Boolean subalgebra B of the projection lattice and a σ-homomorphism E_A from the Borel algebra of the reals to B given by the spectral theorem [21]. This E_A is called the spectral measure of A. Let $f_A : Z \to \mathbb{R} \cup \{\pm\infty\}$ given by

$$f_A(z) = \inf\{\lambda : E_A(-\infty, \lambda] \in z\}$$

Then f_A is continuous. In fact [17] shows that the self-adjoint operators affiliated with the von Neumann algebra generated by B are in bijective correspondence with the continuous extended real valued functions on Z, and that this correspondence preserves functional calculus. Since f_A is continuous, by Proposition 3 $O = (B, f)$ is an observable in the sense defined above. A state s in the usual sense in Hilbert space quantum mechanics gives a σ-additive state on B, and as shown in [10], the descriptions of the probability that O yields a result in a Borel set T when in state s, and the description of the expected value of O when in state s, agree with the usual descriptions in terms of the self-adjoint operator A.

Remark 5 The usual Hilbert space treatment of observables via self-adjoint operators fits within our more operationally motivated treatment of observables. But even in the Hilbert space setting, the operationally motivated approach is more general. The standard Hilbert space treatment effectively chooses observables $O = (B, f)$ where B is a complete subalgebra of the projection lattice and f is not only small Borel measurable, but also continuous. The definition of a self-adjoint operator is quite technically complex, involving intricacies with domain, and it is not physically clear why they are the correct notion.

7 Dynamics

Direct product decompositions illuminate dynamics of a quantum system as well. In this section, we will present the main ideas, and leave many technical details for the reader to find in [14]. We will again frame matters for sets, but they apply equally to decompositions of other types of structures as well.

Definition 18 For a set X, let $\mathrm{Aut}(X)$ be the group of permutations of X. A one-parameter group of automorphisms of X is a group homomorphism $U : \mathbb{R} \to \mathrm{Aut}(X)$. We customarily write U_t for the automorphism $U(t)$.

The idea is that if the element $x \in X$ describes the system at some given time, then for any time $t \in \mathbb{R}$, then after passage of time t, the system will be depicted by the element $U_t(x)$. This provides dynamics for the system. The group homomorphism requirement $U_{s+t} = U_s \circ U_t$ says that letting t units of time pass, and then letting s units of time pass, is equivalent to letting $s + t$ units of time pass. The requirement that U_{-t} is the inverse of U_t amounts to time reversal.

Remark 6 Time reversal is a standard feature of Hilbert space quantum mechanics. Our approach can be adapted by considering the monoid $\mathrm{End}(X)$ of endomorphisms of X and a monoid homomorphism $V : \mathbb{R}^+ \to \mathrm{End}(X)$.

Definition 19 A set with natural frequencies is an algebra $A = (X, (E_t)_{\mathbb{R}})$ consisting of a set X and a family of unary operations on X that satisfy $E_{s+t} = E_s \circ E_t$, and E_0 is the identity. These imply that $E_{-t} = E_t^{-1}$.

The physical idea of a set with natural frequencies is that a physical system has some base "vibration" in a certain specified setting. Mathematically, a set with natural frequencies is simply an algebra in the sense of general algebra [3]. So there are obvious notions of homomorphisms and isomorphisms between sets with natural frequencies. There is also a notion of a finite direct product decomposition, which we spell out in the following.

Definition 20 Let $A = (X, (E_t)_{\mathbb{R}})$ be a set with natural frequencies. A finite direct product decomposition $A \simeq A_1 \times \cdots \times A_n$ is a family $A_i = (X_i, (E_t^i)_{\mathbb{R}})$ for $i = 1, \ldots, n$ of sets with natural frequencies and a bijection $\alpha : X \to X_1 \times \cdots \times X_n$ so that for each $x \in X$, if $\alpha(x) = (x_1, \ldots, x_n)$ then

$$\alpha(E_t x) = (E_t^1 x_1, \ldots, E_t^n x_n)$$

We will often be sloppy with the notation and simply write $E_t x = (E_t^1 x_1, \ldots, E_t^n x_n)$ in the above. All previous results about direct product decompositions of sets were valid for general algebras, and in particular apply to sets with natural frequencies. In particular, we have the following.

Theorem 8 *Let $A = (X, (E_t)_{\mathbb{R}})$ be a set with natural frequencies. Then the set $\mathcal{Q}(A)$ of equivalence classes of binary direct product decompositions of A is an* OMP *under the operations described previously.*

We can then use all other results developed for questions, such as notions of states and observables, and apply them to sets with natural frequencies. We next come to the central idea of this section.

Theorem 9 *Let A be a set with natural frequencies. Suppose $H = (B, f)$ is a finite observable of A where B has n atoms, the scaling f takes values $\lambda_1, \ldots, \lambda_n$ on these atoms, and the n-ary experiment whose questions yield B is given by*

$$h = [A \simeq A_1 \times \cdots \times A_n]$$

Then there is a one-parameter group of automorphisms of A, written E^H, given by

$$E_t^H(x) = (E_{\lambda_1 t}^1 x_1, \ldots, E_{\lambda_n t}^n x_n)$$

The observable H is called the Hamiltonian and represents the energy of the system. The idea is that the factors of the system "vibrate" at a speed proportional to their energy. We require that the dynamics of the system, as given by the one-parameter group U_t is given in this fashion by the energy, a statement written as an equation called the time-independent Schrödinger equation

$$U_t = E_t^H$$

Example 5 Consider the standard Hilbert space setting. Each Hilbert space can be given a natural frequency E_t by setting

$$E_t(v) = e^{-it}v$$

Hilbert space direct product decompositions are given by closed subspaces, and these all respect natural frequencies. So the decompositions of $\mathscr{H}$ as considered being equipped with natural frequencies or not are the same. The standard treatment of time-independent dynamics takes a self-adjoint operator for the energy, called the Hamiltonian, and then gives dynamics by the time-independent Schrödinger equation for Hilbert space quantum mechanics

$$U_t = e^{-iHt}$$

When applied to a self-adjoint operator $H = \lambda_1 P_1 + \cdots + \lambda_n P_n$ that is a finite weighted sum of projectors, this yields exactly the prescription described above because a vector v_i in the range of P_i evolves to $e^{-i\lambda_i t}v_i$. For a Hamiltonian H given by a general unbounded self-adjoint operator, one can show [14] that there is a sequence H_n of finite observables so that for any $v \in \mathscr{H}$ and time t we have

$$E_t^{H_n}(v) \longrightarrow e^{-iHt}\,v$$

in the topology of $\mathscr{H}$.

References

1. Bingham, N. H., & Ostaszewski, A. J. (2010). Normed versus topological groups: Dichotomy and duality. *Dissertationes Mathematicae (Rozprawy Matematyczne), 472*, 138.
2. Birkhoff, G., & von Neumann, J. (1936). The logic of quantum mechanics. *Annals of Mathematics, 37*(4), 823–843.
3. Burris, S., & Sankappanavar, H. P. (1981). A course in universal algebra. In *Graduate texts in mathematics* (Vol. 78). Springer.
4. Dirac, P. A. M. (1958). *The principles of quantum mechanics* (4th ed.). Oxford University Press.
5. Dvurecenskij, A. (1993). *Gleason's theorem and its applications*. Mathematics and Its Applications (East European Series) (Vol. 60). Kluwer Academic Publishers Group, Ister Science Press.
6. Gleason, A. (1957). Measures on the closed subspaces of a Hilbert space. *Journal of Mathematics and Mechanics, Indiana University Mathematics Journal, 6*(4), 885–893.
7. Hannan, T., & Harding, J. (2016). Automorphisms of decompositions. *Mathematica Slovaca, 66*(2), 493–526.
8. Harding, J. (1996). Decompositions in quantum logic. *Transactions of the American Mathematical Society, 348*(5), 1839–1862.
9. Harding, J. (1998). Regularity in quantum logic. *International Journal of Theoretical Physics, 37*(4), 1173–1212.
10. Harding, J. (1999). Axioms of an experimental system. *International Journal of Theoretical Physics, 38*(6), 1643–1675.
11. Harding, J. (2001). States on orthomodular posets of decompositions. *International Journal of Theoretical Physics, 40*(6), 1061–1069.

12. Harding, J. (2006). Orthomodularity of decompositions in a categorical setting. *International Journal of Theoretical Physics, 45*(6), 1117–1127.
13. Harding, J. (2009). A link between quantum logic and categorical quantum mechanics. *International Journal of Theoretical Physics, 48*(3), 769–802.
14. Harding, J. (2017). Dynamics in the decompositions approach to quantum mechanics. *International Journal of Theoretical Physics, 56*(12), 3971–3990.
15. Harding, J. (2018). Wigner's theorem for an infinite set. *Mathematica Slovaca, 68*(5), 1173–1222.
16. Kalmbach, G. (1983). *Orthomodular lattices*. London Mathematical Society Monographs (Vol. 18). Academic Press, Inc.
17. Kadison, R. V., & Ringrose, J. R. (1983). *Fundamentals of the theory of operator algebras. Vol. 1: Elementary theory*. Academic Press.
18. Mackey, G. W. (1963). *The mathematical foundations of quantum mechanics*. A lecture-note volume by W. A. Benjamin, Inc.
19. Neveu, J. (1965). *Mathematical foundations of the calculus of probability*. Holden-Day Inc.
20. Pták, P., & Pulmannová, S. (1991). *Orthomodular structures as quantum logics*. Fundamental Theories of Physics (Vol. 44). Kluwer Academic Publishers Group.
21. Prugovecki, E. (1981). *Quantum mechanics in Hilbert space* (2nd ed.). Academic Press.
22. Stone, M. (1932). On one-parameter unitary groups in Hilbert space. *Annals of Mathematics, 33*(3), 643–648.
23. Varadarajan, V. S. (1985). *Geometry of quantum theory* (2nd ed.). Springer-Verlag.
24. von Neumann, J. (1955). *Mathematical foundations of quantum mechanics*. Princeton University Press.
25. Wigner, E. P. (1959). *Group theory and its applications to the quantum mechanics of the atomic spectra*. Academic Press.

A First Look at Quantum Conditional Events for Economics

John Harding and Hung T. Nguyen

Abstract In view of current interests in borrowing quantum probability notion in quantum mechanics to model uncertainty in economic activities and models, consistent with behavioral economics, we address in this paper an important, but missing, ingredient in *quantum probability logic* for reasoning and inference, namely quantum implication operator as a bona fide mathematical entity, *consistent with quantum conditional probability*, together with associated logical connectives. While our focus is on economics, our approach can be used to develop a similar quantum implication for quantum computing which is based on physics.

Keywords Conditional events · Modus tollens · Probability logic · Quantum conditionals · Quantum implication · Quantum logic · Quantum probability

1 Introduction

In a sense, this paper addresses precisely an "open" research problem mentioned recently in Nguyen [13], namely "to explore the possibility to investigate conditional quantum logic, expending classical logics". We will use the theory of conditional events in probability logic, established in Goodman-Nguyen-Walker [6] (see also [5]) to address the needed problem of having a quantum implication operation, not only for quantum computing and physics, but also for the current interests in borrowing the noncommutative theory of quantum probability in physics to better describe human decision-making and economic models. For an overview of actual interests

J. Harding · H. T. Nguyen (✉)
New Mexico State University, Las Cruces, NM 88003, USA
e-mail: hunguyen@nmsu.edu

J. Harding
e-mail: hardingj@nmsu.edu

H. T. Nguyen
Faculty of Economics, Chiang Mai University, Chiang Mai 50200, Thailand

S. Sriboonchitta et al. (eds.), *Credible Asset Allocation, Optimal Transport Methods, and Related Topics*, Studies in Systems, Decision and Control 429,
https://doi.org/10.1007/978-3-030-97273-8_7

in various fields, following Goodman-Nguyen-Walker's work, see e.g., Milne [12], Nguyen [13].

Back in 1970, Stalnaker [15] proposed the "thesis" that the truth (probability) value of an implication connective, called a conditional "if b then a", should be the conditional probability $P(a|b)$, of course when $P(b) \neq 0$. It was a "thesis" since it makes more sense, in probability logic, to assign conditional probability to an implication of the form "If b then a". But then does that mean $P(b' \vee a) = P(a|b)$? In 1976, Lewis [11] showed that "Probabilities of conditionals are not conditional probabilities" except few trivial cases, where a "conditional" means a material implication.

In other words, conditionals as material implications are not compatible with conditional probabilities: $P(b \to a) \neq P(a|b)$. As a result, if we insist that an implication $b \Longrightarrow a$ should have the truth value as $P(a|b)$, we must define $b \Longrightarrow a$ as an other (bona fide) mathematical entity than $b' \vee a \in \mathcal{A}$ (the Boolean algebra in a probability space $(\Omega, \mathcal{A}, P)$). Lewis's result implied that there is no such an element of $\mathcal{A}$ which can represent $b \Longrightarrow a$ so that $P(b \Longrightarrow a) = P(a|b)$.

So what can be a mathematical entity which can represent $b \Longrightarrow a$, compatible with conditional probability? Clearly, this is not a question that probabilists and statisticians like to ask! This is so because for them, $P(a|b)$ is to represent not the probability of a mathematical object, denoted as $(a|b)$, but is the the probability of a when we change the probability P to the conditional probability measure $P_b : \mathcal{A} \to [0, 1]$, $P_b(a) = \frac{P(a \wedge b)}{P(b)}$, i.e., they never have to consider $P(a|b) = P((a|b))$, with an object $(a|b)$ as an argument of P.

In the 1970's, ruled-based and expert systems took a central stage in engineering research in which rules are of the form "If...Then...", under uncertainty, and it is necessary to be able to combine rules, such a "If a then b" and "If c then d", with probabilistic truth values, of course. Specifically, what is the probability of ["If a then b" and "If c then d"]?

Clearly, probability theory does not have the answer to it since, say, a "conditional event" of the form "If a then b" is even not defined in probability theory, let alone operations between them. Note that, for $b = \Omega$, then of course, $(a|\Omega)$ can be identified with $a \in \mathcal{A}$.

Now, by Lewis' triviality result, it is impossible to search for an element of $\mathcal{A}$ to represent $b \Longrightarrow a$, i.e., $b \Longrightarrow a$ is not a "connective" (a map from $\mathcal{A} \times \mathcal{A} \to \mathcal{A}$). Thus, if there is a bona fide mathematical entity $(a|b)$ such that $P((a|b)) = P(a|b)$, it must be outside of $\mathcal{A}$ (just like we consider complex numbers outside of the real line), i.e., some structure containing $\mathcal{A}$ as a subset. Note right away that if we suceed to do so, we are facing an extended logic: while the set of truth values remains to be $[0, 1]$ the underlying structure $\mathcal{A}$ is changed to another algebra.

In 1988, Goodman and Nguyen [4] proposed a "solution". The implcation $b \Longrightarrow a$ is defined as a "conditional event" which later represented in a simpler form by Walker [17].

It is amazing that, without knowing Goodman-Nguyen's work, in 1989, Redei [14] wrote a paper, just like Stanaker, but for quantum probability logic, with title " Quantum conditional probabilities are not probabilities of quantum conditionals", where quantum conditional means Sasaki hook, and conditional probability is given by Luhder's rule. His result is no surprize since it is exactly the same pattern of classical probability logic: The Sasaki hook is a connective, i.e., an element of the orthmodular lattice L in quantum logic. As we will see, in this paper, a solution for quantum conditional comes simply from the counterpart in probability logic, because, after all, remember quantum probability logic is just a generalization of probability logic (from a Boolean lattice to an orthomodular lattice).

The conclusion of Redei's paper is this: "The logic of quantum statistical inference cannot be described via duality by any reasonable conditional defined in quantum logic", and an open question "How to express the logic of quantum statistical inference in terms of logic of the underlying quantum logic?".

The contribution of this paper is the answer to this question. This paper is structured as follows. In Sect. 2, we review probability logic. In Sect. 3, we review the theory of conditional events in conditional probability logic. In Sect. 4 we review the essentials of quantum logic. Finally, in Sect. 5, we present a solution to the problem of quantum implication consistent with quantum probability.

2 Probability Logic

A logic is a framework for reasoning about a specific set of items, such as propositions in a natural language. Thus, for example, a fuzzy logic is a logic about fuzzy propositions (and not a logic which is fuzzy!). In everyday or in mathematics, we are just concerning with propositions which are either true (T) of false (F) (and nothing else like in fuzzy logic where there are partial membership degrees taken as partial truth values). Thus, by logic, we mean Aristotle's two-valued logic (or classical logic), later formulated in the framework of Boolean algebra by Boole [2]. Besides basic logical connectives $\wedge$, $\vee$, and $(.)'$ in a Boolean algebra $\mathcal{A}$ (say, of subsets of some set Ω), to conduct deductive reasoning, we need rules of deduction (such as modus ponens, modus tollens) which are based upon some implication connective to be established.

The implication connective needed is called a "conditional" to represent the statement "b implies a", or "if b then a", for $a, b \in \mathcal{A}$, and is denoted as $b \Longrightarrow a$. Being a connective means that the "object"$b \Longrightarrow a$ is an element of $\mathcal{A}$. Now by examining the possible truth values of a, b, we can figure out the truth table of $b \Longrightarrow a$, which happens to coincide with the truth table of the element $b' \vee a \in \mathcal{A}$, so that the "conditional" $(b \Longrightarrow a) = b' \vee a$, and is called the "material implication" in classical, two-valued logic.

Using the material implication we conduct reasoning in life or in mathematics (such as establishing mathematical proofs, by induction, and by contradiction, but not by intimidation!). For example, the modus tollens holds in classical logic: If we have $b \Longrightarrow a$ and a'(true), then we can conclude that b' (is true). This is so because we have

$$(b \Longrightarrow a) \wedge a' = (b' \vee a) \wedge a' = a' \wedge b'$$

so that if $t(b \Longrightarrow a) = T$, and $t(a') = T$, then $t(a' \wedge b') = T$, implying that $t(b') = T$ since $t(a') = T$.

In fact, the validity of Modus Tollens in classical logic follows simply by the fact that $(b \Longrightarrow a)$ is equivalent to $a' \Longrightarrow b'$, since

$$a' \Longrightarrow b' = (a')' \vee b' = a \vee b' = b' \vee a$$

Now, reasoning under uncertainty requires a different kind of logic. When uncertainty is modeled as a (classical) probability space $(\Omega, \mathcal{A}, P)$, we are talking about probability logic (see e.g., Hailperin [4]) which is a multivalued logic, but the structure $\mathcal{A}$ remains the same as in classical logic. The truth value of an event $a \in \mathcal{A}$ is its probability $P(a) \in [0, 1]$ (under a given P). Of course, in view of the properties of P, the probability logic is no longer truth functional.

Let's see if the modus tollens is valid (or not) in probability logic, in the sense of classical logic, i.e., if $P(b \Longrightarrow a) = P(b' \vee a) = 1$, and $P(a) = 0$, whether $P(b) = 0$?.

We have

$$b' \vee a = b' \vee (a \wedge (b \vee b')) = b' \vee [(a \wedge b) \vee (a \wedge b')] =$$

$$[b' \vee (a \wedge b)] \vee [b' \wedge a] = b' \vee (a \wedge b)$$

with $b' \wedge (a \wedge b) = \varnothing$.

i.e.,

$$P(b' \vee a) = P(b') + P(a \wedge b)$$

Thus, when $P(a) = 0$, we have $P(a \wedge b) = 0$, so that $P(b') = P(b' \vee a) = 1$, implying that $P(b) = 0$. Yes, the modus tollens holds in probability logic in the classical sense *when we use the material implication (of classical logic) to represent "If b then a"*.

Several remarks are in order here. First, as mentioned in Hailperin [8], that G. Boole made a mistake (In Example 7, p. 284, of his book [2]): "Unfortunately he confuses the probability of a conditional (probability of:if b then a) with conditional probability (probability of a given b) and solves not the stated problem but one involving conditional probabilities". As we will see in the next section on conditional probability logic, Boole's mistake is a *good mistake* (remember the Last Fermat's Theorem?).

Secondly, and this is also interesting (!), in the recent "in defense" of the "logic" underlying the popular approach to statistical hypothesis testing using p-values (Remember that hypothesis testing is formulated as a conditional statement), the "argument" is this. To "soften" the Modus Tollens (of classical logic) in the face of probability uncertainty, in the hope of getting a "statistical proof", Holmes [10] stated: "Statisticians are willing to pay some chance of error to extract knowledge" using induction as follows. Given $(b \Longrightarrow a)$, then the existence of a small ϵ such that $P(a) < \epsilon$ tells us that b is probably not true.This translates into an inference which suggests that if we observe data X, which is very unlikely if b is true(written $P(X|b) < \epsilon$), then b is not "plausible".

Don't we see the "confusion" again between "the probability of a (material) conditional and conditional probability"? How fuzzy it is ! It could be a fuzzy argument for something, but definitely not a basis for "statistical proof" unless we low the level of what we use to call a (mathematical) proof, or by "statistical proof" we mean something different than a "proof".

The "good mistake" of Boole turns out to the right thing to do in probability logic: the truth value of an implication (conditional) "If b then a" in probability logic should be the conditional probability $P(a|b)$, of course when $P(b) \neq 0$. This is later called the Stalnaker's thesis, Stalnaker [15]. But then "If b then a" is not the material implication of classical logic anymore, since $P(b' \vee a) \neq P(a|b)$, except few trivial cases (Lewis's triviality results, Lewis [11]), as it can be seen easily as follows.

Assuming that $P(b) \neq 0$, we have

$$P(b' \vee a) = P(b') + P(a \wedge b) = P(b') + P(a|b)P(b) =$$

$$P(b') + P(a|b)[1 - P(b')] = P(a|b) + P(b')[1 - P(a|b)] =$$

$$P(a|b) + P(b')P(a'|b)$$

Thus, $P(b' \vee a) > P(a|b)$ unless $P(b') = 0$, or $P(a'b) = 0$, rather trivial cases.

In fact, Lewis' trivial result is formulated as follows, which puts to rest any attemp to search for other binary operations on $\mathcal{A}$ whose truth values are conditional probabilities.

Theorem *Let $\mathcal{A}$ be a Boolean ring with more than four elements. Then there is no binary operation $\diamond$ on $\mathcal{A}$ such that for all probability measure P on $\mathcal{A}$, and for all $a, b \in \mathcal{A}$ with $P(b) \neq 0$, $P(a \diamond b) = P(a|b)$.*

For a proof, see Goodman-Nguyen-Walker [6].

Remark It is well-known that, outside of classical two-valued logic, material implication is not a "meaningful" implication to use. That is why other implication operators are considered to replace it. So far, untill 1988, the situation is this. As far as implication operations whose truth values can be equated with conditional probabilities (i.e., consistent with conditional probabilities), there is no efforts to search for a *bona fide mathematical entity*, say, $\rhd$, defined on $\mathcal{A} \times \mathcal{A}$, so that $P(a \rhd b) = P(a|b)$. Instead, that implication is only considered "verbally", i.e., at the "metalanguage" level! Perhaps, the reason is this. In one hand, probabilists and statisticians deal with conditional probability measures, and with conditional models, so they do not think they need a bona fide mathematical object to represent "if... then..." implication in their underlying logic. It is "interesting" to note that, because of conditional models that statisticians ignore the notion of copulas until 1990 when they finally realized that there are lots of realistic situations that they do not have conditional models available to them! On the other hand, any new mathematical notions should be justified by their usefulness. Besides "mathematical interests", what will be the usefulness or the need to consider an "if...then..." implication (consistent with conditional probability) as a mathematical object (rather just leave it at the metalanguage)? One clear answer is the emerging of expert rule-based systems in the 1970's, where there is the need to combine rules to reach decisions, and roughly speaking, for example, we need to assess the probability of the combined rule "if b then a" and "if c then d". And in this paper, we will provide another "usefulness" of $a \rhd b$, namely its extension to *quantum conditional probability logic*.

The morale is this. The notion of "if...then..." implication, as a bona fide mathematical entity, consistent with conditional probability, which we will call *conditional events*, should not be suffered the previous fate of copulas!

Now, within the context of probability logic, without the possible existence of an implication operation $\rhd$, let's take a closer look at how deduction rules for reasoning are used when truth values of $\rhd$ are taken as conditional probabilities.

We have said earlier that, for example, in probability logic, the modus tollens is valid if we use material implication of two-valued logic to represent "if...then..." implication. But, nobody uses material implication in probability logic. So what is the situation? Of course, people use $\rhd$ at the metalanguage level, and focus on their truth values as conditional probability. Thus, the problem takes the form, for modus tollens : Given $P(a|b)$ and $P(b')$, compute $P(a')$ to figure out whether if $P(a|b) = 1$, and $P(a) = 0$, can we get $P(b = 0$? Well, that never happens since, the conditions $P(a|b) = 1$, and $P(a) = 0$ are incompatibe, they cannot happen at the same time (to borrow "quantum mechanics language"!). See also Yalcin [18]. Thus, only hope is to consider softer versions of modus tollens. This was exactly what Wagner [16] has done, whose results are as follows.

Given $P(a|b) = x$ and $P(a') = y$, for $x, y \in [0, 1]$, compute $P(b')$.

Note that, when considering $P(a|b)$, we must assume $P(b) > 0$ (implying that $P(b') < 1$). Thus, when $x = 1$, whatever $P(a')$ is, we will never get $P(b') = 1$ to say that b is False.

If $y = 1$, i.e., a is False, then necessarily $x = 0$, since $P(a|b) \leq \frac{P(a)}{P(b)}$, then : If $b \rhd a$ is False, and a is False, then $P(b') \in (0, 1)$.

In fact, we do not get a modus tollens (exactly). A soft version of modelus tollens could be: If $x \approx 1$, and $y \approx 1$, then $P(b') \approx 1$.

For example (See below): for $x = 1$, $y = 0.99$, then $P(b') \in [0.99, 1)$. This is so since for $x = 1$, $y \in [0, 1)$, we have $y \leq P(b') < 1$.

Here are Wagner's results:

(i) If $x, y \in (0, 1)$ then

$$\max\{\frac{1-x-y}{1-x}, \frac{x+y-1}{x}\} \leq P(b') < 1$$

(ii) If $x = 0$, and $0 \leq y < 1$, then

$$1 - y \leq P(b') < 1$$

(iii) If $x = 1$, and $0 \leq y < 1$, then

$$y \leq P(b') < 1$$

3 Conditional Probability Logic

By conditional probability logic we mean the probability logic in which probability of "conditionals" are taken to be conditional probability. Adams [1] called it "logic of conditionals" since not only the "items" we are reasoning about are conditionals, but mainly because, to be a logic, there should have logical connectives between them. So, a conditional probability logic is a logic where the "propositions" are conditionals and their truth values are conditional probabilities. Of course, it remains to define what is a "conditional" in such a logic.

The core of a conditional probability logic is the derivation of bona fide mathematical entities, denoted as $(a|b)$, called *conditional events* for $a, b \in \mathcal{A}$ (a Boolean ring), such that if P is a probability measure on $\mathcal{A}$, we can assign, without ambiguity, the conditional probability $\frac{P(a \wedge b)}{P(b)}$, when $P(b) \neq 0$, to $(a|b)$. These conditional events live in a larger set than $\mathcal{A}$ (so that Lewis' triviality result does not apply) where logical connectives for them are developed, forming a logic of conditionals.

The first work towards this goal was Goodman and Nguyen [7] in 1988. A complete theory was done in 1991 (see Goodman, Nguyen and Walker [6]). Since we are going to extend the theory to the quantum setting, and since, without any comparison to usefulnees, like copulas, it seems that the theory (of conditional events) is not quite "familiar" to probabilists, statisticians, econometricians, and much more (!), we will review its basics shortly.

But before it, it is interesting to look at another approach to conditional events (Goodman and Nguyen [7]).

Again, let $(\Omega, \mathcal{A}, P)$ be a probability space. Lewis' triviality result prevents the search for an object $(a|b)$ in $\mathcal{A}$, for $a, b \in \mathcal{A}$, such that $P((a|b)) = P(a|b)$ when $P(b) \neq 0$. But, like the extension of the field of real numbers to the field of complex numbers, it does not prevent us from enlarging the $\sigma-$field $\mathcal{A}$ to another $\sigma-$field to house $(a|b)$, so that conditional events are "generalized events" in a larger $\sigma-$ field.

The construction goes as follows. Let Ω^* be the infinite (but countable) product space $\prod_{n \geq 1} \Omega_n$, where $\Omega_n = \Omega$, for all $n \geq 1$. Let $\mathcal{A}^*$ be the usual product $\sigma-$ field on Ω^*, i.e., the $\sigma-$ field of subsets of Ω^* generated by cylinders $a_1 \times a_2 \times ... \times a_n \times \Omega \times \Omega \times ....$, for $n \geq 1$, where $a_1, a_2, ..., a_n \in \mathcal{A}$.

Let $\alpha(.) : \mathcal{A} \times \mathcal{A} \to \mathcal{A}^*$ be the map

$$\alpha(a, b) = \cup_{n\geq 0}(b' \times b' \times ... \times b'(n \text{ times}) \times (a \wedge b))$$

where $\cup$ denotes union among subsets of Ω^*, and $(a \wedge b)$ is the standard shorthand for $(a \wedge b) \times \Omega \times \Omega \times .....$ which is a subset of Ω^*, and for $n = 0$, the first term is simply $a \wedge b$.

Finally, let P^* be the infinite product probability measure on $\mathcal{A}^*$ with identical components P, i.e., P^* is constructed from

$$P^*(a_1 \times a_2 \times ... \times a_n \times \Omega \times \Omega \times ....) = \prod_{n\geq 1} P(a_n)$$

Now, observe that, as subsets of Ω^*, the subsets $b' \times b' \times ... \times b'(n \text{ times}) \times (a \wedge b)$, for $n \geq 0$, are pairwise disjoint, so that, assuming $P(b) \neq 0$,

$$\begin{aligned} P^*(\alpha(a, b)) &= P^*[\cup_{n\geq 0}(b' \times b' \times ... \times b'(n \text{ times}) \times (a \wedge b)] \\ &= \sum_{n=0}^{\infty} P^*[(b' \times b' \times ... \times b'(n \text{ times}) \times (a \wedge b)] \\ &= \sum_{n=0}^{\infty} P(a \wedge b)[P(b')]^n = P(a \wedge b) \sum_{n=0}^{\infty} [P(b')]^n \\ &= \frac{P(a \wedge b)}{1 - P(b')} = \frac{P(a \wedge b)}{P(b)} = P(a|b) \end{aligned}$$

Thus, there is a "canonical" probability space $(\Omega^*, \mathcal{A}^*, P^*)$ extending $(\Omega, \mathcal{A}, P)$, and a map $\alpha(.) : \mathcal{A} \times \mathcal{A} \to \mathcal{A}^*$, such that $P^*[\alpha(a, b)] = P(a|b)$ when $P(b) \neq 0$.

The mathematical entity $\alpha(a, b) \in \mathcal{A}^*$ is an event in the $\sigma-$ field $\mathcal{A}^*$, playing the role of a "conditional event" whose P^*- probability truth value concides with $P(a|b)$.

Note that the "conditional event"

$$\alpha(a, b) = \cup_{n\geq 0}(b' \times b' \times ... \times b'(n \text{ times}) \times (a \wedge b))$$

is "measure-free" in the sense that we define it without involving any probability measure on the measurable space $(\Omega, \mathcal{A})$. However, it is consistent (or compatible) with probability evaluation, i.e., for any probability measure P on $(\Omega, \mathcal{A})$, the probability truth value of $\alpha(a, b)$ is $P(a|b)$. Thus, for example, in Adams' logic of conditionals, Adams [1] , the metalanguage can be specified to a mathematical language!

Remark In a "second" look at quantum conditional events, we will explore the possible quantum version of the above probability space technique, say, by using Fock space or countable tensor product of Hilbert spaces.

We return now to a "non-Boolean" approach to conditional events which is more suitable for extending to quantum logic. For more details, see Goodman and Nguyen [4], Goodman et al. [6], and Walker [17].

The problem is this. Let $(\Omega, \mathcal{A})$ be a measurable space. For $a, b \in \mathcal{A}$, the "conditional event", denoted as $(a|b)$, represents "a given b", or "if b then a". Note that we will proceed to define mathematically such a conditional even (not just put it in the metalanguage), without involving any probability measure on $\mathcal{A}$. That is why the concept of conditional events is termed "measure-free conditioning", although we intend to do this consistently with probability, i.e., whatever a probability measure P is put on the measurable space $(\Omega, \mathcal{A})$, we have $P((a|b)) = P(a|b) = \frac{P(a \wedge b)}{P(b)}$, when $P(b) \neq 0$.

The set of (ordinary) events $\mathcal{A}$ is a Boolean algebra with usual operations $\wedge$, $\vee$, $(.)'$. It is convenient to view $\mathcal{A}(., +)$ as a Boolean ring, i.e., a ring with unit 1 (as Ω), zero 0 (as $\varnothing$) in which every element is idempotent ($a = a^2$), and where $ab = a \wedge b$, $a \vee b = a + b + ab$, and $a' = 1 + a$, $a \leq b$ means $ab = a$.

Our approach to defining $(a|b)$ is axiomatic. We seek a map $f(.)$ defined on $\mathcal{A} \times \mathcal{A}$ to some unknown set S satisfying some desirable properties capturing basic aspects of conditioning compatible with conditional probability evaluations (this is somewhat similar to defining tensor product for composite quantum systems).

Since S is unknown, we work on the domain $\mathcal{A} \times \mathcal{A}$ of $f(.)$, examining the partition on it induced by the equivalence relation: $(a, b) \approx (c, d)$ if $f(a, b) = f(c, d)$. A version of $f(.)$ is then taken by assigning to (a, b) its equivalence class. Without going into technical details, our strategy leads to $f(a, b) = a + \mathcal{A}b'$, a coset of the Boolean ring $\mathcal{A}$. Thus, the range S of such $f(.)$ is $\mathcal{A}|\mathcal{A} = \cup_{b \in \mathcal{A}} \mathcal{A}/\mathcal{A}b'$, where $\mathcal{A}/\mathcal{A}b'$ denotes the quotient ring $\mathcal{A}$ with respect to the principlal ideal $\mathcal{A}b'$. The space $\mathcal{A}|\mathcal{A}$ of conditional events so defined is a conditional extension of (classical) logic.

There are several equivalent representations of conditional events. For example, $f(a, b) = (ab, b)$. It is so because $f(a, b) = f(c, d)$ iff $ab = cd$ and $b = d$. In this equivalent representation, the conditional event $(a|b)$ is the pair (a, b) with $a \leq b$, so that $\mathcal{A}|\mathcal{A}$ is $\{(a, b) : a, b \in \mathcal{A}, a \leq b\}$. With this representation, it is easy to see that if we assign $P(a|b)$ to $f(a, b)$, then it is well defined, since if $(a|b) = (c|d)$, i.e., $(ab, b) = (cd, d)$ then $ab = cd$, and $b = d$ and hence $P(a|b) = P(c|d)$.

Another equivalent representation is this. $(a|b)$ is a "closed interval" in the Boolean ring $\mathcal{A}$. Indeed, a closed interval in $\mathcal{A}$ is of the form $[a, b] = \{x \in \mathcal{A} : a \leq x \leq b\}$, and we have $(a|b) = a + \mathcal{A}b' = [ab, b' \vee a]$, and $[a, b] = (a|b' \vee a)$. Note that $b' \vee a$ is the material implication in classical logic.

The space of conditional events $\mathcal{A}|\mathcal{A}$ is $\{(a, b) : a, b \in \mathcal{A}, a \leq b\}$ is a non-Boolean structure of conditional probability logic. Its basic connectives are

$$(a|b)' = (a'|b)$$

$$(a|b) \wedge (c|d) = (ac|a'b \vee c'd \vee bd)$$

$$(a|b) \vee (c|d) = (a \vee c|ab \vee cd \vee bd)$$

$$(a|b) \leq (c|d) \ \text{ iff } \ (a|b) \wedge (c|d) = (a|b)$$

4 Quantum Logic

This is a two-valued logic (True or False) extending Boolean logic in the sense of extending the Boolean structure $(\mathcal{A}, \varnothing, \Omega, \subseteq, \cap, \cup, (.)')$ to an orthomodular lattice $L(0, 1, \leq, \wedge, \vee, (.)')$ $((.)' = \perp)$ of projections on a (complex) Hilbert space H.

An element $a \in L$ is a (orthogonal) projection on H, i.e., a linear map from H to H (an operator) which is self-adjoint and idempotent. It is a projection from H to a closed subspace $R(a)$,the range of a. L is the set of "quantum events" generalizing Boolean events, i.e., $\mathcal{A} \subseteq L$.

L is a poset with partial order relation $\leq$ defined as $a \leq b$ when $b - a$ is a positive operator. L is bounded with 0, 1 as the projections 0, 1, respectively. With respect to $\leq$, L is a lattice with $a \wedge b$ being the projection corresponding to $R(a) \cap R(b)$, and $a \vee b$ being the projection correpoding the the smallest closed subspace containing $R(a) \cup R(b)$. The orthocomplement $(.)'$ is $a' = 1 - a$.

Unlike $\mathcal{A}$, the lattice L is not distributive, in general. Recall that a lattice L is distributive if every triple $\{a, b, c\}$ is distributive in the sense that every substitution of a, b, c into the following equations is satisfied by L:

$$x \wedge (y \vee z) = (x \wedge y) \vee (x \wedge z)$$

$$x \vee (y \wedge z) = (x \vee y) \wedge (x \vee z)$$

Indeed, for example, let $x_1 = x_2 + x_3$ where $x_2, x_3 \in H$ are unit vectors with $< x_2, x_3 > \ > 0$, and the a_j, $j = 1, 2, 3$ are projections onto the one-dimensional subspaces $\mathbb{C}x_j$ (generated by x_j, $j = 1, 2, 3$), then

$$p_1 \wedge (p_2 \vee p_3) = p_1 \neq 0 = (p_1 \wedge p_2) \vee (p_1 \wedge p_3)$$

Now, as operator, the elements of L have "composition" operation: $a \circ b$, denoted for simplicity ab, is the operator $(a \circ b)(x) = a(b(x))$. Also, as operators, they might be commuting (compatible), i.e., $ab = ba$ or not (not commuting, not compatible), i.e., $ab \neq ba$. Note that $ab \in L$ if and only if a, b are compatible. L is distributive when restrictive to compatible projections.

In the language of L, two projections $a, b \in L$ are compatible (commuting) if and only if

$$a = (a \wedge b) \vee (a \wedge b')$$

Since L is a orthomodular lattice, the above compatibility relation, denoted as aCb, is in fact symmetric.

It follows from the Foulis-Holland theorem (namely: $\{a, b, c\}$ is distributive when aCb and aCc) that $a \leq b$ implies that aCb. For example, for any $a, b \in L$, $a \wedge b$ and a are compatible. Note that the maximal compatible subsets of L are Boolean algebras, called blocks, and that L is built from these "classical" Boolean algebras "pasted" together.

Consider the connective $\supset$ (Sasaki hook) on L: $a \supset b = a' \vee (a \wedge b)$. If a, b are compatible, then $a \supset b = a' \vee b$, i.e., the material implcation on a Boolean logic.

Indeed, if aCb, then by symmetry, we have bCa, so that $b = (b \wedge a) \vee (b \wedge a')$. Thus,

$$a' \vee b = a' \vee [(b \wedge a) \vee (b \wedge a')] =$$

$$[a' \vee (b \wedge a')] \vee (a \wedge b)] = a' \vee (a \wedge b)$$

Remark Orthomodularity of an ortholattice is equivalent to having $a \leq b$ imply that a, b belong to a Boolean subalgebra. So in effect, that all information of an orthomodular lattice is contained in its Boolean subalgebras.

In economics or other phenomena uncertainty arises mainly because of lack of complete information. When uncertainty is modeled as (Kolmogorov) probability, we are led to considering probability logic in which, while the underlying structure remains a Boolean algebra $\mathcal{A}$, the set of possible truth values is extended to the unit interval $[0, 1]$.

In quantum mechanics, the randomness is intrinsic, and hence it is not only natural, but it is a must to consider "quantum probability logic" for reasoning.

Now, the Boolean structure $\mathcal{A}$, on which classical probability theory is based, is extended to the orthomodular lattice L of projections on a Hilbert space H, replacing a classical probability space $(\Omega, \mathcal{A}, P)$ by a "*quantum probability space*" (H, L, ρ), we need to spell out the notions of quantum probability and quantum conditional probability in order to investigate a quantum probability logic.

When L plays the role of "quantum events", probabilities are attached to these new mathematical entities. When the "state" of randomness is represented by a "density matrix" ρ which is a positive operator with unit trace, the natural extension of classical probability of $a \in L$, under ρ, is $\Pr(a) = tr(\rho a)$ (Noting that the trace operator is commutative), taken to be the quantum probability of the quantum event a.

Remark By abuse of language, a "state" is taken to be an element $\psi \in H$ with unit norm. Then $P(a) = ||a\psi||^2$.

How about the notion of quantum conditional probability? Well, it should be some reasonable extension of classical conditional probability!

In classical probability theory, for a given $b \in \mathcal{A}$ with $P(b) \neq 0$ (i.e., under the state P, it is possible that the event b occurs), there is a unique functional, denoted as $P_b(.)$, mapping $\mathcal{A}$ to $[0, 1]$ such that $P_b(.)$ is a probability, and for any $a \leq b$, we have

$P_b(a) = \frac{P(a)}{P(b)}$. That unique functional is given by $P_b(a) = \frac{P(a \wedge b)}{P(b)}$ and is interpreyed as the conditional probability of a given b.

Thus, to defined a quantum analog, we could try to propose a functional which agrees with the classical case for commuting projections. This program is carried out as follows in the context of (H, L, ρ), where the quantum probability measure on L, under ρ, is $P_\rho(a) = tr(\rho a)$.

For $b \in L$ with $P_\rho(b) \neq 0$, define $P_\rho(.|b)$ on L such that it is a quantum probability measure on L, and if $a \leq b$ then $P_\rho(.|b) = \frac{P_\rho(a)}{P_\rho(b)}$.

This rule can be extended uniquely to L, known as Luders' rule of conditioning, as follows.

The extension $P_\rho(.|b)$ to all L, which is unique, and for all $a \in L$, is given by the density operator $\rho_b = \frac{b\rho b}{tr(b\rho b)}$, so that

$$P_\rho(a|b) = tr(\rho_b a) = \frac{tr(b\rho ba)}{tr(b\rho b)}$$

In terms of state ψ, it is written as

$$P_\psi(a|b) = P_{\psi_b}(a) =< \psi_b, a\psi_b >$$

where $\psi_b = \frac{b\psi}{||b\psi||}$.

Note that quantum conditional probabilities defined by Luders' rule reduces to classical conditional probabilities for compatible events. Indeed, let a, b are commuting projections and u is a unit vector of H. The $P_u(a|b)$ is the probability that if you start in the state for u, perform b and get a yes answer, then when you perform a you will get a yes answer. Performing b and getting a yes answer puts you in state $v = \frac{bu}{||bu||}$. So $P_u(a|b) = P_v(a)$. But this then is equal to $||av||^2$. But then

$$P_u(a|b) = ||av||^2 = ||a\frac{bu}{||bu||}||^2 = \frac{||abu||^2}{||bu||^2} = \frac{P_u(a \wedge b)}{P_u(b)}$$

It is amazing that in 1989, Redei [14] established a similar result as Lewis which is not surprizing since after all quantum logic structure is a generalization of Boolean logic, namely "quantum conditional probabilities are not probabilities of quantum conditional", where quantum conditional refers to Sasaki hook, and conditional probability refers to Luders' rule above. The conclusion of Redei's paper is this. It is impossible to describe *quantum statistical inference* in terms of (two-valued) quantum logic, i.e., with the structure of an orthomodular lattice L, in which Sasaki hook is considered (completely generalizing the structure and implication connective of Boolean logic), it is impossible to address the topic of "quantum Probability Logic" because of the lack of a notion of "quantum conditional" compatible with quantum conditional probability. See also Coecke and Smets [3]. For a recent related work on quantum implication, see Younes and Schmitt [18].

5 Quantum Conditional Events (Quantum Implication)

We are going to examine quantum probability logic whose truth values are quantum probabilities on the orthomodular lattice L of projections on a Hilbert space H. In a parallel way with classical probability logic, evoking the fact that the popular Sasaki hook (called a quantum conditional), the countertpart of material implication in classical logic, and which is a connective on L, is not consistent with quantum conditional probability (in Luder's sense), i.e., $P(b \supset a) \neq P(a|b)$, we search for an quantum implication $\rhd$ acting on $L \times L$, but not a connective, such that $P(b \rhd a) = P(a|b)$, simply by mimicking the work in classical probability logic. Specifically, we are going to extend the notion of (measure-free) conditional events to that of quantum conditional events.

Recall that the lifting of the material implication $b \to a = b' \vee a$ from Boolean $\mathcal{A}$ to the lattice L, to get a similar implication on L, is not an option since it does not satisfy the Birkhoff-von Neumann's requirement, namely $b \supset a = 1$ if and only if $b \leq a$ (it is true on $\mathcal{A}$). But, because of distributivity of $\mathcal{A}$, we have

$$\begin{aligned} b' \vee a &= b' \vee [a \wedge (b' \vee b)] = b' \vee [(a \wedge b') \vee (b \wedge a)] \\ &= [b' \vee (a \wedge b')] \vee (b \wedge a) = b' \vee (a \wedge b) \end{aligned}$$

i.e., we can write the material implication on $\mathcal{A}$ also as $b \to a = b' \vee (a \wedge b)$.

In the non distributive lattice L, $b' \vee a \neq b' \vee (a \wedge b)$, unless a and b are compatible (commuting). Now, for any a, b, compatible or not, we always have a and $a \wedge b$ compatible, so that $b' \vee (a \wedge b) = b' \vee a$. Thus, the Sasaki hook, as defined by $b \supset a = b' \vee (a \wedge b)$ for any $a, b \in L$, does generalize material implication of Boolean logic.

But then the question is: what is the truth value we should assign to $b \supset a$? Is it $P(b' \vee (a \wedge b))$?

Of course not, because in the special case of probability logic, where classical probability is a special case of quantum probability, and Boolean lattice is a special case of an orthomodular lattice, the accepted Stanaker's thesis is to assign conditional probabilities as probability truth values of conditonals.

Of course, no problem with the material implication: $b \to a = b' \vee a$ in Boolean logic since it is defined precisely in terms of truth values! In two-valued logic, as a truth functional logic, $b \to a$ is false when b is true but a is false, otherwise it is true. Note that the Modus Tollens is valid: $b \to a$ is true, and a is false imply b is false.

But how about the Sasaki hook in quantum logic? Well, the problem occurs in classical probability logic too with material "connective" when truth values are in $[0, 1]$.

Since quantum logic is a generalization of probability logic, we could extend what has been done in probability logic to the quantum setting.

For events a, b we consider the meaning of $(a|b)$ in the quantum setting. Here are some crucial facts, much different from the classical case, are vital. First, a property does not exist until it is measured. An electron is neither spin up nor spin down until we measure it. We are not saying that we do not know whether it is spin up or spin down until we measure, we can prove that these things have no meaning until we measure. Second, a measurement of one property can destroy a property we know a system to have, and can affect the probability of measurements of other properties.

There are two meanings for quantum conditional probability, one seems suitable for economics which is not physics.

(i) The *operational* interpretation of $(a|b)$ is the event that a measurement for a yields an affirmative answer given that a measurement for b is conducted first and found to give an affirmative answer. The probability of $(a|b)$ when in state ψ is given by (assuming that it is possible for projection b to occur when in state ψ)

$$P^{\psi}_{op}(a|b) = \frac{||ab\psi||^2}{||b\psi||^2}$$

The explication is this. We begin in state ψ. If a measurement is made for b and it reveals that b holds, then the new state is $b\psi$ normalized. So the new state is $\chi = \frac{b\psi}{||b\psi||}$. Then the probability that event a is found to occur when measuring on a system in state ψ is $||a\chi||^2$.

(ii) The *formalistic* interpretation of $(a|b)$ is the probability that $a \wedge b$ occurs given that b occurs. Note, $a \wedge b, b$ are compatible events, so they can be measured for simultaneously. Then

$$P^{\psi}_{form}(a|b) = \frac{||(a \wedge b)\psi||^2}{||b\psi||^2}$$

provided, of course, that it is possible that b occurs when in state ψ.

Note that in the formalistic setting, we have $(a|b) = (a \wedge b|b)$ since its formalistic interpretation is the probability that $a \wedge b$ occurs given that b occurs, and this is the same as the probability that $a \wedge b \wedge b$ occurs given b occurs. This is just as in the classical conditional setting.

Recall the following process in the classical case of a Boolean algebra $\mathcal{A}$. Denote by $\mathcal{C}(\mathcal{A})$ the set of conditional events $(a|b)$ where $a \leq b$. To treat this set of conditional events, and to put structure on it, it is convenient to work with a construction from lattice theory. Let $\mathcal{A}^{[2]} = \{(a, b) \in \mathcal{A}^{[2]} : a \leq b\}$. The map $\Phi : \mathcal{C}(\mathcal{A}) \to \mathcal{A}^{[2]}$, defined by $\Phi(a|b) = (a, a \vee b')$ is a bijection and its inverse is given by $\Phi^{-1}(c, d) = (c, c \vee d')$. Define operations $\wedge, \vee$ on $\mathcal{A}^{[2]}$ componentwise, so that

$$(a, b) \wedge (c, d) = (a \wedge v, b \wedge d) \qquad (a, b) \vee (c, d) = (a \vee v, b \vee d)$$

and define the operation $* : (a, b)^* = (b', a')$. Then define operations on $\mathcal{C}(\mathcal{A})$ via the bijection Φ.

We can mimic the above process exactly for the quantum case leading to quantum conditionals consistent with quantum probability in the formalistic interpretation.

For the orthomodular lattice L of projections on a Hilbert space H, let $\mathcal{C}(L)$ be the set of conditional events $(a|b)$ where $a \leq b$. We can treat formalistic quantum conditional in a lattice theoretic way through a similar construction. Let $L^{[2]} = \{(a, b) \in L^2 : a \leq b\}$. The map $\Phi : \mathcal{C}(L) \to L^{[2]}$, defined by $\Phi(a, b) = (a, a \vee b)$ is a bijection whose inverse is $\Phi^{-1}(c, d) = (c, c \vee d')$. Put operations $\wedge, \vee$ on $L^{[2]}$ componentwise, and define $*$ by $(a, b)^* = (b', a')$. and transfer these operations to $\mathcal{C}(L)$ via this bijection. Note that a projection lattice L is built by gluing together its Boolean algebras of commuting projections. In effect, Boolean subalgebras $\mathcal{A}$ of L correspond to collections of events that are compatible, so behave classically together.

Remarks

(i) If $\mathcal{A}$ is a Boolean subalgebra of L, then $\mathcal{A}^{[2]}$ is a subalgebra of $L^{[2}$, and therefore this algebra $\mathcal{C}(\mathcal{A})$ of classical conditional events is a subalgebra of the algebra $\mathcal{C}(L)$ of formal quantum events.

(ii) In treating formal conditionals, we take the shortcut of considering only those $(a|b)$ with $a \leq b$. But we could have worked with any events a, b to form $(a|b)$. In this case, we treat it as the conditional $(a \wedge b|b)$. Then in moving to $L^{[2}$ we transform it to $(a \wedge b, (a \wedge b) \vee b')$. So we see the Sasaki hook $b \supset a = b' \vee (b \wedge a)$ appears. Namely, the mapping from all conditionals $(a|b)$, without the assumption that $a \leq b$, to $L^{[2}$ is given by $(a|b) \to (a \wedge b, b \supset a)$.

References

1. Adams, E. (1975). *The logic of conditionals*. D. Reidel.
2. Boole, G. (1984). *The laws of thought*. Macmillan.
3. Coecke, B., & Smets, S. (1994). the Sasaki hook is not a (static) implication connective but induces a backward (in time) dynamic one that assigns causes. *International Journal of Theoretical Physics, 43*, 1705–1736.
4. Goodman, I. R., & Nguyen, H. T. (1988). Conditional objects and the modeling of uncertainties. In M. M. Gupta & T. Yamaka (Eds.), *Fuzzy computing, theory, hardware and applications* (pp. 119–138). North Holland.
5. Goodman, I. R., Gupta, M. M., Nguyen, H. T., & Rogers, G. S. (Eds.). (1991). *Conditional logic in expert systems*. North Holland.
6. Goodman, I. R., Nguyen, H. T., & Walker, E. A. (1991). *Conditional inference and logic for intelligent systems*. North Holland.
7. Goodman, I. R., & Nguyen, H. T. (1994). A theory of conditional information for probabilistic inference in intelligent systems: II-product space approach. *Informatiuon Sciences, 76*, 13–42.
8. Hailperin, T. (1975). Probability logic. *Notre Dame Journal Formal Logic, 25*(3), 198–212.
9. Hardegree, G. M. (1975). Stalnaker conditionals and quantum logic. *Journal of Philosophical Logic, 4*(4), 399–421.
10. Holmes, S. (2018). Statistical proof? The problem of irreproducibility. *Bulletin of the American Mathematical Society, 55*(1), 31–55.

11. Lewis, D. (1976). Probabilities of conditionals and conditional probabilities. *The Philosophical Review, 85*, 297–315.
12. Milne, P. (2004). Bruno de Finetti and the logic of conditional events. *British Journal for the Philosophy of Science, 48*, 195–232.
13. Nguyen, H. T., et al. (2020). Conditional event algebras: The state-of-the-art. In O. Koscheleva (Ed.), *Beyond traditional probabilistic data processing techniques* (pp. 545–555). Springer.
14. Redei, M. (1989). Quantum conditional probabilities are not probabilities of quantum conditional. *Physics Letter A, 139*(7), 287–290.
15. Stalnaker, R. C. (1970). Probability and conditionals. *Philosophy of Science, 37*, 64–89.
16. Wagner, C. G. (2004). Modus tollens probabilized. *British Journal for the Philosophy of Science, 55*, 747–753.
17. Walker, E. A., et al. (1991). A simple look at conditional events. In I. R. Goodman (Ed.), *Conditional logic in expert systems* (pp. 101–114). North Holland.
18. Yalcin, S. (2012). A counterexample to modus tollens. *Journal of Philosophical Logic, 41*, 1001–1024.
19. Younes, Y., & Schmitt, I. (2019). On quantum implication, quantum. *Machine Intelligence, 1*, 53–63.

Quantum-Like Modeling: Projection Postulate and Quantum Nonlocality

Andrei Khrennikov

Abstract Applications of the mathematical apparatus of quantum theory to decision making, economics, and finances should be accommodated with the foundations of this theory. Quantum nonlocality (spooky action at a distance) is one of the most exiting features of modern quantum theory. This is one of the main challenges in justification of applications of the quantum formalism to humans. May be consideration of spooky action at a distance is acceptable for micro-systems, say photons. However, it is difficult to imagine such spooky action for humans. The aim of this note is demystification of quantum nonlocality via coupling of the "nonlocal structure" of the Lüders' projection postulate with probabilistic conditioning. The role of the quantum state interpretation, statistical versus individual, is emphasized. By using the statistical interpretation for the Lüders' state update, we are able to work with entangled states without appealing to the ambiguous notion of quantum nonlocality.

Keywords Demystification of quantum nonlocality · Spooky action at a distance · Quantum state update · Projection postulate

1 Introduction

Applications of the mathematical apparatus and methodology of quantum theory to decision making, economics, and finances as well as game theory (see, e.g., [1–19]) should be accommodated with the foundations of this theory. Quantum nonlocality (Einstein's spooky action at a distance [20]) is one of the most exiting features of modern quantum theory. This is one of the main challenges in justification of applications of the quantum formalism to humans.

We elevate the role of the quantum update of probability (based on the Lüders' projection postulate [21]) for compound systems (consisting of a few agents, markets, states, corporations, or financial markets). In particular, we demystify "quantum nonlocality" and entanglement. This demystification is impossible without taking

A. Khrennikov (✉)
Linnaeus University, Växjö, SE 35195, Sweden
e-mail: Andrei.Khrennikov@lnu.se

S. Sriboonchitta et al. (eds.), *Credible Asset Allocation, Optimal Transport Methods, and Related Topics*, Studies in Systems, Decision and Control 429,
https://doi.org/10.1007/978-3-030-97273-8_8

into account the quantum state interpretations. In quantum mechanics endowed with the individual interpretation of a quantum state, the Lüders' projection postulate [21] leads to spooky action at a distance. In principle, such nonlocality is still acceptable for micro-systems. Nonlocal action at a distance is still acceptable (although very exotic) for electrons or photons. But, such a nonlocal behavior would be really surprising for humans. On the other hand, in quantum mechanics endowed with the statistical interpretation, the Lüders' projection postulate [21] can be interpreted as local probabilistic conditioning.

2 Quantum Conditioning: Lüders' Postulate Combined with Born's Rule

Consider an observable A represented by a Hermitian operator $\hat{A}$ acting in Hilbert state space H. This is the simplest mathematical model of quantum measurements [22]. Nowadays, in quantum physics, especially quantum information theory, there is used a more general model of measurements based on quantum instruments [23–25] and theory of open quantum systems (see [9, 10, 12–14] for applications to quantum-like modeling). In this model, quantum observables are represented by positive operator valued measures (POVMs), see [26] for quantum-like studies).

In this note, we restrict our considerations to pure quantum states given by normalized vectors $|\psi\rangle$ of H. The spectral expansion of an observable with purely discrete spectrum has the form:

$$\hat{A} = \sum_x x \, \hat{\pi}^A(x), \tag{1}$$

where $\pi^A(x)$ is the projector on the eigensubspace H_x consisting of eigenvectors with eigenvalue x.

Each measurement is preceded by preparation of systems in some concrete quantum state $|\psi\rangle$. For observables which can be mathematically represented by Hermitian operators (cf. with theory of quantum instruments [23–25]), the post-measurement state is always again the pure state given by normalized projection:

$$|\psi_{A=x}\rangle = \frac{\hat{\pi}^A(x)|\psi\rangle}{\|\hat{\pi}^A(x)\,|\psi\rangle\|}. \tag{2}$$

This is the projection postulate in Lüders' version [21]. We remark that Lüders' paper was published rather late, in 1950th. However, this form of the projection postulate was widely used since the first days of the formulation of quantum mechanics as a mathematical formalism in Hilbert space, e.g., this form of the projection postulate was used in the famous paper of Einstein, Podolsky, and Rosen [20]. (They did not use the abstract Hilbert space and work in the L_2-space.) This is the good place to point out that originally the mathematical formalization of the projection postulate was done by von Neumann [22] and that von Neumann's and Lüders' formulations coincide only for quantum observables with non-degenerate spectra, i.e., those represented by Hermitian operators that are composed of one dimensional orthogonal projectors. For operators with nondegenerate spectra, von Neumann considered a more complicated procedure of the state update [22]. Generally the von Neumann's state update can transfer pure quantum states into mixed states represented by density operators. The complete mathematical formalization of the von Neumann's state update was performed essentially later in the formalism of theory of quantum instruments. We remark that the difference between the update schemes of von Neumann and Lüders have nontrivial foundational consequences [27], especially for measurements on compound systems - local operators representing such measurements have degenerate spectra.

In this paper, we proceed solely in Lüders framework; here measurement with output $A = x$ induces an update of the initially prepared quantum state:

$$|\psi\rangle \rightarrow |\psi_{A=x}\rangle. \tag{3}$$

It should be stressed that, although projector is a linear operator, the operation of the quantum state update is nonlinear, due to normalization by the norm. The Born's rule (one of axioms of quantum theory) has the form:

$$p(A = x; \psi) = \|\hat{\pi}^A(x)\ |\psi\rangle\|^2. \tag{4}$$

3 Lüders Postulate and State and Probability Updates for Compound Systems

From the first sight, the projection transformation of the state given by (2) has nothing to do with nonlocality. However, by considering a compound system $S = (S_1, S_2)$ we shall obtain the state-update procedure making the impression of instantaneous action at a distance. This is famous "quantum nonlocality" making quantum theory so mystical.

Let quantum state $|\Psi\rangle$ belong to tensor product $H = H_1 \otimes H_2$ of state spaces H_i of systems $S_i, i = 1, 2$. Select some observable A on S_1; it is represented by Hermitian operator $\hat{A}$ given by (1), where $\hat{\pi}^A(x)$ acts in H_1. This observable can be also treated as an observable on compound system S. The latter is represented by projector $\hat{\pi}^A(x) \otimes I$ in H. By getting output $A = x$, we transform the state of S on the basis of Lüders projection postulate:

$$|\Psi_{A=x}\rangle = \hat{\pi}^A(x) \otimes I|\Psi\rangle / \|\hat{\pi}^A(x) \otimes I|\Psi\rangle\|. \tag{5}$$

Consider now an observable B on S_2 and its conditional measurement, under output $A = x$. By Born's rule

$$p(B = y|A = x; \Psi) \equiv p(B = y; \Psi_{A=x}) = \|\hat{\pi}^B(y) \otimes I|\Psi_{A=x}\rangle\|^2. \tag{6}$$

If state $|\Psi\rangle$ is separable, i.e., $|\Psi\rangle = |\psi^{(1)}\rangle \otimes |\psi^{(2)}\rangle$, then

$$p(B = y|A = x; \Psi) = \frac{\|\hat{\pi}^B(y) \otimes \hat{\pi}^A(x)|\Psi\rangle\|^2}{\|\hat{\pi}^A(x)|\psi^{(1)}\rangle\|^2} = \|\hat{\pi}^B(y)|\psi^{(2)}\rangle\|^2. \tag{7}$$

Thus, for such a state, measurement of observable A on S_1 does not change statistics for measurements of observable B on S_2,

$$p(B = y|A = x; \Psi) = p(B = y; \Psi) \tag{8}$$

However, if a state is entangled, then generally

$$p(B = y|A = x; \Psi) \neq p(B = y; \Psi) \tag{9}$$

We remark that any state of S determines the states of its subsystems $S_i, i = 1, 2$; for the initial state $|\Psi\rangle$,

$$\rho^{(1)} = \mathrm{Tr}_{\mathrm{H}_2}|\Psi\rangle\langle\Psi|, \ \rho^{(2)} = \mathrm{Tr}_{\mathrm{H}_1}|\Psi\rangle\langle\Psi|,$$

and, for the post-measurement state $|\Psi_{A=x}\rangle$,

$$\rho^{(1)}_{A=x} = \mathrm{Tr}_{\mathrm{H}_2}|\Psi_{\mathrm{A=x}}\rangle\langle\Psi_{\mathrm{A=x}}|, \ \rho^{(2)}_{\mathrm{A=x}} = \mathrm{Tr}_{\mathrm{H}_1}|\Psi_{\mathrm{A=x}}\rangle\langle\Psi_{\mathrm{A=x}}|.$$

The above considerations can be represented in the form of probabilities with respect to the states of S_2. For separable state $|\Psi\rangle$,

$$\rho^{(2)} = \rho^{(2)}_{A=x} \text{ and } p(B = y; \rho^{(2)}) = p(B = y; \rho^{(2)}_{A=x}), \tag{10}$$

and, for entangled states, generally

$$\rho^{(2)} \neq \rho^{(2)}_{A=x} \text{ and } p(B = y; \rho^{(2)}) \neq p(B = y; \rho^{(2)}_{A=x}), \tag{11}$$

The mathematical formulas presented in this section have to be endowed with some interpretation. As is well known, quantum mechanics suffers of the presence of numerous typically contradicting interpretations. Unfortunately, quantum-like modeling with its applications to humanities and economics suffers of the same interpretation diversity problem. In the next section, we shall discuss two basic interpretations of a quantum state and their consequences to the quantum state update and quantum conditioning.

4 Quantum State: A State of An Individual System?

The statistical interpretation of a quantum state can be formulated (due to Einstein) as the following statement:

SI *Quantum state $|\psi\rangle$ represents statistical features of a large ensemble of identically prepared quantum systems.*

By this interpretation, a quantum state $|\psi\rangle$ is not the "personal state" of an individual quantum system (an electron) or quantum-like system (say a human being). By this interpretation the formulas presented in Sect. 2 describe statistics of measurement outputs for ensembles of systems (physical, cognitive, social).

The main interpretation competing with the statistical interpretation is the individual interpretation of a quantum state. Typically it is associated with the Copenhagen interpretation of quantum mechanics:

II *Quantum state $|\psi\rangle$ is the physical state of the concrete quantum system.*

In quantum-like modeling, **II** implies that we can associate a quantum state and the mathematical formulas of Sect. 2 with an individual, say an agent of some financial market.

5 Lüders Postulate and Quantum Nonlocality

Suppose one uses the individual interpretation of a quantum state. In this framework, the use of the Lüders projection postulate implies so-called *spooky action at a distance*, quantum nonlocality. As an example, consider the simplest entangled state:

$$|\Psi\rangle = (|01\rangle + |10\rangle)/\sqrt{2}, \tag{12}$$

where the vectors labeled as $|0\rangle, |1\rangle$ are eignevectors of Hermitian operators $\hat{A} : H_1 \to H_1$ and $\hat{B} : H_2 \to H_2$, and both H_i are qubit spaces (two dimensional Hilbert spaces).

By measuring observable A on S_1 and getting output $A = 0$, we found the compound system in the state $|\Psi_{A=0}\rangle = |01\rangle$. Thus the state of S_2 instanteneously becomes $|\phi\rangle = |01\rangle$. Since by **II** a quantum state has the meaning of the physical state of the concrete system, this is nothing than action at a distance.

II *implies quantum nonlocality.*

Nonlocality is the most exotic feature of quantum theory (endowed with **II**). Even in physics, this feature has been questioned by many scientists and the debates are continued. However, in quantum physics, theory of micro-systems, one still can accept such mysteries: these micro-creatures are so strange and unusual that they can behave non-locally following spooky action at a distance. In quantum-like modeling, the situation is worse. It is really impossible to accept the presence of spooky action at a distance, at least if one does not follow the para-psychological stream of mind. May be using **SI** does not lead to such exotic as spooky action at a distance? I shall consider this issue in Sect. 7. But, before starting **SI**-analyze of the quantum(-like) state update, we would like to turn to classical probability theory (Kolmogorov's axiomatic [28], 1933) and especially conditioning and the probability update in it.

6 Classical Probability and Bayesian Conditioning

Consider now classical probability theory. Probability space (Kolmogorov, 1933 [28]) is a triple $\mathcal{P} = (\Omega, \mathcal{F}, P)$, where Ω is a set of random parameters, $\mathcal{F}$ collection of subsets of Ω representing events, P is a probability measure.

- States of random systems are represented by probability measures
- Observables are defined as maps from Ω to $\mathbf{R}$ (having some special property).

Let A, B be random variables (representing classical observables) with probability distribution $p_A(x), p_B(y)$ and conditional probability $p(B = y|A = x)$. We recall that the latter is given by the Bayes formula:

$$p(B = y|A = x) = P(B = y, A = x)/P(A = x), \tag{13}$$

for $P(A = x) > 0$. To make classical and quantum notations consistent, we set

$$p_A(x) = p(A = y; P), \, p_B(y) = p(B = y; P),$$
$$p(B = y|A = x) = P(B = y; P_{A=x}),$$

where $P_{A=x}$ is the conditional probability corresponding to the output $A = x$ given by the Bayes formula: for an event E,

$$P_{A=x}(E) = P(E|A = x) \equiv P(E, A = x)/P(A = x).$$

In accordance with Bayesian inference, by getting output $A = x$ of observable A we perform an update of the probability measure (the direct analog of the quantum state update (3)):

$$P \to P_{A=x}. \tag{14}$$

It induces the probability update

$$p(B = y; P) \to P(B = y; P_{A=x}). \tag{15}$$

Now, consider a compound classical random system $S = (S_1, S_2)$. Let systems $S_i, i = 1, 2,$ are characterized by sets of random parameters Ω_i. Then S is characterized by set $\Omega = \Omega_1 \times \Omega_2$, Cartesian product. If random systems are *independent,* then the distribution P of random parameters of S is given by $P = P^{(1)} \otimes P^{(2)}$, where $P_i, i = 1, 2$, are probability measures for S_i and tensor product of measures is defined as

$$P(E_1 \times E_2) = P^{(1)}(E_1)P^{(2)}(E_2).$$

We call such a probability state separable. Separability is just another term for independence of random variables $I_i : \Omega \to \Omega_i$, $I_i(\omega) = \omega_i$. If a probability measure on Ω is not separable, we call it entangled.

Let A and B be observables on random systems S_1 and S_2 given by random variables, $A : \Omega_1 \to \mathbf{R}$ and $B : \Omega_2 \to \mathbf{R}$. They can be treated as observables on S represented by random variables $A(\omega) = A(\omega_1)$, $B(\omega) = B(\omega_2)$. It is easy to see that for a separable measure $P = P^{(1)} \otimes P^{(2)}$,

$$P(B = y; P_{A=x}) = p(B = y; P) \tag{16}$$

cf. with the quantum case (8). So, the state update based on to output $A = x$ does not change probability distribution for B-observable. However, if P is entangled, then generally the probability distribution of B is modified:

$$P(B = y; P_{A=x}) \neq p(B = y; P), \tag{17}$$

cf. with the quantum case (9).

We remark that, similarly to the quantum case, in the classical case state P of S determines states of S_i, as the marginal probabilities:

$$P^{(1)}(E_1) = \int\limits_{E_1 \times \Omega_2} dP(\omega_1, \omega_2), \; P^{(2)}(E_2) = \int\limits_{\Omega_1 \times E_2} dP(\omega_1, \omega_2), \tag{18}$$

for discrete parameter-spaces, integration is reduced to summation. If a state is separable, $P = P^{(1)} \otimes P^{(2)}$, then its marginals coincide with $P^{(1)}$, $P^{(2)}$. Hence,

$$P^{(2)}_{A=x} = P^{(2)} \text{ and } P(B = y; P^{(2)}_{A=x}) = p(B = y; P^{(2)}), \tag{19}$$

where $P^{(2)}_{A=x}$ is the marginal of the updated state $P_{A=x}$; cf. with the quantum case (10). For entangled state P, this equality is violated. Generally

$$P^{(2)}_{A=x} \neq P^{(2)} \text{ and } P(B = y; P^{(2)}_{A=x}) \neq p(B = y; P^{(2)}), \tag{20}$$

cf. with the quantum case (11).

Thus by getting result $A = x$, we "instantaneously" get to know that statistics of B-measurement is changed. We can put this consideration into the spatial scenarios, say random variable A is measured in Växjö and random variable B in Moscow. By getting $A = x$ in Växjö one immediately updates the probability to get the output $B = y$ in Moscow. Thus, even in classical probability "faster-than-light change of statistical correlation" takes place. But, this is not surprising. Let us consider this situation from the viewpoint of the statistical interpretations of classical probability: By the this interpretation, the state update given by (13) means that in the original ensemble of systems with state (=probability distribution) P, observer selects the subensemble with state (conditional probability distribution) $P_{A=x}$. It is not surprising at all that this filtration can change statistics of outcomes for another observable B. Of course, the physical filtration and creation of a new ensemble with probability measure $P_{A=x}$ takes time. But, a theoretical prediction is (practically) instantaneous. It is based on correlations encoded in the original state P.

7 Lüders Update Endowed with the Statistical Interpretation Is Local

SI of a quantum state is based on the statistical interpretation of probability. Considerations of the quantum and classical updates of states and the corresponding probabilities (Sects. 2 and 6) are practically identical (up to symbols). Measurement of quantum observable A with output $A = x$ leads to selection of a subensemble of quantum systems and it is natural that statistics of measurements of another observable B is changed. Physical filtration w.r.t. the concrete output $A = x$ takes time, but the theoretical prediction is (practically) instantaneous. For further reading on quantum nonlocality and probability update, we recommend papers [29, 30].

8 Conclusion

Compound systems represented mathematically in the tensor-product of state spaces of subsystems are considered from the viewpoint of the quantum probability conditioning. We elevated the role of the quantum state's interpretation and discuss the most important interpretations, statistical and individual. The latter implies quantum nonlocality and spooky action at a distance. To understand better the mathematical structure of the quantum probability update for compound systems, we compare it with the classical probability update in Kolmogorov measure-theoretic model. We see amazing parallelism in the probability update schemes: quantum based on the Lüders projection postulate and endowed with the statistical interpretation of the quantum state and classical based on the Bayes' formula endowed with the statistical interpretation of probability. In particular, both updates are local and do not imply any sort of spooky action at a distance. This conclusion is important for quantum-like applications outside of physics, e.g., in social and political sciences and economics.

Acknowledgments

I would like to thank Hung Nguyen for initiating and driving the exciting discussion on the interpretation and the role played by the Lüders' projection postulate in quantum-like modeling.

References

1. Khrennikov, A. (1999). Classical and quantum mechanics on information spaces with applications to cognitive, psychological, social and anomalous phenomena. *Found. Physics, 29*, 1065–1098.
2. Khrennikov, A. (2004). *Information dynamics in cognitive, psychological, social, and anomalous phenomena*. Ser.: Fundamental Theories of Physics. Kluwer.
3. Khrennikov, A. (2006). Quantum-like brain: Interference of minds. *BioSystems, 84*, 225–241.
4. Busemeyer, J. R., Wang, Z., & Townsend, J. T. (2006). Quantum dynamics of human decision making. *J. Math. Psych., 50*, 220–241.
5. Khrennikov, A. (2010). *Ubiquitous quantum structure: From psychology to finances*. Springer.
6. Busemeyer, J. R., Pothos, E. M., Franco, R., & Trueblood, J. (2011). A quantum theoretical explanation for probability judgment errors. *Psych. Rev., 118*, 193–218.
7. Busemeyer, J. R., & Bruza, P. D. (2012). *Quantum models of cognition and decision*. Cambridge University Press.
8. Haven, E., & Khrennikov, A. (2013). *Quantum social science*. Cambridge University Press.
9. Asano, M., Khrennikov, A., Ohya, M., Tanaka, Y., & Yamato, I. (2015). *Quantum adaptivity in biology: From genetics to cognition*. Springer.
10. Khrennikov, A. (2015). Quantum-like model of unconscious-conscious dynamics. *Frontiers in Psychology,*, 6, art. 997.
11. Khrennikov, A. (2015). Towards information lasers. *Entropy, 17*(10), 6969–6994.
12. Asano, M., Basieva, I., Khrennikov, A., & Yamato, I. (2017). A model of differentiation in quantum bioinformatics. *Progress in Biophysics and Molecular Biology, 130*, Part A, 88–98.
13. Asano, M., Basieva, I., Khrennikov, A., Ohya, M., & Tanaka, Y. (2017). A quantum-like model of selection behavior. *Journal of Mathematical Psychology, 78*, 2–12.
14. Khrennikova, P., Haven, E., & Khrennikov, A. (2014). An application of the theory of open quantum systems to model the dynamics of party governance in the US Political System. *International Journal of Theoretical Physics, 53*(4), 1346–1360.
15. Khrennikova, P., & Haven, E. (2016). Instability of political preferences and the role of mass-media: a dynamical representation in a quantum framework. *Philosophical Transactions of the Royal Society A, 374*, 20150106.
16. Conte, E., Khrennikov, A., Todarello, O., & Federici, A. (2008). A preliminary experimental verification on the possibility of Bell inequality violation in mental states. *Neuroquantology, 6*(3), 214–221.
17. Asano, M., Khrennikov, A., Ohya, M., Tanaka, Y., & Yamato, I. (2014). Violation of contextual generalization of the Leggett-Garg inequality for recognition of ambiguous figures. *Physica Scripta, T163*, 014006.
18. Khrennikova, P. Quantum-like model of subjective expected utility: A survey of applications to finance. In V. Kreinovich, N. Thach, N. Trung, & D. Van Thanh (Eds.), *Beyond traditional probabilistic methods in economics. ECONVN 2019. Studies in Computational Intelligence* (Vol. 809). Springer.
19. Khrennikov, A., & Haven, E. (2020). Quantum-like modeling: From economics to social laser. *Asian Journal of Economics and Banking, 4*(1), 87–99.

20. Einstein, A., Podolsky, B., & Rosen, N. (1935). Can quantum-mechanical description of physical reality be considered complete? *Physical Review, 47*, 777–780.
21. Lüders, G. (1951). Über die Zustandsänderung durch den Messprozess. *Annalen der Physik (Leipzig), 8*, 322–328.
22. Von Neumann, J. (1955). *Mathematical foundations of quantum mechanics*. Princeton University Press.
23. Davies, E. B. (1976). *Quantum theory of open systems*. Academic Press.
24. Ozawa, M. (2006). *Quantum perfect correlations. An. Phys., 321*, 744–769.
25. Ozawa, M. (2016). Probabilistic interpretation of quantum theory. *New Generation Computing, 34*, 125–152.
26. Khrennikov, A., & Basieva, I. (2014). Quantum model for psychological measurements: From the projection postulate to interference of mental observables represented as positive operator valued measures. *NeuroQuantology, 12*, 324–336. https://arxiv.org/abs/1405.1269.
27. Khrennikov, A. Yu. (2008). The role of von Neumann and Lüders postulates in the Einstein, Podolsky, and Rosen considerations: Comparing measurements with degenerate and nondegenerate spectra *Journal of Mathematical Physics, 49*, N5, art. no. 052102.
28. Kolmolgoroff, A. N. (1933). *Grundbegriffe der Wahrscheinlichkeitsrechnung*. Springer.
29. Khrennikov, A. (2017). After Bell. *Fortschritte der Physik (Progress in Physics), 65*, 1600014.
30. Khrennikov, A. (2019). Get rid of nonlocality from quantum physics. *Entropy, 21*(8), 806.

New Paradigm of Economic Thinking Under Uncertainty

Polina Khrennikova

Abstract Early 'paradoxical' findings in decision theory, such as Allais and Ellsberg paradoxes, revealed an existence of preference reversals and non-neutral attitudes to subjective risks that cannot be mathematically depicted with the aid of the formula of total probability. In many decision tasks agents' decision making is at variance with the classical probability theory, underpinning the standard economic models of rational thinking. In this paper we evaluate in more detail the decision making contexts pertaining to non-consequential reasoning, and how these instances of non-classical preferences can be described and quantified with the aid of alternative frameworks based on quantum probability theory. We also evaluate the implications of asset prices and their deviations from fundamental values that can lead to emergence of excess volatitility. Finally, we consider the major challenges to a unified *quantum-like*framework of thinking and deciding under uncertainty and risk and elaborate on some successful solutions.

Keywords Contextual decision making · Non-classical beliefs · Quantum Probability · Behavioural asset pricing

1 Introduction

The application of quantum probability-based models to the phenomena outside of quantum physics itself is no longer perceived as something exotic, despite the novelty of this interdisciplinary field. The domain of applications includes a wide range of decision-making problems in economics, cognitive science, politics and finance. Many of these findings focus on the violation of the classical probabilistic scheme and the axioms of Boolean logic. The best-known effects that violate the classical representation of observables as random variables in the joint probability space are the conjunction and disjunction errors. Non-additivity of beliefs and the conjunction

P. Khrennikova (✉)
School of Business, University of Leicester, Brookfield Campus, Leicester LE21RQ, UK
e-mail: pk228@leicester.ac.uk

S. Sriboonchitta et al. (eds.), *Credible Asset Allocation, Optimal Transport Methods, and Related Topics*, Studies in Systems, Decision and Control 429,
https://doi.org/10.1007/978-3-030-97273-8_9

errors exhibit themselves in violations of rational reasoning in the Savage Sure Thing Principle [33], and mathematically, as a deviation from rules of classical probability theory [28].

Decision theories under uncertainty and risk are the key building blocks in modern economic and finance models, as well as in public choice theory. In recent decades, starting from the advent of behavioural science and its entrance into the traditional domains of economics and finance, expected utility theories encountered a wave of experimental studies that challenged their axiomatic foundations. One of them is the rational mode of information processing and decision formation, grounded in the canonical formulation of Kolmogorovian probability theory.

Contextual behaviour and non-classical reasoning were initially explained with the aid of an array of biases and external frames that bring a deviation from normative reasoning, see, eg. [14]. Some prominent models of preference formation, such as "Prospect Theory", due to [13] (see also [35]) was devised by relaxing the additivity of objective and subjective probabilities and allowing for the existence of reference-dependent risk attitudes.

An alternative approach to human reasoning and decision making, inspired by the mathematics and probability theory of quantum physics, was adopted after a number of 'paradoxes' and biases were revealed in lab experiments and real-life decision-making. Quantum probability applications to date have provided some accurate explanatory and descriptive accounts of non-consequential reasoning, with the potential to serve as a predictive instrument of human reasoning under uncertainty and risk.

Quantum probability models relax the additivity and distributivity axioms of classical probability theory. Quantum-like models can capture non-additivity and contextuality in agents' prior beliefs across many uncertain states. Quantum probability also enables accommodating the non-Bayesian formation of posterior beliefs. There is by now, a large body of literature that takes advantage of quantum-like models, to describe various decision making contexts in psychology, politics, economics and finance, and to accommodate the emergent probabilistic fallacies, cf. [1, 6, 11, 12, 18, 19, 31, 40]. There are also many applications of quantum frameworks that aim to incorporate the dynamics of human beliefs and preferences as a result of an interaction with our mental observables and external environment, see. e.g., quantum Markovian framework of voters' preference formation [24], and a dynamical framework of asset pricing under uncertainty [23]. A more broad-based evaluation of the empirical data from behavioural experiments with a detailed discussion of the application of quantum models to decision making and financial phenomena can be found in the monographs [4, 7, 11, 18].[1]

Many of these applications focus specifically on the decision-making situations, where agents tend to employ non-classical information processing in ambiguous decision making states. The 'non-classicality' of human reasoning can be well captured through the geometric properties of Hilbert space and the decision-making projectors

[1] Some of the pioneering works on quantum(-like) probabilistic modeling of cognition [15–17], were devised more than 20 years ago.

therein. The key distinguishing feature of quantum probability that makes it a more apt framework to model decision making under uncertainty is non- distributivity and non-commutativity.

In the following sections, we will discuss the current advances of quantum probability-based models and consider the challenges in modelling human preferences and beliefs in various decision-making contexts. In Sect. 2, we review the major findings on non-consequential reasoning and discuss how the generalised quantum-like calculus can accommodate such decision-making patterns. In Part 3 we discuss how quantum-like models can serve as a remedy to preference formation in ambiguous states. In Part 4, we examine, how quantum models can describe contextual behaviour, most easily detected through 'order effects' in individual question-answers and task evaluations. We also evaluate in Sect. 5, how the dynamics of traders' non-classical beliefs can affect the long-term behaviour of financial asset prices. Finally, we also take a more critical stance on the applicability of the quantum formalism to human reasoning and decision making, to consider some challenges with respect to the usage of the calculus of the conventional quantum theory to decision making statistics. Some proposed solutions are reviewed in Sect. 6. In Sect. 7 we conclude and offer some suggestions for the future refinement of the existing quantum probability based frameworks.

2 Non-consequential Reasoning in Economics and Finance

Nowadays, people are constantly faced with more complex decision-making tasks. The surrounding environment contains multiple layers of ambiguity regarding future economic, financial and political outcomes. Some of the recent rare events (some of these events were never observed before), such as BREXIT, the outbreak of COVID-19 and its impact upon economies and financial markets, have shown that it is increasingly challenging to contemplate consequences related to an uncertain future. One of the key factors behind non-consequential reasoning in economic, financial and routine day-to-day decisions is attributed to non-Bayesian processing of information, whereby decision-makers cannot form conditional preferences over outcomes. Consequentialism is grounded in people's ability to think through a hypothetical decision tree. It is at the heart of a celebrated principle of rational reasoning under uncertainty, the "Savage Sure Thing Principle," named after Leonard Savage [33]. We shall discuss the essence of this principle.

2.1 Savage Sure Thing Principle

The Sure Thing Principle (STP) is concerned with the elimination from consideration of the common outcomes in given states of the world. This principle assumes subjective uncertainty associated with these states and can be considered as an equivalent to the independence axiom of expected utility theory under objective risk. In a nutshell, this axiom allows establishing a choice preference in the condition of uncertainty, if the decision-maker knows that an act will produce a more desirable outcome in any state of the world. Savage exemplified how STP works in practice, with the aid of a widely cited story of a businessman, who contemplates purchasing a property before the Presidential elections in the US. The businessman is uncertain, whether the Republican Party or the Democrats will win the election campaign. He decides that he will buy the property in both states (considering his subjective utility associated with a purchase in each state of the world). STP can be considered as a manifestation of consequential reasoning, as individuals consider consequences of different acts given a set of states of the world and some subjective beliefs about these states. A central premise of consequential reasoning is that beliefs obey the axioms of classical probability theory (CP), including the usage of the Bayes formula for updating of these beliefs. It is also important to note, following [9], that the subjective probability estimates associated with some state B and a complimentary state, B' can be expressing an exogenous risk (the uncertainty about the state of nature, such as economic states), as well as a strategic risk, involving uncertainty about the choice of a strategic opponent, which is characteristic for game-theoretic settings (e.g., a competitor starts a price war). Both of these risks are subjective risks that require one to make personal estimates based on their prior beliefs.

In many decision making experiments, entailing both monetary and non-monetary outcomes, the collected decision making statistics did not obey the above stated principle (e.g. [7, 9, 22, 24, 31, 34, 38]) and recently [12]. It was shown that a choice A probability (frequency), without any certain information given on an event B, and its negation, B', was very often below the conditional choice frequencies $A|B$ and $A|B'$, as well as below the total probability (frequency) of A, expressed via the disjunction of these conditional choice outcomes, given by the formula of total probability [28].[2]

A common feature of the above findings on "disjunction effect" is that an inaction is usually preferred to an action, when the states of the world are unknown. These findings apply to experiments, such as "Two-Stage Gambling Task" and its variations, where probabilities are given exogenously [12, 38], but also to decision making tasks, where no probabilities of the states of the world are known, see, e.g., "Hawaii Vacation" experiment in [34]. In "Prisoner's Dilemma" type experiments, a lack of evaluation of counter-party's actions leads to a higher rate of cooperation, despite of this strategy being irrational from a game-theoretic perspective, see [34], and a further analysis in [9, 31]. In the "Hawaii Vacation" experiment the willingness to purchase a ticket to Hawaii was very low among participants, who were in the condition of uncertainty with respect to the outcome of an exam that they took. Only 7% of the participants were willing to purchase the vacation package on the spot, when exam

[2] This means that in the formula of total probability, there is a super-additivity of disjunctions, where $p(A) < p(A \cap B) \cup p(A \cap B')$. We note that in some settings a sub-additivity of disjunctions can be present, which can be interpreted as a form of 'uncertainty attraction'. Such pattern is observed in "Prisoners Dilemma" type experiments, e.g. [9, 34].

outcome (pass, fail) was unknown, versus an average of 55.5%, who were keen to go on a vacation, both after a pass and a fail grade. We can witness that a lack of information about state probabilities makes the decision-making process even more 'non-classical' from the viewpoint of CP.

3 Expectation Formation Under Uncertainty

How do we make decisions, when our beliefs are vague and the exact probabilities of events in question are not attainable?

Standard finance frameworks operate under the assumptions that agents are forming homogeneous subjective beliefs, following the rules of CP theory. This is the essence of preference formation under Subjective Expected Utility Theory (SEUT) conceived by Savage [33]. The subjective beliefs of agents are not observable per se but can be inferred from their preferences. Some studies aim to elicit these beliefs to ascertain, whether they obey the *linearity* condition of classical probability theory see, e.g. [37].

Empirical evidence in behavioural finance studies and psychological experiments show that individuals usually dislike, vague information, or ambiguity and prefer known risks. This behavioural pattern is known as "Ellsberg-type" ambiguity aversion [10]. Attitudes towards ambiguity are not universal, and people can have heterogeneous ambiguity attitudes, for instance, when the experienced events shift their attitude towards some kinds of uncertainties. We can mention the study [32], that detects a more rare phenomenon of 'ambiguity seeking', because of a shift in an individual reference point, caused by earlier gains or losses. There are also other contextual factors, such as emotions [2], that can come into play showing that the agents' ambiguity attitude is not fixed and deviates from additivity of CP theory. In this regard, it becomes of a great significance to further explore the settings that can pertain to instances of non-classical information processing resulting in non-consequential reasoning.[3]

3.1 *Emergence of Co-existing Beliefs*

There is now an established field of quantum physics that inspired quantum probability models that can deal with ambiguous attitudes and their evolution just to mention some representative works [1, 6, 7, 11, 18, 24, 31, 40, 41].

[3] There are also some successful frameworks in behavioural economics that capture the non-additivity of beliefs over objective and subjective risks, cf. [13, 37]. As noted in [7, 40] and recently in [3], models based on quantum probability have a core distinguishing feature in describing and also quantifying 'vague' beliefs.

In quantum probability (QP) calculus an introduction of non-commuting observables enables to capture non-jointly distributed random variables. One can speak about the impossibility of joint measurement of these types of observables. Instead of a classical probability measure with respect to some random variables A and B one considers the interference of their complex probability amplitudes. Non-consequential reasoning can be explained mathematically, as a manifestation of the non-additivity and non-commutativity of probabilistic beliefs and preferences. Different beliefs and preferences can coexist in our mental states at the same time, a phenomenon coined in quantum physics as a superposition state. One can capture non-additivity of human probabilistic beliefs with the aid of a generalised formula of total probability [39]. In this formula interference of the complex probability amplitudes can give raise to either a sub-additivity or super-additivity of probabilistic beliefs.

$$P(A=\alpha)=\sum_{\beta}P(B=\beta)P(A=\alpha|B=\beta)+2\sum_{\beta_1<\beta_2}\cos\theta_{\beta_1\beta_2}\sqrt{P(B=\beta_1)P(A=\alpha|B=\beta_1)P(B=\beta_2)P(A=\alpha|B=\beta_2)} \quad (1)$$

The interference term on the right-hand side of the quantum formula of total probability can serve as a parameter, quantifying the deviation of beliefs from objective probabilities (both, prior and posterior). The size of $\cos\theta$ can vary depending on, the (i) initial cognitive state of the decision-maker; (ii) type of a random variable (observable) that is measured. QP can account for the complementarity of beliefs concerning future states of the world in economic and financial decisions. Complementarity in beliefs suggests an impossibility to *simultaneously* posses firm estimates about the realisation of different states of the world.

4 Non-commutativity and Order Effects

Another widely observed phenomenon in cognitive studies and beyond is the deviation from the commutativity principle of preferences and beliefs. Such non-commutativity, known as *order effect*, emerges when several question-observables are measured after each other. Order effect can be attributed to features of human cognition, whereby new information or question reveals some preferences that the decision-maker was unable to establish in a condition of uncertainty. Formally, for some random variables corresponding to questions A, B, a joint probability distribution does not hold, when $P(A=\alpha, B=\beta)\neq P(B=\beta, A=\alpha)$. Such types of contextual preferences have been successfully modelled via non-commuting observables, corresponding to the random variables, A and B (e.g. [7, 41]).

5 Asset Prices and Agents' Non-classical Information Processing

The above instances of non-classical information processing can result in deviations of prices of financial assets from their fundamental values. Robert Shiller [36], and other scholars in behavioural finance noted that price volatility on the stock market is closely followed by investor confidence. One of the major indexes that monitors the link between investor confidence and stock market performance is the "Yale School of Management Stock Market Confidence Index" [42]. As noted by the index creators: *"Stock prices are of course determined by supply and demand, and there are numerous factors that affect these, fundamental factors, legal, tax-related, demographic, technological, international, as well as other psychological factors related to attention, regret, anchoring, and availability. Indexes of stock market confidence can only play a supportive role in trying to understand market events."* [Yale School of Management, Stock Market Confidence Indexes]. We can witness that investors' expectations can rely on the previous trends in the stock market, creating complex 'feedback loops' that can yield bubbles and periods of high volatility, e.g. [5]. As a result, a 'deeper' form of uncertainty about the future price movements can emerge on the finance market. In the following we briefly illustrate a quantum dynamical model of asset price formation under non-classical beliefs of traders.

5.1 *Quantum Markovian Asset Pricing Model: Some Key Features*

The beliefs of investors can be detrimental in affecting the price-dynamics of financial assets and subsequently the realised returns. Since the expectations of traders are directly affecting asset prices, we can talk about some form of "financial bath" that shapes the behaviour of traders and is reflected in the long-term formation of asset prices. Below, we follow the model formulation devised in [23]. Technically, the prices are depicted as numbers of quantum particles in specific states in a finance analogue of the Fock space (that is a more general construction of the Hilbert space, allowing for superposition of multiple financial assets and their price states). The appropriate size of each asset price state space can be derived empirically from the historically observed prices. The quantum notion of *superposition* allows for coexisting beliefs of traders about asset price realisation. For instance, let us assume AN example of two assets. Asset $\mathcal{A}_1$ can have some possible price outcomes, $k = 0, 1, 2...$, and and asset $\mathcal{A}_2$ can have some other possible price configurations, $m = 0, 1, 2....$[4] A superposition of agents' beliefs about the future asset prices, restricted to two price states, can be denoted as: $\psi = c_1|k_1m_1\rangle + c_2|k_1m_2\rangle + c_3|k_2m_1\rangle + c_4|k_2m_2\rangle$. Such

[4] The prices of financial assets can be expressed in terms of some monetary unit of the finance market that we deal with (e.g., \$) given by natural numbers, $n = 1, 2, ...$ If the price of a financial asset is less than one dollar, then it is identified with a discrete level, $n = 0$.

formalisation allows to depict a *deeper uncertainty* related to the future prices of assets than understood by the classical probability distribution of price realisations. In a more general formulation, we can consider N financial assets $\mathcal{A}_j : j = 1, 2, ...,$ where N in the asset state space.

Remark 1 A state space of a single financial asset $\mathcal{A}_j$ is an infinite dimensional Hilbert space H_j with the basis given by basis-vectors: $|n_j\rangle, n_j = 0, 1, 2, ...j$. Since all price levels are possible in theory, such an infinite state space formalisation overcomes the assumptions on probability distributions and addresses the problem of fat tails and infinite variances. Due to agents' uncertainty about the future prices and returns, this space contains superpositions of the assets' price states in the form:

$$|\psi\rangle = \sum_{n_j} c_{n_j} |n_j\rangle, \tag{2}$$

where c_{n_j} are complex numbers, $\sum_{n_j} |c_{n_j}|^2 = 1$. Here $|c_{n_j}|^2$ give the probabilities P_{n_j} that for that asset we observe the price n_j.

Remark 2 State space for a multi-asset price configuration of N assets is depicted as the *tensor product* of the state spaces H_j :

$$H = \otimes_{j=1}^{N} H_j.$$

Its vectors have the form:

$$|\Psi\rangle = \sum_{n} C_n |n\rangle, \tag{3}$$

where $n = n_1...n_N$ and $\sum_n |C_n|^2 = 1$.

The proposed quantum model can be potentially applicable for a preparation of a long-term probabilistic prognosis of multi-asset price configuration. The model can offer an investor, or a financial analyst insights about asset price distributions at some future point in time. Theoretically, the quantum dynamics of asset price fluctuations can continue to infinity. When interacting with a "bath" of investors' expectations, the probability fluctuations associated with asset prices are dampened, and an equilibrium state of possible price distributions is attained. The so called "Gorini-Kossakowski-Sudarshan-Lindblad" (GKSL)[5] model can be used as a form Markovian approximation of the assets' state prices interacting with bath of the expectations and beliefs of market participants.

The Markovian approximation of the quantum master equation (GKSL) has the form:

$$\frac{d\rho}{dt}(t) = -\frac{i}{\gamma}[\mathcal{H}, \rho(t)] + L(\rho(t)), \tag{4}$$

[5] For a more detailed treatment and mathematical details, see [29].

where $\mathcal{H}$ is a Hermitian operator acting in H and L is a linear operator (also known as the "Lindblad term") acting in the space of linear operators $B(H)$ in H. Technically, the operator $\mathcal{H}$ represents the state dynamics in the absence of impact of the outer environment. The Lindblad term, L, contains the impact of the outer environment or the 'financial bath' that feeds into the asset price dynamics. The financial bath serves as a mathematical proxy of the impact of traders' expectations and the emerging financial news.

6 Contextual Beliefs and Preferences: What Are the Key 'Paradoxes' of Quantum-Like Models?

The introduction of quantum-like models was not free from paradoxes, since not all psychological effects could be nicely accommodated with the aid of 'standard' quantum probability models. In other words, decision making statistics exhibits neither quantum nor classical features as shown in [7, 18, 31]. Hence, a generalization using the mathematics from quantum information theory was proposed in order to come as close as possible to a 'unified' framework of human contextual beliefs and preferences.

6.1 Non-double Stochasticity Problem

One of the major caveats with the above reviewed experimental findings on disjunction effect is the following. While the cognitive statistics is violating the rules of classical probability theory, it is also not obeying a so-called *double-stochasticity* constraint. Double-stochasticity implies a unit additivity of the transition probabilities in the transition probability matrix [18, 31]. This feature of cognitive statistics constraints the application of standard quantum physical formalism based on the orthogonal Hermitian projectors.

To be able to fully accommodate the decision making tasks in a two-dimensional Hilbert space, a class of more general projectors, relaxing the orthogonality constraint of projector-operators, "POVMs" (Positive operator valued measures) can be used following their application in quantum information theory [8]. The authors in [12, 25] employ such types of POVMs, to accommodate preference reversals in voting statistics and investment choice.[6]

[6] Some authors approached non-additivity of transition probabilities, by increasing the dimension of the state-space representation of the observables, see [31]. In these models degenerate spectra can be used to represent eigenvalues by corresponding bases that are given as subspaces of larger dimensions.

6.2 Order Effects and Response Replicability

We have already discussed in Sect. 4 that order effects can be faithfully represented with the aid of non-commuting observables that are given by conventional Hermitian operators. At the same time, there is another feature of individual responses or preferences, coined "response replicability". The same response would be usually repeated by an individual if being asked the same question twice. This is knowns as the "A-A" type response replicability. Also, question surveys show that answers and preferences would be naturally stable over short time horizons and in the absence of new information. Hence, if questions are asked in an order A-B-A, the answer to the question A should not change. This brings a caveat in using conventional quantum probability to capture this second type of response replicability, and at the same time to accommodate order effects of the type "A-B" and "B-A". One can say that there exists a 'paradox', pertaining to the usage of non-commuting Hermitian operators to capture preference reversals. This is due to the impossibility to jointly accommodate both psychological effects as firstly highlighted in [20]. This 'no-go' result was at variance with the key feature of quantum probability, namely, the principle of non-commutativity. This problem was further researched in [21], and a more inclusive probabilistic framework inspired by a generalized quantum measurement theory (based on usage of non-projective operators) was recently proposed in [30].[7] Of course, these generalisations add some complexity and additional free parameters to the quantum framework, which is the cost one has to pay for an " all-inclusive" model of contextual decision making.

6.3 What If the Interference Magnitude Is Too High?

Finally, we would like to mention another possible problem related to the usage of conventional quantum probability framework. This problem emerges when the decision-making statistics exhibits non-additivity of a very high magnitude and hence, cannot be described with the aid of complex probability amplitudes in a Hilbert space. For instance, if the parameter, $\cos\theta$, defined in Eq. (1) is not bounded by ± 1, the usage of complex numbers is not possible. Authors in [18] proposed a solution to this problem, by introducing a hyperbolic analogue of the interference coefficient, where the probability amplitudes are given by hyperbolic, rather than complex numbers. An interference of a very high magnitude ($\cos\theta > 3$) was detected in the survey statistics on voting preferences and a hyperbolic state space was introduced to capture this effect [22].

[7] The authors in [30] employ a slightly different mathematics, to accommodate both of these psychological effects in a single operational framework of contextual preferences. In a nutshell, the distinguishing feature of this approach allows for an additional influence of a 'mental environment' upon the cognitive state of a decision-maker.

7 Future Directions and Concluding Remarks

Quantum probability models are a good analogue of classical probability-based decision frameworks and generalisations thereof. The key advantage of quantum-like models is due to their already existing axiomatic and mathematical apparatus that was originally describing the uncertainty associated with a systems' behaviour in quantum physics. Vast evidence from studies in psychology, behavioural economics and finance shows that people's thinking and reasoning might be at variance with classical symbolic logic and Kolmogorovian probability theory. We can also witness that the non-classicality of beliefs can depend on various objective factors (such as the number of random variables to consider, familiarity with the decision making task, previous decisions, etc.), as well as on the mental environment shaped by our emotions. Emotions are indeed playing an important role in our reasoning and behaviour, see e.g., [2, 26]. These findings on emotion-induced ambiguity and risk attitudes motivate for further research in this field to evaluate the impact of our emotions upon our mental states to capture it with the aid of generalised probability models, such as quantum Markvian equation. The richness of contextuality of human beliefs and preferences might often go beyond the axiomatics of standard quantum formalism. We have endeavoured discuss some of the key caveats related to a straightforward application of quantum probability and the solutions offered so far.

References

1. Aerts, D., Haven, E., & Sozzo, S. (2018). A proposal to extend expected utility in a quantum probabilistic framework. *Economic Theory, 65*, 1079–1109.
2. Baillon, A., Koellinger, P. D., & Treffers, T. (2016). Sadder but wiser: The effects of emotional states on ambiguity attitudes. *Journal of Economic Psychology, 53*, 67–82.
3. Atmanspacher, H., Basieva, I., Busemeyer, J. R., Khrennikov, A., Pothos, E. M., Shiffrin, R., & Wang, Z. (2020). What are the appropriate axioms of rationality for reasoning under uncertainty and resource constrained systems? *Behavioral and Brain Sciences, 43*, e2.
4. Baaquie, B. E. (2004). *Path integrals and Hamiltonians for options and interest rates*. Cambridge University Press.
5. Barberis, N., Shleifer, A., & Vishy, R. (1998). A model of investors sentiment. *Journal of Financial Economics, 49*, 307–343.
6. Busemeyer, J. R., Pothos, E. M., Franco, R., & Trueblood, J. S. (2011). A quantum theoretical explanation for probability judgement errors. *Psychological Review, 118*(2), 193–218.
7. Busemeyer, J. R., & Bruza, P. D. (2012). *Quantum models of cognition and decision*. Cambridge University Press.
8. Busch, P., Grabowski, M., & Lahti, P. (1995). *Operational quantum physics*. Springer.
9. Croson, R. (1999). The disjunction effect and reason-based choice. *Organizational Behavior and Human Decision Processes, 80*(2), 118–133.
10. Ellsberg, D. (1961). Risk, ambiguity and the Savage axioms. *Quarterly Journal of Economics, 75*, 643–669.
11. Haven, E., & Khrennikov, A. (2013). *Quantum Social Science*. Cambridge University Press.
12. Haven, E., & Khrennikova, P. (2018). A quantum-probabilistic paradigm: Non-consequential reasoning and state dependence in investment choice. *Journal of Mathematical Economics, 78*, 186–197.

13. Kahneman, D., & Tversky, A. (1979). Prospect theory: An analysis of decision under risk. *Econometrica, 47*, 263–291.
14. Kahneman, D. (2011). *Thinking, fast and slow*. Penguin Books.
15. Khrennikov, A. (1999). Classical and quantum mechanics on information spaces with applications to cognitive, psychological, social and anomalous phenomena. *Foundations of Physics, 29*, 1065–1098.
16. Khrennikov, A. (2004). On quantum-like probabilistic structure of mental information. *Open Systems and Information Dynamics, 11*(3), 267–275.
17. Khrennikov, A. (2004). *Information dynamics in cognitive, psychological, social, and anomalous phenomena*. Fundamental theories of physics: Kluwer.
18. Khrennikov, A. (2010). *Ubiquitous quantum structure*. Springer.
19. Khrennikov, A., & Haven, E. (2009). Quantum mechanics and violations of the sure-thing principle: the use of probability interference and other concepts. *Journal of Mathematical Psychology, 53*, 378–388.
20. Khrennikov, A., Basieva, I., Dzhafarov, E. N., & Busemeyer, J. R. (2014). Quantum models for psychological measurements: an unsolved problem. *PLoS ONE*, 9(10).
21. Khrennikov, A., & Basieva, I. (2014). Quantum model for psychological measurement: From the projection postulate to interference of mental observables represented as Positive Operator Valued Measures. *NeuroQuantology, 12*(3), 324–336.
22. Khrennikova, P. (2015). Quantum like modelling of the non-separability of voters' preferences in the U.S. Political System. *Lecture Notes in Computer Science, 8951*, 196–209.
23. Khrennikova, P. (2016). Application of quantum master equation for long term prognosis of asset prices. *Physica A: Statistical Mechanics and its Applications, 450*, 253–263.
24. Khrennikova, P., & Haven, E. (2015). Instability of political preferences and the role of mass-media: A dynamical representation in a quantum framework. *Philosophical Transactions of the Royal Society A, 374*, 20150106.
25. Khrennikova, P., & Haven, E. (2017). Quantum generalised observables framework for psychological data: A case of preference reversals in US elections. *Philosophical Transactions of the Royal Society A, 375*, 20160391.
26. Kliger, D., & Levy, O. (2003). Mood-induced variation in risk preferences. *The Journal of Economic Behavior and Organization, 52*(4), 573–584.
27. Knight, F. H. (1921). *Risk, Uncertainty, and Profit*. Houghton Mifflin Company.
28. Kolmogorov A. N. (1933). *Grundbegriffe der Warscheinlichkeitsrechnung*. Springer (English translation: 1956 *Foundations of the probability theory*. Chelsea Publishing Company).
29. Ohya, M., & Volovich, I. (2011). *Mathematical foundations of quantum information and computation and its application to nano- and bio systems*. Springer.
30. Ozawa, M., & Khrennikov, A. (2019). Application of theory of quantum instruments to psychology: combination of question order effect with response replicability effect. *Entropy, 22*(1), 37. https://doi.org/10.3390/e22010037
31. Pothos, E., & Busemeyer, J. (2009). A quantum probability explanation for violations of rational decision theory. *Proceedings of the Royal Society London B, 276*, 2171–2178.
32. Roca, M., Hogarth, A. J., & Maule, A. J. (2006). Ambiguity seeking as a result of the status quo bias. *Journal of Risk and Uncertainty, 32*, 175–194.
33. Savage, L. J. (1954). *The foundations of statistics*. Wiley.
34. Shafir, E., & Tversky, A. (1992). Thinking through uncertainty: Non-consequential reasoning and choice. *Cognitive Psychology, 24*, 449–474.
35. Shafir, E. (1994). Uncertainty and the difficulty of thinking through disjunctions. *Cognition, 49*, 11–36.
36. Shiller, R. (2000). *Irrational exuberance*. Princeton University Press.
37. Tversky, A., & Fox, G. (1995). Weighting risk and uncertainty. *Psychological Review, 102*(2), 269–283.
38. Tversky, A., & Shafir, E. (1992). The disjunction effect in choice under uncertainty. *Psychological Science, 3*(5), 305–309.

39. von Neumann J. (1932). *Mathematische Grundlagen der Quantenmechanik*. Springer. (English translation: 1955 *Mathematical Foundations of Quantum Mechanics*. Princeton University Press).
40. Wang, Z., Busemeyer, J., Atmanspacher, H., & Pothos, E. (2013). The potential for using quantum theory to build models of cognition. *Topics in Cognitive Science, 5*, 672–688.
41. White, L. C., & Pothos, E. M. (2020). The cost of asking: How evaluations bias subsequent judgements. *Decision, 7*(14), 259–286.
42. Yale. (2021). Yale School of Management Stock Market Confidence Indexes, *International Center for Finance*, Yale. https://som.yale.edu/faculty-research/our-centers-initiatives/international-center-finance/research-initiatives. Accessed September 25, 2021.

Reward for Good Performance Works Better Than Punishment for Mistakes: Economic Explanation

Olga Kosheleva, Julio Urenda, and Vladik Kreinovich

Abstract How should we stimulate people to make them perform better? How should we stimulate students to make them study better? Many experiments have shown that reward for good performance works better than punishment for mistakes. In this paper, we provide a possible theoretical explanation for this empirical fact.

1 Formulation of the Problem

Reward versus punishment: an important economic problem. One of the most important issues in economics is how to best stimulate people's productivity, what is the best combination of reward and punishment that makes people perform better.

This problem is ubiquitous. This problem arises not only in economics, it appears everywhere. How do we stimulate students to study better? How do we stimulate our own kids to behave better?

Empirical fact. In his famous book [3] summarizing his research, the Nobelist Daniel Kahneman—one of the fathers of behavioral economics—cites numerous researches that all confirm that reward for good performance, in general, works better than punishment for mistakes; see, e.g., p. 175.

But why? Like many other facts from behavioral economics, this empirical fact does not have a convincing theoretical explanation.

What we do in this paper. In this paper, we provide a theoretical explanation for this empirical phenomenon.

O. Kosheleva · J. Urenda · V. Kreinovich (✉)
University of Texas at El Paso, El Paso, Texas 79968, USA
e-mail: vladik@utep.edu

O. Kosheleva
e-mail: Olgak@utep.edu

J. Urenda
e-mail: jcurenda@utep.edu

S. Sriboonchitta et al. (eds.), *Credible Asset Allocation, Optimal Transport Methods, and Related Topics*, Studies in Systems, Decision and Control 429,
https://doi.org/10.1007/978-3-030-97273-8_10

2 Analysis of the Problem

What people want. People spend some efforts e, and, based on results of these efforts, they get some reward $r(e)$. In the first approximation, we can say that the overall gain is the reward minus the efforts, i.e., the difference

$$r(e) - e. \tag{1}$$

A natural economic idea is that every person wants to maximize his/her gain, i.e., maximize the difference (1).

How to proceed. In view of the formula (1), to explain why rewards work better than punishments, we need to analyze what are the reward functions $r(e)$ corresponding to these two different reward strategies. Similarly to our derivation of the formula (1), in our analysis, we will use simplified "first approximation" models, i.e., models providing a good qualitative understanding of the situation.

What reward function corresponds to rewarding good performance. Crudely speaking, rewarding good performance means that:

- if the performance is not good, i.e., if the effort e is smaller than the smallest needed effort e_0, there is practically no reward:

$$r(e) = r_+ \tag{1}$$

  for some $r_+ \approx 0$;
- on the other hand, the more effort the person uses, the larger the reward; in other words, every effort beyond e_0 is proportionally rewarded, i.e.,

$$r(e) = r_+ + c_+ \cdot (e - e_0), \tag{2}$$

  for some constant c_+.

The constant c_+ depends on the units used for measuring effort and reward: one unit of effort corresponds to c_+ units of reward. The two formulas:

- the formula (1) corresponding to the case $e \le e_0$, and
- the formula (2) corresponding to the case $e \ge e_0$,

can be combined into a single formula

$$r(e) = r_+ + \max(0, c_+ \cdot (e - e_0)) = r_+ + c_+ \cdot \max(0, e - e_0). \tag{3}$$

This dependence has the following form:

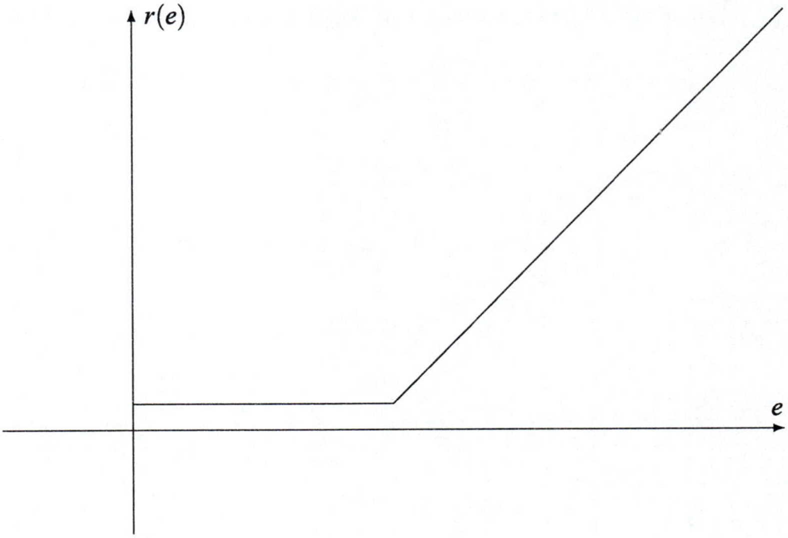

What can we say about this function. It is easy to see that the function (2) is *convex* in the sense that for all $e' < e''$ and for each $\alpha \in [0, 1]$, we have

$$r(\alpha \cdot e' + (1-\alpha) \cdot e'') \le \alpha \cdot r(e') + (1-\alpha) \cdot r(e'').$$

What reward function corresponds to punishing for mistakes. Crudely speaking, punishing for mistakes means that:

- if the performance is good, i.e., if the effort e is larger than or equal to the smallest needed effort e_0, then there is no punishment, i.e., the reward remains the same:

$$r(e) = r_- \tag{4}$$

  for some constant r_-;
- on the other hand, the fewer effort the person uses, the most mistakes he/she makes, so the larger the punishment and the smaller the resulting reward; in other words, every effort below e_0 is proportionally penalized, i.e.,

$$r(e) = r_- - c_- \cdot (e_0 - e), \tag{5}$$

  for some constant c_-.

The constant c_- depends on the units used for measuring effort and reward: one unit of effort corresponds to c_- units of reward. The two formulas:

- the formula (4) corresponding to the case $e \ge e_0$, and
- the formula (5) corresponding to the case $e \le e_0$,

can be combined into a single formula This dependence has the following form:

$$r(e) = r_- - c_- \cdot \max(0, e_0 - e) = r_- + c_- \cdot \min(0, e - e_0). \tag{6}$$

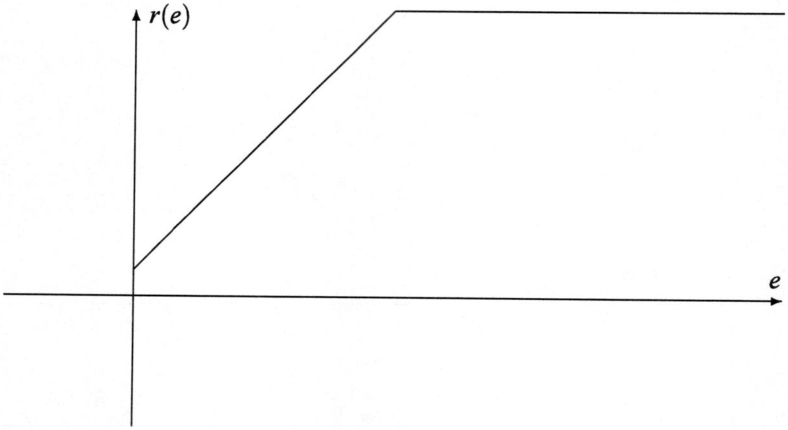

What can we say about this function. It is easy to see that this function is *concave* in the sense that for all $E' < E''$ and for each $\alpha \in [0, 1]$, we have

$$r(\alpha \cdot e' + (1 - \alpha) \cdot e'') \geq \alpha \cdot r(e') + (1 - \alpha) \cdot r(e'').$$

Now, we are ready to present the desired explanation.

3 Our Explanation

Known properties of convex and concave functions: reminder. It is known (see, e.g., [5]) that:

- every linear function is both convex and concave;
- the sum of two convex functions is convex, and
- the sum of two concave functions is concave.

In particular, the linear function $f(e) = -e$ is both convex and concave, thus:

- when the function $r(e)$ is convex, the sum $r(e) + (-e) = r(e) - e$ is also convex; and
- when the function $r(e)$ is concave, the sum $r(e) + (-e) = r(e) - e$ is also concave.

It is also known that:

- for a convex function, the maximum on an interval is always attained at one of this interval's endpoints, while

- for a concave function, its maximum on an interval is always attained at some point inside the interval.

Resulting explanation. As we have mentioned earlier, a person selects the effort e_0 for which the expression $r(e) - e$ attains its largest possible value.

Of course, people's abilities are not unbounded, there are certain limits within which we can apply the efforts. Thus, possible value of the effort e are located within some interval $[\underline{e}, \overline{e}]$. Thus:

- When we reward for good performance, the corresponding function $r(e)$ is convex, thus the difference $r(e) - e$ is convex, and therefore, the selected value e_0 coincides either with $\underline{e}$ or with $\overline{e}$. Thus, if we dismiss the case $e_0 = \underline{e}$ (when the reward is so small that it is not worth spending any effort), we conclude that $e_0 = \overline{e}$, i.e., the person selects the largest possible effort—which is exactly what we wanted to achieve.
- On the other hand, when we punish for mistakes, the corresponding function $r(e)$ is concave, thus the difference $r(e) - e$ is concave, and therefore, the selected value e_0 is always located inside the interval $[\underline{e}, \overline{e}]$: $e_0 < \overline{e}$. Thus, the person will not select the largest possible effort—which is exactly what we wanted to avoid.

This indeed explains why rewarding for good performance works better than punishment for mistakes.

Comments

- What if we have both reward for good performance and punishment for mistakes, i.e., $r(e) = \text{const} + c_+ \cdot \max(0, e - e_0) + c_- \cdot \min(0, e - e_0)$? In this case, for $c_+ > c_-$, the function is still convex, i.e., we still get a very good performance, but if $c_- > c_+$, the function becomes concave, and the performance suffers. Thus, to get good results, reward must be larger than punishment.
- It is interesting to observe that the optimal rewarding function

$$r(e) = r_+ + c_+ \cdot \max(0, e - e_0),$$

in effect, coincides (modulo linear transformations of input and output) with the empirically efficient "rectified linear" activation function $r(e) = \max(0, e)$ used in deep learning; see, e.g., [1, 2, 4]. So, not only people learn better when we use this function—computers learn better too!

Acknowledgements This work was supported in part by the US National Science Foundation grants 1623190 (A Model of Change for Preparing a New Generation for Professional Practice in Computer Science) and HRD-1242122 (Cyber-ShARE Center of Excellence). It was also supported by the program of the development of the Scientific-Educational Mathematical Center of Volga Federal District No. 075-02-2020-1478.

References

1. Fuentes, O., Parra, J., Anthony, E., & Kreinovich, V. (2020). Why rectified linear neurons are efficient: A possible theoretical explanations. In O. Kosheleva, S. Shary, G. Xiang, & R. Zapatrin (Eds.), *Beyond traditional probabilistic data processing techniques: Interval, fuzzy, etc. Methods and their applications* (pp. 603–613). Springer.
2. Goodfellow, I., Bengio, Y., & Courville, A. (2016). *Deep leaning*. MIT Press.
3. Kahneman, D. (2011). *Thinking, fast and slow*. Straus, and Giroux: Farrar.
4. Kreinovich, V., & Kosheleva, O. (2020). Deep learning (partly) demystified. In *Proceedings of the 2020 4th International Conference on Intelligent Systems, Metaheuristics & Swarm Intelligence ISMSI'2020*, Thimpu, Bhutan, April 18–19, 2020.
5. Rockafeller, R. T. (1970). *Convex analysis*. Princeton University Press.

The Conjunction Fallacy in Quantum Decision Theory

Tatyana Kovalenko and Didier Sornette

Abstract The conjunction fallacy is a renowned violation of classical probability laws, which is persistently observed among decision makers. Within Quantum decision theory (QDT), such deviations are the manifestation of interference between decision modes of a given choice prospect. We propose a novel QDT interpretation of the conjunction fallacy, which cures some inconsistencies of a previous treatment, and incorporates the latest developments of QDT, in particular the representation of a decision-maker's state of mind with a statistical operator. Rather than focusing on the interference between choice options, our new interpretation identifies the origin of uncertainty and interference between decision modes to an entangled state of mind, whose structure determines the representation of prospects. Thus, on par with prospects, the state of mind can be a source of uncertainty and lead to interference effects, resulting in characteristic behavioral patterns. We present the first in-depth QDT-based analysis of an empirical study (the touchstone experimental investigations of Shafir et al. [6]), which enables a data-driven exploration of its underlying theoretical construct. We link typicality judgements to probability amplitudes of the decision modes in the state of mind, and quantify the level of uncertainty and the relative contributions of prospect's interfering modes to its probability judgement. This enables inferences about the key QDT interference "attraction" q-factor with respect to different types of prospects—compatible versus incompatible. We propose a novel empirically motivated "QDT indeterminacy (or uncertainty) principle," as a fundamental limit of the precision with which certain sets of prospects can be simultaneously known (or assessed) by a decision maker, or elicited by an experimental procedure. For any type of prospects, we observe a general tendency for the q-factor to converge to the same negative range $q \in (-0.3, -0.1)$ in the presence of high uncertainty, which motivates the hypothesis of an universal "aversion" q. The "aver-

T. Kovalenko · D. Sornette (✉)
Department of Management, Technology and Economics, ETH Zurich, Scheuchzerstrasse 7, 8092 Zurich, Switzerland
e-mail: dsornette@ethz.ch

D. Sornette
Swiss Finance Institute, University of Geneva, 40 blvd. Du Pont d'Arve, 1211 Geneva 4, CH, Switzerland

S. Sriboonchitta et al. (eds.), *Credible Asset Allocation, Optimal Transport Methods, and Related Topics*, Studies in Systems, Decision and Control 429,
https://doi.org/10.1007/978-3-030-97273-8_11

sion" q is independent of the (un-)attractiveness of a prospect under more certain conditions, which is the main difference with the previously considered "QDT quarter law". The universal "aversion" q substantiates the previously proposed "QDT uncertainty aversion principle" and clarifies its domain of application. The universal "aversion" q provides a theoretical basis for modelling different risk attitudes, such as aversions to uncertainty, to risk or to losses.

Keywords Quantum decision theory · Conjunction fallacy · Interference · Indeterminacy principle · Universal aversion

1 Introduction

The conjunction fallacy is a well-known behavioral pattern, when the probability for a conjunction category (AB, e.g. "feminist bankteller") is judged larger than for its constituents (e.g. B, "bankteller"). Although such probability judgement violates the axiomatic probability theory, it is nevertheless consistently observed among decision makers under different experimental setups (indirect, direct-subtle and direct-transparent tests). Several plausible explanations were proposed, such as fallacious representativeness and availability heuristics in the original study [8], or "socially rational" semantic inferences about the meaning of 'probability' [4]. Alternative approaches using the quantum formalism were also applied to tackle the problem. The quantum judgment model suggests that a choice is made by the sequential projection of a decision maker's belief state onto prospects subspaces, resulting in transition probabilities that are based on Lüder's rule [2]. However, the resultant 'question order effect' was criticized for its inability to model double conjunction fallacies. Moreover, empirical analysis confirmed the advantage of another approach to explain the conjunction fallacy, based on modelling 'states of conceptual entities' and 'emergence effects' [1]. Importantly, it can be shown rigorously that Lüder's rule for calculating the probability of consecutive measurements cannot be used for calculating quantum joint probabilities, in particular for non-commuting prospects. The rigorous derivation of quantum joint probabilities for arbitrary prospects, whether commuting or non-commuting, has been proposed within an 'emergence-type' Quantum decision theory (QDT) in [14], which in addition indicates the differences between QDT and other quantum approaches.

The first interpretation of the conjunction fallacy within QDT was suggested in [13]. It exploits the general QDT relation, $p = f + q$, where the probability p that a prospect is chosen is decomposed into the sum of two factors, the utility f and attraction q of that prospect. For a binary choice (between two options), the so called "QDT quarter law" predicts an average amplitude (absolute value) of the attraction factor q to be $\frac{1}{4}$. This top-level non-parametric approach was applied to the experimental results from [6], and the aggregate empirical data was shown to be in agreement with the prediction of the "QDT quarter law".

While acknowledging this first interpretation of the conjunction fallacy within QDT, in particular for its simplicity and absence of adjustable parameters, our detailed analysis reveals some of its weaknesses, i.e. inconsistencies in the underlying assumptions and limited usage of the empirical data. In particular, in [13], the conjunction fallacy is explained by a non-attractiveness of a prospect with a single constituent category B (e.g. "bankteller"). This effect is introduced by imposing a negative interference with a secondary feature A (e.g. "feminist") from a conjunction AB (e.g. "feminist bankteller"). However, the secondary feature is present exclusively in the conjunction category AB. In indirect tests, as in [6], a decision maker is exposed to only one of the judged categories at a time (either a conjunction, or its constituent). Thus, the negative interference with the secondary feature A should not appear in deliberations concerning a single constituent category B.

In [16], a general procedure to introduce uncertainty in decision making was proposed by incorporating an observable, which represents the belief and disbelief of a decision maker about a task setup and a relevant criterion of choice. Indeed, any actual choice situation can be associated with some (irreducible) degree of uncertainty. However, this approach may explain only the occurrence of interference, but not its amplitude and the corresponding sign of the effect, i.e. (non-)attractiveness of a particular choice prospect. Thus, the conjunction fallacy remains unexplained within previous proposals using QDT.

Applying the quantum formalism to decision making and transforming its theoretical construct into an operational tool is challenging. A situation of choice, which includes a decision maker and choice options, should be characterized adequately in terms of the corresponding operators and state vectors, including: (a) a decision-maker's state of mind ρ (for pure states, $\rho = |\psi\rangle\langle\psi|$) and (b) prospects $|\pi_i\rangle$. First, the choice of a basis for the representation of these vectors, i.e. a set of elementary prospects $|e_j\rangle$, is not trivial. For example, in a given experimental setup, a prospect can be postulated via certain action modes, however some of them can themselves turn out to be quantum superpositions. In addition, the state of mind can include interfering elementary prospects that are not originating from the prospects in consideration (i.e. choice options), but from other sources: (i) endogenous (a decision maker's personality traits, experience, beliefs, and so on) and (ii) exogenous (framing, environment, and so on). Second, revealing and estimating the coefficients of the linear decomposition of the state of mind and of prospects over the basis (i.e. elementary prospects) is a subtle task. These coefficients are probability amplitudes, and their squared values correspond to probabilities (i.e. relative weights) of elementary prospects (i.e. decision modes) in the state of mind. The difficulty is that invasive elicitation methods can affect the state of mind of the decision maker and alter probability amplitudes.

These challenges of applying QDT influenced previous explorative directions. The preceding research either checked the agreement between empirical data and the top-level QDT relation $p = f + q$ [15], or sought to construct a sound parameterization of the utility f-factor and interference q-factor, based on "classical" decision theories, such as Expected value [3], Expected utility theory and (stochastic) Cumulative prospect theory [7, 9].

In this paper, we present the first in-depth QDT-based analysis of an empirical dataset, and derive observed prospects' probabilities by quantifying a decision-maker's state of mind and choice options. We focus on the experiment that was reported in [6], and previously analyzed within QDT in [13]. We propose a novel QDT interpretation of the conjunction fallacy, which cures some inconsistencies of this previous treatment, and incorporates the latest developments of QDT, in particular the representation of a decision-maker's state of mind with a statistical operator [16, 18]. Our main contribution is to put the state of mind at the centre of the evaluation of the level of uncertainty that influences prospects' representations, resulting in interference effects. Our novel QDT interpretation of the conjunction fallacy is based on the following propositions:

1. An *interference between the (pre-prepared) state of mind and a prospect* is proposed to be the mechanism of the conjunction fallacy. The descriptions of subjects (instances I) and some of the associated categories (usually a secondary feature A in a conjunction AB) share common characteristics. After the exposition to an instance I, the state of mind of a decision maker is intentionally influenced (framed) by incepting into it specific elementary prospects, which then interfere with resembling choice options and modify the prospects' probabilities. This mechanism is in line with the representativeness and availability heuristics that were suggested for the interpretation of the conjunction fallacy in the original study [8]. QDT provides formal tools to quantify (and predict) such effects.
2. In order to calculate the probabilities of the prospects (explicitly, e.g. for the theoretical formulation of situations of choices, or implicitly, e.g. in real life situations), both prospects and state of mind should be represented with the same set of elementary prospects. Thus, in general, prospects and state of mind are *mutually dependent* in the granularity of their decompositions (i.e., they share the same basis of elementary prospects). In practice, depending on the constituents (elementary prospects) of a state of mind, the decision modes of the considered prospects, as formulated in a subsequent situation of choice, may require further decomposition, i.e. the presented modes in a given experiment may be in a *quantum superposition* themselves, or be a tensor product of several more specific (detailed) elementary prospects. In other words, the choice of the elementary prospects that form a basis *is context dependent*, and may differ from the explicitly presented (e.g. in an experimental setup) formulation.

In summary, the present study differs from previous ones by first reinserting the state of mind at the core of the formalism, as it was initially defined in the axiomatic construction of [12, 13]. This allows us to clarify the interference mechanism of QDT, based on the quantification of the decision maker's state of mind and of the considered prospects. This is achieved by linking typicality judgements and probability amplitudes of decision modes in the state of mind. We decompose the problem into several (extreme) cases, with minimum and maximum interference, as well as with singular or distributed weights of interfering modes in the state of mind. This allows us to estimate the level of uncertainty and the relative contributions of prospect's decision modes to its probability judgement. This level of granularity enables the analysis of broader and more nuanced datasets, when interference effects are less pronounced (for example, for prospects with compatible categories). It also opens the possibility of better characterising interdependencies and the relative importance of QDT elements, and finally makes the theory operational.

The organisation of the article is as follows. Section 2 summarises the analysed experimental set-up and its main results. The detailed presentation of the experimental results from [6] are reproduced in Appendix 1, These experimental results will be compared with the prediction of our novel QDT interpretation in Sect. 4. A self-contained recapitulation of the previous approach to the conjunction fallacy developed by [13] and dissection of its weaknesses can be found in Appendix 2. It can be consulted as a forerunner of the novel formulation, presented in Sect. 3 or can be put aside as it is not needed for the comprehension of the present theory. Section 4 compares the detailed experimental results in the two main set-ups with the predictions of our new QDT interpretation of the conjunction fallacy. Section 5 builds on previous sections and results to suggest a modification of the previous concept of an "attraction factor" q into a universal "aversion factor" q. Section 6 concludes. Appendix 3 shows that the necessary conditions for the emergence of the QDT attraction factor are verified in the experimental set-up. Appendix 4 presents additional results of the analysis.

2 Brief Description of the Analyzed Experimental Setup

The present study analyses empirical results reported in [6]. This section provides a brief description of the experimental setup and its main findings.

During the experiment, 110 decision makers were exposed to 14 instances I (description of a subject) followed by one of the four categories: (1) a compatible conjunction $AB(c)$; (2) its constituent $B(c)$; (3) an incompatible conjunction $AB(i)$; (4) its constituent $B(i)$. The compatibility (c) or incompatibility (i) type was attributed by the experimenters based on their qualitative evaluation of a number of shared properties between conjunction elements A and B.

An example of an instance I is the description of a subject: "Linda was a philosophy major. She is bright and concerned with issues of discrimination and social justice."

The corresponding four categories are:

1. feminist teacher, $AB(c)$;
2. teacher, $B(c)$.
3. feminist bankteller, $AB(i)$;
4. bankteller, $B(i)$;

One group of participants (54 decision makers) was asked to make judgements about the typicality (typ_{exp}) of an instance I in each category. Another group (56 decision makers) provided their judgements about the probability (p_{exp}) that an instance I belonged to the corresponding category. Importantly, 14 instances were presented with each of the categories independently, i.e. an instance and one category at a time, and presented in mixed order. When sampling the four different categories for a given instance, these four pairs were presented randomly and were separated by pairs involving other instances.

Figure 1 illustrates the experimental setup [6].

The conjunction effect was calculated by subtracting the typicality judgement of the instance with respect to a constituent category B from the corresponding typicality judgement with respect to the conjunction AB. The conjunction effect exists when this difference is positive. Similarly, the conjunction fallacy is qualified when the difference between the probability judgements for a conjunction AB and its constituent B is positive. Note that typicality and probability judgements were performed by two distinct groups of participants, which may lead to additional discrepancies between an intended theoretical interpretation (Sect. 3) and the empirical values of prospects' probabilities.

The experimental results from [6] are reproduced in Appendix 1 Table 8, and their aggregate statistic is provided in Table 1. Averaged among participants, the conjunction effect and conjunction fallacy (with positive differences larger then 1%) were reported for 10 out of 14 instances, when coupled with *compatible* categories. The conjunction effect and conjunction fallacy were observed to be stronger and confirmed for all 14 instances, when combined with *incompatible* categories.

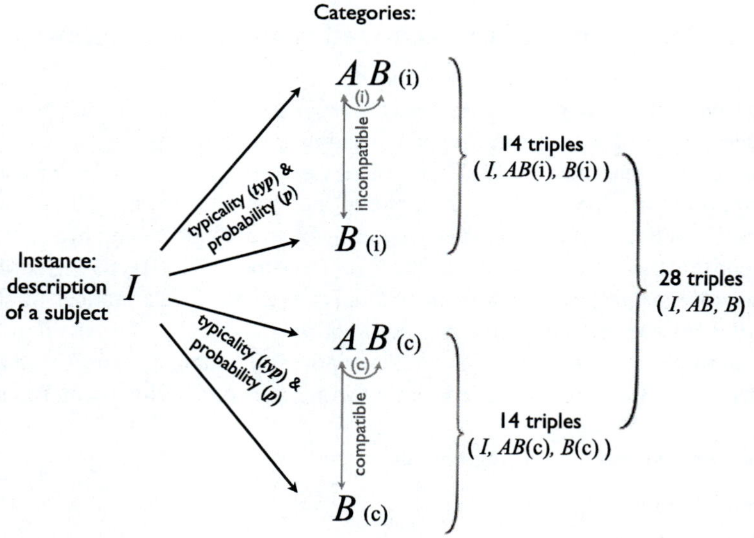

Fig. 1 Experimental setup from [6]. Participants were asked to make typicality and probability judgments about 14 instances I (description of a subject) with respect to four categories: (1) a compatible conjunction ($AB(c)$), (2) its constituent $B(c)$; (3) an incompatible conjunction ($AB(i)$), and (4) its constituent $B(i)$. The compatibility (c) and incompatibility (i) characteristic were attributed by experimenters based on a qualitative assessment of a number of shared properties between conjunction elements (A and B). Judgments were made by two distinct groups: 54 decision makers for typicality, and 56 for probability

Table 1 Aggregate statistics of the experimental results from [6], with a total of 28 triples that combine 14 instances I with categories AB and B, either incompatible (i) or compatible (c)

Triples type	Conjunction effect				Conjunction fallacy				Correlation between conjunction effects and fallacies*
	$typ(AB)$	$typ(B)$	$typ(AB) - typ(B)$		$p(AB)$	$p(B)$	$p(AB) - p(B)$		
	μ	μ	μ	(σ)	μ	μ	μ	(σ)	
(I, AB, B), including:	0.520	0.387	0.133	(0.115)	0.456	0.379	0.078	(0.092)	0.83 (p < 0.01, N=28)
$(I, AB(i), B(i))$	0.371	0.167	**0.204**	(0.063)	0.346	0.220	**0.126**	(0.064)	0.81 ($p < 0.05$, $N = 14$)
$(I, AB(c), B(c))$	0.670	0.606	**0.063**	(0.113)	0.567	0.537	**0.030**	(0.093)	0.58 ($p < 0.05$, $N = 14$)

Sample means μ of typicality (typ) and probability (p) judgments for the 4 categories with respect to I are reported. Sample means μ and sample standard deviations σ for the sizes of the conjunction effect and conjunction fallacy, as well as the correlation between them, are provided. Reproduced from [6]Statistical significance at a level α is obtained when p-value <= α. Significance level α is often set to 0.05, 0.01, 0.001 or lower

Significant positive correlation between the magnitudes of the conjunction effects and fallacies was reported. Note that the correlation is lower for the pairs involving incompatible categories.

For the full description of the experiment, corresponding instances and categories, we refer to the original study [6].

As mention in Introduction, the first interpretation of the conjunction fallacy within QDT was suggested in [13]. A summary of the previous interpretation and our thorough analysis of its problematic issues are presented in Appendix 2. This analysis allows one to understand the evolution of QDT and the contribution that the novel present interpretation of the conjunction fallacy brings to settle the inconsistencies and develop the QDT approach.

3 Novel Theoretical Reinterpretation of the Conjunction Fallacy with QDT

The mean idea to explain the conjunction fallacy is that decision modes of prospects and of a decision-maker's state of mind are mutually related. The probability judgement, i.e. prospect probability, is influenced by the interfering decision modes of the state of mind, which were intentionally incepted by a specific description of a subject (e.g., an instance I as a framing tool). The description of a subject is represented as an uncertain union, due to the incompleteness of the description (a subject is briefly characterized by just a few features) and the large uncertainty in the evolution of the mentioned characteristics. In typical experiments investigating the conjunction fallacy, the decision maker is not confronted with a task of specifying a definitive set of a subjects' characteristics. These attributes remain undefined and are represented by an intermediate inconclusive event with interfering modes.

In the following subsections, we first introduce QDT concepts to characterize a decision-maker and possible events (decisions). Then we describe a decision making process with a two-step methodology and demonstrate the origin of interference effect that can yield the conjunction fallacy.

3.1 *The Strategic Decision-Maker State*

In QDT, a decision maker is characterized by a strategic decision-maker state, which is in general represented by a statistical operator $\hat{\rho}$ on a decision-maker space of mind $\mathcal{H}$. A space of mind is a Hilbert space that is spanned by a set of orthogonal basic states $\{|e\rangle\}$—all admissible events or decisions, which are considered by a decision maker. A strategic decision-maker state $\hat{\rho}$ defines the probabilities of prospective decisions to be taken. It reflects individuality (e.g. persistent personality traits) interconnected with a surrounding (e.g. memory, experience, social influence, as well as fleeting impressions) in the context of a specific choice situation.

A strategic decision-maker state evolves over time. This means that the representation basis as well as coefficients of decomposition are time-dependent. Basic personality traits and fundamental values are relatively stable and undergo a gradual transformation. Weights of these inherent individual characteristics are predominant for choices involving important outcomes, such as life-determining events that require thorough deliberation. However, most of everyday choices, as well as experimental setups, are concerned with relatively minor outcomes and are subjected to a time constraint. Thus, in many situations, simplified rules, i.e. heuristics, are employed to make decisions. Depending on the task at hand, only a few factors come to the fore (e.g. more recent, frequent or typical features) and substantially determine the choice. Practically, this suggests that, in certain cases, a strategic decision-maker state $\hat{\rho}$ can be reduced to a few contextually dependent dimensions (decision modes), or (and) even represented by a pure state $\hat{\rho} = |\psi\rangle\langle\psi|$.

Thus, the description of a choice situation (e.g. a general experimental setup, the formulation of options) is decisive for the basis composition and the resultant prospect probability. Measurable deviations in choice can be generated by intentional manipulations with a task context and produce specific behavioral patterns, such as framing or anchoring effects.

3.2 Decision Modes of the Experiment

At the beginning of the experiment, before any category for a probability judgment is formulated, a decision maker is pre-exposed to an instance that describes a subject by a combination of characteristics. Let us introduce an observable I with a set of possible values:

$$I = \{I_1, I_2\} = \biguplus_k I_k \qquad (k = 1, 2), \tag{1}$$

where I_1 reflects characteristics that are attributed to a subject from a description, while I_2 includes characteristics that do not fit the image presented in an instance. In other words, I_1 and I_2 are defined as complementary, mutually exclusive and exhaustive, characteristics: $I_2 = not\, I_1$. Thus, specific features are highlighted in the mind of a decision maker, and contrasted to others. This classification facilitates certain operations, such as recognition of a subject and its comparison with other subjects, by matching their characteristics. However, a high degree of uncertainty in the attribution of features to a subject remains due to the brevity and fuzziness of a description. Firstly, the classification can be ambiguous for a characteristic that is resembling other qualities of a subject and is typical of him/her, but is not directly mentioned in the presented description. Secondly, the strength and evolution of individual inclinations can be obscure. For example, early interests can be developed to a professional level or dissipated over time. Until the related uncertainty is not resolved in the mind of a decision maker by an observable decision, i.e. a specific value I_k has not been chosen, I is considered to be an inconclusive (or operationally uncertain) event, and the set (1) is an uncertain union [16].

The state vector $|I\rangle$ corresponding to the observable I is thus a linear combination

$$|I\rangle = \gamma_1|I_1\rangle + \gamma_2|I_2\rangle \tag{2}$$

of mode states $|I_k\rangle$ (k=1,2) with probability weights $|\gamma_k|^2$. The set of vectors $\{|I_1\rangle, |I_2\rangle\}$ forms an orthonormal basis

$$\langle I_k||I_l\rangle = \delta_{kl}, \text{ where} \delta_{kl} \text{is the Kronecker delta,} \tag{3}$$

and its linear span is a Hilbert space $\mathcal{H}_I$

$$\mathcal{H}_I \equiv span\{|I_1>, |I_2>\}. \tag{4}$$

For an uncertain union (1), a state vector (2) generates an operator $\hat{P}_I$, which forms an operator algebra $\mathcal{I} \equiv \{\hat{P}_{I_k}\}$ and is expressed as

$$\begin{aligned}\biguplus_k I_k \quad \rightarrow \quad \hat{P}_I \; &\equiv \; |I\rangle\langle I| = \sum_k |\gamma_k|^2 |I_k\rangle\langle I_k| + \sum_{k\neq l} \gamma_k\gamma_l^* |I_k\rangle\langle I_l| \\ &= \sum_k |\gamma_k|^2 \hat{P}_{I_k} + \sum_{k\neq l} \gamma_k\gamma_l^* |I_k\rangle\langle I_l| \quad k, l \in \{1, 2\}. \end{aligned} \tag{5}$$

In the last summand of (5), the term $|I_k\rangle\langle I_l|$, for $k \neq l$, does not belong to the operator algebra $\mathcal{I}$. In other words, the operator $\hat{P}_I$ cannot be represented as a linear combination of the elements from its algebra $\{\hat{P}_{I_k}\}$. Thus, the operator $\hat{P}_I$ is called entangled, and the modes I_k of the corresponding inconclusive event I are interfering with each other [16].

In indirect tests, the description of a subject (an instance I) is followed by one of the categories, either a conjunction $A\&B$, or its constituent B. Let us introduce two observables AB and B with the sets of their possible values:

$$AB = \{(AB)_1, (AB)_2\} = \bigcup_i (AB)_i \quad (i = 1, 2) \text{ and} \tag{6}$$

$$B = \{B_1, B_2\} = \bigcup_i B_i \quad (i = 1, 2), \tag{7}$$

where $(AB)_1$ (resp., B_1) represents characteristics of a conjunction (resp., constituent) category, or attribution of these features to a subject when making a judgement about him/her. Then $(AB)_2$ (resp., B_2) is a complementary subevent (mode), when a conjunction (resp., constituent) category is considered to be absent, or relevant features are not attributed to a subject.

In the course of the experiment, a participant makes an explicit probability judgment about attribution of a conjunction category or its constituent to a described subject. Thus, both events—AB and B—are operationally testable, i.e. each of the observables takes a concrete value from a corresponding set, either (6) or (7). This consideration underpins two assumptions: (i) sets of observable values are represented by standard unions; (ii) a conjunction is treated as a single category AB, i.e. a tensor product of two constituent categories, such that $AB = A \otimes B$.

For both observables AB and B, we put into correspondence the state vectors $|AB\rangle$ and $|B\rangle$ that are decomposed onto orthonormal bases as follows:

$$|AB\rangle = \alpha_1|(AB)_1\rangle + \alpha_2|(AB)_2\rangle, \quad \langle (AB)_i||(AB)_j\rangle = \delta_{ij}, \tag{8}$$

$$|B\rangle = \beta_1|B_1\rangle + \beta_2|B_2\rangle, \quad \langle B_i||B_j\rangle = \delta_{ij}, \tag{9}$$

with respective probability weights $|\alpha_i|^2$ and $|\beta_i|^2$ (i=1,2). The vectors of the bases—$\{|(AB)_i\rangle\}$ and $\{|B_i\rangle\}$—are spanning sets of the corresponding Hilbert spaces $\mathcal{H}_{AB}$ and $\mathcal{H}_B$:

$$\mathcal{H}_{AB} \equiv span\{|(AB)_1\rangle, |(AB)_2\rangle\}, \tag{10}$$

$$\mathcal{H}_B \equiv span\{|B_1\rangle, |B_2\rangle\}. \tag{11}$$

The operators $\hat{P}_{AB}$ and $\hat{P}_B$ are generated by the respective state vectors (8) and (9):

$$\bigcup_i (AB)_i \;\rightarrow\; \hat{P}_{AB} \equiv \sum_i |\alpha_i|^2 \hat{P}_{(AB)_i} = \sum_i |\alpha_i|^2 |(AB)_i\rangle\langle (AB)_i| \;\; i = 1, 2. \tag{12}$$

$$\bigcup_i B_i \;\rightarrow\; \hat{P}_B \equiv \sum_i |\beta_i|^2 \hat{P}_{B_i} = \sum_i |\beta_i|^2 |B_i\rangle\langle B_i| \quad i = 1, 2. \tag{13}$$

Note that, because the operators $\hat{P}_{AB}$ and $\hat{P}_B$ correspond to operationally testable events, they can be decomposed into elements of their operator algebras $\mathcal{AB} = \{\hat{P}_{(AB)_i}\}$ and $\mathcal{B} = \{\hat{P}_{B_i}\}$, as in (12) and (13), i.e. they are not entangled. Thus, decision modes within each of the corresponding operationally testable events AB and B are not interfering [16].

In a typical quantum measurement, an experiment with a physical system consists of two phases: preparation of a system state and measurement of an observable. Decision making can also be described as a two-step process: first, *deliberation* about objectives, desires, choice alternatives, constraints, and so on, and, second, *taking a decision* by adopting a certain choice option. Within QDT, this translates into, (i) an initial preparation and evolution of a strategic decision-maker state, influenced by the context of the choice situation, and (ii) a convergence to and observation of a concrete event (a decision) out of a set of possible basic decision modes. A probabilistic interpretation implies that an event (a decision) occurs with a certain probability that can be predicted from the state of mind of the decision maker.

In the following Sect. 3.3, these two phases of a decision making process are described with two types of representation: in the first Part 3.3.1 a channel representation is applied, which is more suitable to follow the evolution of a strategic decision-maker state during an initial preparation phase; the second Part 3.3.2 uses a composite state representation, which is convenient for representing the measurement, i.e., the observation of a decision.

3.3 *General Representation of a Strategic Decision-Maker State of Mind with Statistical Operators*

3.3.1 Preparation: Evolution of a Space of Mind and of a Strategic Decision-Maker State

In the first step of a preparation phase, a decision maker is exposed to the description of a subject (an instance I). An initial space of mind $\mathcal{H}_M$ is enlarged to a Hilbert-space tensor product $\mathcal{H}_{IM}$

$$\mathcal{H}_M \rightarrow \mathcal{H}_{IM} \equiv \mathcal{H}_I \otimes \mathcal{H}_M. \tag{14}$$

The evolution over the time interval $[t_0; t_1]$ of a strategic decision-maker state $\hat{\rho}_M(t_0)$, influenced by an introduced partial state $\hat{\rho}_I(t_0)$, is represented with the entangling channel

$$C_1 : \hat{\rho}_I(t_0) \otimes \hat{\rho}_M(t_0) \rightarrow \hat{\rho}_{IM}(t_1). \tag{15}$$

Between intentional external inputs, the space of mind $\mathcal{H}_{IM}$ is assumed to be unchanged. However, the strategic state of a decision-maker can evolve due to internal processes of deliberation as well as the influence of the environment in the time interval $[t_1; t_2]$. This is captured by a unitary evolution operator $\hat{U}$:

$$C_2 : \hat{\rho}_{IM}(t_1) \rightarrow \hat{\rho}_{IM}(t_2) = \hat{U}(t_2 - t_1)\hat{\rho}_{IM}(t_1)\hat{U}^{\dagger}(t_2 - t_1) \tag{16}$$

The introduction of a new category over the time interval $[t_2; t_3]$—a conjunction $A\&B$ (resp., its constituent B)—expands the space of mind

$$\mathcal{H}_{IM} \rightarrow \mathcal{H}_{ABIM} \equiv \mathcal{H}_{AB} \otimes \mathcal{H}_{IM}$$
$$(\text{resp.}, \mathcal{H}_{IM} \rightarrow \mathcal{H}_{BIM} \equiv \mathcal{H}_{B} \otimes \mathcal{H}_{IM}), \tag{17}$$

and a decision-maker state is further entangled through the channel

$$C_3 : \hat{\rho}_{AB}(t_2) \otimes \hat{\rho}_{IM}(t_2) \rightarrow \hat{\rho}_{ABIM}(t_3)$$
$$(\text{resp.}, C3 : \hat{\rho}_{B}(t_2) \otimes \hat{\rho}_{IM}(t_2) \rightarrow \hat{\rho}_{BIM}(t_3)). \tag{18}$$

An intermediate evolution over the time interval $[t_3; t_4]$ with a unitary operator $\hat{U}$ completes the preparation of a strategic decision-maker state, which is entangled with a conjunction category AB (resp., its constituent B):

$$C_4 : \hat{\rho}_{ABIM}(t_3) \rightarrow \hat{\rho}_{ABIM}(t_4) = \hat{U}(t_4 - t_3)\hat{\rho}_{ABIM}(t_3)\hat{U}^{\dagger}(t_4 - t_3)$$
$$(\text{resp.}, C_4 : \hat{\rho}_{BIM}(t_3) \rightarrow \hat{\rho}_{BIM}(t_4) = \hat{U}(t_4 - t_3)\hat{\rho}_{BIM}(t_3)\hat{U}^{\dagger}(t_4 - t_3). \tag{19}$$

The attribution to a subject (described in an instance I) of a conjunction category $A\&B$ (resp., its constituent B) is an operationally testable event, which is associated with an explicit decision (similar to a measurement of a quantum system) and can be observed with a certain probability. In general, the procedure of making a decision (measurement) can be described by partially disentangling channels that lead to a separation of $\hat{\rho}_I$ and $\hat{\rho}_{AB}$ (resp., $\hat{\rho}_B$) from the total decision-maker state $\hat{\rho}_{ABIM}$ (resp., $\hat{\rho}_{BIM}$). This description is realistic, but involves several additional entangling-disentangling channels.
However, as demonstrated in [18], the channel-state duality established by the Choi-Jamiolkowski isomorphism allows one to equivalently represent a measurement by the introduction of a composite state. Thus, for the second phase of the decision making processes (revelation of a decision), a channel representation is conveniently substituted by a composite state representation.

3.3.2 Measurement: Decision Under Uncertainty (Composite Prospect and Its Probability)

Within QDT, taking a decision (adopting an option) is equivalent to a transition of the strategic state of a decision-maker to a state corresponding to the chosen option.

In the analyzed experimental setup, a probability judgment is made with respect to the attribution of a conjunction category AB (resp., its constituent B) to a subject (from an instance I). The corresponding choice option is a composite prospect of the form $AB \otimes I$ (resp., $B \otimes I$). Taking into account that a judgment about the absence of a feature, i.e. $(AB)_2$ (resp., B_2), is not investigated, a choice prospect includes only a decision mode $(AB)_1$ (resp., B_1) and is formulated as follows:

$$\pi_{(AB)_1 I} = (AB)_1 \otimes \biguplus_{k=1,2} I_k,$$
$$(\text{resp.}, \pi_{B_1 I} = B_1 \otimes \biguplus_{k=1,2} I_k), \tag{20}$$

with a corresponding prospect state $|\pi_{(AB)_1 I}\rangle$ (resp., $|\pi_{B_1 I}\rangle$):

$$|\pi_{(AB)_1 I}\rangle = |(AB)_1\rangle \otimes (\gamma_1|I_1\rangle + \gamma_2|I_2\rangle) = \sum_{k=1,2} \gamma_k|(AB)_1 I_k\rangle$$
$$(\text{resp.}, |\pi_{B_1 I}\rangle = |B_1\rangle \otimes (\gamma_1|I_1\rangle + \gamma_2|I_2\rangle) = \sum_{k=1,2} \gamma_k|B_1 I_k\rangle). \tag{21}$$

The basic decision modes of the prospect state $\{|(AB)_1 I_k\rangle\}$ (resp.,$\{|B_1 I_k\rangle\}$) are vectors of an orthogonal basis of the space of mind $\mathcal{H}_{ABIM}$ (resp., $\mathcal{H}_{BIM}$) given by (17).

A composite prospect state $\pi_{(AB)_1 I}$ (resp., $\pi_{B_1 I}$) generates a prospect operator

$$\begin{aligned}\pi_{(AB)_1 I} \;&\rightarrow\; \hat{P}_{(AB)_1 I} \equiv |\pi_{(AB)_1 I}\rangle\langle\pi_{(AB)_1 I}| \\ &= \sum_k |\gamma_k|^2 |(AB)_1 I_k\rangle\langle(AB)_1 I_k| + \sum_{k\neq l} \gamma_k\gamma_l^* |(AB)_1 I_k\rangle\langle(AB)_1 I_l| \\ &= \sum_k |\gamma_k|^2 \hat{P}_{(AB)_1} \otimes \hat{P}_{I_k} + \sum_{k\neq l} \gamma_k\gamma_l^* \hat{P}_{(AB)_1} \otimes |I_k\rangle\langle I_l| \;\; k, l \in \{1, 2\}\end{aligned} \tag{22}$$

$$\begin{aligned}(\text{resp.}, \pi_{B_1 I} \;&\rightarrow\; \hat{P}_{B_1 I} \equiv |\pi_{B_1 I}\rangle\langle\pi_{B_1 I}| \\ &= \sum_k |\gamma_k|^2 |B_1 I_k\rangle\langle B_1 I_k| + \sum_{k\neq l} \gamma_k\gamma_l^* |B_1 I_k\rangle\langle B_1 I_l| \\ &= \sum_k |\gamma_k|^2 \hat{P}_{B_1} \otimes \hat{P}_{I_k} + \sum_{k\neq l} \gamma_k\gamma_l^* \hat{P}_{B_1} \otimes |I_k\rangle\langle I_l| \;\; k, l \in \{1, 2\}).\end{aligned} \tag{23}$$

The probability that a decision maker in a strategic state of mind $\hat{\rho}_{ABIM}$ (resp., $\hat{\rho}_{BIM}$) will choose the prospect $\pi_{(AB)_1 I}$ (resp., $\pi_{B_1 I}$) is given by

$$p(\pi_{(AB)_1 I}) = \mathrm{Tr}\hat{\rho}_{ABIM} \hat{P}_{(AB)_1 I}$$
$$(\text{resp.}, p(\pi_{B_1 I}) = \mathrm{Tr}\hat{\rho}_{BIM} \hat{P}_{B_1 I}) \tag{24}$$

Its explicit form reads

$$p(\pi_{(AB)_1 I}) = \sum_{kl} \gamma_k \gamma_l^* \langle (AB)_1 I_l | \hat{\rho}_{ABIM} | (AB)_1 I_k \rangle$$
$$(\text{resp., } p(\pi_{B_1 I}) = \sum_{kl} \gamma_k \gamma_l^* \langle B_1 I_l | \hat{\rho}_{ABIM} | B_1 I_k \rangle), \tag{25}$$

where $k, l \in \{1, 2\}$, which can be expressed as the sum of two parts

$$p(\pi_{(AB)_1 I}) = f(\pi_{(AB)_1 I}) + q(\pi_{(AB)_1 I}), \tag{26}$$

where the diagonal term, i.e. utility factor, reads

$$f(\pi_{(AB)_1 I}) = \sum_{k} |\gamma_k|^2 \langle (AB)_1 I_k | \hat{\rho}_{ABIM} | (AB)_1 I_k \rangle$$
$$(\text{resp., } f(\pi_{B_1 I}) = \sum_{k} |\gamma_k|^2 \langle B_1 I_k | \hat{\rho}_{ABIM} | B_1 I_k \rangle \tag{27}$$

and the off-diagonal term, called "attraction factor" by [13], takes the form

$$q(\pi_{(AB)_1 I}) = \sum_{k \neq l} \gamma_k \gamma_l^* \langle (AB)_1 I_l | \hat{\rho}_{ABIM} | (AB)_1 I_k \rangle$$
$$(\text{resp., } q(\pi_{B_1 I}) = \sum_{k \neq l} \gamma_k \gamma_l^* \langle B_1 I_l | \hat{\rho}_{ABIM} | B_1 I_k \rangle \tag{28}$$

By construction, p and f are probabilities that take values in [0, 1]. [13] proved the "alternation law", which states that the sum of the q-factors (which reads $q(\pi_{(AB)_1 I}) + q(\pi_{(AB)_2 I})$ according to (28) in the present binary case) is identically zero. Thus, in such binary decisions involving the presence or absence of a trait, it is sufficient to discuss the q-factor of just one of the alternatives, the other one being by construction of equal amplitude and opposite sign. Moreover, using non-informative prior assumptions, the typical amplitude of the q-factor for binary choices can be shown to equal 0.25, which is referred to as the "quarter law" [13] (see [17] for generalisation to multiple choices beyond binary ones). Finally, from the structure (26), the contraints $p, f \in [0, 1]$ and the alternation law, [9] showed that the q-factor obeys the additional constraint

$$|q(\pi_{(AB)_1 I})| \leq \min[f(\pi_{(AB)_1 I}), 1 - f(\pi_{(AB)_1 I})]. \tag{29}$$

Interferences between decision modes, which is captured by a non-zero attraction factor (28), have a profound influence on the probability of a prospect to be chosen and may even reverse a decision-maker's preference (in a sense of changin the most probable choice option). As shown in Appendix 3, the proposed QDT interpretation of the experimental setup complies with the necessary conditions for the appearance of the attraction factor.

To explain the conjunction fallacy, i.e. $p(\pi_{(AB)_1 I}) > p(\pi_{B_1 I})$, one should analyze in-depth the values of the coefficients γ in (25), (27) and (28). For this, we use a simplified representation of the strategic decision-maker state in terms of a pure state.

3.4 Simplified Representation of a Strategic Decision-Maker State as a Pure State

3.4.1 Preparation: Entanglement of a Pure State of Mind

As already been mentioned in Sect. 3.1, in certain choice situations, it can be sufficiently realistic and operationally convenient to represent a strategic decision-maker state $\hat{\rho}$ in a simplified form, i.e. as a pure state $\hat{\rho} = |\psi\rangle\langle\psi|$. For example, such situations may involve time restrictions, when a thorough deliberation is not possible, or specific setups, when limited attention resources are focused sharply on a quasi-isolated task at hand. To some degree, both of these features can be attributed to short laboratory experiments as reported in [6], which is analyzed here. Thus, in this section a decision maker is characterized by a state vector $|\psi\rangle$, and is assumed to be isolated from external and internal influences that are not explicitly formulated in the experiment.

The decision modes of the experiment remain the same as formulated in Sect. 3.2.

We assume that, in the first phase of the experiment (period $[t_0; t_1]$), a decision maker concentrates her full attention on the description of a subject (an instance I). Thus, his/her mind space $\mathcal{H}_M$ converges initially to $\mathcal{H}_I$

$$\mathcal{H}_M \;\rightarrow\; \mathcal{H}_I \;\equiv\; span\{|I_k\rangle\},\; k \in 1, 2. \tag{30}$$

The corresponding focused state of mind $|\psi_I\rangle$ can be represented in the Hilbert space $\mathcal{H}_I$ as a linear combination of elementary state vectors from the basis $\{|I_k\rangle\}$, $k = 1, 2$:

$$|\psi_I\rangle(t_1) = \zeta_1|I_1\rangle + \;\zeta_2|I_2\rangle. \tag{31}$$

The time-dependent coefficients $|\zeta_k|^2$ in (31) play a role similar to the weights $|\gamma_k|^2$ of the mode states $|I_k\rangle$ in (2). However, the state of mind vector is assumed to be normalized, which implies an additional constraint on the coefficients $|\zeta_k|^2$:

$$\langle\psi_I||\psi_I\rangle := 1 \Rightarrow \sum_k |\zeta_k|^2 = 1 \;\; (k = 1, 2). \tag{32}$$

Thus, in this case, $|\zeta_k|^2$ and $|\gamma_k|^2$ are proportional, but not necessarily equivalent to each other.

In the course of the experiment (period $[t_1; t_2]$), the description of a subject (an instance I) is followed by a new category—a conjunction $A\&B$ (resp., its constituent

B). Due to the exposition to the category, the decision-maker's mind space $\mathcal{H}_{\mathcal{I}}$ expands and can be represented as the Hilbert-space tensor product

$$\begin{aligned} \mathcal{H}_I &\to \mathcal{H}_{ABI} \equiv \mathcal{H}_{AB} \otimes \mathcal{H}_I \equiv span\{|(AB)_i I_k\rangle\},\ i,k \in 1,2 \\ (\text{resp.}, \mathcal{H}_I &\to \mathcal{H}_{BI} \equiv \mathcal{H}_B \otimes \mathcal{H}_I \equiv span\{|B_i I_k\rangle\},\ i,k \in 1,2). \end{aligned} \tag{33}$$

The expanded state of mind $|\psi_{ABI}\rangle$ (resp., $|\psi_{BI}\rangle$) is a linear combination of elementary prospects $\{|(AB)_i I_k\rangle\}$ (resp., $\{|B_i I_k\rangle\}$):

$$\begin{aligned} |\psi_{ABI}\rangle &= \zeta_{11}|(AB)_1 I_1\rangle + \zeta_{12}|(AB)_1 I_2\rangle + \zeta_{21}|(AB)_2 I_1\rangle + \zeta_{22}|(AB)_2 I_2\rangle \\ (\text{resp.}, |\psi_{BI}\rangle &= \kappa_{11}|B_1 I_1\rangle + \kappa_{12}|B_1 I_2\rangle + \kappa_{21}|B_2 I_1\rangle + \kappa_{22}|B_2 I_2\rangle). \end{aligned} \tag{34}$$

The elementary prospects form an orthonormalised basis, thus squared coefficients ζ sum to 1:

$$\begin{aligned} \langle\psi_{ABI}||\psi_{ABI}\rangle &:= 1 \ \Rightarrow\ \sum_{i,k} |\zeta_{ik}|^2 = 1,\ \ i,k \in \{1,2\} \\ (\text{resp.}, \langle\psi_{BI}||\psi_{BI}\rangle &:= 1 \ \Rightarrow\ \sum_{i,k} |\kappa_{ik}|^2 = 1,\ \ i,k \in \{1,2\}). \end{aligned} \tag{35}$$

Taking into account the definitions of decision modes of Sect. 3.2, the interpretation of the elementary prospects (decision modes) may help to reverse-engineer the probability amplitudes ζ (resp., κ), as follows.

- $|(AB)_1 I_1\rangle$ (resp., $|B_1 I_1\rangle$) implies an attribution of the features of a category AB (resp., B) to a subject that simultaneously possesses features from the description I. Thus, this vector can be associated with compatible features of both, a judged category AB (resp., B) and an instant I. If category and description of a subject share many common features (i.e. AB (resp., B) and I are *compatible*), then one could expect larger values of the probability amplitude ζ_{11} (resp., κ_{11}) of the corresponding elementary prospect, and consequently higher judged typicality/probability of AB (resp., B) with respect to I.
- $|(AB)_1 I_2\rangle$ and $|(AB)_2 I_1\rangle$ (resp., $|B_1 I_2\rangle$ and $|B_2 I_1\rangle$) reflect modes of simultaneous attribution to a subject of either (i) features associated with a judged category AB (resp., B) and features complementary to (i.e. disassociated with) a subject's initial description I; or vice versa (ii) features complementary to (i.e. disassociated with) a judged category AB (resp., B) and features that are compliant with an instance I. Thus, a stronger dissimilarity between a judged category and the description of a subject should increase the probability amplitudes ζ_{ik} (resp., κ_{ik}) ($i \neq k$), which are associated with *incompatibility* of AB (resp., B) and I.
- $|(AB)_2 I_2\rangle$ (resp., $|B_2 I_2\rangle$) represents a state when a subject is endowed with *complementary* features of both a judged category AB (resp., B) and an instance I. This mode corresponds to the judgement of a decision maker that neither a description from an instance I, nor features associated with a category AB (resp., B), can be attributed to the subject under consideration.

Here, we propose that, when a decision maker is focused on an experimental task—i.e. is exposed to the description of a subject I, a category AB (resp., B) and attempts to judge typicality/probability of the latter—he/she concentrates almost exclusively on compatible and incompatible features of the two. Because the state of mind is focused on the presence of specific characteristics, the possibility of the considered subject to deviate from both the description of an instance I and a judged category AB (resp., B), is largely disregarded. Thus, in the notation used above, we suggest that $\zeta_{22} \to 0$ (resp., $\kappa_{22} \to 0$). This assumption evokes employing representativeness and availability heuristics, while neglecting a base rate in judgements.

Taking into account, as discussed above, that $\zeta_{22} \to 0$ (resp., $\kappa_{22} \to 0$) and other coefficients ζ_{ik} (resp, κ_{ik}) ($i, k \in 1, 2$) are non-zero, the state of mind (34) becomes

$$|\psi_{ABI}\rangle = \zeta_{11}|(AB)_1 I_1\rangle + \zeta_{12}|(AB)_1 I_2\rangle + \zeta_{21}|(AB)_2 I_1\rangle$$
$$(\text{resp., } |\psi_{BI}\rangle = \kappa_{11}|B_1 I_1\rangle + \kappa_{12}|B_1 I_2\rangle + \kappa_{21}|B_2 I_1\rangle). \quad (36)$$

Note that (36) is an *entangled* state of mind, because it cannot be represented in a separable form, as the tensor product of the elementary (intention) states,

$$\begin{aligned} |\psi_{ABI}\rangle &= |AB\rangle \otimes |I\rangle = \alpha_1\gamma_1|(AB)_1 I_1\rangle + \alpha_1\gamma_2|(AB)_1 I_2\rangle \\ &\quad + \alpha_2\gamma_1|(AB)_2 I_1\rangle + \alpha_2\gamma_2|(AB)_2 I_2\rangle \\ (\text{resp., } |\psi_{BI}\rangle &= |B\rangle \otimes |I\rangle = \beta_1\gamma_1|B_1 I_1\rangle + \beta_1\gamma_2|B_1 I_2\rangle \\ &\quad + \beta_2\gamma_1|B_2 I_1\rangle + \beta_2\gamma_2|B_2 I_2\rangle), \end{aligned} \quad (37)$$

in the sense that there exists no set of parameters $\{\alpha_1, \alpha_2, \gamma_1, \gamma_2\}$ (resp., $\{\beta_1, \beta_2, \gamma_1, \gamma_2\}$) that can be matched to the set of parameters $\{\zeta_{11}, \zeta_{12}, \zeta_{21}, \zeta_{22} = 0\}$ (resp., $\{\kappa_{11}, \kappa_{12}, \kappa_{21}, \kappa_{22} = 0\}$).

Thus, prior to taking a decision, i.e. a probability judgement, a decision maker is in the entangled state of mind (36). The coefficients $|\zeta_{11}|^2$, $|\zeta_{12}|^2$, $|\zeta_{21}|^2$ (resp., $|\kappa_{11}|^2$, $|\kappa_{12}|^2$, $|\kappa_{21}|^2$) of this decision maker's state of mind are associated with a degree of compatibility of I and AB (resp., B), or with a range of shared properties between the description of a subject and a category. Herewith, $|\zeta_{11}|^2$ (resp., $|\kappa_{11}|^2$) reflects the commonality of the two, while $|\zeta_{12}|^2$ and $|\zeta_{21}|^2$ (resp., $|\kappa_{12}|^2$ and $|\kappa_{21}|^2$) quantify their distinctive unshared properties. Thus, we propose that typicality judgements, which were revealed during the experiment, can be used to quantify the coefficients of the state of mind. Taking into account equation (35), this leads us to parameterise them under the form

$$\begin{cases} |\zeta_{11}|^2 = typ(AB) \\ |\zeta_{12}|^2 = \omega(1 - typ(AB)) \\ |\zeta_{21}|^2 = (1-\omega)(1 - typ(AB)) \end{cases} \quad \left(\text{resp., } \begin{cases} |\kappa_{11}|^2 = typ(B) \\ |\kappa_{12}|^2 = \omega(1 - typ(B)) \\ |\kappa_{21}|^2 = (1-\omega)(1 - typ(B)) \end{cases} \right) \quad (38)$$

where $\omega \in [0, 1]$ is a weighting coefficient between two decision modes, which represents the level of incompatibility of I and AB (resp., B). The coefficient $typ(AB)$ (resp., $typ(B)$) is the value of typicality associated to AB given I, as defined by typ_{exp} in Sect. 2. Note that, in the analyzed experimental data set, typicality and probability judgements were performed by two distinct groups of participants, which may lead to additional discrepancies between theoretical and empirical values of the prospects' probabilities.

3.4.2 Measurement: Operational Prospects Probabilities

A choice option is formulated as a composite prospect similar to (20) of Sect. 3.3.2

$$\pi_{(AB)_1 I} = (AB)_1 \otimes \biguplus_{k=1,2} I_k,$$
$$(\text{resp.}, \pi_{B_1 I} = B_1 \otimes \biguplus_{k=1,2} I_k), \tag{39}$$

with a corresponding prospect state (21)

$$|\pi_{(AB)_1 I}\rangle = \gamma_1 |(AB)_1 I_1\rangle + \gamma_2 |(AB)_1 I_2\rangle$$
$$(\text{resp.}, |\pi_{B_1 I}\rangle = \gamma_1 |B_1 I_1\rangle + \gamma_2 |B_1 I_2\rangle), \tag{40}$$

which generates a prospect operator (22) (resp., (23))

$$\begin{aligned}\pi_{(AB)_1 I} \to \hat{P}_{(AB)_1 I} &\equiv |\pi_{(AB)_1 I}\rangle\langle\pi_{(AB)_1 I}| \\ &= \sum_k |\gamma_k|^2 |(AB)_1 I_k\rangle\langle(AB)_1 I_k| + \sum_{k\neq l} \gamma_k \gamma_l^* |(AB)_1 I_k\rangle\langle(AB)_1 I_l| \\ &= \sum_k |\gamma_k|^2 \hat{P}_{(AB)_1} \otimes \hat{P}_{I_k} + \sum_{k\neq l} \gamma_k \gamma_l^* \hat{P}_{(AB)_1} \otimes |I_k\rangle\langle I_l| \;\; k, l \in \{1, 2\}\end{aligned} \tag{41}$$

$$\begin{aligned}&(\text{resp.}, \pi_{B_1 I} \to \hat{P}_{B_1 I} \equiv |\pi_{B_1 I}\rangle\langle\pi_{B_1 I}| \\ &= \sum_k |\gamma_k|^2 |B_1 I_k\rangle\langle B_1 I_k| + \sum_{k\neq l} \gamma_k \gamma_l^* |B_1 I_k\rangle\langle B_1 I_l| \\ &= \sum_k |\gamma_k|^2 \hat{P}_{B_1} \otimes \hat{P}_{I_k} + \sum_{k\neq l} \gamma_k \gamma_l^* \hat{P}_{B_1} \otimes |I_k\rangle\langle I_l| \;\; k, l \in \{1, 2\}).\end{aligned} \tag{42}$$

The uncertainty related to the description of a subject I, which is captured by an uncertain union, again results in interfering modes $|I_k\rangle\langle I_l|$ $(k, l \in \{1, 2\})$.

The probability that a decision maker in a state of mind $|\psi_{ABI}\rangle$ (resp., $|\psi_{BI}\rangle$) (37) will choose a prospect $\pi_{(AB)_1 I}$ (resp., $\pi_{B_1 I}$) is given by

$$p(\pi_{(AB)_1 I}) = \langle\psi_{ABI}|\hat{P}_{(AB)_1 I}|\psi_{ABI}\rangle = |\langle\pi_{(AB)_1 I}||\psi_{ABI}\rangle|^2$$
$$(\text{resp., } p(\pi_{B_1 I}) = \langle\psi_{BI}|\hat{P}_{B_1 I}|\psi_{ABI}\rangle = |\langle\pi_{B_1 I}||\psi_{BI}\rangle|^2) \tag{43}$$

In an explicit form, we have

$$p(\pi_{(AB)_1 I}) = |\gamma_1\zeta_{11}|^2 + |\gamma_2\zeta_{12}|^2 + 2Re(\zeta_{11}^*\gamma_1\gamma_2^*\zeta_{12})$$
$$(\text{resp., } p(\pi_{B_1 I}) = |\gamma_1\kappa_{11}|^2 + |\gamma_2\kappa_{12}|^2 + 2Re(\kappa_{11}^*\gamma_1\gamma_2^*\kappa_{12})), \tag{44}$$

and it is comprised of two parts: (i) the diagonal term, i.e. utility factor:

$$f(\pi_{(AB)_1 I}) = f(\pi_{(AB)_1 I_1}) + f(\pi_{(AB)_1 I_2}) = |\gamma_1\zeta_{11}|^2 + |\gamma_2\zeta_{12}|^2$$
$$(\text{resp., } f(\pi_{B_1 I}) = f(\pi_{B_1 I_1}) + f(\pi_{B_1 I_2}) = |\gamma_1\kappa_{11}|^2 + |\gamma_2\kappa_{12}|^2), \tag{45}$$

and (ii) the off-diagonal term, i.e. attraction factor:

$$q(\pi_{(AB)_1 I}) = 2Re(\zeta_{11}^*\gamma_1\gamma_2^*\zeta_{12})$$
$$(\text{resp., } q(\pi_{B_1 I}) = 2Re(\kappa_{11}^*\gamma_1\gamma_2^*\kappa_{12})). \tag{46}$$

The uncertainty angle can be defined as [13]

$$\Delta(\pi_{(AB)_1 I}) \equiv \arg(\zeta_{11}^*\gamma_1\gamma_2^*\zeta_{12})$$
$$(\text{resp., } \Delta(\pi_{B_1 I}) \equiv \arg(\kappa_{11}^*\gamma_1\gamma_2^*\kappa_{12})). \tag{47}$$

The uncertainty angle quantifies the "wedge" between the interfering decision modes of a prospect in consideration. In the proposed formulation, for a conjunctive category, the interfering modes are $|(AB)_1 I_1\rangle$ and $|(AB)_1 I_2\rangle$ and, for its constituent category, they are $|B_1 I_1\rangle$ and $|B_1 I_2\rangle$.

The introduction of the uncertainty angle (47) allows us to rewrite the attraction factor (46) as

$$\begin{aligned} q(\pi_{(AB)_1 I}) &= 2\sqrt{f(\pi_{(AB)_1 I_1}) f(\pi_{(AB)_1 I_2})}\ \cos\Delta(\pi_{(AB)_1 I}) \\ &= 2\sqrt{|\gamma_1\zeta_{11}|^2|\gamma_2\zeta_{12}|^2}\ \cos\Delta(\pi_{(AB)_1 I}) \\ (\text{resp., } q(\pi_{B_1 I}) &= 2\sqrt{f(\pi_{B_1 I_1}) f(\pi_{B_1 I_2})}\ \cos\Delta(\pi_{B_1 I}) \\ &= 2\sqrt{|\gamma_1\kappa_{11}|^2|\gamma_2\kappa_{12}|^2}\ \cos\Delta(\pi_{B_1 I})). \end{aligned} \tag{48}$$

To summarize, we propose that the judged probability p_{exp} that an instance I belongs to a conjunction category (resp., constituent category) is equal to the probability of a prospect $\pi_{(AB)_1 I}$ (resp., $\pi_{B_1 I}$). According to equations (44) and (48), we have

$$\begin{aligned}
p_{\exp}(AB) &= p\left(\pi_{(AB)_1 I}\right) = |\gamma_1\zeta_{11}|^2 + |\gamma_2\zeta_{12}|^2 \\
&\quad +2\sqrt{|\gamma_1\zeta_{11}|^2|\gamma_2\zeta_{12}|^2}\,\cos\Delta(\pi_{(AB)_1 I}) \\
(\text{resp., } p_{\exp}(B) &= p(\pi_{B_1 I}) = |\gamma_1\kappa_{11}|^2 + |\gamma_2\kappa_{12}|^2 \\
&\quad +2\sqrt{|\gamma_1\kappa_{11}|^2|\gamma_2\kappa_{12}|^2}\,\cos\Delta(\pi_{B_1 I})).
\end{aligned} \tag{49}$$

Note that one of the modes of a decision maker's state (36) that represents an incompatibility of I and AB (resp., B), i.e. $|(AB)_2 I_1\rangle$ (resp., $|B_2 I_1\rangle$), is absent in the choice prospect (21) and, consequently, in the probability formulation of that prospect (49). Thus, in the focused mind state, when a decision maker is facing only one choice prospect, an incompatibility feature is (mostly) affecting a decision via the decision mode $|(AB)_1 I_2\rangle$ (resp., $|B_1 I_2\rangle$). We account for this by assigning a larger weight to $|(AB)_1 I_2\rangle$ (resp., $|B_1 I_2\rangle$), i.e. $\omega \in [0.5, 1]$ in (38). Thus, the coefficients of (49) are parameterised by

$$\begin{cases} |\zeta_{11}|^2 = typ(AB) \\ |\zeta_{12}|^2 = \omega(1 - typ(AB)) \\ \omega \in [0.5, 1] \end{cases} \left(\text{resp.,} \begin{cases} |\kappa_{11}|^2 = typ(B) \\ |\kappa_{12}|^2 = \omega(1 - typ(B)) \\ \omega \in [0.5, 1] \end{cases} \right) \tag{50}$$

The weight coefficient ω directly impacts the amplitudes of the decision modes. It also influences the uncertainty angle between them via the normalisation condition (32), since a change of amplitude of one component of a normalized vector amounts to a rotation.

The prospect probabilities, i.e. the probability judgements $p_{\exp}(AB)$ and $p_{\exp}(B)$ (49), are by definition real positive numbers $\Re_{\geq 0}$ in the interval [0, 1], and for complementary events sum to 1. Given the orthonormality of the basis (35), the squared probability amplitudes should also belong to $\Re_{\geq 0}$ in the interval [0, 1] and sum to 1 within a state of mind, or within a prospect. In the suggested analytical approach, for the coefficients $|\zeta_{ij}|^2$ and $|\kappa_{ij}|^2$, $i, j \in \{1, 2\}$ of the state of mind, this condition is respected by their direct link to typicality judgements $typ(AB)$ and $typ(B)$, which were made by participants on the same scale from 0 to 1. In the subsequent analysis, special care should be taken to ensure that the coefficients $|\gamma_i|^2$, $i, j \in \{1, 2\}$ of a prospect, as well as their sum, belong to $\Re_{\geq 0}$ in the interval [0, 1].

4 Empirical Analysis and Explanation of the Conjunction Fallacy

In this section, we apply the formulation of section (3.4) to the experimental results reported in [6]. We exploit equations (49) and (50) from two perspectives by (1) quantifying the uncertainty factor ($\cos\Delta(\pi)$) and (2) estimating the relative influence of interfering decision modes (coefficients $|\gamma_i|^2$).

4.1 First Interpretation of the Conjunction Fallacy: Quantifying the Uncertainty Factor and Uncertainty Angle

Here, we focus on quantifying the uncertainty factor $\cos\Delta(\pi)$ associated with different types of categories: $AB(i)$, $B(i)$, $AB(c)$, $B(c)$. To allow for a meaningful comparison with experiments, we need to reduce the number of degrees of freedom contained in equations (49) and (50). We note that the coefficients γ_i, $i \in \{1, 2\}$ embody the uncertainty of an instance I, which can be related to the briefness of the description of a subject and the imprecise time evolution of the described personal characteristics (captured by the uncertain union (1)). Thus, with no prior information, we can assume equal weights for attributing to a subject either characteristics directly mentioned in the description (i.e. I_1), or any other characteristics that are complementary to it (i.e. I_2). This amounts to imposing

$$|\gamma_1|^2 = |\gamma_2|^2 = 0.5. \tag{51}$$

Taking into account Eqs. (49)–(51), the attraction factor reduces to

$$\begin{aligned} q(\pi_{(AB)_1 I}) &= p_{\text{exp}}(AB) - 0.5\left[typ(AB) + \omega(1 - typ(AB))\right] \\ (\text{resp.}, q(\pi_{B_1 I}) &= p_{\text{exp}}(B) - 0.5\left[typ(B) + \omega(1 - typ(B))\right]), \end{aligned} \tag{52}$$

and the uncertainty factor reads

$$\begin{aligned} \cos\Delta(\pi_{(AB)_1 I}) &= \frac{p_{\text{exp}}(AB) - 0.5\left[typ(AB) + \omega(1 - typ(AB))\right]}{\sqrt{typ(AB)\,\omega(1 - typ(AB))}} \\ \left(\text{resp.,}\quad \cos\Delta(\pi_{B_1 I}) \right.&= \left.\frac{p_{\text{exp}}(B) - 0.5\left[typ(B) + \omega(1 - typ(B))\right]}{\sqrt{typ(B)\,\omega(1 - typ(B))}}\right). \end{aligned} \tag{53}$$

Based on the experimental results aggregated over all participants, i.e. the average probability judgement (p_{exp}) of an instance I to belong to one of the four categories and corresponding typicality judgements (typ_{exp}), the attraction factor q is presented in Table 2 for each prospect. The corresponding uncertainty factors, i.e. cosines of the uncertainty angles, are shown in Table 3. Both tables include results for the two boundary values $\omega = 1$ and $\omega = 0.5$. Values of $\cos\Delta(\pi)$ are constrained by the admissible range of the cosine function $[-1, +1]$. As shown in Table 3, all but one (for $\omega = 1$: instance 12, $B(i)$) cosine values are found within this range. This indirectly supports the proposed analytical formulation.

An important observation from Tables 2 and 3 is that for both quantities—q and $\cos\Delta(\pi)$—the absolute difference of their average values between the prospect with conjunction AB and the prospect with constituent B is large for incompatible (i) categories and is small for compatible (c) categories. This signals the higher dissimilarity within a (i) pair in comparison with a (c) pair pertaining to the same

Table 2 Attraction factor $q(\pi)$ by category aggregated over all participants over the 14 instances presented in [6]

Instance I	Attraction factor $q(\pi)$							
	For $\omega = 1$, for each category				**For** $\omega = 0.5$, for each category			
	$B(i)$	$AB(i)$	$B(c)$	$AB(c)$	$B(i)$	$AB(i)$	$B(c)$	$AB(c)$
1	−0.259	−0.099	0.033	0.101	−0.075	0.041	0.123	0.168
2	−0.327	−0.226	0.123	0.207	−0.108	−0.061	0.207	0.264
3	−0.340	−0.274	−0.145	−0.156	−0.134	−0.087	0.032	−0.017
4	−0.234	−0.133	0.098	0.186	−0.017	0.033	0.192	0.257
5	−0.298	−0.231	−0.016	0.052	−0.078	−0.053	0.063	0.120
6	−0.306	−0.218	0.183	0.050	−0.084	−0.030	0.264	0.149
7	−0.348	−0.123	0.044	0.078	−0.138	0.030	0.094	0.135
8	−0.312	−0.248	−0.093	0.016	−0.114	−0.094	0.057	0.125
9	−0.190	−0.029	−0.088	0.059	0.011	0.089	0.028	0.136
10	−0.328	−0.186	0.142	0.184	−0.114	−0.030	0.225	0.266
11	−0.185	0.080	−0.048	0.019	−0.006	0.190	0.067	0.082
12	−0.369	−0.251	0.207	0.009	−0.139	−0.086	0.262	0.115
13	−0.320	−0.161	−0.047	0.003	−0.099	0.016	0.075	0.086
14	−0.108	−0.061	0.129	0.133	0.085	0.083	0.212	0.213
μ	−0.280	−0.154	0.037	0.067	−0.072	0.003	0.136	0.150
σ	0.075	0.101	0.112	0.094	0.066	0.082	0.087	0.079

The four categories are: an incompatible conjunction $AB(i)$ and its constituent $B(i)$, a compatible conjunction $AB(c)$ and its constituent $B(c)$. Sample mean μ and sample standard deviation σ are provided

Table 3 Uncertainty factor (cosine of the uncertainty angle cos $\Delta(\pi)$) by category aggregated over all participants over the 14 instances presented in [6]

Instance I	Cosine of uncertainty angle cos $\Delta(\pi)$							
	For $\omega = 1$, for each category				**For** $\omega = 0.5$, for each category			
	$B(i)$	$AB(i)$	$B(c)$	$AB(c)$	$B(i)$	$AB(i)$	$B(c)$	$AB(c)$
1	−0.588	−0.199	0.069	0.228	−0.241	0.118	0.362	0.536
2	−0.989	−0.477	0.260	0.494	−0.463	−0.182	0.620	0.890
3	−0.891	−0.631	−0.318	−0.314	−0.497	−0.283	0.098	−0.049
4	−0.689	−0.281	0.202	0.412	−0.072	0.098	0.561	0.806
5	−0.914	−0.510	−0.034	0.117	−0.339	−0.166	0.190	0.381
6	−0.978	−0.504	0.391	0.102	−0.377	−0.099	0.798	0.431
7	−0.947	−0.252	0.110	0.186	−0.532	0.086	0.333	0.455
8	−0.769	−0.510	−0.190	0.032	−0.397	−0.273	0.165	0.356
9	−0.480	−0.058	−0.177	0.128	0.040	0.253	0.078	0.416
10	−0.934	−0.384	0.302	0.392	−0.459	−0.087	0.676	0.801
11	−0.411	0.161	−0.096	0.044	−0.017	0.542	0.190	0.266
12	−1(−1.376)	−0.530	0.499	0.018	−0.730	−0.257	0.894	0.330
13	−0.999	−0.354	−0.094	0.006	−0.437	0.051	0.212	0.258
14	−0.257	−0.123	0.274	0.285	0.285	0.238	0.636	0.646
μ	−0.775(−0.801)	−0.332	0.086	0.152	−0.303	0.003	0.415	0.466
σ	0.249 (0.292)	0.221	0.245	0.206	0.272	0.240	0.275	0.254

The four categories are: an incompatible conjunction $AB(i)$ and its constituent $B(i)$, a compatible conjunction $AB(c)$ and its constituent $B(c)$. Sample mean μ and sample standard deviation σ are provided. Values of cos $\Delta(\pi)$ are constrained by the admissible range of the cosine function $[-1, +1]$; unadjusted values are in brackets instance i. This holds for both values of $\omega \in \{1, 0.5\}$. The case $\omega = 1$ seems to be more plausible, since it recovers closely the observed amplitudes of the conjunction fallacy and the conjunction effect, which were reported in Table 1.

Thus, first, consider the case $\omega = 1$. Table 3 shows that the absolute value of the uncertainty factor $|\cos \Delta(\pi)|$ is larger for the pairs of I with *incompatible* (i) categories. Noticeably, in all such pairs, the amplitude of both negative $\cos \Delta(\pi)$ and attraction factor q (Table 2) are larger for a constituent $B(i)$ of an incompatible conjunction. All the categories $B(i)$ in the experiment were formulated such that the corresponding instance I would be atypical of that category (average $typ_{\text{exp}}(B(i)) = 0.167$). According to the experimental setup, the incompatibility of a conjunction category $AB(i)$ indicates that the included categories A and B share only a few common features, i.e. A and B are incompatible between each other. Thus, in the typicality and probability judgements concerning the conjunction $AB(i)$, the category A may partially compensate for the "repulsion" between I and $B(i)$, which is supported by the larger value of the average $typ_{\text{exp}}(AB(i)) = 0.371 > typ_{\text{exp}}(B(i))$. This consequently leads to a higher, though still negative, attraction factor for $AB(i)$, compared to its constituent $B(i)$.

Continuing the case $\omega = 1$, for compatible categories $AB(c)$ and $B(c)$, the uncertainty factors $\cos \Delta(\pi)$ are positive, but much smaller in amplitude than for incompatible pairs. Overall, for compatible categories, there is a slight positive attraction effect. For 12 (resp., 11) out of 14 instances, $\cos \Delta(\pi)$ (resp., q) is larger for a compatible conjunction $AB(c)$ than for a related constituent $B(c)$. Thus, the typicality of $B(c)$ (average $typ_{\text{exp}}(B(c)) = 0.606$) is enhanced by a compatible feature A in conjunction $AB(c)$ (average $typ_{\text{exp}}(AB(c)) = 0.670$), and increases the positive attraction and probability judgements for the latter.

A change in ω from 1 to 0.5 amounts to a rotation. As expected, for the smaller value $\omega = 0.5$, the absolute value of the attraction factor q and of the uncertainty factor increases in comparison with $\omega = 1$. This means that, in order to explain empirical data, a higher uncertainty factor is needed to compensate for a smaller amplitude ζ_{12} (resp., κ_{12}) of the incompatible decision mode $|(AB)_1 I_2\rangle$ (resp., $|B_1 I_2\rangle$) in a decision makers' state of mind (see Eq. 50). The main distinction of the case with $\omega = 0.5$ is that the positive values of q and $\cos \Delta(\pi)$ for the compatible (c) categories become larger, and even increase in magnitude the corresponding values for incompatible (i) pairs.

Figure 2 shows the average over 14 instances I of the uncertainty angle $\Delta(\pi)$ for the four categories and their shift when ω changes from 1 (left subplot) to 0.5 (right subplot). This intuitive representation reveals the differences between incompatible and compatible categories with respect to the associated uncertainty factors. In particular, for the case $\omega = 1$, $\cos \Delta(\pi)$ (and resulting attraction factor q) is slightly positive for compatible (c) categories, and profoundly negative for incompatible (i) ones. Herewith, the average negative $\cos \Delta(\pi)$ is almost twice larger for a constituent $B(i)$ compared with a conjunction $AB(i)$. This leads to a high unattractiveness of the latter, and low probability (judgement) of the corresponding prospect. This constitutes one of the plausible perspectives to interpret the conjunction fallacy. For the

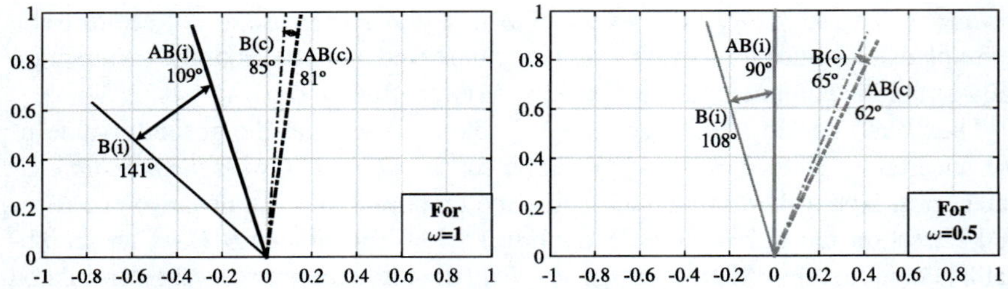

Fig. 2 Average over 14 instances I of the uncertainty angle $\Delta(\pi)$ with $\omega = 1$ (left subplot, black) or $\omega = 0.5$ (right subplot, grey), for each of the four categories: an incompatible conjunction $AB(i)$ (bold solid line), and its constituent $B(i)$ (solid line), a compatible conjunction $AB(c)$ (bold dash-dotted line), and its constituent $B(c)$ (dash-dotted line). The conjunction fallacy is associated with the higher dissimilarity within the pair of incompatible (i) prospects, than within compatible (c) pairs (as indicated by arrows). For the main considered case $\omega = 1$ (left subplot), the uncertainty factor $\cos\Delta(\pi)$ (and resulting attraction factor q) is slightly positive for (c) categories, and profoundly negative for (i) categories. The average negative $\cos\Delta(\pi)$ is almost twice larger in amplitude for a constituent $B(i)$ compared with a conjunction $AB(i)$, suggesting an explanation of the conjunction fallacy

case $\omega = 0.5$, the absolute values of $\cos\Delta(\pi)$ increase, but the same mechanism (i.e. higher dissimilarity between incompatible (i) prospects) explains the observed conjunction fallacy.

4.2 *Second Interpretation of the Conjunction Fallacy: Estimating the Relative Influence of Interfering Decision Modes ($|\gamma_1|^2$ and $|\gamma_2|^2$)*

We now analyze Eq. (49) from a different perspective, namely as a system of two coupled equations. The analysis is conducted for each of the 28 triples defined as follows. A compatible triple $(I, AB(c), B(c))$ consists in an instance I combined with two compatible categories $AB(c)$ and $B(c)$. An incompatible triple $(I, AB(i), B(i))$ consists in an instance I combined with two incompatible categories $AB(i)$ and $B(i)$. Using the 14 instances I studied in [6], we thus have 28 triples to consider.

Recall that the uncertainty, which originates from the indeterminacy of a subject's description I, was introduced as an uncertain union (1). We assume that this uncertainty has a similar influence onto the two prospects that are associated with the categories (AB and B) of the same type (compatible or incompatible). This assumption amounts to imposing that the coefficients γ_1 and γ_2 are the same within each triple (I, AB, B).

Again, to allow for a meaningful comparison with experiments, we need to reduce the number of degrees of freedom contained in Eqs. (49) and (50). We thus consider two extreme cases: minimum (vanishing) interference and maximum interference.

- Minimum interference ($q_{\min} = 0$) is achieved for $\cos\Delta(\pi) = 0$, i.e. when the interfering decision modes are orthogonal ($\Delta(\pi) = 90°$). Solving simultaneously both equations in (49), we obtain:

$$\begin{cases} |\gamma_1|^2 = \frac{p_{\text{exp}}(AB) - |\gamma_2|^2 |\zeta_{12}|^2}{|\zeta_{11}|^2} \\ |\gamma_2|^2 = \frac{p_{\text{exp}}(B)|\zeta_{11}|^2 - p_{\text{exp}}(AB)|\kappa_{11}|^2}{|\zeta_{11}|^2|\kappa_{12}|^2 - |\zeta_{12}|^2|\kappa_{11}|^2} \end{cases}$$

Taking into account (50), this leads to

$$\begin{cases} |\gamma_1|^2 = \frac{p_{\text{exp}}(AB) - |\gamma_2|^2 \omega(1 - typ(AB))}{typ(AB)} \\ |\gamma_2|^2 = \frac{1}{\omega} \frac{p_{\text{exp}}(B) typ(AB) - p_{\text{exp}}(AB) typ(B)}{typ(AB) - typ(B)} \end{cases} \tag{54}$$

- Maximum interference occurs when $|\cos\Delta(\pi)| = 1$, i.e. when the interfering decision modes are collinear. For maximum positive (resp., negative) interference, when the uncertainty angle between interfering decision modes $\Delta(\pi) = 0°$ and $\cos\Delta(\pi) = 1$ (resp., $\Delta(\pi) = 180°$ and $\cos\Delta(\pi) = -1$), Eq. (49) give for $|\gamma^+|$ (resp., $|\gamma^-|$):

$$\begin{cases} |\gamma_1^{\pm}| = \frac{\sqrt{p_{\text{exp}}(AB)} \mp |\gamma_2^{\pm}||\zeta_{12}|}{|\zeta_{11}|} \\ |\gamma_2^{\pm}| = \pm \frac{\sqrt{p_{\text{exp}}(B)}|\zeta_{11}| - \sqrt{p_{\text{exp}}(AB)}|\kappa_{11}|}{|\zeta_{11}||\kappa_{12}| - |\zeta_{12}||\kappa_{11}|} \end{cases}$$

Taking into account (50), this leads to

$$\begin{cases} |\gamma_1^{\pm}| = \frac{\sqrt{p_{\text{exp}}(AB)} \mp |\gamma_2^{\pm}|\sqrt{\omega}\sqrt{(1 - typ(AB))}}{\sqrt{typ(AB)}} \\ |\gamma_2^{\pm}| = \pm \frac{1}{\sqrt{\omega}} \frac{\sqrt{p_{\text{exp}}(B) typ(AB)} - \sqrt{p_{\text{exp}}(AB) typ(B)}}{\sqrt{typ(AB)(1 - typ(B))} - \sqrt{(1 - typ(AB)) typ(B)}} \end{cases} \tag{55}$$

Note that $|\gamma_2^+| = -|\gamma_2^-|$ and $|\gamma_1^+| = |\gamma_1^-|$. Interestingly, in the expressions of $|\gamma_1|$, $\sqrt{\omega}$ cancels out when the value of $|\gamma_2|$ is substituted into it, implying that $|\gamma_1|$ does not depend on ω. Moreover, for positive and negative maximum interferences, the squared coefficients are equal: $|\gamma_2^+|^2 = |\gamma_2^-|^2$ and $|\gamma_1^+|^2 = |\gamma_1^-|^2$. These squared coefficients are the mean objects of interest since they are interpreted as the weights of the corresponding decision modes of a prospect.

For all 28 triples (I, AB, B), the squared coefficients $\{|\gamma_i|^2, i \in \{1, 2\}\}$ were calculated for the cases of both minimum (zero) and maximum interference, and for the two values $\omega = 1$ and $\omega = 0.5$. The values of the squared probability amplitudes $|\gamma_i|^2$, $i \in \{1, 2\}$, and their sums, for 28 triples (I, AB, B) and all considered conditions (minimum and maximum interference, and $\omega = \{1, 0.5\}$), are reported in Appendix 4 (Tables 9 and 10). Raw values, prior to the adjustments mentioned at the end of this subsection (concerning small deviations from the allowed window and outliers), are provided in brackets.

Figure 3 presents the complementary cumulative distribution function (CCDF) of these coefficients and their sum over the set of the 28 triples. Note the predominance of coefficient $|\gamma_1|^2$, which characterizes the compatibility (typicality) between a described subject (I) and categories in a triple $(AB$ and $B)$. Thus, $|\gamma_1|^2$ is the major contribution within a prospect that influences the probability for that prospect to be chosen (i.e. probability judgment). Coefficient $|\gamma_2|^2$, which is associated with the incompatibility (atypicality) between a considered subject (I) and the triple's categories $(AB$ and $B)$, plays a secondary role. It is inversely proportional to $\omega \in \{1, 0.5\}$, as was shown analytically in Eq. (55). This dependence has a simple explanation. *Within an entangled state of mind* (36), if the atypicality factor is not loaded purely onto the mode $|AB_1I_2\rangle$ ($|B_1I_2\rangle$), but is alos distributed onto an additional mode $|AB_2I_1\rangle$ ($|B_2I_1\rangle$) via decreasing ω from 1 to 0.5, then the observed prospect probabilities can be explained only by a proportional increase of the influence of the original atypicality modes $|AB_1I_2\rangle$ and $|B_1I_2\rangle$ *in the prospect.*

Figure 3 shows that, for each triple, the calculated values of $|\gamma_i|^2$ and of their sum $|\gamma_1|^2 + |\gamma_2|^2$ are found mostly inside their allowed region [0, 1]. Thus, the condition that constrains the squared probability amplitudes of a prospect and their sum to be real positive numbers $\Re_{\geq 0}$ in the interval [0, 1], which was introduced at the end of Sect. 3.4, is in general satisfied. This is especially true for the maximum interference case (lower subplot of Fig. 3), as well as for the minimum interference case with $\omega = 1$. For the minimum interference case (upper subplot), in particular for $\omega = 0.5$, most of the values that fall outside the allowed interval [0, 1] remain close to it. Taken together, the above observations support overall the proposed analytical approach as applied to the experiments of [6].

Concerning the cases that lead to failures to fall in the allowed interval [0, 1], two groups should be distinguished.

- Small departures from [0, 1] can be considered insignificant, because they could arise from minor inconsequencial causes, such as the unavoidable simplifications and approximations in the parameterisation of the theoretical framework, imperfections in the experimental design, erroneous answers ("noise") of decision makers, and so on. Thus, in the subsequent analysis, these small deviations are adjusted to satisfy the constraint $\Re_{\geq 0} \in [0, 1]$, i.e. slightly negative values are replaced by 0, and when $|\gamma_1|^2 + |\gamma_2|^2 > 1$, we normalise $|\gamma_1|^2$ and $|\gamma_2|^2$ correspondingly so that the sum becomes 1.
- Coefficient and/or sum values that are far away from the allowed region are considered "outliers." They pose a challenge to our theoretical framework. They should not be rejected by analysed carefully as their origin may point to new insights. This is the purpose of the next subsection.

4.3 Outliers Detection and Analysis

Several approaches can be followed to identify outliers. The simple visual analysis (heuristic approach) of Fig. 3 suggests extending the allowed region from [0, 1] to [−0.25, 1.25] to account for noise and imperfections in the experiments and in the theoretical parameterisation. This extended allowed region contains most of the values that are outside [0, 1], even for the case of minimum interference with $\omega = 0.5$.

Another approach consists in using a robust statistical approach based on the interquartile range (IQR). In this way, boundaries to identify outliers could be determined by a multiple M of the IQR, which is subtracted from (resp., added to) the first quartile Q_1 (resp., third quartile Q_3). A desired property of these boundaries

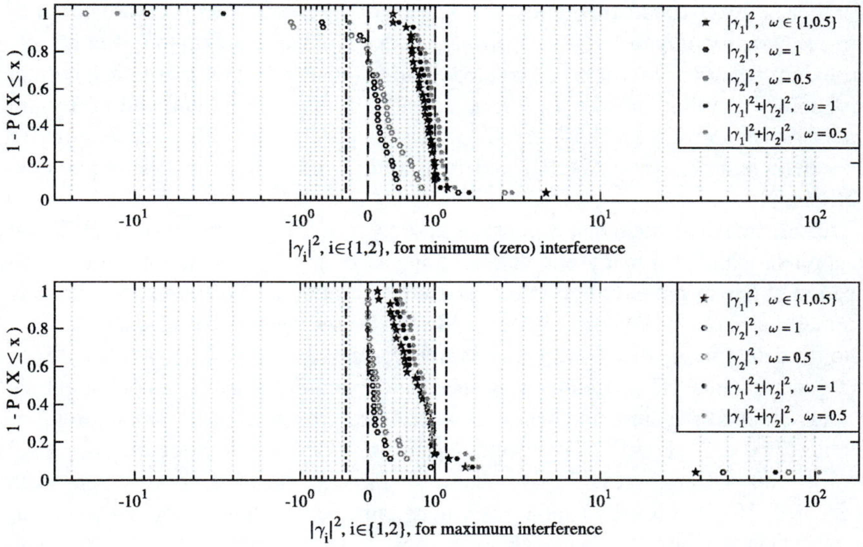

Fig. 3 Complementary cumulative distribution function (CCDF) of squared probability amplitudes $\{|\gamma_i|^2, i \in \{1, 2\}\}$, for minimum (upper panel) and maximum interference (lower panel) cases, over the set of the 28 triples described in the text. The coefficients $\{|\gamma_i|^2, i \in \{1, 2\}\}$ are the probability weights of interfering decision modes within a prospect, and are calculated for all 28 triples (I, AB, B), including 14 incompatible (i) and 14 compatible (c). Note the predominance of coefficient $|\gamma_1|^2$ (star), which is associated with the typicality between a described subject (I) and two categories in the triple (AB and B). Coefficient $|\gamma_2|^2$ (empty circle), which quantifies the atypicality between a considered subject (I) and the triple's categories (AB and B), is inversely proportional to $\omega \in \{1, 0.5\}$, i.e. to the weight of the prospect's atypicality decision mode in the entangled state of mind (36). Most of the observed values of coefficients $|\gamma_i|^2$ and their sum $|\gamma_1|^2 + |\gamma_2|^2$ (filled black circle for $\omega = 1$ and filled grey circle for $\omega = 0.5$) are within their allowed region [0, 1] (dashed vertical lines), in particular for the maximum interference case. The extended region [−0.25, 1.25] (dash-dotted vertical lines) captures values in the vicinity of the allowed region, especially relevant for the case of minimum interference. A logarithmic transformation that is symmetric with respect to the origin on the x-axis has been performed with the Matlab *symlog* function, which was created by R. Perrotta, and based on [10]

would be that they cover and enlarge the allowed range of the data. However, due to the particularities of the analyzed data set, this desired property cannot be achieved with the same M for all types of coefficients. In particular, the distributions of $|\gamma_2|^2$ are centered closer to 0, while the distributions of $|\gamma_1|^2$ and $|\gamma_1|^2 + |\gamma_2|^2$ gravitate towards 1. This leads to unintuitive boundaries that are skewed to the negative side for $|\gamma_2|^2$, and at the same time skewed to the right side (above 1) for other coefficients. Despite this shortcoming, we use the IQR-based method to determine median value of multiplier M, among all types of γ coefficients, which is implied by the extended allowed region $[-0.25, 1.25]$ introduced from the visual analysis of Fig. 3. The implied median $M = 1.5$, which is a widely used multiplier's value with the IQR method, applied to normally distributed datasets. This finding indirectly supports the extended allowed region $[-0.25, 1.25]$ of the heuristic approach.

A better grounded approach to identify outliers requires formal statistical testing. For this, an approximate theoretical distribution of the coefficients γ should be determined. Quantile-quantile (Q-Q) plots allow for graphical comparison of empirical probability distributions of the calculated raw coefficients with the theoretical normal probability distribution (Fig. 4). For all considered cases (minimum and maximum interference, and $\omega \in \{1, 0.5\}$), the Q-Q plots support the normality assumption for the coefficients $\{|\gamma_i|^2, i \in \{1, 2\}\}$ and their sums, and moreover expose potential outliers.

As a formal statistical test, a generalized (extreme Studentized deviate) ESD test is applied, which is a many-outlier detection procedure. It is suitable for the identification of one or more outliers in a univariate dataset that follows an approximately normal distribution. The generalized ESD test was shown to be adequately accurate for detecting up to 10 outliers in samples as small as 25 [5]. To detect outliers, the generalized ESD procedure was repeated separately—for each coefficient γ_i, $i \in \{1, 2\}$, and their sum, and under each condition (minimum and maximum interference, and $\omega \in \{1, 0.5\}$). The results of the two-sided test with significance level $\alpha^{GESD} = 0.001$ are reported in Appendix 4, Table 11. In total, four outliers were identified: 10 (c), 7 (c), 5(c) and 6 (c), where names correspond to an index number of an instance I and a type of categories AB, B—compatible (c) or incompatible (i). The same outliers appear under all considered conditions of interference and values of ω, though the number of outliers differs. One condition, maximum interference with $\omega = 1$ stands out, as it includes the least number of outliers (two), which provides additional support to the validity of this parameterisation.

The identified outliers have been marked by an asterisk in Tables 9 and 10 of Appendix 4. If, for a given triple (I, AB, B), at least one of the coefficients $|\gamma_i|^2$, $i \in \{1, 2\}$, or their sum, is identify as an outlier, than all triple's coefficients for that condition are marked also as outliers, and excluded from the general dataset for further calculations. However, outliers should not be thrown away. Their separate analysis can provide useful insights about the origins of the abnormal values of the coefficients and inform on the limits of the proposed analytical approach.

We have identified a number of outliers based on the above analysis of the probability amplitudes $|\gamma_1|^2$ and $|\gamma_2|^2$ and their sums. Are there also outliers directly observable in the distribution of the empirical conjunction effect? To address this question,

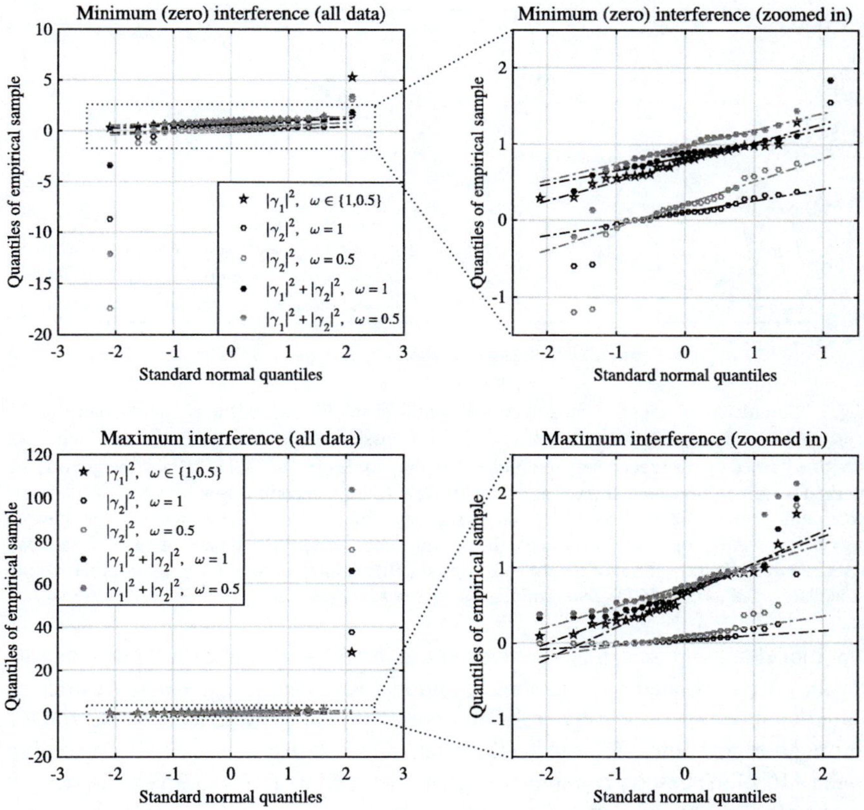

Fig. 4 Quantile-quantile (Q-Q) plots that compare empirical probability distributions of the calculated row coefficients (i.e squared probability amplitudes) $|\gamma_1|^2$ and $|\gamma_2|^2$ and their sums, to the theoretical normal probability distribution. For both—minimum (zero) and maximum interference cases (resp., upper and lower subplots), and both $\omega \in \{1, 0.5\}$, the Q-Q plots support the normality assumption for coefficients $|\gamma_1|^2$ and $|\gamma_2|^2$ and their sums. Moreover, potential outliers are exposed. Plots on the left side present the whole datasets, while plots on the right side zoom them in, for better appreciation of central part of the distributions

Fig. 5 shows the distribution of the absolute values $|typ_{\text{exp}}(AB) - typ_{\text{exp}}(B)|$ of the conjunction effect, which can be well approximated by a normal probability distribution $N(\mu = 0.1550, \sigma = 0.0064)$. For this sample, no outliers were identified with a formal generalized ESD test. Note that most of the tail values, both left and right side, lay above the single Gaussian model (solid line), while values in the center of the sample are found below the theoretical line. These systematic deviations motivate the search for an improved approximation model. Thus, the single Gaussian (G) model was compared with Gaussian mixture (GM) models. Several mixtures with different proportions of constituent normal distributions are drawn in Fig. 5, and results of their multiple-criteria testing against a single Gaussian model is presented in Table 4. Though the log-likelihood objective function is found maximum (or minimum for the negative log-likelihood) under the (unrestricted) Gaussian mix-

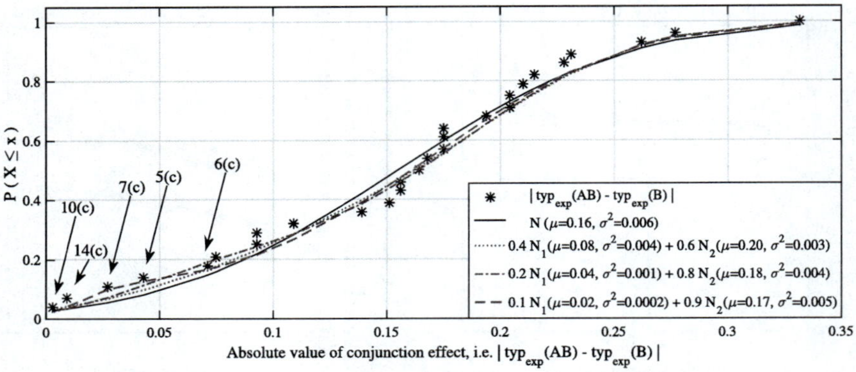

Fig. 5 Cumulative distribution function (CDF) of the absolute value of the conjunction effect, i.e. $|typ_{\text{exp}}(AB) - typ_{\text{exp}}(B)|$, best fitted by a single Gaussian model. All the outliers, which were identified based on prospects' squared probability amplitudes $|\gamma_i|^2$ ($i \in \{1, 2\}$) and their sums, are marked by arrows and concentrated at the far left tail, where the conjunction effect $\to 0$. The only exception—14 (c)—is not an outlier, but its corresponding γ coefficients exceeded the allowed region and were subject to adjustment, thus confirming the specific nature of the left tail. We associate this effect to a limit of the discriminating ability (of a decision maker, or an experimental procedure), and term it *"QDT indeterminacy (uncertainty) principle"*

ture model assumptions, this improvement of the fit is not sufficient to compensate for the cost associated with the three additional parameters. The single (restricted) Gaussian model performs better for all criteria of quality of fit, such as the Akaike information criterion (AIC) and the Bayesian information criterion (BIC). According to the AIC of the best performing mixture model (GM: $0.1N_1 + 0.9N_2$) with respect to a single Gaussian model, the former is only 0.28 times as probable as the Gaussian model to minimize the information loss. In addition, the log-likelihood ratio test of nested hypothesis [11] does not reject the single Gaussian model (p-value = 0.33).

Thus, on the one hand, we demonstrate clearly the existence of outliers in the set of probability amplitudes $|\gamma_1|^2$ and $|\gamma_2|^2$ and their sums, while no outlier exists in the set of conjunction effect amplitudes. This apparent contradiction is resolved by the following observation that sheds light on a potential reason of the appearance of the former outliers. In Fig. 5, which presents the cumulative distribution function (CDF) of the absolute value $|typ_{\text{exp}}(AB) - typ_{\text{exp}}(B)|$ of the conjunction effect, one can observe that all outliers identified by the above analysis of the probability amplitudes $|\gamma_1|^2$ and $|\gamma_2|^2$ and their sums are found to be concentrated at the far left tail of the distribution, and are associated with the lowest magnitudes of the observed conjunction effect. This means that the abnormal values arise in the situation when typicality judgements of the two categories in a triple (AB and B) are very close to each other. On the one hand, this can be understood analytically from Eqs. (54) and (55). For instance, in Eq. (54), $|\gamma_2|^2$ shots up dramatically as the difference in typicality judgements becomes very small: $(typ_{\text{exp}}(AB) - typ_{\text{exp}}(B)) \to 0 \implies |\gamma_2|^2 \to \infty$. At first sight, this property could be considered as a limitation of the proposed analytical formulation. On the other hand, if the theoretical formulation is adequate for most of the choice situations, its inapplicability to certain prospects (judgments) with

Table 4 Selection of the best fitting model for the empirical distribution of the absolute value of the conjunction effect, i.e. $|typ_{\exp}(AB) - typ_{\exp}(B)|$

Model selection		Negative log-likelihood	AIC	Relative likelihood, $e^{\frac{AIC_G - AIC_{GM}}{2}}$	BIC	Log-likelihood ratio test for nested models, *p*-value*
Single Gaussian (G) model	N	−30.92	**−57.85**	**1.00**	**−55.18**	+
Gaussian mixture (GM) model	$0.1N_1$+$0.9N_2$	**−32.65**	−55.30	0.28	−48.64	0.33
	$0.2N_1$+$0.8N_2$	−32.14	−54.28	0.17	−47.61	0.49
	$0.4N_1$+$0.6N_2$	−31.37	−52.75	0.08	−46.09	0.83

A single Gaussian (G) model is tested against several Gaussian mixture (GM) models, with different proportions of constituent normal distributions, and is found to perform best. Figure 5 illustrates the fitsStatistical significance at a level α is obtained when p-value $<= \alpha$. Significance level α is often set to 0.05, 0.01, 0.001 or lower extremely small differentiating characteristics may reveal a limitation of a more fundamental nature: a ceiling in discriminating abilities, either of a decision maker, or of an experimental procedure. Some differences may simply be *too small to notice*. This can be termed the *"QDT indeterminacy (uncertainty) principle"*, as representing a fundamental limit to the precision with which certain pairs (sets) of prospects can be simultaneously known (assessed) by a decision maker, or elicited by an experimental procedure. The formulation of the "QDT indeterminacy (uncertainty) principle", and its characterization, e.g. as a special regime, is proposed as an important and promising research direction. However, larger empirical datasets under various conditions should be analyzed before arriving to a conclusive understanding.

Note that, in the proposed analytical formulation, typicality judgements (typ) are directly linked to the weights ζ and κ of decision modes *in the state of mind* of a decision maker (see Eq. 50). In turn, the coefficients γ are associated to the uncertainty of interfering decision modes *within a prospect*, which is captured by an uncertain union (see Eq. 1). The interconnection between the two sources of indeterminacy (uncertainty), and towards the considered "QDT indeterminacy (uncertainty) principle", calls for further investigation. However, the current analysis provides some first evidence for this novel QDT proposition.

4.4 Return to the Second Interpretation of the Conjunction Fallacy: Analysis of the Adjusted Coefficients $|\gamma_1|^2$ and $|\gamma_2|^2$

As a result of the detailed analysis of the coefficients γ, their raw values were adjusted as explained in the previous subsection: small deviations were corrected to satisfy the constraint $\Re_{\geq 0} \in [0, 1]$, and outliers were excluded from further calculations. Adjusted (and raw) values are reported in Appendix 4 (Tables 9 and 10). The adjusted coefficients γ aggregated over the triples (I, AB, B) of different types (i) or (c) are included in Table 5.

The first notable observation from Table 5 is that the sample means of the sum $|\gamma_1|^2 + |\gamma_2|^2$, for the two decision modes that make up a prospect, are close to 1, especially for the minimum interference case. Thus, for this case, the prospect states are found approximately normalized, which is a characteristic property of a decision-maker's state of mind. Within QDT, a decision consists in the transition of a current state of mind into a new state, which is equivalent to the chosen prospect. During the transition, the state vector of a chosen prospect is being normalized, so that it corresponds to the new state of mind. The results of our analysis show that the prospects in the deliberation phase, when interference is minimal, are already close to being normalized. In other words, even before the transition (decision) occurs, prospects resemble a normalized state of mind.

As expected from derivations (54)–(55), for both interference cases (minimum and maximum interference), changing ω from 1 to 0.5 increases the squared coefficients

Table 5 Aggregated adjusted squared probability amplitudes $|\gamma_i|^2$, $i \in \{1, 2\}$, and their sums for the minimum (zero) interference and maximum interference, as defined in Sect. 4.2

	Minimum (zero) interference, $\cos\Delta(\pi) = 0$						Maximum interference, $\lvert\cos\Delta(\pi)\rvert = 1$					
For $\omega = 1$	$\lvert\gamma_1\rvert^2$		$\lvert\gamma_2\rvert^2$		$\sum_{i=1,2} \lvert\gamma_i\rvert^2$		$\lvert\gamma_1\rvert^2$		$\lvert\gamma_2\rvert^2$		$\sum_{i=1,2} \lvert\gamma_i\rvert^2$	
	μ	σ	μ	σ	μ	σ	μ	σ	μ	σ	μ	σ
24 triples (I, AB, B), incl.	**0.74**	0.18	0.13	0.12	**0.87**	0.12	**0.62**	0.28	0.05	0.06	**0.68**	0.24
14 triples $(I, AB(i), B(i))$	**0.73**	0.17	0.11	0.09	**0.84**	0.14	**0.55**	0.28	0.05	0.05	**0.60**	0.24
10 triples $(I, AB(c), B(c))$	**0.74**	0.20	0.16	0.14	**0.90**	0.09	**0.71**	0.26	0.06	0.08	**0.77**	0.21
For $\omega = 0.5$	$\lvert\gamma_1\rvert^2$		$\lvert\gamma_2\rvert^2$		$\sum_{i=1,2} \lvert\gamma_i\rvert^2$		$\lvert\gamma_1\rvert^2$		$\lvert\gamma_2\rvert^2$		$\sum_{i=1,2} \lvert\gamma_i\rvert^2$	
	μ	σ	μ	σ	μ	σ	μ	σ	μ	σ	μ	σ
24 triples (I, AB, B), incl.	**0.70**	0.19	0.23	0.20	**0.93**	0.09	**0.60**	0.28	0.10	0.13	**0.70**	0.19
14 triples $(I, AB(i), B(i))$	**0.71**	0.18	0.21	0.16	**0.91**	0.10	**0.55**	0.28	0.10	0.11	**0.65**	0.21
10 triples $(I, AB(c), B(c))$	**0.69**	0.22	0.28	0.25	**0.97**	0.05	**0.68**	0.27	0.10	0.16	**0.77**	0.14

The sample mean μ and sample standard deviation σ for triples (I, AB, B) of different types—incompatible (i) and compatible (c)—are reported. The coefficients $|\gamma_1|^2$ and $|\gamma_2|^2$ represent the weights of the interfering decision modes $|(AB)_1 I_1\rangle$ and $|(AB)_1 I_2\rangle$ within a prospect, i.e. the probability judgement of an instance I to belong to a category (AB or B).

$|\gamma_2|^2$ by a factor of 2. As was previously mentioned, when the influence of the atypicality factor in the mind of a decision maker is redistributed evenly from one decision mode $|(AB)_1 I_2\rangle$ onto two modes $|(AB)_1 I_2\rangle$ and $|(AB)_2 I_1\rangle$, the squared probability amplitude of the former atypicality mode $|(AB)_1 I_2\rangle$ in the prospect is raised twofold in order to explain the empirical values of the prospects' probabilities (probability judgements).

The most interesting insight from Table 5 is a manifestation of triples with distinct category types—incompatible (i) and compatible (c)—in the two different interference cases. For a moment, consider condition $\omega = 1$, which required smaller coefficient adjustments and, thus, can be perceived as a more reliable parameterisation. When zero interference between decision modes is imposed, the means of the squared coefficients $|\gamma_1|^2$, as well as the means of $|\gamma_1|^2 + |\gamma_2|^2$, are found similar for all triples: there are no difference between triples, which involve either (i) or (c) categories. In contrast, when maximum interference is assumed, the difference between (i) and (c) categories becomes more evident. For triples $(I, AB(i), B(i))$, the means of $|\gamma_1|^2$ and of $|\gamma_1|^2 + |\gamma_2|^2$ drop visibly. For triples $(I, AB(c), B(c))$, the average of $|\gamma_1|^2$ remains essentially unchanged (it decreases by just 0.03 in absolute value), and the mean of $|\gamma_1|^2 + |\gamma_2|^2$, while moderately decreasing, remains distinctly higher than for the alternative (i) category.

To formally ascertain the above observations, two-sample t-tests for equal means, without assumption of equal variances, were conducted. The hypotheses are formulated as follows:

- H_0: the coefficients $|\gamma_1|^2$ (or $|\gamma_1|^2+|\gamma_2|^2$) for triples $(I, AB(i), B(i))$ and $(I, AB(c), B(c))$ come from independent random samples from normal distributions with *equal means*;
- H_1: the coefficients $|\gamma_1|^2$ (or $|\gamma_1|^2+|\gamma_2|^2$) for triples $(I, AB(i), B(i))$ and $(I, AB(c), B(c))$ come from distributions with *unequal means*;
- H_1^* (a modified one-side hypothesis): the mean of coefficients $|\gamma_1|^2$ (or $|\gamma_1|^2 + |\gamma_2|^2$) for $(I, AB(i), B(i))$ *is smaller* than the corresponding mean for $(I, AB(c), B(c))$.

Results of the t-tests for equal means are summarized in Table 6. For $\omega = 1$, the previous observation is in general confirmed. Under minimum interference condition, H_0 of equal means of $|\gamma_1|^2$ (or $|\gamma_1|^2+|\gamma_2|^2$) for triples $(I, AB(i), B(i))$ and $(I, AB(c), B(c))$ cannot be rejected, with p-values as high as 0.907, and the lowest p-value being 0.127. At the same time, for the maximum interference case, the p-values of H_0 are 4–6 times smaller than for the minimum interference case and, for $|\gamma_1|^2+|\gamma_2|^2$, the p-values of H_0 decrease to p-value=0.03, when pitting H_0 against H_1^*.

For $\omega = 0.5$, the tendency is similar for the means of $|\gamma_1|^2$: H_0 cannot be rejected, but the p-value for the maximum interference case is almost 4 times lower than for the minimum interference case, signaling an increasing discriminability between (i) and (c) types. The means of $|\gamma_1|^2 + |\gamma_2|^2$ under both conditions—minimum and maximum interference—can be considered unequal with reasonable significance levels: H_0 can be rejected with p-value=0.10 (resp., 0.05) in favor of H_1 (resp., H_1^*).

Table 6 Results of two-sample t-tests for equal means (without assumption of equal variances) between two types of triples

	Discriminability of triple's types—(i) and (c)—within each interference case							
	$\lvert\gamma_1\rvert^2$				$\lvert\gamma_1\rvert^2 + \lvert\gamma_2\rvert^2$			
	Minimum (zero) interference		Maximum interference		Minimum (zero) interference		Maximum interference	
For $\omega = 1$	μ	p-value	μ	p-value	μ	p-value	μ	p-value
$(I, AB(i), B(i))$	0.73	**0.907 (0.454)**	0.55	**0.133 (0.067)**	0.84	**0.253 (0.127)**	0.60	**0.062 (0.030)**
$(I, AB(c), B(c))$	0.74		0.71		0.90		0.77	
For $\omega = 0.5$	μ	p-value	μ	p-value	μ	p-value	μ	p-value
$(I, AB(i), B(i))$	0.71	0.871 (0.565)	0.55	0.268 (0.134)	0.91	0.100 (0.050)	0.65	0.103 (0.051)
$(I, AB(c), B(c))$	0.69		0.68		0.97		0.77	

H_0: means of $|\gamma_1|^2$ (or $|\gamma_1|^2 + |\gamma_2|^2$) for both types of triples $(I, AB(i), B(i))$ and $(I, AB(c), B(c))$ are *equal*; H_1: means of $|\gamma_1|^2$ (or $|\gamma_1|^2 + |\gamma_2|^2$) for triples $(I, AB(i), B(i))$ and $(I, AB(c), B(c))$ are ***unequal***; H_1^* (a modified one-side hypothesis, reported in brackets): mean of $|\gamma_1|^2$ (or $|\gamma_1|^2 + |\gamma_2|^2$) for $(I, AB(i), B(i))$ *is less* than the mean for $(I, AB(c), B(c))$. Though reasonable significance levels Statistical significance at a level α is obtained when p-value$<= \alpha$. Significance level α is often set to 0.05, 0.01, 0.001 or lower are achieved only for $|\gamma_1|^2 + |\gamma_2|^2$, the general tendency is confirmed: in the case of minimum interference, the two types of triples—(i) and (c)—are indistinguishable, while for maximum interference (i) and (c) triples manifest themselves differently

However, the condition $\omega = 0.5$ should be analyzed with caution, as many of the raw calculated coefficients required adjustments, which may lead to some distorsions.

Table 7 provides results of a similar two-sample t-test, but now conducted not between two types of triplets—$(I, AB(i), B(i))$ and $(I, AB(c), B(c))$—but rather within each type. The test investigates the susceptibility of triple's types—(i) and (c)—to a change of interference, from the minimum to the maximum interference case. Analyzing $|\gamma_1|^2$, for triples with compatible (c) categories, the hypothesis H_0 of equal means is convincingly not rejected while, for triples with incompatible (i) categories, H_0 can be rejected at a significant level (with maximum p-value= 0.02). Thus, triples $(I, AB(i), B(i))$ are distinguished by their susceptibility to a change of interference.

To summarize the main observation on aggregated coefficients, under two interference conditions—minimum (zero) and maximum—triples with incompatible (i) categories behave differently, than triples with compatible (c) categories. Vice versa, the comparison of the coefficients' dynamics under different interference cases may help to classify prospects, e.g. (i) versus (c) type.

In our above analyses of different conditions—minimum and maximum interference, $\omega \in \{1, 0.5\}$—the same empirical data is used and the general analytical formulation of prospects and decision maker's state of mind are kept unchanged. Our analysis thus reveals how the experimental data *could* be explained under each condition, and whether and to what degree the influence of different factors (decision modes and interference between them) varies. In this context, a decrease of the weight $|\gamma_1|^2$ of the predominant (typicality) decision mode (and the subsequent decrease of the sum $|\gamma_1|^2 + |\gamma_2|^2$), which is observed for the maximum interference case (relative to zero interference), is associated with the increased influence of the QDT interference factor q under this condition. Recall that the attraction factor q reflects the interference between decision modes of a prospect which, after exposure to the prospect, are also present in the state of mind. In addition, the q-factor may include interferences contributed by other elementary prospects of the state of mind, which are originated, not from the prospect in consideration but, from other sources. The latter corresponds to the general QDT formulation of a state of mind as a statistical operator, or would require the identification of additional elementary prospects within the current formulation in terms of a pure state of mind. However, whatever the source, interference effects are condensed into the QDT attraction q-factor.

The analysis of this section shows that the decrease of $|\gamma_1|^2$, followed by a growing effect of the q-factor, is mostly observed for the triples with incompatible (i) categories, and not for compatible (c) type. Thus, the observed probabilities (probability judgements) of prospects with (c) categories can often be explained with a relatively small q-factor. In contrast, in order to explain the probabilities of prospects with (i) categories, the q-factor plays a substantial role. Given the findings in Sect. 4.1, triples $(I, AB(i), B(i))$ are characterized by a negative attraction factor q. Now an additional hypothesis can be formulated: with the decrease of the coefficient $|\gamma_1|^2$ of the predominant (typicality) decision mode, the magnitude of the negative attraction factor q is expected to increase. In other words, an increase $|\gamma_1|^2$ should lead to an increase of q.

Table 7 Results of two-sample t-tests for equal means (without assumption of equal variances) within each triple type, under changing interference conditions

	Susceptibility of triple's types—(i) and (c)—to a change of interference							
	$\lvert\gamma_1\rvert^2$				$\lvert\gamma_1\rvert^2 + \lvert\gamma_2\rvert^2$			
	$(I, AB(i), B(i))$		$(I, AB(c), B(c))$		$(I, AB(i), B(i))$		$(I, AB(c), B(c))$	
For $\omega = 1$	μ	p-value	μ	p-value	μ	p-value	μ	p-value
Interference case								
minimum (zero)	0.73	**0.048 (0.024)**	0.74	**0.757 (0.379)**	0.84	0.003 (0.002)	0.90	0.080 (0.040)
maximum	0.55		0.71		0.60		0.77	
For $\omega = 0.5$	μ	p-value	μ	p-value	μ	p-value	μ	p-value
Interference case								
minimum (zero)	0.71	**0.089 (0.044)**	0.69	**0.892 (0.446)**	0.91	0.0006 (0.0003)	0.97	0.002 (0.001)
maximum	0.55		0.68		0.65		0.77	

H_0: means of $|\gamma_1|^2$ (or $|\gamma_1|^2 + |\gamma_2|^2$) within one type of triples—either $(I, AB(i), B(i))$, or $(I, AB(c), B(c))$—are *equal*; H_1: means of $|\gamma_1|^2$ (or $|\gamma_1|^2 + |\gamma_2|^2$) within one type of triples—either $(I, AB(i), B(i))$, or $(I, AB(c), B(c))$—are *unequal*;H_1^* (a modified one-side hypothesis, reported in brackets): mean of $|\gamma_1|^2$ (or $|\gamma_1|^2 + |\gamma_2|^2$) for the case of maximum interference *is smaller* than for the case of minimum interference. The table shows that triples $(I, AB(i), B(i))$ can be distinguished by their susceptibility to a change of interferenceStatistical significance at a level α is obtained when p-value $<= \alpha$. Significance level α is often set to 0.05, 0.01, 0.001 or lower

To test this hypothesis, the adjusted coefficients $|\gamma_1|^2$ for the case of maximum interference are compared with the attraction factor q from Table 2. Figure 6 illustrates this dependence. Values of $|\gamma_1|^2$ and q are analyzed for the case $\omega = 1$. Firstly, this case is considered to be more reliable. It leads to fewer outliers among γ coefficients, see Table 10 in Appendix 4. In addition, as was mentioned in Sect. 4.1, the case $\omega = 1$ seems to be more plausible, since it recovers closely the observed amplitudes of the conjunction fallacy and the conjunction effect, which were reported in Table 1. Secondly, for the value $\omega = 1$, the attraction factor q from Table 2, i.e. from Eq. 52, reduces to a simple form

$$q(\pi_{(AB)_1 I}) = p_{\exp}(AB) - 0.5$$
$$(\text{resp., } q(\pi_{B_1 I}) = p_{\exp}(B) - 0.5). \quad (56)$$

Indeed, the attraction factor from Eq. (56) can be more generally understood as a refinement of a prospect probability (probability judgement) from a simple toss-like guess (50/50) by incorporating within the q-factor the additional information about a subject (I) and one of four categories ($AB(i)$, $B(i)$, $AB(c)$, $B(c)$).

Figure 6 confirms the hypothesis. As the predominant (typicality) decision mode is characterized by a larger coefficient $|\gamma_1|^2$ (≥ 0.5), i.e. the decision (probability judgement) is determined, to a larger degree, by a single major factor, the QDT

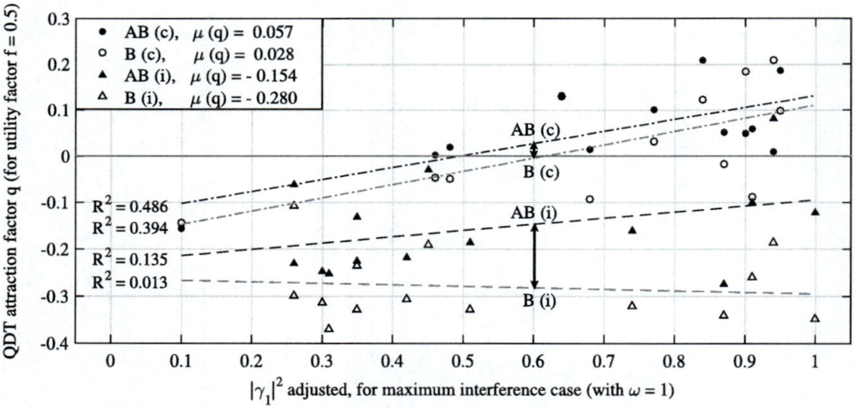

Fig. 6 Dependence between the QDT attraction factor q and the average (over participants) squared probability amplitudes $|\gamma_1|^2$ of a predominant (typicality) decision mode $|(AB)_1 I_1\rangle$ of a prospect. A prospect is a probability judgement of an instance I (a subject) to belong to one of the four categories: an incompatible conjunction $AB(i)$ (filled triangles), or its constituent $B(i)$ (empty triangles), a compatible conjunction $AB(c)$ (filled circles), or its constituent $B(c)$ (empty circles). For $|\gamma_1|^2$, the case of maximum interference between typicality $|(AB)_1 I_1\rangle$ (resp. $|B_1 I_1\rangle$) and atypicality $|(AB)_1 I_2\rangle$ (resp., $|B_1 I_2\rangle$) decision modes is demonstrated. Values of q and $|\gamma_1|^2$ are presented for the case $\omega = 1$. For each category type, sample mean of the attraction factor $\mu(q)$ is provided in the inset. A typical distance between attraction factors for prospects of the same type—either (c), or (i)—is indicated by arrows. The large q distance for incompatible (i) pairs, with $q(B(i)) << q(AB(i))$, illustrates strong repulsion towards $B(i)$ category and explains the conjunction fallacy, which is observed for (i) prospects

attraction factor q varies in a wide range (-0.4, 0.3). However, when the influence of the predominant decision mode decreases and $|\gamma_1|^2 < 0.5$, i.e. the uncertainty of a decision increases, than the magnitude of the negative attraction q increases. The next section elaborates this observation.

As also vividly illustrated by Fig. 6, in the region of higher certainty (right side), triples with compatible categories $(I, AB(c), B(c))$ are characterized by a moderate positive attraction factor. The gradual increase of uncertainty flips the corresponding q-factor to the negative side, however the absolute magnitude of the interference effect remains relatively small. This confirms our aggregate analysis, and demonstrates the mechanism in a finer manner.

Triples with incompatible categories $(I, AB(i), B(i))$ has been shown (Table 7) to exhibit a significant decrease of their $|\gamma_1|^2$ coefficients, as well as sums of the coefficients, when going from the minimum to the maximum interference condition, leading to growing negative interferences and q-factors. With increased uncertainty (Fig. 6, left side), the negative attraction factor of prospects with $AB(i)$ categories further decreases. However, within a conjunction, category A is typical of an instance I, and provides a compensating (positive) effect on the overall negative q-factor. This keeps the q-factor of a prospect within the conjunction category $AB(i)$ larger than for its constituent prospect with a single incompatible $B(i)$ category.

The negative q-factors of prospects within the incompatible constituent category $B(i)$ consistently exhibit the largest amplitude (average $q(B(i)) = -0.28$), of the order of and even exceeding the prediction of the QDT "quarter law" (see Introduction and Appendix 1). This raises interesting questions, to be explored in the future, regarding possible lower values of the q-factor, the existence of a lower barrier, and its insensitivity to a change of the uncertainty level.

This completes the second interpretation of conjunction fallacy.

5 From "Attraction" q to Universal (Uncertainty, Risk and Loss) "Aversion" q

A remarkable insight from Fig. 6 is that, as the uncertainty of a prospect increases, all linear regressions, although relatively noisy (maximum $R^2 = 0.49$), converge to the range of $q \in (-0.3, -0.1)$. This convergence is observed for all prospects, regardless of their type. A negative attraction factor for prospects with incompatible (i) categories could be expected. But the q-factor of prospects with compatible (c) categories also converges to the same *negative* range. We propose to refer to this universal convergence value of the attraction factor (equiv., convergence range $(-0.3, -0.1)$) as the "aversion" q.

The value of this universal "aversion" q resembles the QDT "quarter law" (see Introduction and Appendix 1), which applies to a choice situation with two prospects and predicts an average (among participants or repetitions) absolute value $|q| = 1/4$. The quarter law suggests an identical average value q in both directions—positive and negative—and independently of uncertainty (or of the average uncertainty level). In contrast, the observed "aversion" q is the result of a general tendency observed for *any type of prospect* to converge to the same *negative* range at *high uncertainty* levels and independently of the (un-)attractiveness of a prospect under more certain conditions.

The universal "aversion" q provides a theoretical and empirical background for the use of a general QDT-based "uncertainty aversion principle". According to [13], this principle suggested that the uncertain prospect alternative is associated with the most negative attraction factor. This was shown to lead to reasonable agreements with top-level aggregated empirical results. However, no justification of such an assumption was provided, and the choice of a prospect with higher uncertainty was quite arbitrary. The current empirical analysis reveals a much more complicated and subtle picture than previously assumed. The universal convergence of the QDT attraction factor to $q \in (-0.3, -0.1)$ that we have documented here is shown to be associated with increased uncertainty, and thus resembles closely the well-known *uncertainty aversion*. It actually can be understood as an interpretation and mechanism of uncertainty aversion within the QDT framework, which in addition provides quantitative predictions, e.g. within the convergence range $q \in (-0.3, -0.1)$.

Going one step further, different types of aversions—uncertainty, risk, loss aversion—which have been thoroughly discussed in "classical" decision theories, may have the same core repulsion mechanism that manifests itself (slightly differently) under different conditions (more/less uncertainty and gain/loss domain - e.g. along horizontal and vertical axes of Fig. 6 respectively). Attraction to certain risky prospects—risk-seeking behavior—can be understood as a positive interference q-factor under specific choice conditions. Scattered evidence of the possibility to quantify different risk attitudes with the QDT attraction factor q were provided in previous studies:

- evidence of larger risk aversion of females compared to males [3],
- confirmation of a gender risk aversion effect, and evidence of a "safe q" that is based on standard deviation risk measure [7],
- integration of loss aversion with "large loss" q [9].

For the first time, the current study may be able to provide a common theoretical basis for modelling different risk attitudes with the QDT attraction factor. The various well-documented risk preferences may originate from the same interference mechanism and be modelled with universal principles: an irreducible "QDT indeterminacy (uncertainty) principle", and the universal "aversion" q .

6 Conclusions

In this article, we have taken a fresh look at a classical example of behavioral patterns, the conjunction fallacy, which turned out to be insightful. Our reinterpretation of the conjunction fallacy within Quantum decision theory (QDT) clarified the distinction between two origins of interfering elementary prospects (decision modes):

- different prospects associated with a choice situation (which is at the core of previous QDT developments);
- the state of mind that characterizes the decision maker (background, knowledge, experience, psychological traits, feeling, emotions, and so on).

This distinction does not contradict, but rather aims at building on top and complementing previous QDT developments. We clarify that interfering decision modes may be present (incepted, framed) in the *state of mind*, prior (and during) to the occurrence of a choice situation, and those interfering elementary prospects may *affect* the representation of the choice prospects (options). Thus, not only choice options influence the constituents (elementary prospects) of a state of mind, as previously considered, but the reverse dependence should also be taking into account. The second source of interference can be useful to understand many of the observed behavioral patterns and cognitive biases, such as: representativeness and availability heuristics, which were previously invoked to explain the conjunction fallacy [8], as well as anchoring, attentional bias, belief bias, confirmation bias, framing effect, optimism and pessimism biases, etc. The different cognitive biases should thus been classified according to the two sources of interference.

In this article, the conjunction fallacy was reinterpreted and quantitatively analysed from two perspectives involving either

- the uncertainty factor (uncertainty angles)—Fig. 2; or
- the probability amplitudes of interfering decision modes—Fig. 6.

For the first time, an in-depth quantitative analysis within QDT was performed. Previous studies involved top-level aggregate experimental results, exploiting the most general QDT relation: $p = f + q$. They either checked agreement of data with that relation, or sought for parametrization of f and q, based on "classical" decision theories. The current study aimed at exploring the interference mechanism of QDT, based on the quantification of two fundamental elements: a decision maker's state of mind and the prospects under consideration. The quantification of the decision maker's state of mind was achieved by introducing a link between typicality judgements and probability amplitudes of decision modes in the state of mind. Dividing the problem into several (extreme) cases—minimum and maximum interference, as well as varying the weights of interfering modes in the state of mind—allowed us to estimate the uncertainty and relative contributions of prospect's decision modes to their probability judgement. This level of granularity enabled the analysis of broader and more detained datasets, where interference effects are less profound (for example, data on compatible prospects). This also opens the possibility of discovering the interdependencies and dynamics of QDT elements, and of making the theory operational.

Our explanation of the conjunction fallacy remains squarely based on the core idea of QDT, that uncertainty is the source and modulating factor of the interference (q) between decision modes, which affects the probability of a prospect to be chosen (i.e. a probability judgement). We showed that the interference factor q is present for all types of prospects, conjunctions (both compatible and incompatible) and their constituents. However, it manifests itself differently. For example, prospects with compatible (c) categories are found less susceptible to a change of interference conditions (minimum or maximum interference), and the coefficients of interfering decision modes vary within a smaller range. For this type of prospects, the q-factor is more likely to be characterized by a smaller absolute amplitude, which may switch between positive and negative values depending on the level of uncertainty (i.e. relative contribution of the predominant decision mode). For prospects with incompatible (i) categories, interference effects are more easily detected, because they are more likely to be negative for any uncertainty level. Thus, they are usually characterized by a larger absolute q-factor. Within an incompatible conjunction $AB(i)$, we found that a category A, which is formulated such that it is typical of an instance I, then provides a compensating (positive) effect on the overall negative q-factor of the prospect. In contrast, for a prospect with an incompatible constituent category $B(i)$, the negative q-factor has the largest amplitude, with an average $q(B(i)) = -0.28$, which is at the order and even exceeds the prediction of the QDT "quarter law" in previous studies. Issues concerning the lowest possible values of the q-factor, the possible existence of a lower "barrier", and the insensitivity of such prospects to a change of the uncertainty level, are important future research directions.

Based on our detailed empirical analysis, the novel principle of a "QDT indeterminacy (uncertainty)" has been introduced, as being the fundamental limit to the precision with which certain pairs (sets) of prospects can be simultaneously known (assessed) by a decision maker, or elicited by an experimental procedure.

The observed general tendency, for *any type of prospect*, of the QDT attraction factor to converge to the same *negative* range $q \in (-0.3, -0.1)$ for *high uncertainty* levels motivated the introduction of an universal "aversion" q. The "aversion" q was found essentially independent of the (un-)attractiveness of a prospect under more certain conditions. This is contrast with the QDT "quarter law" previously introduced [13, 17]. We have provided supporting evidence for the *QDT uncertainty aversion principle* and clarified its application.

For the first time, the present study may be able to provide a theoretical common ground for modelling different risk attitudes—uncertainty-, risk-, loss-aversion—with the QDT attraction factor. These various behavioral characteristics are well documented by "classical" decision theories in different choice conditions. We suggest that they may originate from the same interference mechanism and explained with universal principles: the irreducible "QDT indeterminacy (uncertainty) principle" and the universal "aversion" q.

Acknowledgements The authors are grateful to V.I. Yukalov for useful feedbacks on the final manuscript. We appreciate comments from S. Sozzo, T. Veloz and other participants of the Symposium "Worlds of Entanglement", which was organized by the Centre Leo Apostel for Interdisciplinary Studies (CLEA) of the Free University of Brussels (VUB), Brussels, Belgium. This work was partially supported by the Swiss National Foundation, under grant 105218_159461 for the project on "Quantum Decision Theory".

Appendices

1 Appendix 1: Experimental Results from [6]

See Appendix Table 8.

Table 8 Each of the 14 instances I (description of a subject) was presented to decision makers on separate occasions with one of 4 categories: either incompatible (i) conjunction and its constituent, or compatible (c) conjunction and its constituent

Instance I	Type	Conjunctive category AB		Constituent category B		Conjunction effect	Conjunction fallacy
		$typ_{exp}(AB)$	$p_{exp}(AB)$	$typ_{exp}(B)$	$p_{exp}(B)$	$typ_{exp}(AB) - typ_{exp}(B)$	$p_{exp}(AB) - p_{exp}(B)$
1	i	0.439	0.401	0.264	0.241	0.175	0.160
	c	0.733	0.601	0.64	0.533	0.093	0.068
2	i	0.340	0.274	0.125	0.173	0.215	0.101
	c	0.773	0.707	0.664	0.623	0.109	0.084
3	i	0.252	0.226	0.177	0.160	0.075	0.066
	c	0.445	0.344	0.294	0.355	0.151	−0.011
4	i	0.337	0.367	0.133	0.266	0.204	0.101
	c	0.716	0.686	0.623	0.598	0.093	0.088
5	i	0.289	0.269	0.121	0.202	0.168	0.067
	c	0.729	0.552	0.686	0.484	0.043	0.068
6	i	0.249	0.282	0.110	0.194	0.139	0.088
	c	0.604	0.550	0.675	0.683	−0.071	−0.133
7	i	0.389	0.377	0.161	0.152	0.228	0.225
	c	0.772	0.578	0.799	0.544	−0.027	0.034
8	i	0.383	0.252	0.208	0.188	0.175	0.064
	c	0.564	0.516	0.400	0.407	0.164	0.109
9	i	0.527	0.471	0.195	0.310	0.332	0.161
	c	0.694	0.559	0.538	0.412	0.156	0.147
10	i	0.375	0.314	0.144	0.172	0.231	0.142
	c	0.671	0.684	0.668	0.642	0.003	0.042
11	i	0.559	0.580	0.282	0.315	0.277	0.265
	c	0.750	0.519	0.540	0.452	0.210	0.067
12	i	0.340	0.249	0.078	0.131	0.262	0.118
	c	0.575	0.509	0.779	0.707	−0.204	−0.198
13	i	0.291	0.339	0.116	0.180	0.175	0.159
	c	0.668	0.503	0.512	0.453	0.156	0.050
14	i	0.423	0.439	0.229	0.392	0.194	0.047
	c	0.679	0.633	0.670	0.629	0.009	0.004

110 participants made judgements about typicality (typ_{exp}) of an instance I in each category (54 participants), and probability (p_{exp}) that an instance I belonged to the corresponding category (distinct 56 participants). Difference between typicality (resp., probability) judgements with respect to a conjunction (AB) and its constituent (B) are referred to as conjunction effect (resp., conjunction fallacy)

2 Appendix 2: Previous Interpretation of the Conjunction Fallacy Within Quantum Decision Theory

Summary of the Previous Interpretation

The previous interpretation of the conjunction fallacy within Quantum decision theory (QDT), based on the experimental results of [6], was proposed in [13]. The current section condenses the arguments of that interpretation, and provides citations from the original.[1] Parts of sentences in italic correspond to pieces of text extracted from [13]. Within these quotes, we indicate within parentheses when we had to adapt the text to make it understandable with our present conventions.

Consider the following two intentions. *One intention, with just one representation, is "to decide whether the (su)bject* (which is described in an instance I) *has the feature* B*." The second intention is "to decide about the secondary feature" which has two representations, when one decides whether "the (su)bject has the special characteristic"* (A_1) *or "the (su)bject does not have this characteristic"* (A_2). Thus, according to the notation (B = "bankteller", A_1 = "feminist", A_2 = "non-feminist"), prospects are formulated as follows:

- for conjunction AB (e.g. $AB(i)$ "feminist bankteller"):

$$\pi_1 = BA_1, \text{ which is represented by the state } |\pi_1\rangle = a_{11}|BA_1\rangle, \tag{57}$$

- for its constituent B (e.g. $B(i)$ "bankteller"):

$$\pi_2 = BA = B(A_1 + A_2), \text{ which is represented by the state}$$
$$|\pi_2\rangle = a_{21}|BA_1\rangle + a_{22}|BA_2\rangle. \tag{58}$$

The following general scheme is applied to calculate the probability of the prospect with a constituent category B:

$$\begin{aligned} p(\pi_2) = p(BA) &= p(BA_1) + p(BA_2) + q(BA) \\ &= p(B|A_1)p(A_1) + p(B|A_2)p(A_2) + q(BA) \end{aligned} \tag{59}$$

This is a typical situation where a decision is taken under uncertainty. The uncertainty-aversion principle requires that the interference term $q(BA)$ *should be negative* ($q(BA) < 0$).

For the set of compatible (c) *pairs of characteristics, it turned out that the average probabilities were* $p(BA) = 0.537$ *and* $p(BA_1) = 0.567$*, with statistical errors of 20%. Hence, within this accuracy,* $p(BA)$ *and* $p(BA_1)$ *coincide and no conjunction*

[1] To unify notation, B is used for a characteristic that appears in both (i) a conjunction and (ii) a constituent categories (e.g. "bankteller"), and A—for a secondary characteristic (e.g. "feminist") that occurs only in a conjunction. This notation corresponds to [6] and replaces the corresponding symbols that were used in [13], where A = "bankteller", X_1 = "feminist" and X_2 = "non-feminist".

fallacy arises for compatible characteristics. From the view point of QDT, this is easily interpreted as due to the lack of uncertainty: since the features B and A_1 are similar to each other, one almost certainly yielding the other, there is no uncertainty in deciding, hence, no interference, and, consequently, no conjunction fallacy.

For incompatible (i) categories, *the simplest and most natural mathematical embodiment of the property of "incompatibility" is to take the probabilities of possessing B, under the condition of either having or not having A_1, as equal, that is,* $p(B|A_j) = 0.5$. *For these incompatible (i) categories, equation (59) reduces to*

$$p(BA) = \frac{1}{2} + q(BA). \tag{60}$$

For incompatible (i) categories, the average values of the reported probabilities are $p(BA) = 0.220$ and $p(BA_1) = 0.346$ [6].

Given the observed values of $p(BA)$ for each of the 14 constituents of incompatible categories (i) [6] and Eq. (60), the observed interference terms are found fluctuating around a mean of -0.28, with a standard deviation of ± 0.06:

$$q(BA) = -0.28 \pm 0.06. \tag{61}$$

The conjunction error is found to be:

$$\varepsilon(BA_1) \equiv p(BA_1) - p(BA) = 0.126. \tag{62}$$

From Eq. (59), the average value of $p(BA_2)$ is equal to 0.154. In addition, the proposed assumption that $p(B|A_j) = 0.5$ leads to $p(A_1) = p(BA_1)/0.5 = 0.692$, and similarly $p(A_2) = p(BA_2)/0.5 = 0.308$.

[13] conclude: *QDT interprets the conjunction effect as due to the uncertainty underlying the decision, which leads to the appearance of the intention interferences. The interference of intentions is caused by the hesitation whether, under the given primary feature (B), the (su)bject possesses the secondary feature (A_1) or does not have it (A_2). The term $q(BA)$ is negative, reflecting the effect of deciding under uncertainty (according to the uncertainty-aversion principle). Quantitatively, we observe that the amplitude $|q(BA)|$ is in agreement with the QDT interference-quarter law.*

Weaknesses of the Previous Interpretation

As summarised in the previous Sect. 2, the interpretation of the conjunction fallacy in [13] rests on two assumptions:

1. the formulation of a prospect for a constituent category B such that it includes uncertainty about a secondary feature ($A_1 + A_2$), i.e. Eq. (58);
2. the independence of incompatible features, which underlies Eq. (60).

The current section analyses these two assumptions and demonstrates the existence of some inconsistencies.

Formulation of the Prospect for a Constituent Category (B): Intention Concerning a Primary Feature and Undetermined Sign of q (Eqs. 57–59)

For a secondary characteristic (e.g. "feminist"), the uncertainty about its presence (i.e. an undecided attribution of this feature to the subject from an instance I) is represented as a *composite* action A, which is a sum of two action modes A_1 ("the subject *has* a secondary feature") and A_2 ("the subject *does not have* a secondary feature"). However, for a primary characteristic (e.g. "bankteller"), a *simple* action with one action mode B ("*to decide* whether the subject has a primary feature") is suggested. This formulation of intention B as an *active decision* concerning a primary feature is necessary to justify the negative sign of the attraction factor q for a corresponding prospect with a constituent category (B), if a level of uncertainty about a secondary feature ($A_1 + A_2$) is introduced in this prospect, such that:

$$\pi_2 = B(A_1 + A_2)\ , \text{i.e. } |\pi_2\rangle = a_{21}|B \otimes A_1\rangle + a_{22}|B \otimes A_2\rangle, \tag{63}$$

where the symbol $\otimes$ represents the tensor product operator (see [13] for details).

Thus, under an assumption of passivity in the presence of uncertainty, it is proposed that making a *decision* concerning a primary feature B (e.g. "bankteller") is not attractive, i.e. $q(\pi_2) \leq 0$.

The following two arguments reveal an inconsistency in the above formulation of a constituent intention B.

First, it is natural to assume that, for a primary characteristic, an intention complementary to B should exist, which is denoted for simplicity *not* B and stands for "*not to decide* whether the subject has a primary feature"). Action *not* B reflects an undecided attribution of a primary feature and, similar to action A for a secondary feature, can be presented as a sum of two action modes: B_1 ("the subject has a primary feature") and B_2 ("the subject does not have a primary feature"). Continuing this analogy, in formulating the prospects for both categories, i.e. for a conjunction and its constituent, action mode B_1 should be used to represent a consideration (attribution) of a primary feature (e.g. "bankteller"). However, the introduction in [13] of action B as described above remains unclear. In fact, an interpretation of action B as an *active decision* ("*to decide* whether the object has a primary feature") seems unrealistic, as in the experiment participants were exposed to the predefined categories and *had to* judge the corresponding probabilities, i.e. participants were not asked to decide whether to make the judgement or not.

Second, if the intention B represents a single action mode of possessing a primary feature that is equivalent to B_1 ("the subject *has* a primary feature"), then a complementary action mode B_2 ("the subject *does not have* a primary characteristic) exists and also requires an *active decision* of a decision maker about the possession (or absence) of a primary feature in the subject. Thus, the sign of the attraction factor q for a constituent category cannot be determined, even when assuming the presence of uncertainty concerning a secondary feature.

These two inconsistencies can thus be summarised as follows:

- an intention concerning a primary feature B (e.g. "bankteller") should be formulated similarly to an intention about a secondary feature A (e.g. "feminist"), in the form of "the subject *has* a feature";
- the sign of the attraction factor q for a constituent category cannot be determined, even when assuming the presence of uncertainty with respect to a secondary feature in this prospect.

Formulation of the Prospect for a Constituent Category (B): Uncertainty About a Secondary Feature (Eq. 58)

In order to fully understand the assumptions underlying the QDT interpretation of [13], it is important to note that the experiments that have investigated the conjunction fallacy have been performed under several distinct treatments, which introduce subtle but important differences for their theoretical interpretation. The following three main classes of experiment treatment have been used.

1. *Indirect* tests, when subjects were exposed to a description of a subject (an instance I) and only one of the categories (either conjunction AB, or its constituent B) at a time. In this setup, the judgements about the probability of an instance I with respect to each category—AB or B—were made *separately* and were not juxtaposed. For example, in [8], an indirect between-subjects comparison was conducted, when the probability of the conjunction was evaluated by one group and the probability of its constituent was evaluated by another group. In [6], judgments of probabilities for each category—AB or B—were performed separately, but by the same decision makers.
2. *Direct-subtle* tests, when, following an instance I, participants are exposed to both a conjunction and its constituent category, but the inclusion relation is not made apparent. For example, in [8], the two categories of interest are shown simultaneously, but are camouflaged among five additional filler items.
3. *Direct-transparent* tests, when an instance I, a conjunction and its constituent are presented together to highlight the connection between the categories.

In the current QDT treatment [13], the conjunction fallacy, i.e. $p(AB) > p(B)$, is explained by the negative attraction factor of a prospect with constituent category B. For a negative q to appear, a judgement about the prospect with a constituent B is *assumed* to be influenced by the existence of a level of *uncertainty* about the presence of a secondary feature A. In other words, a judgment about a primary feature B is saddled with an added degree of uncertainty about the attribution of a secondary feature A, even when absent in the judged category B (Eq. 58).

However, in the *indirect* test design as presented in [6] and analyzed in [13], the judgement about the probability of an instance I with respect to a constituent category B was made *separately*, without exposition to a conjunction category AB. In this indirect setup, it is not obvious that a secondary feature, which is present only in a conjunction, has any influence on a judgment about a constituent category. Thus, there is no evidence that a secondary feature should be included in the formulation of the QDT prospect for a constituent category B (e.g. "bankteller"), which instead can be simply represented by

$$\pi_2 = B \text{ , i.e. } |\pi_2\rangle = a_{21}|B\rangle, \tag{64}$$

instead of Eq. 58.

Importantly, for *indirect* tests, with this formulation of a prospect for a constituent B, there is no uncertainty about the presence of a secondary feature A in a constituent category. Thus, the QDT uncertainty aversion principle ought not to be invoked to explain the observed conjunction fallacy, and another mechanism is required.

For the *direct* test designs, which allow for a direct comparison of a conjunction and a constituent categories, the formulation of prospects proposed in [13] is more plausible. It could be expected though that a more profound manifestation of uncertainty for a constituent category, which is associated with the presence of a secondary feature, would increase the absolute value of a negative attraction factor of this prospect ($|\pi_2\rangle$), amplifying the conjunction fallacy. However, the opposite results are observed in experiments [8]. Probably, the most convincing evidence was obtained in a *direct-transparent* test, where the probability judgment about an instance I (the description of Linda) was made with respect to the following two categories:

1. Linda is a bank teller *whether or not* she is active in the feminist movement: $B(i)$ (versus $B(A_1 + A_2)$ in [13]);
2. Linda is a bank teller and is active in the feminist movement: $AB(i)$ (versus BA_1 [13]).

In this example, the degree of uncertainty about the presence of a secondary "feminist" feature $(A_1 + A_2)$ is made explicit in a constituent category $B(i)$, which provides a good match for the formulation of a prospect with a constituent in [13]. However, contrary to what one could expect from the formulation of [13], the portion of decision makers who committed the conjunction fallacy dropped from above 80% (observed in both indirect and direct tests) to 57% (for the direct-transparent test) [8]. This finding signals that the recognition by decision makers of a level of uncertainty about a secondary feature in a constituent category does not make this prospect less attractive, but rather emphasizes the inclusive relation between the two categories (conjunction and its constituent) and facilitates the correct application of the conjunction rule.

Furthermore, [8] outlined that the *representativeness heuristic* may be at the heart of the persistent conjunction fallacy. Even when provided with a valid and clear explanation of the inclusion of a conjunction category into a constituent, the majority of subjects choose to stick to an "emotional" resemblance argument.

This suggests an alternative QDT mechanism for the explanation of the conjunction fallacy: rather than a negative attraction of a constituent category due to the uncertainty of a secondary feature, the key ingredient is a *higher attraction to a conjunction prospect*, if a secondary feature is compatible with the description of an instance I.

Independence of (Incompatible-Type) Prospects (Eq. 60)

Equation (60), which is a key result in [13], is based on two underlying assumptions:

- the "incompatibility" of constituents in a conjunction category is treated as leading to their "independence", i.e. the probability of possessing a primary characteristic B is assumed to be independent from having a secondary characteristic A_j, yielding $p(B|A_1) = p(B|A_2) = p(B)$;
- with no prior information, it is assumed that $p(B) = 0.5$.

A first general criticism can be raised about the assumption that the existence of uncertainty about an independent category would lead to a negative attraction factor when deciding about another independent category. If this was the case, any decision would then be associated with $q < 0$, as it is impossible to create a completely certain environment in our complex uncertain world, as there are always many variables that remain uncertain around us.

Secondly, in [13], the "compatibility" of categories is associated with an absence of uncertainty, i.e. the possession of one of the compatible characteristics yields the other one, which implies positive correlation close to 1. In this context, "incompatibility" is more likely to imply negative correlation close to -1, rather than independence with correlation close to 0.

Thirdly, an assumption that a high degree of "compatibility" (or "incompatibility") is associated with very low uncertainty, and thus leads to $q \to 0$, has to be tested. For example, it is plausible that the subjective estimation of a high "compatibility" ("incompatibility") of categories may increase the 'subjective' confidence of a decision maker, which could be reflected in a high positive (negative) value of the attraction factor, and make a choice deviate from an 'objective' judgement. Empirical evidence should be gathered to support this hypothesis.

We are thus led to suggest two alternative propositions replacing the two assumptions of [13] discussed above:

- Independent intentions do not interfere. Thus, the existence of uncertainty about one intention does not influence the probability that another intention will be realized, if these intentions are independent.
- Equation (60) requires revision: if (incompatible) intentions are treated as independent, they should not interfere and $q = 0$; if (incompatible) intentions interfere and $q \neq 0$, then $p(B|A_1) = p(B|A_2) = p(B) = 0.5$ can not be assumed.

Partial Use of Data

[13] did not make use of experimental data on *compatible* conjunctions and their constituents, and on *typicality* judgements. Most importantly, the description of a subject—instance I—is a key element of the experiment, which consists in framing participants prior the choice (judgement). However, this was ignored in many theoretical interpretations, including [13].

Synthesis

The previous interpretation of the conjunction fallacy within QDT [13] aimed at explaining it for the most clearcut cases of one type of incompatible prospects with a constituent category $B(i)$. Agreement between partial empirical data and the general QDT relation $p{=}f{+}q$ was obtained. However, the needed underlaying assumptions have been shown to be unsubstantiated.

In particular, the representation of prospects, which is needed to justify the application of the uncertainty-aversion principle and the corresponding negative sign of the attraction factor q, leads to serious inconsistencies. The interference effects argued to occur for a single constituent category B, as formulated in [13], have a shaky foundation. As discussed above in details, within an indirect test setup [6], the inclusion of uncertainty about a secondary feature A from a conjunction category AB should not be relevant to a separate judgement regarding B. Another essential assumption about the independence of incompatible categories, upon which the proposed interpretation rests, is arbitrary. Importantly, the influence of framing, i.e. the pre-exposure of decision makers to an instance I (the description of a subject), is disregarded. Last but not least, the available empirical data is used only partially (just for one out of four judged categories).

Since the attempt of [13], the theoretical construct of QDT has been significantly enriched. We use this opportunity to 'cure' the above mentioned weaknesses, to propose a genuine rationalisation and quantitative explanation of the conjunction fallacy within QDT, which allows us to further explore the limits of QDT.

3 Appendix 3: Verification of the Necessary Conditions for the Emergence of the QDT Attraction Factor

The necessary conditions for the attraction factor to be non-zero are: (a) entanglement in a strategic decision-maker state, and (b) entanglement of a prospect, i.e. a decision is to be made under uncertainty. Concerning the former condition (a), a strategic decision-maker state can be separable (not entangled) only if the measurements of observables are not temporally correlated. Section 3.3.1 demonstrates the evolution of a strategic decision-maker state through a sequence of channels, which produces an entangled state $\hat{\rho}_{ABIM}$ (resp., $\hat{\rho}_{BIM}$) just prior to a decision. For the second condition (b), to determine whether prospect $\pi_{(AB)_1 I}$ from (20) is entangled, we investigate the separability of the corresponding operator $\hat{P}_{(AB)_1 I}$ (22), as proposed in [16, 18]. For this, we introduce two Hilbert-Schmidt spaces below. The first one is defined by

$$\widetilde{\mathcal{AB}} \equiv \{\mathcal{AB}, \mathcal{H}_{AB}, \sigma_{AB}\} \tag{65}$$

where $\mathcal{AB} = \{\hat{P}_{(AB)_i}\}$ is an operator algebra (or an algebra of local observables), acting on the Hilbert space $\mathcal{H}_{AB}$, while σ_{AB} is the scalar product $\sigma_{AB} : \mathcal{AB} \times \mathcal{AB} \rightarrow$

$\mathbb{C}$ that is defined as

$$\sigma_{AB} : (\hat{P}_{(AB)_1}, \hat{P}_{(AB)_2}) = \mathrm{Tr}_{AB} \hat{P}^{\dagger}_{(AB)_1} \hat{P}_{(AB)_2} \tag{66}$$

and generates the Hilbert-Schmidt norm $||\hat{P}_{(AB)}|| \equiv \sqrt{(\hat{P}_{(AB)_i}, \hat{P}_{(AB)_i})}$, $i \in 1, 2$.
Similarly, for the second Hilbert-Schmidt space:

$$\widetilde{\mathcal{I}} \equiv \{\mathcal{I}, \mathcal{H}_I, \sigma_I\} \tag{67}$$

where $\mathcal{I} = \{\hat{P}_{I_i}\}$ is an operator algebra (or an algebra of local observables) on the Hilbert space $\mathcal{H}_I$, the scalar product is a map $\sigma_I : \mathcal{I} \times \mathcal{I} \to \mathbb{C}$ that is defined as

$$\sigma_I : (\hat{P}_{I_1}, \hat{P}_{I_2}) = \mathrm{Tr}_I \hat{P}^{\dagger}_{I_1} \hat{P}_{I_2} \tag{68}$$

and generates the Hilbert-Schmidt norm $||\hat{P}_I|| \equiv \sqrt{(\hat{P}_{I_i}, \hat{P}_{I_i})}$, $i \in 1, 2$.
Now, a composite Hilbert-Schmidt space can be introduced as a tensor-product space

$$\widetilde{\mathcal{AB}} \otimes \widetilde{\mathcal{I}} = \{\mathcal{AB}, \mathcal{H}_{AB}, \sigma_{AB}\} \otimes \{\mathcal{I}, \mathcal{H}_I, \sigma_I\}. \tag{69}$$

An operator acting on this composite Hilbert-Schmidt space $\widetilde{\mathcal{AB}} \otimes \widetilde{\mathcal{I}}$ is called *separable* (or disentangled) if and only if it can be represented as a sum

$$\sum_i \hat{P}_{(AB)_i} \otimes \hat{P}_{I_i} \quad (\hat{P}_{(AB)_i} \in \widetilde{\mathcal{AB}}, \ \ \hat{P}_{I_i} \in \widetilde{\mathcal{I}}) \tag{70}$$

Importantly, the operator $\hat{P}_{(AB)_1 I}$ cannot be represented in the separable form (70), because the last term $|I_k\rangle\langle I_l|$ in (22) does not pertain to $\widetilde{\mathcal{I}}$. Thus, we conclude that the corresponding composite prospect $\pi_{(AB)_1 I}$ in (20) is *entangled*.
Following the same procedure, the operator $\hat{P}_{B_1 I}$ in (23) is non-separable, i.e. it cannot be represented as a sum

$$\sum_i \hat{P}_{B_i} \otimes \hat{P}_{I_i} \quad (\hat{P}_{B_i} \in \widetilde{\mathcal{B}}, \ \ \hat{P}_{I_i} \in \widetilde{\mathcal{I}}) \tag{71}$$

and the related composite prospect $\pi_{B_1 I}$ in (20) is *entangled* as well.

Thus, the necessary conditions for the emergence of the QDT attraction factor in (28), i.e. $q(\pi_{(AB)_1 I}) \neq 0$ and $q(\pi_{B_1 I}) \neq 0$, are satisfied.

4 Appendix 4: Squared Probability Amplitudes $|\gamma_i|^2$, $i \in \{1, 2\}$, of Prospects' Decision Modes

See Appendix Tables 9, 10 and 11.

Table 9 Adjusted coefficients $|\gamma_i|^2$, $i \in \{1, 2\}$, and their sums, for the Minimum interference case

Minimum (zero) interference, $\cos\Delta(\pi) = 0$

I	Type (AB, B)	$\omega = 1$			$\omega = 0.5$		
		$\lvert\gamma_1\rvert^2$	$\lvert\gamma_2\rvert^2$	$\sum_{i=1,2} \lvert\gamma_i\rvert^2$	$\lvert\gamma_1\rvert^2$	$\lvert\gamma_2\rvert^2$	$\sum_{i=1,2} \lvert\gamma_i\rvert^2$
1	i	0.91	0.00	0.91	0.91	0.00	0.91
	c	0.80	0.07	0.86	0.80	0.13	0.93
2	i	0.58	0.11	0.70	0.58	0.23	0.81
	c	0.88	0.11	0.99	0.80 (0.88)	0.20 (0.22)	1.00 (1.10)
3	i	0.88	0.00	0.89	0.88	0.01	0.89
	c	0.30	0.38	0.68	0.29 (0.30)	0.71 (0.75)	1.00 (1.06)
4	i	0.70	0.20	0.90	0.63 (0.70)	0.37 (0.40)	1.00 (1.10)
	c	0.95	0.01	0.96	0.95	0.02	0.97
5	i	0.55	0.15	0.71	0.55	0.31	0.86
	c	−(0.98)	−(−0.60)*	−(0.38)	−(0.98)	−(−1.20)*	−(−0.22)
6	i	0.76	0.12	0.88	0.75 (0.76)	0.25	1.00 (1.01)
	c	−(1.29)	−(−0.58)*	−(0.71)	−(1.29)	−(−1.16)*	−(0.13)
7	i	0.98	0.00(−0.01)	0.98 (0.97)	0.98	0.00(−0.01)	0.98 (0.97)
	c	−(0.29)	−(1.55)*	−(1.84)*	−(0.29)	−(3.10)*	−(3.39)*
8	i	0.48	0.11	0.59	0.48	0.22	0.70
	c	0.81	0.14	0.95	0.74 (0.81)	0.26 (0.28)	1.00 (1.09)
9	i	0.70	0.22	0.92	0.62 (0.70)	0.38 (0.43)	1.00 (1.13)
	c	0.85	0.00(−0.09)	0.85 (0.75)	0.85	0.00(−0.19)	0.85 (0.66)
10	i	0.70	0.08	0.78	0.70	0.17	0.87
	c	−(5.29)*	−(−8.71)*	−(−3.42)*	−(5.29)*	−(−17.42)*	−(−12.13)*
11	i	0.96 (1.00)	0.04 (0.05)	1.00 (1.05)	0.92 (1.00)	0.08 (0.09)	1.00 (1.09)
	c	0.60	0.28	0.88	0.52 (0.60)	0.48 (0.56)	1.00 (1.16)
12	i	0.55	0.10	0.64	0.55	0.19	0.74
	c	0.92	0.00(−0.05)	0.92 (0.87)	0.92	0.00 (-0.10)	0.92 (0.82)
13	i	0.93 (0.98)	0.07	1.00 (1.06)	0.87 (0.98)	0.13 (0.15)	1.00 (1.13)
	c	0.61	0.29	0.90	0.51 (0.61)	0.49 (0.58)	1.00 (1.19)
14	i	0.58	0.34	0.92	0.46 (0.58)	0.54 (0.67)	1.00 (1.25)
	c	0.70 (0.78)	0.30 (0.33)	1.00 (1.11)	0.54 (0.78)	0.46 (0.66)	1.00 (1.44)
μ	i	0.73 (0.74)	0.11	0.84 (0.85)	0.71 (0.74)	0.21 (0.22)	0.91 (0.96)
	c	0.74 (1.10)	0.16 (−0.49)	0.90 (0.60)	0.69 (1.10)	0.28 (−0.98)	0.97 (0.11)
	All	0.74 (0.92)	0.13 (−0.19)	0.87 (0.73)	0.70 (0.92)	0.23 (−0.38)	0.93 (0.54)
σ	i	0.17 (0.18)	0.09	0.14 (0.15)	0.18	0.16 (0.19)	0.10 (0.16)
	c	0.20 (1.24)	0.14 (2.42)	0.09 (1.20)	0.22 (1.24)	0.25 (4.84)	0.05 (3.61)
	All	0.18 (0.89)	0.12 (1.71)	0.12 (0.85)	0.19 (0.89)	0.20 (3.41)	0.09 (2.55)
IQR	All	0.30 (0.34)	0.17 (0.20)	0.12 (0.24)	0.33 (0.34)	0.30 (0.40)	0.11 (0.29)

The coefficients are calculated for all 28 triples of an instance I and categories AB and B, including 14 incompatible (i) and 14 compatible (c). Adjustments are: exclusion of outliers (marked by a dash); and corrections to satisfy the constraint $\Re_{\geq 0} \in [0, 1]$, i.e. replacement of negative values by 0, and renormalization of $|\gamma_1|^2$ and $|\gamma_2|^2$, if their sum exceeds 1. Identified by the general ESD test, outlier values are marked by an asterisk *. Unadjusted values are in brackets

Table 10 Adjusted coefficients $|\gamma_i|^2$, $i \in \{1, 2\}$, and their sums for the maximum interference case

Maximum interference, $\cos\Delta(\pi) = 1$							
I	Type (AB, B)	$\omega = 1$			$\omega = 0.5$		
		$\lvert\gamma_1\rvert^2$	$\lvert\gamma_2\rvert^2 2$	$\sum_{i=1,2} \lvert\gamma_i\rvert^2$	$\lvert\gamma_1\rvert^2$	$\lvert\gamma_2\rvert^2$	$\sum_{i=1,2} \lvert\gamma_i\rvert^2$
1	i	0.91	0.00	0.91	0.91	0.00	0.91
	c	0.77	0.00	0.77	0.77	0.00	0.77
2	i	0.35	0.05	0.39	0.35	0.10	0.44
	c	0.84	0.01	0.85	0.84	0.01	0.85
3	i	0.87	0.00	0.87	0.87	0.00	0.87
	c	0.10	0.26	0.36	0.10	0.51	0.61
4	i	0.35	0.10	0.45	0.35	0.21	0.56
	c	0.95	0.00	0.95	0.95	0.00	0.95
5	i	0.26	0.08	0.34	0.26	0.17	0.43
	c	0.87 (1.31)	0.13 (0.20)	1.00 (1.52)	−(1.31)	−(0.41)	−(1.72)*
6	i	0.42	0.06	0.48	0.42	0.11	0.54
	c	0.90 (1.73)	0.10 (0.20)	1.00 (1.93)	−(1.73)	−(0.40)	−(2.13)*
7	i	1.00	0.00	1.00	1.00	0.00	1.00
	c	−(0.12)	−(0.92)*	−(1.04)	−(0.12)	−(1.84)*	−(1.96)*
8	i	0.30	0.04	0.35	0.30	0.08	0.39
	c	0.68	0.02	0.70	0.68	0.05	0.72
9	i	0.45	0.08	0.53	0.45	0.17	0.62
	c	0.91	0.01	0.92	0.91	0.01	0.92
10	i	0.51	0.02	0.54	0.51	0.05	0.56
	c	−(28.26)*	−(37.83)*	−(66.09)*	−(28.26)*	−(75.66)*	−(103.92)*
11	i	0.94	0.00	0.94	0.94	0.01	0.95
	c	0.48	0.06	0.54	0.48	0.12	0.60
12	i	0.31	0.05	0.36	0.31	0.09	0.40
	c	0.94	0.00	0.95	0.94	0.00	0.95
13	i	0.74	0.02	0.76	0.74	0.04	0.78
	c	0.46	0.07	0.53	0.46	0.14	0.60
14	i	0.26	0.19	0.45	0.26	0.38	0.64
	c	0.64	0.06	0.70	0.64	0.11	0.76
μ	i	0.55	0.05	0.60	0.55	0.10	0.65
	c	0.71 (2.73)	0.06 (2.83)	0.77 (5.56)	0.68 (2.73)	0.10 (5.66)	0.77 (8.39)
	All	0.62 (1.64)	0.05 (1.44)	0.68 (3.08)	0.60 (1.64)	0.10 (2.88)	0.70 (4.52)
σ	i	0.28	0.05	0.24	0.28	0.11	0.21
	c	0.26 (7.36)	0.08 (10.08)	0.21 (17.43)	0.27 (7.36)	0.16 (20.15)	0.14 (27.50)
	All	0.28 (5.23)	0.06 (7.13)	0.24 (12.35)	0.28 (5.23)	0.13 (14.27)	0.19 (19.49)
IQR	All	0.52 (0.57)	0.08 (0.09)	0.46 (0.47)	0.53 (0.57)	0.12 (0.17)	0.32 (0.36)

The coefficients are calculated for all 28 triples of an instance I and categories AB and B, including 14 incompatible (i) and 14 compatible (c). Adjustments are: exclusion of outliers (marked by dash); and corrections to satisfy the constraint $\Re_{\geq 0} \in [0, 1]$, i.e. replacement of negative values by 0, and renormalization of $|\gamma_1|^2$ and $|\gamma_2|^2$, if their sum exceeds 1. Identified by a general ESD test, outlier values are marked by an asterisk *. Unadjusted values are in brackets

Table 11 Results of the generalized (extreme Studentized deviate) ESD test[a] to detect $i = 1...k$ outliers in samples of γ coefficients, approximated by the normal distribution

Minimum (zero) interference, $\cos \Delta(\pi) = 0$														
$\|\gamma_1\|^2, \omega \in \{1, 0.5\}$			$\|\gamma_2\|^2, \omega = 1$			$\|\gamma_2\|^2, \omega = 0.5$			$\sum_{i=1,2} \|\gamma_i\|^2, \omega = 1$			$\sum_{i=1,2} \|\gamma_i\|^2, \omega = 0.5$		
R_i	λ_i	Outlier	R_i	λ_i	Outlier	R_i	λ_i	Outlier	R_i	λ_i	Outlier	R_i	λ_i	Outlier
4.939	3.464	5.29	4.990	3.464	−8.71	4.990	3.464	−17.42	4.885	3.464	−3.42	4.973	3.464	−12.13
2.360	3.441		3.924	3.441	1.55	3.927	3.441	3.10	3.838	3.441	1.84	4.108	3.441	3.39
2.189	3.416		2.918	3.416	−0.60	2.924	3.416	−1.20	2.844	3.416		3.359	3.416	
2.438	3.390		3.594	3.390	−0.58	3.605	3.390	−1.16	2.011	3.390		3.306	3.390	
I_{out}: 10(c)			I_{out}: 10(c), 7(c), 5(c), 6(c)			I_{out}: 10(c), 7(c), 5(c), 6(c)			I_{out}: 10(c), 7(c)			I_{out}: 10(c), 7(c)		
Maximum interference, $\cos \Delta(\pi) = 1$														
$\|\gamma_1\|^2, \omega \in \{1, 0.5\}$			$\|\gamma_2\|^2, \omega = 1$			$\|\gamma_2\|^2, \omega = 0.5$			$\sum_{i=1,2} \|\gamma_i\|^2, \omega = 1$			$\sum_{i=1,2} \|\gamma_i\|^2, \omega = 0.5$		
R_i	λ_i	Outlier	R_i	λ_i	Outlier	R_i	λ_i	Outlier	R_i	λ_i	Outlier	R_i	λ_i	Outlier
5.090	3.464	28.26	5.101	3.464	37.83	5.101	3.464	75.66	5.100	3.464	66.09	5.101	3.464	103.92
2.855	3.441		4.589	3.441	0.92	4.589	3.441	1.84	3.214	3.441		2.939	3.441	2.13
2.212	3.416		2.719	3.416		2.646	3.416		2.846	3.416		3.229	3.416	1.96
1.674	3.390		2.366	3.390		2.410	3.390		1.557	3.390		3.511	3.390	1.72
I_{out}: 10(c)			I_{out}: 10(c), 7(c)			I_{out}: 10(c), 7(c)			I_{out}: 10(c)			I_{out}: 10(c), 6(c), 7(c), 5(c)		

[a] A generalized ESD test is done with Matlab *gesd* function, which was created by F.A. Alcaraz Garcia, and based on [5].
The test statistic $R_i = (max_i|x_i - \overline{x}|)/s$, where $\overline{x}$ is the sample mean and s is the sample standard deviation, is provided. The critical value λ_i is calculated based on the t-distribution. Two-sided test was performed with significance level $\alpha^{GESD} = 0.001$, for each coefficient $|\gamma_i|^2, i \in \{1, 2\}$, and their sum, and under each condition (minimum and maximum interference, and $\omega \in \{1, 0.5\}$). The values of the detected outliers are presented, and the corresponding triples (I, AB, B) are listed in I_{out} by an index number of an instance I and a type of categories AB, B—compatible (c) or incompatible (i)

References

1. Aerts, D., Aerts Arguëlles, J., Beltran, L., Beltran, L., Sassoli de Bianchi, M., Sozzo, S., & Veloz, T. (2017). Testing quantum models of conjunction fallacy on the world wide web. *International Journal of Theoretical Physics.*
2. Busemeyer, J. R., Pothos, E. M., Franco, R., & Trueblood, J. S. (2011). A quantum theoretical explanation for probability judgment errors. *Psychological review, 118*(2), 193.
3. Favre, M., Wittwer, A., Heinimann, H. R., Yukalov, V. I., & Sornette, D. (2016). Quantum decision theory in simple risky choices. *PLoS ONE, 12*(11), 2016.
4. Hertwig, R., & Gigerenzer, G. (1999). The 'conjunction fallacy' revisited: How intelligent inferences look like reasoning errors. *Journal of behavioral decision making, 12*(4), 275.
5. Rosner, B. (1983). Percentage points for a generalized esd many-outlier procedure. *Technometrics, 25*(2), 165–172.
6. Shafir, E. B., Smith, E. E., & Osherson, D. N. (1990). Typicality and reasoning fallacies. *Memory & Cognition, 18*(3), 229–239.
7. Siffert, M., Kovalenko, T., & Sornette, D. (2017). *Calibration of quantum decision theory, analysis of parameters, gender difference and risk aversion.* Working Paper, ETH Zurich.
8. Tversky, A., & Kahneman, D. (1983). Extensional versus intuitive reasoning: The conjunction fallacy in probability judgment. *Psychological Review, 90*(4), 293.
9. Vincent, S., Kovalenko, T., Yukalov, V. I., & Sornette, D. (2016).*Calibration of quantum decision theory, aversion to large losses and predictability of probabilistic choices*. Research paper no. 16-31, Swiss Finance Institute.
10. Webber, J. B. W. (2013). A bi-symmetric log transformation for wide-range data. *Measurement Science & Technology, 2*(24), 2013.
11. Wilks, S. S. (1938). The large-sample distribution of the likelihood ratio for testing composite hypotheses. *The Annals of Mathematical Statistics, 9*, 60–62.
12. Yukalov, V. I., & Sornette, D. (2008). Quantum decision theory as quantum theory of measurement. *Physics Letters A, 372*(46), 6867–6871.
13. Yukalov, V. I., & Sornette, D. (2009). Processing information in quantum decision theory. *Entropy, 11*(4), 1073–1120.
14. Yukalov, V. I., & Sornette, D. (2013). Quantum probabilities of composite events in quantum measurements with multimode states. *Laser Physics, 23*(10), 105502.
15. Yukalov, V. I., & Sornette, D. (2014). Manipulating decision making of typical agents. *IEEE Transactions on Systems, Man, and Cybernetics: Systems, 44*(9), 1155–1168.
16. Yukalov, V. I., & Sornette, D. (2015). Positive operator-valued measures in quantum decision theory. Revised Selected PapersIn Harald Atmanspacher, Claudia Bergomi, Thomas Filk, & Kirsty Kitto (Eds.), *Quantum Interaction: 8th International Conference, QI 2014, Filzbach, Switzerland, June 30–July 3, 2014* (pp. 146–161). Springer International Publishing.
17. Yukalov, V. I., & Sornette, D. (2016a). Inconclusive quantum measurements and decisions under uncertainty. *Frontiers in Physics, 4*(12), 1–9. https://doi.org/10.3389/fphy.2016.00012
18. Yukalov, V. I., & Sornette, D. (2016b). Quantum probability and quantum decision-making. *Philosophical Transactions of the Royal Society A, 374*(2058), 20150100.

Predicting (Economic) Trends: Why Signature Method in Machine Learning

Vladik Kreinovich and Chon Van Le

Abstract In many practical situations, we can predict the trend—i.e., how the system will change—but we cannot predict the exact timing of this change: this timing may depend on many unpredictable factors. For example, we may be sure that the economy will recover, but how fast it will recover may depend on the status of the pandemic, on the weather-affected agriculture input, etc. In such trend predictions, one of the most efficient methods is signature method, which is based on applying machine learning techniques to several special characteristics of the corresponding time series. In this paper, we provide an explanation for the empirical success of the signature method.

1 What Is Signature Method

Predictions are important. Prediction is one of the main objective of science, and economic predictions are one of the main objectives of econometrics. We want to predict what will happen to economy if we do not interfere. If we do not like this prediction, we need to decide what action to take to improve the economy—and for that, we need to be able to predict what will happen if we undertake different actions.

In some cases, we can predict the trend but not the timing. In some cases, we are able to predict exactly what will happen at different future moments of time—e.g., what will be the GDP next year. In other cases, there are too many factors affecting the situation: for example, in agriculture, a lot depends on current weather patterns. In many such situations, it is possible to predict the trend—but not the timing.

For example, we may predict that under an appropriate fiscal policy, the economy will improve, that at some future moment of time, the GDP will grow by 20% and

V. Kreinovich (✉)
University of Texas at El Paso, 500 W. University, El Paso, TX 79968, USA
e-mail: vladik@utep.edu

C. V. Le
International University of Ho-Chi-Minh City, Ho-Chi-Minh City, Vietnam
e-mail: lvchon@hcmiu.edu.vn

S. Sriboonchitta et al. (eds.), *Credible Asset Allocation, Optimal Transport Methods, and Related Topics*, Studies in Systems, Decision and Control 429,
https://doi.org/10.1007/978-3-030-97273-8_12

the unemployment will decrease to half of the current value—but we cannot predict whether this will happen in 3 years or in 6 years. In such situations, what is important for prediction is also not when exactly different events happened in the past—the exact timing is too much affected by random events to be useful—but rather what was the state of economy at different moments in the past. What is important, e.g., is that when the crops decreased by 30%, the unemployment grew by 20%—this show how the country's economy depends on its agriculture sector—but it does not matter whether this happened 3 or 4 years ago.

Let us describe this situation in precise terms. The state of the economy at each moment of time t can be characterized by the values of several characteristics $x = (x_1, \ldots, x_n)$ at this moment of time. In these terms, what we know is how, in the past, the state of the economy changed, i.e., what were the values $x_i(t)$ of all these characteristics for all the moments of time t starting from the moment T_0 when we started recording these values to the current moment T. Based on this information, we want to predict how the state $x(t)$ will change in the future, for moments $t > T$.

This prediction should not depend on the exact duration of each state, only on the general trend. In other words, we should get the same prediction based on the actual values $x(t)$ and on the values $X(t) = x(\tau(t))$ for any increasing function $\tau(t)$.

- The processes may have been slower than they actually were—in this case, we may have $\tau(t) = c \cdot t$ for some $c < 1$.
- The processes may have been faster than they actually were—in this case, we may have $\tau(t) = c \cdot t$ for some $c > 1$.
- The processes may have been slower at some periods of time and faster and other—in this case, the dependence $\tau(t)$ is nonlinear.

In all these cases, whether we use the original records $x(t)$ or re-scaled records $X(t) = x(\tau(t))$, we should get the exact same predicted trend.

Additional requirement: predictions often depend only on changes, not on the initial state. Another reasonable assumption is that the trend's predictions should depend only on the relative changes, not on the actual initial state. For example, it is important to know that the GDP declined by 20% and the unemployment increased by 80%, but it does not matter that much whether we talk about a big country with a large population and large GDP or a smaller one with smaller population and a smaller GDP.

In precise terms, this means that what is important is not the actual values $x_i(t)$, but rather the ratios $x_i(t)/x_i(T_0)$ describing how these values changed. In other words, the predictions should remain the same whether we use the values $x_i(t)$ or the values $c_i \cdot v_i(t)$ for some constants c_i. In many economic situations, it is convenient to use logarithms $v_i(t) = \ln(x_i(t))$ of the actual values. The logarithm $\ln(c_i \cdot x_i(t))$ of each re-scaled value is equal to the sum $\ln(x_i(t)) + \ln(c_i) = v_i + C_i$, where $C_i \stackrel{\text{def}}{=} \ln(c_i)$. In these terms, we should predict the same trends whether we use the original dependence $v_i(t)$ or the re-scaled dependence $v_i(t) + C_i$.

Signature method: a brief description. In such situations, it turned out to be very efficient to replace the original description $v_i(t)$ with the so-called *signature*, i.e., with the sequence of the values

$$s_{i_1} = \int \dot{v}_{i_1}(t_1)\, \mathrm{d}t_1,$$

where, as usual, $\dot{v}_i(t)$ indicates the derivative,

$$s_{i_1,i_2} = \int\limits_{T_0 \le t_1 \le t_2 \le T} \dot{v}_{i_1}(t_1) \cdot \dot{v}_{i_2}(t_2)\, \mathrm{d}t_1\, \mathrm{d}t_2,$$

$$\ldots$$

$$s_{i_1,\ldots,i_k} = \int\limits_{T_0 \le t_1 \le \cdots \le t_k \le T} \dot{v}_{i_1}(t_1) \cdot \ldots \cdot \dot{v}_{i_k}(t_k)\, \mathrm{d}t_1 \ldots \mathrm{d}t_k, \tag{1}$$

$$\ldots$$

This idea is especially useful when we use machine learning: trend predictions based on the signature are much more accurate than if we apply deep learning to the actual record $v_i(t)$; see, e.g., [1–20].

Known fact: signature has the desired invariance properties. If we add the same constant vector C_i to all the values $v_i(t)$, the derivatives will not change, and thus, the signature values will not change. Hence, the signature values remain the same whether we use the original dependence $v_i(t)$ or the re-scaled dependence $v_i(t) + C_i$.

Similarly, since $\dot{v}_i(t) \cdot \mathrm{d}t_i = \mathrm{d}v_i(t_i)$, each signature value can be represented in the following equivalent form

$$s_{i_1,\ldots,i_k} = \int\limits_{T_0 \le t_1 \le \cdots \le t_k \le T} \mathrm{d}v_{i_1}(t_1) \ldots \mathrm{d}v_{i_k}(t_k). \tag{2}$$

From this expression, it is clear that this value does not change if we re-scale time, i.e., replace the original dependence $v_i(t)$ with the re-scaled dependence $v_i(\tau(t))$.

But why signature? A natural question is: why signature and not other characteristics? In this paper, we provide a possible explanation of why it makes sense to use signature.

2 Why Signature: An Explanation

What we want: reminder. We want to find characteristics depending on all the values $v_i(t)$ for all i and t, characteristics that will be invariant under re-scaling of time $t \mapsto \tau(t)$ and under re-scaling of the values $v_i(t) \mapsto v_i(t) + C_i$.

How to achieve independence with respect to re-scaling of values $v_i(t)$. Independence on re-scaling of values can be achieved if we only consider dependence on the derivatives $\dot{v}_i(t)$.

Indeed, derivatives do not change under such re-scaling. Vice versa, once we know the derivatives, we can reconstruct the differences

$$v_i(t) - v_i(T_0) = \int_{T_0}^{t} \dot{v}_i(s)\,ds$$

and thus, indeed reconstruct the value $v_i(t)$ modulo such re-scaling.

Resulting reformulation of the problem. So, to make sure that our characteristics do not change under re-scaling of values $v_i(t)$, it makes sense to consider characteristics depending on all the values $\dot{v}_i(t)$ for all i and t, characteristics that will be invariant under re-scaling of time $t \mapsto \tau(t)$.

How can we describe general characteristics. Most dependencies are smooth. There are seemingly non-smooth processes like phase transition, but in reality, they are smooth too: just the time scale becomes different. Similarly in economics, most characteristics smoothly change with time. Sometimes the changes are fast and this seem abrupt and discontinuous, but in reality, they are reasonably smooth.

In general, a sufficiently smooth function $b = f(a_1, \ldots, a_n)$ of n inputs a_i can be described by its Taylor series:

$$b = b_0 + \sum_{i_1=1}^{n} b_{i_1} \cdot a_{i_1} + \sum_{i_1=1}^{n}\sum_{i_2=1}^{n} b_{i_1,i_2} \cdot a_{i_1} \cdot a_{i_2} + \cdots$$
$$+ \sum_{i_1=1}^{n} \cdots \sum_{i_k=1}^{n} b_{i_1,\ldots,i_k} \cdot a_{i_1} \cdot \ldots \cdot a_{i_k} + \cdots$$

In our case, the unknowns a_i are the values $\dot{v}_i(t)$ corresponding to different values of i and t. Theoretically, there are infinitely many moments of time t. However, of course, in practice, we only have values $v_i(t)$ corresponding to finitely many moments of time $t_1 < \cdots < t_m$. In this case, as approximations to derivatives, we have finite differences

$$\dot{v}_i(t_\ell) \approx \delta v_i(t_\ell) \stackrel{\text{def}}{=} \frac{v_i(t_{\ell+1}) - v_i(t_\ell)}{\Delta t_\ell}, \tag{3}$$

where we denoted

$$\Delta t_\ell \stackrel{\text{def}}{=} t_{\ell+1} - t_\ell. \tag{4}$$

By definition of the derivative, when Δt_ℓ tends to 0, the finite difference $\delta v_i(t_\ell)$ tends to the derivative. Thus, when the differences Δt_ℓ are sufficiently small—i.e., when the moments t_ℓ are sufficiently close to each other—we can safely assume, for all practical purposes, that the differences $\delta v_i(t_\ell)$ are equal to the corresponding derivatives.

In terms of these variables $\delta v_i(t_\ell)$ corresponding to difference values of i and ℓ, the general Taylor series expansion of a characteristic

$$s\,(\delta v_1(t_1), \ldots, \delta v_1(t_m), \delta v_2(t_1), \ldots)$$

takes the form

$$s = S_0 + \sum_{i_1=1}^{n} S_{i_1} + \sum_{i_1=1}^{n}\sum_{i_2=1}^{n} S_{i_1,i_2} + \cdots + \sum_{i_1=1}^{n} \ldots \sum_{i_k=1}^{n} S_{i_1,\ldots,i_k} + \cdots, \tag{5}$$

where we denoted

$$S_{i_1,\ldots,i_k} \stackrel{\text{def}}{=} \sum_{\ell_1=1}^{m} \ldots \sum_{\ell_k=1}^{m} b_{i_1,\ldots,i_k,\ell_1,\ldots,\ell_k} \cdot \delta v_{i_1}(t_{\ell_1}) \cdot \ldots \cdot \delta v_{i_k}(t_{\ell_k}). \tag{6}$$

Dividing and multiplying each terms in the sum by the product $\Delta t_{\ell_1} \cdot \ldots \cdot \Delta t_{\ell_k}$, we conclude that

$$S_{i_1,\ldots,i_k} = \sum_{\ell_1=1}^{m} \ldots \sum_{\ell_k=1}^{m} B_{i_1,\ldots,i_k}(t_{\ell_1}, \ldots, t_{\ell_k}) \cdot \delta v_{i_1}(t_{\ell_1}) \cdot \ldots \delta v_{i_k}(t_{\ell_k}) \cdot \Delta t_{\ell_1} \cdot \ldots \cdot \Delta t_{\ell_k}, \tag{7}$$

where we denoted

$$B_{i_1,\ldots,i_k}(t_{\ell_1}, \ldots, t_{\ell_k}) \stackrel{\text{def}}{=} \frac{b_{i_1,\ldots,i_k,\ell_1,\ldots,\ell_k}}{\Delta t_{\ell_1} \cdot \ldots \cdot \Delta t_{\ell_k}}. \tag{8}$$

Taking into account that, for the practical purposes, the differences $\delta v_{i_k}(t_{\ell_k})$ are equal to the corresponding derivatives $\dot{v}_i(t_\ell)$, we conclude that

$$S_{i_1,\ldots,i_k} = \sum_{\ell_1=1}^{m} \ldots \sum_{\ell_k=1}^{m} B_{i_1,\ldots,i_k}(t_{\ell_1}, \ldots, t_{\ell_k}) \cdot \dot{v}_{i_1}(t_{\ell_1}) \cdot \ldots \dot{v}_{i_k}(t_{\ell_k}) \cdot \Delta t_{\ell_1} \cdot \ldots \cdot \Delta t_{\ell_k}. \tag{9}$$

The expression (9) is nothing else but the integral sum for the integral

$$I_{i_1,\ldots,i_k} = \int\limits_{t_1=T_0}^{T} \ldots \int\limits_{t_k=T_0}^{T} B_{i_1,\ldots,i_k}(t_1,\ldots,t_k) \cdot \dot{v}_{i_1}(t_1) \cdot \ldots \dot{v}_{i_k}(t_k)\,\mathrm{d}t_1 \ldots \mathrm{d}t_k. \quad (10)$$

When the differences Δt_ℓ are small, for practical purposes, the integral sum $S_{i_1,\ldots,i_k}$ is equal to the integral $I_{i_1,\ldots,i_k}$.

So, we conclude that a generic representation of a characteristic s has the form

$$s = S_0 + \sum_{i_1=1}^{n} I_{i_1} + \sum_{i_1=1}^{n}\sum_{i_2=1}^{n} I_{i_1,i_2} + \cdots + \sum_{i_1=1}^{n} \ldots \sum_{i_k=1}^{n} I_{i_1,\ldots,i_k} + \cdots, \quad (11)$$

where the integrals $I_{i_1,\ldots,i_k}$ are determined by the formula (10).

When is the general expression (11) invariant under re-scaling of time? To come up with a description of all possible invariant characteristics, we need to find out which expressions (11) are invariant under re-scaling of time $t \mapsto \tau(t)$.

To find out which expressions (11) are thus invariant, let us first take into account—as we did in the previous section—that $\dot{v}_i(t) \cdot \mathrm{d}t_i = \mathrm{d}v_i(t)$. Then, the expression (10) takes the equivalent form

$$I_{i_1,\ldots,i_k} = \int\limits_{t_1=T_0}^{T} \ldots \int\limits_{t_k=T_0}^{T} B_{i_1,\ldots,i_k}(t_1,\ldots,t_k)\,\mathrm{d}v_{i_1}(t_1) \ldots \mathrm{d}v_{i_k}(t_k). \quad (12)$$

If we re-scale time, we get an expression

$$I^{\tau}_{i_1,\ldots,i_k} = \int\limits_{t_1=T_0}^{T} \ldots \int\limits_{t_k=T_0}^{T} B_{i_1,\ldots,i_k}(\tau(t_1),\ldots,\tau(t_k))\,\mathrm{d}v_{i_1}(t_1) \ldots \mathrm{d}v_{i_k}(t_k). \quad (13)$$

Invariance means that we should have $I_{i_1,\ldots,i_k} = I^{\tau}_{i_1,\ldots,i_k}$ for all possible functions $\dot{v}_i(t)$. This means that the coefficients $B_{i_1,\ldots,i_k}$ should be the same in both cases, i.e., that we should have

$$B_{i_1,\ldots,i_k}(t_1,\ldots,t_k) = B_{i_1,\ldots,i_k}(\tau(t_1),\ldots,\tau(t_k)) \quad (14)$$

for all possible increasing functions $\tau(t)$.

One can easily check that every two tuples $t_1 < \cdots < t_k$ and $t'_1 < \cdots < t'_k$ can be obtained from each other by some increasing function. Thus, for all such tuples, the value $B_{i_1,\ldots,i_k}(t_1,\ldots,t_k)$ is the same; we will denote it by $\mathbf{B}_{i_1,\ldots,i_k}$.

Similarly, for any other ordering of the moments $t_{\pi(1)} < \cdots < t_{\pi(k)}$ corresponding for any permutation $\pi : \{1,\ldots,k\} \mapsto \{1,\ldots,k\}$, the value $B_{i_1,\ldots,i_k}(t_1,\ldots,t_k)$ depends only on this permutation and is the same for all tuples $(t_1,\ldots,t_k)$ for which this ordering is true. We will denote this common value by $\mathbf{B}_{i_{\pi(1)},\ldots,i_{\pi(k)}}$.

The whole domain of all possible tuples $(t_1, \ldots, t_k)$—over which the integral (10) is computed—can be divided into sub-domains corresponding to different orders between t_i. For example, for $k = 2$, we divide the domain of all the pairs (t_1, t_2) into two sub-domains:

- the set of all the pairs for which $t_1 < t_2$ that corresponds to the identity permutation $\pi(i) = i$, and
- the set of all the pairs for which $t_2 < t_1$ that corresponds to swap $\pi(1) = 2$ and $\pi(2) = 1$.

On the sub-domain $D_{k,\pi}$ corresponding to the identity permutation π, the value $B_{i_1,\ldots,i_k}(t_1, \ldots, t_k)$ is a constant $\mathbf{B}_{i_1,\ldots,i_k}$, so the integral over this sub-domain has the form

$$\int\limits_{D_{k,\pi}} B_{i_1,\ldots,i_k}(t_1, \ldots, t_k) \cdot \dot{v}_{i_1}(t_1) \cdot \ldots \dot{v}_{i_k}(t_k)\, dt_1 \ldots dt_k$$
$$= \mathbf{B}_{i_1,\ldots,i_k} \cdot \int\limits_{T_0<t_1<\cdots<t_k<T} dv_1(t_1) \ldots dv_{i_k}(t_k). \tag{15}$$

As usual, in the integration, the integral over measure-0 parts corresponding to possible equalities such as $t_1 = T_0$, $t_1 = t_2$, etc. is 0, so we can say that

$$\int\limits_{D_{k,\pi}} B_{i_1,\ldots,i_k}(t_1, \ldots, t_k) \cdot \dot{v}_{i_1}(t_1) \cdot \ldots \dot{v}_{i_k}(t_k)\, dt_1 \ldots dt_k$$
$$= \mathbf{B}_{i_1,\ldots,i_k} \cdot \int\limits_{T_0\le t_1\le\cdots\le t_k\le T} dv_1(t_1) \ldots dv_{i_k}(t_k). \tag{16}$$

The integral in the right-hand side is exactly one of the signature values $s_{i_1,\ldots,i_k}$, so we get

$$\int\limits_{D_{k,\pi}} B_{i_1,\ldots,i_k}(t_1, \ldots, t_k) \cdot \dot{v}_{i_1}(t_1) \cdot \ldots \dot{v}_{i_k}(t_k)\, dt_1 \ldots dt_k = \mathbf{B}_{i_1,\ldots,i_k} \cdot s_{i_1,\ldots,i_k}. \tag{17}$$

The integral $I_{i_1,\ldots,i_k}$ over the whole set of tuples $(t_1, \ldots, t_k)$ is equal to the sum of the integrals over all sub-domains corresponding to different permutations, so we get

$$I_{i_1,\ldots,i_k} = \sum_{\pi} \mathbf{B}_{i_{\pi(1)},\ldots,i_{\pi(k)}} \cdot s_{i_{\pi(1)},\ldots,i_{\pi(k)}}. \tag{18}$$

So, in the invariant case, the general expression (11) for a characteristic takes the form

$$s = S_0 + \sum_{i_1=1}^{n} \mathbf{B}_{i_1} \cdot s_{i_1} + \sum_{i_1=1}^{n} \sum_{i_2=1}^{n} \mathbf{B}_{i_1,i_2} \cdot s_{i_1,s_2} + \cdots + \sum_{i_1=1}^{n} \cdots \sum_{i_k=1}^{n} \mathbf{B}_{i_1,\ldots,i_k} \cdot s_{i_1,\ldots,i_k} + \cdots, \tag{19}$$

i.e., it is a linear combination of the signature values.

Conclusion. It was known that signature values are invariant. What we have shown is that any other invariant characteristic is nothing else but a linear combination of signature values. In this sense, signature values is all that we can extract from the data, they provide full information about the inputs.

This explains why signature values are so successful—since they provide full information about the input.

Acknowledgements This work was supported in part by the National Science Foundation grants 1623190 (A Model of Change for Preparing a New Generation for Professional Practice in Computer Science), and HRD-1834620 and HRD-2034030 (CAHSI Includes), and by the AT&T Fellowship in Information Technology.

It was also supported by the program of the development of the Scientific-Educational Mathematical Center of Volga Federal District No. 075-02-2020-1478, and by a grant from the Hungarian National Research, Development and Innovation Office (NRDI).

References

1. Amendola, C., Friz, P., & Sturmfels, B. (2018). *Varieties of signature tensors*. arXiv preprint arXiv:1804.08325
2. Arribas, I. P. (2018). *Derivatives pricing using signature payoffs*. arXiv preprint arXiv:1809.09466
3. Arribas, I. P., Goodwin, G. M., Geddes, J. R., Lyons, T., & Saunders, K. E. (2018). A signature-based machine learning model for distinguishing bipolar disorder and borderline personality disorder. *Translational Psychiatry, 8*(1), Paper 274. https://doi.org/10.1038/s41398-018-0334-0
4. Boedihardjo, H., Geng, X., Lyons, T., & Yang, D. (2016). The signature of a rough path: Uniqueness. *Advances in Mathematics, 293*, 720–737. https://doi.org/10.1016/j.aim.2016.02.011
5. Boedihardjo, H., Ni, H., & Qian, Z. (2014). Uniqueness of signature for simple curves. *Journal of Functional Analysis, 267*(6), 1778–1806.
6. Chen, K. T. (1958). Integration of paths—A faithful representation of paths by noncommutative formal power series. *Transactions of the American Mathematical Society, 89*(2), 395–407. https://doi.org/10.2307/1993193
7. Chevyrev, I., & Kormilitzin, A. (2016). *A primer on the signature method in machine learning*. arXiv preprint arXiv:1603.03788
8. Field, J., Gyurkó, L. G., Kontkowski, M., & Lyons, T. (2014). *Extracting information from the signature of a financial data stream*. arXiv preprint arXiv:1307.7244
9. Geng, X. (2015). *Reconstruction for the signature of a rough path*. arXiv preprint arXiv:1508.06890
10. Hambly, B., & Lyons, T. (2010). Uniqueness for the signature of a path of bounded variation and the reduced path group. *Annals of Mathematics, 171*, 109–167. https://doi.org/10.4007/annals.2010.171.109
11. Kormilitzin, A. *The signature method in machine learning*. https://github.com/kormilitzin/
12. Lyons, T. (2014). *Rough paths, signatures and the modelling of functions on streams*. arXiv preprint arXiv:14054537

13. Lyons, T., & Qian, Z. (2007). *System control and rough paths*. Oxford University Press.
14. Lyons, T., & Xu, W. (2015). *Hyperbolic development and inversion of signature*. arXiv preprint arXiv:1507.00286
15. Lyons, T., & Xu, W. (2015). *Inverting the signature of a path*. arXiv preprint arXiv:1406.7833
16. Moore, P. J., Lyons, T. J., & Gallacher, J. (2019). Using path signatures to predict a diagnosis of Alzheimer's disease. *PLoS ONE, 14*(9), Paper e0222212. https://doi.org/10.1371/journal.pone.0222212
17. Ni, H. (2015). *A multi-dimensional stream and its signature representation*. arXiv preprint arXiv:1509.03346
18. Pfeffer, M., Seigal, A., & Sturmfels, B. (2018). *Learning paths from signature tensors*. arXiv preprint arXiv:1809.01588
19. Suguira, N., & Hosoda, S. (2020). Machine learning technique using the signature method for automated quality control of Argo profiles. *Earth and Space Science, 7*(9), Paper e2019EA001019. https://doi.org/10.1029/2019EA001019
20. Yang, W., Lyons, T., Ni, H., Schmid, C., & Jin, L. (2019). *Developing the path signature methodology and its application to landmark-based human action recognition*. arXiv preprint arXiv:1707.03993

Why Geometric Progression in Selecting the LASSO Parameter: A Theoretical Explanation

William Kubin, Yi Xie, Laxman Bokati, Vladik Kreinovich, and Kittawit Autchariyapanitkul

Abstract In situations when we know which inputs are relevant, the least squares method is often the best way to solve linear regression problems. However, in many practical situations, we do not know beforehand which inputs are relevant and which are not. In such situations, a 1-parameter modification of the least squares method known as LASSO leads to more adequate results. To use LASSO, we need to determine the value of the LASSO parameter that best fits the given data. In practice, this parameter is determined by trying all the values from some discrete set. It has been empirically shown that this selection works the best if we try values from a geometric progression. In this paper, we provide a theoretical explanation for this empirical fact.

1 Formulation of the Problem

Need for regression. In many real-life situations, we know that the quantity y is uniquely determined by the quantities $x_1, \ldots, x_n$, but we do not know the exact formula for this dependence. For example, in physics, we know that the aerodynamic resistance increases with the body's velocity, but we often do not know how exactly. In economics, we may know that a change in tax rate influences the economic growth, but we often do not know how exactly.

W. Kubin · Y. Xie · L. Bokati · V. Kreinovich (✉)
Computational Science Program, University of Texas at El Paso, El Paso, TX 79968, USA
e-mail: vladik@utep.edu

W. Kubin
e-mail: wkubin@miners.utep.edu

Y. Xie
e-mail: yxie3@miners.utep.edu

L. Bokati
e-mail: lbokati@miners.utep.edu

K. Autchariyapanitkul
Maejo University, Chiang Mai, Thailand
e-mail: kittawit_a@mju.ac.th

S. Sriboonchitta et al. (eds.), *Credible Asset Allocation, Optimal Transport Methods, and Related Topics*, Studies in Systems, Decision and Control 429,
https://doi.org/10.1007/978-3-030-97273-8_13

In all such cases, we need to find the dependence $y = f(x_1, \ldots, x_n)$ between several quantities based on the available data, i.e., based on the previous observations $(x_{k1}, \ldots, x_{kn}, y_k)$ in each of which we know both the values x_{ki} of the input quantities x_i and the value y_k of the output quantity y. In statistics, determining the dependence from the data is known as *regression.*

Need for linear regression. In most cases, the desired dependence is smooth—and usually, it can even be expanded in Taylor series; see, e.g., [1, 5]. In many practical situations, the range of the input variables is small, i.e., we have $x_i \approx x_i^{(0)}$ for some values $x_i^{(0)}$. In such situations, after we expand the desired dependence in Taylor series, we can safely ignore terms which are quadratic or of higher order with respect to the differences $x_i - x_i^{(0)}$ and only keep terms which are linear in terms of these differences:

$$y = f(x_1, \ldots, x_n) = c_0 + \sum_{i=1}^{n} a_i \cdot \left(x_i - x_i^{(0)}\right),$$

where $c_0 \stackrel{\text{def}}{=} f\left(x_1^{(0)}, \ldots, x_n^{(0)}\right)$ and $a_i \stackrel{\text{def}}{=} \dfrac{\partial f}{\partial x_i}_{|x_i = x_i^{(0)}}$. This expression can be simplified into a general linear expression:

$$y = a_0 + \sum_{i=1}^{n} a_i \cdot x_i, \tag{1}$$

where $a_0 \stackrel{\text{def}}{=} c_0 - \sum_{i=1}^{n} a_i \cdot x_i^{(0)}$.

In practice, measurements are never absolutely precise, so when we plug in the actually measured values x_{ki} and y_i, we will only get an approximate equality:

$$y_k \approx a_0 + \sum_{i=1}^{m} a_i \cdot x_{ki}. \tag{2}$$

Thus, the problem of finding the desired dependence can be reformulated as follows:

- given the values y_k and x_{ki},
- find the coefficients a_i for which the property (2) holds for all k.

The usual least squares approach. We want each left-and side y_k of the approximate equality (2) to be close to the corresponding right-hand side. In other words, we want the tuple $(y_1, \ldots, y_K)$ consisting of all the left-hand sides to be close to a similar tuple formed by the right-sides

$$\left(\sum_{i=1}^{m} a_i \cdot x_{1i}, \ldots, \sum_{i=1}^{m} a_i \cdot x_{Ki}\right).$$

It is reasonable to select the values a_i for which the distance between these two tuples is the smallest possible. Minimizing the distance is equivalent to minimizing the square of this distance, i.e., the expression

$$\sum_{k=1}^{K}\left(y_k - \left(a_0 + \sum_{i=1}^{m} a_i \cdot x_{ki}\right)\right)^2. \tag{3}$$

This minimization is know as the *Least Squares method*. This is the most widely used method for processing data. The corresponding values a_i can be easily found if we differentiate the quadratic expression (3) with respect to each of the unknowns a_i and then equate the corresponding linear expressions to 0. Then, we get an easy-to-solve systems of linear equations.

Comment. The above heuristic idea becomes well-justified when we consider the case when the measurement errors are normally distributed with 0 mean and the same standard deviation σ. This indeed happens in many situations when the measuring instrument's bias has been carefully eliminated, and most major sources of measurement errors have been removed. In such situations, the resulting measurement error is a joint effect of many similarly small error components. For such joint effects, the Central Limit Theorem states that the resulting distribution is close to Gaussian (= normal); see, e.g., [4]. Once we know the probability distributions, a natural idea is to select the most probable values a_i, i.e., the values for which the probability to observe the values y_k is the largest. For normal distributions, this idea leads exactly to the least squares method.

Need to go beyond least squares. When we know that all the inputs x_i are essential to predict the value y of the desired quantity, the least squares method works reasonably well. The problem is that in practice, we often do not know which inputs x_i are relevant and which are not. As a result, to be on the safe side, we include as many inputs as possible, perfectly understanding that many of them will turn out to be irrelevant.

If all the measurements were exact, this would not be a problem: for irrelevant inputs x_i, we would get $a_i = 0$, and the resulting formula would be the desired one. However, because of the measurement errors, we do not get exactly 0s. Moreover, the more such irrelevant variables we add, the more non-zero "noise" terms $a_i \cdot x_i$ we will have, and the larger will be their sum—negatively affecting the accuracy of the formula (3) and thus, of the resulting desired (non-zero) coefficients a_i.

LASSO method. Since we know that many coefficients will be 0, a natural idea is, instead of considering all possible tuples $a \stackrel{\text{def}}{=} (a_0, a_1, \ldots, a_n)$, to only consider tuples for which a bounded number of coefficients is 0, i.e., for which $\|a\|_0 \le B$ for some constant b, where $\|a\|_0$ (known as the ℓ_0-norm) denotes the number of non-zero coefficients in a tuple.

The problem with this natural idea is that the resulting optimization problem becomes NP-hard, which means, crudely speaking, that no feasible algorithm is possible that would always solve all the instances of this problem. A usual way to solve such problem is by replacing the ℓ_0-norm with an ℓ_1-norm $\sum_{i=0}^{n} |a_i|$ which is convex and for which, therefore, the optimization problem is easier to solve. So, instead of solving the problem of unconditionally minimizing the expression (3), we minimize this expression under the constraint $\sum_{i=0}^{n} |a_i| \le B$ for some constant B. This minimum can be attained when we have strict inequality or when the constraint becomes an equality. If the constraint is a strict inequality, then we have a local minimum of (3), which, for quadratic functions, is exactly the global minimum that we try to avoid. Thus, to avoid using least squares, we must consider the case when the constraint becomes an equality $\sum_{i=0}^{n} |a_i| = B$.

According to the Lagrange multiplier method, minimizing a function under an equality-type constraint is equivalent, for an appropriate value of a parameter λ, to unconstraint minimization of the linear combination of the original objective function and the constraint, i.e., to minimizing the expression

$$\sum_{k=1}^{K} \left(y_k - \left(a_0 + \sum_{i=1}^{m} a_i \cdot x_{ki} \right) \right)^2 + \lambda \cdot \sum_{i=0}^{n} |a_i|. \quad (4)$$

This minimization is known as the *Least Absolute Shrinkage and Selection Operator* method—*LASSO method*, for short; see, e.g., [3, 6].

How the LASSO parameter λ is selected: main idea. The success of the LASSO method depends on what value λ we select. When λ is close to 0, we retain all the problems of the usual least squares method. When λ is too large, the λ-term dominates, so we select the values $a_i = 0$, which do not provide any good description of the desired dependence.

In different situations, different values λ will work best. The more irrelevant inputs we have, the more important it is to deviate form the least squares, and thus, the larger the parameter λ—that describes this deviation—should be. We rarely know beforehand which inputs are relevant—this is the whole problem—so we do now know beforehand what value λ we should use. The best value λ needs to be decided based on the data.

A usual way of testing any dependence is by randomly dividing the data into a (larger) training set and a (smaller) testing set. We use the training set to find the value of the desired parameters (in our case, the parameters a_i), and then we use the testing set to gauge how good is the model. As usual with the methods using randomization, to get more reliable results, we can repeat this procedure several times, and make sure that the results are good for all cases,

In precise terms, we select several training subsets $S_1, \ldots, S_m \subseteq \{1, \ldots, K\}$. For each of these subsets S_j, we find the values $a_{ij}(\lambda)$ that minimize the functional

$$\sum_{k \in S_j} \left(y_k - \left(a_0 + \sum_{i=1}^{m} a_i \cdot x_{ki} \right) \right)^2 + \lambda \cdot \sum_{i=0}^{n} |a_i|. \quad (5)$$

We can then compute the overall inaccuracy, as

$$\Delta(\lambda) \stackrel{\text{def}}{=} \sum_{j=1}^{m} \left(\sum_{k \notin S_j} \left(y_k - \left(a_{j0}(\lambda) + \sum_{i=1}^{m} a_{ji}(\lambda) \cdot x_{ki} \right) \right)^2 \right). \quad (6)$$

We then select the value λ for which the corresponding inaccuracy is the smallest possible.

How the LASSO parameter λ is selected: details. In the ideal world, we should be able to try all possible real values λ. However, there are infinitely many real numbers, and in practice, we can only test finitely many of them. Which set of values λ should we choose?

It turned out that empirically, the best results are obtained of we use the values λ that form a geometric progression $\lambda_n = c_0 \cdot q^n$. Of course, a geometric progression also has infinitely many values, but we do not need to test all of them: usually, as λ increases from 0, the value $\Delta(\lambda)$ first decreases then increases again, so it is enough to catch a moment when this value starts increasing.

Natural question and what we do in this paper. A natural question is: why geometric progression works best? In this paper, we provide a theoretical explanation for this empirical fact.

2 Our Result

What do we want? At first glance, the answer to this question is straightforward: want to select a discrete set of values, i.e., a set

$$S = \{\dots < \lambda_n < \lambda_{n+1} < \dots\}.$$

However, a deeper analysis shows that the answer is not so simple. Indeed, what we are interested in is the dependence between the quantities y and x_i. However, what we have to deal with is not the quantities themselves, but their numerical values, and the numerical values depend on what unit we choose for measuring these quantities. For example:

- a person who is 1.7 m high is also 170 cm high,
- an April 2020 price of 2 US dollars is the same as the price of $2 \cdot 23{,}500 = 47{,}000$ Vietnam Dong, etc.

In most cases, the choice of the units is rather arbitrary. It is therefore reasonable to require that the results of data processing—when converted to original units—should not depend on which units we originally used. And hereby lies a problem. Suppose that we keep the same units for x_i but change a measuring unit for y to a one which is α times smaller. In this case, the new numerical values of y become α times larger: $y \to y' = \alpha \cdot y$. To properly capture these new values, we need to increase the original values a_i by the same factor, i.e., replace the values a_i with the new values $a_i' = \alpha \cdot a_i$. In terms of these new values, the minimized expression (4) takes the form

$$\sum_{k=1}^{K}\left(y_k' - \left(a_0' + \sum_{i=1}^{m} a_i' \cdot x_{ki}\right)\right)^2 + \lambda \cdot \sum_{i=0}^{n} |a_i'|,$$

i.e., taking into account that $y_k' = \alpha \cdot y_k$ and $a_i' = \alpha \cdot a_i$, the form

$$\alpha^2 \cdot \sum_{k=1}^{K}\left(y_k - \left(a_0 + \sum_{i=1}^{m} a_i \cdot x_{ki}\right)\right)^2 + \alpha \cdot \lambda \cdot \sum_{i=0}^{n} |a_i|.$$

Minimizing an expression is the same as minimizing α^{-2} times this expression, i.e., the modified expression

$$\sum_{k=1}^{K}\left(y_k - \left(a_0 + \sum_{i=1}^{m} a_i \cdot x_{ki}\right)\right)^2 + \alpha^{-1} \cdot \lambda \cdot \sum_{i=0}^{n} |a_i|.$$

This new expression is similar to the original minimized expression (4), but with a new value of the LASSO parameter $\lambda' = \alpha^{-1} \cdot \lambda$.

What this says is that when we change the measuring units, the values of λ are also re-scaled—i.e., multiplied by a constant. What was the set $\{\lambda_n\}$ in the old units becomes the re-scaled set $\{\alpha^{-1} \cdot \lambda_n\}$ in the new units. Since this is, in effect, the same set but corresponding to different measuring units, we cannot say that one of these sets is better than the other, they clearly have the same quality.

So, we cannot choose a single set S, we must choose a family of sets $\{c \cdot S\}_c$, where the notation $c \cdot S$ means $\{c \cdot \lambda : \lambda \in S\}$.

Natural uniqueness requirement. Eventually, we need to select some set S. As we have just explained, we cannot select one set a priori, since with every set S, a set $c \cdot S$ also has the same quality. To fix a unique set, we can, e.g., fix one of the values $\lambda \in S$. Let us require that with this fixture, we will be end up with a unique optimal set S. This means, in particular, that, if we select a real number $\lambda \in S$, then the only set $c \cdot S$ that contains this number will be the same set S.

Let us describe this requirement in precise terms.

Definition 1 By a *discrete set*, we mean a subset S of the set $\mathbb{R}^+$ of all positive real numbers for which, for every $\lambda \in S$, there exists an $\varepsilon > 0$ such that for every other element $\lambda' \in S$, we have $|\lambda - \lambda'| > \varepsilon$.

Comment. For such sets, for each element λ, if there are larger elements, then there is the "next" element—i.e., the smallest element which is larger than λ. Similarly, if there are smaller elements, then there exists the "previous" element—i.e., the largest element which is smaller than λ. Thus, such sets have the form

$$\{\cdots < \lambda_{n-1} < \lambda_n < \lambda_n < \cdots\}$$

Notation. *For each set S and for each number $c > 0$, by $c \cdot S$, we mean the set*

$$\{c \cdot \lambda : \lambda \in S\}.$$

Definition 2 We say that a discrete set S is—*uniquely determined* if for every $\lambda \in S$ and $c > 0$, if $\lambda \in c \cdot S$, then $c \cdot S = S$.

Proposition 1 *A set S is uniquely determined if and only if it is a geometric progression, i.e., if and only if it has the form*

$$S = \{c_0 \cdot q^n : n = \ldots, -2, -1, 0, 1, 2, \ldots\}$$

for some c_0 and q.

Discussion. This results explains why geometric progression is used to select the LASSO parameter λ.

Proof It is easy to prove that every geometric progression is uniquely determined. Indeed, if for $\lambda = c_0 \cdot q^n$, we have $\lambda \in c \cdot S$, this means that $\lambda = c \cdot c_0 \cdot q^m$ for some m, i.e., $c_0 \cdot q^n = c \cdot c_0 \cdot q^m$. Dividing both sides by $c_0 \cdot q^m$, we conclude that $c = q^{n-m}$ for some integer $n - m$. Let us show that in this case, $c \cdot S = S$. Indeed, each element x of the set $c \cdot S$ has the form $x = c \cdot c_0 \cdot q^k$ for some integer k. Substituting $c = q^{n-m}$ into this formula, we conclude that $x = c_0 \cdot q^{k+(n-m)}$, i.e., that $x \in S$. Similarly, we can prove that if $x \in S$, then $x \in c \cdot S$.

Vice versa, let us assume that the set S is uniquely determined. Let us pick any element $\lambda \in S$ and denote it by λ_0. The next element we will denote by λ_1, the next to next by λ_2, etc. Similarly, the element previous to λ_0 will be denoted by λ_{-1}, previous to previous by λ_{-2}, etc. Thus,

$$S = \{\ldots, \lambda_{-2}, \lambda_{-1}, \lambda_0, \lambda_1, \lambda_2, \ldots\}.$$

Clearly, $\lambda_1 \in S$, and for $q \stackrel{\text{def}}{=} \lambda_1/\lambda_0$, we have $\lambda_1 \in q \cdot S$—since $\lambda_1 = (\lambda_1/\lambda_0) \cdot \lambda_0 = q \cdot \lambda_0$ for $\lambda_0 \in S$. Since the set S is uniquely determined, this implies that $q \cdot S = S$. Since

$$S = \{\ldots, \lambda_{-2}, \lambda_{-1}, \lambda_0, \lambda_1, \lambda_2, \ldots\},$$

we have

$$q \cdot S = \{\ldots, q \cdot \lambda_{-2}, q \cdot \lambda_{-1}, q \cdot \lambda_0, q \cdot \lambda_1, q \cdot \lambda_2, \ldots\}.$$

The sets S and $q \cdot S$ coincide. We know that $q \cdot \lambda_0 = \lambda_1$. Thus, the element next to $q \cdot \lambda_0$ in the set $q \cdot S$—i.e., the element $c \cdot \lambda_1$—must be equal to the element which is next to λ_1 in the set S, i.e., to the element λ_2: $\lambda_2 = q \cdot \lambda_1$. For next to next elements, we get $\lambda_3 = q \cdot \lambda_2$ and, in general, we get $\lambda_{n+1} = q \cdot \lambda_n$ for all n—which is exactly the definition of a geometric progression.

The proposition is proven.

Comment. Similar arguments can explain why in machine learning methods such as deep learning (see, e.g., [2])—which usually use the gradient method

$$x_{i+1} = x_i - \lambda_i \cdot \frac{\partial J}{\partial x_i}$$

to find the minimum of an appropriate objective function J, empirically the best strategy for selecting λ_i also follows approximately a geometric progression: e.g., some algorithms use:

- $\lambda_i = 0.1$ for the first ten iterations,
- $\lambda_i = 0.01$ for the next ten iterations,
- $\lambda_i = 0.001$ for the next ten iterations, etc.

Indeed, in this case, similarly, re-scaling of J is equivalent to re-scaling of λ, and thus, we need to have a family of sequences $\{c \cdot \lambda_i\}$ corresponding to different $c > 0$. A natural uniqueness requirement—as we have shown—leads to the geometric progression.

Acknowledgements This work was supported in part by the US National Science Foundation grants 1623190 (A Model of Change for Preparing a New Generation for Professional Practice in Computer Science) and HRD-1242122 (Cyber-ShARE Center of Excellence). It was also supported by the program of the development of the Scientific-Educational Mathematical Center of Volga Federal District No. 075-02-2020-1478.

References

1. Feynman, R., Leighton, R., & Sands, M. (2005). *The Feynman lectures on physics*. Addison Wesley.
2. Goodfellow, I., Bengio, Y., & Courville, A. (2016). *Deep learning*. MIT Press.
3. Hastie, T., Tibshirani, R., & Wainwright, M. (2015). *Statistical learning with sparsity: The LASSO and generalizations*. Chapman & Hall/CRC.
4. Sheskin, D. J. (2011). *Handbook of parametric and nonparametric statistical procedures*. Chapman and Hall/CRC.
5. Thorne, K. S., & Blandford, R. D. (2017). *Modern classical physics: Optics, fluids, plasmas, elasticity, relativity, and statistical physics*. Princeton University Press.
6. Tibshirani, R. (1996). Regression shrinkage and selection via the LASSO. *Journal of the Royal Statistical Society, 58*(1), 267–288.

How to Train A-to-B and B-to-A Neural Networks So That the Resulting Transformations Are (Almost) Exact Inverses

Paravee Maneejuk, Torben Peters, Claus Brenner, and Vladik Kreinovich

Abstract In many practical situations, there exist several representations, each of which is convenient for some operations, and many data processing algorithms involve transforming back and forth between these representations. Many such transformations are computationally time-consuming when performed exactly. So, taking into account that input data is usually only 1–10% accurate anyway, it makes sense to replace time-consuming exact transformations with faster approximate ones. One of the natural ways to get a fast-computing approximation to a transformation is to train the corresponding neural network. The problem is that if we train A-to-B and B-to-A networks separately, the resulting approximate transformations are only approximately inverse to each other. As a result, each time we transform back and forth, we add new approximation error—and the accumulated error may become significant. In this paper, we show how we can avoid this accumulation. Specifically, we show how to train A-to-B and B-to-A neural networks so that the resulting transformations are (almost) exact inverses.

P. Maneejuk
Faculty of Economics, Chiang Mai University, Chiang Mai, Thailand

T. Peters · C. Brenner
Institute of Cartography and Geoinformatics, Leibniz University of Hannover, Hannover, Germany
e-mail: peters@ikg.uni-hannover.de

C. Brenner
e-mail: Claus.Brenner@ikg.uni-hannover.de

V. Kreinovich (✉)
Department of Computer Science, University of Texas at El Paso, El Paso, TX 79968, USA
e-mail: vladik@utep.edu

S. Sriboonchitta et al. (eds.), *Credible Asset Allocation, Optimal Transport Methods, and Related Topics*, Studies in Systems, Decision and Control 429,
https://doi.org/10.1007/978-3-030-97273-8_14

1 Formulation of the Problem

Need for A-to-B and B-to-A transformations. In many practical problems, there are two (or more) different representations of a state, so that:

- some operations are easier to perform in one representation, while
- other operations are easier to perform in a different representation.

A well-known historical case is the use of logarithms in a slide rule. Normally, a positive real number a is represented by two points at a distance x (or at a distance proportional to x). In this representation, it is easy to perform additions and subtractions. However, to perform multiplication or division, it is more convenient to represent each number x in a logarithmic scale, as the interval of width $\ln(x)$. In this case, e.g., multiplication $a, b \to a \cdot b$ can be efficiently performed as follow:

- first, we transform both inputs a and b into the logarithmic scale, computing the values $a' = \ln(a)$ and $b' = \ln(b)$;
- then, we add the results a' and b' of this transformation, thus computing

$$c' = a' + b';$$

- finally, we apply the inverse transformation to the value c', i.e., find c for which $\ln(c) = c'$ (this c is, of course, equal to $\exp(c')$.

One can easily see that $c = \exp(c') = \exp(a' + b') = \exp(a') \cdot \exp(b') = a \cdot b$, so we indeed get the desired product.

Computing the ration a/b is similar, the only difference is that instead of adding a' and b', we compute their difference $c' = a' - b'$.

Such situations are ubiquitous, let us just name a few cases:

- In fluid mechanics, there are two alternative representations of dynamics: Euler and Lagrange representations. In the Euler representation, we describe how the quantities like density and velocity depend on time and on spatial coordinates. In this representation, when a particle moves, its coordinates change. In the Lagrange approach, we "tag" the moving particles so that when a particle moves, its coordinates remain the same—but, e.g., the distance between particles changes.
- In quantum physics, we can have the Schroedinger representation, in which the state of the systems changes, and we can have the Heisenberg representation, in which the state of the particle remains the same, but the operators corresponding to physical quantities (such as coordinates or momentum) change with time.
- In optics, sometimes it is more convenient to represent light as particles, and in other problems, it is more convenient to represent it as a wave.
- In cartography, some ways of representing the Earth surface by a map provide better description of angles, others better description of areas, etc.

- In signal processing, sometimes it is more convenient to describe how the signal changes with time, e.g., when we want to compute the largest possible deviation from the ideal signal. On the other hand, for filtering (and for data processing in general), it is often more efficient to apply the Fourier transform and thus, use the corresponding frequency representation.

Need for machine learning. Sometimes, the transformations are straightforward—e.g., the transformation between different maps of the same area can usually be described by explicit formulas. However, in many practical situations, the exact implementation of the corresponding transformations can be very time consuming. A good example of such transformations are transformations between Euler and Lagrange coordinates: these transformations require solving a complex system of partial differential equations. This is especially important for time-critical applications, where we need to finish computations before a deadline—e.g., for predicting tomorrow's weather.

In such situations, a natural way to drastically decrease computation time is to take into account that in practice, the values of the quantities come from measurements and are, thus only known with some reasonable accuracy—usually, around 1–10%. Thus, there is no need to compute the answer with 10 or 13 digits after the period. It makes sense to replace the original time-consuming practically exact computations with faster approximate ones, that would provide an answer with the corresponding accuracy.

An efficient way to come with such an approximation is to use machine learning. In this approach, several times, we run the original exact model on different inputs, and then use the corresponding results to train a machine learning algorithm—e.g., a neural network [2, 3]. This training may take some time, but once we freeze the weights, neural-network computations become very fast. This idea has been efficiently applied to many real-life problems, and it indeed allows us to drastically reduce computation time.

Traditional methodology of using machine learning and its limitations. When we have two different representations—let us denote them A and B—we need both A-to-B and B-to-A transformations. In line with the above idea, it is reasonable to replace both transformations by appropriately trained neural networks. For this purpose, we start, e.g., with a large number of different states $a_1, \ldots, a_n$ in the A-representation, and we use the exact A-to-B algorithm to find the corresponding B-states $b_1, \ldots, b_n$. Then:

- we train the A-to-B neural network on patterns (a_i, b_i) in which a_i is the input, and b_i is the desired output, and
- we train the B-to-A neural network on patterns (b_i, a_i) in which b_i is the input, and a_i is the desired output.

The main limitation of this scheme is that a neural network provides only an approximation to the actual transformation. It is Ok if we apply the neural network only once: in this case, if we can select the approximations to be more accurate that the measurement accuracy, the resulting inaccuracy will be negligible in comparison with the measurement inaccuracy.

However, in many data processing algorithm, we need to constantly switch between different representations. For example, in signal and image processing, we often have an iterative algorithm that switches all the time between the time and frequency domains. In this case, if we replace each exact transformation with an approximate one, every time we apply a transformation, we add an extra approximation error. When we apply A-to-B and B-to-A transformations many time, the resulting errors accumulate, and we may end up with a very inaccurate result.

What we need and what we do in this paper. To avoid the above situations, it is desirable to make sure that the corresponding A-to-B and B-to-A transformations are (almost) exactly inverses: if we first apply the A-to-B neural network to some input state a and apply the B-to-A neural network to the resulting state b, we should get the state a back.

The need for this inversion is especially important in economic and financial applications. In such applications, A-to-B and B-to-A transformations may describe options from classes A and B that customers perceive as equivalent ones. In this case, if a trader follows a neural network computations, and as a result of applying first A-to-B and then B-to-A transformations, get a state a' which should be equivalent to a but is actually slightly different—e.g., slightly better than a, we get an undesirable *arbitrage* phenomenon, when a trader can earn huge amounts of money by exploiting this seemingly minor difference.

In this paper, we describe how we can train A-to-B and B-to-A neural networks so that the resulting transformations are (almost) exact inverses.

Comment. An alternative solution is described in [1, 4], where the authors propose to restrict ourselves by *invertible* neural networks, i.e. networks in which each layer can be inverted. If this is how we train the A-to-B network, then, by simply inverting each layer, we will indeed get a B-to-A neural network for which the resulting transformations are (almost) exact inverses. However, the restriction to invertible neural networks may make the training of a neural network less efficient—e.g., less accurate or taking more computation time, since the current successes of neural networks are based on non-invertible neural networks. With this possibility in mind, we believe that it is better not to restrict the type of neural networks.

2 Our Proposal

First stage is similar. On the first stage of our proposal, we train the A-to-B network the same way as usual. Namely:

- we start with a large number of different states $a_1, \ldots, a_n$ in the A-representation,
- we use the exact A-to-B algorithm to find the corresponding B-states $b_1, \ldots, b_n$, and then

- we train the A-to-B neural network on patterns (a_i, b_i) in which a_i is the input and b_i is the desired output.

Second stage is different. On the second stage, we train the B-to-A network. The main difference from the usual approach is that, to train this network, in addition to the sample (b_i, a_i) obtained on the first two sub-stages of the first stage, we also use other patterns. To be precise, here is what we suggest:

- Once the A-to-B network is trained, we generate more examples of A-states

$$a_{n+1}, \ldots, a_N \ (N \gg n).$$

- To each of these new examples a_j, we apply the A-to-B network and record the corresponding B-state b_j. Since the A-to-B network is much faster than the exact A-to-B transformation that we approximating, during the same time as the second sub-stage of the first stage, we can process much more examples ($N \gg n$).
- Finally, to train the B-to-A network, we use *both* the patterns (b_i, a_i) generated on the first stage and the newly generated patterns (b_j, a_j).

Why it works. In the original method, the A-to-B and B-to-A networks are exact inverses only on n patterns (a_i, b_i). On all other inputs $a \neq a_i$, if we first apply the A-to-B network and then the B-to-A network, we, in general, do not get the same original state back. To be more precise, the closer a to one of a_i, the closer the result of the back-and-forth transformation to a. The larger n, the denser are the states a_i in the class of all possible A-states and thus, in general, the smaller the distance from a to the nearest point a_i and the closer the back-and-forth result to the original state a.

In our proposed approach, the A-to-B and B-to-A networks are exact inverses on $N \gg n$ patterns (a_j, b_j). Since $N \gg n$, the new states a_j are placed much denser in the class of all possible A-states. Thus, in general, the distance from an A-state a to the nearest new state a_j is much smaller that the distance from a to the nearest original state a_i. Thus, for the newly trained networks, for a generic A-state a, the result of applying the back-and-forth to a is mich closer to the original state a than for the original neural network—which is exactly what we wanted.

Comment. In our description, we started with the A-to-B transformation. Alternatively, we could start with the B-to-A transformation and then apply the new idea to the A-to-B transformation. Which transformation to start with depends on the relative computational complexity of the original exact A-to-B and B-to-A transformations: we should start with the one which is faster:

- if, in general, the exact A-to-B transformation is faster, we start with the A-to-B transformation (as in the above description);
- on the other hand, if, in general, the exact B-to-A transformation is faster, we should start with the B-to-A transformation on the first stage, and only use the A-to-B transformation on the second stage.

3 What If We Have More Than Two Different Representations?

Formulation of the problem. In some practical situations, we have more than two different representations $A^{(1)}, \ldots, A^{(K)}$, $K > 2$. In such situations, we need to be able to perform a transformation between each pair, so we need transformations $A^{(k)} \to A^{(k')}$ for each pair $k \neq k'$.

How this problem is solved now.

- We start with a large number of different states $a_1^{(1)}, \ldots, a_n^{(1)}$, e.g., in the $A^{(1)}$-representation.
- For each of these states $a_i^{(1)}$ and for each representation $k > 1$, we use the exact $A^{(1)}$-to-$A^{(k)}$ algorithm to find the corresponding $A^{(k)}$-states $a_1^{(k)}, \ldots, a_n^{(k)}$.
- Then, for each pair $k \neq k'$, we train the $A^{(k)}$-to-$A^{(k')}$ neural network on patterns

$$\left(a_i^{(k)}, a_i^{(k')}\right), \quad i = 1, \ldots, n.$$

This process has the same limitation and in the case of two representations ($K = 2$): if we apply several neural networks and get back to the same representation that we started with, the resulting state may be different. How can we make this result closer to the original stage?

Proposed new algorithm: first stage. Similar to the case $K = 2$, the first stage of the algorithm is similar to what we do in the existing scheme:

- We start with a large number of different states $a_1^{(1)}, \ldots, a_n^{(1)}$, e.g., in the $A^{(1)}$-representation.
- For each of these states $a_i^{(1)}$ and for each representation $k > 1$, use the exact $A^{(1)}$-to-$A^{(k)}$ algorithm to find the corresponding $A^{(k)}$-states $a_1^{(k)}, \ldots, a_n^{(k)}$.
- Then, for each $k > 1$, we train the $A^{(1)}$-to-$A^{(k)}$ neural network on patterns

$$\left(a_i^{(1)}, a_i^{(k)}\right), \quad i = 1, \ldots, n.$$

Proposed new algorithm: second stage. On the second stage, we do the following:

- We generate more examples of $A^{(1)}$-states $a_{n+1}^{(1)}, \ldots, a_N^{(1)}$ ($N \gg n$).
- To each of these new examples $a_j^{(1)}$, for each $k > 1$, we apply the $A^{(1)}$-to-$A^{(k)}$ network and record the corresponding $A^{(k)}$-states $a_j^{(k)}$. Since the $A^{(1)}$-to-$A^{(k)}$ neural network is much faster than the exact $A^{(1)}$-to-$A^{(k)}$ transformation that we approximating, during the same time as the second sub-stage of the first stage, we can process much more examples $N \gg n$.

- Finally, for all $k > 1$ and $k' \neq k$, to train the $A^{(k)}$-to-$A^{(k')}$ network, we use *both* the patterns $\left(a_i^{(k)}, a_i^{(k')}\right)$ generated on the first stage and the newly generated patterns $\left(a_j^{(k)}, a_j^{(k')}\right)$.

Acknowledgements The first author is grateful for the financial support of the Center of Excellence in Econometrics, Chiang Mai University, Thailand.

This work was also supported by the German Research Foundation (DFG) as a part of the Research Training Group i.c.sens (grant GRK2159), by the Institutes of Cartography and Geoinformatics and of Geodesy of the Leibniz University of Hannover, and the US National Science Foundation grants 1623190 (A Model of Change for Preparing a New Generation for Professional Practice in Computer Science) and HRD-1242122 (Cyber-ShARE Center of Excellence). It was also supported by the program of the development of the Scientific-Educational Mathematical Center of Volga Federal District No. 075-02-2020-1478.

This paper was written when V. Kreinovich was visiting the Leibniz University of Hannover.

References

1. Ardizzone, L., Kruse, J., Wirkert, S., Rahner, D., Pellegrini, E. W., Klessen, R. S., Maier-Hein, L., Rother, C., & Köthe, U. (2019, May 6–9). Analyzing inverse problems with invertible neural networks. In *Proceedings of the Seventh International Conference on Learning Representations ICLR'2019*, New Orleans, Louisiana.
2. Bishop, C. M. (2006). *Pattern recognition and machine learning*. Springer.
3. Goodfellow, I., Bengio, Y., & Courville, A. (2016). *Deep learning*. MIT Press.
4. Jacobsen, J.-H., Smeulders, A., & Oyallon, E. (2018, April 30–May 3). i-RevNet: Deep invertible networks. In *Proceedings of the Sixth International Conference on Learning Representations ICLR'2018*, Vancouver, Canada.

Use Cases of Quantum Optimization for Finance

Samuel Mugel, Enrique Lizaso, and Román Orús

Abstract In this paper we briefly review two recent use-cases of quantum optimization algorithms applied to hard problems in finance and economy. Specifically, we discuss the prediction of financial crashes as well as dynamic portfolio optimization. We comment on the different types of quantum strategies to carry on these optimizations, such as those based on quantum annealers, universal gate-based quantum processors, and quantum-inspired Tensor Networks.

1 Introduction

Some computational problems have no efficient classical solution. An important example is searching for an item in a disordered list. For a list of size N, this problem can be solved in $O(N)$ operations on a standard (*classical*) computer. If you are searching Google's database for a given keyword, this can take a truly phenomenal amount of time.

In some cases, quantum computing can solve intractable problems efficiently. A spectacular example is Grover's algorithm, which can find an item in a disordered

S. Mugel (✉)
Multiverse Computing, Banting Institute, Toronto, ON M5G 1L5, Canada
e-mail: sam.mugel@multiversecomputing.com

E. Lizaso · R. Orús
Multiverse Computing, Parque Cientifico y Tecnologico de Gipuzkoa, Paseo de Miramón 170, 20014 Donostia-San Sebastián, Spain
e-mail: enrique.lizaso@multiversecomputing.com

R. Orús
e-mail: roman.orus@multiversecomputing.com

R. Orús
Donostia International Physics Center, Paseo Manuel de Lardizabal 4, 20018 Donostia-San Sebastián, Spain

Ikerbasque Foundation for Science, Maria Diaz de Haro 3, 48013 Bilbao, Spain

S. Sriboonchitta et al. (eds.), *Credible Asset Allocation, Optimal Transport Methods, and Related Topics*, Studies in Systems, Decision and Control 429,
https://doi.org/10.1007/978-3-030-97273-8_15

list in $O(\sqrt{N})$ computations [1]. There is no way to devise a classical algorithm with this performance.

The key to Grover's speedup is superposition. Quantum systems can be in several states at once. A quantum bit (*qubit*), for instance, can be in a superposition of states $|0\rangle$ and $|1\rangle$. If we perform an operation on a system which is in many states at once, the operation is applied simultaneously to each state. Quantum computing therefore implements parallel computing natively on a massive scale.

At its core, a computation is a physical process. Quantum computing is the idea that quantum processes can also be used to perform calculations. The laws which govern quantum objects are different from the physical laws which are important for classical computing. This means that computational resources are available in quantum computing that don't exist in classical computing. There are a vast amount of examples where a computational advantage can be obtained from these resources [2].

There exist many problems of practical interest which cannot be solved efficiently on a classical computer. Many high value problems can be found in the field of finance. Option pricing, for instance, is usually solved through Monte Carlo due to the inherently stochastic nature of the variables involved. It is common for these computations to take upwards of 24 h. Problems like credit scoring, financial threat detection, and customer identification usually employ complex machine learning models. Each of these can cost millions of dollars to train. There also exists a myriad of financial optimization problems of great monetary value. These include investment optimization, liquidity management, and macro-economics problems such as financial crash prediction. These problems are complex because financial institutions or assets can be correlated in extremely complex ways. Optimization problems are often the bottleneck of financial calculations.

There is an intense research effort underway to efficiently solve high value financial problems using quantum computing. This field has already demonstrated spectacular applications of quantum computing of commercial value, several of which are reviewed in this paper. For a complete description of verticals of quantum computing in finance, see Ref. [3].

In this article, we focus on intractable financial optimization problems, and review some examples of how these can be solved efficiently using quantum computing. In Sect. 2, we describe an algorithm to find the stable state of a financial network using a present day quantum computer. This result can be used for financial crash prediction. In Sect. 3, we find the optimal portfolio of investment, chosen from 52 assets over a four year period. This work relied on present day quantum computers and original quantum inspired methods.

2 Predicting Financial Crashes

We live in a world where financial institutions hold vast portfolios, which often include shares in other financial institutions. In this situation, *could a particular transaction lead to a financial crash*?

This is an incredibly complex question. It's so complex that traditional supercomputers are incapable of tackling it, even for simple toy models. Even given complete knowledge of all the assets and cross-holdings in a simple network of 20 institutions, it would take more time than the age of the universe (13.7 billion years!) to compute the effect of a perturbation.

At present, we mainly rely on empirical or statistical tools to answer this question. It is not clear that these methods can systematically and reliably predict financial crashes because indicators of a crisis generally fail at predicting the next crisis. Our failure to prevent these events is directly responsible for economic crises and their devastating consequences.

While the situation might seem dire, quantum computing has shown it could tackle these types of problems efficiently, both in theory and in practice. In particular, quantum computing has proved extremely successful at tackling complex financial problems [3].

The work presented in this section builds upon Ref. [4]. A physical implementation of this work was presented in [5].

2.1 Quantum Annealing

In Ref. [4], we developed a quantum algorithm for predicting financial crashes on a quantum annealer. These quantum processors solve quadratic integer valued optimization problems, which can be intractable for classical computers.

To solve a problem on a quantum annealer, we map the problem's cost function to an idealized quantum system, the Ising model. The Ising model analogous to our problem is simulated on the quantum annealer. The annealing procedure drives the simulated material to its most stable state, which can then be read off [6].

Here, the cost function is the stability equation of the financial network. Its most stable state—or *ground state*—is the network's equilibrium configuration. When the network is perturbed, we can measure the new equilibrium of the system and ask: has a financial crash occurred to reach this configuration?

2.2 Financial Model

We base our study on a model for financial networks originally proposed by Elliott et al. [7]. Their model predicts the equilibrium *market value* of an institution, which is defined as its non-inflated value to the economy. The vector of equilibrium market values $\vec{v}$ satisfies the financial stability condition:

$$\vec{v} = A\left(D\vec{p} - \vec{b}(\vec{v}, \vec{p})\right). \quad (1)$$

From this equation, we can see that the market value of an institution is determined by two terms. First, it is determined by the price p_j of each asset j, scaled with the asset ownership D. This matrix has terms D_{ij}, the share of asset j held by the ith institution. The asset prices are related to the institutions' market values through the dependency matrix A, which accounts for institutions' cross ownership and debt contracts between institutions.

The second term $\vec{b}(\vec{v}, \vec{p})$ is a discontinuous quantity. It has value $b_i = 0$ when v_i is above a particular value v_i^c. For $v_i^c < v_i^c$, the ith institution suffers an extra drop in value b_i. This term captures sociological effects, such as investors' loss of faith, and the loss of capital incurred when a company goes into liquidation.

It is straightforward to map Eq. (1) to an optimization problem:

$$\vec{v} = \text{argmin}_{\vec{v}} \left\{\vec{v} - A\left[D\vec{p} - \vec{b}(\vec{v}, \vec{p})\right]\right\}^2. \quad (2)$$

Because Eq. (2) is a discontinuous equation, it is extremely difficult to minimize.

2.3 Formulation as a QUBO

To solve Eq. (2) on a quantum annealer, we encode the market values $\vec{v}$ on our quantum processor. One way to do this is to use a binary encoding, such that institution i's market value is given by:

$$v_i = \sum_q 2^q x_{iq}, \quad (3)$$

where $x_{iq} \in \{0, 1\}$ is the readout value of the qth qubit assigned to institution i.

All the details of the financial network are encoded as interactions between qubits. These are captured in the Hamiltonian H:

$$H = \left\{\hat{\vec{v}} - A\left[D\vec{p} - \vec{b}(\hat{\vec{v}}, \vec{p})\right]\right\}^2. \quad (4)$$

Note that we have defined the operator $\hat{\vec{v}}$ by quantizing Eq. (2). This means we have replaced the value variables $\vec{v}$ by operators—mathematical objects which act on the quantum processor's qubits and determine their state (0, 1, or some superposition of these states).

In Ref. [5], we implemented Eq. (4) on a physical quantum annealer. Because all present day quantum annealers can only solve quadratic optimization problems—

QUBOs—we had to approximate Eq. (4). We used ancilla qubits to simulate the complex, multi-qubit interactions between the logical qubits. In general, we can approximate the lowest energy state of an interaction between k qubits by using k ancillary qubits [8].

The next step is to enter the approximate H into the quantum annealer, which finds its most stable state. It is this state that encodes the values at equilibrium of all assets in the financial network. The rest is easy: we calculate the values of all institutions, and ensure that none of them drops below their critical value. In the future, this could allow us to proceed responsibly, ensuring that none of our actions can cause financial havoc.

3 Portfolio Optimization

Investors are always on the lookout for the best investment opportunities. In general, this is a complex sequence of buys and sells. What makes this problem hard is that every buy or sell of shares has cost. This means that the optimal investment at time t is highly dependent on your past investment trajectory.

Excellent tools exist, which exploit the cost function structure to tackle large optimization problems. This is for instance the case of gradient descent methods. Unfortunately, financial optimization problems are in general non-convex, meaning they do not have a simple mathematical structure. This makes them very hard to solve using traditional computing methods.

What makes our problem even harder is this: we only allow our investor to invest in large bundles. This is typically the case for exchange-traded funds (ETF) shares. Time can also be discrete. You may, for instance, choose to buy and sell assets only once a day. This helps avoid solutions with many short transactions, as these generally lead to highly fluctuating (i.e.: risky) portfolios.

This means the variables we are optimizing—the amount of assets you own—are integer variables. The fact these are not continuous means we cannot use standard calculus tools, making it almost impossible to tackle this problem using classical computers. Clearly this is a high value problem where new, adapted optimizers could prove disruptive.

In Ref. [9], we described a way to tackle this problem using quantum and quantum inspired methods. These tools allowed us to find the optimal investment trajectory spanning 52 assets and four years of data. This study is important because it is one of the few applications of quantum computing to finance of commercial value.

Financial optimization is an active field of study, which has given rise to many exciting studies, which include Refs. [10–13].

3.1 *Optimizers' Overview*

The solver treats the objective function like a black box. The solver gives the box a portfolio, the box returns its score. The aim for the solver is to repeat this process until the best portfolio is found. A good solver repeats the process as few times as possible.

Financial optimization problems are difficult because there is an almost infinite number of portfolios you could build. However not all of these are interesting to us: most portfolios have mediocre performance.

3.1.1 Quantum Annealing

One optimization method we implemented is quantum annealing, which we already described in Sect. 2.1. We used it to explore only portfolios which are close to optimal, while ignoring the ones which are of no interest to us. This is analogous to the way water spreads over land, exploring all the valleys but none of the peaks.

3.1.2 Variational Quantum Eigensolver (VQE)

While solving an optimization problem, your solver must jump from one solution to another. Quantum computers have an advantage here because, in the quantum world, there exist many more ways to go from one solution to the next. While these intermediate portfolios have no physical significance, they can give the cost function a simpler (convex) structure. This allows us to use popular optimization tools such as gradient descent. This is the idea behind the Variational Quantum Eigensolver (VQE) algorithm.

The quantum processor (QPU) is necessary in the VQE algorithm because an exponential number of parameters are needed to describe the intermediate portfolios. This makes it very expensive for a classical computer to calculate the value of the cost function. We therefore differ this task to the quantum processor. The classical computer is used to iteratively prepare portfolios, and find the lowest cost solution.

3.1.3 Tensor Networks

In some cases, a smart use of classical computers can perform just as well as a quantum computer. By extending the space of possible portfolios, we can find shortcuts to the optimal solution.

However, not all of these new intermediates are important to find the shortcut. Tensor networks are a family of algorithms which simulate quantum mechanics on a classical computer, focusing only on the subspace of relevant portfolios.

Using these algorithms, we were able to build a solver which can compete with—and in some cases outperform!—our quantum solver.

3.2 *The Financial Model*

According to Modern Portfolio Theory, a portfolio's risk can be estimated from its volatility, which is related to its covariance. The optimal investment can be found by maximizing the portfolio's profit, penalized with risk [14]. The ratio of risk to returns is the *Sharpe ratio*, which is our metric for comparing investments.

The optimal amount ω_{tn} of asset n held at time t can then be found by minimizing:

$$H_0 = \sum_t -\mu_t^T \omega_t + \frac{\gamma}{2} \omega_t^T \Sigma_t \omega_t + \lambda(\Delta\omega_t)^2. \quad (5)$$

Here, μ_t are the returns at time t, Σ_t is the covariance matrix, and λ is the matrix of transaction costs. The *risk aversion* γ is the portfolios penalty for risk. This determines the amount of risk an investor is willing to take.

The ω_t vector should be normalized, meaning that 100% of the available budget is invested at any time. We enforced this by penalizing portfolios which did not respect this constraint. The cost function we optimized was therefore:

$$H = H_0 + \sum_t \rho \left(\sum_n \omega_{tn} - 1 \right)^2, \quad (6)$$

where ρ is a hyperparameter of the model.

Note that Eq. (6) is already in quadratic form. Using the procedure described in Sect. 2.3, it is straightforward to map this to an Ising Hamiltonian.

3.3 *Dimensional Reduction Algorithms*

The largest quantum computer on Earth only has 5000 noisy quantum bits (for reference, a standard laptop has billions of bits). How can we fit such a large problem on such a small chip?

3.3.1 Asset Clustering

One way to reduce the problem's size is to group assets which move in similar ways. The key intuition here is that investing in highly correlated assets does not significantly lower the portfolios risk. A rough description of the optimal portfolio can be gained by grouping similar assets as indices. Once an initial guess is obtained, it can be used as a starting point for a finer portfolio search.

This is not only useful when resources are scarce. Even the best solvers struggle when the space of candidate portfolios is vast. Eliminating uninteresting portfolio options early reduces the risk of getting stuck with a suboptimal solution.

3.3.2 Problem Fragmentation

In general, you can't just invest in the most profitable shares every day: the costs of buying and selling stocks will likely outweigh your profits. You can expect, however, the best investment portfolio for one month to be some combination of each day's near optimal investments. Because quantum processors are samplers, it is easy for them to give us many good portfolios. These can then be combined using classical computers (which is computationally cheap).

This is an example of a hybrid algorithm. We fragmented our problem into two bits: (1) find several good investments for today; (2) combine them into a portfolio. We gave the quantum computer small, hard jobs. This means its small number of quantum bits wasn't an issue. We brought all these results together using a classical computer. This makes the most of its huge computational resources, without being bogged down by its slow algorithms.

3.4 Results

Using quantum computers and tensor networks, we were able to find the best portfolio among 10^{382} candidates (see Table 1)—many, many times more than the number of

Table 1 Specifics of the different datasets used for benchmarking the different algorithms and hardware platforms

	XS	S	M	L	XL	XXL
N_a	3	4	4	8	8	8
N_t	2	5	7	17	29	53
N_q	1	1	1	2	2	3
N_{qubits}	6	20	28	272	464	1272
N_{states}	64	$O(10^6)$	$O(10^8)$	$O(10^{81})$	$O(10^{139})$	$O(10^{382})$
K	2	3	3	5	10	15
K'	1	1	1	3	3	7

Risk aversion for all datasets is fixed to $\gamma = 1$. Time steps N_t are measured in business months (i.e., not considering weekends due to closure of some markets). N_a is the number of assets, N_q the number of bits per asset, N_{qubits} the total number of qubits, N_{states} the total number of sates in the portfolio space, K the total amount of investment, and K' the maximum investment per asset

Table 2 Sharpe ratios obtained using the different optimization algorithms considered in Ref. [9]

Method	XS	S	M	L	XL	XXL
VQE	3.59	–	–	–	–	–
Exhaustive	6.31	8.90	–	–	–	–
VQE constrained	6.31	6.04	4.81	–	–	–
Gekko	5.98	8.90	8.39	15.83	20.76	–
D-wave hybrid	5.98	8.90	8.39	7.47	9.70	12.16
Tensor networks	5.98	8.90	9.54	16.36	15.77	15.83

The dataset parameters are described in Table 1

atoms in the Universe! This problem is completely out of reach of standard computers. Our final portfolio boasted a staggering 68% return on investment over a 4 year period. A complete benchmarking of the solutions, as compared by the Sharpe ratio, is shown in Table 2.

4 Conclusions

In this paper we have sketched two basic applications of quantum optimization for financial problems, namely, financial crash prediction, and dynamic portfolio optimization. These examples show that real business value can be derived from present day quantum computers. This is particularly true for the portfolio optimization case, where we found the best investment portfolio by optimizing over 52 assets and four years of data.

There exist many more verticals of quantum optimization in finance. It is our belief that the most interesting applications have yet to be discovered. It is time for financial institutions to revisit bottleneck problems. Quantum computing has already proved it can break down computational barriers, and is redefining what we consider *intractable*.

Acknowledgements Thanks to Christophe Jurczak, Pedro Luis Uriarte, Pedro Muñoz-Baroja, Creative Destruction Lab, BIC-Gipuzkoa, DIPC, Ikerbasque, and Basque Government for constant support. We extend special thanks to our collaborators Francesco Benfenati, Beñat Mencia Uranga, and Samuel Palmer, for stimulating discussions and interesting ideas.

References

1. Grover, L. K. (1996). A fast quantum mechanical algorithm for database search. In *Proceedings of the Twenty-Eighth Annual ACM Symposium on Theory of Computing—STOC '96* (pp. 212–219).
2. Nielsen, M. A., & Chuang, I. L. (2010). *Quantum computation and quantum information* (10th Anniversary ed.). Cambridge University Press.

3. Orús, R., Mugel, S., & Lizaso, E. (2019). Quantum computing for finance: Overview and prospects. *Reviews in Physics, 4*, 100028.
4. Orús, R., Mugel, S., & Lizaso, E. (2019). Forecasting financial crashes with quantum computing. *Physical Review A, 99*, 060301.
5. Ding, Y., Lamata, L., Sanz, M., Martín-Guerrero, J. D., Lizaso, E., Mugel, S., Chen, X., Orús, R., & Solano, E. (2019). Towards prediction of financial crashes with a D-wave quantum computer (pp. 1–8).
6. Kadowaki, T., & Nishimori, H. (1998). Quantum annealing in the transverse Ising model. *Physical Review E, 58,* 5355–5363.
7. Elliott, M., Golub, B., & Jackson, M. O. (2014). *Financial networks and contagion* (Vol. 104).
8. Chancellor, N., Zohren, S., & Warburton, P. A. (2017). Circuit design for multi-body interactions in superconducting quantum annealing systems with applications to a scalable architecture. *npj Quantum Information,* 1–6.
9. Mugel, S., Kuchkovsky, C., Sanchez, E., Fernandez-Lorenzo, S., Luis-Hita, J., Lizaso, E., & Orus, R. (2020). Dynamic portfolio optimization with real datasets using quantum processors and quantum-inspired tensor networks (pp. 1–11).
10. Rosenberg, G., Haghnegahdar, P., Goddard, P., Carr, P., Wu, K., & López De Prado, M. (2016). Solving the optimal trading trajectory problem using a quantum annealer. *IEEE Journal of Selected Topics in Signal Processing, 10,* 1053–1060.
11. Elsokkary, N., Khan, F. S., Torre, D. L., Humble, T. S., & Gottlieb, J. (2017). *Financial portfolio management using D-wave's quantum optimizer: The case of Abu Dhabi securities exchange* (Vol. 2, pp. 1–4). IEEE.
12. Grant, E., Humble, T., & Stump, B. (2020). Benchmarking quantum annealing controls with portfolio optimization.
13. Cohen, J., Khan, A., & Alexander, C. (2020). Portfolio optimization of 40 stocks using the Dwave quantum annealer (pp. 1–16).
14. Markowitz, H. (1952). Portfolio selection. *The Journal of Finance, 7,* 77.

Classical Optical Modelling of Social Sciences in a Bohr–Kantian Framework

Sudip Patra and Partha Ghose

Abstract There is a recent surge of interest in proposing quantum-like models in cognition and social sciences at large. In the current paper we attempt to contribute to that growing body of literature. We distinguish our framework from other notable works like Qbism and CBD (contextuality by default) by grounding it in the larger Bohr–Kantian philosophical and conceptual paradigm. Since our modelling techniques are motivated by the classical optical model (COM), there is a simpler geometric interpretation, with tools like Poincaré sphere representation of cognitive states and a novel Pancharatnam–Berry phase, which helps visualizing deviations from classical measure theoretic predictions of decision making.

Keywords Bohr–Kantian · COM · Poincaré sphere · Pancharatnam–Berry phase · FTP · POVM

1 Introduction

Niels Bohr, one of the founding fathers of quantum mechanics, was strongly influenced by the eighteenth century German philosopher Immanuel Kant. Much of Kant's philosophy can be seen as an attempt to provide satisfactory philosophical grounds for the objective basis of Newtonian mechanics against Hume's scepticism. Bohr used a Kantian approach, with subtle modifications, to do the same for quantum mechanics. His approach must be sharply distinguished from what has come to be popularly known as the Copenhagen interpretation of quantum mechanics. This term was used by physicists *opposed* to Bohr's idea of complementarity and contextuality,

S. Patra (✉)
Center for Complexity Economics, Applied Spirituality and Public Policy (CEASP),
O P Jindal Global University, Sonipat, Haryana 131001, India
e-mail: spatra@jgu.edu.in

P. Ghose
Tagore Centre for Natural Sciences and Philosophy,
Rabindra Tirtha, New Town, Kolkata 700156, India

S. Sriboonchitta et al. (eds.), *Credible Asset Allocation, Optimal Transport Methods, and Related Topics*, Studies in Systems, Decision and Control 429,
https://doi.org/10.1007/978-3-030-97273-8_16

to identify what they saw as the common features of the Bohr–Heisenberg interpretation as it emerged in the late 1920s.

It is important to point out here a fundamental difference between Bohr and Heisenberg. After Heisenberg came up with his uncertainty or indeterminacy relations in 1927, the inalienable and uncontrollable 'disturbance' of a system caused by an experimental probe emerged as the real novel feature of quantum physics. Bohr was not happy with this situation. After the 1935 EPR paper in which the possibility of a *mechanical disturbance* of a measurement on one member of an entangled pair of systems on the other was eliminated, Bohr adopted the stance that *the state of the measuring device and the state of the object cannot be separated from each other during a measurement and constitute a dynamical whole* or an *unanalyzable whole*. Most physicists sided with Bohr and declared Einstein an 'old hat' without actually understanding Bohr or following him later in their works.

Whereas Heisenberg looked to the formalism of quantum mechanics to arrive at his famous uncertainty principle or indeterminacy relation, Bohr preferred to analyze the role of experimental arrangements of a classical type with definite outcomes, exemplified by the double-slit experiment. As a result, he saw Heisenberg's relations as formal or mathematical expressions of the general notion that our understanding of atomic phenomena must build on complementary descriptions emerging from incompatible experimental arrangements of a classical type, i.e. a type not subject to Heisenbergs indeterminacy relations. He justified this by the necessity to communicate in an unambiguous way the results of scientific experiments. Such unambiguity is guaranteed by reliance on a shared language, perhaps refined by the terminology of classical physics.

In the case of the double-slit experiment, the basic apparatus is classically describable, and further, it is impossible to have one or both slits open at the same time. They are physically *mutually exclusive* arrangements, giving rise to complementary particle-like and wave-like *phenomena* which together give the complete description of the system. Bohr's position was:

> Consequently, evidence obtained under different experimental conditions cannot be comprehended within a single picture, but must be regarded as complementary in the sense that only the totality of the phenomena exhausts the possible information about the objects.

It follows from this that, unlike in classical physics, *observation in quantum physics is context-dependent*. This is the essence of Bohr's view of contextuality and complementarity, the two being intimately connected to each other.

What led Bohr initially to emphasize the importance of the principle of complementarity in physics was the existence of the indivisible quantum of action h.

> This crucial point ... implies the impossibility of any sharp separation between the behaviour of atomic objects and the interaction with the measuring instruments which serve to define the conditions under which the phenomena appear. In fact, the individuality of the typical quantum effects finds its proper expression in the circumstance that any attempt of subdividing the phenomena will demand a change in the experimental arrangement introducing new possibilities of interaction between objects and measuring instruments which in principle cannot be controlled.

Later, however, Bohr extended the principle to other areas of human knowledge such as psychology in which the quantum of action can hardly be expected to play any fundamental role [1]. He drew attention to the close analogy between observations of atomic phenomena and those in human psychology. He wrote:

> In introspection it is clearly impossible to distinguish sharply between the phenomena themselves and their conscious perception, and although we may often speak of lending our attention to some particular aspect of a psychical experiment, it will appear on closer examination that we really have to do, in such cases, with mutually exclusive situations. ... By such a comparison it is, of course, in no way intended to suggest any closer relation between atomic physics and psychology, but merely to stress an epistemological argument common to both fields, and thus to encourage us to see how far the solution of the relatively simple physical problems may be helpful in clarifying the more intricate psychological questions with which human life confronts us, ...

It is clear from this that to Bohr the principle of complementarity provided the epistemological basis for a comprehensive understanding of both physical and psychological phenomena witnessed in mutually exclusive contexts.

It is also clear from the quote above that in the social sciences *complementarity*, and by implication *contextuality*, has been known for a long time, quite independent of quantum theory. It is therefore tempting to investigate whether contextuality in the social sciences can be understood in a similar fashion to that in quantum physics. Now, it is possible in quantum physics to separate the measuring device from the system under investigation, as was done by Heisenberg and his followers, and continues to be done by the majority of quantum physicists today as well as by those engaged in 'quantum modelling' of social sciences [2–4]. In Bohr's view such a separation is purely *epistemological*: the cut is arbitrary and has no ontological significance. This is more intrinsically so in the social sciences where *the subjective state of an individual under investigation is inaccessible to the 'other'*. Hence, it is impossible, in principle, for the 'other' to detect any change in it directly or indirectly. It is a primary phenomenon, a first-person experience whose intrinsic nature cannot be observed by any other external agent. Under the reasonable assumption that there are many such individuals with subjective first-person experience, what can be learnt and described in an inter-subjectively objective manner is what is revealed by the answers (responses in general) to specific questions (stimuli) given by these individuals in specific contexts. The contexts and questions together with the internal mental states give rise to *phenomenal* mental states that are observable.

When one talks about quantum mechanics, one usually has nonrelativistic particle mechanics in mind. What distinguishes the quantum mechanics of particles from the classical mechanics of particles is its ability to account for (i) interference phenomena, (ii) contextuality and complementarity and (iii) entanglement. This is achieved by describing a quantum state as a vector in a Hilbert space rather than by a point in phase space. However, there is another branch of classical physics that is just as important as classical mechanics, namely classical optics, or more generally, Maxwell's electromagnetic theory of fields. In this theory, too, states are represented by vectors in a Hilbert space. Interference is a well known phenomenon in classical optics (in fact, historically the very empirical basis of a wave theory as opposed to a corpuscular theory of light), and recently it has come to light that even entanglement occurs in classical polarization optics [5–10]. This blurs the boundary between classical and quantum physics, and much of what typically characterized quantum physics is now known to occur in classical optics as well. The difference between the two is very subtle and may lie in the difference between classical and quantum superpositions, leading to differences in the power of information processing by classical optical and quantum mechanical methods [11].

The close mathematical resemblance between classical optics and quantum mechanics and the successes of 'quantum modelling' of the social sciences prompt us to explore 'classical optical modelling' (COM) of social science phenomena as an alternative and perhaps complementary way of understanding them in an intuitive fashion. In this task we will be guided by a Bohr–Kantian epistemological framework with emphasis on contextuality and complementarity, and also a subjective interpretation of probabilities suitably modified to take account of the 'common knowledge' and the 'zero priors' problems [12]. Most importantly, we will see that it automatically leads to a new prediction, namely a Pancharatnam–Berry phase for a cyclic set of questions in a given context [13, 14].

Our approach can be summarized by the use of the following key elements:

1. The mathematical techniques of classical polarization optics.
2. The Bohr–Kant epistemology of 'unanalyzable wholeness' as explained above.
3. The Lüders rule, which preserves pure states, for updation purposes [15].
4. The subjective interpretation of probability.
5. The formula for total probability (FTP) with the interference term included [16].

We will call this *Classical Optical Modelling* (COM). It provides a simple, intuitive and geometric picture.

2 The Schematic Framework

Assumptions

1. A mental state is a normalized vector in a Hilbert space H.
   This guarantees superposition and interference. This is common with 'quantum modelling'. Where it differs from quantum modelling is that such a state is defined only by the context c the subject is in when a question q is asked, and all questions of 'disturbances' to the internal inaccessible subjective state $|\psi\rangle$ are bracketed (in the sense of Saussere). The pair (c, q) defines the *transcendental condition* (in the sense of Kant) or *epistemological precondition* for assigning a mental state to a subject. The word *transcendental* is to be understood in the Kantian sense as distinct from *transcendent*. The former is an epistemological precondition for experience whereas the latter exceeds experience. It must be emphasized that in a Bohr–Kantian framework there is no collapse of the state and no measurement problem.

The normalization of the states, however, makes it possible to assign a probability to pure states, and the corresponding formula for total probability (FTP) has an interference term.

To summarize, if $|\psi\rangle$ is the ontic state (the unknowable *noumenon* x), we label a mental state as $|\psi\rangle_q^c$ which is an *unanalyzable whole*.

2. The answers are random variables R_q^c (not necessarily dichotomous). Hence, if either c or q varies, the answer varies. This ensures *contextuality*.
3. If c and c' are incompatible, one has *complementarity*. The incompatibility is to be understood in the sense of *mutual exclusiveness* of the contexts, like in the quantum double-slit experiment with one or both slits open. They cannot coexist. The conditions are *incompatible*, but they are not represented by non-commuting operators. In terms of incompatible contexts c and c' and the same question q, one can cite the example of the question 'What did you eat?' in two different contexts c = Alice ate a banana and c' = Alice ate an apple. Alice's answers will be different in the two cases. This is not surprising if we remember that her phenomenal mental states are given by $|\psi\rangle_q^c$ and $|\psi\rangle_q^{c'}$ with c and c' incompatible. Hence, $R_q^c \neq R_q^{c'}$. This is the COM representation of Bohr's *Complementarity Principle* in social sciences.

Applications

QOE (question order effect) follows when $q = (q_1, q_2)$ is different from $q' = (q_2, q_1)$ so that $R_q^c \neq R_{q'}^c$ without requiring non-commuting observables representing q and q'. No non-commuting observables like $[\hat{p}, \hat{q}] = -i\hbar$ are involved.
RRE (replicable response to the first and last question q_1 in the series $q_1 - q_2 - q_1$ with q_2 incompatible with q_1) also follows since the response $R_{q_1}^c$ depends only on c and q_1.

3 The Basic Mathematical Model (COM)

Let us now give an explicit mathematical representation of the theory outlined above. A simple choice is the use of the Poincaré sphere. It was developed by the French mathematician Henri Poincaré to represent polarization states of light by points on the surface of a unit sphere. In our case it will be tacitly assumed that these points represent all possible responses to all possible questions, or all possible combinations of all possible concepts, and so on, depending on the context. This is a reasonable assumption in the absence of any a priori reason for excluding certain parts of the sphere from consideration, though, in practice, it may turn out to be so in certain cases.

Defining the canonical basis states by

$$|0\rangle = \begin{pmatrix} 1 \\ 0 \end{pmatrix}, \quad |1\rangle = \begin{pmatrix} 0 \\ 1 \end{pmatrix}, \tag{1}$$

one can write a general normalized state as

$$|\psi\rangle = \cos\theta|0\rangle + e^{i\phi}\sin\theta|1\rangle \tag{2}$$

where $-\pi/2 \leq 2\theta \leq +\pi/2$ and $0 \leq 2\phi \leq 2\pi$ are the polar (measured from the XY plane) and azimuthal angles. Since the choice of basis is arbitrary and can be changed, resulting in a different representation of the state, it will be convenient to denote a particular choice of basis as representing a specific experimental context c. (In optics this is done by choosing an optical element like a polarizer to filter out a definite state of polarization from unpolarized light.) It follows that in the canonical basis defined above, the point $(2\theta, 2\phi)$ represents the response to a particular question q in the context c. With this identification, a mental state can be written as the normalized vector

$$|\psi\rangle_q^c = \cos\theta|0\rangle + e^{i\phi}\sin\theta|1\rangle \tag{3}$$

Another mental state corresponding to a different question q' in the same context will be written as

$$|\psi\rangle_{q'}^c = \cos\theta'|0\rangle + e^{i\phi'}\sin\theta'|1\rangle \tag{4}$$

which corresponds to a different point $(2\theta', 2\phi')$ on the Poincaré sphere. These two states are clearly different.

Let us now choose the basis

$$|+\rangle = \frac{1}{\sqrt{2}}\left[|0\rangle + |1\rangle\right] \tag{5}$$

$$|-\rangle = \frac{1}{\sqrt{2}}\left[|0\rangle - |1\rangle\right], \tag{6}$$

which would correspond to a different context c' (different experimental condition) which is incompatible with c (i.e. c and c' are mutually exclusive experimental arrangements). Then the normalized state produced would be

$$|\psi\rangle_q^{c'} = \alpha|+\rangle + \beta|-\rangle \tag{7}$$

with

$$\alpha = \frac{1}{\sqrt{2}}\left[\cos\theta - e^{i\phi}\sin\theta\right], \tag{8}$$

$$\beta = \frac{1}{\sqrt{2}}\left[\cos\theta + e^{i\phi}\sin\theta\right]. \tag{9}$$

The states described thus far are called 'pure states'. They can be represented by the density matrix $\rho = |\psi\rangle_q^c\langle\psi|_q^c$ (with the properties $\rho^2 = \rho$, $\mathrm{Tr}\rho = 1$) which has off-diagonal elements representing interference effects. These interference effects are well known in classical optics, but not in classical particle mechanics. They also occur in quantum mechanics where they represent non-classical effects. It is well known that 'classical probabilities' cannot be defined for such pure states. Nevertheless, it *is* possible to define a total probability even for pure states by including an interference term. In quantum modelling such probabilities are called 'quantum-like probabilities' [16].

Classical probabilities *can* however be defined for 'mixed states' because they do not have off-diagonal elements. They are defined in general by

$$\hat{\rho} = \sum_{k=1}^{N} p_k |\psi_k\rangle\langle\psi_k| \tag{10}$$

where $\{|\psi_k\rangle\}$ are a set of pure states, $\hat{\rho}^2 \neq \hat{\rho}$, $\mathrm{Tr}\hat{\rho} = 1$ so that $\sum_{k=1}^{N} p_k = 1$. The coefficients p_k can therefore be interpreted as probabilities. Hence, the mixed states corresponding to the states (3) and (7) are

$$\hat{\rho}_q^c = \cos^\theta |0\rangle_q^c \langle 0|_q^c + \sin^2\theta |1\rangle_q^c \langle 1|_q^c, \tag{11}$$

$$\hat{\rho}_q^{c'} = \frac{1}{2}\left[|\alpha|^2|+\rangle\langle+| + |\beta|^2|-\rangle\langle-|\right]. \tag{12}$$

In quantum optics the state (11) would correspond to a photon horizontally polarized with probability $\cos^2\theta$ or vertically polarized with probability $\sin^2\theta$, and the state (12) to a photon right-circularly polarized with probability $|\alpha|^2$ or left-circularly polarized with probability $|\beta|^2$ (hence the notation $+$ and $-$). In quantum physics a measurement invariably reduces a pure state to a mixed state, enabling the use of classical probabilities.

A generalized measurement in quantum mechanics can be described by POVMs (Positive Valued Operator Measures), a collection of positive operators $E_i \geq 0$ satisfying the conditions $\sum_i E_i = \mathbb{I}$. Such a measurement is denoted by $M = \{E_i\}$. Each E_i is associated with an outcome of the measurement and since $E_i \geq 0$, it has the decomposition $E_i = M_i^\dagger M_i$. For a state ρ (pure or mixed), the probability of obtaining the result associated with E_i is $\mathrm{Tr}(E_i\rho)$. The post measurement state after obtaining the result i is

$$\rho \rightarrow \frac{M_i \rho M_i^\dagger}{\mathrm{Tr}(E_i\rho)}. \tag{13}$$

By Neumark's dilation theorem a POVM can be obtained from projective measurements (PVMs) acting in a larger Hilbert space [17].

Lüders Rule or Ansatz

To avoid using standard measurement theory and in keeping with the Bohr–Kantian framework, in COM a state updation will be implemented by using the Lüders rule or *ansatz* [15]. It is important to bear in mind what Lüders wrote in his paper about measurement:

> It may not be redundant to emphasize that statements on the change of state due to measurement do not arise out of quantum theory itself through the inclusion of the measurement apparatus in the Schrödinger equation. Measurement, *an act of cognizance*, adds an element not already contained in the formulation of quantum theory. (Italics inserted)

In quantum mechanics if an observable $A = \sum_i a_i P_i$ with non-degenerate eigenvalues a_i and projectors P_i is measured on a system in a state ρ, then the Lüders rule states that

$$\rho \to \rho'_k = \frac{P_k \rho P_k}{\mathrm{Tr} P_k \rho} \tag{14}$$

on the condition that the result a_k was obtained. In the extreme case of $P_k = \mathbb{I}$, the identity operator, this gives $\rho \to \rho$, which is what one would expect: there is no state change and a pure state remains a pure state. This is where the Lüders rule differs from a von Neumann projection which, even in the case of the identity operator, would result in a mixed state $\hat{\rho} = w \sum_i |i\rangle\langle i|$ with the same weight w for all states. The Lüders rule describes a change from a general state ρ (pure or mixed) to a specific pure state ρ'_k in Hilbert space ($\rho \to \rho'_k, \rho'^2_k = \rho'_k$), and can be generalized to any Hilbert space theory, including classical polarization optics and COM.

As regards the state of the ensemble after the measurement, one may consider either the separate subensemble in which a certain measured value, e.g. a_k, occurs (measurement followed by selection of a_k), or one may consider the complete ensemble again combined after the measurement process (measurement followed by aggregation). In the latter case one obtains the mixed state

$$\rho' = \sum_k \rho'_k = \sum_k P_k \rho P_k. \tag{15}$$

This looks mathematically identical to a von Neumann projection. It is in this sense, in our opinion, that Lüders wrote, 'The ansatz presented here is essentially identical to the "reduction of the wave function".'

In classical polarization optics *pure states are regularly produced and observed* and plotted on a Poincaré sphere. Measurements per se do not cause any change of state. However, it is possible to change the polarization of a state by passing the light through optical elements like polarization rotators. In cases where the optical elements change a general polarization state (like elliptical, which are the points above and below the Equator of the Poincaré sphere) to a specific state (like, say, right-circular polarization, which is the North Pole), they implement the Lüders rule.

We will now show how COM gives simple intuitive explanations in certain areas of the social sciences like, for example, the *Conjunction Fallacy*, the Question Order Effect (QOE), the Replicable Response Effect (RRE), entanglement in visual perception, etc. In these cases, *the novel insight that COM offers is that of contextuality and complementarity inherent in the mutual exclusiveness of different basis choices*. It also makes a striking new prediction, a Pancharatnam–Berry phase for a cyclic set of questions, as we will see.

Another promising area for COM is the modelling of multiple perceptions produced by ambiguous figures like the Necker cube.

There are other areas also like finance and economics where generalized measurements and QBism are known to play important roles. It will be worthwhile investigating such areas using COM instead.

4 Applications in the Social Sciences

4.1 QOE and RRE

Let us now consider the Question Order Effect (QOE) [18, 19]. This involves two questions (q, q') asked in reverse order in a given context c, i.e. if (q', q) signifies asking question q first and then q', then (q, q') implies the reverse order in the same context. Now, we have seen that the response to a question in a given context c is represented on the Poincaré sphere by a point $(2\theta, 2\phi)$. Hence, the two answers corresponding to the two questions will be represented by the two vectors (3) and (4). It is clear that asking the questions in the order (q', q) can be formally represented by a rotation of vector (3) to vector (4), whereas asking them in the reverse order (q, q') can be formally represented by a rotation of the vector (4) to vector (3). The end results in the two cases are obviously different. This provides a simple geometric and intuitive understanding of QOE.

If there are three successive questions $q_1 - q_2 - q_1$ such that q_1 and q_2 are incompatible, it has been found that the response to the third question q_1 is the same as the response to it when it was asked the first time. This is known as the 'Replicable Response Effect' (RRE) [20]. In conventional quantum modelling this poses a problem because the questions are represented by operators and $[\hat{O}(q_1), \hat{O}(q_2)] \neq 0$. In the COM framework this poses no problem because questions are not represented by operators and an observable answer depends only on the question asked in a certain context (a point on the Poincare sphere), independent of what precedes it.

4.2 Guppy Effect

The 'Guppy Effect' arises in situations where combinations of concepts are involved [21]. While guppy is a very typical example of a pet-fish, it is neither a very typical example of a pet nor of a fish. This psychological behaviour is unexpected in any classical theory. It implies *overextension* and *underextension* of membership weights of items, and this cannot be explained by any classical model based on set theory. A guppy can, however, be represented by a point $(2\theta, 2\phi)$ on a Poincaré sphere whose points represent combinations of concepts and whose basis vectors are chosen such that one of them is a typical pet like a dog and the other a typical non-pet like a frog. So one possibility is $|0\rangle = |\text{dog}\rangle$ and $|1\rangle = |\text{frog}\rangle$. Then

$$|\text{guppy}\rangle = \cos\theta|0\rangle + e^{i\phi}\sin\theta|1\rangle \tag{16}$$

for some choice of the angles $(2\theta, 2\phi)$. This is a pure state described by the density matrix $\rho = |\text{guppy}\rangle\langle\text{guppy}|$ which has off-diagonal elements which represent interference effects and are 'non-classical' in the sense of classical probability theory.

4.3 Conjunction Fallacy: Linda Problem

The most celebrated example of this fallacy originated with Tversky and Kahneman [22]. The information given about Linda is that she is 31 years old, single, outspoken, and very bright. She majored in philosophy. As a student, she was deeply concerned with issues of discrimination and social justice, and also participated in anti-nuclear demonstrations. Question: Which is more probable: (A) Linda is a bank teller? (B) Linda is a bank teller and is active in the feminist movement? The majority of those who were asked this question chose option B. However, the classical probability of two events occurring together (in 'conjunction') is always less than or equal to the probability of either one occurring alone:

$$P(A\,and,\,B) \le P(A), \quad P(A\,and\,B) \le P(B)$$

Classical theories are based on set theory, so probabilities will always be expressed mathematically by a measure on sets:

$$\mu(A \cup B) \le \mu(A) + \mu(B)$$

$$\mu(A \cap B) \le \mu(A), \;\; \mu(A \cap B) \le \mu(B)$$

This problem has a ready solution if one represents combinations of concepts by points on a Poincarë sphere, as in the case of the Guppy Effect. In this case if the basis vectors are chosen such that, say $|0\rangle = |\text{teacher}\rangle$ and $|1\rangle = |\text{Marxist}\rangle$, then

$$|\text{Linda}\rangle = \cos\theta|0\rangle + e^{i\phi}\sin\theta|1\rangle \tag{17}$$

for some choice of the angles $(2\theta, 2\phi)$. This is a pure state like guppy that violates classical probability rules.

4.4 Entanglement in Visual Perception

A very interesting psychological experiment was carried out in 2011 by the Brussels group in which human participants performed joint measurements [23]. The system was formed by two distinct concepts $Animal$ and $Acts$, which were combined in a specific sentence $The Animal Acts$. This sentence carries a certain meaning, which corresponds to the specification of a state of the bipartite system formed by the two concepts $Animal$ and $Acts$. In the Brussels operational-realistic approach to cognition, concepts are treated as entities that can be in different states and be subjected to measurements performed by cognitive entities sensitive to their meaning, like human minds. The four joint measurements AB, AB', $A'B$ and $A'B'$ are defined as follows. Measurement AB consisted in participants jointly selecting a good example for the concept *Animal* from the two possibilities $A_1 = Horse$ and $A_2 = Bear$, and a good example for the concept *Acts* from the two possibilities $B_1 = Growls$ and $B_2 = Whinnies$. Hence, the four outcomes of AB were: $The\ Horse\ Growls$, $The\ Horse\ Whinnies$, $The\ Bear\ Growls$ and $The\ Bear\ Whinnies$. The experimental probabilities obtained were

$$p(A_1, B_1) = \frac{4}{81},\ p(A_1, B_2) = \frac{51}{81},\ p(A_2, B_1) = \frac{21}{81},\ p(A_2, B_2) = \frac{5}{81} \tag{18}$$

The joint measurement AB' was carried out in the same way, but the exemplars of *Acts* were changed to the two possibilities $B'_1 = Snorts$ and $B'_2 = Meows$. Hence, the four outcomes of AB' were: $The\ Horse\ Snorts$, $The\ Horse\ Meows$, $The\ Bear\ Snorts$ and $The\ Bear\ Meows$. The experimental probabilities obtained were:

$$p(A_1, B'_1) = \frac{48}{81},\ p(A_1, B'_2) = \frac{2}{81},\ p(A_2, B'_1) = \frac{24}{81},\ p(A_2, B'_2) = \frac{57}{81} \tag{19}$$

Similarly, the joint measurement $A'B$ considered $Growls$ and $Whinnies$ for the two exemplars of $Acts$, as in the AB experiment, but this time $A'_1 = Tiger$ and $A'_2 = Cat$, for the two possible choices for $Animal$. Hence, the four outcomes were: $The\ Tiger\ Growls$, $The\ Tiger\ Whinnies$, $The\ Cat\ Growls$ and $The\ Cat\ Whinnies$, and the experimental probabilities were:

$$p(A'_1, B_1) = \frac{63}{81},\ p(A'_1, B_2) = \frac{7}{81},\ p(A'_2, B_1) = \frac{7}{81},\ p(A'_2, B_2) = \frac{4}{81} \tag{20}$$

Finally, in the joint measurement $A'B'$ the four outcomes were $The\ Tiger\ Snorts$, $The\ Tiger\ Meows$, $The\ Cat\ Snorts$ and $The\ Cat\ Meows$, and the experimental probabilities were:

$$p(A'_1, B'_1) = \frac{12}{81},\ p(A'_1, B'_2) = \frac{7}{81},\ p(A'_2, B'_1) = \frac{8}{81},\ p(A'_2, B'_2) = \frac{54}{81} \tag{21}$$

Hence, the correlation functions are

$$\begin{aligned}
E(A, B) &= p(A_1, B_1) - p(A_1, B_2) - p(A_2, B_1) + p(A_2, B_2) = -\frac{31}{81} \\
E(A, B') &= p(A_1, B'_1) - p(A_1, B'_2) - p(A_2, B'_1) + p(A_2, B'_2) = \frac{29}{81} \\
E(A', B) &= p(A'_1, B_1) - p(A'_1, B_2) - p(A'_2, B_1) + p(A'_2, B_2) = \frac{53}{81} \\
E(A', B') &= p(A'_1, B'_1) - p(A'_1, B_2) - p(A'_2, B'_1) + p(A'_2, B'_2) = \frac{51}{81}
\end{aligned} \tag{22}$$

Therefore, considering the quantity

$$S = -E(A, B) + E(A, B') + E(A', B) + E(A', B'), \tag{23}$$

defined by Clauser–Horne–Shimony–Holt (CHSH) which satisfies the bounds $-2 \leq S \leq +2$ for product states, they obtained the value

$$S = 2 + \frac{34}{81} > 2 \tag{24}$$

which violates the CHSH bound and signals entanglement in visual perception.

We will now give a COM of this entanglement. As we have seen, the points on a Poincaré sphere can represent the set of all possible combinations of concepts. Let us now consider the Bell state in the standard basis c,

$$|\Psi^+\rangle^c = \frac{1}{\sqrt{2}} \left[|0\rangle_1 \otimes |0\rangle_2 + |1\rangle_1 \otimes |1\rangle_2 \right], \tag{25}$$

which is a state in a generalized Poincaré sphere in the tensor product Hilbert space $H_1 \otimes H_2$ [24–28]. In the $\pm$ basis c' it can be written as

$$|\Psi^+\rangle^{c'} = \frac{1}{\sqrt{2}} \left[|+\rangle_1 \otimes |+\rangle_2 + |-\rangle_1 \otimes |-\rangle_2 \right] \tag{26}$$

Let us now identify $|0\rangle$ with an $Animal$ and $|1\rangle$ with $Action$. Let $|0\rangle_1 = Horse$, $|0\rangle_2 = Bear$, $|1\rangle_1 = Whinnies$, $|1\rangle_2 = Growls$, $|+\rangle_1 = Tiger$, $|+\rangle_2 = Cat$, $|-\rangle_1 = Snorts$, $|-\rangle_2 = Meows$. Thus, only the statements $Horse\, Whinnies$, $Bear\, Growls$, $Tiger\, Snorts$ and $Cat\, Meows$ would have equal probabilities, and all other cross terms like $Horse\, Growls$, $Tiger\, Meows$, etc. vanish, which is approximately supported by the data. However, the data also shows that the larger probabilities are not all equal. For example, the probability $p(A_1, B_2)$ $(Horse\, Whinnies) \approx 2p(A_2, B_1)(Bear\, Growls)$, and the probability $p(A'_2, B'_2)$ $(Cat\, Meows) \approx 5p(A'_1, B'_1)(Tiger\, Snorts)$. Also, the experimental value $S = 2.42 < 2\sqrt{2}$, the Tsirelson bound [29]. Therefore, the cognitive state involved in the 2011 Brussels study was certainly an entangled state but not one that was maximally entangled like a Bell state.

This opens up the possibility of further psychological experiments involving spatial separation.

Spatially Separable Bell States in Classical Optics

There is some controversy in the literature as to whether the non-separability that has been observed in classical light should be called 'entanglement' because entanglement is a term coined by Schrödinger specifically to express a purely quantum non-separability that persists in spite of arbitrary spatial separation of the sub-systems [30]. It is indeed true that all instances of non-separability proposed and observed with classical light so far are of a local nature, involving different modes of the same optical vortex beam. It is therefore necessary to investigate whether, indeed, Bell states with arbitrary spatial separation are possible in classical polarization optics, and if so, what they imply.

To describe polarized light it is most convenient to use normalized Jones vectors

$$|J\rangle = \frac{1}{\sqrt{\langle J|J\rangle}} \begin{pmatrix} E_x \\ E_y \end{pmatrix} \tag{27}$$

with $E_x = A_0\hat{e}_x \exp(i\phi_x)$ and $E_y = A_0\hat{e}_y \exp(i\phi_y)$ the complex transverse electric fields, $\hat{e}_x$ and $\hat{e}_y$ the unit polarization vectors, and $\langle J|J\rangle = |E_x|^2 + |E_y|^2 = A_0^2$, the intensity I_0. This is a column vector in a 2-dimensional Hilbert space of polarization states. The Jones vector for horizontally (x axis) polarized light is

$$|h\rangle \equiv |0\rangle = \begin{pmatrix} 1 \\ 0 \end{pmatrix}, \tag{28}$$

and that of vertically polarized (y axis) light is

$$|v\rangle \equiv |1\rangle = \begin{pmatrix} 0 \\ 1 \end{pmatrix}. \tag{29}$$

Jones vectors for other linear polarization states can be written as linear combinations of these basis vectors. Mathematically, these states are exact analogs of qubits in quantum mechanics, and can be manipulated in analogous manner by using classical optical emulations of quantum logic gates like the Hadamard gate

$$H = \frac{1}{\sqrt{2}} \begin{pmatrix} 1 & 1 \\ 1 & -1 \end{pmatrix}, \tag{30}$$

the XOR gate

$$XOR = \begin{pmatrix} 0 & 1 \\ 1 & 0 \end{pmatrix}, \tag{31}$$

and the CNOT gate

$$CNOT = \begin{pmatrix} 1 & 0 & 0 & 0 \\ 0 & 1 & 0 & 0 \\ 0 & 0 & 0 & 1 \\ 0 & 0 & 1 & 0 \end{pmatrix}. \tag{32}$$

Let us consider two identical monochromatic sources S_a and S_b emitting *coherent* trains of horizontally polarized classical light pulses in opposite directions. The sources could consist of identical *cw* laser beams sliced into coherent pulses by two optical choppers acting in unison. Let the a train pass through a Hadamard gate (a $\lambda/2$ waveplate oriented at $\pi/4$ to the horizontal). Then the stationary states of the pulse trains are

$$\frac{1}{\sqrt{2}}\psi(\mathbf{x}_a)[|0\rangle_a + |1\rangle_a] \tag{33}$$

$$\text{and} \quad \psi(\mathbf{x}_b)|0\rangle_b, \tag{34}$$

$$\text{with} \quad \mathbf{x}_a + \mathbf{x}_b = 0, \quad \psi(\mathbf{x}_a) \cap \psi(\mathbf{x}_b) = \emptyset, \tag{35}$$

where $\psi(\mathbf{x}_a)$ and $\psi(\mathbf{x}_b)$ are the spatial wave functions. The two qubit state is therefore

$$\frac{1}{\sqrt{2}}\psi(\mathbf{x}_a)\psi(\mathbf{x}_b)[|0\rangle_a \otimes |0\rangle_b + |1\rangle_a \otimes |0\rangle_b]. \tag{36}$$

One must note that the classical state (33) is plane polarized at $+45^0$, and hence it has simultaneous projections on both the basis states $|0\rangle_a$ and $|1\rangle_a$ with equal amplitude, whereas a quantum state is fundamentally different—it is projected either to the state $|0\rangle_a$ or to the state $|1\rangle_a$ with equal probability. Hence, although the normalized classical Jones vectors are mathematically analogous to qubits, their physical significance is quite different.

Let the a train (33) be rotated to the state $|0\rangle_a$ or $|1\rangle_a$ randomly. Every time a $|0\rangle_a$ state is produced, nothing is done to the corresponding $|0\rangle_b$ pulse, but every time a $|1\rangle_a$ state is produced, a polarization flipper acts on the corresponding $|0\rangle_b$ pulse and converts it to a $|1\rangle_b$ pulse. Since these transformations are unitary and the pulses are coherent, the final state is

$$|\Phi^+\rangle = \frac{1}{\sqrt{2}}\psi(\mathbf{x}_a)\psi(\mathbf{x}_b)[|0\rangle_a \otimes |0\rangle_b + |1\rangle_a \otimes |1\rangle_b]. \tag{37}$$

The polarization rotations on the a train together with the polarization flipper on the b train emulate a CNOT gate to produce this classical analog of an EPR-Bohm state in which the Bell state in polarization space is independent of the spatial separation between the pulses a and b.

Like the state (33) the state (37) is also only mathematically analogous to a quantum mechanical EPR-Bohm state, and its physical significance is quite different. However, observationally, a measurement of the polarization on either side (a, b) would yield vertical and horizontal states with 50% probability, and the state is rotationally invariant, like a quantum entangled state.

Now, consider a separable state like (36) and joint polarization measurements on the two sides a and b at angles $(\theta_a, \theta_{a'})$ and $(\theta_b, \theta_{b'})$. Instead of local realism, assume only *noncontextuality* which is the requirement that the result of a measurement is predetermined and not affected by how the value is measured, i.e. not affected by previous or simultaneous measurement of any other compatible or co-measureable observable. Treating the normalized classical Jones vectors mathematically as qubits, one can then quite easily derive the CHSH inequality [31, 32]

$$-2 \leq E(\theta_a, \theta_b) - E(\theta_a, \theta_{b'}) + E(\theta_{a'}, \theta_b) + E\theta_{a'}, \theta_{b'}) \leq +2 \tag{38}$$

where $E(\theta_a, \theta_b)$ are expectation values. [The outcomes need not be ± 1 as is usually assumed. It is sufficient that they lie between ± 1 which is guaranteed by Malus' Law in classical polarization optics with normalized Jones vectors [33].] If the nonseparable state (37) is used, one gets

$$E(\theta_a, \theta_b) = \cos(\theta_a - \theta_b), \tag{39}$$

and for a specific set of values $\theta_a = \pi/2, \theta_{a'} = 0, \theta_b = \pi/4, \theta_{b'} = -\pi/4$ one gets $S = 2\sqrt{2}$ which violates the CHSH inequality. Hence, the use of the word 'entanglement' in classical polarization optics is justified.

The original derivations of the Bell and CHSH inequalities involved the use of local hidden variables (local realism). Hence, in such a framework a CHSH violation would imply the breakdown of 'local realism', i.e. locality + realism. However, the impression gained ground that the violation implied nonlocality alone. That can be justified only if one further invokes the projection postulate and regards projection as an ontic phenomenon. However, the projection postulate has serious problems [34–36]. As we have seen, the CHSH inequality can also be derived using *noncontextuality* alone, without any reference to local hidden variables. Hence, a CHSH violation in classical polarization optics with spatially separated beams would imply a violation of noncontextuality, not of locality [37, 38], implying an important change in the very notion of classical realism.

4.5 Pancharatnam–Berry Phase in Social Science: A New Prediction

We have seen how QOE and RRE can be simply described in the COM framework. We saw that the answers to questions are represented by points on the Poincaré sphere, and that these points can be connected by a rotation of the associated vectors, the end point of the vectors tracing out a curve on the sphere. These rotations can generate an overall phase which is normally unobservable because observables are always expressed in terms of $|||\psi\rangle||^2$. However, whenever the trajectory on the Poincaré sphere, corresponding to variations of the parameters (θ, ϕ), traces out a closed loop enclosing a non-vanishing surface area, a non-trivial geometric phase γ is generated. It is defined by half the solid angle Ω subtended by the closed loop at the centre of the Poincaré sphere:

$$\gamma = \frac{1}{2}\Omega.$$

[13, 14]. A sequence of questions like $(q_1 - q_2 - q_3 - q_1)$ whose answers trace out a closed loop is required. This is a novel prediction of the theory, and in principle can be tested.

4.6 Ambiguous Figures

There are a number of examples of ambiguous figures or images that produce bistable perceptions, like the Necker cube, the Rubin vase, My Wife and My Mother-in-Law, Jastrow's duck-rabbit, etc. Once one has had both perceptions, the perceptions spontaneously reverse back and forth while one keeps looking at the stimulus (the visual information, the ambiguous figure) which remains the same, and it is impossible to have both perceptions at the same time. The flipping back and forth of the perceptions is called 'reversal'. This is clearly an instance of complementarity, and the reversals point to something extremely interesting and important from the point of view of COM. We have seen above the difference between pure and mixed states. The transition from a pure to a mixed state is non-unitary and irreversible. Hence, it is impossible to understand reversal using quantum modelling which inherently uses quantum measurement. On the other hand, in the COM framework one can understand it as spontaneous reversal of incompatible basis choices by the brain. The brain launches a basis search and selects the bases that produce the two perceptions whenever they arise. The reversal time depends on the efficiency of the search engine used by the brain.

Of course, what exactly happens in the brain during biperception is something for cognitive neuroscience and electrophysiology to figure out, and there are competing theories (top-down vs. bottom-up) [39]. But for our purpose, they are 'hidden variable' models that lie outside our purview. No matter what theory is eventually proven to be correct, our modelling will remain unaffected.

5 The Role of Noise in Cognitive Systems

Among many striking features of cognitive systems (for example, contextuality which we have based our framework on) is 'noise'. There may be many sources of noise, either due to faulty framing of experiments (like faulty framing of questions/questionnaire posed to agents), or due to the inability of agents to provide sharp responses (yes or no, 0 or 1 etc.) [40]. To take care of 'noise', Positive Operator Valued Measures (POVM) have been used in the cognitive modelling of the choice behaviours of agents under uncertainty. This is a very helpful tool in describing agents' behaviour in case of uncertainty in financial markets since many interesting results, like order effects, can be explained.

Noise in the system has an important interpretation in the decision theory literature. For example, say due to some noise in the final choice action, or due to some error, the agent, rather than choosing the optimal, chooses a wrong option. Now, such actions can be represented by positive operators rather than by more restrictive projection operators. One important example would be the sub-optimal choice of portfolios by agents, which then can have significant impacts for mispricing of assets. In real cognitive experiments, or, in choice making by agents even under full information, other errors can occur. For example, if there are two mutually exhaustive choices A and B, then there may be agents who would rather prefer A but instead choose B and vice versa. In such cases choices cannot be represented by typical orthonormal projection operators. Rather, one would need some operators, say E_A whose expectation value in the state $|A\rangle$ is close to 1 but with an error, and similarly for E_B. In the $\{|A\rangle, |B\rangle\}$ basis such operators can be termed as positive operators. Positive operators such as these are (a) complete: $E_A + E_B = \mathbb{I}$, and (b) positive $\langle\psi|E_i|\psi\rangle \geq 0,\ \forall|\psi\rangle \in H$, which also means, for any general state (including a mixed state like ρ), $\mathrm{Tr}(\rho E_i) \in [0, 1]$, and $\sum_{A,B} \mathrm{Tr}\rho E_i = 1$.

Positive operators have more general properties than projection operators. For example, if E_i is a positive operator, then one can write it (by no means uniquely) as $E_i = M_i^\dagger M_i$, where $M_i^\dagger$ is the Hermitian conjugate of M_i, such that for all such observations $\sum_i M_i^\dagger M_i = \sum_i E_i = \mathbb{I}$ where $\mathbb{I}$ is the identity operator. Again, M_i can be given a square matrix representation. For example, if ϵ is the noise in the system, then

$$M_A = \begin{pmatrix} \sqrt{1-\epsilon} & 0 \\ 0 & \sqrt{\epsilon} \end{pmatrix}, \quad M_B = \begin{pmatrix} \sqrt{\epsilon} & 0 \\ 0 & \sqrt{1-\epsilon} \end{pmatrix}. \tag{40}$$

There are several interesting properties of positive operators [40] such as: they are non-repeatable ($E_i^2 \neq E_i$); they are not unique; they are used when the basis elements in the Hilbert space of the model are not orthogonal; and they are used when there are more responses than there are basis states. This last property can be used in decision making models with noise in the system.

The probability that a measurement described by E_i gives a positive answer is then given by

$$p(i) = \mathrm{Tr}(\mathrm{E_i}\rho), \tag{41}$$

and the analogue of the collapse postulate in terms of POVMs is simply that if a measurement of the POVM $\{E_i\}$ yields the outcome i, then a pure state ρ collapses to

$$\rho' = \frac{M_i \rho M_i^\dagger}{\mathrm{Tr}(E_i \rho)}. \tag{42}$$

Notice that the state ρ' is not a pure state because $\rho'^2 \neq \rho'$, which is characteristic of a von Neumann type measurement which turns a pure state into a mixed state.

Now consider the Lüders rule (14) for a non-ideal measurement. In the simple case of two mutually exhaustive choices A and B with $p(A) \simeq p(B)$ one can write $E_A = (1-\epsilon)P_A + \epsilon P_B$, $E_B = \epsilon P_A + (1-\epsilon)P_B$ with $P_A = |A\rangle\langle A|$, $P_B = |B\rangle\langle B|$, $0 \leq \epsilon \leq 1$. Then,

$$\langle A|E_A|A\rangle = 1-\epsilon, \ \ \langle B|E_A|B\rangle = \epsilon, \tag{43}$$

$$\langle A|E_B|A\rangle = \epsilon, \ \ \langle B|E_B|B\rangle = 1-\epsilon, \tag{44}$$

$$\langle A|E_A|B\rangle = 0, \quad \text{etc.} \tag{45}$$

Hence, the rule changes to

$$\rho \to \rho' = \frac{E_k \rho E_k}{\mathrm{Tr} E_k \rho} \simeq (1 - 2\epsilon) \frac{P_k \rho P_k}{\mathrm{Tr} P_k \rho} + \mathcal{O}(\epsilon^2), \quad k = A,\ B \tag{46}$$

This implies that $\rho'^2 = (1 - 4\epsilon)\rho' + \mathcal{O}(\epsilon^2)$ which signals a small and calculable departure from a pure state, representing noise in the system.

6 Conclusions: Some Reflections

Quantum-like cognitive modellings do have a probability updating scheme via Born's rule, which would apply to the updating of cognitive states. Here a pertinent question would be: why not then use Bayesian updating only? There are some problems with the pure Bayesian paradigm, which have been emphasized in the quantum-like modelling paradigm:

1. The standard Bayesian updating formula suffers from the 'zero prior' problem [12], which means if the prior for any event is very close to 0, then no amount of belief updating would generate significant posterior probabilities. Using Born's rule would avoid the same. The QBist line of thinking is also similar in this case. After all, the Bayesian framework is based on classical Boolean logic.
2. Recently Khrennikov has achieved a breakthrough [41]. It has been shown that use of the quantum-like frame work rather than the pure Bayesian updating framework generates a more expanded form of the 'common knowledge' theorem, with deep implications in economics and finance. Robert Aumann in his seminal paper proposed that given common priors and a common knowledge about posteriors of an event, Bayesian rational agents (who update according to Bayes' rule) cannot 'agree to disagree' any further on probabilities of any event occurring [42]. Now this assumption is central to many asset pricing models/general equilibrium models in economics. However, it has been shown that if agents have a quantum-like framework (may be QBist or Växjö or COM) for updating, then even if other conditions hold, agents can still agree to disagree. This also means under special conditions one could recover Aumann's theorem. We believe this has very deep implications for economics and social science as a whole.
3. Overall, a quantum-like framework or a COM framework can resolve many 'anomalies' like the question order effect and replicable response effect which the Bayesian paradigm cannot.

QBism, Växjö and COM: Implications for FTP?
A cognitive state, as we have defined earlier, is as an *unanalysable whole*, which has contextuality embedded in its description. It can be given a purely subjective epistemological interpetation as in QBism. The QBist method provides a subjectivist description to the state and uses Born's rule for further updating of states and evaluating probabilities. There have been multiple versions of QBism, one distinctive

version being the Vaxjo version, which came up from the discussions since 2001 [43]. According to the QBist approach, quantum states can be represented by density operators ρ in Hilbert spaces which are finite dimensional, and measurements are described by POVMs M_j, where j signifies an individual agent's measurement experience. Hence, the individual probability, similar to degree of belief (which is widely used in social sciences like finance) of the outcome is given by $p(M_j) = \text{Tr}\rho M_j$. QBists view this rule to be a generalization of the classical probability update. Hence we will first briefly explore the *Jeffrey Conditioning* framework for belief updates in the classical set up.

According to the famous Bayesian framework of updating of subjective degrees of beliefs, if H is an hypothesis to be tested conditioned on the set of observed data E, then

$$p(H|E) = \frac{p(H\&E)}{p(E)} \tag{47}$$

where $1 > p(E) > 0$. This expression can be further simplified as

$$p(H|E) = \left[\frac{p(H)}{p(E)}\right] p(E|H). \tag{48}$$

The commutativity property of Boolean logic ($E\&H = H\&E$) is assumed in writing this, and also there is the so called 'zero prior' trap of Bayesian logic, which means if the prior probability for H is already 0 or near 0, then any amount of updating will not generate any significantly large posterior. However, while using subjective probability updating, if agents update according to Born's mechanism or rule, such a trap can be avoided. This point is important because social sciences, mainly financial market scenarios, exhibit rare events or so called 'black swans' which mean large updates from near 0 prior probabilities or degree of beliefs.

There can be two types of updating of subjective degrees of beliefs, simple conditioning and a more pragmatic Jeffrey Conditioning. In the former case, if the agent is absolutely certain of E happening, then the posterior belief $q(H) = p(H|E)$. However, this might seem to be an ideal case, and hence in a more pragmatic scenario, if the posterior belief about E is updated as $q(E) > 0$, then we arrive at a formula for total probability, FTP:

$$FTP = q(H) = q(E)p(H|E) + (1 - q(E))p(H|\bar{E}), \tag{49}$$

where $\bar{E}$ signifies $\neg E$. Based on this FTP, one can formulate the ratio of odds, which means, ratio of probabilities of H happening and H not happening, and then use the rule that if that ratio > 1, then the collected data E can be treated as evidence of H happening.

One of the main contributions of the quantum-like and QBist frameworks in decision making theory is the generalization of the above FTP so as to incorporate 'perturbative' terms, which might explain deviations of real decision making behaviours, than prescribed by the classical probability framework or Game theory framework.

We refer here to two forms of FTP recently suggested, QBism FTP and Växjö FTP [43], and we then discuss how the present framework can contribute to the topic. Though Bayesianism is mentioned in the generalised formulations of FTP, the only commonality is the framework of subjective belief updating, and the specific forms of FTP need not refer at all to the Bayesian updating formulas which are based on the Boolean logical set up. In the quantum-like framework, one can work with a complete set of hypotheses $\{H_j\}$ which are represented by POVMs. In the case of QBism, if the condition SIC-POVM, i.e. symmetric informationally complete POVM, is valid, and d is the dimensionality of the Hilbert space set up, then

$$q(H_j) = \sum_i \left[(d+1)q(E_i) - \frac{1}{d} \right] p(H_j|E_i). \tag{50}$$

On the other hand, in the Växjö set up, which is a statistical-realist-contextual set up, the observables are represented by Neumann–Lüders type operators with discrete spectral decompositions, and the FTP formula (for simplicity for bivariate observables) can be written down as follows. It is convenient to use the following representation of POVMs: $M_j = V_j^* V_j$, where $V_j : H \to H$ are linear operators. A POVM can be considered as a random observable. Take any set of labels $\alpha_1, ..., \alpha_m$, e.g. for $m = 2, \alpha_1 = yes, \alpha_2 = no$. Then the corresponding observable takes these values (for systems in the state ρ) with the probabilities $p(a_j) = p\rho(a_j) = \mathrm{Tr}\rho M_j = \mathrm{Tr} V_j \rho V_j^*$. We are also interested in the post-measurement states. Suppose the state ρ was given, a generalized observable was measured and the value a_j was obtained. Then the output state after this measurement has the form

$$\rho_j = \frac{V_j \rho V_j^*}{\mathrm{Tr}(V_j \rho V_j^*)} \tag{51}$$

Order Effects and Modified FTP

Both order effects and interference terms in FTP can be demonstrated using POVMs. Consider two generalized observables a and b corresponding to POVMs $M_a = V_j^* V_j$ and $M_b = W_j^* W_j$, where $V_j = V(\alpha_j)$ and $W_j = W(\beta_j)$ correspond to the values α_j and β_j. If the state ρ is given, the probabilities of observations of values α_j and β_j have the form

$$p(\alpha) = \mathrm{Tr}\rho M_a(\alpha) = \mathrm{Tr} V(\alpha)\rho V^*(\alpha), \tag{52}$$

$$p(\beta) = \mathrm{Tr}\rho M_b(\beta) = \mathrm{Tr} W(\beta)\rho W^*(\beta). \tag{53}$$

Now we consider two consecutive measurements: first the a-measurement and then the b-measurement. If in the first measurement the value $a = \alpha$ was obtained, then the initial state ρ was transformed into the state

$$\rho_{a=\alpha} = \frac{V(\alpha)\rho V^*(\alpha)}{\mathrm{Tr}\,(V(\alpha)\rho V^*(\alpha))} \tag{54}$$

For the consecutive b-measurement, the probability to obtain the value $b = \beta$ is given by

$$p(\beta|\alpha) = \mathrm{Tr} a(\alpha) M b(\beta) = \frac{\mathrm{Tr} W(\beta) V(\alpha) \rho V^*(\alpha) W^*(\beta)}{(\mathrm{Tr} V(\alpha) \rho V^*(\alpha))} \tag{55}$$

This is the conditional probability to obtain the result $b = \beta$ under the condition of the result $a = \alpha$. We set $p(\alpha, \beta) = pa(\alpha) p(\beta|\alpha)$. Since operators need not commute, $p(\alpha, \beta) \neq p(\beta, \alpha)$. We recall that, for two classical random variables a and b which can be represented in the Kolmogorov measure-theoretic approach, the formula of total probability (FTP) has the form $pb(\beta) = \sum pa(\alpha) p(\beta|\alpha)$. Further, we restrict our consideration to the case of dichotomous variables, $\alpha = \alpha_1, \alpha_2$ and $\beta = \beta_1, \beta_2$.

In the presence of the interference term the FTP for general non-pure states given by density operators and generalized quantum observables given by two (dichotomous) PVOMs is

$$pb(\beta) = pa(\alpha_1) p(\beta|\alpha_1) + pa(\alpha_2) p(\beta|\alpha_2) + 2\lambda \sqrt{pa(\alpha_1) p(\beta|\alpha_1) pa(\alpha_2) p(\beta|\alpha_2)}, \tag{56}$$

or by using ordered joint probabilities

$$pb(\beta) = p(\alpha_1, \beta) + p(\alpha_2, \beta) + 2\lambda\beta \sqrt{p(\alpha_1, \beta) p(\alpha_2, \beta)}. \tag{57}$$

The coefficient λ of the interference term has the form

$$\lambda = \frac{\mathrm{Tr} \rho W^*(\beta) V^*(\alpha_i) V(\alpha_i) W(\beta) V^*(\alpha_i) W^*(\beta) W(\beta) V(\alpha_i)}{2\sqrt{pa(\alpha_1) p(\beta|\alpha_1) pa(\alpha_2) p(\beta|\alpha_2)}} \tag{58}$$

Now, introduce the parameters

$$\gamma_{\alpha\beta} = \frac{\mathrm{Tr}(\rho W^*(\beta) V^*(\alpha) V(\alpha) W(\beta))}{\mathrm{Tr}(\rho V^*(\alpha) W^*(\beta) W(\beta) V(\alpha))} = \frac{p(\beta, \alpha)}{p(\alpha, \beta)} \tag{59}$$

This parameter is equal to the ratio of the ordered joint probabilities of the same outcome, but in the reverse order, namely, ‘b then a’ or ‘a then b’. Then the interference term is given by

$$\lambda = \frac{1}{2}\left(\sqrt{\frac{p(\alpha_1, \beta)}{p(\alpha_2, \beta)}}(\gamma_{\alpha_1\beta} - 1) + \sqrt{\frac{p(\alpha_2, \beta)}{p(\alpha_2, \beta)}}(\gamma_{\alpha_2\beta} - 1)\right) \tag{60}$$

In principle, this coefficient can be larger than one. Hence, it cannot be represented as $\lambda = \cos\theta$ for some angle (‘phase’) θ [16]. However, if POVMs M_a and M_b are, in fact, spectral decompositions of Hermitian operators, then the coefficients of interference are always less than one, i.e., one can find phases θ.

In the case of COM, the unit radius of the Poincaré sphere will act as the substitute for the Born rule in quantum mechanics. We have seen that a general state corresponding to a point $(2\theta, 2\phi)$ on the sphere is given by Eq. (3) in the standard basis c. The coefficients of the basis states can therefore be interpreted as probability amplitudes. Hence, considering another state $(2\theta, 2\phi')$ on the sphere and the state corresponding to it, and forming their inner product, one gets the FTP

$$FTP^{c}_{COM} = p_0 + p_1 + 2\sqrt{p_1 p_2}\cos\delta \tag{61}$$

where $p_0 = \cos^2\theta$, $p_1 = \sin^2\theta$ and $\delta = \phi - \phi'$. Since $-1 < \cos\delta < +1$, the total probability can be both larger than and less than the classical probability $p_0 + p_1$, giving rise to *overextension* and *underextension* (see Guppy Effect, Sect. 4.2). This FTP is clearly contextual as it varies from one basis to another. For example, in the $\pm$ basis,

$$FTP^{c'}_{COM} = p_+ + p_- + 2\sqrt{p_+ p_-}\cos\delta \tag{62}$$

where $p_+ = |\alpha|^2$, $p_- = |\beta|^2$ and α, β are given by Eqs. (8) and (9).

FTP: Implications in Social Sciences
We would like to conclude with some observations on implications of FTP in social sciences, specifically in financial decision making under contexts like uncertainty or ambiguity [44]. The FTP formulation can play an important, and yet, not fully recognized role in resolving some long standing confusions regarding mechanisms through which market observables change. A market can be perceived as a network of agents, such that we can provide a finite dimensional Hilbert space description to the state of the market as a whole. Such a state can then undergo updating based on a similar description to what we have provided in the earlier sections.

For example, in neoclassical finance, there is a long-standing debate on the exact mechanism through which agents' beliefs about asset prices impact asset returns. As we have discussed earlier, there can be both pure state to pure state and mixed state updating of beliefs. Now, as we have seen, in the latter case we have a classical ensemble of beliefs (probability distribution over possible states) with no interference terms. Such a mixed state can be analogous to the standard dispersion of belief description in finance [45], which in neoclassical finance has been studied statistically. However, if there is a resultant pure state of the market as a whole with interference terms (as described for the bivariate observables in the earlier section), there is no analogous description in the standard finance literature since the standard decision-making models are based on classical Boolean Logic. In such a case the impact of the phase factor in the FTP would be of crucial importance to observables like asset prices.

Recently, secondary market data has been used to reconstruct phase factors for game theoretic scenarios involving investors [46]. In the particular framework presented here, we hope to exploit the geometric representation of states (say market states as a whole) on the Poincaré sphere to derive phase factors.

References

1. Bohr, N. (1958). *Atomic physics and human knowledge*. Wiley. Reprinted as *The philosophical writings of Niels Bohr* (Vol. II). Ox Bow Press.
2. Khrennikov, A. (2010). *Ubiquitous quantum structure: From psychology to finance*. Springer.
3. Haven, E., & Khrennikov, A. (2013). *Quantum social science*. Cambridge University Press.
4. Aerts, D. (2009). Quantum structure in cognition. *Journal of Mathematical Psychology, 53*, 314–348.
5. Spreeuw, R. J. C. (1998). A classical analogy of entanglement. *Foundations of Physics, 28*, 361–374.
6. Spreeuw, R. J. C. (2001). Classical wave-optics analogy of quantum information processing. *Physical Review A, 63*, 062302.
7. Ghose, P., & Samal, M. K. (2001). EPR type nonlocality in classical electrodynamics. arXiv:quant-ph/0111119
8. Ghose, P., & Mukherjee, A. (2014). Entanglement in classical optics. *Reviews in Theoretical Science, 2*, 1–14.
9. Aiello, A., Töppel, F., Marquardt, C., Giacobino, E., & Leuchs, G. (2015). Quantum-like nonseparable structures in optical beams. *New Journal of Physics, 17*, 043024.
10. Qian, X.-F., Little, B., Howell, J. C., & Eberly, J. H. (2015). Shifting the quantum-classical boundary: Theory and experiment for statistically classical optical fields. *Optica, 2*(7), 611–615.
11. Khrennikov, A. (2019). Roots of quantum computational supremacy: Superposition, entanglement, or complementarity? https://doi.org/10.20944/preprints201912.0006.v1
12. Basieva, I., Pothos, E., Trueblood, J., Khrennikov, A., & Busemeyer, J. (2017). Quantum probability updating from zero prior (by-passing Cromwell's rule). *Journal of Mathematical Psychology, 77*, 58–69.
13. Pancharatnam, S. (1956). Generalized theory of interference, and its applications. Part I. Coherent pencils. *Proceedings of the Indian Academy of Sciences-Section A, 44*(5), 247–262.
14. Berry, M. V. (1984). Quantal phase factors accompanying adiabatic changes. *Proceedings of the Royal Society of London A: Mathematical and Physical Sciences, 392*(1802), 45–57.
15. Lüders, G. (1951). Über die Zustandsänderung durch den Meßprozeß. *Annalen der Physik, 8*, 322–328. https://doi.org/10.1002/andp.200610207. For an English translation: Concerning the state-change due to the measurement process. *Annals of Physics (Leipzig), 15*, 663–670 (2006), see quant-ph/0403007v2
16. Khrennikov, A. (2006). A formula of total probability with interference term and the Hilbert space representation of the contextual Kolmogorovian model. arXiv:math/0609197 [math.PR].
17. Peres, A. (1993). *Quantum theory: Concepts and methods*. Kluwer Academic Publishers.
18. Wang, Z., & Busemeyer, J. R. (2013). A quantum question order model supported by empirical tests of an a priori and precise prediction. *Topics in Cognitive Sciences, 5*, 689–710.
19. Trueblood, J. S., & Busemeyer, J. R. (2011). A quantum probability account of order effects in inference. *Cognitive Science, 35*, 1518–1552.
20. Khrennikov, A., Basieva, I., Dzhafarov, E. N., & Busemeyer, J. R. (2014). Quantum models for psychological measurements: An unsolved problem. *PLoS ONE*.
21. Osherson, D. N., & Smith, E. E. (1981). On the adequacy of prototype theory as a theory of concepts. *Cognition, 9*, 35–58.
22. Tversky, A., & Kahneman, D. (1983). Extension versus intuitive reasoning: The conjunction fallacy in probability judgment. *Psychological Review, 90*(4), 293–315. https://doi.org/10.1037/0033-295X.90.4.293
23. Aerts, D., Aerts Arguelles, J., Beltran, L., Geriente, S., Sassoli de Bianchi, M., Sozzo, S., & Veloz, T. (2019). Quantum entanglement in physical and cognitive systems: A conceptual analysis and a general representation. *The European Physical Journal Plus, 134*, 493.
24. Tilma, T., Byrd, M. S., & Sudarshan, E. C. G. (2002). A parametrization of bipartite systems based on SU(4) Euler angles. *Journal of Physics A: Mathematical and General, 35*, 10445–10466.

25. Havel, T. F., & Doran, C. J. L. (2004). A Bloch-sphere-type model for two qubits in the geometric algebra of a 6D Euclidean vector space. In *2nd Conference on Quantum Information and Computation. Proceedings of SPIE—The International Society for Optical Engineering* (Vol. 5436, pp. 93–106). arXiv:quantph/0403136
26. Mosseri, R., & Dandoloff, R. (2001). Geometry of entangled states, Bloch spheres and Hopf fibrations. *Journal of Physics A: Mathematical and General, 34*, 10243–10252 (2001). arXiv:quantph/0108137
27. Mosseri, R. (2006). Two-qubit and three-qubit geometry and Hopf fibrations. In *Topology in condensed matter*. Springer Series in Solid-State Sciences (Vol. 150, pp. 187–203). arXiv:quant-ph/0310053
28. Wie, C.-R. (2014). Bloch sphere model for two-qubit pure states. arXiv:1403.8069
29. Tsirelson, B. S. (1980). Quantum generalizations of Bell's inequality. *Letters in Mathematical Physics, 4*(2), 93–100.
30. Paneru, D., Cohen, E., Fickler, R., Boyd, R. W., & Karimi, E. arXiv:1911.02201 [quant-ph] and references therein.
31. Clauser, J. F., Horne, M. A., Shimony, A., & Holt, R. A. (1969). *Physical Review Letters, 23*(15), 8804.
32. Bell, J. S. (1964). *Physics, 1*(3), 195–200. Reproduced as Ch. 2 of Bell, J. S. (2004). *Speakable and unspeakable in quantum mechanics*. Cambridge University Press.
33. Ghose, P., & Mukherjee, A. (2014). *Advanced Science, Engineering and Medicine, 6*, 246–251.
34. Everett, H. (1957). *Reviews of Modern Physics, 29*(3), 454–462.
35. Bell, J. S. (2004). Against measurement (Chap. 23). In *Speakable and unspeakable in quantum mechanics*. Cambridge University Press.
36. Mermin, N. D. (2014). *Nature, 507*(7493), 421–423.
37. Khrennikov, A. (2020). Quantum versus classical entanglement: Eliminating the issue of quantum nonlocality. *Foundations of Physics*. https://doi.org/10.1007/s10701-020-00319-7. arXiv:1909.00267v1 [quant-ph].
38. Khrennikov, A. (2020). Two faced Janus of quantum nonlocality. arXiv: 2001.02977v1 [quant-ph].
39. Ramachandran, D. R., & Ramachandran, V. S. (2008). Ambiguities & perception: What uncertainty tells us about the brain. *Scientific American Special Editions, 18*(2s), 56–59. https://doi.org/10.1038/scientificamerican0508-56sp
40. Yearsley, J. M. (2016). Advanced tools and concepts for quantum cognition: A tutorial. *Journal of Mathematical Psychology, 78*, 24–39. https://doi.org/10.1016/j.jmp.2016.07.005
41. Khrennikov, A., & Basieva, I. (2014). Possibility to agree on disagree from quantum information and decision making. *Journal of Mathematical Psychology, 62–63*, 1–15.
42. Aumann, R. J. (1976). Agreeing on disagree. *Annals of Statistics, 4*, 1236–1239.
43. Khrennikov, A. (2016). Quantum Bayesianism as the basis of general theory of decision-making. *Philosophical Transactions of the Royal Society A, 374*, 20150245. https://doi.org/10.1098/rsta.2015.0245
44. Khrennikova, P., & Patra, S. (2019). Asset trading under non-classical ambiguity and heterogeneous beliefs. *Physica A, 521*, 562–577.
45. Chatterjee, S., John, K., & Yan, A. (2012). Takeovers and divergence of investor opinion. *The Review of Financial Studies, 25*, 227–277.
46. Haven, E., & Khrennikova, P. (2018). A quantum-probabilistic paradigm: Non-consequential reasoning and state dependence in investment choice. *Journal of Mathematical Economics, 78*(C), 186–197.

Classical Optical Modelling of the 'Prisoner's Dilemma' Game

Sudip Patra and Partha Ghose

Abstract Though there is a strong body of literature on 'pure quantum games' where quantum entanglement needs to be introduced in the modelling for Pareto Superior equilibria, unattainable in classical games, we demonstrate in the current paper a particular version of the game involving three strategies in which the physical setting of classical optics makes a superior strategy possible even without entanglement/nonseparability. Non-zero sum cooperation games with superior equilibria can also be designed using a new type of nonlocal classical nonseparability, the advantage of such designs being ease of implementation and robustness of entanglement, unlike fragile quantum entanglement due to decoherence. We demonstrate the foundational set up of classical optical modelling of game theory, which is conceptually different from quantum modelling, by using the basic Prisoners' Dilemma game under full information.

Keywords Classical optical modelling (COM) · Entanglement · Prisoner's Dilemma (PD) · Pareto superiority · Joint probabilities · Non-zero-sum games

1 Introduction

It was John von Neuman and Oskar Morgenstern [1] who first developed the mathematical theory of games in the form known to economists, social scientists and biologists. Over the past seven decades the framework has been deepened and generalized. The paradigmatic game that has found widespread applications in economy, psychology, ecology and biology is the Prisoners' Dilemma which falls in the category of 'non-zero sum games' where—in contrast to 'zero-sum games'—the two

S. Patra (✉)
O P Jindal Global University: CEASP, Sonipat, Haryana 131001, India
e-mail: spatra@jgu.edu.in

P. Ghose
Tagore Centre for Natural Sciences and Philosophy, Rabindra Tirtha, New Town, Kolkata 700156, India

S. Sriboonchitta et al. (eds.), *Credible Asset Allocation, Optimal Transport Methods, and Related Topics*, Studies in Systems, Decision and Control 429,
https://doi.org/10.1007/978-3-030-97273-8_17

players are not in strict opposition to each other, but may rather benefit from mutual cooperation. Recently, some of these games have been 'quantized' [2, 3]. Quantum games differ from their classical counterparts in three main ways: they admit (i) superposition of initial states, (ii) entanglement of initial states and (iii) superposition of strategies to be used on the initial states.

Prisoners' Dilemma (PD) has become paradigmatic in another sense. In this non-zero sum game, if we start from the assumption that individual players are typically rational and utility maximizing, where the utility function is typically individual, and rationality is 'common knowledge', then instead of a Pareto Superior cooperation strategy, the players are forced to choose an inferior defection strategy. This is a tragic consequence of rationality itself, and of late a repeated game or evolutionary Prisoners' Dilemma game has been proposed [4] where cooperation might emerge based on 'discount rates' of individual players (we elaborate on this point later in the paper).

Now the advantage of a quantum-like modification of the game is that, there can be additional quantum strategies. Some authors propose a 'super operator' or a 'super strategy' which would converge the game to a Pareto Superior cooperative equilibrium, thus demonstrating 'quantum supremacy' [5]. Hence, if such superior strategies are played or are available to the players, then a good cooperative equilibrium can be established in the first round itself, rather than in extensions.

We think there is another advantage of such an alternative modelling, namely, it is not necessary to modify the underlying utility function structure (as in the behavioural economics and finance literature) by, for example, introducing 'altruism' or 'inequity-aversion' [6]. Rather, here the players are still 'rational' in the neoclassical sense of the word, but the major modifications are in the strategic profiles, 'entangled' or 'correlated' game structure, and decision framework (probability computation using a modified formula for total probability (FTP) and updating of beliefs when required of the players).

It is interesting to note that classical polarization optics, as opposed to classical particle mechanics, offers similar advantages to games because of the occurrence of superpositions and entanglement [7–12]. In fact, classical entanglement/nonseparability has the advantage of being robust, unlike quantum entanglement which is fragile in the presence of noise. Based on this, an alternative *Classical Optical Model* (COM) of social sciences has recently been developed [13] which has striking similarities with 'quantum models' [14–16] but is *conceptually quite different*. It has the advantage of being simple, geometric and intuitive, and is partly based on the epistemological approach of Bohr. It does not use the full mathematical apparatus of quantum mechanics as in 'quantum modelling' but only the Hilbert space structure of classical optics that captures superposition and entanglement. In particular, as is well known, *measurement plays no special role in classical optics*.

The possibility, however, remains that there may be cases in the social sciences which are truly quantum-like and cannot be captured by COM. It would be instructive to look for them.

Philosophical Background
Bohr adopted a definite Kantian viewpoint with subtle differences. Much of Kant's philosophy can be seen as an attempt to arrive at satisfactory philosophical grounds for the objective basis of Newtonian mechanics against Hume's scepticism. Bohr used a similar Kantian approach, albeit with subtle modifications, to do the same for quantum mechanics. His approach must be sharply distinguished from the so called *Copenhagen interpretation* of quantum mechanics. This term was coined by physicists *opposed* to Bohr's idea of complementarity and contextuality, to identify what they saw as the common features of the Bohr-Heisenberg interpretation as it emerged in the late 1920s.

There is a fundamental difference between Bohr and Heisenberg. Heisenberg's uncertainty or indeterminacy relations are based on the inalienable and uncontrollable 'disturbance' of a system caused by an experimental probe. Bohr was not happy with this situation. After the 1935 EPR paper which eliminated the possibility of a *mechanical disturbance* of a measurement on one member of an entangled pair of systems on the other, Bohr adopted the stance that *the state of the measuring device and the state of the object cannot be separated from each other during a measurement and constitute a dynamical whole* or an *unanalyzable whole*. He saw Heisenberg's relations as formal or mathematical expressions of the general notion that our understanding of atomic phenomena must be based on *complementary* descriptions of incompatible experimental arrangements of a classical type. He justified this by the necessity to communicate the results of scientific experiments in an unambiguous way. Such unambiguity is guaranteed by reliance on a shared language refined by the terminology of classical physics.

In a double-slit experiment, for example, the basic apparatus is classically describable, and further, it is impossible to have one or both slits open at the same time. They are physically *mutually exclusive* arrangements, giving rise to *complementary* particle-like and wave-like *phenomena* which together give the complete description of the system. Bohr's position was:

> Consequently, evidence obtained under different experimental conditions cannot be comprehended within a single picture, but must be regarded as complementary in the sense that only the totality of the phenomena exhausts the possible information about the objects.

It follows from this that, unlike in classical physics, *observation in quantum physics is context-dependent*. This is the essence of Bohr's view of contextuality and complementarity, the two being intimately connected to each other.

Later Bohr extended the principle to other areas of human knowledge such as psychology in which the quantum of action can hardly be expected to play any fundamental role [17]. He drew attention to the close analogy between observations of atomic phenomena and those in human psychology. He wrote:

> In introspection it is clearly impossible to distinguish sharply between the phenomena themselves and their conscious perception, and although we may often speak of lending our attention to some particular aspect of a psychical experiment, it will appear on closer examination that we really have to do, in such cases, with mutually exclusive situations. ... By such a comparison it is, of course, in no way intended to suggest any closer relation between

atomic physics and psychology, but merely to stress an epistemological argument common to both fields, and thus to encourage us to see how far the solution of the relatively simple physical problems may be helpful in clarifying the more intricate psychological questions with which human life confronts us, ...

This makes it clear that to Bohr the principle of complementarity provided the *epistemological basis* for a comprehensive understanding of both physical and psychological phenomena witnessed in mutually exclusive contexts.

It is also clear from the above quote that in the social sciences too *complementarity*, and by implication *contextuality*, has been known for a long time, quite independent of quantum theory. It is therefore tempting to investigate whether contextuality in the social sciences can be understood in a similar fashion to that in quantum physics. Now, it is possible in quantum physics to separate the measuring device from the system under investigation, as was done by Heisenberg and his followers, and continues to be done by those engaged in 'quantum modelling' of social sciences. In Bohr's view such a separation is purely *epistemological*: the cut is arbitrary and has no ontological significance. This is more intrinsically so in the social sciences where *the subjective state of an individual under investigation is inaccessible to the 'other'*. Hence, it is impossible, in principle, for the 'other' to detect any change in it directly or indirectly. It is a primary phenomenon, a first-person experience whose intrinsic nature cannot be observed by any other external agent. Under the reasonable assumption that there are many such individuals with subjective first-person experience, what can be learnt and described in an inter-subjectively objective manner is what is revealed by the answers (responses in general) to specific questions (stimuli) given by these individuals in specific contexts. The contexts and questions together with the internal mental states give rise to *phenomenal* mental states that are observable.

Quantum mechanics is essentially particle mechanics. What distinguishes quantum mechanics from classical particle mechanics is its ability to account for (i) interference phenomena, (ii) contextuality and complementarity and (iii) entanglement. This is achieved by describing a quantum state as a vector in a Hilbert space rather than by a point in phase space. However, there exists another branch of classical physics, namely classical optics, or more generally, Maxwell's electromagnetic theory of fields in which also states are represented by vectors in a Hilbert space. Interference is a well known phenomenon in classical optics, and recently it has been demonstrated that even entanglement occurs in classical polarization optics [7–12, 18]. This blurs the boundary between classical and quantum physics, and much of what typically characterized quantum physics is now known to occur in classical optics as well. The close mathematical resemblance between classical optics and quantum mechanics and the successes of 'quantum modelling' of the social sciences prompt us to explore Classical Optical Modelling (COM) of social science phenomena as *an alternative way* of understanding them in an intuitive fashion.

Hence overall, our framework would provide: (i) a simple geometric representation of the PD game with Pareto Superior outcomes for 'classical optical' entanglement-like scenario which can be in reality sustained (unlike the very fragile pure quantum entanglement where the no-signaling theorem further prohibits utilizing such entanglement for communication purposes), and (ii) a formula for total probability (FTP) which can be used to demonstrate the deviant behaviours of 'rational' players from the Nash Equilibrium (NE) [19] under scenarios of uncertainty.

The use of the Formula for Total Probability (FTP) needs to be emphasized here. It has been observed that under real life scenarios, for example, when players are in an ambiguity or uncertainty scenario, the moves exhibited by them do not follow dominant Nash Equilibrium, though exact reasons for this deviant behavior may well be behavioral and complex [15]. FTP computed on the basis of Born's rule can provide a description of such deviations due to the presence of additional perturbative terms in the probability computations also known as 'interference' terms. Again, such interference terms would be absent in classical measure theory-based computations. In the recent development of COM of decision making or cognition [13] a modified formula for total probability has been developed. However, such a formula can be utilized in case of 'sequential' move games, rather than in 'simultaneous' move games, which is the Prisoners' Dilemma in normal form, as will be discussed. In PD games therefore we will use joint probabilities rather than conditional probabilities. In future versions we would expand on 'sequential' move cooperative non-zero sum games.

We will now develop a COM of the Prisoners' Dilemma Game. We emphasize that *COM is not 'quantum modelling'*, the only commonality between COM and quantum modelling being the use of states in Hilbert space. We believe it is a useful alternative way of modelling problems in the social sciences.

2 COM of Prisoners' Dilemma Game

Our starting point will be two players/prisoners Alice (A) and Bob (B) who independently decide whether they choose to defect (strategy D) or cooperate (strategy C). Depending on the decision they take, each player receives a certain pay-off (Table 1). The objective of each player is to maximize his or her individual pay-off. The catch of the dilemma is that D happens to be the 'dominant strategy', i.e. the optimal move for an individual player regardless of how the other players act. Each player rationally and independently chooses to defect, although by doing so, they do worse than if they had both decided to cooperate (see pay-off Table). This is because D is the *Nash Equilibrium* in which none of the players sees any advantage in choosing to act otherwise. This is the 'classical game'. We wish to develop a 'non-classical' game with a richer structure of possible strategies by utilizing superposition and entanglement in classical optics rather than quantum mechanical superposition and entanglement as in quantum modelling [3].

In analogy to Eisert et al. [3], we propose a physical model which consists of (a) sources of two polarized classical light beams, one for each player, (b) a set of optical instruments which enable each player to manipulate his or her own light beam in a strategic manner, and (c) a measurement device (polarization analyzers) which determines the players' pay-off according to the polarization states of the light beams detected. All of these are assumed to be perfectly known to both Alice and Bob.

Table 1 Pay-off table

	Bob: C	Bob: D
Alice: C	(3,3)	(0,5)
Alice: D	(5,0)	(1.1)

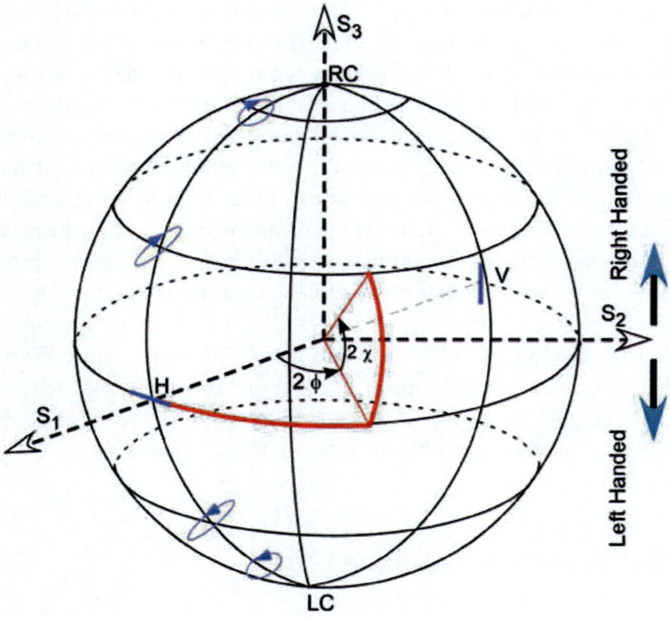

Fig. 1 The Poincaré Sphere; (read θ for χ)

To the possible outcomes of the classical strategies C and D are assigned two vectors $|C\rangle$ and $|D\rangle$ which span the basis of the 2-dimensional Hilbert space $\mathcal{H}$ of a polarized light beam. The state of the game at each instant is described by a vector in the tensor product space $\mathcal{H}_{\mathcal{A}} \otimes \mathcal{H}_{\mathcal{B}}$ of the two light beams spanned by the basis $|C\rangle_A \otimes |C\rangle_B$, $|C\rangle_A \otimes |D\rangle_B$, $|D\rangle_A \otimes |C\rangle_B$ and $|D\rangle_A \otimes |D\rangle_B$.

Let us identify $|C\rangle := |0\rangle$ and $|D\rangle := |1\rangle$ as the standard basis vectors of a Poincaré sphere (Fig. 1). They correspond to horizontal (H) and vertical (V) polarization states of light. Let the optical states with Alice and Bob in this basis be

$$|\psi\rangle_i = \cos\theta_i |C\rangle_i + e^{i\phi_i} \sin\theta_i |D\rangle_i , \tag{1}$$

where $i = A, B$ and $-\pi/2 \leq 2\theta_i \leq +\pi/2$ and $0 \leq 2\phi_i \leq 2\pi$. These states are the points $(2\theta, 2\phi)$ on the Poincaré sphere. The 'strategic space' is the family of unitary matrices $U(2)$ [20]

$$\hat{U}(\theta, \phi) = e^{i\phi} \begin{pmatrix} r & it^* \\ it & r^* \end{pmatrix} \tag{2}$$

where ϕ is any phase and (r, t) are parameters satisfying the condition $|r|^2 + |t|^2 = 1$. Using the parametrization $r = \cos\theta e^{-i\phi}$, $t = \sin\theta e^{-i\phi}$, this can be written in the form

$$\hat{U}(\theta, \phi) = \begin{pmatrix} \cos\theta & i\sin\theta \\ ie^{2i\phi}\sin\theta & e^{2i\phi}\cos\theta \end{pmatrix}. \tag{3}$$

One readily sees that

$$\hat{U}(0, 0) = \begin{pmatrix} 1 & 0 \\ 0 & 1 \end{pmatrix} = \sigma_0 = \mathbb{I}, \tag{4}$$

$$-i\hat{U}\left(\frac{\pi}{2}, \pi\right) = \begin{pmatrix} 0 & 1 \\ 1 & 0 \end{pmatrix} = \sigma_1, \tag{5}$$

$$-\hat{U}\left(\frac{\pi}{2}, \frac{\pi}{2}\right) = \begin{pmatrix} 0 & -i \\ i & 0 \end{pmatrix} = \sigma_2, \tag{6}$$

$$\hat{U}\left(0, \frac{\pi}{2}\right) = \begin{pmatrix} 1 & 0 \\ 0 & -1 \end{pmatrix} = \sigma_3, \tag{7}$$

where $\sigma_i\,(i = 1, 2, 3)$ are the Pauli matrices which are generators of the group $SU(2)$. They satisfy the algebra

$$[\sigma_i, \sigma_j] = 2i\epsilon_{ijk}\sigma_k. \tag{8}$$

Since these matrices form a closed set of non-trivial operators, one can associate at most three independent strategies with them which may be realized using beam splitters [20]. We shall make the following identifications: 'cooperation $\hat{C} = \sigma_0$ and 'defect' $\hat{D} = \sigma_1$.

Let us note in passing here that

$$\frac{1}{\sqrt{2}}[\sigma_3 + \sigma_1] = \frac{1}{\sqrt{2}} \begin{pmatrix} 1 & 1 \\ 1 & -1 \end{pmatrix} := \hat{H} \tag{9}$$

which is the Hadamard gate, and that the unitary operator $\hat{U}(\theta, \phi)$ can also be written in the form

$$\hat{U}(\theta, \phi) = e^{i\phi}\exp(-i\theta\vec{\sigma}.\vec{n}) \tag{10}$$

which is a rotation by angle 2θ about an arbitrary direction $\vec{n}$ of the Poincaré sphere.

It must be emphasized here that the use of unitary transformations and the Pauli matrices is quite legitimate within a purely classical Hilbert space theory. The algebra (8) of the Pauli matrices is a property of the non-Abelian Lie group $SU(2)$, and Planck's constant does not enter into it.

Denoting Alice's and Bob's strategies by $\hat{U}_A$ and $\hat{U}_B$ respectively, their joint strategy will be $\hat{U}_A \otimes \hat{U}_B$, each one of them acting only on his/her light beam. If the initial state in the tensor product space is $|\psi_0\rangle = |\psi\rangle_A \otimes |\psi\rangle_B$ where the states $|\psi\rangle_A$ and $|\psi\rangle_B$ are given by (1), then the final state is $|\psi_f\rangle = \hat{U}_A \otimes \hat{U}_B|\psi_0\rangle$. The subsequent detection of this state by two polarization analyzers (e.g. polarizing beam splitters) will yield particular outcomes, CC, CD etc., depending on the strategies adopted by Alice and Bob. In the standard version of the Prisoners' Paradox game the pay-off is returned according to the corresponding entry of the pay-off matrix (Table 1). Let us consider Alice's and Bob's pay-offs in general,

$$\$_A = rP_{CC} + pP_{DD} + tP_{DC} + sP_{CD}, \tag{11}$$

$$\$_B = rP_{CC} + pP_{DD} + sP_{DC} + tP_{CD} \tag{12}$$

where $P_{\lambda\lambda'} = |\langle\lambda\lambda'|\psi_f\rangle|^2$ is the joint probability that the output ports record λ and λ' polarization states. The symbols (r, p, s, t) stand for 'reward', 'punishment', 'temptation' and 'sucker's pay-off', the numerical values being those given in Table 1 (r = 3, p = 1, s = 0, t = 5).

Table 1 the first entry in the parentheses is Alice's pay-off and the second entry that of Bob in the standard Prisoners' Paradox game. Now,

$$\begin{aligned}|\psi_f\rangle &= \hat{U}_A \otimes \hat{U}_B|\psi_0\rangle \\ &= \cos\theta_A \cos\theta_B \hat{U}_A|C\rangle_A \otimes \hat{U}_B|C\rangle_B + e^{i\phi_B}\cos\theta_A \sin\theta_B \hat{U}_A|C\rangle_A \otimes \hat{U}_B|D\rangle_B \\ &+ e^{i\phi_A}\sin\theta_A \cos\theta_B \hat{U}_A|D\rangle_A \otimes \hat{U}_B|C\rangle_B \\ &+ e^{i(\phi_A+\phi_B)}\sin\theta_A \sin\theta_B \hat{U}_A|D\rangle_A \otimes \hat{U}_B|D\rangle_B. \end{aligned} \tag{13}$$

This shows a richer structure of strategies than in the classical game. If both Alice and Bob choose strategy $\hat{C}$, a simple calculation gives

$$\begin{aligned} P_{CC} &= \cos^2\theta_A \cos^2\theta_B, \;\; P_{DD} = \sin^2\theta_A \sin^2\theta_B, \\ P_{CD} &= \cos^2\theta_A \sin^2\theta_B, \;\; P_{DC} = \sin^2\theta_A \cos^2\theta_B. \end{aligned} \tag{14}$$

On the other hand, if they both choose strategy $\hat{D}$,

$$\begin{aligned} P_{CC} &= \sin^2\theta_A \sin^2\theta_B, \;\; P_{DD} = \cos^2\theta_A \cos^2\theta_B, \\ P_{CD} &= \sin^2\theta_A \cos^2\theta_B, \;\; P_{DC} = \cos^2\theta_A \sin^2\theta_B. \end{aligned} \tag{15}$$

Since they are both rational and knowledgeable about polarization optics (which is 'common knowledge'), given the pay-off table (Table 1), each one of them will choose his/her initial state to be $|C\rangle$ so that $\theta_A = \theta_B = 0$ and opt for the $\hat{D}$ strategy so that $P_{DD} = 1$ and the rest of the probabilities are zero. This will make the Nash Equilibrium DD the only possible outcome. This corresponds to the players playing 'pure strategies' in classical games. Thus, the classical game is included in the 'non-classical' extension.

Table 2 Extended pay-off table

	Bob: C	Bob: D	Bob: L
Alice: C	(3,3)	(0,5)	(2,2)
Alice: D	(5,0)	(1.1)	(2,2)
Alice: L	(2,2)	(2,2)	(2,2)

Extended Prisoners' Dilemma

Now, Alice and Bob are not restricted to the choices $\hat{C}$ and $\hat{D}$ only in classical games. They can also choose a third strategy $\hat{L}$. It is clear from the pay-off Table 2 that $\hat{L}$ is the dominant strategy. Hence, a new Nash Equilibrium emerges which is LL. The pay-offs are now given by

$$\$_A = rP_{CC} + pP_{DD} + tP_{DC} + sP_{CD} + lP_{LL}, \tag{16}$$

$$\$_B = rP_{CC} + pP_{DD} + sP_{DC} + tP_{CD} + lP_{LL}. \tag{17}$$

This is an extended Prisoners' Dilemma game with three strategies [21]. Thus, in addition to 'cooperate' and 'defect', the players can also 'abstain', obtaining the 'loner's pay-off' L. The value of L must be set such that: (a) it is not greater than r, otherwise the advantage of not playing will be sufficiently large to ensure that players will always abstain, and (b) it is greater than s, otherwise there are no benefits to abstaining. Thus, the values of L can be set in the range $[s, r]$. We choose $l = 2$. It turns out that LL is also Pareto Optimal. So, there is no dilemma any more.

Table 2 if l denotes the 'loner's' pay-off, the condition $t > r > l > p > s$ is satisfied in this table [21].

A COM of this 3-strategy game can be developed using the group SO(3) (whose representations are also representations of SU(2)) to build the 'strategy space'. The 3×3 matrices s_x, s_y, s_z which are generators of SO(3) satisfy the algebra

$$[s_x, s_y] = is_z, \text{ cyclic} \tag{18}$$

These generators represent rotations $R_x(\theta) = e^{i\theta s_x}$, $R_y(\theta) = e^{i\theta s_y}$, $R_z(\theta) = e^{i\theta s_z}$ about the x, y, z axes. The most general rotation is $R(\phi, \theta, \psi)$ where ϕ, θ, ψ are Euler angles. Let the three unit basis vectors in the (x, y, z) frame be denoted by $|C\rangle, |D\rangle, |L\rangle$, and let the initial states with Alice and Bob be

$$|\psi\rangle_i = \alpha_i|C\rangle_i + \beta_i|D\rangle_i + \gamma_i|L\rangle_i \tag{19}$$

with $\alpha_i^2 + \beta_i^2 + \gamma_i^2 = 1, i = A, B$. Then the initial state $|\psi_0\rangle = |\psi\rangle_A \otimes |\psi\rangle_B$ and the final state is $|\psi_f\rangle = R_A(\phi_A, \theta_A, \psi_A) \otimes R_B(\phi_B, \theta_B, \psi_B)|\psi_0\rangle$. By choosing the initial states and applying the appropriate rotations, Alice and Bob can always ensure that the final state is LL.

The game can be played using the $l = \pm 1$ and $l = 0$ orbital angular momentum states of classical light with components corresponding respectively to $|C\rangle, |D\rangle, |L\rangle$. Solutions of the paraxial approximation to the Helmholtz equation

$$\left(\nabla^2 + k^2\right) \psi(\vec{r}) = 0 \tag{20}$$

are Laguerre-Gauss polynomials $\psi_{pl}(\rho, \phi, z) = u_{pl}(\rho, z)e^{-il\phi}$ where (ρ, ϕ, z) are cylindrical coordinates, p is a non-negative whole number and l is an integer. They represent Laguerre-Gaussian beams of order (p, l) which can be readily produced using holographic plates [22]. When a Gaussian beam passes through the holographic plate, in addition to the transmitted Gaussian beam with $l = 0$, two Laguerre-Gaussian beams of higher order with opposite phases are created. This is a consequence of angular momentum conservation. These two beams travel at an angle with respect to the original Gaussian beam and can be easily distinguished.

Such beams carry a well-defined orbital angular momentum. Since the dependence on ϕ is only in the phase, the complex amplitude of Laguerre-Gaussian beams is cylindrically symmetric about the beam axis. It causes the wavefronts to tilt helically around the z-axis. The sign of l determines the handedness of the helix, and the intensities are the same for beams with l and $-l$ while their phases are opposite. One can identify the $l = \pm 1$ states with $|C\rangle$ and $|D\rangle$ and the $l = 0$ state with $|L\rangle$, and use standard optical elements like beam splitters, mirrors etc. to manipulate the beams to correspond to 'cooperate', 'defect' and 'abstain'.

What is interesting is that this has been achieved with product states. It will therefore be very interesting to investigate what additional results, if any, can be obtained if entanglement is taken into account in COM.

3 Entanglement and the Prisoners' Dilemma

It is now well known that inter-system nonseparability/entanglement does occur in classical polarization optics though quantum entanglement, which is inter-system and nonlocal, may have certain advantages over such classical entanglement [23]. We now extend the idea of nonseparability in classical polarization optics from the local to the nonlocal case. Consider two identical monochromatic sources S_A and S_B emitting *coherent* trains of horizontally polarized classical light pulses for Alice and Bob. Then the initial state is $|C\rangle_A \otimes |C\rangle_B$. Let Alice apply the Hadamard gate $\hat{H}$ (Eq. 9) on her state. That will produce the state

$$|\Psi\rangle = \frac{1}{\sqrt{2}}\left[|C\rangle_A + |D\rangle_A\right] \otimes |C\rangle_B = \frac{1}{\sqrt{2}}\left[|C\rangle_A \otimes |C\rangle_B + |D\rangle_A \otimes |C\rangle_B\right]. \tag{21}$$

Let a polarization filter be placed in Alice's train of pulses that selects $|C\rangle_A$ and $|D\rangle_A$ randomly, each 50% of the time, and let Bob apply $\hat{D} = \sigma_1$ on his state only whenever he finds Alice's pulse to be $|D\rangle_A$. That will generate the maximally entangled state

$$|\Psi^+\rangle_0 = \frac{1}{\sqrt{2}}[|C\rangle_A \otimes |C\rangle_B + |D\rangle_A \otimes |D\rangle_B]. \tag{22}$$

Then

$$|\Psi_f\rangle = \frac{1}{\sqrt{2}}[\hat{U}_A|C\rangle_A \otimes \hat{U}_B|C\rangle_B + \hat{U}_A|D\rangle_A \otimes \hat{U}_B|D\rangle_B]. \tag{23}$$

If Bob chooses $\hat{D}$, the best response for Alice is (a) $\hat{D}$ or (b) $\hat{L}$, as will be explained shortly. But, if Bob chooses $\hat{L}$, Alice's best response is $\hat{L}$. Similarly, if Alice chooses $\hat{D}$, Bob's best response is (a) $\hat{D}$ or (b) $\hat{L}$, but if Alice chooses $\hat{L}$, Bob's best response is $\hat{L}$. The reason is that in case (a) $|\Psi^+\rangle_0$ is preserved, and in case (b) the state changes to

$$|\Psi_f^-\rangle = \frac{e^{i\pi}}{\sqrt{2}}[|C\rangle_A \otimes |C\rangle_B - |D\rangle_A \otimes |D\rangle_B]. \tag{24}$$

Maximal entanglement is preserved in both cases. However, now DD is no longer a Nash Equilibrium—the new NE in dominant strategies is LL (Table 2).

Thus, when entanglement is present, a new possibility arises in the PD game without requiring extension to a third strategy. And the new equilibrium is Pareto Optimal, which resolves the dilemma.

It turns out that all the strategies preserve the 'degree of entanglement' measure called 'concurrence'. In our case a general tensor product state between Alice's and Bob's beams would be

$$|\Psi\rangle = \alpha_1|0\rangle_A \otimes |0\rangle_B + \alpha_2|0\rangle_A \otimes |1\rangle_B + \alpha_3|1\rangle_A \otimes |0\rangle_B + \alpha_4|1\rangle_A \otimes |1\rangle_B \tag{25}$$

with $\sum_i |\alpha_i|^2 = 1$. Following Ref. [24], one can write the state as

$$|\Psi\rangle = c|\Psi_e\rangle + \sqrt{1-c^2}e^{i\chi}|\Psi_f\rangle \tag{26}$$

where $|\Psi_e\rangle$ is a maximally entangled state, $|\Psi_f\rangle$ is a factorizable state, and $c^2 = 2|\alpha_1\alpha_4 - \alpha_2\alpha_3|$ is the concurrence measure with $c \in [0, 1]$. It is straightforward to verify that c is preserved by all the three strategies. In fact, this follows generally from the fact that c is invariant under local unitary transformations, and all the strategies are unitary operators.

Here, however, we would like to maintain caution in claiming that the Pareto Superior equilibrium obtained by (LL) is a cooperation strategy in the standard sense, since both players refrain from playing at that point, which is nevertheless a Pareto improvement for sure.

Superposition of strategies and FTP

Till now we have considered a pure strategy game, allowing for entanglement as described above. However, it has been demonstrated [15] that rational players do not follow standard rules of the game when contexts change, for example from full information to uncertainty. In our case when Alice and Bob face uncertainty (which may be either related to incomplete knowledge of strategy profiles of players or the criterion of rationality of each other's context), as contrasted to a full information scenario. Hence, under uncertainty (in generation of an action potential by a neuron [25]) the 'mental states' of players, i.e. their beliefs about strategy states of the other players, may be described by a superposition of states. For example, for two possible choices $|0\rangle'_A$ or $|1\rangle'_A$ available to Alice, Bob's state before a real move would be

$$\begin{aligned} |\phi\rangle_B &= \cos\chi|0\rangle'_B + e^{i\phi}\sin\chi|1\rangle'_B \\ &:= \alpha|0\rangle'_B + \beta|1\rangle'_B \end{aligned} \tag{27}$$

We assume that the two states $|0\rangle'$ and $|1\rangle'$ are orthogonal and represent the quiescent and firing bases. The amplitudes $\alpha = \cos\chi$ and $\beta = e^{i\phi}\sin\chi$ represent potentialities for a neuron to be quiescent or firing. This is a pure state which is fundamentally different from an ensemble of states. A pure state captures uncertainty. It can also be represented by a density matrix $\rho_B = |\phi\rangle_B\langle\phi|_B$. Since this particular pure state of Bob corresponds to Alice's states $|0\rangle'_A$ or $|1\rangle'_A$, Alice's state is either

$$|\phi_0\rangle_A = \alpha|0\rangle'_A \otimes |0\rangle'_B + \beta|0\rangle'_A \otimes |1\rangle'_B, \tag{28}$$

which is a product state, or

$$|\phi_1\rangle_A = \alpha|1\rangle'_A \otimes |0\rangle'_B + \beta|1\rangle'_A \otimes |1\rangle'_B \tag{29}$$

which is also a product state.

Now, since in general it is possible to define a 'mental state' by a density matrix, let us define a general mental state by

$$\Theta = \sum_{i,j} x_{i,j}|\phi_i\rangle\langle\phi_j|, \tag{30}$$

where (i, j) have values 0 and 1 [25]. Hence the initial mental states of Alice are

$$\Theta_{0A} = |\phi_0\rangle_A\langle\phi_0|_A, \tag{31}$$

$$\text{or } \Theta_{1A} = |\phi_1\rangle_A\langle\phi_1|_A \tag{32}$$

with $|\phi_0\rangle_A$ and $|\phi_1\rangle_A$ given by (28), (29). These mental states would evolve over time in a non-Markovian way, which would be the dynamics of the model. One can measure the probabilities of Alice choosing the states (28) and (29) as

$$P_{0A} = \text{Tr}\Theta_{0A} = \text{Tr}|\phi_0\rangle_A\langle\phi_0|_A, \tag{33}$$

$$P_{1A} = \text{Tr}\Theta_{1A} = \text{Tr}|\phi_1\rangle_A\langle\phi_1|_A. \tag{34}$$

Such total probabilities contain interference terms due to the presence of superpositions embedded in them, and thus will differ significantly from classical probability-based computations. This is where the formula for total probability (FTP) developed in [13] comes in, as we will show in detail elsewhere.

4 Discussions and Conclusions

Distinguishing between correlated equilibrium games and entangled games, Aumann [26] initiated a strong literature on correlated equilibrium for classical games. Correlated equilibria always exist, and are achieved via a third party or a mediator. A trusted mediator can select strategies from joint probability distributions and 'privately' send such information to players, who then, based on such information, can maximize their payoffs, resulting in a Pareto superior equilibrium relative to standard Nash Equilibrium games.

However, the challenge has been to establish such superior outcomes without the intervention of any mediator or, in effect, any communication between players, which is the novelty of entangled games. Here the players can achieve such superior outcomes without the necessity to communicate.

In 'quantum correlated' games [27] players with full classical information can have access to a quantum entangled state, i.e. entangled qubits that are disjoint or spatially separated, and perform local operations on their qubits. No interaction is needed between the players. The novelty of our COM framework here is that similar outcomes can be accomplished utilizing *classical optical devices*, without any intervention of a mediator.

The quantum Prisoner's Dilemma has grown into an interesting exploration in itself [5]. Some authors [28] have suggested that any quantum game (where players have access to quantum strategies such as, for example, a super cooperation strategy as in [5], through quantum devices like quantum GATES) can be viewed as state preparation, transition and general measurement of the outcome state.

Pheonix et al. [28] also envisage quantum games as *reverse engineered*, as for example, by choosing the optimal state first and then designing a playable game. However, general measurements of the outcome state can be fundamentally different from the classical game case, since in classical games the final output state (equilibrium state) and the pay-offs converge, while if a quantum game is designed via suitable instruments (i.e. it is playable), then there is no guarantee that the output state when measured would be an eigenstate of the measurement operator. Hence, in effect, the output state informs about an expectation of pay-offs.

However, since maximal quantum entanglement, which is a necessary condition for superior cooperation outcomes, is fragile in nature due to decoherence, overall we may conjecture that compared to any classical PD game, or more broadly a non-zero sum game of cooperation, one may achieve a superior equilibrium outcome if and only if entanglement is guaranteed to be stable in the presence of suitable operator representations of strategies.

The advantage of COM over pure quantum game modelling has already been pointed by us in the introduction section. Here we further observe that in classical game theory in a single-shot game with no further rounds, the pay-off matrix is given, and hence empirically joint probabilities which help forming the expected pay-offs can either be computed by observing a large ensemble of PDs played by identical players, or in a repeated game scenario, by using Bayesian updating. However, in our formulation the probabilities would be computed using the 'square of the amplitude rule', according to foundational theorems in Hilbert space theory.

The classical entanglement, which is a well-accepted feature of classical optics and is stable, can be utilized instead of quantum entanglement to design such a game. While this will suffice for most cases, it is possible that quantum entanglement will have superiority in special cases, the quintessentially quantum ones.

Returning to Auman's work, we would like to add that although it is generally true that correlated mixed strategy classical games improve equilibrium outcomes [29], since we need mediator correlation devices or agents to correlate the actions of players, there can be further inefficiencies due to noise in signaling communication either due to manipulations of such devices, or information asymmetry problems among participants. This is where quantum games with entanglement can be more cost effective. Further, since Aumann's seminal paper, correlated equilibrium has been associated with 'cost of disobedience' among players for a non-zero-sum game. Hence, such equilibria can be termed as 'costly correlated equilibria' (CCE). It has recently been shown [30] that for 'mixed strategy' games a set of CCE increases for even arbitrarily small deviations in cost as imposed by the mediator, but for pure strategy games (for example PD games in finite normal form), CCE can only improve when the cost is sufficiently high, making it difficult to achieve. In contrast a quantum PD game makes such 'costly' mediation redundant. Hence, using entanglement to achieve a Pareto improved solution at a lesser cost is preferable to costly correlated games.

Here we would like to emphasize that, in contrast to quantum games, our design using classical entanglement is achievable at much less cost and is also environmentally robust.

After this work was completed, we came across two papers which, to the best of our knowledge, are the first examples of the application of nonseparability between spin and orbital degrees of freedom of laser modes to game theory [31, 32]. While this is encouraging from our point of view, there are some fundamental differences with COM. Nonseparability (entanglement) is essential to quantum games, but that is not the case in COM in which, as shown in Sect. 2 above, superior strategies arise even without nonseparability. Furthermore, the type of nonseparability used in the experiments is intra-system and local. Though nonlocality is not essential for quantum games, it is a more striking illustration of the quantum advantage. In our COM model of the prisoners dilemma with entanglement/nonseparability (Sect. 3) we have taken care to introduce inter-system and nonlocal entanglement to parallel the quantum case.

References

1. von Neumann, J., & Morgenstern, O. (1944). *The theory of games and economic behavior*. Princeton University Press
2. Meyer, D. A. (1999). Quantum strategies. *Physical Review Letters, 82*(5), 1052.
3. Eisert, J., Wilkens, M., & Lewenstein, M. (1999). Quantum games and quantum strategies. *Physical Review Letters, 83*(15), 3077.
4. Axelrod, R., & Dion, D. (1988). Classical Optical Modelling of the 'Prisoner's Dilemma' Game. *Science, 242*, 138590.
5. Li, A., & Yong, X. (2014). Entanglement guarantees emergence of cooperation in quantum prisoner's dilemma games on networks. *Scientific Reports, 4*, 6286.
6. Fehr, E., & Schmidt, K. M. (2003). Theories of fairness and reciprocity: Evidence and economic applications. In M. Dewatripont, L. P. Hansen, & S. Turnovski (Eds.), *Advances in economic theory, Eighth World Congress of the Econometric Society* (Vol. 1, pp. 208–257). Cambridge University Press.
7. Spreeuw, R. J. (1998). A classical analogy of entanglement. *Foundations of Physics, 28*(3), 361–374.
8. Spreeuw, R. J. (2001). Classical wave-optics analogy of quantum-information processing. *Physical Review A, 63*(6), 062302.
9. Ghose, P., & Samal, M. K. (2001). arXiv:quant-ph/0111119.
10. Ghose, P., & Mukherjee, A. (2014). Entanglement in classical optics. *Reviews in Theoretical Science, 2*, 1–14.
11. Aiello, A., Toppel, F., Marquardt, C., Giacobino, E., & Leuchs, G. (2015). Quantum–like nonseparable structures in optical beams. *New Journal of Physics (NJP), 17*, 043024.
12. Qian, X.-F., Little, B., Howell, J. C., & Eberly, J. H. (2015). Entanglement limits duality and vice versa. *Optica, 2*(7), 611–615
13. Patra, S., & Ghose, P. (2020). Classical Optical Modelling of Social Sciences in a Bohr-Kantian Framework. (preprint).
14. Khrennikov, A. (2010). *Ubiquitous quantum structure: From psychology to finance*. Springer.
15. Haven, E., & Khrennikov, A. (2013). *Quantum social science*. Cambridge University Press.
16. Aerts, D. (2009). Quantum structure in cognition. *Journal of Mathematical Psychology, 53*(5), 314–348.
17. Bohr, N. (1958). *Atomic physics and human knowledge*. Wiley. Reprinted as The Philosophical Writings of Niels Bohr (Vol. II). Ox Bow Press.
18. Borges, C. V. S., Hor-Meyll, M., Huguenin, J. A. O., & Khoury, A. Z. (2010). Bell-like inequality for the spin-orbit separability of a laser beam. *Physical Review A, 82*(3), 033833 and references therein.
19. Myerson, R. B. (1991). *Game theory: An analysis of conflict*. MIT Press.
20. Zeilinger, A., Bernstein, H. J., & Horne, M. A. (1994). Information transfer with two-state two-particle quantum systems. *Journal of Modern Optics, 41*(12), 2375–2384.
21. Cardinot, M., O'Riordan, C., & Griffith, J. (2016). The optional Prisoner's Dilemma in a spatial environment: Coevolving game strategy and link weights, in *Proceedings of the 8th International Joint Conference on Computational Intelligence* (IJCCI 2016) (Vol. 1, pp. 86–93). ECTA.
22. Allen, L., Beijersbergen, M. W., Spreeuw, R. J. C., & Woerdman, J. P. (1992). Orbital angular momentum of light and the transformation of Laguerre-Gaussian laser modes. *Physical Review A, 45*(11), 8185–8189.
23. Khrennikov, A. (2020). *Foundations of Physics*. https://doi.org/10.1007/s10701-020-00319-7; arXiv:1909.00267v1 [quant-ph]. To link to this article: https://doi.org/10.1080/09500349414552211.
24. Abouraddy, A. F., Saleh, B. E., Sergienko, A. V., & Teich, M. C. (2001). Degree of entanglement for two qubits. *Physical Review A, 64*(5), 050101.
25. Khrennikov, A., & Asano, M. (2020). A quantum-like model of information processing in the brain. *Applied Sciences, 10*(2), 707.

26. Aumann, R. J. (1974). Subjectivity and correlation in randomized strategies. *Journal of mathematical Economics, 1*(1), 67–96.
27. Deckelbaum, A. (2011). arXiv:1101.3380 [quant-ph].
28. Phoenix, S., Khan, F., & Teklu, B. (2020). Preferences in quantum games. *Physics Letters A, 384*(15), 126299.
29. Szopa, M. (2021). Pareto efficiency of mixed quantum strategy equilibria. *Entropy, 23*, 506. https://doi.org/10.3390/e23050506
30. Ozdogan, A., & Saglam, I. (2021). Correlated Equilibrium Under Costly Disobedience. MRPA archive, March 2020.
31. Pinheiro, A. R. C., Souza, C. E. R., Caetano, D. P., Huguenin, J. A. O., Schmidt, A. G. M., & Khoury, A. Z. (2013). Vector vortex implementation of a quantum game. *Journal of the Optical Society of America B, 30*, 3210–3214.
32. Balthazar, W. F., Passos, M. H. M., Schmidt, A. G. M., Caetano, D. P., & Huguenin, J. A. O. (2015). Experimental realization of the quantum duel game using linear optical circuits. *Journal of Physics B: Atomic, Molecular and Optical Physics, 48*, 165505.

The Probability of Being Better or Worse Off, and by How Much, Depending on Experimental Conditions with Skew Normal Populations

Tingting Tong, Tonghui Wang, David Trafimow, and Cong Wang

Abstract Much research in the economics and social sciences is based on non-meaningful units. A consequence is that substantive researchers often have difficulty in interpreting their data. If there is a one-point difference between the means in two conditions on, say, subjective well-being, it need not be clear how to interpret that difference. In addition, many researchers have criticized the typical conversion to standardized effect size indices, expressed as standard deviation units, such as Cohen's d. In contrast, we take a completely different approach that features conversion into a probability that a randomly selected person from one population will score higher on the dependent variable than a randomly selected person from the other population. Because most distributions are skewed, the mathematics to be presented fall under the large umbrella of skew normal distributions rather than under the smaller umbrella of normal distributions. Two real data examples are given for illustration of our main results.

1 Introduction

A researcher randomly assigns people to an experimental condition designed to increase subjective well-being or a control condition. Consistent with expectations, mean subjective well-being in the experimental condition exceeds mean subjective well-being in the control condition, with a medium effect size (Cohen's $d = 0.50$). How should the researcher interpret the finding?

T. Tong · T. Wang (✉) · D. Trafimow
New Mexico State University, Las Cruces, NM 88003, USA
e-mail: twang@nmsu.edu

T. Tong
e-mail: tongtt@nmsu.edu

D. Trafimow
e-mail: dtrafimo@nmsu.edu

C. Wang
University of Nebraska Omaha, Omaha, NE 68182, USA
e-mail: congwang@unomaha.edu

S. Sriboonchitta et al. (eds.), *Credible Asset Allocation, Optimal Transport Methods, and Related Topics*, Studies in Systems, Decision and Control 429,
https://doi.org/10.1007/978-3-030-97273-8_18

At present, it is not necessarily clear how to interpret the finding. Although it might be reasonable to interpret higher scores as indicating higher subjective well-being than lower scores, it is not clear what a point of difference on the subjective well-being scale actually means. And although the effect size indicates that the difference in means is half a standard deviation unit, it also need not be clear whether that is impressive, unimpressive, or somewhere between, though [7] famously characterized it as a medium size effect. The difficulties in interpreting existing effect sizes indices have been well-documented (e.g., [4, 5, 9, 14, 15, 17, 18]).

As a quick example, consider again the subjective well-being experiment. Consistent with researchers' hopes, the mean score on the subjective well-being scale is higher for the experimental than control condition, with a medium effect size. How should the researchers interpret the finding given the lack of clarity about what subjective well-being scores actually mean. For example, what does it mean for a patient to circle a "2" on the subjective well-being scale? An important problem is that there is no way to know how much more likely a person randomly selected from the experimental group is to experience more subjective well-being than a person randomly selected from the control group.

But suppose it were possible to express the effect as a probability. With respect to subjective well-being, where higher scores are considered better than lower scores, suppose it were possible to express the difference in conditions as a probability that a randomly selected person from the experimental population is better off than a randomly selected person from the control population. Such an expression of effect size as a probability would have a clear interpretation, in contrast to a difference in means, Cohen's d, or other current ways of expressing effect sizes, whether standardized or not.

To our knowledge, there is only one other article that has considered this issue. In a praiseworthy article, Demidenko [8] assumed normally distributed populations and provided relevant mathematics for determining the probability that a randomly selected person from one normally distributed population would exceed a randomly selected person from another normally distributed population. Although we consider this progress, the progress is limited because very few populations are normally distributed, and there have been several systematic demonstrations that most distributions are skewed (e.g., [3, 11, 16]). It would be an advance to determine how to calculate the probability that a randomly selected person from a skew normal population scores higher than a randomly selected person from another skew normal population. Furthermore, as there may be differences in the parameters of the two skew normal populations, the mathematics should be sufficiently flexible to allow for those differences. Therefore, our goal is to provide the relevant mathematics, with illustrative figures and implications. An important implication is that it is easy to go away assuming normality if there is even slight skew normality.

2 Skew Normal Distributions

Before exploring the implications of skew normal probability and cumulative distribution functions, it is worthwhile to attempt a quick explanation of how skew normal distributions compare to normal distributions, to render the following discussion more intuitive. Whereas normal distributions have two parameters, mean μ and standard deviation σ; skew normal distributions have three parameters. The location ξ replaces the mean, the scale ω replaces the standard deviation, and there is a shape (or skewness) parameter α. When the shape parameter equals 0 the distribution is normal, and the location equals the mean and the scale equals the standard deviation. But when the shape parameter does not equal 0, the location does not equal the mean and the scale does not equal the standard deviation. In symbols, when $\alpha = 0$, $\xi = \mu$ and $\omega = \sigma$; but when $\alpha \neq 0$, $\xi \neq \mu$ and $\omega \neq \sigma$. The family of normal distributions is a subset of the family of skew normal distributions. Thus, the family of skew normal distributions is more generally applicable than the family of normal distributions.

Let us now derive the probability that a randomly chosen person from one skew normal population has a higher score on the dependent variable than a randomly chosen person from another skew normal population. We will assume dependent populations at any level of correlation between -1 and +1, with independent populations (correlation = 0) as a special case.

Definition 1 ([2]) A random vector $\mathbf{X} = (X_1, \cdots, X_k)'$ is said to have an k-dimensional multivariate skew normal distribution with the vector of location parameter $\boldsymbol{\mu} = (\mu_1, \ldots, \mu_k)' \in \Re^k$, the scale parameter of the positive definite matrix Σ, and the vector of skewness (shape) parameters $\boldsymbol{\alpha} = (\alpha_1, \ldots, \alpha_k)'$, denoted as $\boldsymbol{X} \sim SN_k(\boldsymbol{\mu}, \Sigma, \boldsymbol{\alpha})$, if its density function (pdf) is given by

$$f_{\mathbf{x}}(\mathbf{x}) = 2\phi_k(\mathbf{x}; \boldsymbol{\mu}, \Sigma)\Phi\left(\boldsymbol{\alpha}'\Sigma^{-1/2}(\mathbf{x} - \boldsymbol{\mu})\right), \tag{1}$$

where $\phi_k(\mathbf{x}; \boldsymbol{\mu}, \Sigma)$ is the density of the k-dimensional multivariate normal distribution $N_k(\boldsymbol{\mu}, \Sigma)$ with mean $\boldsymbol{\mu}$ and covariance matrix Σ, and $\Phi(z)$ is the cumulative distribution function (cdf) of the standard normal random variable $Z \sim N(0, 1)$. Here $\mathbf{x}'$ is the transpose of the the vector $\mathbf{x} \in \Re^{\mathbf{k}}$.

2.1 *Calculation of* $\delta_{ab} = P(Z > 0)$ *in Independent Distributions*

As we mentioned above, Let $X \sim SN(\xi_1, \omega_1, \alpha_1)$, $Y \sim SN(\xi_2, \omega_2, \alpha_2)$ and we assume that X and Y are independent. We want to obtain the distribution of the linear combination, $Z = X + aY + b$, of X and Y for specified $a \in \Re$ and $b \in \Re$ and then calculate $\delta_{ab} = P(Z > 0)$. We need to find the distribution of Z first, which is given below.

Theorem 1 *Let* $X \sim SN(\xi_1, \omega_1^2, \alpha_1)$, $Y \sim SN(\xi_2, \omega_2^2, \alpha_2)$. *First we assume that two skew normal populations are independent. Then the probability density function (pdf) of* $Z = X + aY + b$ *is*

$$f_Z(z) = c\phi(z; \nu, \tau^2)\Phi_2[B(z-\nu); \mathbf{0}_2, \Delta], \tag{2}$$

where

$$c = \Phi_2^{-1}(\mathbf{0}_2; \mathbf{0}_2, \Delta + \tau^2 BB'), \quad \nu = \xi_1 + a\xi_2 + b, \quad \tau^2 = \omega_1^2 + a^2\omega_2^2, \quad B = \mathbf{b}'/\tau^2,$$

$$\mathbf{0}_2 = \begin{pmatrix} 0 \\ 0 \end{pmatrix}, \qquad \Delta = \begin{pmatrix} 1+\alpha_1^2 & 0 \\ 0 & 1+\alpha_2^2 \end{pmatrix} - \frac{\mathbf{bb}'}{\tau^2}, \qquad \mathbf{b} = \begin{pmatrix} \alpha_1\omega_1 \\ a\alpha_2\omega_2 \end{pmatrix}.$$

Here $\phi(z; \nu, \tau^2)$ *is the pdf of the normal distribution* $N(\nu, \tau^2)$ *with mean* ν *and variance* τ^2, *and* $\Phi_2(\mathbf{u}; \mathbf{0}_2, \Delta)$ *is the cumulative distribution function (cdf) of the bivariate normal distribution with mean vector* $\mathbf{0}_2$ *and covariance* Δ.

Proof The joint pdf of $X \sim SN(\xi_1, \omega_1, \alpha_1)$ and $Y \sim SN(\xi_2, \omega_2, \alpha_2)$ is

$$f(x, y) = 4\phi(x; \xi_1, \omega_1^2)\phi(y; \xi_2, \omega_2^2)\Phi\left(\alpha_1\frac{x-\xi_1}{\omega_1}\right)\Phi\left(\alpha_2\frac{y-\xi_2}{\omega_2}\right).$$

Let $Z = X + aY + b$, $Y = Y$, then $X = Z - aY - b$, $Y = Y$ and it is easy to see that the Jacobian J of this transformation is 1. Thus, we obtain the pdf of Z as

$$f_Z(z) = \int_{-\infty}^{\infty} f_X(z - ay - b) f_Y(y) dy,$$

which is

$$\begin{aligned} f_Z(z) = 4 &\int_{-\infty}^{\infty} \phi(z - ay - b; \xi_1, \omega_1^2)\phi(y; \xi_2, \omega_2^2)\Phi\left(\alpha_1\frac{z-ay-b-\xi_1}{\omega_1}\right) \\ &\Phi\left(\alpha_2\frac{y-\xi_2}{\omega_2}\right) = \frac{2}{\pi\omega_1\omega_2}\int_{-\infty}^{\infty} e^{-\left[\frac{(z-ay-b-\xi_1)^2}{2\omega_1^2} + \frac{(y-\xi_2)^2}{2\omega_2^2}\right]} \\ &\Phi_2\left[\left(\alpha_1\frac{z-ay-b-\xi_1}{\omega_1}, \alpha_2\frac{y-\xi_2}{\omega_2}\right)'; \mathbf{0}, I_2\right] dy \\ &= \frac{4}{\omega_*}\phi(z; \xi_1 + a\xi_2 + b, \omega_1^2 + a^2\omega_2^2)\int_{-\infty}^{\infty} \phi(y; c, \omega_*^2) \\ &\times \Phi_2\left[\left(\alpha_1\frac{z-ay-b-\xi_1}{\omega_1}, \alpha_2\frac{y-\xi_2}{\omega_2}\right)'; \mathbf{0}, I_2\right] dy, \end{aligned} \tag{3}$$

where

$$\omega_* = \frac{\omega_1 \omega_2}{\sqrt{\omega_1^2 + a^2\omega_2^2}} \quad \text{and} \quad c = \frac{a\omega_2^2(z - b - \xi_1) + \omega_1^2 \xi_2^2}{\omega_1^2 + a^2\omega_2^2}.$$

Let $T = \frac{y-c}{\omega_*}$. Then Eq. (3) will be

$$f_Z(z) = 4\phi(z; \xi_1 + a\xi_2 + b, \omega_1^2 + a^2\omega_2^2)\times$$
$$E_T\left\{\Phi_2\left[\left(\alpha_1 \frac{z - a(t\omega_* + c) - \xi_1 - b}{\omega_1}, \alpha_2 \frac{t\omega_* + a + c - \xi_2}{\omega_2}\right)'; \mathbf{0}, I_2\right]\right\} \tag{4}$$

with $T \sim N(0, 1)$. If we denote $X_1 = \alpha_1 \frac{z-a(t\omega_*+c)-\xi_1-b}{\omega_1}$ and $X_2 = \alpha_2 \frac{t\omega_*+a+c-\xi_2}{\omega_2}$, then $E_T(\dot{})$ in Eq. (4) can be simplified as

$$\begin{aligned} E_T\left[\Phi_2(X_1,\ X_2)', \mathbf{0}, I_2\right] &= E_T[P(U_1 \le X_1,\ U_2 \le X_2 | X_1, X_2)] \\ &= P(U_1 \le X_1,\ U_2 \le X_2) \\ &= P(U_1 - X_1 \le 0,\ U_2 - X_2 \le 0), \end{aligned} \tag{5}$$

where U_1, U_2 and T are independent standard normal random variables. Note that $E(U_i - X_i) = -E(X_i)$ and $Var(U_i - X_i) = 1 + Var(X_i)$ for $i = 1,\ 2$. Additionally,

$$Cov(U_1 - X_1,\ U_2 - X_2) = Cov(X_1,\ X_2) = \frac{-a\alpha_1\alpha_2\omega_*^2}{\omega_1\omega_2}.$$

Therefore, the joint distribution of $(U_1 - X_1,\ U_2 - X_2)'$ is

$$(U_1 - X_1,\ U_2 - X_2)' \sim N_2(\boldsymbol{\mu},\ \Sigma_J),$$

where

$$\boldsymbol{\mu} = \begin{pmatrix} -\alpha_1 \frac{z-ac-\xi_1-b}{\omega_1} \\ -\alpha_2 \frac{c-\xi_2}{\omega_2} \end{pmatrix} \quad \text{and} \quad \Sigma_J = \begin{pmatrix} 1 + \alpha_1^2 \frac{a^2\omega_*^2}{\omega_1^2} & -\frac{a\alpha_1\alpha_2\omega_*^2}{\omega_1\omega_2} \\ -\frac{a\alpha_1\alpha_2\omega_*^2}{\omega_1\omega_2} & 1 + \alpha_2^2 \frac{\omega_*^2}{\omega_2^2} \end{pmatrix}.$$

Then Eq. (5) can be written to be

$$E_T\left[\Phi_2(X_1,\ X_2)', \mathbf{0}, I_2\right] = \Phi_2(-\boldsymbol{\mu}; \mathbf{0}, \Sigma_J),$$

and therefore, Eq. (3) will be

$$f_{1,2}(z) = 4\phi(z; \xi_1 + a\xi_2 + b, \omega_1^2 + a^2\omega_2^2)\Phi_2[\boldsymbol{B}(z - (\xi_1 + a\xi_2 + b)); \mathbf{0}_2, \Sigma_J], \tag{6}$$

where $\boldsymbol{B} = \left(\frac{\alpha_1\omega_1}{\omega_1^2+a^2\omega_2^2}, \frac{a\alpha_2\omega_2}{\omega_1^2+a^2\omega_2^2}\right)'$.

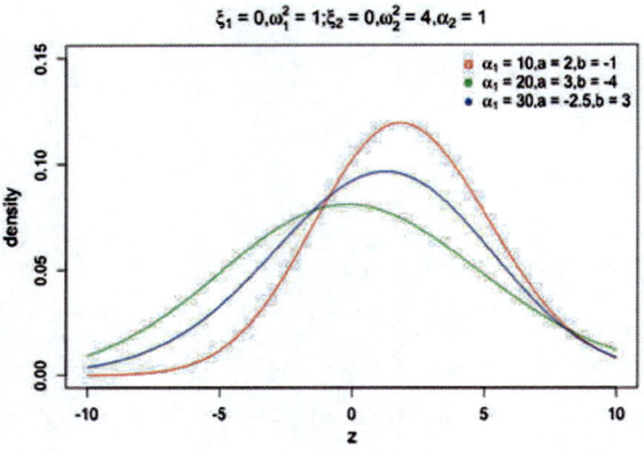

Fig. 1 The density of Z when location parameters of both populations equal 0, the scale parameters for first and second population equal 1 and 2 respectively. The shape parameters for second population are 1 and for first population equal 10, 20, 30 with $a = 2, b = -1$, $a = 3, b = -4$ and $a = -2.5, b = 3$. The corresponding probability $P(Z > 0)$ is 0.725462, 0.50538 and 0.60111, respectively

Note that the pdf of Z given in Eq. (2) is the special case of the closed skew normal density. For details of closed skew normal distributions, see [10, 22].

Comment. For the case where $\alpha_1 = \alpha_2 = 0$, we know that $Z \sim N(\xi_1 - \xi_2, \omega_1^2 + a^2\omega_2^2)$, the normal distribution with mean $\xi_1 - \xi_2$ and variance $\omega_1^2 + a^2\omega_2^2$. Assume that $\omega_1 = \omega_2 = \omega$ and $a = -1$, then

$$Z^* = \frac{Z}{\omega} \sim N\left(\frac{\theta}{\sqrt{2}}, , 1\right),$$

where $\theta = (\xi_1 - \xi_2)/\omega$ is the Cohen's effect size. Thus, based on a sample of size n from this population, we can make inferences on θ. Therefore Theorem 1 can be used for the inference of Cohen's effect size θ under skew normal settings.

Now we can compute the probability $P(Z > 0)$, which is given by

$$\delta_{ab} = Pr(Z > 0) = \int_0^\infty c\phi(z; \nu, \tau^2)\Phi_2[B(z - \nu); \mathbf{0}_2, \Delta]dz. \tag{7}$$

Density curves of Z for different parameters and their corresponding $P(Z > 0)$ are given in Figs. 1 and 2, respectively. These values can be instantiated into the equations presented earlier, but there is an easier way too. We provide a freely available online calculator at https://probab.shinyapps.io/inde_prob/.

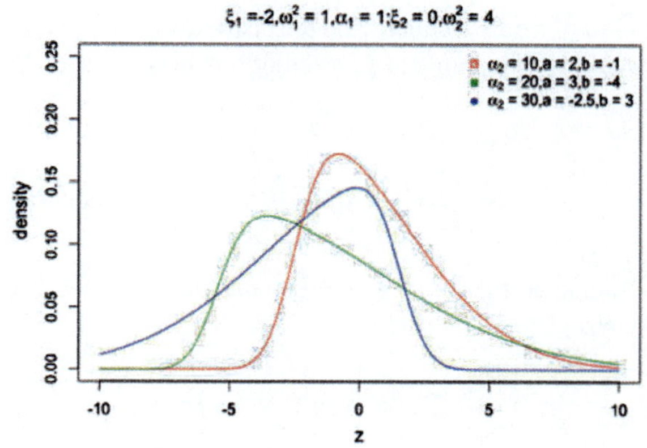

Fig. 2 The density of Z when location parameters for first and second population equal –2 and 0 respectively, the scale parameters for first and second population equal 1 and 2 respectively. The shape parameters for first population are 1 and for second population equal 10, 20 and 30 with $a = 2, b = -1, a = 3, b = -4$ and $a = -2.5, b = 3$. The corresponding probability $P(Z > 0)$ is 0.997147, 0.809657 and 0.727792, respectively

2.2 *Calculation of $\delta_{\rho ab} = P(U > 0)$ in Dependent Distributions*

Now we consider the bivariate skew normal random vector $\mathbf{X} = (X_1, X_2)' \sim SN_2(\boldsymbol{\xi}, \Sigma, \boldsymbol{\alpha})$, with location parameter $\boldsymbol{\xi}$, scale parameter Σ, and skewness parameter $\boldsymbol{\alpha}$ given by

$$\boldsymbol{\xi} = \begin{pmatrix} \xi_1 \\ \xi_2 \end{pmatrix}, \quad \Sigma = \begin{pmatrix} \omega_1^2 & \rho\omega_1\omega_2 \\ \rho\omega_1\omega_2 & \omega_2^2 \end{pmatrix}, \quad \boldsymbol{\alpha} = \begin{pmatrix} \alpha_1 \\ \alpha_2 \end{pmatrix}.$$

Then we have the following result and its proof is given in Appendix.

Theorem 2 *Let $\mathbf{X} \sim SN_2(\boldsymbol{\xi}, \Sigma, \boldsymbol{\alpha})$ given above and consider $U = X_1 + aX_2 + b$ with $a, b \in \Re$. Then the pdf of U is*

$$\begin{aligned} f_U(u) &= 2\phi(u; \xi_1 + a\xi_2 + b, \omega_1^2 + a^2\omega_2^2 + 2a\rho\omega_1\omega_2) \\ &\quad \times \Phi\left[\alpha_* \frac{(u - \xi_1 - a\xi_2 - b)}{(\omega_1^2 + a^2\omega_2^2 + 2a\rho\omega_1\omega_2)^{1/2}}\right], \end{aligned} \tag{8}$$

where

$$\alpha_* = \frac{\alpha_1 d_1 + \alpha_2 d_2}{\sqrt{(1 - \alpha_1^2 - \alpha_2^2)k^2\sigma^2 - (\alpha_1 d_1 + \alpha_2 d_2)^2}}$$

with $d_1 = \omega_1^2 + \omega_1\omega_2(\sqrt{1-\rho^2} + a\rho)$, $d_2 = a\omega_2^2 + \omega_1\omega_2(a\sqrt{1-\rho^2} - \rho)$, $k = \sqrt{\omega_1^2 + \omega_2^2 + 2\sqrt{1-\rho^2}\omega_1\omega_2}$, *and* $\sigma^2 = \omega_1^2 + a^2\omega_2^2 + 2a\rho\omega_1\omega_2$.

Proof Let $\boldsymbol{c} = (1, a)'$, we are trying to find the distribution of $U = X_1 + aX_2 + b = \boldsymbol{c}'\boldsymbol{X} + b$. First we derive the moment generating function (mgf) of U:

$$\begin{aligned} M_U(t) &= E[\exp(t\boldsymbol{c}'\boldsymbol{X} + tb)] \\ &= 2\exp\left\{t(\boldsymbol{c}'\boldsymbol{\xi} + b) + \frac{1}{2}t^2\boldsymbol{c}'\Sigma\boldsymbol{c}\right\}\Phi(\boldsymbol{\delta}'\Sigma^{1/2}\boldsymbol{c}t). \end{aligned} \tag{9}$$

It is easy to obtain that $\boldsymbol{c}'\boldsymbol{\xi} + b = \xi_1 + a\xi_2 + b$ and $\sigma^2 = \boldsymbol{c}'\Sigma\boldsymbol{c} = \omega_1^2 + a^2\omega_2^2 + 2a\rho\omega_1\omega_2$. From Wang et al. [21],

$$\Sigma^{1/2} = \begin{bmatrix} \frac{\omega_1^2+\sqrt{1-\rho^2}\omega_1\omega_2}{k} & \frac{\rho\omega_1\omega_2}{k} \\ \frac{\rho\omega_1\omega_2}{k} & \frac{\omega_2^2+\sqrt{1-\rho^2}\omega_1\omega_2}{k} \end{bmatrix},$$

where

$$k = \sqrt{\omega_1^2 + \omega_2^2 + 2\sqrt{1-\rho^2}\omega_1\omega_2}.$$

Thus, $\delta_* = \boldsymbol{\delta}'\Sigma^{1/2}\boldsymbol{c}(\boldsymbol{c}'\Sigma\boldsymbol{c})^{-1/2}$. By Eq. (1), we obtain that $U \sim SN(\xi_1 + a\xi_2 + b, \sigma^2, \alpha_*)$ with

$$\begin{aligned} \alpha_* &= \frac{\delta_*}{\sqrt{1-\delta_*^2}} \\ &= \frac{\alpha_1 d_1 + \alpha_2 d_2}{\sqrt{(1-\alpha_1^2-\alpha_2^2)k^2\sigma^2 - (\alpha_1 d_1 + \alpha_2 d_2)^2}}, \end{aligned} \tag{10}$$

where $d_1 = \omega_1^2 + \omega_1\omega_2(\sqrt{1-\rho^2} + a\rho)$ and $d_2 = a\omega_2^2 + \omega_1\omega_2(a\sqrt{1-\rho^2} + \rho)$. Then the density of U is given by

$$f_U(u) = 2\phi(u; \boldsymbol{c}'\boldsymbol{\xi} + b, \sigma^2)\Phi\left[\alpha_*\frac{(u - \boldsymbol{b}'\boldsymbol{\xi} - b)}{\sigma}\right] \tag{11}$$

$$\begin{aligned} &= 2\phi(u; \xi_1 + a\xi_2 + b, \omega_1^2 + a^2\omega_2^2 + 2a\rho\omega_1\omega_2) \\ &\times \Phi\left[\alpha_*\frac{(u - \xi_1 - a\xi_2 - b)}{(\omega_1^2 + a^2\omega_2^2 + 2a\rho\omega_1\omega_2)^{1/2}}\right]. \end{aligned} \tag{12}$$

The probability $\delta_{\rho ab} = P(U > 0)$ is given by

$$\delta_{\rho ab} = P(U > 0) = \int_0^\infty f_U(u)du. \tag{13}$$

Density curves of U for different parameters and their corresponding probability $P(U > 0)$ are given in Figs. 3 and 4, respectively. For example, in Fig. 3, we can see

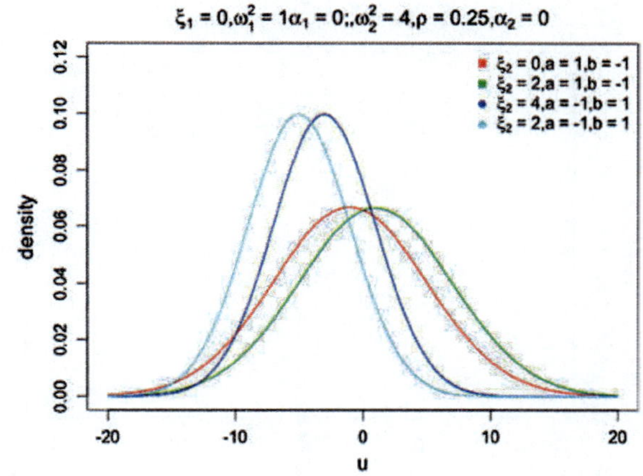

Fig. 3 The density of U when location parameters of the first population equal 0, the scale parameters for first and second population equal 1 and 2 respectively. The shape parameters for both populations equal 0 and the location parameters of the the second population equal 0, 2, 4, 6 with $a = 1, b = -1$ and $a = -1, b = 1$. The corresponding probability $P(U > 0)$ is 0.433816, 0.566184, 0.226627 and 0.10565, respectively

that the density curves are affected by the location parameter $\xi_2 = 0, 2, 4, 2$ with other parameters specified, and $\rho = 0.25$, together with their corresponding probabilities $P(U > 0) = 0.433816, 0.566184, 0.226627, 0.10565$, respectively. We provide a different online calculator, still freely available online at https://probab.shinyapps.io/ProbU/.

2.3 *Estimation of* $\delta_{ab} = P(Z > 0)$ *and* $\delta_{\rho ab} = P(U > 0)$

The above two theorems imply that if the location, scale and shape parameters are known in two populations, one can calculate the probabilities $\delta_{ab} = P(Z > 0)$ and $\delta_{\rho ab} = P(U > 0)$. But in most situations in real life the population parameters are not known. Thus estimating δ_{ab} and $\delta_{\rho ab}$ is necessary to be studied. The parametric estimation of $P(X > Y)$ for normal distributions in the context of probabilistic environmental risk assignment was discussed by Jacobs et al. (2015). Here, we consider the method of moment estimator (MME) and maximum likelihood estimator (MLE). The estimators obtained in this two ways are denoted by $\hat{\delta}_{ab}^{MME}$ and $\hat{\delta}_{ab}^{MLE}$ respectively.

Method of moment estimator (MME) Equation (14), below, relates location to mean and scale to standard deviation.

$$\xi = \mu - \sqrt{\frac{2}{\pi}}\delta\omega \quad \text{and} \quad \omega^2 = \sigma^2\left(1 - \frac{2}{\pi}\delta^2\right)^{-1}. \tag{14}$$

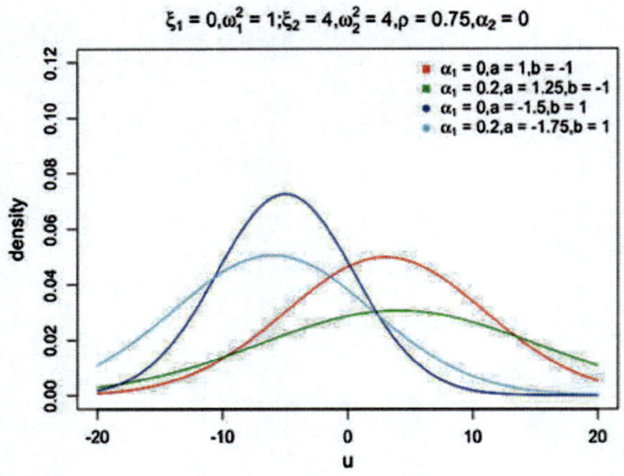

Fig. 4 r of U when location parameters of the first and second population equal 0 and 4, the scale parameters for first and second population equal 1 and 2 respectively. The shape parameter for the second populations equal 0 and for the first population equal 0, 0.2, 0, 0.2 with $a = 1$, $b = -1$, $a = 1.25$, $b = -1$, $a = -1.5$, $b = 1$ and $a = -1.75$, $b = 1$. The corresponding probability $P(U > 0)$ is 0.64617, 0.666848, 0.181651 and 0.224808, respectively

where $\delta = \alpha/\sqrt{1-\alpha^2}$. To obtain parameter estimators from samples, it is necessary to obtain an estimate of delta δ, which is the moment estimate $\hat{\delta}$ in Eq. (15) below.

$$|\hat{\delta}| = \sqrt{\frac{\pi}{2}\frac{|\hat{\gamma}_1|^{\frac{2}{3}}}{|\hat{\gamma}_1|^{\frac{2}{3}} + ((4-\pi)/2)^{\frac{2}{3}}}} \tag{15}$$

where $\hat{\gamma}_1$ is the sample skewness and the sign of $\hat{\delta}$ is the same as the sign of $\hat{\gamma}_1$. In turn, it is easy to obtain an estimate of the shape parameter α using Eq. (16):

$$\hat{\alpha} = \frac{\hat{\delta}}{\sqrt{1-\hat{\delta}^2}}. \tag{16}$$

Rewriting Eq. (14) in terms of sample estimates, as opposed to population parameters, renders Eq. (17).

$$\hat{\xi} = \bar{X} - \sqrt{\frac{2}{\pi}}\hat{\delta}\hat{\omega} \quad \text{and} \quad \hat{\omega}^2 = S^2\left(1 - \frac{2}{\pi}\hat{\delta}^2\right)^{-1}. \tag{17}$$

Then we can obtain the $\hat{\delta}_{ab}^{MME}$ by substituting the MMEs of ξ_1,ξ_2,ω_1,ω_2,α_1 and α_2 in Eq. 7.

Maximum likelihood estimator (MME) The log-likelihood function for the skew normal distribution is given by

$$\log(L) = n\log(\frac{2}{\omega}) - \frac{n}{2}\log(2\pi) - \frac{1}{2}\sum_{i=1}^{n}\left(\frac{x_i-\xi}{\omega}\right)^2 + \sum_{i=1}^{n} ln\left(\Phi(\alpha\frac{x_i-\xi}{\omega})\right) \tag{18}$$

To derive the MLEs of the parameters in skew normal distribution, we take the partial derivatives of the lnL functions with respect to parameters of interest and equal them to zero. Then the corresponding MLEs are obtained by solving the following equations:

$$\begin{aligned}
\frac{lnL}{\xi} &= \sum_{i=1}^{n}\left(\frac{x_i-\xi}{\omega}\right) - \alpha\sum_{i=1}^{n}\frac{\phi(\alpha\frac{x_i-\xi}{\omega})}{\Phi(\alpha\frac{x_i-\xi}{\omega})} = 0,\\
\frac{lnL}{\omega} &= -n + \sum_{i=1}^{n}\left(\frac{x_i-\xi}{\omega}\right)^2 - \alpha\sum_{i=1}^{n}\frac{\phi(\alpha\frac{x_i-\xi}{\omega})}{\Phi(\alpha\frac{x_i-\xi}{\omega})}\left(\frac{x_i-\xi}{\omega}\right) = 0,\\
\frac{lnL}{\alpha} &= \sum_{i=1}^{n}\frac{\phi(\alpha\frac{x_i-\xi}{\omega})}{\Phi(\alpha\frac{x_i-\xi}{\omega})}\left(\frac{x_i-\xi}{\omega}\right) = 0.
\end{aligned} \tag{19}$$

By the invariance property of MLEs, we obtain $\hat{\delta}_{ab}^{MLE}$ by substituting the MLEs of ξ_1,ξ_2,ω_1,ω_2,α_1 and α_2 in Eq. (7).

To derive $\hat{\delta}_{\rho ab}^{MME}$ and $\hat{\delta}_{\rho ab}^{MLE}$ under matched data setting, instead of estimating ξ_1,ξ_2,ω_1,ω_2,α_1 and α_2 we estimate $\xi_* = \xi_1 + a\xi_2 + b$, $\sigma^2 = \mathbf{c}'\Sigma\mathbf{c} = \omega_1^2 + a^2\omega_2^2 + 2a\rho\omega_1\omega_2$ and

$$\begin{aligned}
\alpha_* &= \frac{\delta_*}{\sqrt{1-\delta_*^2}} \\
&= \frac{\alpha_1 d_1 + \alpha_2 d_2}{\sqrt{(1-\alpha_1^2-\alpha_2^2)k^2\sigma^2 - (\alpha_1 d_1 + \alpha_2 d_2)^2}}
\end{aligned} \tag{20}$$

by using the same methods as in the independent setting. Then by substituting the MMEs and MLEs of ξ_*, σ^2 and α_* in Eq. (13), we can get $\hat{\delta}_{\rho ab}^{MME}$ and $\hat{\delta}_{\rho ab}^{MLE}$.

3 Implications

The foregoing mathematics imply a variety of potential ways to advance the social sciences. These are explored below with a focus on independent samples from true experiments where participants are randomly assigned to conditions.

3.1 Converting Nonmeaningful Units to Meaningful Probabilities

The foregoing mathematics imply that for any two populations where the locations, scales, and shapes are known, it is possible to convert to a meaningful effect size with respect to probability. The researcher can know the probability that a randomly selected person from one population will receive a higher score on the dependent variable than a randomly selected person from the other population. Thus, even if there are nonmeaningful units, such as subjective well-being units, attitude units, pain units, and so on; these can be converted into probabilities. Or if the population parameters are not known, as is usually the case, it is possible to calculate estimators based on sample mean, standard deviation, and skewness.

To obtain estimators, it is merely necessary to use one's favorite statistical program to obtain the mean, standard deviation, and skewness for both samples. Although there are multiple ways to calculate skewness, that will not be a present focus.

Let us return to the subjective well-being experiment but with made-up sample data. Suppose that the means in the experimental and control groups are 0.56 and 0, respectively; the standard deviations are 0.83 and 1.00, respectively; and the skewness values are 0.137 and 0, respectively. Using the foregoing equations implies that the location is 0 and the scale is 1 in both conditions; and the shape is 1 in the experimental condition and 0 in the control condition, respectively: Cohen's d is 0.62. Thus, going by normal estimators, there is a slightly better than medium size effect whereas going by skew normal estimators, there is no effect. The present mathematics implies that the best estimate of the probability that a randomly selected person from the experimental group will have more subjective well-being than a randomly selected person from the control group is 0.67.

It is interesting to parse the implications of the different effect sizes. The effect size based on normal estimators is 0.62, which implies that the treatment worked but the effect size based on skew normal estimators is 0, which implies that the treatment did not work. However, even using skew normal estimators, we also have seen that a randomly selected person from the experimental condition has an estimated probability of 0.67 of doing better than a randomly selected person from the control condition, which suggests that perhaps the experiment worked after all, even according to skew normal estimators. In a therapy situation, what likely will matter to the patient is the probability that he or she will be better off undergoing the therapy. In that case, the probability-based effect size is the one that should matter most. But if the experiment had been performed to test a hypothesis that the manipulation is supposed to shift the experimental distribution relative to the control distribution, then the location-based effect size contradicts the hypothesis. The fact that the locations are the same in both conditions indicates that the manipulation did not shift the experimental group distribution relative to the control group distribution. In essence, whether the manipulation "worked" depends on the researcher's goals. In a therapeutic context, the probability-based effect size arguably is best but in a hypothesis testing context, the location-based effect size arguably is best. The present example demonstrates the value of researchers using more than one kind of effect size.

3.2 *Equal Locations and Scales, but Not Shapes*

It is instructive to see just how much of a difference the shape parameter can make with respect to the proposed probability-based effect size, even when two populations do not differ at all with respect to location and scale. We created Fig. 5 for illustration. In Fig. 5, the probability that a randomly selected person from the first population will obtain a higher score on the dependent variable than a randomly selected person from the second population is expressed along the vertical axis as a function of the shape parameter for the first population (shape = 0, 1, 2, 3, or 4) and the shape parameter for the second population (shape = -4, -3, -2, -1, or 0). The reason the shape parameter was set at zero or at positive values for the first population, whereas it was set a 0 or at negative values for the second population, was to dramatize the effects that shape can have, even in a figure where locations and scales are the same in both conditions.

When both shapes are set at 0, then because the locations and scales were set at the same values, the two population distributions are identical. Thus, the probability expressed along the vertical axis is 0.5, which is the lowest point illustrated by Fig. 5. Following the bottom most curve in Fig. 5, note that increasing the shape parameter in the positive direction causes the probability to increase. The other curves in Fig. 5 show that the probability is maximized by increasing the shape parameter of the first population in the positive direction and increasing the shape parameter of the second population in the negative direction. Thus, even with equal locations and scales, it is possible to have extremely high probabilities. A dramatic example is when the shape parameter is 4 for the first population and -4 for the second population, with a resultant probability of 0.99999.

3.3 *Unequal Locations and Shapes, with Equal Scales*

Figure 5 showed the dramatic effects that distribution shapes can have, even when locations are equal. Figure 5 shows even more dramatic effects when locations are not equal. Figure 6 is similar to Fig. 5, but with an important exception. Specifically, although the scales equal 1 for both populations, the location is 1 for the first population and 0 for the second population. At first glance, Fig. 6 may seem similar to Fig. 5, but this is not so because the values along the vertical axis start at 0.50 in Fig. 5 but at 0.75 in Fig. 6. Thus, the difference in location strongly influences the probability expressed along the vertical axis, as does the shapes of the distributions.

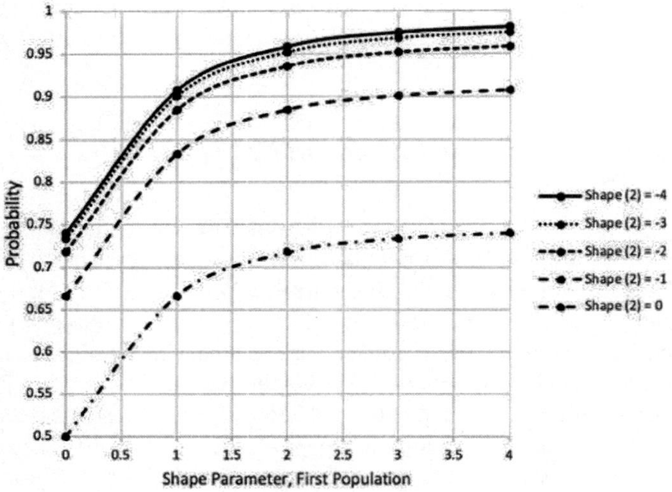

Fig. 5 The probability that a randomly selected person from the first population will obtain a higher score on the dependent variable than a randomly selected person from the second population is expressed along the vertical axis as a function of the shape parameter for the first population (shape = 0, 1, 2, 3, or 4) and the shape parameter for the second population (shape = –4, –3, –2, –1, or 0). The locations and scales of the two populations are equal

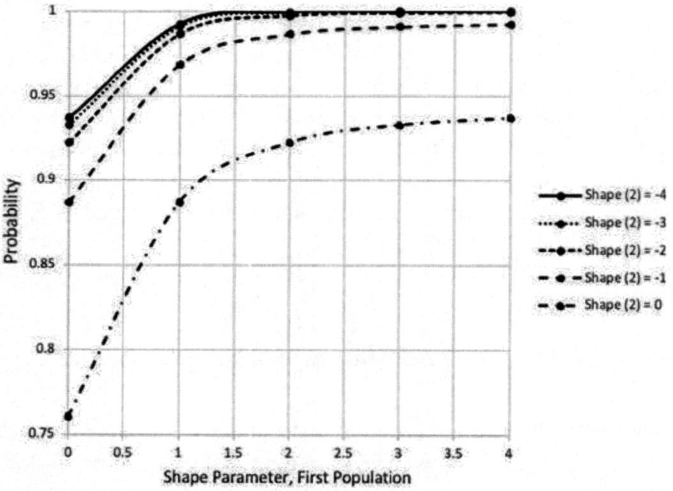

Fig. 6 The probability that a randomly selected person from the first population will obtain a higher score on the dependent variable than a randomly selected person from the second population is expressed along the vertical axis as a function of the shape parameter for the first population (shape = 0, 1, 2, 3, or 4) and the shape parameter for the second population (shape = –4, –3, –2, –1, or 0). The location of the first population is 1 and the location of the second population is zero, but the scale is 1 for both populations

3.4 The Interaction of Scale and Shape

The effects of scale sizes are more subtle. Thus, instead of having a figure analogous to Figs. 5 and 6, with a single graph, Fig. 7 includes four of them where the shape parameter for the second population is 0 (first panel), –1 (second panel), –2 (third panel), or –3 (fourth panel). Let us commence with the first panel, where the probability is expressed along the vertical axis as a function of the scale multiple along the horizontal axis, with different curves for different scale parameters for the first population. The scale multiple needs a short explanation. Because the scale parameter was always set to 1 for the second population, setting the scale parameter at 1 too for the first population implies that the scale is the same for both populations, so the multiple is 1. But setting the scale parameter at 2, 3, or 4 for the first population implies that this scale parameter is twice, thrice, or four times that of the scale parameter for the second population. Hence, the multiple ranges from 1 to 4. The first panel of Fig. 7, where the second population is normally distributed, indicates that as the scale multiple increases, the probability decreases. Thus, from the point of view of maximizing the probability, it is an advantage to minimize variation in the experimental condition. In contrast, the other panels of Fig. 7 suggest a more complicated conclusion. That is, when the shape parameter for the first population is large, increasing the scale multiple decreases the probability, consistent with the first panel. But when the shape parameter for the first population is small (e.g., 0), then increasing the scale multiple increases the probability. Hence, whether it is good or bad to increase the scale multiple depends on the shapes of the distributions involved.

3.5 Wrongly Assuming Normality

Although professional statisticians often bemoan that substantive researchers are too prone to assume that they have normally distributed populations, the fact of the matter is that substantive researchers routinely assume this. From the typical significance testing standpoint, the assumption is traditionally justified via the Central Limit Theorem. But from the present perspective the Central Limit Theorem does not get the job done because the issue is to have the most precise estimate of the probability that it is possible to obtain. And as Fig. 8 illustrates, assuming normality when there is skewness can cause the resultant probability calculation to be quite wrong. Figure 8 was created under similar conditions as Fig. 5 in that the locations and scales of the two populations were set at equal values. In the event that both shapes equal 0, so that the two populations have exactly the same distribution, the probability equals 0.5, as we have seen before. Because all the other values in Fig. 8 are larger than 0.5, it was a simple matter to subtract 0.5 from all probabilities to render the degree of wrongness in probability assuming normality when there actually is skew normality. For this reason, the point at the origin of Fig. 8, where the probability equals 0.50, is actually the correct value. That is, subtracting the correct value from itself renders

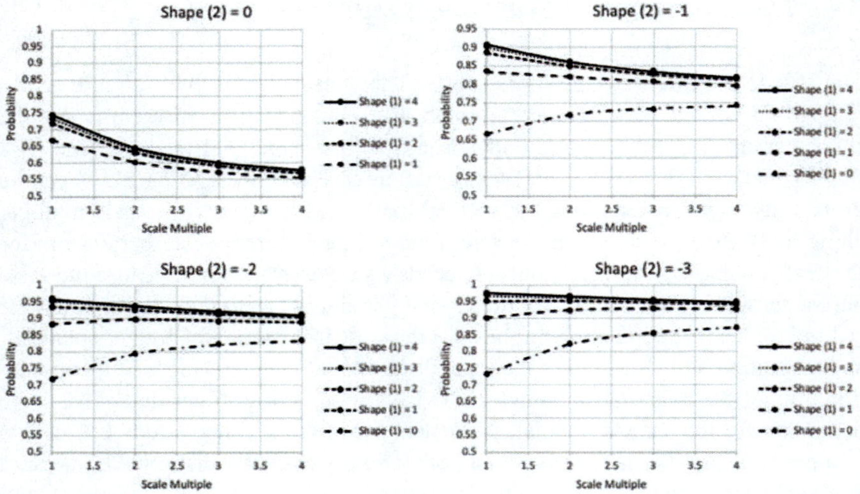

Fig. 7 The probability that a randomly selected person from the first population will obtain a higher score on the dependent variable than a randomly selected person from the second population is expressed along the vertical axis as a function of the shape parameter for the first population (shape = 0, 1, 2, 3, or 4) and the multiple by which the scale for the first population exceeds the scale for the second population (1, 2, 3, or 4). The different panels give different values for the shape parameter for the second population (shape = –3, –2, –1, or 0)

a value of 0, which is the smallest value for wrongness in probability that Fig. 8 exhibits. Figure 8 shows that as the difference in shape across the two distributions increases, so does the degree of wrongness caused by assuming normality.

3.6 The Role of Dependence

Although the foregoing mathematical work is sufficiently general to apply to dependent populations, with independent populations as a special case, our focus has thus far been on independent populations. To explore the issue of dependent populations, Fig. 9 expresses probability along the vertical axis as a function of the shape parameter of the first population along the horizontal axis, keeping the locations and shapes equal across populations, and keeping the shape parameter of the second population at zero. The unique feature of Fig. 9 is the different curves representing dependence at the level of $\rho = 0, 0.25$, or 0.75. We did not include a curve for $\rho = 0.50$ because it overlapped so much with the $\rho = 0.75$ curve that it is difficult to see in a figure. As Fig. 9 illustrates, the degree of dependence matters; but not very much, as the curves are not very far apart from each other. The greatest deviation is when $\rho = 0.25$, with the curves for when $\rho = 0$ or 0.75 quite close to each other.

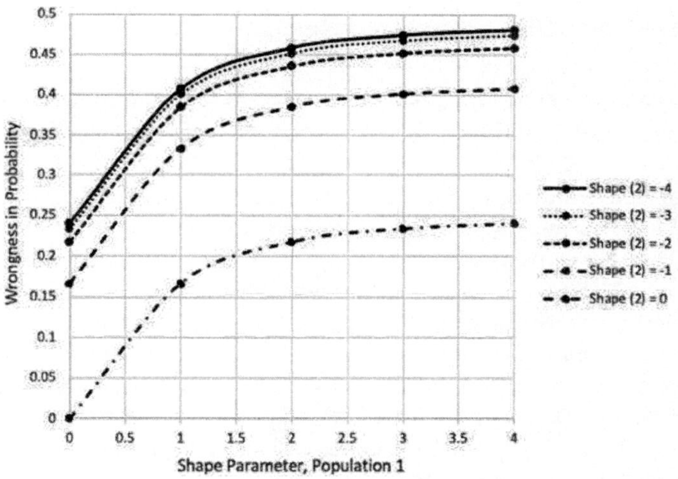

Fig. 8 Degree of wrongness in probability caused by assuming normality when there is skew normality is expressed along the vertical axis as a function of the shape parameter for the first population (shape = 0, 1, 2,, 3, or 4) and the shape parameter for the second population (shape = –4, –3, –2, –1, or 0). The locations and scales of the two populations are equal

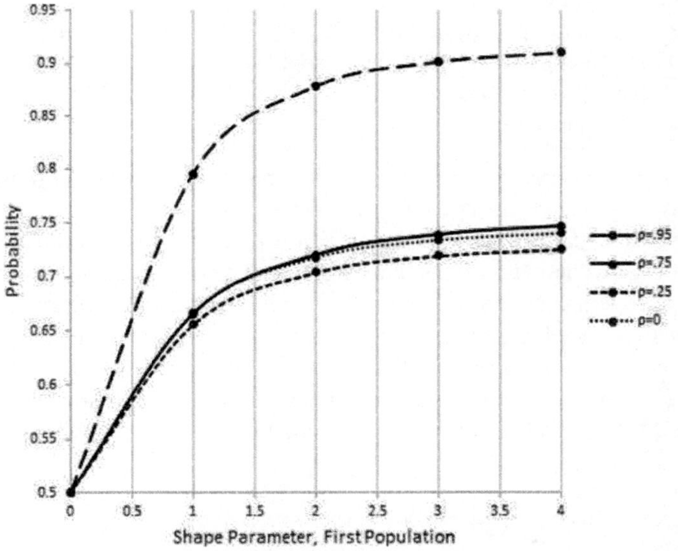

Fig. 9 The probability that a randomly selected person from the first population will obtain a higher score on the dependent variable than a randomly selected person from the second population is expressed along the vertical axis as a function of the shape parameter for the first population (shape = 0, 1, 2, 3, or 4) and the correlation between populations (0, 0.25, 0.75, or 0.95). The location of the both populations have the same locations and scales, and the shape parameter for the second population was set at zero

4 Monte Carlo Simulation and Real Data Examples

In this section, we provide the Monte Carlo (MC) simulation study following the guidelines provided in Burton et al. [6] and two real data examples to support our main results.

The simulation study is performed in R. We use the rsn function from package 'sn' (version 2.0.0) to sample from a skew normal distribution. The MLEs of $P(Z > 0)$ and $P(U > 0)$ are derived by carrying out $q = 1000$ times simulations for different sample sizes n = 50, 100, 300, 500. Here we let the corresponding MMEs be the initial values for the parameters to be optimized over when performing the MLEs in program. The estimation results of $\delta_{ab} = P(Z > 0)$ for an arbitrary set of parameters $\xi_1 = 2, w_1^2 = 1, \alpha_1 = 5, \xi_2 = 0, w_2^2 = 4, \alpha_2 = 1, a = -1, b = 0$ are shown in Table 1. The absolute bias, and standard deviations of estimators are also presented in the same table. A freely available online calculator for estimation is give here: http://feic.shinyapps.io/estim_inde.

Table 2 shows the estimation results of $\delta_{\rho ab} = P(U > 0)$ and the corresponding absolute bias and standard deviations of estimators with the values of parameters $\xi_1 = 0, w_1^2 = 1, \alpha_1 = 2, \xi_2 = 2, w_2^2 = 2, \alpha_2 = 1, \rho = 0.25, a = -1, b = 0$. A freely available online calculator for estimation is give here: https://probdiffgamma.shinyapps.io/estimator_dep/.

To illustrate how to calculate the probability that a randomly selected person in one group will score higher on the dependent variable than a randomly selected person in the other group, we provide an real data example for independent and dependent sample cases, respectively.

Example 1 For independent case, we consider Salary data (in 2011) from science (astronomy, biology, chemistry and biochemistry, computer science, geology, mathematical science and physics) and non-science (creative media, criminal justice, anthropology, art, communications studies, geography, philosophy, sociology, music, theater arts, government, journalism and mass communications, English, languages and linguistics, sociology, history) departments at College of Arts and Sciences, New Mexico State University are publicly available https://riograndefoundation.org/downloads/rgf_pr_nmsu.pdf and have distributions that are approximately skew normal (see Figs. 10 and 11 for histograms). Using the method of moment estimation, we obtain that

$$\text{science}: \quad X \sim SN(4.451107, (2.686328)^2, 1.306512)$$

and

$$\text{non-science}: \quad Y \sim SN(3.892328, (2.536494)^2, 2.199257).$$

with unit of 10, 000. That is, the estimated parameters for the salaries for faculty in science departments are \$44, 511.07, \$(26, 863.28)^2, and 1.306512 for location, scale, and shape, respectively. The estimated parameters for non-science departments are \$38, 923.28, \$(25, 364.94)^2, and 2.199257 for location, scale, and shape, respec-

Table 1 Results of estimations of $\delta_{ab} = P(Z > 0)$ with sample size $n = 50, 100, 300, 500$

n = 50	ξ_1	w_1^2	α_1	ξ_2	w_2^2	α_2	δ_{ab}
True value	2	1	5	0	4	1	0.827214
MLE	2.026176	0.958448	26.522025	0.180594	3.998715	2.272122	0.793338
|Bias|	0.026176	0.041552	21.522025	0.180594	0.001285	1.272122	0.033876
Sd.MLE	0.133713	0.140849	55.315524	0.810161	0.425222	15.410818	3.745321
n = 100	ξ_1	w_1^2	α_1	ξ_2	w_2^2	α_2	δ_{ab}
True value	2	1	5	0	4	1	0.827214
MLE	2.005763	0.992181	8.968298	0.149149	3.990593	1.037311	0.80729
|Bias|	0.005763	0.007819	3.968298	0.149149	0.009407	1.037311	0.019924
Sd.MLE	0.084613	0.097681	36.707371	0.695285	0.341006	0.948792	1.004219
n = 300	ξ_1	w_1^2	α_1	ξ_2	w_2^2	α_2	δ_{ab}
True value	2	1	5	0	4	1	0.827214
MLE	2.002854	0.995511	5.291312	0.134164	3.890977	0.943318	0.815191
|Bias|	0.002854	0.004489	0.291312	0.134164	0.109023	0.056682	0.012023
Sd.MLE	0.042161	0.050586	1.258475	0.498855	0.233332	0.530859	0.432133
n = 500	ξ_1	w_1^2	α_1	ξ_2	w_2^2	α_2	δ_{ab}
True value	2	1	5	0	4	1	0.827214
MLE	2.002686	0.99369	5.129285	0.118159	3.862159	0.927141	0.819216
|Bias|	0.002686	0.00631	0.129285	0.118159	0.137841	0.072859	0.007998
Sd.MLE	0.033373	0.039716	0.942239	0.395079	0.187855	0.415904	0.176534

tively. By typing in the parameter estimates for location, scale, and shape for both groups, and the calculator renders a probability of 0.56 for the example. That is, a randomly selected person from a science department has approximately a 55% chance of registering a higher salary than a randomly selected person from a non-science department. We also provide the density curve of $Z = X - Y$, see Fig. 12.

Example 2 For matched case, we consider the data from [19] to compare the market values and the selling prices of 45 homes from the same region. Let D be the values of sale price minus market values, we obtain the estimated distribution of D is $SN(-1.1457, 6.0164^2, 2.7646)$ by MME. The histogram and estimated density curve of D are given in Fig. 13. The pattern of the points of the scatter plot shown in Fig. 14 shows a positive linear relationship between this two variables. After calculation, we get the correlation coefficient is $\rho = 0.95$. Using the method of moment estimation, we obtain the MMEs of ξ_*, σ^2 and α_* given in Eq. (20) are $\hat{\xi_*} = 1.3967$, $\hat{\sigma}^2 = 272.9404$ and $\hat{\alpha_*} = 1.2267$. Then by substituting the MMEs and the $\rho = 0.95$ into Eq. (13), we get $\delta_{\rho ab} = P(U > 0) = 0.7747$. That is, a randomly selected observation from sale price set has approximately a 77% chance of registering a higher value than a randomly observation from the market value set.

Table 2 Results of estimations of $\delta_{\rho ab} = P(U > 0)$ with sample size $n = 50, 100, 300, 500$

n = 50	ξ_*	σ^2	α_*	$\delta_{\rho ab}$
True value	–2	2.292893	0.115073	0.215947
MLE	–2.781363	2.634482	2.218410	0.290509
|Bias|	0.781363	0.341589	2.103337	0.074562
Sd.MLE	1.104951	0.518234	19.050117	0.689110
n = 100	ξ_*	σ^2	α_*	$\delta_{\rho ab}$
True value	–2	2.292893	0.115073	0.215947
MLE	–2.673824	2.557147	0.626441	0.244653
|Bias|	0.673824	0.264254	0.511368	0.028706
Sd.MLE	0.925449	0.404528	0.75068	0.069355
n = 300	ξ_*	σ^2	α_*	$\delta_{\rho ab}$
True value	–2	2.29289	0.115073	0.215947
MLE	–2.544261	2.483328	0.478602	0.233782
|Bias|	0.544261	0.190438	0.363529	0.017835
Sd.MLE	0.689628	0.267692	0.468418	0.0475246
n = 500	ξ_*	σ^2	α_*	$\delta_{\rho ab}$
True value	–2	2.292893	0.115073	0.215947
MLE	–2.493209	2.456621	0.435145	0.230785
|Bias|	0.493209	0.163728	0.320072	0.014838
Sd.MLE	0.634892	0.227552	0.414563	0.042566

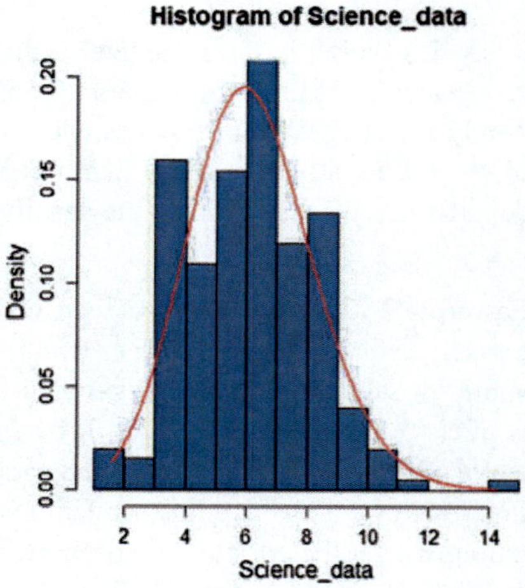

Fig. 10 Frequency histogram of the salary data (with sample size 201) from Science department with fitted skew normal density curve

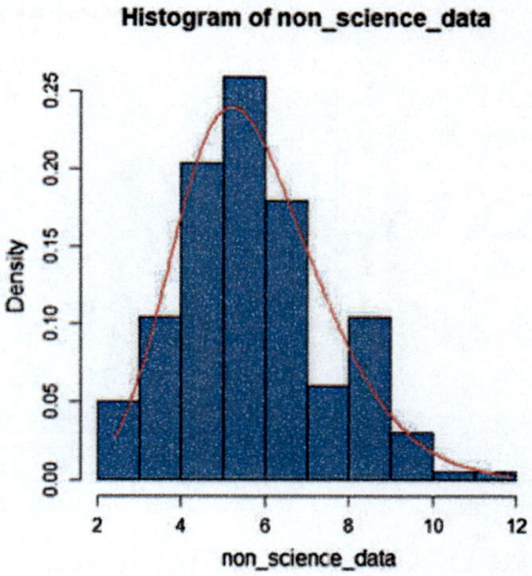

Fig. 11 Frequency histogram of the salary data (with sample size 201) from non-science department with fitted skew-normal density curve

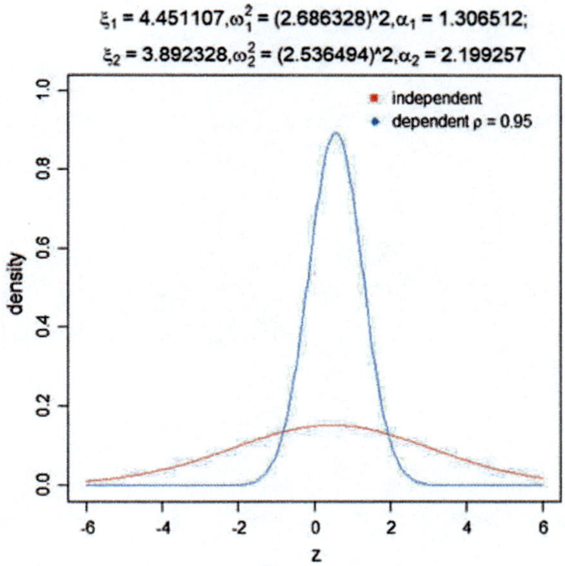

Fig. 12 The density curve of Z (in independent case) and density curve of D (in dependent case) of Example 1 and Example 2, respectively

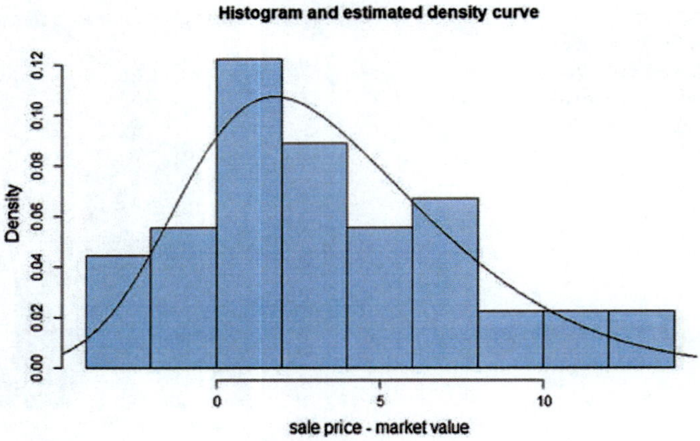

Fig. 13 Histogram and estimated density curve of D

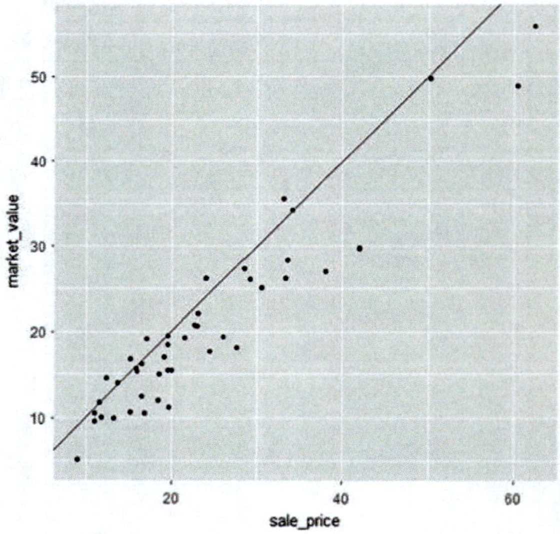

Fig. 14 Scatter plot between the market value Y and sales price X

5 Discussion

As related earlier, an important problem in the social sciences is the issue of meaningful units. For example, it is not clear what an attitude unit, a depression unit, or a pain unit is. Therefore, the meaning of a difference in means or a difference in locations across populations (or across estimators of population parameters) is often unclear. Furthermore, typical effect size indices, such as Cohen's d, give effect sizes in standard deviation units but it often is not clear how to interpret these either. Our solution was to find a way to convert skew normal parameters (or estimators) into the probability that a randomly selected person from one population would score higher on the dependent variable than a randomly selected person from the other population. This solution has several advantages, though we do not recommend that researchers dispense with other effect size indices.

First, converting skew normal parameters (or estimators) to a probability imparts a clear meaning that provides a solution to the problem of nonmeaningful units.

Second, converting to a probability enables comparisons and contrasts easier across research studies that may include populations with very different shape parameters. Prior to the present work, existing effect size indices were not suitable.

Third, in our view, social science researchers routinely fail to take distribution shapes into account. This may be due to the lessons in many graduate statistics classes that the Central Limit Theorem means that distribution shapes are irrelevant. Of course, the Central Limit Theorem does not say this; but that is the message that all too often comes across. And then there also is the ubiquitous advice to researchers that if they have non-normal distributions, they should perform a data transformation to normalize (see [20] for a list of data transformations and critical discussion). In contrast, the present work, especially Fig. 5, shows that shape parameters can have impressive effects on probability even when the location and scale parameters are the same across populations. And it is worthwhile to emphasize this point. Most researchers do not even consider shape, and typical effect size indices such as Cohen's *d* also do not consider shape. Figure 5 shows that not considering shape can lead to quite serious errors, as does Fig. 8.

Fourth, although shape is of blatant importance, it is important to take differences in locations and scales into account too. Figures 6 and 7 illustrate this importance.

Fifth, it is convenient that, as Fig. 9 illustrates, the issue of dependence does not matter very much. A researcher who wrongly assumes complete independence will nevertheless arrive at a probability that is reasonably accurate. Thus, researchers who do not know the degree of dependence between their populations need not be discommoded by the lack of knowledge.

Sixth, there also is the issue of anticipating future work. Although our umbrella of skew normal distributions is much more general than the typical umbrella of normal distributions, distributions nevertheless need not be skew normal anymore than they need be normal. As future research expands the present reasoning to handle more distribution families, we see probability as providing a common currency that will eventually allow comparisons even across studies with different distribution families (e.g., exponential rather versus skew normal).

In conclusion, the present work, with its emphasis on probability, simultaneously addresses the issue of imposing clarification when the substantive researcher is plagued with nonmeaningful units but also of providing a new type of standardized effect size that is not expressed in problematic standard deviation units. We hope and expect that mathematicians and statisticians will expand the present reasoning to many distribution families. As such expansions continue to be accomplished, it will become increasingly possible to express differences between conditions as meaningful probabilities across substantive topics, research paradigms, and populations of interest.

References

1. Azzalini, A. (1985). A class of distributions which includes the normal ones. *Scandinavian Journal of Statistics, 12*(2), 171–178.
2. Azzalini, A., & Valle, A. D. (1996). The multivariate skew normal distribution. *Biometrika, 83*(4), 715–726.
3. Blanca, M. J., Arnau, J., Lopez-Montiel, D., Bono, R., & Bendayan, R. (2013). Skewness and kurtosis in real data samples. *Methodology: European Journal of Research Methods for the Behavioral and Social Sciences, 9*(2), 78–84.
4. Baguley, T. (2009). Standardized or simple effect size: What should be reported? *British Journal of Psychology, 100*(Pt 3), 603–617.
5. Bobko, P., Roth, P. L., & Bobko, C. (2001). Correcting the effect size of d for range restriction and unreliability. *Organizational Research Methods, 4*(1), 46–61.
6. Burton, A., Altman, D. G., Royston, P., & Holder, R. L. (2006). The design of simulation studies in medical statistics. *Statistics in Medicine, 25*, 4279–4292.
7. Cohen, J. (1988). *Statistical power analysis for the behavioral sciences* (2nd ed.). Erlbaum.
8. Demidenko, E. (1996). The p-value you can't buy. *The American Statistician, 70*(1), 33–38.
9. Gillett, R. (2003). The metric comparability of meta-analytic effect-size estimators from factorial designs. *Psychological Methods, 8*(4), 419–433.
10. Gupta, A. K., Gonzalez-Farias, G., & Domínguez-Molina, A. A. (2004). A multivariate skew normal distribution. *Journal of Multivariate Analysis, 89*, 181–190.
11. Ho, A. D., & Yu, C. (2015). Descriptive statistics for modern test score distributions: Skewness, kurtosis, discreteness, and ceiling effects. *Educational and Psychological Measurement, 75*(3), 365–388.
12. Jacobs, R., Bekker, A. A., Van der Voet, H., & Ter Braak, C. J. (2015). Parametric estimation of P (X> Y) for normal distributions in the context of probabilistic environmental risk assessment. *PeerJ, 3*, e1164.
13. Joanes, D. N., & Gill, C. A. (1998). Comparing measures of sample skewness and kurtosis. *Journal of the Royal Statistical Society (Series D): The Statistician, 47*(1), 183–189.
14. Kirk, R. (1996). Practical significance: A concept whose time has come. *Educational and Psychological Measurement, 56*(5), 746–759.
15. McGrath, R. E., & Meyer, G. J. (2006). When effect sizes disagree: The case of *r* and *d*. *Psychological Methods, 11*(4), 386–401.
16. Micceri, T. (1989). The unicorn, the normal curve, and other improbable creatures. *Psychological Bulletin, 105*(1), 156–166.
17. Olejnik, S., & Algina, J. (2000). Measures of effect size for comparative studies: Applications, interpretations, and limitations. *Contemporary Educational Psychology, 25*(3), 241–286.
18. Olejnik, S., & Algina, J. (2003). Generalized eta and omega squared statistics: Measures of effect size for some common research designs. *Psychological Methods, 8*(4), 434–447.
19. Sharpe, N. R., De Veaux, R. D., Velleman, P. F., Wright, D., & Bock, D. E. (2010). *Business statistics*. Addison Wesley.
20. Trafimow, D., Wang, T., & Wang, C. (2019). From a sampling precision perspective, skewness is a friend and not an enemy!. *Educational and Psychological Measurement, 79*(1), 129–150.
21. Wang, C., Wang, T., Trafimow, D., & Myuz, H. A. (2019). Necessary sample size for specified closeness and confidence of matched data under the skew normal setting. Communications in Statistics-Simulation and Computation.
22. Zhu, X., Li, B., Wang, T., & Gupta, A. K. (2019). Sampling distributions of skew normal populations associated with closed skew normal distributions. *Random Operators and Stochastic Equations, 27*(2), 75–87.

A Priori Procedure (APP) for Estimating the Scale Parameter in Gamma Populations for Known Shape

Cong Wang, Tonghui Wang, David Trafimow, and Zhaohui Xu

Abstract There is a growing statistical literature pertaining to how researchers can make specifications for precision and confidence, with a priori procedure (APP) equations providing the necessary sample size to meet those specifications. However, an important limitation of that literature is that there are no APP equations pertaining to estimating parameters of gamma distributions. The present work addresses the lack by proposing new APP equations by which researchers can determine the necessary sample size to meet specifications for precision and confidence with respect to estimating the scale parameter in gamma distributions. Our work extends to the ratio of scales in two independent gamma distributions too.

Keywords A priori procedure · Gamma population · Necessary sample size · Confidence level · Precision

1 Introduction

Our goal is to extend the a priori procedure to the family of gamma distributions. Consequently, this introduction contains a brief description of the a priori procedure and an explanation of why an extension to the family of gamma distributions is desirable. Subsequent sections provide the relevant mathematics and implications.

C. Wang
Department of Mathematics, University of Nebraska Omaha, Omaha, NE, USA
e-mail: congwang@unomaha.edu

T. Wang (✉)
Department of Mathematical Sciences, New Mexico State University, Las Cruces, NM, USA
e-mail: twang@nmsu.edu

D. Trafimow
Department of Psychology, New Mexico State University, Las Cruces, NM, USA
e-mail: dtrafimo@nmsu.edu

Z. Xu
Dongfang College, Zhejiang University of Finance and Economy, Hangzhou, China
e-mail: 20020073@zufedfc.edu.cn

S. Sriboonchitta et al. (eds.), *Credible Asset Allocation, Optimal Transport Methods, and Related Topics*, Studies in Systems, Decision and Control 429,
https://doi.org/10.1007/978-3-030-97273-8_19

Trafimow [10] designed the a priori procedure (APP) as an alternative to the null hypothesis significance testing procedure or traditional confidence intervals. Whereas p-values and traditional confidence intervals are computed based on obtained data, the APP is pre-data and independent of what the researcher finds. The goal is to find the sample sizes necessary to ensure that the sample statistics to be obtained are within particular ranges of corresponding population parameters with known probabilities. There have already been several extensions of this work, such as Trafimow and MacDonald [11]; and Trafimow et al. [8, 9].

There is an important difference between typical confidence intervals based on sample data and a priori confidence intervals not based on data. To gain a basic understanding of the concept of a priori confidence intervals, consider the simple case where there is a single group randomly sampled from a normally distributed population. The researcher is interested in determining the sample size n necessary to obtain a sample mean within some distance f in standard deviation units, from the population mean, with a specified probability. Trafimow [10] provides an accessible derivation of a simple equation:

$$n = \left(\frac{z_{(1-c)/2}}{f}\right)^2,$$

where f is the fraction of a standard deviation the researcher defines as sufficiently "close", $z_{(1-c)/2}$ is the standard z-score that corresponds to the desired probability of being close, and n is the minimum sample size necessary to meet specifications for closeness and confidence.

For example, suppose the researcher desires to have a $c = 95\%$ probability of obtaining a sample mean within $f = 0.1$ of a standard deviation of the population mean. Then the necessary sample size would be $n = \left(\frac{1.96}{0.1}\right)^2 = 384.16$ or 385 rounded to the nearest higher whole number. Wang et al. [13, 15] expanded the APP to render possible for researchers to find the sample sizes needed to meet specifications for precision and confidence pertaining to differences in means in matched or independent samples. More recently, the APP has been extended to address binary data in normal population, where the statistic of interest is not a mean but rather a correlation coefficient and see Wang et al. [12] for more details.

Although much APP work has been completed, there are no extensions to the family of gamma distributions. Here is why such an extension is desirable. The family of gamma distributions is a two-parameter family of continuous probability distributions. This family includes the well-known families of exponential distributions, Erlang distributions, and chi-squared distributions as special cases. The family of gamma distributions is widely used and plays an important role in the reliability field and survival analysis; therefore, a successful procedure to facilitate estimation of its parameters would be a useful contribution. The maximum likelihood estimation for its parameters are discussed by many researchers such as Harter and Moore [5], Bowman et al. [2], Bowman and Shenton [1] and Johnson et al. [7]. Hwang and Huang [6] provided new moment estimators of parameters of the gamma distribution using its characterization. Beside that, bias of the maximum likelihood estimators of the two-parameter gamma distribution were revisited by Giles and Feng [4].

The paper is organized as follows. Properties of Gamma distributions are reviewed briefly in Sect. 2. In Sect. 3, we derive the APP equations for the scale parameter in gamma distributions as well as the ratio of scales in two independent gamma distributions. Numerical results, as well as the simulation results, are illustrated in Sect. 4.

2 Properties of the Gamma Distribution

The Gamma distribution is widely used in engineering, science, and business, to model continuous variables that are always positive and have skewed distributions. In order to construct the APP for estimating the scale (or the ratio of scales) parameter, we need to discuss the properties of the gamma distribution.

Definition 2.1 A random variable X that is gamma distributed with shape k and scale θ is denoted by $\Gamma(k, \theta)$ and the probability density function using the shape-scale parametrization is

$$f(x) = \frac{1}{\Gamma(k)\theta^k} x^{k-1} \exp\left\{-\frac{x}{\theta}\right\}, \tag{1}$$

where $k, \theta, x \in (0, \infty)$.

The moment generating function of $X \sim \Gamma(k, \theta)$ is

$$M_X(t) \equiv E(e^{tX}) = (1 - \theta t)^{-k}, \quad t < \frac{1}{\theta} \tag{2}$$

and the mean and variance of X are

$$E(X) = k\theta \quad \text{and} \quad Var(X) = k\theta^2, \tag{3}$$

respectively.

For the sampling distribution of the sample mean, we have the following result.

Proposition 2.1 *Let $X_1, X_2, \ldots, X_n$ be a random sample from the gamma distribution and its density function is given in 1. Then we have the following results:*

(i) The sample mean $\bar{X}$ is of gamma distributed, and

$$\bar{X} \sim \Gamma\left(nk, \frac{\theta}{n}\right).$$

(ii) The mean and variance of $\bar{X}$ are, respectively,

$$E(\bar{X}) = k\theta \quad and \quad Var(\bar{X}) = \frac{k\theta^2}{n}.$$

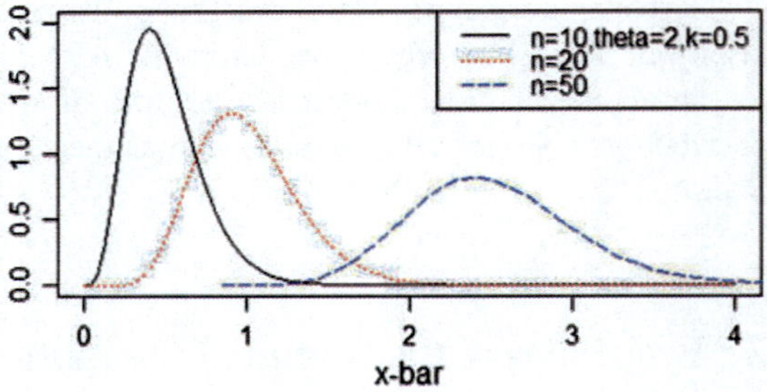

Fig. 1 Density curves of $\bar{X}$ for values of $n = 10, 20, 50$ when $k = 0.5, \theta = 2$

(iii) *The estimator* $\frac{\bar{X}}{k}$ *for known k is the best unbiased estimator for* θ.

Proof The moment generating function $\bar{X}$ is

$$M_{\bar{X}}(t) = E[\exp\{\bar{X}t\}] = E\left[\exp\left\{\sum_{i=1}^{n} X_i \frac{t}{n}\right\}\right] = \left(1 - \frac{\theta}{n}t\right)^{-kn}.$$

It i.e. easy to obtain (ii) from 3.
For (iii), it is clear that $Var(\bar{X}/k) = \frac{\theta^2}{kn}$. By Definition 2.1,

$$\log f(x) = -\log\Gamma(k) - k\log\theta + (k-1)\log x - \frac{x}{\theta}$$

and

$$\frac{\partial \log f(x)}{\partial x} = -\frac{k}{\theta} + \frac{x}{\theta^2},$$

from which we can obtain that

$$nE\left[\left(\frac{\partial \log f(X)}{\partial x}\right)^2\right] = Var\left(\frac{\partial \log f(X)}{\partial x}\right) + \left[E\left(\frac{\partial \log f(X)}{\partial x}\right)\right]^2 = \frac{nk}{\theta^2}$$

and the Cramér-Rao lower bound is $\frac{\theta^2}{nk}$ which is equal to $Var\left(\frac{\bar{X}}{k}\right)$. □

The density curves of $\bar{X}$ for various parameters k and θ with different sample sizes are given in Figs. 1 and 2.

For estimating the ratio of scale parameters in two independent gamma populations, we have the following result.

Proposition 2.2 *Let* $X_1, X_2, \ldots, X_{n_1}$ *be a random sample from* $\Gamma(k_1, \theta_1)$ *and* $Y_1, Y_2, \ldots, Y_{n_2}$ *be a random sample from* $\Gamma(k_2, \theta_2)$. *Assume that two gamma populations are independent. Consider the random variable R given by* $R = \frac{\bar{X}}{\bar{Y}}$, *where* $\bar{X}$ *and* $\bar{Y}$ *are sample means. Then we have the following results.*

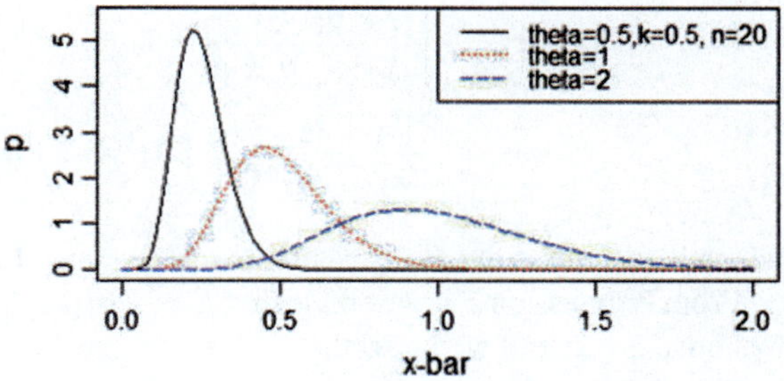

Fig. 2 Density curves of $\bar{X}$ for values of $\theta = 0.5, 1, 2$ when $k = 0.5$ and $n = 20$

(i) The distribution of $\frac{k_2\theta_2}{k_1\theta_1} R$ *is the F-distributed with degrees of freedoms* $2n_1k_1$ *and* $2n_2k_2$.

(ii) The mean and variance of R are, respectively,

$$E(R) = \frac{n_2k_1}{n_2k_2 - 1}\frac{\theta_1}{\theta_2} \quad \text{and} \quad Var(R) = \frac{n_2^2k_1(n_1k_1 + n_2k_2 - 1)}{n_1(n_2k_2 - 1)^2(n_2k_2 - 2)}\frac{\theta_1^2}{\theta_2^2}.$$

(iii) The Cramer-Rao lower bound for $\hat{R} = \frac{n_2k_2-1}{n_2k_1} R$ *is given by*

$$L_C = \left(\frac{1}{n_1k_1} + \frac{1}{n_2k_2}\right)\frac{\theta_1^2}{\theta_2^2}$$

Proof For (i), we will first show that if $X \sim \Gamma(k_1, \theta_1)$ and $Y \sim \Gamma(k_2, \theta_2)$, where X and Y are independent, then $T = \frac{k_2\theta_2 X}{k_1\theta_1 Y}$ has F-distribution with degrees of freedom $2k_1$ and $2k_2$. Let $f_{X,Y}$, f_X, and f_Y be, respectively, the densities of (X, Y), X, and Y. Then the density of $Z = \frac{X}{Y}$ is

$$f_Z(z) = \int_0^\infty y f_{X,Y}(yz, y)dy = \int_0^\infty y f_X(yz) f_Y(y)dy$$
$$= \frac{\Gamma(k_1 + k_2)}{\Gamma(k_1)\Gamma(k_2)}\left(\frac{\theta_1}{\theta_2}\right)^{-k_1} z^{k_1-1}\left(\frac{\theta_2}{\theta_1}z + 1\right)^{-(k_1+k_2)}.$$

It is well known that the mean and variance of Z are

$$E(Z) = \frac{k_1}{k_2 - 1}\frac{\theta_1}{\theta_2} \quad \text{and} \quad Var(Z) = \frac{k_1(k_1 + k_2 - 1)}{(k_2 - 1)^2(k_2 - 2)}\frac{\theta_1^2}{\theta_2^2},$$

respectively. Also it is easy to see that the density of T is

$$f_T(t) = f_Z\left(\frac{k_1\theta_1}{k_2\theta_2}t\right)\frac{k_1\theta_1}{k_2\theta_2}$$
$$= \frac{\Gamma(k_1+k_2)}{\Gamma(k_1)\Gamma(k_2)}\left(\frac{k_1}{k_2}\right)^{k_1} t^{k_1-1}\left(1+\frac{k_1}{k_2}t\right)^{-(k_1+k_2)},$$

which is the pdf of the F-distribution with degrees of freedom $2k_1$ and $2k_2$.

From Proposition 2.1, we obtain that $\bar{X} \sim \Gamma(n_1k_1, \frac{\theta_1}{n_1})$ and $\bar{Y} \sim \Gamma(n_2k_2, \frac{\theta_2}{n_2})$. Thus by above argument, we know that

$$\frac{k_2\theta_2}{k_1\theta_1}R \sim F_{2n_1k_1,2n_2k_2},$$

the F-distribution with $2n_1k_1$ and $2n_2k_2$ degrees of freedoms. Part (ii) follows immediately from the mean and variance of Z given above.

For (iii), $\hat{R}$ is an unbiased estimator of θ_1/θ_2. Note that the joint density of $X_1, \ldots, X_{n_1}$ and $Y_1, \ldots, Y_{n_2}$ is

$$f_{X,Y}(\mathbf{x}, \mathbf{y}) = \prod_{i=1}^{n_1}\frac{x_i^{k_1-1}}{\Gamma(k_1)\theta_1^{k_1}}\exp\left\{-\frac{x_i}{\theta_1}\right\}\prod_{j=1}^{n_2}\frac{y_j^{k_2-1}}{\Gamma(k_2)\theta_2^{k_2}}\exp\left\{-\frac{y_j}{\theta_2}\right\}.$$

Thus the log-likelyhood function is

$$\log f_{X,Y}(\mathbf{x}, \mathbf{y}) = -n_1\left(\log\Gamma(k_1) + k_1\log\theta_1\right) + (k_1-1)\sum_{i=1}^{n_1}\log x_i - \sum_{i=1}^{n_1}\frac{x_i}{\theta_1}$$
$$-n_2\left(\log\Gamma(k_2) + k_2\log\theta_2\right) + (k_2-1)\sum_{j=1}^{n_2}\log y_j - \sum_{j=1}^{n_2}\frac{y_j}{\theta_2}.$$

Assume that k_1 and k_2 are known and let $\boldsymbol{\theta} = (\theta_1, \theta_2)'$. Then we obtain

$$\frac{\partial \log f_{X,Y}(\mathbf{x}, \mathbf{y})}{\partial \boldsymbol{\theta}} = \begin{bmatrix} -\frac{n_1k_1}{\theta_1} + \frac{\sum_{i=1}^{n_1}x_i}{\theta_1^2} \\ -\frac{n_2k_2}{\theta_2} + \frac{\sum_{j=1}^{n_2}y_j}{\theta_2^2} \end{bmatrix}.$$

Therefore,

$$\frac{\partial^2 \log f_{X,Y}(\mathbf{x}, \mathbf{y})}{\partial \boldsymbol{\theta}\partial \boldsymbol{\theta}'} = \begin{bmatrix} \frac{n_1k_1}{\theta_1^2} - \frac{2\sum_{i=1}^{n_1}x_i}{\theta_1^3} & 0 \\ 0 & \frac{n_2k_2}{\theta_2^2} - \frac{2\sum_{j=1}^{n_2}y_j}{\theta_2^3} \end{bmatrix},$$

and

$$I(\boldsymbol{\theta}) = -E\left(\frac{\partial^2 \log f_{X,Y}(\mathbf{x}, \mathbf{y})}{\partial \boldsymbol{\theta}\partial \boldsymbol{\theta}'}\right) = \begin{bmatrix} \frac{n_1k_1}{\theta_1^2} & 0 \\ 0 & \frac{n_2k_2}{\theta_2^2} \end{bmatrix}.$$

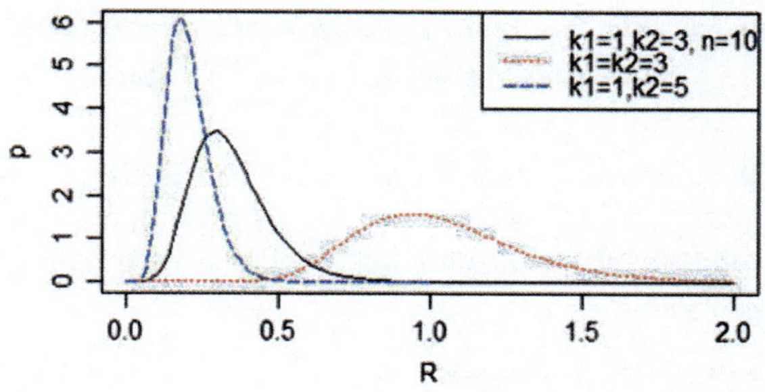

Fig. 3 Density curves of R for different values of $k_1 = 1, 3$ and $k_2 = 3, 5$ when $n = 10$

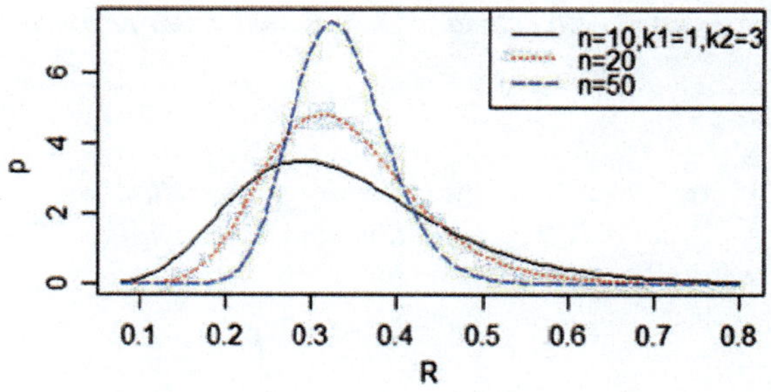

Fig. 4 Density curves of R for different values $n = 10, 20, 50$ when $k_1 = 1$ and $k_2 = 3$

Consequently, the Cramer-Rao lower bound L_C is

$$L_C = \left(\frac{\partial\phi(\boldsymbol{\theta})}{\partial\boldsymbol{\theta}}\right)' I(\boldsymbol{\theta})^{-1}\left(\frac{\partial\phi(\boldsymbol{\theta})}{\partial\boldsymbol{\theta}}\right) = \left(\frac{1}{n_1k_1} + \frac{1}{n_2k_2}\right)\frac{\theta_1^2}{\theta_2^2}$$

if we denote $\phi(\boldsymbol{\theta}) = \frac{\theta_1}{\theta_2}$. □

Remark 2.1 In Proposition 2.2, it is easy to check that $Var(\hat{R}) = \frac{n_1k_1+n_2k_2-1}{n_1k_1(n_2k_2-2)}\frac{\theta_1^2}{\theta_2^2}$ by (ii), that is, $Var(\hat{R}) > L_C$ where L_C is given in (iii).

The following graphs (Figs. 3 and 4) show the density curves of R for different values of k_1, k_2 and $n = n_1 = n_2$ when $\theta_1 = \theta_2$.

3 The APP for Estimating the Scale Parameter in One Sample and the Ratio of the Scales in Two Independent Samples

In this section, we will set up the APP procedures for estimating the scale parameter θ by a random sample from the gamma distribution with shape k and scale θ, and the ratio of scales in two independent gamma populations.

3.1 The Necessary Sample Size for Estimating the Scale Parameter θ with Known Shape k

In order to determine the necessary sample size n to be $c \times 100\%$ confident for the given precision, we consider the distribution of the maximum likelihood estimator estimator, $\hat{\theta} = \bar{X}/k$ of θ, given in Proposition 2.1 for known shape k. Note that $\hat{\theta}$ is also unbiased of θ,

Theorem 3.1 *Suppose that $X_1, X_2, \ldots, X_n$ is a random sample from $\Gamma(k, \theta)$. Let c be the confidence level and f be the precision which are specified such that the error associated with estimator $\bar{X}$ is $E = f\sigma$. More specifically, if*

$$P[f_1\sigma \leq \hat{\theta} - \theta \leq f_2\sigma] = c, \tag{4}$$

where f_1 and f_2 are the left and right precision which are restricted by $max\{|f_1|, f_2\} \leq f$, and σ is the population standard deviation. Then the minimum sample size n required can be obtained by

$$\int_L^U f(z)dz = c \tag{5}$$

such that $U - L$ is minimized, where

$$L = \sqrt{n}kf_1, \qquad U = \sqrt{n}kf_2,$$

and

$$f(z) = \frac{(kn)^{kn/2}}{\Gamma(kn)} \left(z + \sqrt{n/k}\right)^{kn-1} \exp\left[-\sqrt{kn}(z + \sqrt{n/k})\right].$$

Proof Let $Z = \frac{\hat{\theta}-\theta}{\sigma_{\hat{\theta}}}$. Then $Z = \frac{\bar{X}}{\theta\sqrt{k/n}} - \sqrt{kn}$, and Eq. 4 can be rewritten as

$$P\left[\sqrt{n}kf_1 \leq Z \leq \sqrt{n}kf_2\right]$$

since $\sigma = \sqrt{k}\theta$, that is,

$$\int_L^U f(z)dz = c,$$

where $L = \sqrt{n}kf_1$, $U = \sqrt{n}kf_2$ and $f(z)$ is the pdf of Z.
Note that by(i) in Proposition 2.1, $\bar{X} \sim \Gamma(nk, \theta/n)$. Then we have

$$T \equiv \frac{\bar{X}}{\theta\sqrt{k/n}} \sim \Gamma(kn, 1/\sqrt{kn}),$$

and the pdf of Z is given by

$$f(z) = f_T(z + \sqrt{kn})$$
$$= \frac{(kn)^{kn/2}}{\Gamma(kn)} \left(z + \sqrt{kn}\right)^{kn-1} \exp\left[-\sqrt{kn}(z + \sqrt{kn})\right].$$

Remark 3.1 The specified values of n, f_1 and f_2 are obtained simultaneously, given that f and the $c \times 100\%$ confidence interval have been specified. The corresponding confidence interval for θ based on Theorem 3.1 is

$$\left[\frac{\bar{X}}{k(1 + f_2\sqrt{k})}, \frac{\bar{X}}{k(1 + f_1\sqrt{k})}\right].$$

3.2 *The Necessary Sample Size for Estimating the Ratio of Scales in Two Independent Gamma Random Sample for Known Shapes*

In this part, the APP procedures will be set up for estimating the ratio for the scales by two random samples from two independent gamma distributions with shapes k_1, k_2 and scales θ_1, θ_2. Similarly as in one sample case, we assume that the shapes are known. To determine the necessary sample size to be $c \times 100\%$ confident for the given precision, we consider the distribution of the unbiased estimator $\hat{R} \equiv \frac{n_2k_2-1}{n_2k_1} R$ of $r \equiv \frac{\theta_1}{\theta_2}$ given in Proposition 2.2. Note that $F_{n,m_2} \leq F_{n,m_1}$ for $m_1 \leq m_2$ where F_{n,m_1} is the the critical point of F distribution with numerator degrees of freedom n and denominator degrees of freedom m_1. Therefore, without loss of generality, our APP is applied to finding the minimum value of m_1 and m_2, see Wang et al. [14] for more details. Then we have the following result.

Theorem 3.2 *Suppose that $X_1, X_2, \ldots, X_{n_1}$ and $Y_1, Y_2, \ldots, Y_{n_2}$ are two random samples from gamma distributions with shapes k_1 and k_2 and scales θ_1 and θ_2. Let c be the confidence level and f be the precision which are specified such that the error associated with the unbiased estimator $\hat{R}$ of r is $E = f\sigma_{X/Y}$. More specifically, if*

$$P\left[f_1\sigma_{X/Y} \leq \hat{R} - r \leq f_2\sigma_{X/Y}\right] = c, \tag{6}$$

where $\sigma_{X/Y}$ is the standard deviation of the ratio of two independent random variable X and Y with $X \sim \Gamma(k_1, \theta_1)$ and $Y \sim \Gamma(k_2, \theta_2)$, f_1 and f_2 are the left and right precision which are restricted by $\max\{|f_1|, f_2\} \leq f$. Then the sample size n where $n = \min\{n_1, n_2\}$ required can be obtained by

$$\int_{L}^{U} f(t)dt = c \tag{7}$$

such that $U - L$ is minimized, where $f(t)$ is the pdf of the F-distribution with degrees of freedom $2nk_1$ and $2nk_2$,

$$L = b(1 + af_1) \quad and \quad U = b(1 + af_2),$$

where

$$a = \sqrt{\frac{k_1(k_1 + k_2 - 1)}{(k_2 - 1)^2(k_2 - 2)}} \quad and \quad b = \frac{nk_1}{nk_2 - 1}.$$

Proof From Proposition 2.2, $\frac{k_2\theta_2}{k_1\theta_1} R \sim F_{2nk_1,2nk_2}$, then $\frac{nk_2\theta_2}{(nk_2-1)\theta_1}\hat{R} \sim F_{2nk_1,2nk_2}$. Thus, we can rewrite Eq. 6 to

$$P\left[\frac{nk_1}{(nk_2 - 1)r}\left(r + f_1\sigma_{X/Y}\right) \le \frac{nk_1}{(nk_2 - 1)r}\hat{R} \le \frac{nk_1}{(nk_2 - 1)r}\left(r + f_2\sigma_{X/Y}\right)\right] = c.$$

By the proof of (i) in Proposition 2.2,

$$Var(X/Y) = \frac{k_1(k_1 + k_2 - 1)}{(k_2 - 1)^2(k_2 - 2)} r^2,$$

which implies $\sigma_{X/Y} = a^2r^2$ if we denote $a = \sqrt{\frac{k_1(k_1+k_2-1)}{(k_2-1)^2(k_2-2)}}$. Thus,

$$P\left[b(1 + af_1) \le T \le b(1 + af_2)\right] = c,$$

where $b = \frac{nk_2}{nk_2-1}$ and $T = b\frac{\hat{R}}{r}$, and Eq. 7 holds.

Remark 3.2 The specified values of n, f_1 and f_2 are obtained simultaneously, given that f and the $c \times 100\%$ confidence interval have been specified. The corresponding confidence interval for θ based on Theorem 3.2 is

$$\left[\frac{\hat{R}}{1 + af_2}, \frac{\hat{R}}{1 + af_1}\right].$$

4 Simulation Results for Necessary Sample Sizes

In this section, we provide simulation results of the necessary sample sizes for specific precision values f and confidence levels c given in (5) and (6) of Theorems 3.1 and

Table 1 The value of sample size n, left precision f_1 and right precision f_2 under different f for the given $c = 0.95, 0.9$ and $k = 1$

f	c	n	f_1	f_2
0.15	0.95	176	−0.1462	0.1495
	0.9	121	−0.1500	0.1484
0.2	0.95	99	−0.1935	0.1999
	0.9	68	−0.2000	0.1973
0.3	0.95	45	−0.2847	0.2973
	0.9	31	−0.2955	0.2895
0.4	0.95	25	−0.3760	0.3992
	0.9	17	−0.3965	0.3861
0.5	0.95	16	−0.4613	0.4993
	0.9	11	−0.4884	0.4730

Table 2 The value of sample size n, left precision f_1 and right precision f_2 under different f for the given $c = 0.95, 0.9, k_1 = 1$ and $k_2 = 3$

f	c	n	f_1	f_2
0.2	0.95	177	−0.1930	0.1992
	0.9	122	−0.1995	0.1966
0.3	0.95	79	−0.2859	0.2999
	0.9	54	−0.2992	0.2930
0.4	0.95	45	−0.3741	0.3991
	0.9	30	−0.3994	0.3889
0.5	0.95	29	−0.4596	0.4989
	0.9	19	−0.4973	0.4817

3.2, respectively. Here we consider $c = 95\%, 90\%$ with different values of f, and the corresponding necessary sample sizes are given in Tables 1 and 2.

Using the Monte Carlo simulations, we account relative frequency for different values of θ and r (the ratio of θ_1 and θ_2). Table 3 shows the result for the relative frequency of 90% and 95% confidence intervals for $f = 0.2, 0.3, 0.4, 0.5, \theta = 1, 2, 5$ and $k = 1$ corresponding to Table 1, and the relative frequencies of 90% and 95% confidence intervals for different r values corresponding to Table 2 are shown in Table 4. All results are illustrated with a number of simulation runs $M = 10000$.

5 A Real Data Application

We provide a real data application to illustrate the necessary sample sizes under our APP. The data set is on the ages of best actors and actresses (in years) of which is listed

Table 3 The relative frequencies for confidence intervals of θ with confidence level $c = 0.95, 0.9$, different values of precision f and θ for $k = 1$

f	c	$\theta = 1$	$\theta = 2$	$\theta = 5$
0.2	0.95	0.9493	0.9499	0.9515
	0.9	0.8922	0.9016	0.9001
0.3	0.95	0.9502	0.9509	0.9506
	0.9	0.9014	0.9007	0.9002
0.4	0.95	9481	0.9511	0.9501
	0.9	0.8984	0.9022	0.9004
0.5	0.95	0.9561	0.9515	0.9533
	0.9	0.8987	0.8991	0.9000

Table 4 The relative frequencies for confidence intervals of r with confidence level $c = 0.95, 0.9$, different precision f and r for $k_1 = 1, k_2 = 3$

f	c	$r = 1$	$r = 2$	$r = 5$
0.2	0.95	0.9537	0.9512	0.9524
	0.9	0.9092	0.9013	0.9002
0.3	0.95	0.9504	0.9534	0.9495
	0.9	0.9034	0.9027	0.8921
0.4	0.95	9527	0.9533	0.9542
	0.9	0.9011	0.9030	0.9024
0.5	0.95	0.9497	0.9502	0.9484
	0.9	0.8990	0.8999	0.8971

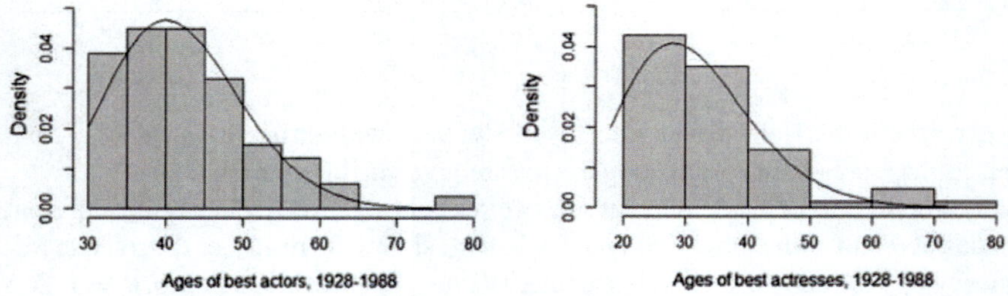

Fig. 5 The histograms of the ages of Oscar-winning best actors and actresses given in appendix and its corresponding gamma curves

Appendix [3]. The moment estimates $\hat{k}_A = 1.7774$, $\hat{k}_B = 3.3335$, $\hat{\theta}_A = 24.5107$ and $\hat{\theta}_B = 10.4091$ (here we keep four decimal places) based on the data. The histograms of the data curved by $\Gamma(1.7774, 24.5107)$ and $\Gamma(3.3335, 10.4091)$ are shown in Fig. 5.

Now if we want to construct the 90% confidence interval for θ_A and θ_B with given $k_A = 1.7774$, $k_B = 3.3335$ and $f = 0.15$. From Eq. 5, the necessary sample sizes

$n_A = 56$ for Data (A) and $n_B = 17$ for Data (B). Randomly select a sample of size $n_A = 56$ from Data (A), the sample mean $\bar{X}_A = 43.5893$. Then the 95% confidence interval of θ_A is [20.4224, 30.3570], and it is clear that the moment estimate $\hat{\theta} = 24.5107$ is included. Similarly, random choose a sample with size $n_B = 17$ from Data (B) and the sample mean $\bar{X}_B = 34.0588$. Thus, the 95% confidence interval of θ_B is [8.1527, 13.8645]. Since the $\hat{R} = 2.3116$, then the necessary sample size for $c = 0.95$ and $f = 0.15$ is 44 and the corresponding confidence interval of $\frac{\theta_A}{\theta_B}$ is [1.3231, 2.8753].

6 Discussion

Although there is a growing APP literature, reviewed in the introduction; that literature does not include a way for researchers to estimate necessary sample sizes to meet specifications for precision and confidence under gamma distribution settings. The present work addresses that lack, both with respect to the scale parameter for a single gamma distributed population and with respect to ratios when there are two gamma distributed populations. We also provided tables and a free and publicly available computer program so that researchers should be able to perform the calculations, regardless of their level of mathematical sophistication. Finally, we reported an example with real data so that researchers can better understand how to apply the APP to data from gamma distributed populations. Future research, along a similar vein, might pertain to estimating the shape parameter of gamma distributed populations.

7 Appendix

Here is the data set which is used in the section of example which represent the ages of best actors (Data A) and actresses (Data B).

Data A:
30, 31, 32, 32, 33, 33, 34, 34, 35, 35, 35, 35, 37, 37, 38, 38, 38, 38, 38, 38, 39, 39, 40, 40, 40, 40, 41, 41, 41, 41, 42, 42, 43, 43, 43, 43, 43, 44, 44, 45, 46, 46, 47, 47, 48, 48, 48, 49, 49, 49, 51, 51, 52, 53, 55, 56, 56, 56, 60, 61, 62, 76

Data B:
21, 22, 24, 24, 24, 24, 24, 25, 26, 26, 26, 26, 26, 26, 26, 27, 27, 27, 28, 28, 29, 29, 29, 30, 30, 30, 30, 31, 31, 32, 33, 33, 33, 34, 34, 34, 34, 34, 34, 34, 35, 35, 36, 37, 37, 38, 38, 38, 40, 41, 41, 41, 41, 41, 42, 45, 48, 49, 60, 61, 61, 62, 74

References

1. Bowman, K., & Shenton, L. R. (1988). *Properties of estimators for the gamma distribution*. Marcel Dekker.
2. Bowman, K., Shenton, L. R., & Lam, H. K. (1987). Simulation and estimation problems associated with the three-parameter gamma distribution. *Communications in Statistics—Simulation and Computation,16*, 1147–1188.
3. Brown, R., & Davis, G. (1990). Ages of Oscar-winning best actors and actresses. *The Mathematics Teacher, 83*(2), 96–102 (7 pages)
4. Giles, D. E. & Feng, H. (2009). Bias of the maximum likelihood estimators of the two-parameter gamma distribution revisited. *Econometrics Working Paper EWP0908*. ISSN 1485-6441
5. Harter, H. L., & Moore, A. H. (1965). Maximum-likelihood estimation of the parameters of the gamma and Weibull populations from complete and from censored samples. *Technometrics, 7*, 639–643.
6. Hwang, T., & Huang, P. (2002). On new moment estimation of parameters of the gamma distribution using its characterization. *Annals of the Institute of Statistical Mathematics, 54*(4), 840–847.
7. Johnson, R. W., Kliche, D. V., & Smith, P. L. (2014). Maximum likelihood estimation of gamma parameters for coarsely binned and truncated raindrop size data. *Quarterly Journal of the Royal Meteorological Society, 140*(1245–1256), 2014. https://doi.org/10.1002/qj.2209.
8. Trafimow, D., Wang, T., & Wang, C. (2018). From a sampling precision perspective, skewness is a friend and not an enemy! In *Educational and Psychological Measurement*, pp. 1–22. https://doi.org/10.1177/0013164418764801.
9. Trafimow, D., Wang, C., & Wang, T. (2019). Making the a prior procedure work for differences between means. *Educational and Psychological Measurement,1–13*, https://doi.org/10.1177/0013164419847509.
10. Trafimow, D. (2017). Using the coefficient of confidence to make the philosophical switch from a posteriori to a priori inferential statistics. *Educational and Psychological Measurement, 77*(5), 831–854.
11. Trafimow, D., & MacDonald, J. A. (2017). Performing inferential statistics prior to data collection. *Educational and Psychological Measurement, 77*(2), 204–219.
12. Wang, C., Wang, T., Li, B., Zhu, X., & Ma, Y. (2020). Extending a priori procedure (APP) to address correlation coefficients. In N. Trung, N. Thach, and V. Kreinovich (Eds.), *Data science for financial econometrics*. Springer.
13. Wang, C., Wang, T., Trafimow, D., & Chen, J. (2019). Extending a priori procedure to two independent samples under skew normal setting. *Asian Journal of Economics and Banking, 03*(02), 29–40.
14. Wang, C., Wang, T., Trafimow, D., & Talordphop, K. (2020). Extending the a priori procedure to one-way analysis of variance model with skew normal random effects. *Asian Journal of Economics and Banking, 4*(2), 77–90.
15. Wang, C., Wang, T., Trafimow, D., & Myuz, H. A. (2019). Necessary sample size for specified closeness and confidence of matched data under the skew normal setting. *Communications in Statistics-Simulation and Computation*. https://doi.org/10.1080/03610918.2019.1661473.

Practical Applications

Testing CAPM Using Markov Switching Models: Application to ASEAN-6 Stock Markets

Pichayakone Rakpho, Woraphon Yamaka, and Songsak Sriboonchitta

Abstract In this study, six finance stocks traded in FTSE ASEAN 40 index (ASEAN market) are tested and analysed using the two-state Markov Switching Capital Asset Pricing Model (MS-CAPM). We consider eight MS-CAPM specifications including MSI-CAPM, MSIH-CAPM, MSIB-CAPM, MSIBH-CAPM, MSH-CAPM, MSB-CAPM, MSBH-CAPM and linear CAPM and the best specification of CAPM. We find that MSB-CAPM model (with switching beta risk parameter) is the best fit model for Vietnam, Indonesia, Malaysia and Singapore stock returns, MSBH-CAPM model (with switching beta parameter and heteroskedasticity) for Philippines stock return and MSIBH-CAPM model (with switching intercept term, beta parameter and heteroskedasticity) for Thailand stock return. These results confirm the heterogeneous structure of CAPM using Markov switching models.

Keywords CAPM · Markov switching · ASEAN-6 stock markets

1 Introduction

Greater financial market integration has widened the opportunity and offered more and diverse choices for present day investors to invest in different stock markets. Association of Southeast Asian Nations (ASEAN) stock market, and the official ASEAN Exchanges, is one of the attractive capital markets in the world consequential to the implementation of the agreement on abolishment of capital inflow barrier and foreign exchange restriction among member countries. In recent years, ASEAN has been considerably interesting to investors who expect the higher return. Although, this market is attractive to investors, but it is inevitably risky due to market volatility. Therefore, the decision of investor must rely on the risk and return of the stock. To assess stock risk and return, in this study, we consider applying the capital asset

P. Rakpho · W. Yamaka (✉) · S. Sriboonchitta
Center of Excellence in Econometrics, Faculty of Economics, Chiang Mai University, Chiang Mai, Thailand
e-mail: woraphon.econ@gmail.com

S. Sriboonchitta et al. (eds.), *Credible Asset Allocation, Optimal Transport Methods, and Related Topics*, Studies in Systems, Decision and Control 429,
https://doi.org/10.1007/978-3-030-97273-8_20

pricing model (CAPM). This approach is very well-known and extensively used for measuring risk in financial analysis. The estimated parameter, known as beta risk, reflects the sensitivity of asset return to the market returns. However, Pastpipatkul, Maneejuk and Sriboonchitta [9] argued the unconditional beta which is constant over time is clearly an unreasonable assumption since the investors' horizon can be more than one period, and their risk aversion/taking behaviour as well as market conditions may change from one period to another. This implies a time-varying risk and risk premium [17].

The movements of returns are obviously different overtime depending on the regimes of the markets or economic conditions at that period, therefore, the Markov-switching (MS) model of Hamilton [1], also known as the regime switching model, is considered to measure the beta risk in different economic conditions. This model has been applied to CAPM in many studies, for examples, Gu [5], Pastpipatkul, Maneejuk, Sriboonchitta [9], Rakpho, Yamaka and Sriboonchitta [11], Vendrame, Guermat and Tucker [17], Thamprasert, Pastpipatkul, and Yamaka [15]. However, none of them addressed the parameter change in the model. In the general MS specifications, all the parameters are regime-dependent. However, in the empirical applications, it is preferable to consider only some parameters as dependent on the state as this may offer a lot of alternatives for the MS specification (see [6, 8]). Therefore, this study proposes a more general form of MS model which allows better flexibility for capturing almost all possible structures of CAPM which have not been studied in earlier literature. In the literature, some studies, such as Huang [4], Skrinjaric [13], Prukumpai [10], and Vendrame, Guermat and Tucker [17] conducted the MS-CAPM to investigate the beta risk of stock markets, However, none of them considered other specifications of MS-CAPM and compared various MS-CAPM specifications. To the best of our knowledge, we are the first attempt in testing and comparing the best MS-CAPM specification.

This study aims to test the efficacy of MS-CAPM in various specifications for ASEAN stock market. Specifically, investors use the models to investigate the relationship between the expected return and the degree of risk in ASEAN-6 financial markets, consisting of Thailand, the Philippines, Indonesia, Malaysia, Singapore and Vietnam whose shares are traded in the ASEAN exchanges. This type of models is selected for the main reason that it allows us to capture the differences of systematic risks of ASEAN-6 financial markets in the period of high and low market volatility. Systematic risks are measured using two-state Markov Switching Model for ASEAN-6 financial markets in the period of high and low volatility by alternative specifications of MS models which allow shifts in mean, intercept, coefficient, and variance across regimes.

The organization of this paper is as follows. Section 2 describes the methodologies used in the model. Section 3 exhibits data description. Section 4 employs the proposed model to investigate the relationship between the expected return and the degree of risk in ASEAN-6 financial markets. Finally, we provide concluding remarks in Sect. 5.

2 Methodology

2.1 CAPM Model

Sharpe [12] put forward the first Capital Asset Pricing Model (CAPM). The CAMP is designed for examining the relationship between expected asset return and its risk. The idea is the trade-off compensation of risk and return while comparing to the market return, or how the asset return deviates from group. By setting expected asset return equal to risk-free asset return equal to risk-free asset return plus a multiplication of market risk premium and systematic risk [15]. The model equation can be expressed as follows

$$r_{it} = \alpha_i + \beta_i (r_{mt}) + \varepsilon_{it}, \quad (1)$$

where $r_{it} = R_{it} - R_{ft}$ and $r_{mt} = R_{mt} - R_{ft}$, α and β are parameter. R_{it} is the return on asset i. R_{mt} is the market return. R_{ft} is the risk free asset return at time t. The error term ε_{it} is assumed to be i.i.d. $N\left(0, \sigma^2\right)$. The beta coefficient indicates relation between risk of selected asset and the risk of the whole market, or in another implication is an indicator for systematic risk of an asset.

2.2 MS-CAPM Model

Gu [5] developed a regime-switching CAPM model which can be written in the general form as follows:

$$r_{it} = \alpha_{i,s_t} + \beta_{i,s_t} r_{mt} + \varepsilon_{t,s_t}, \quad (2)$$

where $r_{it} = R_{it} - R_{ft}$ and $r_{mt} = R_{mt} - R_{ft}$ indicates excess return on asset and on the market portfolio at time t. α_{i,s_t} and β_{i,s_t} denote regime dependent intercept and beta risk parameter, respectively. $\varepsilon_{t,s_t} \sim N\left(0, \sigma^2_{s_t}\right)$ and the unobserved state variable s_t is first order Markov switching process and $\sigma^2_{s_t}$ is the regime dependent variance. Given the information available at time $t-1$, the conditional probability of observing the state s_t is given as follows:

$$\begin{aligned} P\left[s_t = 1 \,|s_{t-1} = 1\right] &= p \\ P\left[s_t = 0 \,|s_{t-1} = 1\right] &= 1 - p \\ P\left[s_t = 0 \,|s_{t-1} = 0\right] &= q \\ P\left[s_t = 1 \,|s_{t-1} = 0\right] &= 1 - q \end{aligned} \quad (3)$$

where p and q are the fixed transition probabilities of being in low or high volatility regime, respectively. In this study, some parameters are conditioned on the state of the Markov chain, while the other parameters are regime-invariant. In the general MS-CAPM specification, all the parameters are regime-dependent, see Eq. 1.

Table 1 Types of MS-CAPM models

		MS	
		α_i varying	α_i invariant
β_i invariant	σ^2 invariant	MSI-CAPM	CAPM (linear)
	σ^2 varying	MSIH-CAPM	MSH-CAPM
β_i varying	σ^2 invariant	MSIB-CAPM	MSB-CAPM
	σ^2 varying	MSIBH-CAPM	MSBH-CAPM

However, in the empirical applications, it is preferable to consider only some parameters as dependent on the state s_t as this may offer a lot of alternatives for the MS-CAPM specification.

Here, we establish a common notation to provide simplicity in expressing various specifications of MS-CAPM in which some parameters are subject to shifts with the varying state. In other words, the model allows intercept, beta risk, and variance to either vary or not vary across regimes. In order to establish a unique notation for each model, we specify the following regime-dependent parameters with the general MS term (Table 1):

I Markov-switching intercept term (α)
B Markov-switching beta risk parameter (β)
H Markov-switching heteroskedasticity (σ)

MS-CAPM specifications In this study, we proposed eight MS-CAPM specifications with two regimes, as follows:

1. Markov-switching intercept, beta parameter and heteroscedasticity: MSIBH-CAPM

$$r_{it} = \alpha_{i,s_t} + \beta_{i,s_t} r_{mt,s_t} + \varepsilon_{t,s_t}, \quad \varepsilon_{t,s_t} \sim N(0, \sigma_{s_t}^2);\, s_t = 1, 2$$

2. Markov-switching intercept and beta parameter: MSIB-CAPM

$$r_{it} = \alpha_{i,s_t} + \beta_{i,s_t} r_{mt} + \varepsilon_t \quad , \varepsilon_t \sim N(0, \sigma^2);\, s_t = 1, 2$$

3. Markov-switching intercept and heteroscedasticity: MSIH-CAPM

$$r_{it} = \alpha_{i,s_t} + \beta_i r_{mt} + \varepsilon_t \quad , \varepsilon_{t,s_t} \sim N(0, \sigma_{s_t}^2);\, s_t = 1, 2$$

4. Markov-switching intercept: MSI-CAPM

$$r_{it} = \alpha_{i,s_t} + \beta_i r_{mt} + \varepsilon_t \quad , \varepsilon_t \sim N(0, \sigma^2);\, s_t = 1, 2$$

5. Markov-switching beta parameter: MSB-CAPM

Table 2 Descriptive Statistics

	FTSE40	INDONESIA	MALAYSIA	PHILIPPINES	SINGAPORE	THAILAND	VIETNAM
Mean	0.0000	0.0003	0.0001	0.0002	0.0000	0.0002	–0.0004
Median	0.0002	0.0004	0.0000	0.0000	0.0000	0.0000	0.0000
Maximum	0.0697	0.0762	0.0425	0.0936	0.0753	0.0891	0.0474
Minimum	–0.0816	–0.1095	–0.0997	–0.1308	–0.0869	–0.1256	–0.0866
Std. Dev.	0.0112	0.013	0.0071	0.01221	0.011	0.0133	0.0156
Skewness	–0.263	–0.645	–1.1593	–0.746	–0.1511	–0.4934	–0.2782
Kurtosis	8.6241	11.9245	19.9538	12.6295	10.2597	11.5944	4.5756
Jarque-Bera	3872.8	9869.3	35539.85	11525.09	6408.03	9083.62	338.93
MBF-JB	0.0000	0.0000	0.0000	0.0000	0.0000	0.0000	0.0000
Unit root test	–49.946	–48.8297	–49.1787	–48.4922	–52.5943	–53.4897	–44.2598
MBF-Unit root	0.0000	0.0000	0.0000	0.0000	0.0000	0.0000	0.0000

Note MBF is Minimum Bayes factor which computed by using formulas from [14]

$$r_{it} = \alpha_i + \beta_{i,s_t} r_{mt,s_t} + \varepsilon_t \quad, \varepsilon_t \sim N(0, \sigma^2);\ s_t = 1, 2$$

6. Markov-switching heteroscedasticity: MSH-CAPM

$$r_{it} = \alpha_i + \beta_i r_{mt} + \varepsilon_t \quad, \varepsilon_{t,s_t} \sim N(0, \sigma_{s_t}^2);\ s_t = 1, 2$$

7. Markov-switching beta parameter and heteroscedasticity: MSBH-CAPM

$$r_{it} = \alpha_i + \beta_{i,s_t} r_{mt,s_t} + \varepsilon_t \quad, \varepsilon_{t,s_t} \sim N(0, \sigma_{s_t}^2);\ s_t = 1, 2$$

8. CAPM (linear)

$$r_{it} = \alpha_i + \beta_i r_{mt} + \varepsilon_t$$

To compare the various types of models with the above specifications, for choosing from the different MS specifications, we use the Akaike information criterion (AIC) and Bayesian information criterion (BIC).

3 Data

The data used are the daily prices of the overall ASEAN market (FTSE ASEAN 40) and ASEAN-6 financial stocks consisting of Thailand (SET50), Philippines (PSEi), Indonesia (JCI), Malaysia (FBMKLCI), Singapore (STI) and Vietnam (TFVTTU). Here, we assume risk free is equal to zero. The daily time series data covering period from June 2007 to August, 2018 are used. All data are collected from Bloomberg database. We then transform these time series variables into log-return rate and the movement of these returns are plotted in Fig. 1. Extremely high positive and negative returns are evident from September 2008 onward, and have continued well into 2009.

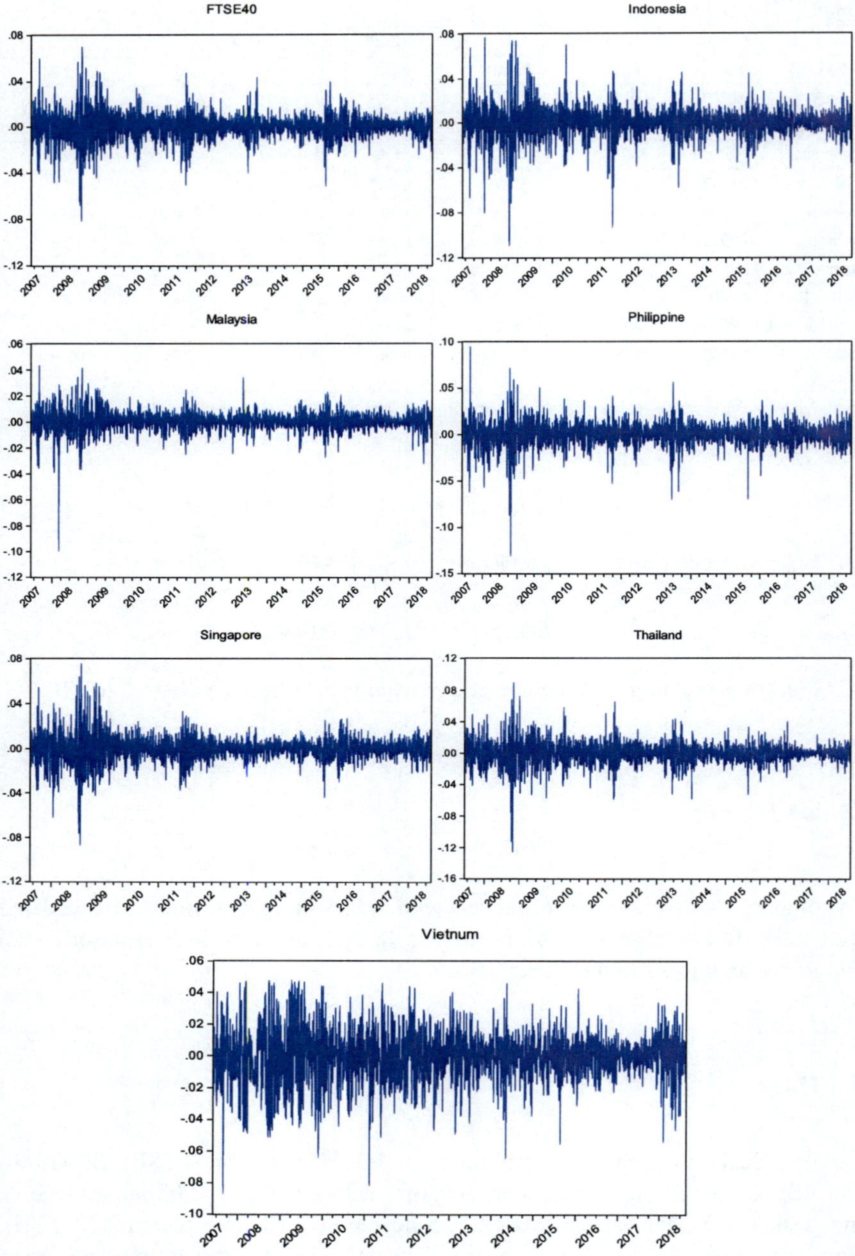

Fig. 1 Daily returns of ASAEN market stock and ASEAN-6 financial stocks from 2007 to 2018

These return series also show volatility under uncertainty situation particularly as occurred during the financial crisis of 2007–2009. The descriptive statistics are provided in Table 2. We observe that ASEAN-6 stocks and ASEAN market stock are not normally distributed. Minimum Bayes factor (MBF) values indicate that the returns are decisive not normally distributed. Unit root test is also conducted to investigate the stationarity of the data and the result shows that all returns are stationary with decisive evidence. Consequently, these variables can be used to estimate in the MS-CAPM models in the further step. Note that the interpretation of MBF is referred to Vovk [16].

4 Empirical Results

4.1 *Model Selection and Results*

In this study, we make a model comparison among various specifications of two regime used to estimate eight models in the MS-CAPM models for ASEAN-6 stock market (Vietnam, Thailand, Indonesia, Malaysia, Singapore and Philippines) using the Akaike Information Criterion (AIC) and Bayesian Information Criterion (BIC). According to the results in Table 3, it is found that MSB-CAPM model (with switching beta parameter) is the best fit for four country stock returns namely Vietnam, Indonesia, Malaysia, and Singapore. MSBH-CAPM model (with switching beta parameter and heteroskedasticity) is the best fit for Philippines stock return, and MSIBH-CAPM model (with switching intercept term, beta parameter, and heteroskedasticity) is the best fit for Thailand stock return because these models provide the minimum AIC and BIC. Also, this evidences confirm the variability of the beta risk across market conditions.

The results of MS-CAPM models are given in Tables 4, 5 and 6. In each table, it shows beta parameter in low and high volatile periods. The beta risks are calculated for two regimes, $\beta_{s_t=1}$ and $\beta_{s_t=2}$. Table 4 presents the estimation MSB-CAPM model (with switching beta parameter) of Indonesia, Singapore, Vietnam, Malaysia stock markets. We note that the beta risk is less than one, indicating that the stock return moves less than the ASEAN market return (FTSE ASEAN 40 Index) while the beta risk greater than one, demonstrating that the stock return moves larger than the ASEAN market return. It is found that the beta risk of Indonesia stock is positive beta risk for both regimes whereas the beta risk of regime one is lower than regime two, we thus interpret regime 1 as high volatility or bull regime and regime 2 as low volatility or bear regime. For the beta risk of Singapore, Vietnam, and Malaysia stock markets, they are less than one with decisive evidence. During the low volatile period, the beta parameters of Singapore and Malaysia markets are positive and less than one, whereas the beta of Vietnam stock market is less than zero with weak evidence, indicating that the return of the Vietnam stock market may move in the opposite direction to the market return. The probabilities of switching from regime

Table 3 Model selection

	Vietnam		Thailand	
Models	AIC	BIC	AIC	BIC
MSB	**–16060.4**	**–16024.5**	–18889	–18853.1
MSBH	–10039.9	–9998.04	–18946.1	–18904.2
MSH	–16015.3	–15979.4	582812	582847.9
MSI	–15948.7	–15912.8	–18820.9	–18785
MSIB	–16016.7	–15974.9	–18885.9	–18844
MSIBH	–16011.8	–15964	**–19456.6**	**–19408.8**
MSIH	–15957.9	–15916.1	–19229.8	–19187.9
CAPM linear	–6.165389	–6.161288	–6.806792	–6.809416
	Indonesia		Malaysia	
Models	AIC	BIC	AIC	BIC
MSB	**–19879.2**	**–19843.3**	**–22674.6**	**–22638.7**
MSBH	–19788.9	–19747.1	–21558.1	–21516.3
MSH	582812	582847.9	–22519.4	–22483.5
MSI	–19599.3	–19563.5	–22582.4	–22546.5
MSIB	–19531.6	–19489.7	–21665.9	–21624
MSIBH	–19781.6	–19733.8	–21934.3	–21886.5
MSIH	–1946.46	–1904.62	–9709.07	–9667.23
CAPM linear	–7.032866	–7.028765	–6.852161	–6.848058
	Singapore		Philippines	
Models	AIC	BIC	AIC	BIC
MSB	**–22066.6**	**–22030.8**	–18181.3	–18145.4
MSBH	–21846.7	–21804.8	**–18345.6**	**–18303.7**
MSH	–21864.3	–21828.4	–18134.8	–18098.9
MSI	–21862.8	–21827	–18153.4	–18117.6
MSIB	–20531.7	–20489.9	–18150.8	–18108.9
MSIBH	–20867.7	–20819.9	–18281	–18233.2
MSIH	–2691.01	–2649.17	–10670	–10628.2
CAPM linear	–7.481407	–7.477306	–6.40551	–6.401409

Bold number indicates the lowest AIC and BIC

Table 4 Estimation MSA-CAPM model of four stock markets

Parameter	Indonesia	Singapore	Vietnam	Malaysia
α_i	0.0004(0.0235)	0.0001(0.9954)	–0.0003(0.6621)	0.0001(0.7911)
$\beta_{s_t=1}$	0.5460(0.0000)	0.6960(0.0000)	–0.1061(0.6705)	0.3625(0.0000)
$\beta_{s_t=2}$	1.2362(0.0000)	0.9722(0.0000)	0.6309(0.0000)	0.5224(0.0000)
σ_1	0.0075(0.0000)	0.0053(0.0000)	0.0149(0.0000)	0.0050(0.0000)
P_{11}	0.7800(0.0000)	0.8021(0.0000)	0.7432(0.0000)	0.7991(0.0000)
P_{22}	0.8263(0.0000)	0.7979(0.0000)	0.8339(0.0000)	0.8009(0.0000)

Note () is MBF Minimum Bayes factor (see [14])

Table 5 Estimated MSBH-CAPM model of Philippines stock market

Parameter	Philippines
α_i	0.0003(0.3369)
$\beta_{S_t=1}$	0.4352(0.0000)
$\beta_{S_t=2}$	0.6335(0.9945)
$\sigma_{S_t=1}$	0.0097(0.0000)
$\sigma_{S_t=2}$	0.9970(0.0000)
P_{11}	0.9961(0.0000)
P_{22}	0.7928(0.0000)

Note () is MBF Minimum Bayes factor (see [14])

Table 6 Estimated MSIBH-CAPM model of Thailand stock market

Parameter	Thailand
$\alpha_{S_t=1}$	0.0004 (0.0280)
$\alpha_{S_t=2}$	−0.0006(0.7265)
$\beta_{S_t=1}$	0.7561 (0.0000)
$\beta_{S_t=2}$	0.7572 (0.0000)
$\sigma_{S_t=1}$	0.0064 (0.0000)
$\sigma_{S_t=2}$	0.0167 (0.0000)
P_{11}	0.9585 (0.0000)
P_{22}	0.7806 (0.0000)

Note () is MBF Minimum Bayes factor (see [14])

1 (bear market) and regime 2 (bull market) are shown in this table; we can see that there is high probability of staying in each regime, indicating a high persistence of market condition.

Table 5 presents the estimation results of MSBH-CAPM model (with switching beta parameter and heteroskedasticity) of ASEAN stock markets. We can see that the beta risk indicates a positive return of 0.4352 in bear market with decisive evidence while the bull market shows a positive of 0.6335 with weak evidence. We also observe that the beta risks are less than one indicates that the stock return moves less than the market ASEAN return (FTSE ASEAN 40 Index). The probability of switching from regime 1 to regime 2 is 0.0039, while remaining in regime 1 is 0.9961. On the other hand, the probability of switching from regime 2 to regime 1 is 0.2072, while remaining in regime 2 is 0.7928. This suggests higher probability for staying in the low volatility regime or bear market when compared to high volatility regime or bull market.

Finally, Table 6 presents the estimation results of MSIBH-CAPM model (with switching intercept term, beta parameter and heteroskedasticity) for Thailand stock market. There are similar beta risks in both regimes with decisive evidence and both regimes show beta risks with values less than one, indicating that the stock return moves less than the ASEAN market return (FTSE ASEAN 40 Index). The probability of staying in each regime shows that there is higher probability of staying in regime 1 than regime 2.

Considering structural change, if we misspecify the switching equation structure, the result may be incorrect with the over or under estimated parameter. This paper shows that CAPM of ASEAN-6 stocks do not switch all parameters. Therefore, we gain more accuracy by restricting the structure of the model.

5 Conclusion

Various specifications of CAPM, which measures the risk of the stock returns against the overall market returns, are considered in this study. The non-linear feature or structural change of the stock market is addressed in our study, hence we present the Markov switching CAPM models. In the general MS-CAPM specifications, all the parameters are regime-dependent. However, in the empirical applications, it is preferable to consider only some parameters as dependent on the state as this may offer a lot of alternatives for the MS-CAPM specification. In this study, we introduce eight MS-CAPM specifications with two regimes including MSI-CAPM, MSIH-CAPM, MSIB-CAPM, MSIBH-CAPM, MSH-CAPM, MSB-CAPM, MSBH-CAPM and CAPM (linear) models. The filtered probabilities of the bull and bear regimes are estimated using the Hamilton's filter. The data used are the daily prices of overall ASEAN market (FTSE ASEAN 40 Index) and ASEAN-6 financial stocks consisting of Thailand (SET50), Philippines (PSEi), Indonesia (JCI), Malaysia (FBMKLCI), Singapore (STI) and Vietnam (TFVTTU).

To test and find the best fit specification of the MS models, we employ selection criteria that are AIC and BIC, where the minimum AIC and BIC. The empirical results in this paper reveal that MSB-CAPM model (with switching beta parameter) represents the best fit for Vietnam, Indonesia, Malaysia and Singapore stock returns, MSBH-CAPM model (with switching beta parameter and heteroskedasticity) for Philippines stock return and MSIBH-CAPM model (with switching intercept term, beta parameter and heteroskedasticity) for Thailand stock return. We can say that If switching equation structure is not well defined, the obtained result can be misleading and the risk can be higher than (or lower) the market making investors to have lower (or higher) risk than the market. We show that CAPM for ASEAN-6 stocks do not switch due to the whole parameters in the equation. By restricting the structure of the model to identify the determinants of the structural change will enhance the beta risk estimates.

References

1. Hamilton, J. D. (1989). A new approach to the economic analysis of nonstationary time series and the business cycle. *Econometrica, 57,* 357–384.
2. Hamilton, J. D., & Susmel, R. (1994). Autoregressive conditional heteroskedasticity and changes in regime. *Journal of Econometrics, 64*(1–2), 307–333.
3. Held, L., & Ott, M. (2018). On p-values and Bayes factors. *Annual Review of Statistics and Its Application, 5,* 393–419.
4. Huang, H.-C., & (River). (2003). Tests of regime-switching CAPM under price limits. *International Review of Economics and Finance, 12*(3), 305–326. https://doi.org/10.1016/s1059-0560(03)00013-3
5. Gu, L. (2005). Asymmetric risk loadings in the cross section of stock returns. Available at SSRN 676845.
6. Krolzig, H. M. (2013). Markov-switching vector autoregressions: Modelling, statistical inference, and application to business cycle analysis (Vol. 454). Springer Science and Business Media.
7. Kuan, C. M. (2002). Lecture on the Markov switching model. Institute of Economics Academia Sinica, 1–30.
8. Pastpipatkul, P., Yamaka, W., & Sriboonchitta, S. (2016). Effect of quantitative easing on ASEAN-5 financial markets. In Causal Inference in Econometrics (pp. 525–543). Springer.
9. Pastpipatkul, P., Maneejuk, P., & Sriboonchitta, S. (2017). Markov switching regression with interval data: Application to financial risk via CAPM. *Advanced Science Letters, 23*(11), 10794–10798.
10. Prukumpai, S. (2015). Time-varying industrial portfolio betas under the regime-switching model: Evidence from the stock exchange of Thailand. *Applied Economics Journal, 22*(2), 54–76.
11. Rakpho, P., Yamaka, W., & Sriboonchitta, S. (2018). Which quantile is the most informative? Markov switching quantile model with unknown quantile level. *Journal of Physics: Conference Series, 1053*(1), 012121. (IOP Publishing).
12. Sharpe, W. F. (1964). Capital asset prices: A theory of market equilibrium under conditions of risk. *The Journal of Finance, 19*(3), 425–442.
13. Skrinjaric, T. (2014). Testing for regime-switching CAPM on Zagreb Stock Exchange. *Croatian Operational Research Review,* 119–133.
14. Stern, H. S. (2016). A test by any other name: P values, Bayes factors, and statistical inference. *Multivariate Behavioral Research, 51*(1), 23–29.
15. Thamprasert, K., Pastpipatkul, P., & Yamaka, W. (2018). Interval-valued estimation for the five largest market capitalization stocks in the stock exchange of Thailand by Markov-switching CAPM. In *International Econometric Conference of Vietnam* (pp. 916–925). Springer.
16. Vovk, V. G. (1993). A logic of probability, with application to the foundations of statistics. *Journal of the Royal Statistical Society. Series B (Methodological),* 317–351.
17. Vendrame, V., Guermat, C., & Tucker, J. (2018). A conditional regime switching CAPM. *International Review of Financial Analysis, 56,* 1–11.

A Bayesian Approach to Quantile Regression for Interval-Valued Data: Application to CAPM

Rungrapee Phadkantha, Woraphon Yamaka, and Songsak Sriboonchitta

Abstract The purpose of this paper is to introduce an approach to fitting a quantile regression on interval-valued data. This approach consists of fitting a model on the appropriate point of the interval values. To obtain this point, the convex combination method is applied in the quantile regression. Moreover, we also introduce the Bayesian approach for estimating all the unknown parameters in the model. The approach is illustrated via a simulation study and real data sets. In the real data study, we apply this methodology to measure the beta risk of Thai stock returns through Capital Asset Pricing model. The results show the high performance and accuracy of the Bayesian estimation in both simulated data and real application study.

Keywords Bayesian · Quantile regression · Interval-valued data · CAPM

1 Introduction

Due to an imprecise observation or an inexact measurement of the single-valued data particularly those related to stock or bond market include shares, preferred shares and government bonds etc., the interval-valued data, which contains the maximum and minimum values of the observed data, has been considered in the recent years for replacing the single-valued data. By using the interval-valued data, we can capture more realistic the movement or information of the stock price. In the pioneer work, Billard and Diday [2] presented the first approach to fitting a linear regression model using an interval-valued data set, derived from the midpoints between the lower and upper boundaries of the independent and dependent variables. Lima Neto and De Carvalho [12] improved this approach by presenting a new method based on two

R. Phadkantha · W. Yamaka (✉) · S. Sriboonchitta
Center of Excellence in Econometrics, Faculty of Economics, Chiang Mai University, Chiang Mai, Thailand
e-mail: woraphon.econ@gmail.com

S. Sriboonchitta et al. (eds.), *Credible Asset Allocation, Optimal Transport Methods, and Related Topics*, Studies in Systems, Decision and Control 429,
https://doi.org/10.1007/978-3-030-97273-8_21

linear regression models, the first regression model on the midpoints of the intervals and the second one on the ranges which reconstruct the boundaries of the interval values of the dependent variable in a more efficient manner than the Billard and Diday's method. In a more recent work, Chanaim et al. [4] argued that the mid-point and range methods are not good enough to be the representative of the interval data, thus they introduced convex combination method to get the representative point of the interval-valued data. Instead of restricting the weight of interval-valued data (w) to be 0.5 (mid-point), $\tilde{x} = 0.5(x_{\max}) + 0.5(x_{\min})$, they generalized this weight to be unknown. Hence, the representative point data becomes $\tilde{x} = \mathrm{w}(x_{\max}) + (1 - \mathrm{w})(x_{\min})$, where $\mathrm{w} \in [0, 1]$ is the weight parameter.

In this study, we aim to go beyond the classical linear regression for interval-valued data. We would rather introduce quantile regression model for interval-valued data as there are various drawbacks of the linear regression model. It is well known that this classical regression examines how the conditional mean of the response variable relates to a set of predictors. However, this regression estimate is not robust against outliers in the response measurements. This model is not desirable if the conditional quantile functions are of interest. Thus, the quantile regression of Koenker and Bassett [10] is considered in this study as we can investigate how the median or other quantiles of the response variable relate to the predictors. The study of Allen et al. [1] stated that linear regression is unable to capture the tail behavior of the market, namely extreme market events. Nowadays, after the recent global financial crises, many financial institutions have been getting more cautious to understand the risk of their financial investments under extreme market conditions than the normal market conditions. Thus, the quantile regression for interval-valued data is introduced in this study to deal with these complicated problems.

From the estimation point of view, there is a rich body of literature on quantile regression with the frequentist view however, not much work has been done under the Bayesian framework. Pastpipatkul et al. [13] suggested that Bayesian estimation can escape possible misspecification errors and improve efficiency in estimation due to an identification of a stochastic or deterministic trend for the estimation of parameters. Also, when there are many parameters to be estimated, it is claimed to have more efficiency than the traditional estimators like Least squares and Maximum likelihood. By using the Markov chain Monte Carlo (MCMC) method, it may be used for obtaining the posterior distributions, even in complex situations.

It is therefore more suitable to use quantile regression to quantify the beta risk of the asset returns in Capital Asset Pricing model at given quantile level. We first briefly review how quantile regression works following Koenker and Zhao [11] and Chen et al. [6]. Moreover, we use the Bayesian approach to estimate the parameters in quantile regression. This estimation has attracted increasing attention from many studies ([8, 16]).

The main objective of this study is to propose a Bayesian quantile regression for interval-valued data. To our knowledge, this study is the first attempt to estimate the quantile regression with interval-valued data using Bayesian approach. We then illustrate our model by investigating the CAPM. There have been other recent extensions of the CAPM with interval data, as seen in Tibprasorn et al. [18], Phochanachan et al. [15] and Phadkantha et al. [14]. Nevertheless, their works are restricted in the regression framework.

The remainder of the paper is as follows: Sect. 2 describes the methodology used in this study, Sect. 3 conducts a simulation study to evaluate and illustrate our method, Sect. 4 presents the empirical analysis and employs the proposed model to investigate the risk of Thai stock market through CAPM. Finally, we provide concluding remarks in Sect. 5.

2 Methodology

In this section, we briefly review the center method and convex combination proposed by Billard and Diday [2] and Chanaim et al. [4], respectively. We then present the quantile regression model for interval-valued data.

2.1 Center Method

The center method was firstly applied to the time series model by Billard and Diday [2]. They proposed this method to the linear regression model. The main idea of this method is the estimated slope parameter is based on the center or mid-point of the interval data. Let $y = \left[y_u, y_l\right]$ where y_u denotes the upper bound value of observed y while y_l denotes the lower bound of observed y. To obtain the mid-point value of y, we can derive by

$$y_c = \frac{y_u + y_l}{2}, \tag{1}$$

where y_c is a center price of interval valued data, derived from the sum of the maximum and minimum prices divided by two.

2.2 Convex Combination

Chanaim et al. [5] suggested that the center method may lead to the misspecification problem since the midpoint of the intervals might not be the good representative of the intervals. To overcome this problem, they introduced a convex combination approach

$$y_{cc} = \mathrm{w}y_l + (1-\mathrm{w})y_u \;\; ; \mathrm{w} \in [0,1], \tag{2}$$

where w is the weight parameter of the interval data with values [0,1]. The advantage of this method lies in the flexibility to assign weights in calculating the appropriate value between intervals.

2.3 Quantile Regression with Interval Data

Quantile regression is a statistical method proposed by Koenker and Bassett [10]. It presents the relationship between the dependent and independent variables at the specific quantiles (also called "percentiles"), therefore the estimated parameters change according to the specified quantile level. It is widely used in cases where a study seeks to estimate the different percentiles of a population of interest. This method can be considered as a natural analogue in the regression analysis for providing a thorough and robust data analysis. For a further comprehensive explanation of this methodology, see Buchinsky [3] and Yamaka et al. [20]. As we deal with the interval-valued data, the quantile regression with one covariate model can be simply defined as

$$\mathrm{w}_1(\tau)y_{l,t} + (1-\mathrm{w}_1(\tau))y_{u,t} = \beta(\tau)[w_2(\tau)x_{l,t} + (1-\mathrm{w}_2(\tau))x_{u,t}] + \varepsilon_t(\tau), \;\; t = 1,\ldots,T, \tag{3}$$

where $y_{l,t}$ is the lower bound of dependent variabley_t and $y_{u,t}$ the upper bound of dependent variable y_t. $x_{l,t}$ is the lower bound of dependent variable x_t and $x_{u,t}$ is the upper bound of dependent variable x_t. w is the coefficient of the expected market risk premium and the error term, ε_t has an asymmetric Laplace distribution. Thus, $(0 < \tau < 1)$ conditional quantile of y_t given x'_t is simply as

$$Q_{(\mathrm{w}_1(\tau))y_{l,t}+(1-\mathrm{w}_1(\tau))y_{u,t}}(\tau\,|x_l, x_u) = \left[\mathrm{w}_2(\tau)\,x_{l.t} + (1-\mathrm{w}_2(\tau))x'_{u.t}\right]\beta(\tau), \tag{4}$$

where $\beta(\tau)$ is a vector of coefficients dependent on τ. In the conventional least squares method, the τ-th is defined as any solution for $\hat{\beta}(\tau)$. Thus, the minimization problem is

$$\begin{aligned}\hat{\beta}(\tau) &= \min_{\beta \in R^k} E\left[\rho_\tau(y_{cc,t} - x_{cc,t}\beta(\tau))\right] \\ &= \min_{\beta \in R^k} \sum_{t=1}^{T}\left[\rho_\tau(y_{cc,t} - x_{cc,t}\beta(\tau))\right],\end{aligned} \tag{5}$$

where the loss function is

$$\rho_\tau(\varepsilon(\tau)) = u(\tau - I(\varepsilon(\tau) < 0)). \tag{6}$$

2.4 Bayesian Estimation

The minimization problem in Eq. (5) corresponds to the maximum asymmetric Laplace likelihood [21]. Thus, our Bayesian estimation is based on this likelihood distribution. In this estimation, MCMC method is employed to extract the posterior distributions of unknown parameters $\beta(\tau)$,variance $\sigma(\tau)$ and w(τ) which is given by

$$P(\beta^{\tau}, \mathrm{w}(\tau), \sigma \,|y_{cc}, x_{cc}) = L(\beta(\tau), \mathrm{w}(\tau), \sigma(\tau)\,|y_{cc}, x_{cc})p(\beta(\tau), \mathrm{w}(\tau), \sigma(\tau)), \tag{7}$$

where $L(\beta(\tau), \mathrm{w}(\tau), \sigma(\tau)\,|y_{cc}, x_{cc})$ is the asymmetric Laplace likelihood, $p(\beta(\tau), \sigma(\tau))$ is prior distribution of $\beta(\tau)$and $\sigma(\tau)$. Here, we can choose any prior distribution depending on our belief to produce a proper joint posterior. In this study, the uniform prior is assumed for constructing the posterior distribution. Inference about model parameters then follows conventional Bayesian procedures though Metropolis Hasting (MH) as described Chib et al. [7]. As we apply our model to study the CAPM, thus we specified the prior distribution of our model as in the following:

$$p(\beta(\tau)) \sim U(-1, 1)$$

$$p(\sigma(\tau)) \sim U(0, 1),$$

$$p(w(\tau)) \sim U(0, 1),$$

where $U(\cdot)$ is uniform distribution. The MH iterations for $\theta(\tau) = \{\beta(\tau), \sigma(\tau), \mathrm{w}(\tau)\}$ process can be described as follows:

1. Starting at an initial parameter value for $\theta^0(\tau)$
2. Choosing a new parameter value close to the old value based on proposal function. The proposal distribution employed in the MH algorithm is a normal distribution with mean $\tilde{\theta}$ and covariance (C^j), that is $\theta^j = N(\tilde{\theta}, C^j)$ where $C^j = \sigma_d \operatorname{cov}(\theta^j) + \sigma_d \varepsilon I_d$ after initial period, where σ_d is a parameter that depends on dimension d and ε is a constant term which is very tiny when compared with the size of the likelihood function.
3. Computing the acceptance probability which is calculated by

$$\theta_j = \frac{P(\beta^j(\tau), \mathrm{w}^j(\tau), \sigma^j(\tau), C\,|y_{cc}, x_{cc})}{P(\beta^{j-1}(\tau), \mathrm{w}^{j-1}(\tau), \sigma^{j-1}(\tau), C\,|y_{cc}, x_{cc})}$$

   If $\theta_j \geq 1$ then draw trace $\theta^j = \theta^{j-1}$ else we draw trace θ^j from a proposal distribution.
4. Repeat steps 2–3 for $j = 1, \ldots, J$ in order to obtain samples $\theta^1, \ldots, \theta^J$.

3 Simulation Study

We describe our approach to Bayesian quantile regression for interval-valued data through simulated data. The Monte Carlo simulation is conducted to examine the finite sample performance of our proposed model. To this end, we consider the following model specifications.

Model 1: Center of interval data: $w_1 = 0.5$, $w_2 = 0.5$

$$\left[(0.5)y_t^U + (1-0.5)y_t^L\right] = 2 + 3\left[(0.5)x_t^U + (1-0.5)x_t^L\right] + \varepsilon_t \quad (8)$$

Model 2: Convex combination of interval data: $w_1 = 0.4$, $w_2 = 0.6$

$$\left[(0.4)y_t^U + (1-0.4)y_t^L\right] = 2 + 3\left[(0.6)x_t^U + (1-0.6)x_t^L\right] + \varepsilon_t \quad (9)$$

In this section, we also consider different quantile levels of our proposed model; hence is assumed to be asymmetric Laplace distribution with skew parameter $\tau = (0.20, 0.50, 0.80)$. We simulate 500 and 1000 samples for the Monte Carlo experiment using the specified parameters of each case. For each simulation case, we perform 100 replications for all the three cases. In each simulation, we proceed as follows.

(1) We draw $\varepsilon_t(\tau)$ from the ALD with mean 0, variance 1, and skew parameter $\tau = (0.20, 0.50, 0.80)$.
(2) Random variable from normal distribution with mean 0, variance 1. Then, we generate $x_{u,t}$ according to $x_{l,t} = x_{u,t} - U[0, 2]$. This guarantees that the bounds are not crossing each other.
(3) $x_{cc,t}$ of the intervals have been computed by $x_{cc,t} = (w_2 x_{u,t}) + (1 - w_2)x_{u,t}$. Then $y_{cc,t}$ is generated from $y_{cc,t} = x_{cc,t}\beta(\tau) + \varepsilon_t(\tau)$.
(4) Compute the upper and lower bounds of y_t by $y_{l,t} = y_{cc,t} - U(0, 2)$ and $y_{u,t} = \left(y_{cc,t} - (1 - w_1)y_{cc,t}\right)/w_1$.

To assess the performance of our proposed model and compare our result with 2 model, based on Bias value of each parameter estimate. In this simulation study, we perform $B = 100$ replications. The Bias value can be defined as

$$Bias = \sum_{b=1}^{B}\left(\widehat{\theta}_b - \theta^{true}\right)/B \quad (10)$$

Table 1 displays the summary statistics from 100 replications. It can be seen that the median estimates are close to the true values, indicating that the posterior median obtained by our sampling scheme is reliable. We compare the results of the model specifications and find that the Bias of parameters obtained from the center method are mostly lower than those from the convex combination method when the data is simulated from Model 1. This is not a surprise as the true value of the weight parameters is assumed to be 0.5. However, when we compare the simulation results under Model 2, we can see that the quantile regression with convex combination

Table 1 Simulation study

Bias	Parameter	Model 1				Model 2			
		Center		CC		Center		CC	
		MLE	Bayesian	MLE	Bayesian	MLE	Bayesian	MLE	Bayesian
N=500	$\alpha(0.2)$	2.6381	2.6729	2.0885	1.9461	0.6381	0.6729	0.0347	0.5497
	$\beta(0.2)$	2.8931	2.8299	2.9012	2.8943	0.1069	0.1701	0.0632	0.0081
	$\sigma(0.2)$	1.0726	1.0784	1.013	1.0184	0.0726	0.0784	0.0059	0.0596
	$w_1(0.2)$			0.5978	−0.6666			0.1117	0.5878
	$w_2(0.2)$			0.7353	0.6634			0.1442	0.1458
N=1000	$\alpha(0.2)$	1.9886	1.8772	1.7631	1.8837	0.0114	0.1228	0.1114	0.2255
	$\beta(0.2)$	2.8935	2.8916	3.014	2.9857	0.1065	0.1084	0.002	0.1205
	$\sigma(0.2)$	0.9696	0.9727	1.0195	1.0219	0.0304	0.0273	0.0032	0.05
	$w_1(0.2)$			0.5823	0.4968			0.0201	0.0737
	$w_2(0.2)$			0.5075	0.5359			0.0411	0.0316
N=500	$\alpha(0.5)$	2.4893	2.3993	1.5569	1.5112	0.4893	0.3993	0.4431	0.4888
	$\beta(0.5)$	3.1589	3.1489	2.948	2.941	0.1589	0.1489	0.052	0.059
	$\sigma(0.5)$	1.0245	1.0305	0.9422	0.948	0.0245	0.0305	0.0578	0.052
	$w_1(0.5)$			0.6206	0.5998			0.3206	0.2998
	$w_2(0.5)$			0.3758	0.3429			0.3242	0.3571
N=1000	$\alpha(0.5)$	2.2096	2.1091	2.0443	2.0565	0.6381	0.6729	0.0885	0.0539
	$\beta(0.5)$	2.9498	2.9574	3.1275	3.1065	0.1069	0.1701	0.0988	0.1057
	$\sigma(0.5)$	0.9781	0.9807	1.0449	1.0481	0.0726	0.0784	0.013	0.0184
	$w_1(0.5)$			0.2997	-0.1304			0.2978	0.9666
	$w_2(0.5)$			0.5702	0.5721			0.0353	0.0366
N=500	$\alpha(0.8)$	1.9747	1.8604	2.1143	1.7995	0.0253	0.1396	0.1143	0.2005
	$\beta(0.8)$	2.8764	2.8652	3.2079	3.2466	0.1236	0.1348	0.2079	0.2466
	$\sigma(0.8)$	0.9551	0.9625	0.9608	0.9658	0.0449	0.0375	0.0392	0.0342
	$w_1(0.8)$			0.235	-0.0896			0.065	0.3896
	$w_2(0.8)$			0.7001	0.577			0.0001	0.123
N=1000	$\alpha(0.8)$	1.7888	1.7236	1.5569	1.5112	0.4893	0.3993	0.4431	0.4888
	$\beta(0.8)$	2.8387	2.8049	2.948	2.941	0.1589	0.1489	0.052	0.059
	$\sigma(0.8)$	0.9803	0.9814	0.9422	0.948	0.0245	0.0305	0.0578	0.052
	$w_1(0.8)$			0.6206	0.5998			0.3206	0.2998
	$w_2(0.8)$			0.3758	0.3429			0.3242	0.3571

Note CC—is convex combination

method is superior to quantile regression with center method as the lower Bias values are obtained. According to this simulation study, we can say that our method is reliable and more flexible than the center method.

4 Empirical Analysis

In this study, we use the interval daily data consisting of Krungthai Card (KTC), TISCO Financial Group PCL (TISCO), Airports of Thailand PCL (AOT), Beauty Community PCL (BEAUTY), PTT Public Company Limited (PTT) and SET index of Thailand for the period from 13 December 2012 to 16 July 2018. We have considered the daily minimum and maximum prices and the data were collected from Bloomberg. To preserve the interval format, we calculate daily returns with respect to the previous day weighted average price, that is,

$$r_{m,i} = \frac{P^U_{m,t} - P^A_{t-1}}{P^A_{t-1}}, \frac{P^L_{m,t} - P^A_{t-1}}{P^A_{t-1}} = \left[r^U_{m,i}, r^L_{m,i}\right], \tag{11}$$

$$r_{s,i} = \frac{P^U_{s,t} - P^A_{t-1}}{P^A_{t-1}}, \frac{P^L_{s,t} - P^A_{t-1}}{P^A_{t-1}} = \left[r^U_{s,i}, r^L_{s,i}\right], \tag{12}$$

$$r_{m,i} - r_{f,i} = \left[r^U_{m,i} - r_{f,i}, r^L_{m,i} - r_{f,i}\right], \tag{13}$$

$$r_{s,i} - r_{f,i} = \left[r^U_{s,i} - r_{f,i}, r^L_{s,i} - r_{f,i}\right], \tag{14}$$

where $P^U_{m,i}$, $P^L_{m,i}$, and $P^A_{m,i}$ are maximum, minimum and average price of our intervals respectively. Figure 1 shows our interval returns, and the description of interval returns is shown in Table 1.

4.1 Data Description

The data description is provided in Table 1. In this study, we use Minimum Bayes factor (MBF) as the tool for checking the significant result. This MBF can be considered as an alternative of p-value [9]. If $1 < MBF < 1/3$, $1/3 < MBF < 1/10$, $1/10 < MBF < 1/30$, $1/30 < MBF < 1/100$, $1/100 < MBF < 1/300$ and $MBF < 1/300$, there is a chance that the MBF favors the weak evidence, moderate evidence, substantial evidence, strong evidence, very strong evidence and decisive evidence for $H_1 : \beta \neq 0$ respectively. All data series are stationary as shown by the MBF values. Skewness is asymmetry in a statistical distribution, in which the curve appears distorted or skewed either to the left or to the right. If it is negative skew, the distribution is concentrated on the right. On the contrary, if it is positive skew, the distribution is concentrated on the left. We observe that the returns exhibit either negative or positive skewness, meaning that our returns may not have a normal distribution. Thus, we conduct the Jarque - Bera normality test and it is evident that stock returns reject the null hypothesis of normal distribution. To preserve the interval format, we calculate daily returns with respect to the previous day weighted average price, that is

$$ry_{max,t} = \frac{y_{\max,t} - y_{avg,t-1}}{y_{avg,t-1}} \text{ and } ry_{min,t} = \frac{y_{\min,t} - y_{avg,t-1}}{y_{avg,t-1}}, \tag{15}$$

where $y_{\min,t}$, $y_{\max,t}$, and $y_{avg,t-1}$ are minimum, maximum, and average price of our intervals.

4.2 *Estimation Results*

In this study, we explain the estimation results from our empirical data. We use quantile regression with convex combination and center methods for interval data as presented above to estimate $\alpha(\tau)$,$\beta(\tau)$,$\sigma(\tau)$,$w_1(\tau)$,$w_2(\tau)$. The results are provided in Table 2. The results of center method suggested on the coefficient values $\beta(\tau)$ indicate TISCO stock and AOT stock are less risky compared to the market return as the quantile level gets higher. On the other hand, the coefficient values $\beta(\tau)$of

Table 2 Data description

	Mean	Median	Max	Min	SD	Skewness	Kurtosis	MBF JB	MBF Unit root
KTC_H	0.018	0.013	0.909	−0.894	0.043	−0.254	281.594	0.000	0.000
KTC_L	−0.014	−0.012	0.888	−0.902	0.042	0.058	290.875	0.000	0.000
TISCO_H	0.011	0.01	0.902	−0.07	0.03	19.06	565.502	0.000	0.000
TISCO_L	−0.01	−0.008	0.886	−0.103	0.03	19.463	589.167	0.000	0.000
AOT_H	0.012	0.01	0.902	−0.88	0.039	0.014	383.417	0.000	0.000
AOT_L	−0.01	−0.008	0.884	−0.883	0.039	0.486	398.161	0.000	0.000
BEAUTY_H	0.019	0.014	0.891	−0.262	0.039	8.846	189.54	0.000	0.000
BEAUTY_L	−0.016	−0.013	0.854	−0.421	0.038	6.873	213.207	0.000	0.000
PTT_H	0.01	0.008	0.048	0.001	0.005	1.845	8.905	0.000	0.000
PTT_L	−0.01	−0.008	−0.001	−0.048	0.005	−1.845	8.905	0.000	0.000
SET_H	0.005	0.005	0.901	−0.054	0.027	27.394	916.916	0.000	0.000
SET_L	−0.004	−0.004	0.895	−0.063	0.027	27.14	907.268	0.000	0.000

Note JB—is Jarque-Bera test

KTC stock, BEAUTY stock and PTT stock are more volatile than market as the quantile level gets higher. The results of convex combination method introduced that the coefficient values $\beta(\tau)$ of KTC stock and PTT stock are more volatile than the market as quantile level gets higher. Conversely, the coefficient values $\beta(\tau)$ of TISCO stock, AOT stock and BEAUTY stock are less volatile than the market as quantile level gets higher. The weights of interval value $w(\tau)$ for center method and convex combination are however different, fixed weight at 0.5 from the former but flexible by assigning $w(\tau)$ between [0,1] for the latter.

5 Conclusion

In this study, we introduce a Bayesian quantile regression for interval-valued data. The convex combination method is applied to find the best representative value between the maximum and minimum values of the observed intervals. We conduct the simulation study to examine the performance of the Bayesian estimation and we find that our Bayesian estimation shows the reliable parameter estimates. Also, we compare the conventional center method and convex combination under the quantile regression framework. The results show the superiority of the convex combination and the flexibility of this method over the center method, when the true weight parameter is not 0.5. We then apply our method to study the interval data of Thailand's Stock Exchange market including Krungthai Card (KTC), TISCO Financial Group PCL (TISCO), Airports of Thailand PCL (AOT), Beauty Community PCL (BEAUTY) and PTT Public Company Limited (PTT). These five stocks can be viewed as high market value in the Thai stock market. Considering the difference in method between the center method and the convex combination. The results confirm that the convex has the flexibility to assign weight between the lower and upper bounds of interval. As the center method defines a weight fixed at 0.5 ignoring uptrend or downtrend of economy, the results of this method seem either underestimate and overestimate in every quantile level. For the future, the study should apply interval data to nonlinear regression or other econometric problems.

Parameter	Bayesian center method	Bayesian CC method	Parameter	Bayesian center method	Bayesian CC method
$\alpha_{KTC}(0.2)$	−0.0049	0.0014	$\alpha_{TISCO}(0.2)$	0.0078	−0.0006
$\beta_{KTC}(0.2)$	1.1578	0.6645	$\beta_{TISCO}(0.2)$	1.3001	0.6263
$\sigma_{KTC}(0.2)$	0.0013	0.0054	$\sigma_{TISCO}(0.2)$	0.0021	0.003
$W_1(0.2)$		0.0268	$W_1(0.2)$		0.0972
$W_2(0.2)$		0.5186	$W_2(0.2)$		0.4995
$\alpha_{KTC}(0.5)$	−0.0002	0.0011	$\alpha_{TISCO}(0.5)$	−0.0002	0.0071
$\beta_{KTC}(0.5)$	0.6737	1.1475	$\beta_{TISCO}(0.5)$	0.6022	0.5144
$\sigma_{KTC}(0.5)$	0.0003	0.0073	$\sigma_{TISCO}(0.5)$	0.0022	0.0047
$W_1(0.5)$		0.5201	$W_1(0.5)$		0.0983
$W_2(0.5)$		0.6275	$W_2(0.5)$		0.5001
$\alpha_{KTC}(0.8)$	0.0078	0.002	$\alpha_{TISCO}(0.8)$	0.0041	0.0088
$\beta_{KTC}(0.8)$	1.2997	1.2576	$\beta_{TISCO}(0.8)$	0.5869	0.6877
$\sigma_{KTC}(0.8)$	0.0021	0.005	$\sigma_{TISCO}(0.8)$	0.001	0.0033
$W_1(0.8)$		0.7344	$W_1(0.8)$		0.5171
$W_2(0.8)$		0.6822	$W_2(0.8)$		0.6976
$\alpha_{BEAUTY}(0.2)$	−0.0038	−0.0005	$\alpha_{PTT}(0.2)$	−0.0007	−0.0065
$\beta_{BEAUTY}(0.2)$	1.0504	0.5001	$\beta_{PTT}(0.2)$	0.5839	0.5981
$\sigma_{BEAUTY}(0.2)$	0.002	0.0053	$\sigma_{PTT}(0.2)$	0.0012	0.0038
$W_1(0.2)$		0.096	$W_1(0.2)$		0.3762
$W_2(0.2)$		0.5005	$W_2(0.2)$		0.5311
$\alpha_{BEAUTY}(0.5)$	−0.0034	−0.0043	$\alpha_{PTT}(0.5)$	−0.0007	0.0058
$\beta_{BEAUTY}(0.5)$	1.034	1.0769	$\beta_{PTT}(0.5)$	0.6819	0.9319
$\sigma_{BEAUTY}(0.5)$	0.0011	0.0073	$\sigma_{PTT}(0.5)$	0.001189	0.005
$W_1(0.5)$		0.7555	$W_1(0.5)$		0.0985
$W_2(0.5)$		0.6855	$W_2(0.5)$		0.5003
$\alpha_{BEAUTY}(0.8)$	−0.0018	0.0135	$\alpha_{PTT}(0.8)$	−0.0049	0.0165
$\beta_{BEAUTY}(0.8)$	1.0201	0.4579	$\beta_{PTT}(0.8)$	1.1578	1.0369
$\sigma_{BEAUTY}(0.8)$	0.0002	0.0062	$\sigma_{PTT}(0.8)$	0.0013	0.0041
$W_1(0.8)$		0.5318	$W_1(0.8)$		0.0981
$W_2(0.8)$		0.4794	$W_2(0.8)$		0.4996
$\alpha_{AOT}(0.2)$	0.0341	0.0008			
$\beta_{AOT}(0.2)$	0.5869	0.5297			
$\sigma_{AOT}(0.2)$	0.0045	0.0038			
$W_1(0.2)$		0.098			
$W_2(0.2)$		0.5012			
$\alpha_{AOT}(0.5)$	0.023	0.0008			
$\beta_{AOT}(0.5)$	0.5009	0.5347			
$\sigma_{AOT}(0.5)$	0.0021	0.0051			
$W_1(0.5)$		0.5228			
$W_2(0.5)$		0.6484			
$\alpha_{AOT}(0.8)$	0.0041	0.0093			
$\beta_{AOT}(0.8)$	0.4409	0.5084			
$\sigma_{AOT}(0.8)$	0.001	0.0042			
$W_1(0.8)$		0.5102			
$W_2(0.8)$		0.5377			

Note () is MBF, which is computed by $MBF_{01}(p) = \begin{cases} -\exp(1)p\log p & \text{for } p < 1/\exp(1) \\ 1 & \text{for } p \geq 1/\exp(1) \end{cases}$ [17, 19] and CC is Convex Combination

References

1. Allen, D. E., Singh, A. K., Powell, R. J., McAleer, M., Taylor, J., & Thomas, L. (2012). The volatility-return relationship: Insights from linear and non-linear quantile regressions.
2. Billard, L., & Diday, E. (2000). Regression analysis for interval-valued data. In *Data analysis, classification, and related methods* (pp. 369–374). Springer.
3. Buchinsky, M. (1998). Recent advances in quantile regression models: a practical guideline for empirical research. *Journal of Human Resources*, 88–126.
4. Chanaim, S., Sriboonchitta, S., & Rungruang, C. (2016). A convex combination method for linear regression with interval data. In *International symposium on integrated uncertainty in knowledge modelling and decision making* (pp. 469-480). Springer.
5. Chanaim, S., Khiewngamdee, C., Sriboonchitta, S., & Rungruang, C. (2018). A convex combination method for quantile regression with interval data. In *International econometric conference of Vietnam* (pp. 440-449). Springer.
6. Chen, K. G., Valencia, J. C., Gillet, J. P., Hearing, V. J., & Gottesman, M. M. (2009). Involvement of ABC transporters in melanogenesis and the development of multidrug resistance of melanoma. *Pigment Cell and Melanoma Research, 22*(6), 740–749.
7. Chib, S., & Greenberg, E. (1995). Understanding the metropolis-hastings algorithm. *The American Statistician, 49*(4), 327–335.
8. Gerlach, R. H., Chen, C. W., & Chan, N. Y. (2011). Bayesian time-varying quantile forecasting for value-at-risk in financial markets. *Journal of Business and Economic Statistics, 29*(4), 481–492.
9. Held, L., & Ott, M. (2016). How the maximal evidence of p-values against point null hypotheses depends on sample size. *The American Statistician, 70*(4), 335–341.
10. Koenker, R., & Bassett Jr, G. (1978). Regression quantiles. *Econometrica: Journal of The Econometric Society*, 33–50.
11. Koenker, R., & Zhao, Q. (1996). Conditional quantile estimation and inference for ARCH models. *Econometric Theory, 12*(5), 793–813.
12. Neto, E. D. A. L., & de Carvalho, F. D. A. (2008). Centre and range method for fitting a linear regression model to symbolic interval data. *Computational Statistics and Data Analysis, 52*(3), 1500–1515.
13. Pastpipatkul, P., Maneejuk, P., & Sriboonchitta, S. (2017). Markov switching regression with interval data: Application to financial risk via CAPM. *Advanced Science Letters, 23*(11), 10794–10798.
14. Phadkantha, R., Yamaka, W., & Tansuchat, R. (2018). Analysis of risk, rate of return and dependency of REITs in ASIA with capital asset pricing model. In *International conference of the Thailand econometrics society* (pp. 536-548). Springer.
15. Phochanachan, P., Pastpipatkul, P., Yamaka, W., & Sriboonchitta, S. (2017). Threshold regression for modeling symbolic interval data. *International Journal of Applied Business and Economic Research, 15*(7), 195–207.
16. Reich, B. J., Fuentes, M., & Dunson, D. B. (2011). Bayesian spatial quantile regression. *Journal of the American Statistical Association, 106*(493), 6–20.
17. Sellke, T., Bayarri, M. J., & Berger, J. O. (2001). Calibration of p values for testing precise null hypotheses. *The American Statistician, 55*(1), 62–71.
18. Tibprasorn, P., Khiewngamdee, C., Yamaka, W., & Sriboonchitta, S. (2017). Estimating efficiency of stock return with interval data. In *Robustness in econometrics* (pp. 667–678). Springer.
19. Vovk, V. G. (1993). A logic of probability, with application to the foundations of statistics. *Journal of the Royal Statistical Society: Series B (Methodological), 55*(2), 317–341.
20. Yamaka, W., Autchariyapanitkul, K., Meneejuk, P., & Sriboonchitta, S. (2017). Capital asset pricing model through quantile regression: An entropy approach. *Thai Journal of Mathematics*, 53–65.
21. Yu, K., & Moyeed, R. A. (2001). Bayesian quantile regression. *Statistics and Probability Letters, 54*(4), 437–447.

The Asymmetric Effect of Trade, Financial, and Political Globalization on Economic Development in ASEAN+3

Wilawan Srichaikul, Paravee Maneejuk, and Woraphon Yamaka

Abstract This study investigates the heterogeneous effect of trade globalization, financial globalization and political globalization on economic development in ASEAN+3 countries (Brunei, Cambodia, Laos, Indonesia, Malaysia, Myanmar, the Philippines, Singapore, Thailand, Vietnam, China, Japan and South Korea) over the period 1990–2016 using Panel quantile cointegration regression approach. The results of long run relationship show the asymmetric effect of trade globalization on economic development with a strong positive effect in low economic development countries but a negative effect in high economic development countries. Meanwhile, political globalization has only the positive effect on economic development for both poorer and richer nations, but the coefficient values vary across quantiles. Finally, financial globalization shows the weak contribution to economic development. Furthermore, the short run relationship between economic development and each globalization variable evidently exists with a weak negative speed of adjustment toward equilibrium.

Keywords Trade globalization · Financial globalization · Political globalization · Economic development · ASEAN+3 · Panel quantile cointegration

1 Introduction

Globalization, although multi-faceted with complex definitions, generally refers to the increasing interdependence of world economies and declining barriers to trade, capital flows, labor movements and migration, communications, as well as knowledge and technological transfer. Through these processes, economies become much closer and more interrelated. Yet, globalization is not solely an economic process even if it once was. There is also more substance in economic globalization alone than the

W. Srichaikul · P. Maneejuk (✉) · W. Yamaka
Faculty of Economics, Center of Excellence in Econometrics, Chiang Mai University, Chiang Mai 50200, Thailand
e-mail: mparavee@gmail.com

S. Sriboonchitta et al. (eds.), *Credible Asset Allocation, Optimal Transport Methods, and Related Topics*, Studies in Systems, Decision and Control 429,
https://doi.org/10.1007/978-3-030-97273-8_22

mere flow of goods and capital. More recent measurements of globalization were built on the conceptualization by Keohane and Nye [1] of three different relevant dimensions of globalization: (1) economic: goods, capital and services (2) political: the diffusion of government policies internationally, and (3) social: the spread of ideas, information, images, and people.

The contribution of globalization has been extensively studied for the recent decade. For instance, Afzal [2] examined the impact of globalization on the growth of Pakistan's economy and found that trade openness and financial integration, while having no short-run effect, did have long run relationship with economic growth proxied by GDP. Mutascu and Fleischer [3] found that there are both positive and negative effects of globalization. The positive effects include increased national income through comparative advantage, access to global capital, spread of technology, and opportunities for individuals, spread of human rights. The effects are negative considering the deteriorating stability of global capital markets, loss of cultural integrity, weakening of national economic autonomy, further impoverishment of countries lacking skills and capital, and failure of openness to be managed well by poor countries. Ying et al. [4] found for ASEAN that economic globalization has a significantly positive influence on economic growth while social globalization and political globalization have respectively a negative and a non-significant negative impact. Kilic [5] found that economic and political globalization positively affected the economic growth of developing countries. Deluna and Chelly [6] investigated the relationship among economic growth, financial and trade globalization in the Philippines and found that growth in trade volumes accelerates economic growth, but financial openness has no significant effect on the current value of GDP. Suci et al. [7] investigated the impact of globalization on economic growth of six countries in ASEAN and revealed that economic and political globalization had positive and significant impacts on the economic growth in the region. Furthermore, Zerrin and Dumrul [8] examined and found the positive impacts of "economic" and "social" globalization on economic growth in Turkey.

As we mentioned earlier, all three components of globalization might have contributed to economic development in various countries, especially developing countries which are expediting their openness in recent years. In our investigation, ASEAN + 3 (the Association of South-East Asian Nations plus China, Japan and South Korea) countries are considered. We consider China, Japan and South Korea in our analysis as they are high development countries extending the involvement of ASEAN into the East Asian regional cooperation in particular trade cooperation to make ASEAN+3 become one of the most dynamic economic regions. As a significant subset of the developing world, ASEAN countries are also challenged by opportunities and threats of globalization.

While each component of globalization has both positive and negative implications, there is no clear picture on the relationship between the three components of globalization and economic growth. Samimi and Jenatabadi [9] suggested that the impact of globalization on economic growth of countries could be varied by the set of complementary policies such as improvement in human capital and financial system. In fact, globalization by itself does not increase or decrease economic growth. Under the context of ASEAN+3 and macroeconomics, we are interested in the economic growth implications of the broader and deeper integration of member countries in terms of trade, finance, and politics which resemble the effects of trade, finance, and political globalization that can be asymmetric. However, we have noticed that previous scholarly works in literature did not consider the heterogeneous casual effects of these three globalization and normally employed the linear mean regression model as the tool for their analysis. As mentioned by Ibrahim et al. [10], and Pipitpojanakarn et al. [11, 12], the conditional mean of the linear model is a vital location measure but not enough to fully characterize the conditional distribution. Therefore, the conventional linear model may not be appropriate to study our ASEAN+3 problems.

This study addresses this gap in the literature by using Panel quantile cointegration regression (PQCR) to examine whether there exists asymmetric causal effect of globalization on economic development of ASEAN+3 countries along quantiles of the conditional growth distribution. We explore a range of conditional quantiles in short run and long run, and thereby exposing a variety of forms of conditional heterogeneity. In the model estimation, the fixed effects (FE) specification is used for controlling individual heterogeneity as it offers a more flexible approach to the analysis of panel data than that afforded by the random effects estimation (Kato et al. [13]). The PQCR model has a linear structure but it allows us to explore the full spectrum of the conditional quantiles. Specifically, instead of regressing the "mean" of dependent and independent variables, a PQCR model is estimated in which quantiles of the conditional distribution of the dependent variable is expressed as functions of observed covariates. Therefore, this model allows us for investigating the impact of trade, financial, and political globalization on economic development in ASEAN-3 at given quantile level.

Although previous studies tested the globalization-growth nexus, none examined it for varying degrees of economic development in ASEAN+3 for both short run and long run relationship. In addition, no studies examined all three-dimension globalizations—trade, financial, political—on ASEAN+3 economic development yet. Moreover, as the previous results indicated a ambiguous results about the effects of trade, financial, political on economic development, so we aim to reinvestigate this issue using a PQCR model.

The remainder of the paper is organized as follows. Section 2 describes the methodology; Sect. 3 presents the data used in this study and model specification. In Sect. 4, we present the empirical results. Section 5 presents our conclusions.

2 Methodology

In this paper, we use a fixed effects Panel quantile cointegration regression model to investigate heterogeneous effects of trade globalization, financial globalization and political globalization on economic development. We start briefly describing the approach of the Panel quantile regression with fixed effects specification. Then, the Panel quantile cointegration regression is introduced. Finally, the cointegration test is presented.

2.1 Fixed Effect Panel Quantile Regression

The quantile regression technique was introduced in the seminal paper by Koenker and Bassett [14]. This method is a generalization of median regression analysis to quantile regression. The estimation method for this model depends on the two error components, says individual effects and idiosyncratic error term. More formally, the formulation of a panel data regression model can be expressed by the following equation

$$y_{it} = {x'}_{kit}\beta_k^{\tau} + \alpha_i^{\tau} + u_{it}^{\tau}, \tag{1}$$

where $i = 1, \ldots, N$ is the cross sectional unit, $t = 1, \ldots, T$ is the time index, y_{it} is $NT \times 1$ outcome variable, x_{kit} is $NT \times K$ explanatory variables, β_k^{τ} is $K \times 1$ coefficient parameters at given quantile τ and u_{it}^{τ} is mean zero stationary idiosyncratic error term or residual. The assumption about this error term is that it is usually not correlated with x_{kit}. α_i^{τ} is the individual-specific effect error component at given quantile τ which is allowed to vary across individual countries but is constant over time. That is, the individual effects α_i^{τ} is assumed to be fixed parameters to be estimated. This is called the fixed effects panel data model. The advantages of the fixed effects model are: (1) that it can allow unobserved individual effects to be correlated with explanatory variable x_{it} and (2) that there is no need to specify the conditional density for unknown individual effects.

In this model, we are mainly interested in analyzing β^{τ} under the fixed effects estimation, hence we need to eliminate the influence of individual effects. To do this, we use demean variable method and transform our variables to eliminate this effect. Thus, $\tilde{y}_{it} = y_{it} - (1/N)\sum_{i=1}^{N} y_{it}$ and $\tilde{x}_{kit} = x_{kit} - (1/N)\sum_{i=1}^{N} x_{kit}$. Then, the long run model can be written as

$$Q_{\tilde{y}_{it}}(\tau|\beta, \tilde{x}_{kit}) = \delta_0^{\tau} + {x'}_{kit}\beta_k^{\tau}, \quad i = 1, \ldots, N; t = 1, \ldots, T, \tag{2}$$

where δ_0^{τ} has a pure location shift effect on the conditional quantiles of the response. The effects of the covariates x_{it} are permitted to depend upon the quantile τ of interest.

Koenker [15] treated unobservable fixed effects as parameters to be jointly estimated with the covariate effects for different quantiles.

2.2 Fixed Effects Panel Quantile Regression Cointegration Approach

In this study, we apply the two-step procedures proposed by Engle and Granger [16] to estimate our model. Thus, a panel-based error correction model (short-run model) can be constructed with a one-period lagged error correction and can be specified in Eq. (3).

$$\Delta \ln \tilde{y}_{it} = \varphi_0^{\tau} + \sum_{t=1}^{T-1} \phi_k^{\tau} \Delta \ln \tilde{x}_{kit-j} + \pi_i ECT_{it-1} + \tilde{\varepsilon}_{it} \quad (3)$$

where Δ is the difference operator. φ_0^{τ} is the deterministic component (constant) at given quantile τ. ϕ_k^{τ} and π_i are the parameters of regressors. $ECT_{it-1} = (\tilde{y}_{it-1} - \tilde{x'}_{it-1}\beta^{\tau})$ is the error correction term. $\tilde{\varepsilon}_{it}$ is a stationary random error with a zero mean. Note that is called speed of adjustment, if there exists $-2 < \pi_i < 0$, cointegration is confirmed for Eq. (2).

2.3 Testing Panel-Based Multivariate Regression Model

Since using standard quantile regression techniques with nonstationary series can lead to the spurious regression problem (Harris and Sollis [17]), it is important to test the order of integration of each panel data in Eq. (2). To examine the presence of a unit root for the error u_{it}^{τ} which is estimated from the Eq. (2) (first step), we apply the Im, Pesaran and Shin (IPS) test proposed by Im, Pesaran and Shin [18]. The test is based on the following model.

$$\Delta \tilde{u}_{it}^{\tau} = \gamma z_{it} + \rho_i \tilde{u}_{it-1}^{\tau} + \sum_{j=1}^{P} \omega_j \Delta \tilde{u}_{it-j}^{\tau} + \tilde{v}_{it}, \quad (4)$$

where z_{it} is the deterministic components such as fixed effects or a mixture of fixed effects and heterogeneous time trends. The null hypothesis $(H_0 : \rho_i = 0)$ is that each series in the panel contains a unit root for all i, and the alternative hypothesis $(H_1 : \rho_i < 0$ for at least one $i)$ is that at least one of the individual series in the panel is stationary.

If there are unit roots in the series, tests of cointegration in the panel can be applied to examine if individual series are linked to form an equilibrium relationship spanning in the long-run. The approaches used in this paper are the estimation methods proposed by Pedroni [19] and Kao [20].

Once the variables in a panel data set are cointegrated, the next step is to estimate a cointegration (long-run) vector. The long-run parameters in Eq. (2) and the short-run parameters in Eq. (3) are estimated using the group-means fully modified OLS (FMOLS) and dynamic OLS (DOLS) estimators developed by Pedroni [21, 22]. The FMOLS estimator is a non-parametric approach, while the DOLS estimator is a parametric approach where lagged first-differenced terms are explicitly estimated (Harris and Sollis [17]). Both approaches incorporate corrections for endogeneity and serial correlation. However, Kao and Chiang [23] suggested that FMOLS may be more biased than DOLS. In this paper, therefore, we apply both FMOLS and DOLS to estimate our PQCR model.

3 Data Analysis

3.1 Data Description

In this paper, we investigate the heterogeneous effects of trade globalization, financial globalization and political globalization on economic development using data on ASEAN+3 (Brunei, Cambodia, Laos, Indonesia, Malaysia, Myanmar, the Philippines, Singapore, Thailand, Vietnam, China, Japan and South Korea) over the period 1990–2016. The data are collected from the KOF globalization index and World Bank Indicators. We also consider urbanization for inclusion in the model. Urbanization is measured by urban population in GDP. All variables are transformed into natural log. This study is conducted using an econometric framework as follows:

$$Q_{\ln GDP_{it}}(\tau|\beta, \tilde{x}_{kit}) = \delta_0^\tau + \beta_1^\tau \ln TRADE_{it} + \beta_2^\tau \ln FIN_{it} + \beta_3^\tau \ln POLI_{it} + \beta_4^\tau \ln URB_{it} \tag{5}$$

where the countries are indexed by i and time by t. The descriptions of variables are provided in Table 1.

Table 1 Variable definitions

Variable	Definition	Source
lnGDP	Logarithm of economic growth (GDP, PPP current international $)	World development indicators
lnTRADE	Logarithm of trade globalization	KOF globalization index
lnFIN	Logarithm of financial globalization	KOF globalization index
lnPOLI	Logarithm of political globalization	KOF globalization index
lnURB	Logarithm of urbanization (% of GDP)	World development indicators

Table 2 Summary statistics

Variable	lnGDP	lnTRADE	lnFIN	lnPOLI	lnURB
Mean	26.113	3.992	3.974	4.075	3.798
Median	26.521	4.034	4.05	4.247	3.827
Maximum	30.696	4.575	4.546	4.532	4.605
Minimum	0.000	3.048	2.979	2.749	2.737
Std. Dev.	3.089	0.312	0.33	0.435	0.534
Skewness	−5.126	−0.354	−0.715	−1.093	−0.131
Kurtosis	44.245	2.857	3.068	3.171	1.833
Jarque-Bera (MBF)	26417.09 (0.000)	7.609 (0.000)	29.952 (0.000)	70.334 (0.000)	20.932 (0.000)

Note ( ) *MBF*—is minimum bayes factor (see [24])

3.2 Descriptive Statistics

Table 2 presents the descriptive statistics of the variables of interest. In Table 2, it can be seen that GDP has the highest standard deviation, while POLI has the lowest. The Jarque-Bera test significantly rejects the null hypothesis of normality, demonstrating the non-normality of the unconditional distribution of all the variables.

Before estimating the panel quantile cointegration model, this study checks the stationary properties for all the variables concerned. In this study, the Levin et al. [25] (LLC) unit root test is employed where the best number of lags to be used in each test is selected by the Akaike Information Criterion. Table 3 presents the result of the LLC unit root test, both in level and in first difference. lnTRADE, lnFIN and lnURB fail to reject the null hypothesis of unit roots in the level, but they reject the null hypothesis in first difference with the decisive evidence. Thus, we conclude that lnGDP, lnTRADE, and lnURB variables are integrated in the order of I(1), while lnGDP and lnPOLI are stationary at the level, indicating that the order integration of these two variables are zero, I(0). These results qualify the PQCR model to be an appropriate model for this study.

3.3 Cointegration Tests

To confirm the existence of the cointegration relationship, we test the model considering ten quantiles: 0.1, 0.2, 0.3, 0.4, 0.5, 0.6, 0.7, 0.8, 0.9, and 0.95. Note that the IPS test is conducted to test the stationarity of the residual term obtained from each quantile level model (Eq. 4). Table 4 shows that MBFs are equal to zero, indicating that the null hypotheses are rejected with the decisive evidence. These results confirm the existence of long-run relationship between economic development and globalization for all quantiles.

Table 3 Panel unit root tests

Variable	Levin, Lin & Chu					
	Level			First difference		
	None	Intercept	Intercept and trend	None	Intercept	Intercept and trend
lnGDP	6.742 (0.000)	2.722 (0.024)	61.809 (0.000)	−2.302 (0.070)	45.8836 (0.000)	62.959 (0.000)
lnTRADE	0.885 (0.675)	−4.158 (0.000)	−0.9703 (0.624)	−5.951 (0.000)	3.567 (0.001)	6.380 (0.000)
lnFIN	2.056 (0.120)	−2.829 (0.018)	0.470 (0.895)	−7.479 (0.000)	−6.787 (0.000)	3.297 (0.004)
lnPOLI	4.684 (0.000)	−5.642 (0.000)	−5.839 (0.000)	−7.923 (0.000)	2.530 (0.040)	12.022 (0.000)
lnURB	1.799 (0.197)	−2.853 (0.017)	−1.326 (0.414)	−2.968 (0.012)	−36.410 (0.000)	−34.579 (0.000)

Note ( ) *MBF*—is minimum bayes factor (see [24])

Table 4 Variable definitions

IPS test	Quantile cointegration									
	10th	20th	30th	40th	50th	60th	70th	80th	90th	95th
t-stat (MBF)	−10.81 (0.000)	−10.849 (0.000)	−10.78 (0.000)	−10.614 (0.000)	−10.56 (0.000)	−10.262 (0.000)	−10.087 (0.000)	−9.941 (0.000)	−9.533 (0.000)	−9.094 (0.000)

Note ( ) *MBF*—is minimum bayes factor (see [24])

4 Results

Our empirical results are presented in this section. The result of our long run equation obtained from quantile regression is illustrated in Table 5. We report the 10th, 20th, 30th, 40th, 50th, 60th, 70th, 80th, 90th and 95th quintiles of the conditional concentration's distribution. The lower quantiles, including 10th, 20th, 30th, 40th, and 50th, refer to the countries with lower economic development levels. The higher quantiles, including 60th, 70th, 80th, 90th, and 95th refer to the countries with higher economic development levels.

Trade globalization exhibits both negative and positive effects on economic development of the ASEAN+3 countries. It can be seen that coefficients of trade globalization are positive in the low quantiles, but turn to be negative in the high quantile levels. Note that at the 10th, 20th, 30th, 40th, 50th, 60th, 70th and 80th quantiles, the coefficient of trade globalization conveys weak evidence. The 90th quantile corresponds to moderate evidence of impact while the 95th quantile gets strong evidence. This result indicates that trade globalization exacerbates economy in the high economic development group of ASEAN+3 countries but has positive impact on the economic development in the low economic development countries. Also, we can say that the impact of trade globalization is clearly heterogeneous and strongly asymmetric

Table 5 Panel quantile regression results in the long run

Variable	Quantile cointegration				
	10th	20th	30th	40th	50th
lnTRADE	0.4023 (0.9771)	0.3806 (0.7104)	0.2663 (0.7345)	0.2239 (0.7345)	0.0747 (0.9807)
lnPOLI	3.0966 (0.777)	1.7534 (0.3534)	1.4284 (0.3083)	1.5457 (0.3064)	1.7365 (0.183)
lnFIN	0.0446 (0.9998)	0.1694 (0.9249)	0.2653 (0.6446)	0.2433 (0.7992)	0.2991 (0.7406)
lnURB	1.3362 (0.6886)	1.9482 (0.0073)	2.0641 (0.0001)	2.0388 (0.0005)	1.9364 (0.0009)
Variable	Quantile cointegration				
	60th	70th	80th	90th	95th
lnTRADE	−0.0990 (0.9565)	−0.2402 (0.8146)	−0.3468 (0.5907)	−0.6106 (0.1819)	−0.8229 (0.0047)
lnPOLI	1.8128 (0.1603)	1.6220 (0.1393)	1.5817 (0.1522)	2.0429 (0.0511)	2.5878 (0.0003)
lnFIN	0.1819 (0.8429)	0.0356 (0.9947)	0.0281 (0.9944)	0.0092 (0.9990)	−0.0182 (0.9968)
lnURB	2.1151 (0.0008)	2.2443 (0.0001)	2.2796 (0.0000)	2.088 (0.0003)	1.6435 (0.0001)

Note ( ) *MBF*—is minimum bayes factor (see [24])

because the coefficients of different quantile levels change greatly with respect to different conditional distribution of economic development levels.

It is interesting that the trade globalization renders a negative impact to the high development countries. We expect that the increase of the trade in the high development ASEAN+3 countries may shift their abundant resources away from their competitive sector in order to boost the export sector. Thus, the allocation of resources to produce additional unspecialized products can increase the production costs.

Political globalization apparently has positive effect on economic development for both lower and higher quantiles. Although there is weak evidence at the 10th and 20th quantile levels; for 30th, 40th, 50th, 60th, 70th, 80th quantiles, the coefficient of political globalization conveys moderate evidence. Likewise, political globalization has coefficient value with substantial evidence at the 90th quantile but the highest with decisive evidence at the 95th quantile. The results imply that, as political globalization increases 1%, the level of economic growth increases by 1.4284–2.5878% along the quantile levels. The result indicates that political globalization can affect economic development in both poorer and richer countries in ASEAN+3. In the case of financial globalization, there is weak evidence supporting the effect of this factor on economic development for all quantile levels as the MBF values are quite high and close to one. We expect that the ASEAN+3 countries are independent of the world financial in-flow.

Table 6 Panel quantile cointegration regression results in the short run

Variable	Quantile cointegration				
	10th	20th	30th	40th	50th
ΔlnTRADE	−0.0146 (0.9552)	−0.0118 0.8528)	−0.0064 (0.9645)	0.0042 (0.9744)	0.0069 (0.9283)
ΔlnPOLI	0.0256 (0.7882)	−0.0215 (0.8334)	−0.0427 (0.6630)	0.0136 (0.9709)	0.0408 (0.6055)
ΔlnFIN	-0.0386 (0.6328)	−0.0166 (0.7069)	0.0108 (0.9020)	0.0367 (0.4522)	0.0462 (0.1933)
ΔlnURB	0.9760 (0.3366)	0.5324 (0.5009)	0.4789 (0.6358)	0.4309 (0.6111)	0.4010 (0.5312)
ECT_{it-1}	−0.0030 (0.7827)	−0.0032 (0.8235)	−0.0036 (0.7497)	−0.0040 (0.6747)	−0.0042 (0.7432)
Variable	Quantile cointegration				
	60th	70th	80th	90th	95th
ΔlnTRADE	0.0110 (0.9243)	0.0164 (0.9138)	0.0229 (0.7953)	0.0231 (0.5402)	0.0147 (0.7557)
ΔlnPOLI	0.0311 (0.8200)	0.0408 (0.7203)	0.0228 (0.8081)	0.0456 (0.5314)	0.0521 (0.2784)
ΔlnFIN	0.0524 (0.1119)	0.0334 (0.2703)	0.0215 (0.7040)	−0.0121 (0.8847)	−0.0144 (0.8941)
ΔlnURB	0.3453 (0.7849)	−0.0349 (0.9982)	−0.1761 (0.9459)	−0.1089 (0.9785)	−0.3119 (0.8893)
ECT_{it-1}	−0.0041 (0.6190)	−0.0048 (0.6898)	−0.0052 (0.8842)	−0.0055 (0.8453)	−0.0062 (0.7577)

Note ( ) *MBF*—is minimum bayes factor (see [24])

Finally, the results for the control variable, urbanization, reveal that urbanization has the effect on the economic development in the same direction for low and high quantiles. For 20th quantile the coefficient of urbanization is with very strong evidence. At the 30th, 40th, 50th, 60th, 70th, 80th, 90th and 95th quantiles, urbanization has coefficient values conveying decisive evidence but weak evidence at the 10th. At the 80th quantile, the coefficient value of urbanization is 2.2796 which is the highest among all quantiles. Urbanization plays a vital role in the economic and social fabric of both lower and higher income economies by offering opportunities for education, employment and health services. Education capital determines the ability of a nation to develop new technologies and adopt existing technologies. Health capital can influence economic activity directly through its impact on labor productivity. Urbanization seems to be a key factor in entrepreneurship.

Turning to the analysis of short-term dynamics in Table 6 the findings show that there exists a weakly negative effect of the error correction term on economic growth, Δ ln GDP, indicating the short-run relationship between economic development and each globalization variable for all quantile levels. However, the speed of adjustment to equilibrium is relatively slow.

5 Conclusion

The objective of this paper is to investigate the effects of trade globalization, financial globalization and political globalization on economic development of countries having different development levels within the ASEAN+3. This study utilizes the panel quantile cointegration regression model to achieve our objectives taking into account the unobserved individual and distributional heterogeneity. The quantile cointegration regression analysis obtains a more detailed picture of the influencing factors that affect economic development at different economic levels in short-run and long-run. Moreover, to avoid an omitted-variable bias and obtain more comprehensive and tenable results, urbanization as control variable is included in our model.

The long run investigation discloses many interesting findings. First, the direct effect of trade globalization on economic growth is strongly negative in countries with higher economic development levels but is weakly positive in lower economic development countries. Second, political globalization has a positive direct effect on economic development. Unlike, trade globalization, the effect of political globalization on economic development depends on the existing levels of economic development of the countries (30th, 40th, 50th, 60th, 70th, 80th, 90th and 95th quantiles). The 1 % higher in political globalization has brought about 1.4284–2.5878% increase in economic growth in all quantile levels. Third, this study also confirms the decisive effect of urbanization for both low economic and high economic development countries (20th, 30th, 40th, 50th, 60th, 70th, 80th, 90th and 95th quantiles). Finally, financial globalization has a weak effect on economic development for all quantiles. In addition to the long run relationship, the short run relationship between economic development and globalizations is found for all quantile levels with relatively slow speed of adjustment toward equilibrium is under negative changes.

Based on the aforementioned findings, the following policies are proposed.

(1) The effects of globalization on economic growth not only depend on the economic structure of a country during the globalization process but also can be altered by a series of complementary policies such as the improvement of human capital and the technology. The influence of complementary policies is very important because it helps the country to succeed in the process of globalization. Therefore, the state needs to respond critically through appropriate diplomacy and economic policy instruments, so that the country's economy can be strong and can increase bargaining position in the midst of increasingly tight global competition.

(2) The government should maintain political stability and define its policies consistent with the economy structure of ASEAN+3. Moreover, governments should also pay close attention to the negative impact of trade on economic growth in high economic development countries.

Acknowledgements The authors would like to thank the Dr. Laxmi Worachai for her helpful comments on an earlier version of the paper. This research work was partially supported by Chiang Mai University.

References

1. Keohane, R. O., & Nye, J. S. (2000). Globalization: What's new? What's not?(And so what?). *Uoreign Policy-Washington*, 104–119.
2. Afzal, M. (2007). The impact of globalisation on economic growth of Pakistan. *The Pakistan Development Review*, 723–734 .
3. Mutascu, M., & Fleischer, A. M. (2011). Economic growth and globalization in Romania. *World Applied Sciences Journal, 12*(10), 1691–1697.
4. Ying, Y. H., Chang, K., & Lee, C. H. (2014). The impact of globalization on economic growth. *Romanian Journal of Economic Forecasting, 17*(2), 25–34, e87824.
5. Kilic, C. (2015). Effects of globalization on economic growth: Panel data analysis for developing countries. *Petroleum-Gas University of Ploiesti Bulletin*
6. Deluna Jr, R., & Chelly, A. (2014). Economic growth, financial and trade globalization in the Philippines: A vector autoregressive analysis.
7. Suci, S. C., Asmara, A., & Mulatsih, S. (2016). The impact of globalization on economic growth in ASEAN. *Bisnis & Birokrasi Journal, 22*(2), 79–87, e87824.
8. Zerrin, K., & Dumrul, Y. (2018). The Impact of Globalization on Economic Growth: Empirical Evidence from the Turkey. *International Journal of Economics and Financial Issues, 8*(5), 115, e87824.
9. Samimi, P., & Jenatabadi, H. S. (2014). Globalization and economic growth: Empirical evidence on the role of complementarities. *PloS one, 9*(4), e87824.
10. Ibrahim, S., Mazlina, A. R., Azman-Saini, W. N. W., & Zakaria, M. F. M. (2016). Financial integration-growth nexus: A quantile regression analysis. *Journal of Economic Integration, 31*(3), 531–546.
11. Pipitpojanakarn, V., Maneejuk, P., Yamaka, W., & Sriboonchitta, S. (2016). Analysis of agricultural production in Asia and measurement of technical efficiency using copula-based stochastic frontier quantile model. In *International Symposium on Integrated Uncertainty in Knowledge Modelling and Decision Making* (pp. 701–714). Springer, Cham.
12. Pipitpojanakarn, V., Maneejuk, P., Yamaka, W., & Sriboonchitta, S. (2017). How does economic growth affect the well-being in Asia? *International Journal of Economic Research, 14*(6), 231–244.
13. Kato, K., Galvao, A. F., Jr., & Montes-Rojas, G. V. (2012). Asymptotics for panel quantile regression models with individual effects. *Journal of Econometrics, 170*(1), 76–91.
14. Koenker, R., & Bassett Jr, G. (1978). Regression quantiles. *Econometrica: Journal of the Econometric Society*, 33–50 .
15. Koenker, R. (2004). Quantile regression for longitudinal data. *Journal of Multivariate Analysis, 91*(1), 74–89.
16. Engle, R. F., & Granger, C. W. (1987). Co-integration and error correction: representation, estimation, and testing. *Econometrica: Journal of the Econometric Society*, 251–276.
17. Harris, R., Sollis, R., Harris, H., & Sollis, R. (2003). *Applied time series modelling and forecasting*. Wiley.
18. Im, K. S., Pesaran, M. H., & Shin, Y. (2003). Testing for unit roots in heterogeneous panels. *Journal of Econometrics, 115*(1), 53–74.
19. Pedroni, P. (1999). Critical values for cointegration tests in heterogeneous panels with multiple regressors. *Oxford Bulletin of Economics and Statistics, 61*(S1), 653–670.
20. Kao, C. (1999). Spurious regression and residual-based tests for cointegration in panel data. *Journal of Econometrics, 90*(1), 1–44.
21. Pedroni, P. (2001). Fully modified OLS for heterogeneous cointegrated panels. In *Nonstationary panels, panel cointegration, and dynamic panels* (pp. 93–130). Emerald Group Publishing Limited.
22. Pedroni, P. (2001). Purchasing power parity tests in cointegrated panels. *Review of Economics and Statistics, 83*(4), 727–731.

23. Kao, C., & Chiang, M.-H. (2001). On the estimation and inference of a cointegrated regression in panel data. In B.H. Baltagi, T.B. Fomby, R. Carter Hill (Eds.), *Nonstationary panels, panel cointegration, and dynamic panels* (Vol. 15, pp. 179–222) Emerald Group Publishing Limited. (Advances in Econometrics)
24. Held, L., & Ott, M. (2018). On p-values and Bayes factors. *Annual Review of Statistics and Its Application, 5,* 393–419.
25. Levin, A., Lin, C. F., & Chu, C. S. J. (2002). Unit root tests in panel data: Asymptotic and finite-sample properties. *Journal of Uconometrics, 108*(1), 1–24.

Interdependence of Macroeconomic Factors and Economic Growth in OECD Countries: Evidence Based on a Bayesian Panel VAR Model

Wilawan Srichaikul and Woraphon Yamaka

Abstract This study examines the interdependency between macroeconomics factors, specifically financial development, industrial structure, and gross capital formation and economic growth in OECD countries including Austria, Belgium, Denmark, France, Greece, Iceland, Ireland, Italy, Luxembourg, Norway, the Netherlands, Portugal, United Kingdom, Sweden, Switzerland, Turkey, United States, and Germany. We use panel data of these 18 countries from 2001 to 2016 to examine their macroeconomic interdependency using the Bayesian panel vector autoregression model. Four different priors are used to combine with the likelihood of the model, and the results show that stochastic search variable selection prior is selected as it gives the lowest MSE and MAD. The empirical results show that there exists the interdependence among macroeconomic factors and economic growth in OECD countries.

Keywords Interdependency · Macroeconomics factors · Economic growth · OECD countries, Bayesian Panel VAR

1 Introduction

Economic growth refers to the increase of the capacity of a country's economy in producing goods and services in the particular period, compared to the previous period. Growth is an essential and necessary condition for development, but it is not sufficient as it cannot guarantee development. Thus, many countries are focusing on sustainable economic growth as it guarantees the overall economic development. It brings a better standard of living for the community. There are many factors which contribute simultaneously to maintain the sustainable economic growth of a nation. Therefore, the measurement of a particular country's economic growth is complex as there are so many determinants of growth contributing simultaneously to gross domestic products (GDP).

W. Srichaikul · W. Yamaka (✉)
Faculty of Economics, Center of Excellence in Econometrics, Chiang Mai University, Chiang Mai 50200, Thailand
e-mail: woraphon.econ@gmail.com

S. Sriboonchitta et al. (eds.), *Credible Asset Allocation, Optimal Transport Methods, and Related Topics*, Studies in Systems, Decision and Control 429,
https://doi.org/10.1007/978-3-030-97273-8_23

The relationship between macroeconomic factors and economic growth has been extensively studied for several decades. For instance, Bassanini et al. [1] explained that the financial system has some contributions to economic growth as it provides funding for capital accumulation and helps the diffusion of new technologies. A well-developed financial system is likely to mobilize individuals' savings into profitable large-scale investments, thereby boosting the liquidity in the economy. Levine [2] suggested that financial institutions and markets can enhance economic growth through several channels, i.e., by (i) easing the exchange of goods and services through the provision of payment services and (ii) mobilizing and pooling savings from a large number of investors. Kenourgios and Samitas [3] examined the long-run relationship between finance and economic growth for Poland and concluded that credit to the private sector had been one of the main driving forces of the long-run growth. Hagmayr et al. [4] investigated the finance-growth nexus in four emerging economies of Southeastern Europe and found a positive and significant effect of bond markets and the capital stock on growth. Furthermore, the study of Feng and Long [5] mentioned that the change in the industrial structure is one of the main driving forces of the economic cycle.

From the econometric point of view, Han and Zheng [6] used an unconstrained Vector Autoregressive model and granger causality test as well as Impulse response function and Variance decomposition to investigate the macroeconomic relations. They found that the industrial structure has counterproductive to economic development. In addition, several studies have empirically examined the causal relationship between capital formation and economic growth. Blomstrom et al. [7] and Bose and Haque [8] examined the causal direction between them using the Granger-Sims causality framework. Uneze [9] examined this relationship by using a panel cointegration. They revealed that higher economic growth leads to higher capital formation and that in turn, increases in capital formation results in higher economic growth. From the literature, we found many macroeconomic factors related to economic growth, and most of the studies employed the VAR framework to investigate this issue. This study will mainly investigate the interdependence of macroeconomics factors, including financial development, the industrial structure, gross capital formation, and economic growth in OECD countries. However, the conventional VAR model has some methodological difficulties. It is well known that this model is not suitable for panel data as it does not consider the inherent heterogeneity across countries. Thus, the panel vector autoregression (PVAR) model is considered in this study. However, due to the large number of estimated parameters and the limited number of observations, it has a possibly sparse structure. To deal with this problem, this study uses the Bayesian approach to improve efficiency. The accuracy of the Bayesian estimation relies on the prior selection. This study also estimates the posterior distribution with many different prior distributions and selects the best fit prior distribution considering the lowest mean squared errors (MSE) and mean absolute deviation (MAD). Van de Schoot et al. [10] suggested that Bayesian estimation can provide reliable results when the sample size is small. It is more informative, flexible, and efficient than the Maximum likelihood estimator. The implementation of Bayesian prior on the parameters of the model can reduce the estimation uncertainty and obtain the inference and forecast accurately Pastpipatkul et al. [11] and Kaewsompong et al. [12].

The structure of the rest of this paper is as follows. The next section provides the model used in this study. Section 3 presents data and model specification, Sect. 4 presents the empirical results. Finally, the conclusion is provided in Sect. 5.

2 Methodology

2.1 The PVAR Model

Similar to VAR model, the panel VAR (PVAR) assumes that all variables are treated endogenous variables. However, the model incorporates a cross sectional dimension into the time series dimension in order to construct the panel data. In this study, we consider an unrestricted G-variate PVAR of lag order $P = 1$, which can be written as:

$$y_{it} = A_{i1} y_{1,t-1} + \cdots + A_{ii} y_{i,t-1} + \cdots + A_{iN} y_{N,t-1} + \epsilon_{it}, \tag{1}$$

where the subscript $i = \{1, \ldots, N\}$ is the cross-sectional dimension or number of countries and $t = \{1, \ldots, T\}$ is the time dimension. The y_{it} is an endogenous variable of country i at time t. A_{ij} are $G \times G$ matrices of unknown parameters to be estimated for each $i, j = \{1, \ldots, N\}$. ϵ_{it} is an error term assumed to have a normal distribution $\epsilon_{it} \sim N(0, \Sigma_{it})$, where Σ_{it} is the $G \times G$ positive definite and symmetric matrix of covariance. Following Korobilis [13], we assume that there are dynamic interdependencies between country i and country j for $i \neq j$.

2.2 Posterior Distribution with Different Priors

The posterior distributions depend on the interaction between likelihood and prior. With the influence of prior distribution on the posterior inference, it is necessary to conduct a prior sensitivity analysis, which becomes an important tool in Bayesian statistics. To perform sensitivity analysis on the priors, we estimate the posterior distribution with different prior distributions. If these posteriors are not available analytically, they are usually approximated using the Markov chain Monte Carlo (MCMC) method [14]. As suggested by Korobilis [13], we employed four different priors in this study including Bayesian Factor Clustering and Selection (BFCS) prior [13], Bayesian Mixture Shrinkage (BMS) prior [13], Stochastic Search Variable Selection (SSVS) prior for VAR [15] and Bayesian Semiparametric Shrinkage (BSS) prior [16].

Let Θ be the parameters in the PVAR model. The posterior estimation can be done by multiplying these priors' distribution with the likelihood function using Bayes theorem. Thus, the posterior estimation can be formed as follows.

$$Pr(\Theta|Y_t) = L(\Theta|Y_t) \times Pr(\Theta), \tag{2}$$

where $L(\Theta|Y_t)$ is the likelihood function of Θ given Y_t and $Pr(\Theta)$ is the prior distribution of Θ. Let we define the alternative form of PVAR (Eq. 1) as

$$Y_t = W_t\beta + U_t, \tag{3}$$

where $W_t = I_{NG} \otimes X_t'$, $X_t' = (y_{j,t-1})$ and $\beta = \{A_{i1}y_{1,t-1}, \ldots, A_{ii}y_{i,t-1}, \ldots, A_{iN} y_{N,t-1}\}$,. Y_t and U_t are $NG \times 1$ vectors of the endogenous variables and the random disturbances of the model. Thus, we can write the likelihood function this equation as

$$L(\Theta|Y_t) \propto \left(\prod_t |\Sigma|^{-0.5T} exp\left[-0.5\sum_t (Y_t - W_t\beta)'(\Sigma)^{-1}(Y_t - W_t\beta)\right]\right), \tag{4}$$

Since there are dynamic interdependencies from country j to country i, we only allow β to vary with cross-sectional units. We can obtain the linear structure of β as follows [17].

$$\beta = \Phi\theta + e. \tag{5}$$

Equation (5) is structure specification of the vector of PVAR coefficients β where Φ is a $K \times s$ matrix of predetermined loadings with ones and zeros. Let $K = NG^2$ and θ be a $s \times 1$ lower dimensional factor vector capturing components in the coefficient vector that are common across units, within units, specific variable and lagged coefficients. (See, [17]). The error e is assumed to be normal with mean zero and variance $\Sigma \otimes \sigma^2 I$.

Factoring β as in (5) gives us various advantages. In computation, this approach can reduce the problem of estimating NG^2 coefficients into the one of estimating $s + N + G$ factors characterizing them. In practice, we can transform an over-parametrized panel VAR into a Seemingly unrelated regression (SUR) model, where the regressors are averages of certain right-hand side VAR variables.

Using the panel restrictions described above, this study conducted four priors for the dynamic interdependency restrictions.

2.2.1 Bayesian Factor Clustering and Selection (BFCS) Prior

Korobilis [13] suggested that the coefficient can be zero and a single coefficient β_k is not always clustered with non-zero coefficient. Thus, the parametric BFCS prior is proposed to deal with this culprit and it was modified based on spike and slab priors. Let $k = ij$,the hierarchical structure of this prior becomes

$$\begin{aligned} \beta_k &\sim (1-\kappa_k)\,\delta_0(\beta_k) + \kappa_k \Delta_k, \\ \Delta_k &\sim N\left(\Phi\theta, \Sigma \otimes \sigma^2 I\right), \\ \theta &\sim N(0, c), \\ \kappa_k &\sim Bernoulli(\pi), \\ \pi &\sim Beta\ (1, \varphi). \end{aligned} \tag{6}$$

In this case, the linear structure of β is estimated by

$$\beta = \Gamma \times (\Phi\theta) + e, \tag{7}$$

where $e \sim N(0, \Sigma \otimes 0)$, Γ is a $K \times K$ diagonal matrix with elements $\Gamma_{it} = \kappa_i, i = 1, \ldots, K$. For π, it is a probability of the same coefficient β_k coming from the clustering while $1-\pi$ is the probability that coefficient has prior a point mass at zero.

2.2.2 Bayesian Mixture Shrinkage (BMS) Prior

Korobilis [13] proposed a nonparametric prior in order to allow similar coefficients to shrink into a common prior location, which can be different for different groups of similar coefficients. This prior can gain more information of the common prior locations for homogenous prior and it will not make the posterior sharper when the likelihood function is weak. The BMS prior can be defined as

$$\begin{aligned} \beta_k &\sim N\left(\mu_k, \tau_k^2\right), \\ \mu_k, \tau_k^{-2} &\sim \pi\delta_0(\beta) \times \delta_{1e+10}\left(\tau^{-2}\right) + (1-\pi)\,F, \\ F &\sim DP\,(\theta F_0), \\ F_0 &\sim N(0, \lambda) \times gamma\,(0.5, 0.5), \\ \pi &\sim Beta\ (1, \varphi), \end{aligned} \tag{8}$$

where μ_k is the location or mean hyper-parameter, τ_k^{-2} is variance hyper-parameter, and DP is the Dirichlet process.

2.2.3 Stochastic Search Variable Selection (SSVS) Prior

The SSVS is usually motivated as a method for selecting variables in a PVAR model proposed by George et al. [15]. This study also considers the SSVS prior, which can be interpreted as a hierarchical prior for all of the elements of β_k and Σ.

$$\begin{aligned} \beta_k &\sim (1-\kappa_k)\, N\left(0, \tau_k^2 I\right) + \kappa_k N\left(0, \tau_k^2 I\right), \\ \kappa_k &\sim Bernoulli\,(\pi_k)\,,\; k = ij,\, j \neq i, \\ \Sigma &\sim IW\,(S_i, v_i)\,, \\ \pi_k &\sim Beta\;(1, \varphi)\,, \end{aligned} \tag{9}$$

where Σ is distributed standard Inverse Wishart (IW) with degrees of freedom v_i and scale matrix S_i. π_k is the prior inclusion probability of variable k in country i.

2.2.4 Bayesian Semiparametric Shrinkage (BSS) Prior

Following Korobilis [18], the prior for coefficient of PVAR model β_k is formed as

$$\begin{aligned} \beta_k &\sim \pi\delta_0\,(\beta) + (1-\pi)\, G, \\ G &\sim DP\,(\theta G_0)\,, \\ G_0 &\sim N\left(0, \tau^2\right), \\ \tau^2 &\sim IG\,(0, 0)\,, \\ \pi &\sim Beta\;(1, \varphi)\,, \end{aligned} \tag{10}$$

where the semiparametric selection prior and Dirichlet process (DP) prior are used to derive the simple stick breaking representation of β_k.

3 Data

In this section, we describe the data used in this study. The annual data for 18 OECD countries over the period 2001–2016 is collected. We use OECD countries' data for two reasons. First, the main focus of this study is to understand the effects of macroeconomic factors on GDP in advanced or high-income economics, which are now the center of the world economic activities. Second, most OECD members are high-income economies with a very high Human Development Index (HDI) and are regarded as developed countries, so we can better loosen the effects of macroeconomic factors on GDP. In this study, we consider gross domestic product per capita (GDP_{it}), financial development, domestic credit to the private sector (FIN_{it}), the industrial structure (IND_{it}), and gross capital formation (GCF_{it}) as the variables used. The source of these data is the World Bank database. All variables are transformed into natural log. In this study, PVAR(P) is considered thus the empirical model can be written as.

$$\begin{aligned}
GDP_t &= a_1 + \sum_{i=1}^{p} A_{1i} GDP_{t-i} + \sum_{i=1}^{p} B_{1i} FIN_{t-i} + \sum_{i=1}^{p} C_{1i} GCF_{t-i} + \sum_{i=1}^{p} D_{1i} IND_{t-i} + \varepsilon_{1t} \\
FIN_t &= a_2 + \sum_{i=1}^{p} A_{2i} GDP_{t-i} + \sum_{i=1}^{p} B_{2i} FIN_{t-i} + \sum_{i=1}^{p} C_{2i} GCF_{t-i} + \sum_{i=1}^{p} D_{2i} IND_{t-i} + \varepsilon_{2t} \\
GCF_t &= a_3 + \sum_{i=1}^{p} A_{3i} GDP_{t-i} + \sum_{i=1}^{p} B_{3i} FIN_{t-i} + \sum_{i=1}^{p} C_{3i} GCF_{t-i} + \sum_{i=1}^{p} D_{3i} IND_{t-i} + \varepsilon_{3t} \\
IND_t &= a_4 + \sum_{i=1}^{p} A_{4i} GDP_{t-i} + \sum_{i=1}^{p} B_{4i} FIN_{t-i} + \sum_{i=1}^{p} C_{4i} GCF_{t-i} + \sum_{i=1}^{p} D_{4i} IND_{t-i} + \varepsilon_{4t}.
\end{aligned} \tag{11}$$

The summary of the descriptive statistics is illustrated in Table 1. The results of Jarque-Bera test are evident that the null hypothesis of normality is strongly rejected for all variables, indicating the non-normality of the unconditional distribution of all the variables. We use Minimum Bayes Factor (MBF) to make the statistical inference. The MBF is interpreted following what Goodman [19] labelled intervals. As a result, a MBF between 1-1/3 is considered weak evidence for H_1, 1/3-1/10 moderate evidence for H_1, 1/10-1/30 substantial evidence , 1/30-1/100 strong, 1/100-1/300 very strong, and < 1/300 decisive. The results of the panel unit root test of Levin, Lin and Chu [20] show that GDP_{it}, FIN_{it}, GCF_{it} have decisive evidence and IND_{it} has strong evidence for H_1 thus they are stationary data.

Table 1 Descriptive statistics

Variable	GDP_{it}	FIN_{it}	GCF_{it}	IND_{it}
Mean	10.5821	4.6423	25.4111	3.1994
Median	10.6905	4.6833	25.3862	3.1925
Maximum	11.6888	5.7434	28.9398	3.8008
Minimum	8.0455	2.6487	21.3148	2.4808
Std. Dev.	0.5754	0.4575	1.5654	0.2384
Skewness	−1.2121	−1.3125	−0.2586	−0.2340
Kurtosis	5.7677	6.7489	3.4749	3.7979
Jarque-Bera	162.4451 [0.0000]	251.3481 [0.0000]	5.9174 [0.0000]	10.2671 [0.0000]
Unit root test	−9.46362 [0.0000]	−4.58520 [0.0000]	−8.07188 [0.0000]	−2.77452 [0.0213]

Note [ ] MBF—is minimum bayes factor, computed by $e^{p \log p}$, where p is p-value (see, [21, 22])

4 Empirical Results

To determine the number of lags for our empirical model, we use the Bayesian Information Criterion (BIC) as the measure. Table 2 provides the BIC of each model specification and the results suggest that only the second lag of each variable is needed in the PVAR model.

The accuracy of the Bayesian estimation relies on prior selection. The lowest mean squared errors (MSE) and mean absolute deviation (MAD) are preferred to select the best prior. In this study, we estimate the posterior distribution with four different prior distributions, including Bayesian Factor Clustering and Selection (BFCS), Bayesian Mixture Shrinkage (BMS), Stochastic Search Variable Selection (SSVS), and Bayesian Semiparametric Shrinkage (BSS). The results in Table 3 show that the PVAR(2) model with SSVS prior is suitable for estimating the interdependence of macroeconomic factors and economic growth in OECD countries as the lowest MAD and MSE are presented. Besides, we also compare Bayesian estimation with the frequentist Maximum likelihood estimator(MLE), and the result shows that the PVAR model with SSVS prior outperforms the MLE, indicating the robustness of the Bayesian estimation.

From the models comparison, the Bayesian PVAR(2) with SSVS priors are applied to examine the interdependency of the four factors and estimate their impacts on OECD countries' economic growth. It should be noted that the concept of causality in the VAR model is the Granger Causality, which refers to the effects of lagged

Table 2 Lag length selection

Lag	Model
1	−9.3502
2	−9.7272[a]
3	−9.4887

Note ([a]) denotes Bayesian significant

Table 3 Estimations comparison

Prior	Criteria	Model
BFCS	MAD MSE	0.1456 0.1078
BMS	MAD MSE	0.1501 0.1107
BSS	MAD MSE	0.9113 0.8503
SSVS	MAD MSE	0.1455[a] 0.1077[a]
MLE	MAD MSE	0.7315 0.6831

Note ([a]) indicates the lowest MAD and MSE. MLE is maximum likelihood estimator

Table 4 The Bayesian panel VAR estimation for OECD countries

Lag	GDP_{it}	FIN_{it}	GCF_{it}	IND_{it}
GDP_{it-1}	0.8848 (0.0085) [0.0000]	−0.0089 (0.0079) [0.5290]	−0.0019 (0.0027) [0.7646]	−0.0067 (0.0044) [0.3069]
FIN_{it-1}	0.0089 (0.0074) [0.4794]	0.0045 (0.0036) [0.4698]	0.9891 (0.0455) [0.0000]	−0.0280 (0.0499) [0.8544]
GCF_{it-1}	0.0103 (0.0016) [0.0000]	0.0030 (0.0014) [0.1197]	0.0074 (0.0015) [0.0000]	0.0016 (0.0014) [0.5123]
IND_{it-1}	0.2840 (0.0135) [0.0000]	0.0018 (0.0070) [0.9673]	0.0055 (0.0117) [0.8943]	0.0071 (0.0146) [0.8886]
GDP_{it-2}	−0.0401 (0.0141) [0.0174]	−0.0452 (0.0155) [0.0143]	0.0072 (0.0070) [0.5875]	0.0068 (0.0057) [0.4879]
FIN_{it-2}	−0.0100 (0.0041) [0.0488]	0.0021 (0.0040) [0.8645]	0.0034 (0.0056) [0.8322]	0.0018 (0.0032) [0.8449]
GCF_{it-2}	0.9823 (0.0043) [0.0000]	−0.0096 (0.0044) [0.0963]	0.0021 (0.0014) [0.3168]	0.0033 (0.0013) [0.0490]
IND_{it-2}	0.1005 (0.0594) [0.2386]	0.4126 (0.0460) [0.0000]	0.8995 (0.0308) [0.0000]	0.0031 (0.0228) [0.9908]

Note ( ) is standard error, [ ] is Minimum Bayes factor, computed byby $e^{p \log p}$, where p is p-value (see [21])

variables on the dependent variables. A macroeconomic factor can granger cause the GDP in two ways as direct and indirect effects. For the direct effect, the factors must significantly affect the GDP directly. For the indirect effect, the factors do not directly affect the GDP but affect another factor that affects the GDP.

Table 4 presents the estimated coefficients, standard errors and Minimum Bayes Factors (MBF). From the results, we find a decisive evidence for the relationship among macroeconomic variables and GDP. Concerning our GDP equation, we find a decisive evidence that GDP can react to all lag of macroeconomic variables, except for FIN_{t-1} and IND_{t-2}. There are some economic interpretations. A 1% GDP change in the previous year will cause GDP at the current year to change by 0.8848% in the same direction. Likewise, the change of GCF_{t-1}, and that of IND_{t-1}, also contribute a decisive positive impact to GDP. In the second lag results, we find the decisive, strong, and substantial evidences, respectively, for GCF_{t-2}, GDP_{t-2} and FIN_{t-2} in affecting the GDP in the current period, while IND_{t-2} shows a weak evidence affecting the GDP in current year. The results show that the 1% GDP change two years ago causes GDP at the current year to change 0.0401% in the opposite direction. By 1% FIN_{t-2} change two years ago, GDP at the current time changes

0.0100% in the opposite direction. Furthermore, the 1% GCF change two years ago causes GDP at the current time to change 0.9823% in the same direction.

For other equations, namely FIN_{it}, GCF_{it}, and IND_{it} equations, we turn our attention to investigate the effect of GDP on these macroeconomic variables. The results show a weak evidence for impact of GDP on these macroeconomic factors, except for the effect of GCF_{t-2} on FIN_{it}.

5 Conclusion

In this study, we employ a Panel VAR approach to investigate the relationship between macroeconomic factors and OECD countries' economic growth. These countries are viewed as advanced economies, which are now the center of global economic activities. In the estimation model, as we deal with a large number of parameter estimates and the limited number of observations, this study considers the Bayesian approach as an estimator for the Panel VAR model. However, there are some concerns over this method's accuracy as the prior distribution is generally unknown in the real empirical study, and it is not easy to have an appropriate one. Therefore, the posterior distributions of the Panel VAR model are constructed from various prior distributions.

According to the results, each prior performance relies on the mean squared errors (MSE) and mean absolute deviation (MAD). SSVS priors is then selected on the ground of the lowest MAD and MSE. The best fit prior is also compared with the classical Maximum likelihood estimation, and it is evident that the higher performance of Panel VAR with SSVS prior. The findings from our analysis show that there exists substantial evidence of the effect of macroeconomic factors on the GDP of OECD countries; however, there exists weak evidence of the effect of GDP on macroeconomic factors.

Acknowledgements We acknowledge the partial of the Center of Excellence in Econometrics, Faculty of Economics, Chiang Mai University, Thailand.

References

1. Bassanini, A., Scarpetta, S., & Hemmings, P. (2001). Economic growth: The role of policies and institutions. *Panel Data Evidence from OECD Countries*.
2. Levine, R. (2005). Finance and growth: Theory and evidence. *Handbook of Economic Growth, 1*, 865–934.
3. Kenourgios, D., & Samitas, A. (2007). Financial development and economic growth in a transition economy: Evidence for Poland. *Journal of Financial Decision Making, 3*(1), 35–48.
4. Hagmayr, B., Haiss, P. R., & Sümegi, K. (2007). Financial sector development and economic growth-evidence for southeastern Europe.
5. Feng, Y., Li, D., & Long, Y. (2015). Dynamic relationship of industrial structure change and economic fluctuations: Evidence from Sichuan, China. *International Journal of Economics and Finance, 7*(12), 253.

6. Han, Z., Ma, P., & Zheng, X. (2011). Industrial structure to economic growth influence. In *2011 9th International Conference on Reliability, Maintainability and Safety (ICRMS)* (pp. 450–453). IEEE.
7. Blomstrom, M., Lipsey, R. E., & Zejan, M. (1993). Is fixed investment the key to economic growth? (No. w4436). *National Bureau of Economic Research.*
8. Bose, N., & Haque, M. (2005). Causality between public investment in transport and communication and economic growth.
9. Uneze, E. (2013). The relation between capital formation and economic growth: Evidence from sub-Saharan African countries. *Journal of Economic Policy Reform, 16*(3), 272–286.
10. Van De Schoot, R., Broere, J. J., Perryck, K. H., Zondervan-Zwijnenburg, M., & Van Loey, N. E. (2015). Analyzing small data sets using Bayesian estimation: the case of posttraumatic stress symptoms following mechanical ventilation in burn survivors. *European Journal of Psychotraumatology, 6*(1), 25216.
11. Pastpipatkul, P., Yamaka, W., Wiboonpongse, A., Sriboonchitta, S. (2015). Spillovers of quantitative easing on financial markets of Thailand, Indonesia, and the Philippines. In *International Symposium on Integrated Uncertainty in Knowledge Modelling and Decision Making* (pp. 374–388). Springer.
12. Kaewsompong, N., Sriboonchitta, S., Maneejuk, P., & Yamaka, W. (2016). Relationships among prices of rubber in ASEAN: Bayesian structural VAR model. *Thai Journal of Mathematics*, 101–116.
13. Korobilis, D. (2015). Prior selection for panel vector autoregressions. MPRA Paper 64143, University Library of Munich.
14. Bornn, L., Doucet, A., & Gottardo, R. (2010). An efficient computational approach for prior sensitivity analysis and cross-validation. *Canadian Journal of Statistics, 38*(1), 47–64.
15. George, E., Sun, D., & Ni, S. (2008). Bayesian stochastic search for VAR model restrictions. *Journal of Econometrics, 142*, 553–580.
16. Giannone, D., Lenza, M., & Primiceri, G.E. (2015). Prior selection for vector autoregressions. *Review of Economics and Statistics, 97*(2), 436–451.
17. Canova, F., & Ciccarelli, M. (2013). Panel vector autoregressive models: A survey. *Emerald Group Publishing Limited, 32*, 205–246.
18. Korobilis, D. (2013). Hierarchical Shrinkage priors for dynamic regressions with many predictors. *International Journal of Forecasting, 29*, 43–59.
19. Goodman, S. N. (1999). Toward evidence-based medical statistics. 1: The P value fallacy. *Annals of Internal Medicine, 130*(12), 995–1004.
20. Levin, A., Lin, C. F., & Chu, C. S. J. (2002). Unit root tests in panel data: Asymptotic and finite-sample properties. *Journal of Econometrics, 108*(1), 1–24.
21. Held, L., & Ott, M. (2018). On p-values and Bayes factors. *Annual Review of Statistics and Its Application, 5*, 393–419.
22. Maneejuk, P., & Yamaka, W. (2020). Significance test for linear regression: How to test without P-values? *Journal of Applied Statistics*, 1–19.

Economic Policy Uncertainty and Stock-Bond Correlations: Evidence from the Thailand Market

Nachatchapong Kaewsompong and Terdthiti Chitkasame

Abstract This study examines the effects of economic policy uncertainty (EPU) index shocks on stock-bond correlations in the case of Thailand. The method of research used applied the EPU index as a proxy for economic policy uncertainty in Thailand, and employed a dynamic copula method with the EPU index as an additional exogenous variable to create a conditional correlation. The data set used in this study is the monthly return of stock (SET) and bond (10-year government bond) data from June 2005 to June 2020. The empirical results show that the high EPU index affects the stock-bond correlations. Due to the high level of the EPU results, the investors have no confidence in the economy and paid attention to the economic policy uncertainty and tended to adjust the proportion of holdings in risky assets or stocks and safe-haven assets or bonds to be consistent.

Keywords EPU · Stock exchange of Thailand · Thailand government bond · Dynamic copula

1 Introduction

Concerns about economic policy uncertainty have intensified from the occurrence of the global financial crisis of 2008, the European debt crisis in 2011, and the US-China trade war in 2019. In 2016, Baker et al. [1] developed the economic policy uncertainty (EPU) index as a proxy for movements in policy-related economic uncertainty of the United States. The EPU index shows that the value is high when major events occur for example, the September 11 attacks, especially the failure of Lehman Brothers, and other major battles over fiscal policy. This index represents the concerns of the people in society. The movement of the EPU index is perceived by investors as policy-induced shocks. Several studies suggest that the EPU has co-movements with stock market returns and implied volatility (see [2, 3]).

N. Kaewsompong (✉) · T. Chitkasame
Faculty of Economics, Chiang Mai University, Chiang Mai, Thailand
e-mail: nachat_flysky@gmail.com

S. Sriboonchitta et al. (eds.), *Credible Asset Allocation, Optimal Transport Methods, and Related Topics*, Studies in Systems, Decision and Control 429,
https://doi.org/10.1007/978-3-030-97273-8_24

Over the past few years, there have been many studies that introduced various approaches for measuring the uncertainties in policy on economics, especially in the financial sector. Baker et al. [1] developed the method to get a new economic policy uncertainty (EPU) index, which was based on newspaper coverage frequency, as a proxy for movements in policy-related economic uncertainty over time. They found that the EPU index spiked sharply around the period of major economic or political shocks. Additionally, the data from the newspaper cannot indicate investor's behaviors. Some news related to EPU in the newspaper will not reflect the decision of investors in the stock markets [4]. In the case of Thailand, Luangaram and Sethapramote [5] also followed the method developed by Baker et al. [1] for measuring political uncertainty or political uncertainty index (PU), which was used to analyze the quantifying effects of the economic impacts. However, political uncertainty is one part of the vagueness in the economy, there is no policy-related economic uncertainty that its uncertainty is important as well. Therefore, Punwong et al. [6] consider the keywords involved with economic policy uncertainty investigating the effect on the financial markets. As mentioned in Castelnuovo and Tran [7] and Liemieux and Peterson [8], Consumers always focus on maximizing search utilities; searches continue as long as the marginal benefits of doing so exceeds the marginal cost. Thus, internet searches can be applied to quantify the behavior of people nowadays. Therefore, the EPU index from Punwong et al. [6] in this research was applied.

It is well known that the major assets in the capital market, which are generally popular among investors due to their high liquidity, are bonds and stocks. The link between the stock market and bond yields is a powerful innovation to drive portfolio building and interest rate premiums [9]. The exploration of the stock-bond correlation examines asset class behavior in different economic conditions. Volatility shocks tend to push stocks and bonds in opposite directions [10], which means that in times of high uncertainty this causes the price of the bond market to rise relative to the stock market and their correlation is weaker. In other words, when high-level EPU shocks increase due to their negative shock, investors have less security in the economy and tend to adjust the proportion of holdings [11]. In the case of Thailand, Siddiqui [12] consider the stock-bond correlation in emerging markets, including the Thailand stock market. The result found that the conditional volatility is higher and the negative market momentum increases the correlation of stocks and bonds applying the ADCC-GARCH model.

Many studies have proposed and applied the DCC-GARCH model to estimate the dependence because it widely detects the dynamic correlation, Asgharian et al. [13] Although the dynamic correlation, determined by the DCC-GARCH model with the elliptical distribution assumption (the normal or Student's t distribution), is widespread in measuring the time-varying volatility and correlation, there are several drawbacks of this model. First, it narrowly captures only their linear comovement [14]. Secondly, the estimate of the conditional correlation may be biased due to the poor specification of the multivariate distribution. If the relationship among the variables is asymmetry, the DCC-GARCH model with the elliptical distribution is not effective for estimating this dependence. Thirdly, the marginal distribution of variables can be different and exhibit skewness and fat-tail. Thus, the elliptic marginal distribution based on the DCC-GARCH may fail to capture the volatility of the variables [15]. Thus, this research applied the dynamic copula method, since previous methods used less general correlation models.

This paper aims to investigate the relationship between the EPU and the correlation between securities and bonds using the EPU index of Punwong et al. [6] and a dynamic copula method to create a conditional correlation between stocks and bonds. We treat the EPU index as an additional exogenous variable in the GARCH equations. However, we found a similar tendency to previous studies by Punwong et al. [6]. The findings of this study would be beneficial to investors for the management of portfolios.

The rest of the paper is structured as follows. Section 2 reviews the literature. The methodology used and the scope of the study are described in Sects. 3 and 4. Section 5 discusses the empirical results. The last section concludes the study.

2 Literature Review

In recent years, there have been different approaches capable of measuring uncertainties in economic policy. In 2016, Baker et al. [1] developed the EPU (economic policy uncertainty) index as a proxy for movements in policy related to the economic uncertainty of the United States. EPU index is based on newspaper coverage frequency. They found that the EPU index rose sharply near periods of major events. However, the trend of newspaper readership has decreased rapidly due to the advent of innovation and technology changing the world dynamics from analog to a digital society. Castelnuovo and Tran [7] suggest that internet search is an important resource that can reflect economic policy uncertainty. They developed the uncertainty index for the United States based on real-time data from Google Trends and found that the uncertainty index affects the dynamics of unemployment in the United States. Tobback et al. [16] proposed an alternative methodology applying text mining techniques in articles and newspapers to create the Belgium's EPU index. They found that the level of political uncertainty changes during high-risk turbulence, and that uncertainty can predict changes in bond yields, credit spreads and consumer confidence.

A number of studies have investigated the relationship of the EPU index to various financial variables. Antonakakis et al. [2] found a joint movement between the uncertainties of economic policy on the returns of the stock market and the implied volatility. Brogaard and Dtzel [17] found that economic policy uncertainties have a negative effect on the stock market in Australia. Similarly, a study by Xiong et al. [18] in China and Balcilar et al. [19] in South Korea, Malaysia, and Hong Kong. Li et al. [11] examined the effects of economic policy uncertainty shocks on stock-bond correlations in the United States by applying the DCC model and found that the political uncertainty index had a negative impact on stock-bond correlations. Bartsch [20] studied linkages between economic policy uncertainty and volatility of Dollar-Pound exchange rates employing the GARCH model. The result indicated that EPU relates to exchange rate volatility.

In the case of Thailand, Luangaram and Sethapramote [5] sstudied the impact of political uncertainty on Thailand's economy by creating an index to measure political uncertainty (PU) in Thailand, based on keyword searches in Thai-language newspapers. Evidence suggests that the rise in the EPU index has had significant negative effects on the Thai economy in the short term and economic growth in the long term. Punwong et al. [6] created the EPU index for Thailand based on Baker et al. [1] method and applied the DCC-GARCH model to study the impact of the EPU index on the performance of the Thai stock market. The result shows that the EPU is having a negative impact on the Thai stock market.

3 Methodology

This study uses several bivariate static and dynamic copula models to study the relationship between economic policy uncertainty and stock-bond correlations. In this study, we first employ GARCH (1,1) with skewed student-t distribution and EPU shock to model the marginal distribution for each index return. Patton [21] suggested that an appropriate marginal distribution is required for modelling the copulas. If the marginal distribution is mis-specified, the probability integral transformation may not be uniform (0,1). Also, the standardized residuals may face with the heteroscedasticity and autocorrelation problems. In this section, we present the main methodology employed in this study: the time-varying dependence process is outlined first, and then we describe the estimation of the model.

3.1 GARCH-type Models

The general form of a GARCH (p,q) is defined as

$$\begin{aligned} r_t &= \mu_t + \varepsilon_t \\ \varepsilon_t &= \eta_t \sqrt{h_t} \\ h_t &= \omega + \sum_{i=1}^{p} \alpha_i \varepsilon_{t-i}^2 + \sum_{j=1}^{q} \beta_j \sigma_{t-j}^2, \end{aligned} \tag{1}$$

where r_t is the log-return of an asset at time t. ε_t is residual series at time t. h_t is the conditional variance at time t. η_t is the independent and identically distributed (i.i.d) standardized random variable. Another type of GARCH model considered in this study is the GJR-GARCH model of Glosten et al. [22] which incorporates the impact of good news and bad news in the conditional variance. Therefore, the model is defined as

$$h_t = \omega + \sum_{i=1}^{p} (\alpha_i + \lambda_i I_{t-i})\varepsilon_{t-1}^2 + \sum_{j=1}^{q} \beta_j \sigma_{t-j}^2 + \gamma X_t, \tag{2}$$

where X_t is the EPU index at time t, and

$$I_{t-i} = \begin{cases} 0 \ if \ \varepsilon_{t-i} \geq 0, \\ 1 \ if \ \varepsilon_{t-i} < 0. \end{cases} \tag{3}$$

λ is the leverage effect, $\varepsilon_{it} = h_{it}\eta_{it}$, and η_{it} is i.i.d. of standard innovation. $\omega, \alpha, \beta, \lambda \geq 0$, stationary condition is $\alpha + \gamma \leq 0$. Moreover, we add the EPU index in stock and bond as the exogenous variabale in both GARCH and GJR-GARCH equations. Thus we have, respectively, GARCH-X and GJR-GARCH-X as follows

$$h_t = \omega + \sum_{i=1}^{p} \alpha_i \varepsilon_{t-1}^2 + \sum_{j=1}^{q} \beta_j \sigma_{t-j}^2 + \gamma X_t \tag{4}$$

$$h_t = \omega + \sum_{i=1}^{p} (\alpha_i + \lambda_i I_{t-i})\varepsilon_{t-1}^2 + \sum_{j=1}^{q} \beta_j \sigma_{t-j}^2 + \gamma X_t. \tag{5}$$

In this study, we simply assume the order of p and q as 1 ($p = 1, q = 1$).

3.2 Bivariate Dynamic Copulas

Our interest is to measure the dependence structure between economic policy uncertainty and stock-bond correlations; therefore, bivariate copula specification is considered here. We use 2 copulas; Gaussian and Student-t family. Each of these families describes a different dependence structure between two random variables. For Gaussian and Student-t, they feature asymmetric dependence measure, which is commonly used in many financial works in the literature as the benchmark. The time-varying version of these copula models is briefly discussed in this section.

The bivariate time-varying Gaussian copula is defined by

$$C_{Ga,t}(u_t, v_t \left| \rho_{Ga,t} \right.) = \int_{-\infty}^{\Phi^{-1}(u_t)} \int_{-\infty}^{\Phi^{-1}(v_t)} \frac{1}{2\pi\sqrt{1-\rho_{Ga,t}^2}} \left\{ -(\frac{x_t^2 - 2\rho_{Ga,t}x_t y_t + y_t^2}{2(1-\rho_{Ga,t}^2)} \right\} dxdy. \tag{6}$$

where $\Phi^{-1}(\cdot)$ is the inverse of the standard normal distribution function with the correlation coefficient $(-1 < \rho_{Ga,t} < 1)$. There is no tail dependence for this copula. To allow for time-varying dependence, Patton [21] suggested predicting the time-varying dependence $(\rho_{Ga,t})$ using ARMA (1,1) process, as follows:

$$\rho_{Ga,t} = \Lambda\left(\varpi_0^{Ga} + \varpi_1^{Ga}\rho_{Ga,t-1} + \varpi_2^{Ga}\left(\frac{1}{10}\sum_{j=1}^{1} F_1^{-1}(u_{t-j})F_2^{-1}(v_{t-j})\right)\right) \tag{7}$$

where $\Lambda(z) = (1 - e^{-z})/(1 + e^{-z})$ is the logistic transformation ensuring the $\rho_{Ga,t} \in (-1, 1)$. $\varpi_0^{Ga}, \varpi_1^{Ga}, and \varpi_2^{Ga}$ are the parameters to be estimated. Thus, the ARMA (1, 1) process of Patton [21] is modified to be ARMAX (1,1).

The bivariate Student-t copula is defined by

$$C_{T,t}(u_t, v_t \,|\rho_{T,t}\,, v_t) = \int_{-\infty}^{t_v^{-1}(u_t)}\int_{-\infty}^{t_v^{-1}(v_t)} \frac{1}{2\pi\sqrt{1-\rho_{T,t}^2}} \left\{1 + \frac{x^2 - 2\rho_{T,t}xy + y^2}{v_t(1-\rho_{T,t}^2)}\right\}^{-\frac{v_t+2}{2}} dxdy, \tag{8}$$

where $t_v^{-1}(\cdot)$ is the inverse of the standard student-t distribution function with degrees of freedom. $\rho_{T,t}$ is the time-varying correlation within the interval (−1,1). Like Gaussian copula, this copula has the symmetric dependence structure and presents non-zero tail dependence. The time-varying dependence process of the Student-t copula is, therefore,

$$\rho_{T,t} = \Lambda\left(\begin{array}{c}\varpi_0^T + \varpi_1^T\rho_{T,t-1} + \varpi_2^T\left(\frac{1}{10}\sum_{j=1}^{1} F_1^{-1}(u_{t-j})F_2^{-1}(v_{t-j})\right)\\ +\kappa_1^T GI_{US} + \kappa_2^T GI_O\end{array}\right). \tag{9}$$

Copulas can also be used to characterize the dependence in the tails of the distribution. Two measures of tail dependence related to copulas are known as the upper and lower tail dependence coefficients. They are indeed very helpful for investigating the behavior of the extreme market, say boom and burst. The computation of tail dependence is referred to as Joe [23].

3.3 Estimation of Copulas: The Inference for Margins Method (IFM)

In the estimation of the dynamic copula, this study follows Joe and Xu [24] using the two-step Maximum Likelihood procedure known as inference function for marginal (IFM). Joe and Xu [24] showed that the IFM method has high efficiency of the easily implemented procedure compared to the one-step Maximum Likelihood method. According to Sklar's theorem [25] and Patton [21], we can derive the bivariate dynamic copula density by taking the partial derivative of joint density function to obtain the joint density:

$$\begin{aligned} h_t(x, y \,|W_{t-1}) &= \frac{\partial^2 C_t(F_{1,t}(x|W_{t-1}), F_{2,t}(y|W_{t-1})|G_{t-1})}{\partial F_{1,t}(x|W_{t-1}), \partial F_{2,t}(y|W_{t-1})} \cdot \frac{\partial F_{1,t}(x|W_{t-1})}{\partial x} \cdot \frac{\partial F_{2,t}(y|W_{t-1})}{\partial y} \\ &= c_t(F_{1,t}(x\,|W_{t-1}), F_{2,t}(y\,|W_{t-1})\,|W_{t-1}) \cdot f_{1,t}(x\,|W_{t-1}) \cdot f_{2,t}(y\,|W_{t-1}), \end{aligned} \tag{10}$$

where $c_t(\cdot)$ is the conditional copula density (dynamic copula density). $f_t(\cdot)$ is the univariate marginal distribution. Thus, our full likelihood function for dynamic copula-based GARCH(1,1) can be written as

$$l(\Theta) = c_t(u_t, v_t\,|W_{t-1};\ \Theta_C) \cdot f_{1,t}(x_t\,|W_{t-1};\ \Theta_{G,1}) \cdot f_{2,t}(y_t\,|W_{t-1};\ \Theta_{G,2}), \tag{11}$$

where $\Theta_C = \{\varpi_0, \varpi_1, \varpi_2, \kappa_1, \kappa_2)$ is the vector of the dynamic copula parameter. $\Theta_{G,i} = \{\text{u}, \omega, \alpha, \beta, \text{v}, \lambda)$, $i = 1, 2$, is the vector of the GARCH-skewed-t parameter. Then, the joint conditional log-likelihood function to be maximized is expressed as

$$\begin{aligned} L(\Theta) = \ & \log c_t(u_t, v_t\,|W_{t-1};\ \Theta_C) \\ & + \log f_{1,t}(x_t\,|W_{t-1};\ \Theta_{G1}) \\ & + \log f_{2,t}(y_t\,|W_{t-1};\ \Theta_{G2}). \end{aligned} \tag{12}$$

According to IFM, we can estimate the parameters of GARCH-skewed-t and the parameters of the dynamic copula, separately. Therefore, the parameters for our model are estimated by the following process:

Step 1: Estimate the parameters for the GARCH-skewed-t by maximum likelihood method

$$\widehat{\Theta}_{G1} = \arg\max \sum_{t=1}^{T} \log f_{1,t}(x_t\,|W_{t-1};\ \Theta_{G1}), \tag{13}$$

$$\widehat{\Theta}_{G2} = \arg\max \sum_{t=1}^{T} \log f_{2,t}(y_t\,|W_{t-1};\ \Theta_{G2}). \tag{14}$$

Step 2: Estimate the parameters for the dynamic copula based on the estimated parameters obtained from step 1

$$\widehat{\Theta}_C = \arg\max \sum_{t=1}^{T} \log c_t(u_t, v_t \,|W_{t-1};\ \Theta_C, \widehat{\Theta}_{G1}, \widehat{\Theta}_{G2}). \quad (15)$$

4 Data Description

The data set used in this study is the monthly close of the Thailand Stock Index (SET) and the 10-year government bonds (Bond). The data was retrieved from Thomson Reuter's databases. Research was undertaken using the EPU index from Punwong et al. [6]. the study period is from June 2005 to June 2020. The monthly index of stocks, bonds and EPU has been transformed into rate of return and growth, respectively. The descriptive statistics are provided in Table 1. The mean values are negative value for all variables, the skewness of data return is not equal to zero and display a negative skewness. All feedback data also showed excessive kurtosis, indicating that data returns have the typical characteristics of leptokurtosis and fat-tail. We guarantee that the Thai Stock Market Index (SET) and the 10-year government bond are not normally distributed as the minimum values of the Bayes factor (MBF) of the Jarque-Bera statistic. Indicating the returns are decisive not normally distributed. The unit root test is also conducted to investigate the stationarity of the data and the result shows that all returns and the growth rate of EPU are stationary with decisive evidence. Consequently, these variables can be used to estimate the dynamic copula-type models in a further step.

Table 1 Descriptive statistics

	EPU	SET	Bond
Mean	−0.002	0.039	−0.007
Median	−0.004	0.011	−0.008
Maximum	0.286	0.177	0.166
Minimum	−0.336	−0.302	−0.317
Std. Dev.	0.105	0.059	0.071
Skewness	−0.058	−1.348	−0.487
Kurtosis	2.931	9.268	5.043
Jarque-Bera	0.14	351.023[a]	38.625[a]
MBF unit-root	0.00	0.00	0.00
Observations	181	181	181

Note [a] indicates the strong evidence support the alternative hypothesis by MBF

5 Estimated Results

As several models and distributions are considered in this study. It would be best to compare these models to find the best fit. We note that GARCH-X (1,1), GJR-GARCH-X (1,1) models with three distributions: normal (N), student-t (std.), and skewed student-t (sstd.) including the Gaussian and Student-t copula families are considered in this study. The best fit model is selected based on a minimum of the Bayesian information criterion (BIC). Table 2 provides AIC, BIC, and log-likelihood of each model specification for our two empirical groups of markets. We observe that the dynamics copula-GARCH-X model with skewed student-t distribution (Gaussian family) clearly outperforms other models in both variables where the lowest BIC is obtained.

Furthermore, we compared our select model with the previous model (DCC-GARCH) to verify that the selected model produced results that are closer to realistic or better, both in the case of within and without the EPU effect, considering the BIC value (Table 3). The results present that the dynamic copula with the EPU model has the best BIC value. This confirms that the model selected was more general and more realistic than previous models used (DCC-GARCH).

Table 2 Model selection

	Gussian					
	GARCH-N	GJR-GARCH-N	GARCH-Std	GJR-GARCH-Std	GARCH-Sstd	GJR-GARCH-Sstd
AIC	5.472	4.749	4.852	4.749	4.746	4.587
BIC	15.061	14.495	14.448	14.344	14.041	14.183
Log-likelihood	0.264	0.55	0.574	0.626	0.627	0.706
	Student-T					
AIC	5.647	11.642	11.359	10.953	11.155	11.415
BIC	15.242	30.833	30.55	30.144	30.346	30.606
Log-likelihood	0.177	0.179	0.32	0.523	0.422	0.293

Table 3 Model comparison

	Gussian dynamic copula model (with GARCH-Sst)		DCC-GARCH model (with GARCH-Sst)	
	With EPU effect	Without EPU effect	With EPU effect	Without EPU effect
AIC	4.746	5.342	10.223	12.312
BIC	14.041	16.432	20.242	32.843
Log-likelihood	0.627	0.555	0.157	0.749

In the second step, the univariate GARCH-X (1,1) was employed with the skewed student-t model to model the marginal distributions in this study. The results for the margins are summarized in Table 4. What was found was that the estimated ARCH coefficient (α), GARCH coefficient (β), and the impact of the EPU coefficient (γ)are strong evidence for supporting all variables' returns, thus signifying the presence of the clustering volatility phenomenon in stocks and bonds. In addition, the EPU resulted in the volatility of stock and bond of about 0.002 and 0.009, respectively (as shown in Table 4). In addition, there is strong evidence to support the EPU effect of both stocks and bonds. This indicates that the EPU appears to have a moderate impact on stock and bond volatilities.

Afterward, we considered the conditional variance equations ($\alpha + \beta$), which are the volatility feedback effect decays over time, which are discovered to be less than one for both cases, indicating Thailand's stock and bond returns series satisfy the stationary condition. We remark that the $\alpha + \beta$ is very close to 1. This implies the high volatility persistence of our series. In addition, we note that the degree of volatility persistence of stock (0.961) was the highest, followed by bonds (0.743). This evidence allows us to indicate that the persistence of high volatility in the stock market (SET) is a crucial contributor to the EPU index.

Next, we estimated the GARCH models on the stock market (SET) and bonds. We obtained the standardized residuals from this estimation to evaluate the dynamic copula with the Gaussian family. The parameters of the time-varying dependencies of the chosen copula are also reported in Table 4. The parameter ϖ_0 represents the dependence levels. While, ϖ_1 and ϖ_2 represents the degree of persistence and captures the adjustment in the dependence process, respectively. This indicates that the dependency adjustment process (0.989) has less effect on the conditional correlation than the degree of persistence (-1.043).

Table 5 represents the average of conditional correlations in stock-bonds. The correlation of the mean is 0.004, while the maximum and minimum values are 0.500 and -0.415, respectively. In addition, the standard deviation is 0.072, which indicates the highest fluctuation correlation. Moreover, the stock-bond correlations are a measure of an investor's sentiment, it reflects an investors' concerns. When investors are not confident in the economy, investors will avoid risk and tend to adjust the riskier equity market holding ratio and buy relatively safer bond market, leading to increased correlations between bonds and stocks. On the other hand, if investors are confident in the economy, the correlations between stocks and bonds are weak.

From our results, it is quite surprising that the equity-bond co-movement is weak (0.004). One possible explanation for this result is that investors may be behaving randomly or unreasonably. As mentioned in Lin [26], the irrational investor cannot fully understand and synthesize huge volumes of complex information before making an investment decision.

Next, we separated our dynamic conditional correlations into two sub-periods, including crisis period and normal period, relative to the EPU index. According to the global financial and COVID-19 crisis, economists warned that the economic crisis was likely to be triggered by the reverse yield curve phenomenon in 2008 and the World Health Organization (WHO) announcement of the spread of COVID-19

Table 4 Maximum likelihood estimates of the GARCH-X (1,1) and estimation of dynamic copula model

Gussian						
GARCH-skewed student-t						
Parameter	SET	S.D.	MBF	BOND	S.D.	MBF
ω	0.000	0.000	1	0.000	0.000	1
α_1	0.000	0.046	0.467	0.000	0.002	0.467
β_1	0.961	0.017	0.013	0.743	0.15	0.017
ν_1	0.097	0.067	0.153	0.352	0.024	0.024
γ_1	0.002	0.002	0.023	0.009	0.001	0.019
λ	0.643	0.08	0.025	0.904	2.561	0.154
Shape	3.882	1.128	0.015	7.154	17.679	0.252
$\alpha_1 + \beta_1$	0.961			0.743		
Parameter	Coefficient		S.D.		MBF	
ϖ_0	−0.683		0.74		0.543	
ϖ_1	−1.043		0.76		0.045	
ϖ_2	0.989		1.033		0.054	

Table 5 Summary statistics of conditional correlations of the SET and bond

Correlation	Mean	Max	Min	S.D.
SET-Bond	0.004	0.5	−0.415	0.072

Table 6 Summary statistics on conditional correlations of SET and bonds during the crisis and normal period compared to the EPU index

	Variable	Mean	Max	Min	S.D.
Crisis period					
Subprime	Stock-bond	0.014	0.15	−0.11	0.056
	EPU	116.659	163.544	91.236	13.474
COVID-19	Stock-bond	0.093	0.166	0.038	0.048
	EPU	105.678	122.18	85.288	14.128
Normal period					
Normal	Stock-bond	−0.005	0.134	−0.088	0.042
	EPU	77.064	99.217	59.606	8.507

disease to the pandemic caused in March 2020. The results in Table 6 represent that the mean conditional correlation during the financial crisis and COVID-19 (EPU index greater than 100) has seemed significantly related. During this time, it was found to have a weak relationship at regular intervals between 2011 and the end of 2019. The results indicate that during the crisis period, the value of EPU is above 100.

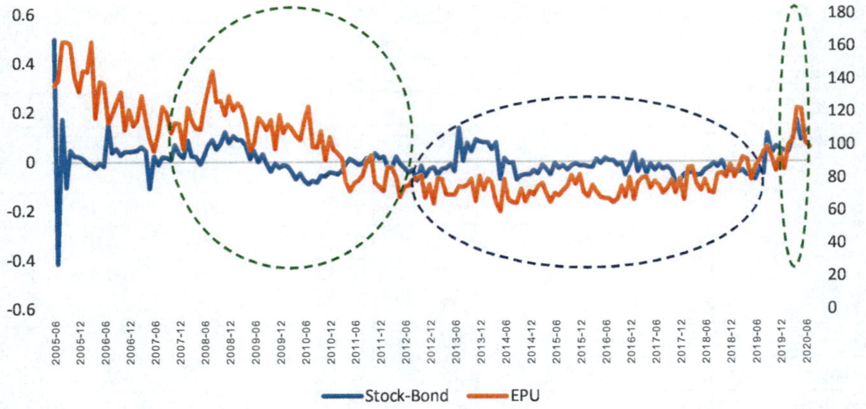

Fig. 1 The relationship between the EPU index and the stock-bond correlations

Figure 1 presents the relationship between the economic policy uncertainty (EPU) index and the stock-bond correlations. Investors view the movement of the EPU index as a policy shock. However, when the EPU index has a low value (less than 100), for example during the years 2011–2018 (as shown in Table 6), the median value of the financial bond correlations is 0. The low EPU index indicates that the uncertainty on economic policy is low. Hence, investors trust the investment environment and invest in a variety of assets, especially risky and high return assets. The correlation between stocks and bonds was at a low level during the period.

While the EPU index has a high value (more than 100), such as during the finical crisis 2006–2009 and the COVID-19 pandemic 2019–2020 (as shown in Table 6), the mean value of the stock-bond correlations is equal to 0.2 and 0.25, respectively. The EPU's high level suggests that economic policy uncertainty is high. Investors lack confidence in the economy and pay attention to economic policy uncertainty and tend to adjust the proportion of holdings in risky assets or stocks and safe-haven assets or bonds to be consistent. Therefore, stocks and bonds are more related. This result is consistent with Fang and Li [27] and Li et al. [11] who conclude that the high EPU index affects correlations between stocks and bonds.

6 Conclusion

This paper examines how policy uncertainty shocks affect the correlations between stocks and bonds in the case of Thailand. The EPU index from Punwong et al. [6] as a proxy for economic policy uncertainty in Thailand and applied a dynamic copula method with the EPU index to establish a conditional correlation. The result shows that if there is a high level of economic policy uncertainty, investors will be aware of the economic policy uncertainty. This leads to a higher correlation between stocks and bonds. In other words, the high EPU index affects the correlations between stocks and bonds, which Fang and Li [27] results confirm. This study provides investors with useful information for managing portfolios. It also includes suggestions for policy makers on how to deal with equity and bond markets.

References

1. Baker, S. R., Bloom, N., & Davis, S. J. (2016). Measuring economic policy uncertainty. *The Quarterly Journal of Economics, 131*(4), 1593–1636.
2. Antonakakis, N., Chatziantoniou, I., & Filis, G. (2013). Dynamic co-movements of stock market returns, implied volatility and policy uncertainty. *Economics Letters, 120*(1), 87–92.
3. Jones, P. M., & Olson, E. (2013). The time-varying correlation between uncertainty, output, and inflation: Evidence from a DCC-GARCH model. *Economics Letters, 118*(1), 33–37.
4. Peiser, W. (2000). Cohort replacement and the downward trend in newspaper readership. *Newspaper Research Journal, 21*(2), 11–23.
5. Luangaram, P., & Sethapramote, Y. (2018). *Economic impacts of Political Uncertainty in Thailand* (No. 86). Puey Ungphakorn Institute for Economic Research.
6. Punwong, S., Kaewsompong, N., & Tansuchat, R. (2021). Impact of economic policy uncertainty on the stock exchange of Thailand: Evidence from the industry-level stock returns in Thailand. In *Behavioral predictive modeling in economics* (pp. 393–406). Springer.
7. Castelnuovo, E., & Tran, T. D. (2017). Google it up! a google trends-based uncertainty index for the united states and Australia. *Economics Letters, 161*, 149–153.
8. Liemieux, J., & Peterson, R. (2011). Purchase decline as a moderator of the effects of price uncertainty on search behavior. *Journal of Economic Psychology, 32*, 33–44.
9. Sueppel, R. (20181). Understanding *the correlation of equity and bond returns*. Retrieved from https://www.sr-sv.com/understanding-equity-bond-return-correlation
10. Ilmanen, A. (2003). Stock-bond correlations. *The Journal of Fixed Income, 13*(2), 55–66.
11. Li, X. M., Zhang, B., & Gao, R. (2015). Economic policy uncertainty shocks and stock-bond correlations: Evidence from the US market. *Economics Letters, 132*, 91–96.
12. Siddiqui, D. A. (2019). Analyzing stock-bond correlation in emerging markets.
13. Asgharian, H., Christiansen, C., & Hou, A. J. (2016). Macro-finance determinants of the long-run stock-bond correlation: The DCC-MIDAS specification. *Journal of Financial Econometries, 14*(3), 617–642.
14. Chang, K. L. (2012). The time-varying and asymmetric dependence between crude oil spot and futures markets: Evidence from the Mixture copula-based ARJI-GARCH model. *Economic Modelling, 29*(6), 2298–2309.
15. Maneejuk, P., & Yamaka, W. (2019). Predicting contagion from the US financial crisis to international stock markets using dynamic copula with google trends. *Mathematics, 7*(11), 1032.
16. Tobback, E., Naudts, H., Daelemans, W., de Fortuny, E. J., & Martens, D. (2018). Belgian economic policy uncertainty index: Improvement through text mining. *International Journal of Forecasting, 34*(2), 355–365.
17. Brogaard, J., & Detzel, A. (2015). The asset-pricing implications of government economic policy uncertainty. *Management Science, 61*(1), 3–18.
18. Xiong, X., Bian, Y., & Shen, D. (2018). The time-varying correlation between policy uncertainty and stock returns: Evidence from China. *Physica A: Statistical Mechanics and its Applications, 499*, 413–419.
19. Balcilar, M., Gupta, R., Kim, W. J., & Kyei, C. (2019). The role of economic policy uncertainties in predicting stock returns and their volatility for Hong Kong, Malaysia and South Korea. *International Review of Economics and Finance, 59*, 150–163.

20. Bartsch, Z. (2019). Economic policy uncertainty and dollar-pound exchange rate return volatility. *Journal of International Money and Finance, 98,* 102067.
21. Patton, A. J. (2006). Estimation of multivariate models for time series of possibly different lengths. *Journal of Applied Econometrics, 21*(2), 147–173.
22. Glosten, L., Jagannathan, R., & Runkle, D. (1993). Relationship between the expected value and volatility of the nominal excess returns on stocks. *Journal of Finance, 48,* 1779–1802.
23. Joe, H. (1997). *Multivariate models and multivariate dependence concepts*. CRC Press.
24. Joe, H., & Xu, J. J. (1996). The estimation method of inference functions for margins for multivariate models.
25. Sklar, M. (1959). Fonctions de repartition an dimensions et leurs marges. *Publ. inst. statist. univ. Paris, 8,* 229–231.
26. Lin, T. C. (2015). Reasonable investor(s). *Boston University Law Review, 95*(2), 461–518.
27. Fang, L., Yu, H., & Li, L. (2017). The effect of economic policy uncertainty on the long-term correlation between US stock and bond markets. *Economic Modelling, 66,* 139–145.

Revisiting the Determinants of Thai Economic Growth: A Mixed Frequency Approach

Rungrapee Phadkantha, Wiranya Puntoon, and Kongliang Zhu

Abstract This paper aims to analyze the factors affecting the economic growth of Thailand. Even though this issue is uncomplicated and has already been discussed by many scholars, it is still important. In this study, we conduct an econometric analysis taking into account various key macro-economic factors that can play a significant role in expediting the Thai economic growth. However, these factors are normally recorded at different frequencies as monthly, quarterly, and yearly data. Thus, we suggest applying the mixed frequency regression approach namely U-MIDAS and R-MIDAS for the investigation using various time series data reported at different frequencies. This study uses mix-frequency variables of quarterly real GDP, and monthly unemployment, tourism, total export, government budget, and direct investment. The results show that the U-MIDAS model is more suitable than R-MIDAS. We find that export and government budget in the previous month shows the decisive evidence of affecting the economic growth in the current quarter, while the tourism in the previous month shows strong evidence of affecting the economic growth.

Keywords Economic growth · MIDAS regression · R-MIDAS and U-MIDAS

1 Introduction

Economic growth is often determined by many factors. It is vital to ask which factors are important for driving the economic growth in the rapidly evolving economic context. Therefore, many scholars have paid increasing attention to investigating the factors determining the recent economic growth (see [1–6]). However, the results

R. Phadkantha · W. Puntoon (✉)
Center of Excellence in Econometrics, Faculty of Economics, Chiang Mai University, Chiang Mai, Thailand
e-mail: wiranya.pun@gmail.com

K. Zhu
Faculty of Economics Khon Kaen University, Khon Kaen, Thailand

S. Sriboonchitta et al. (eds.), *Credible Asset Allocation, Optimal Transport Methods, and Related Topics*, Studies in Systems, Decision and Control 429,
https://doi.org/10.1007/978-3-030-97273-8_25

remain controversial. Hence, in this study, we aim to revisit the key determinants of Thailand's economic growth.

This study attempts to identify the leading factors that affect the economic growth of Thailand and to provide policymakers with useful information for their decision-making and implementation of appropriate policies to deal with the economic dynamism. To these ends, we first explore the recent economic growth literature for a list of some prime movers. Gunduz and Hatemi [7]; Katircioglu [8], and Tang and Tan [9] indicated that the tourism-led growth hypothesis is valid in many countries and they confirmed that tourism is a key factor driving economic growth. Pegkas [10] and Maneejuk et al. [5] confirmed a strong effect of government expenditure. Saget [11] and Fofana [12] attested that unemployment harms economic growth. Foreign direct investment (FDI) is another factor investigated in many studies [13–15]. These studies estimated the effect of FDI on economic growth in many countries and revealed that there is a positive impact of FDI on economic growth. Finally, we also find that export is another important economic factor contributing to economic growth (see, [16–18]). Many factors are operating in the economic environment that will influence the GDP growth rate. However, in this study, we focus on the variables that have empirically shown their strong and significant roles in economic growth according to the literature. Thus, we consider only unemployment, tourism, total export, government budget, and foreign direct investment as the key factors affecting the economic growth of Thailand. Then, the impact of these factors on economic growth can be estimated by the regression model.

However, due to the nature of data collection, the economic statistics are normally recorded at different frequencies, say monthly, quarterly, and yearly. Therefore, it is difficult to implement an econometric model with data having different frequencies. To handle this difficulty, some contributors to the literature in this area advised using the simplistic methods to convert the data into the same frequency (disaggregate low-frequency data to the high-frequency data or aggregate the high-frequency data to the low-frequency data), for example, by disaggregating the quarterly data into daily data. In this way, the influence of different-frequency factors can be easily performed in the linear regression, and its impact on economic growth is quantified.

However, Ghysel et al. [19] demonstrated that both disaggregation and aggregation approaches may lead to a very rough and distorted estimate of the impact of the independent variable on the dependent variable. They introduced the Mixed Data Sampling (MIDAS) Regression model to deal with time-series data sampled at different frequencies, where the relationship between the low-frequency variables and high-frequency variables is applicable to be estimated. Penev et al. [20] mentioned this model imposes an unequal weighting scheme on the high- and low-frequency observations. Hence, we consider using the Mixed frequency approach, namely the Unrestricted (U)-MIDAS and Restricted (R)-MIDAS models for flexible modelling of the impact of unemployment, tourism, total export, government budget, and foreign direct investment on Thailand's economic growth.

The remainder of the paper is organized as follows. Section 2 brierly describes the methodology used in this study. Section 3 presents the data and model specification. In Sect. 4, we present the empirical results. Section 5 presents our conclusions.

2 Methodology

In this section, we briefly explain the models proposed by Ghysel et al. [19] and Foroni and Marcellino [21] namely, the R-MIDAS model and the U-MIDAS model. Moreover, we discuss the case of a single high-frequency indicator for each regression in this study to reduce the curse of dimensionality and to avoid the estimation issue in the MIDAS framework [19].

2.1 The Restricted MIDAS Model (R-MIDAS)

It will be convenient to focus on a mixture of two frequencies, respectively high and low. In terms of notation, $t = 1, \ldots, T$ indexes the low-frequency time unit, and m is the number of times the higher sampling frequency appears in the same basic time unit (assumed fixed for simplicity). In this study, we aim to predict the quarterly real GDP growth using monthly factors as explanatory variables (tourism, government expenditure, unemployment, export, and foreign direct investment). The low-frequency (quarterly) variable will be denoted by y_t^L, where a high-frequency (monthly) series will be denoted by $x_{t-j/m}^H$ where $t - j/m$ is the jth (past) high-frequency period with $j = 0, 1, \ldots$ For a quarter/month mixture, one has $x_t^H, x_{t-1/3}^H, x_{t-2/3}^H$ as the last, second to last, and first months of the quarter t. The parameters of MIDAS regression are adjusted by regressions that involve processes sampled at different frequencies. If the low-frequency variable (y_t^L) is explained by the high-frequency variable (x_t^H), the basic MIDAS model is shown as

$$y_t^L = \beta_0 + \beta_1 (\sum_{j=1}^{K} C(j; \theta_h) x_{t-j/m}^H) + \varepsilon_{t+h} \;\; ; \; t = 1, .., T \tag{1}$$

where β_0 and β_i are the unknown parameters to be estimated, K is the number of monthly data used to predict y_t^L. The parameterization of the lagged coefficient of $C(j; \theta)$ is one of the main features of MIDAS. For the weight function of this study, we choose the normalized exponential Almon lag polynomial weighting function proposed by Gysels et al. [22] that can be specified as

$$C(j; \theta) = \frac{\exp(\theta_1 j + \theta_2 j^2)}{\sum_{i=1}^{K} \exp(\theta_1 i + \theta_2 i^2)} \tag{2}$$

where the parameters θ_1, θ_2 are also part of the estimation problem and can be influenced by the last K values of x_t^H. This weighting specification gives more weight to the most recent observed period and the weights sum up to one as it has been widely used in the MIDAS literature. We use the restricted MIDAS model with exponential Almon weight $K = 3$ and $h = 1$. Then the MIDAS model can be shown as Eq. 3

$$y_t^L = \beta_0 + \beta_1 \left(\frac{\exp(\theta_1 + \theta_2)}{\sum_{i=1}^{K} \exp(\theta_1 i + \theta_2 i^2)} (x_t^H) + \frac{\exp(2\theta_1 + 4\theta_2)}{\sum_{i=1}^{K} \exp(\theta_1 i + \theta_2 i^2)} (x_{t-(1/3)}^H) + \frac{\exp(3\theta_1 + 9\theta_2)}{\sum_{i=1}^{K} \exp(\theta_1 i + \theta_2 i^2)} (x_{t-(2/3)}^H) \right) + \varepsilon_t \tag{3}$$

where y_t^L is the quarterly independent variable at time t which can be explained by the monthly dependent variables x_t^H at time t, β_0, β_i, θ_1 and θ_2 are the unknown parameters to be estimated through the maximize likelihood (ML) technique. However, the number of lags in the regression model does not influence the number of the estimated parameters. Thus, the weight function appears to incorporate historical data of the indicator while maintaining the parsimonious parameter estimation that is one of the key features of the MIDAS approach [19]. Note that this study uses the BIC to select the number of lags for the autoregressive terms and the high-frequency terms.

2.2 *The Unrestricted MIDAS Model (U-MIDAS)*

Kingnetr et al. [23] said that the basic MIDAS model was significant to reduce the coefficients via a specific weighting scheme. But the model needs a strong assumption about the dynamics of the data and its process. To succeed in dealing with this problem, we use the unrestricted MIDAS (U-MIDAS) model proposed by Foroni and Marcellino [21]. The U-MIDAS model can be specified as

$$y_t^L = \beta_0 + \sum_{j=1}^{K} \beta_j x_{t-1-j/m}^H + \varepsilon_t \;\; ; \; t = 1, .., T \tag{4}$$

where y_t^L is a low-frequency data measured quarterly at time t , x_{t-1}^H is a high-frequency data (monthly data). j is the first month of the quarter t, m is the frequency ratio, which is 3 in our case, K is the number of monthly data used to predict y_t^L. From their study, Foroni et al. [24] found that This model specification has the advantage of higher flexibility compared to the functional lag polynomials in the classical MIDAS approach. U-MIDAS can be helpful, if the differences in sampling frequencies are small, for example, monthly-quarterly, which is considered in our study. The quarterly GDP should be related to various monthly economic factors, and a small number of lags is necessary to capture the dynamics. Therefore, the U-MIDAS model is used in this study. Barsoum and Stankiewicz [25], Baumeister et al. [26], and Foroni et al. [24] confirmed that U-MIDAS generally performs better than MIDAS when mixing quarterly and monthly data.

3 Data

This study uses the data recorded at different frequencies, especially quarterly and monthly time-series variables. The quarterly real GDP (low-frequency series) covers 2012-Q2 to 2020-Q1. All other indicators are collected at monthly frequency covering 2012-M4 to 2020-M3, which include government expenditure, foreign direct investment, total export, the number of tourists, and unemployment. The quarterly real GDP and government expenditure were obtained from CEIC Data. The foreign direct investment and total export were collected from the Ministry of Commerce, Thailand. The number of tourists is downloaded from the Immigration Bureau of Thailand. Finally, unemployment was obtained from the Bank of Thailand. To eliminate the long-term stochastic trends or unit-roots, all data transforming into growth rates are typically done.

3.1 Data Description

Table 1 shows the summary statistics of real GDP (GDP), Government budget (G), Tourism (TOUR), Export (EX), Unemployment (UN), and Foreign Direct Investment (FDI). Firstly, this study considers the sign of skewness, which indicates asymmetric distribution. GDP, G, EX and UN have positive skewness meaning that the growth tends to be positive along the sample period. The variable Tour tends to be in a downtrend since its skewness is negative. The kurtosis of all variables is more than 3. Secondly, considering the results of the Jarque-Bera normality test, it is evident that the MBF values are zero for all variables, indicating the non-normal distribution of our data.

Table 1 Descriptive statistics

	GDP (Q)	Government expenditure (M)	Tourism (M)	Export (M)	Unemployment (M)	Foreign direct investment (M)
˙Mean	3.384	0.033	0.003	0.178	0.022	1.001
˙Median	3.233	−0.003	−0.002	0.007	−0.007	−0.213
˙Maximum	15.298	0.961	0.237	9.916	0.838	62.796
˙Minimum	−1.774	−0.42	−0.561	−0.901	−0.292	−26.761
˙Std. Dev.	2.748	0.269	0.131	1.269	0.208	7.837
˙Skewness	2.32	1.202	−1.005	6.8	1.319	5.14
˙Kurtosis	12.468	4.992	6.087	48.65	5.5	44.11
˙Jarque?Bera (JB)	148.226	38.987	54.279	9075.387	52.839	7182.747
˙MBF-JB	0	0	0	0	0	0
Observations	32	96	96	96	96	96

Note This study uses Minimum Bayes factor (MBF) as the tool for checking the significant result, for more detail see Maneejuk and Yamaka [27]. MBF calculated by $MBF = exp(t^2/2)$ ([28, 29]). "Q" and "M" denote quarter and month, respectively

4 Results

4.1 Unit Root Test

Before estimation, to make sure that variables fluctuate around a constant mean, and their variances are constant, the stationarity of the variables should be tested. This study applies the unit root test based on the Augmented Dickey-Fuller (ADF) and Phillips-Perron (PP), and the Minimum Bayes Factor (MBF) as a tool for making a statistical test. In Table 2, the results show there is strong evidence supporting the stationarity of all variables, while MBF values are mostly less than 0.33.

Table 2 Unit root test

Variable	Method	Intercept	Trend and intercept	None
GDP	ADF	−5.2312	−1.9175	−2.3612
		−0.0001	−0.6214	−0.0195
	PP	−3.6359	−0.3537	−2.2918
		−0.0098	−0.0465	−0.0231
G	ADF	−3.3284	−3.4265	−3.4433
		−0.0678	−0.0123	−0.0007
	PP	−16.5112	−17.6783	−14.9081
		0	0	0
TOUR	ADF	−0.3705	−0.8929	−0.9248
		−0.9089	−0.952	−0.3135
	PP	−5.655	−5.6202	−5.7399
		0	0	0
EX	ADF	−4.4309	−4.74935	−12.9061
		−0.0005	−0.0011	0
	PP	−28.8831	−293127	−23.4084
		−0.0001	−0.0001	0
UN	ADF	−10.404	−10.4578	−10.0354
		0	0	0
	PP	−16.694	−18.1262	−15.3189
		0	0	0
FDI	ADF	−10.1208	−10.0668	−9.9995
		0	0	0
	PP	−10.1564	−10.1223	−10.0063
		0	0	0

Note () is the Minimum Bayes Factor (MBF)

4.2 Empirical Results

In this section, the study employs two types of R-MIDAS (the best fit specification of R-MIDAS and R-MIDAS with the same lag selection as U-MIDAS), and U-MIDAS to analyze the Thai economic growth. Since the low-frequency variable is quarterly data and the high-frequency variables are monthly data, the frequency is a mismatch. Therefore, we set m = 3 (quarter to month). The empirical results are shown in Tables 3, 4 and 5.

Prior to interpreting the results, we compare the performance of the three models (two of R-MIDAS and one U-MIDAS). The U-MIDAS performs better than the R-MIDAS models as the lower BIC is obtained. Therefore, the influence of the selected factors on the economic growth of Thailand can be better interpreted based on the results of the U-MIDAS model.

Table 5 shows the results of the U-MIDAS estimation, with the unemployment and government budget variables lag at time $t-2$ while the rest lag at time $t-1$. All variables are significant at different degrees of evidence. The unemployment at month

Table 3 Results from the best fit restricted MIDAS (R-MIDAS) model

	Estimate	Standard error	t value	*p*-value	MBF	Interpretation of MBF
β_0	1.7100	0.2429	7.0400	0	0	Decisive evidence
$\beta_{t-10,UN}$	−12.2972	3.5962	−3.4190	0.0033	0.0029	Decisive evidence
C_{UN}	0.1266	0.0577	2.1930	0.0425	0.0903	Strong evidence
$\beta_{t-13,\ TOUR}$	61.9690	9.0241	6.8670	0	0	Decisive evidence
C_{TOUR}	−0.0173	0.0374	−0.4640	0.6486	0.8979	Weak evidence
$\beta_{t-7,G}$	−13.3120	5.8250	−2.2850	0.0354	0.0735	Strong evidence
C_G	3.4122	4.4057	0.775	0.4493	0.7406	Weak evidence
$\beta_{t-6,\ EX}$	5.5029	2.4834	2.2160	0.0406	0.0858	Strong evidence
C_{EX}	2.6906	3.7934	0.7090	0.4878	0.7778	Weak evidence
$\beta_{t-5,\ FDI}$	−0.0850	0.0276	−3.078	0.0068	0.0088	Strong evidence
C_{FDI}	0.3984	0.2852	1.3970	0.1805	0.3769	Weak evidence
BIC	90.8816					

Table 4 Results from the restricted MIDAS (R-MIDAS) using the lag length suggested by U-MIDAS

	Estimate	Standard error	t value	p-value	MBF	Interpretation of MBF
β_0	2.0590	0.9464	2.1750	0.0412	0.0939	Strong evidence
$\beta_{t-1,\ UN}$	−8.2850	5.7880	−1.431	0.1671	0.3592	Weak evidence
C_{UN}	10.2000	20820	0	0.9996	1.0000	Weak evidence
$\beta_{t-1,\ TOUR}$	10.1100	4.9820	2.0290	0.0554	0.1277	Moderate evidence
C_{TOUR}	0.3409	1.4160	0.2410	0.8121	0.9714	Weak evidence
$\beta_{t-2,G}$	7.9260	8.6940	0.9120	0.3723	0.6598	Weak evidence
C_G	1.9480	0.3987	4.8860	0.0001	0	Decisive evidence
$\beta_{t-1,\ EX}$	3.9620	3.2820	1.2070	0.2407	0.4827	Weak evidence
C_{EX}	0.3107	1.3240	0.2350	0.8168	0.9728	Weak evidence
$\beta_{t-1,\ FDI}$	−0.0613	0.0788	−0.7770	0.4456	0.7394	Weak evidence
C_{FDI}	−7.0790	582.7000	−0.0120	0.9904	0.9999	Weak evidence
BIC	180.0384					

$t-1$ has a positive effect on real GDP while the unemployment at month $t-2$ has a negative effect. The different effects may be caused by government policy at that time. The number of tourists affects economic growth positively and the magnitude is the largest. When the number of tourists at month $t-1$ decreases by 1%, the real GDP will decrease by 7.29%. This indicates that economic growth relies largely on the tourism sector. The main reason is that tourism revenue accounts for around 20% of Thai GDP. Thus, the increase/decrease in the number of tourists would contribute to a great growth/loss of Thai GDP. Tourism is quite important to the Thai economy as it provides jobs and supports businesses, especially the Hotel and Restaurant, and Passenger Transport Business sub-sectors. The government budget shows a positive effect on economic growth. In other words, the increase in government spending is attended by economic growth which also indicates that the budget of government has efficiently been used. Export has a positive and significant impact on economic growth with decisive evidence. It can be said that export will cause economic growth that may depend on the ability of the country to adopt new technology, the economic policy, financial market development, human resource development, infrastructure,

Table 5 Results from the best fit unrestricted MIDAS (U-MIDAS) model

	Estimate	Standard error	t value	*p*-value	MBF	Interpretation of MBF
β_0	2.3552	0.7829	3.0080	0.0061	0.0108	Strong evidence
$\beta_{t-1,UN}$	4.7502	6.1159	0.7770	0.4449	0.7394	Weak evidence
$\beta_{t-2,UN}$	−6.9468	3.5575	−1.9530	0.0626	0.1485	Moderate evidence
$\beta_{t-1,TOUR}$	7.2900	2.5656	2.8420	0.0090	0.0176	Strong evidence
$\beta_{t-1,G}$	4.5626	1.9640	2.3230	0.0290	0.0673	Substantial evidence
$\beta_{t-2,G}$	4.1076	1.0438	3.9350	0.0006	0.0004	Decisive evidence
$\beta_{t-1,EX}$	0.3883	0.0443	8.7660	0.0000	0.0000	Decisive evidence
$\beta_{t-1,FDI}$	−0.0744	0.0551	−1.3500	0.1895	0.4020	Weak evidence
BIC	168.88					

etc. Finally, similar to Ang [30], the influence of foreign direct investment on the economic growth of Thailand is negative. This result is quite surprising. The possible reason is that the FDI recipient country may fear foreign ownership of domestic firms. According to Schoors et al. [31], when domestic firms are not so productive compared to the foreign ones, it could force the local productive firms to leave the market and results in structural unemployment and income of the country.

5 Conclusion

The main objective of this paper is to analyze the determinants of economic growth of Thailand. However, the data needed for this analysis may be reported at different frequencies causing the estimation difficulty. To deal with this problem, our study suggests the use of one of the novel models called Mixed Data Sampling (MIDAS) Regression to deal with time-series data sampled at different frequencies. Different from the equally weighted assignment in the converting process (aggregating and disaggregating data), this model imposes an unequal weighting scheme on the high-frequency and low-frequency observations. In this study, we consider quarterly real GDP as the low-frequency observation and monthly unemployment, tourism, total export, government budget, and direct investment as the high-frequency observation.

The results show that the U-MIDAS model is more suitable for investigating the impact of economic factors on Thailand's economic growth. According to the results obtained from the U-Midas model, we find strong evidence supporting the impact of unemployment, tourism, total export, government budget, and direct investment on the Thai economy. Specifically, export and government budget at the previous month shows decisive evidence while tourism at the previous month shows strong evidence supporting the impact on the Thai economy. For the unemployment factor, we observe the different sign of effects on the Thai economy, but the evidence is not quite strong. Finally, foreign direct investment is found to have a weak impact on Thailand's economy.

Acknowledgements The authors would like to thank Dr. Laxmi Worachai for her helpful comments on an earlier version of the paper. The authors are also grateful for the financial support offered by the Center of Excellence in Econometrics, Chiang Mai University, Thailand.

References

1. Anbao, T., & Danhua, Z. (2011). A comparative analysis on the factors promoting China's economic growth based on demand. *Energy Procedia, 5,* 1388–1393.
2. Brida, J. G., Gómez, D. M., & Segarra, V. (2020). On the empirical relationship between tourism and economic growth. *Tourism Management,81*, 104131.
3. Chinnakum, W., Sriboonchitta, S., & Pastpipatkul, P. (2013). Factors affecting economic output in developed countries: A copula approach to sample selection with panel data. *International Journal of Approximate Reasoning, 54*(6), 809–824.
4. Divya, K. H., & Devi, V. R. (2014). A study on predictors of GDP: Early signals. *Procedia Economics and Finance, 11,* 375–382.
5. Maneejuk, P., Yamaka, W., & Sriboonchitta, S. (2019). Does the Kuznets curve exist in Thailand? A two decades' perspective (1993-2015). *Annals of Operations Research,* 1–32.
6. Pastpipatkul, P., Maneejuk, P., & Sriboonchitta, S. (2017). Testing the validity of economic growth theories using copula-based seemingly unrelated quantile kink regression. In *Robustness in Econometrics* (pp. 523–541). Springer.
7. Gunduz, L., & Hatemi-J, A. (2005). Is the tourism-led growth hypothesis valid for Turkey? *Applied Economics Letters, 12*(8), 499–504.
8. Katircioglu, S. T. (2009). Revisiting the tourism-led-growth hypothesis for Turkey using the bounds test and Johansen approach for cointegration. *Tourism Management, 30*(1), 17–20.
9. Tang, C. F., & Tan, E. C. (2013). How stable is the tourism-led growth hypothesis in Malaysia? Evidence from disaggregated tourism markets. *Tourism Management, 37,* 52–57.
10. Pegkas, P. (2018). The effect of government debt and other determinants on economic growth: The Greek experience. *Economies, 6*(1), 10.
11. Saget, C. (2000). Can the level of employment be explained by GDP growth in transition countries? (Theory versus the quality of data). *Labour, 14*(4), 623–643.
12. Fofana, N. Z. F. (2001). Employment and economic growth in the Cote d'Ivoire: An analysis of structural determinants. *African Development Review, 13*(1), 98–113.
13. Alfaro, L., Chanda, A., Kalemli-Ozcan, S., & Sayek, S. (2004). FDI and economic growth: The role of local financial markets. *Journal of International Economics, 64*(1), 89–112.
14. Nair-Reichert, U., & Weinhold, D. (2001). Causality tests for cross-country panels: A New look at FDI and economic growth in developing countries. *Oxford Bulletin of Economics and Statistics, 63*(2), 153–171.

15. Pegkas, P. (2015). The impact of FDI on economic growth in Eurozone countries. *The Journal of Economic Asymmetries, 12*(2), 124–132.
16. Tang, C. F., & Abosedra, S. (2019). Logistics performance, exports, and growth: Evidence from Asian economies. *Research in Transportation Economics, 78,* 100743.
17. Tremblay, R. (1990). The "export-import" effect and economic growth. *North American Review of Economics and Finance, 1*(2), 241–252.
18. Tyler, W. G. (1981). Growth and export expansion in developing countries: Some empirical evidence. *Journal of Development Economics, 9*(1), 121–130.
19. Ghysels, E., Santa-Clara, P., & Valkanov, R. (2004). The MIDAS touch: Mixed data sampling regression models.
20. Penev, S., Leonte, D., Lazarov, Z., & Mann, R. A. (2014). Applications of MIDAS regression in analysing trends in water quality. *Journal of Hydrology, 511,* 151–159.
21. Foroni, C., Ghysels, E., & Marcellino, M. (2013). Mixed-frequency vector autoregressive models. *Advances in Econometrics*.
22. Ghysels, E., Sinko, A., & Valkanov, R. (2007). MIDAS regressions: Further results and new directions. *Econometric Reviews, 26*(1), 53–90.
23. Kingnetr, N., Tungtrakul, T., & Sriboonchitta, S. (2017). Forecasting GDP growth in Thailand with different leading indicators using MIDAS regression models. In *Robustness in econometrics* (pp. 511–521). Springer.
24. Foroni, C., Marcellino, M., & Schumacher, C. (2015). Unrestricted mixed data sampling (MIDAS): MIDAS regressions with unrestricted lag polynomials. *Journal of the Royal Statistical Society. Series A (Statistics in Society,)* 57–82.
25. Barsoum, F., & Stankiewicz, S. (2015). Forecasting GDP growth using mixed-frequency models with switching regimes. *International Journal of Forecasting, 31*(1), 33–50.
26. Baumeister, C., & Hamilton, J. D. (2015). Sign restrictions, structural vector autoregressions, and useful prior information. *Econometrica, 83*(5), 1963–1999.
27. Maneejuk, P., & Yamaka, W. (2020). Significance test for linear regression: how to test without P-values? *Journal of Applied Statistics*, 1–19.
28. Sellke, T., Bayarri, M. J., & Berger, J. O. (2001). Calibration of ρ values for testing precise null hypotheses. *The American Statistician, 55*(1), 62–71.
29. Vovk, V. G. (1993). A logic of probability, with application to the foundations of statistics. *Journal of the Royal Statistical Society: Series B (Methodological), 55*(2), 317–341.
30. Ang, J. B. (2009). Foreign direct investment and its impact on the Thai economy: The role of financial development. *Journal of Economics and Finance, 33*(3), 316–323.
31. Schoors, K., & van der Tol, B. (2001). The productivity effect of foreign ownership on domestic firms in Hungary. In *EAE Conference in Philadelphia*, PA.

A New Approach for Estimating Probability Density Function with Fuzzy Data

Hoa Le, Uyen Pham, and Pham The Bao

Abstract "Data Is the New Oil of the Digital Economy". With this quotation, how to explore data becomes keyword in economic. However, due to the fuzziness of the environment restriction on human and other resources, the data values may be imprecise or collected only in the form of intervals. That leads to the fuzziness of data. Especially, data in financial application may appear as fuzzy numbers such as in stock price prediction. There are several research on fuzzy lifetime data and other reliability problems but not on fuzzy financial data. In this paper, we focus on financial fuzzy data and propose a estimating method for fuzzy probability density functions. The method is expressed as an algorithm running on a simulated data set and a real data set of stock price.

Keywords Probability density function · Fuzzy probability density fun ction · Empirical density function · $\alpha-$level functions · Bayesian statistics · Application in financial analysis

1 Introduction

In "The Economists" published on May 6, 2017, the topic "The world's most valuable resource is no longer oil, but data" put a giant impression to people about the role of data. Fortunately, if data appear as precise numbers, a variability of research methods has been widely studied from estimating to forecasting, as well as probability density functions.

H. Le · U. Pham (✉)
Faculty of Economic Mathematics, University of Economics and Law,
Vietnam National University of Ho Chi Minh City, Quarter 3, Linh Xuan Ward, Thu Duc District, Ho Chi Minh City, Vietnam
e-mail: uyenph@uel.edu.vn

P. T. Bao
Information Science Faculty, Sai Gon University, 273, An Duong Vuong St., Ward 3, District 5, Ho Chi Minh City, Vietnam

S. Sriboonchitta et al. (eds.), *Credible Asset Allocation, Optimal Transport Methods, and Related Topics*, Studies in Systems, Decision and Control 429,
https://doi.org/10.1007/978-3-030-97273-8_26

However, due to the fuzziness of the environment restriction on measurements and other resources, the data values may be imprecise or collected only in the form of intervals [1–4]. That leads to the fuzziness of data. Specifically, in financial applications, the impact of Covid-19 has a great influence on investors, Cash flow tends to flow into many different investment channels: stock, real estate, gold ...Nevertheless, the stock market is still a favorite investment channel. Evidently, US stock market has had a series of impressive increasing days or the Vietnamese market has received many positive assessments on investment worthiness level. Moreover, the amount of investment accounts have increased sharply in order to catch the trend of Vietnamese stock market which show some extremely spectacular during the last few months. In particular, the stock price in each trading session will fluctuate continuously, influenced by the supply and demand of investors. Remarkably, the Vietnamese stock prices are strongly influenced by the buying and selling actions of foreign investors and limited by a specified range based on the closing price of the previous session per session. This closing price should be between floor and ceiling prices for each day and is used to create the floor and ceiling prices for the next day by a dispersion of 7 or 10%. Therefore, in order to give decision on investing stock for a certain day, the investor have access to data under the form of an interval and therefore, we could consider the predicted stock price per session to be a fuzzy number in research to help investors decide when to buy or sell to take profit or stop loss.

Scientifically, fuzzy sets and fuzzy density distribution have been in interest of many researchers [5–10]. One of the most popular and widely used methods is the Bayesian Point Estimation Method to obtain estimation in Rayleigh distribution [7], exponential [1, 2] or normal, Weibull [4, 11, 12]. Alternatively, the Reliability Function is applied in non-parametric case [13]. However, these studies focus mainly on lifetime data and reliability problems [2, 13, 14]. To the best of our knowledge, a study on fuzzy density probability with fuzzy financial data has not been carried.

The paper presents a method of estimating fuzzy density probability given fuzzy numbers in interval form. The effectiveness of the algorithm is illustrated with a simulated data based on Monte Carlo method and a real Vietnamese stock price data set.

In the next section, we will introduce main notations and definitions on fuzzy probability density functions and state the main theorem on existence of a probability density function with fuzzy data. Then, we present some applications using simulated data and empirical data.

2 Estimating the Fuzzy Probability Density Function

2.1 Fuzzy Numbers and Membership Functions

Definition 1 [8] A fuzzy number x^* is determined by its so-called membership function $\xi(.)$ which is a real function of one real variable x obeying the following:

i. $\xi : \mathcal{R} \to [0, 1]$.
ii. $\forall \alpha \in (0, 1]$ the so-called $\alpha-$cut

$$C_\alpha(x^*) := \{x \in \mathcal{R} : \xi(x) \geq \alpha\},$$

is a finite union of compact intervals, $[a_{\alpha,j}, b_{\alpha,j}]$, i.e.

$$C_\alpha(x^*) = \cup_{j=1}^{k_\alpha}[a_{\alpha,j}, b_{\alpha,j}] \neq \emptyset.$$

iii. The support of $\xi(.)$, defined by

$$supp[\xi(.)] := \{x \in \mathcal{R} : \xi(x) > 0\}$$

is bounded.

The set of all fuzzy numbers is denoted by $\mathcal{F}(\mathcal{R})$.

Lemma 1 *[8] For the membership function $\xi(.)$ of a fuzzy number x^* the following holds true:*

$$\xi(x) = \max\{\alpha \times I_{C_\alpha(x^*)}(x) : \alpha \in [0, 1]\}, \forall x \in \mathcal{R},$$

where $I_{C_\alpha(x^*)}(x) = I_{x:\xi(x)\geq\alpha}(x) = \begin{cases} 1, \textit{for } x : \xi(x) \geq \alpha \\ 0, \textit{elsewhere} \end{cases}$

Definition 2 [8] A fuzzy number is called a fuzzy interval if all its $\alpha-$cuts are non-empty closed bounded intervals. The set of all fuzzy intervals is denoted by $\mathcal{F}_I(\mathcal{R})$.

Let x_1^* and x_2^* be two fuzzy numbers with corresponding membership functions $\xi_1(.)$ and $\xi_2(.)$. Let $C_\alpha(x_1^*) = [a_{\alpha,1}, b_{\alpha,1}]$ and $C_\alpha(x_2^*) = [a_{\alpha,2}, b_{\alpha,2}]$, $\forall \alpha \in (0, 1]$ then the $\alpha-$cut of the fuzzy sum $x_1^* \oplus x_2^*$ is given by [8]:

$$C_\alpha(x_1^* \oplus x_2^*) = [a_{\alpha,1} + a_{\alpha,2}, b_{\alpha,1} + b_{\alpha,2}], \forall \alpha \in (0, 1].$$

2.2 *Fuzzy Valued Functions*

Fuzzy valued functions $f^*(.)$ are functions whose range of values are fuzzy intervals. i.e.

$$f^* : M \to \mathcal{F}_I(\mathcal{R})$$

where M denotes the observation space of a random quantity.

Definition 3 [8] Let $f^*(.)$ be a fuzzy valued functions with domain M, and $x \in M$.

Denoting the $\alpha-$cuts of $f^*(x)$ by:

$$C_\alpha\left[f^*(x)\right] = \left[f_\alpha^L(x), f_\alpha^U(x)\right], \forall\alpha \in (0, 1],$$

for variable x the real valued functions $f_\alpha^L(.)$ and $f_\alpha^U(.)$ are called $\alpha-$ level functions.

Definition 4 [8] Let $f^*(.)$ be a fuzzy valued function defined on a measure space $(M, \mathcal{A}, \mu)$. If all $\alpha-$level functions $f_\alpha^L(.)$ and $f_\alpha^U(.)$ are integrable with finite integrals $\int f_\alpha^L(x)d\mu(x)$ and $\int f_\alpha^U(x)d\mu(x)$, then the generalized integral $\int_M f^*(x)d\mu(x)$ is the fuzzy number y^* whose $\alpha-$cuts $C_\alpha(y^*)$ are defined by:

$$C_\alpha(y^*) = \left[\int_M f_\alpha^L(x)d\mu(x), \int_M f_\alpha^U(x)d\mu(x)\right], \forall\alpha \in (0, 1].$$

The membership function of the fuzzy integral y^* is given by:

$$\xi_{y^*}(x) = \sup\left\{\alpha \times I_{[y_\alpha^L, y_\alpha^U]}(x) : \alpha \in [0, 1]\right\}, \forall x \in \mathcal{R},$$

where $y_\alpha^L = \int_M f_\alpha^L(x)d\mu(x)$ and $y_\alpha^U = \int_M f_\alpha^U(x)d\mu(x)$. The resulting fuzzy integral is denoted by $(f) \int_M f^*(x)d\mu(x)$.

2.3 The Fuzzy Probability Density Function

Definition 5 [8] A fuzzy-valued function $f^* : M \to \mathcal{F}_I(\mathcal{R})$ defined on a measure space $(M, \mathcal{A}, \mu)$ is called a *fuzzy probability density function* if it fulfills the following conditions:

i. The integral $y^* = (f) \int_M f^*(x)d\mu(x)$ exists.
ii. $1 \in C_1(y^*)$ and $supp(y^*) \subseteq (0, +\infty)$.
iii. There exists a probability density function $f : M \to [0; +\infty)$ satisfies

$$f_1^L(x) \le f(x) \le f_1^U(x), \tag{1}$$

where $x \in M$.

The property iii. in Definition 5 is helpful in determining fuzzy density function fitting real fuzzy data based on theory of density functions.

2.4 Estimating the Fuzzy Probability Density Function based on Histogram Method

Definition 6 [15] Suppose that a fuzzy sample $x_1^*, x_2^*, \ldots, x_n^*$ from a population with unknown probability density function $f_X(.)$ in the domain of $[a, b]$, is observed. Let $\alpha-$cut of $x_{\alpha,i}^* = \left[\underline{x}_{\alpha,i}, \overline{x}_{\alpha,i}\right], i = 1, 2, \ldots, n$. For partitioning $a = t_0, t_1, \ldots, t_{m-1}, t_m = b$ the fuzzy histogram is defined by the fuzzy values $\hat{f}_H(t)$ whose $\alpha-$ cuts $\hat{f}_{\alpha,H}(t)$ are given by:

$$\begin{cases} \hat{f}_{\alpha,H}(t) = c_{\alpha,j}, & for\; t_j \le t \le t_{j+1}, j = 0, 1, \ldots, m-1, \\ \hat{f}_{\alpha,H}(b) = c_{\alpha,m-1}, & \\ \hat{f}_{\alpha,H}(t) = 0, & otherwise, \end{cases} \tag{2}$$

where

$$c_{\alpha,j} = \frac{1}{n(t_{j+1} - t_j)} \sum_{i=1}^{n} \left[\inf\left(\mathcal{I}_{[t_j,t_{j+1})}(y); y \in x_{\alpha,i}^*\right), \sup\left(\mathcal{I}_{[t_j,t_{j+1})}(y); y \in x_{\alpha,i}^*\right)\right],$$

$$\inf \mathcal{I}_{[t_j,t_{j+1})}(y) = \begin{cases} 1 \; for\; t_j \le y < t_{j+1} \\ 0 \; otherwise, \forall y \in x_{\alpha,i}^* \end{cases},$$

$$\sup \mathcal{I}_{[t_j,t_{j+1})}(y) = \begin{cases} 0 \; for\; y < t_j \; or \; y \ge t_{j+1} \\ 1 \; otherwise, \forall y \in x_{\alpha,i}^* \end{cases}.$$

in which $\mathcal{I}_{[t_j,t_{j+1})}(y) = \begin{cases} 1 \; for\; t_j \le y < t_{j+1} \\ 0 \; otherwise, \forall y \in \mathcal{R}, j = 0, 1, \ldots, m-1. \end{cases}$

Based on the fuzzy histogram, Arefi, M. constructed a fuzzy probability density function, in which each of $[t_j, t_{j+1})$ is replaced by $[t - h, t + h)$ through the following definition:

Definition 7 [15] Let $x_{\alpha,i}^*, i = 1, 2, \ldots, n$ be the $\alpha-$cuts of the fuzzy numbers $x_i^*, i = 1, 2, \ldots, n$. The $\alpha-$cuts of the fuzzy probability density function with empirical cumulative distribution function method are defined as:

$$\hat{f}_{\alpha,n}(t) = \frac{1}{2nh_n} \sum_{i=1}^{n} \left[\inf\left(\mathcal{I}_{(t-h_n,t+h_n]}(y); y \in x_{\alpha,i}^*\right), \sup\left(\mathcal{I}_{(t-h_n,t+h_n]}(y); y \in x_{\alpha,i}^*\right)\right], \tag{3}$$

where

$$\mathcal{I}_{(t-h_n,t+h_n]}(y) = \begin{cases} 1, & for\; t - h_n < y \le t + h_n, \\ 0, & otherwise, \forall y \in \mathcal{R}. \end{cases}$$

Theorem 1 *[15] Consider the fuzzy empirical cumulative distribution function method in Definition 7*

i. *For every $y_i \in x^*_{\alpha,i}$, $i = 1, 2, \ldots, n$, the expectation of $\hat{f}^L_\alpha$ (or $\hat{f}^U_\alpha$) converges to $f(t)$, if $h_n \to 0$ as $n \to \infty$.*
ii. *For every $y_i \in x^*_{\alpha,i}$, $i = 1, 2, \ldots, n$, the variance of $\hat{f}^L_\alpha$ (or $\hat{f}^U_\alpha$) converges to 0, if $h_n \to 0$ as $n \to \infty$.*

After estimating the $\alpha-$level functions [8, 15], we propose a new theorem, which indicates the existence of a density curve between $\alpha-$level functions as Theorem 2.

Theorem 2 *There exists a function $g : [\min \underline{x}_{\alpha,i}, \max \overline{x}_{\alpha,i}] \to [0, +\infty)$ and $\int_{[\min \underline{x}_{\alpha,i}, \max \overline{x}_{\alpha,i}]} g(t)dt = 1$ is defined as:*

$$\hat{f}^L_\alpha(t) \le g(t) \le \hat{f}^U_\alpha(t), \tag{4}$$

where $\hat{f}^L_\alpha(t), \hat{f}^U_\alpha(t)$ is determined by Definition 7 with the conditions $\forall t \in [\min \underline{x}_{\alpha,i}, \max \overline{x}_{\alpha,i}]$, for all $i = 1, 2, \ldots, n$.
Additionally, the function g is an estimation of the probability density function between two $\alpha-$level functions.

Proof i. First, we prove $\begin{cases} \int_{[\min \underline{x}_{\alpha,i}, \max \overline{x}_{\alpha,i}]} \hat{f}^L_\alpha(t)dt \le 1 \\ \int_{[\min \underline{x}_{\alpha,i}, \max \overline{x}_{\alpha,i}]} \hat{f}^U_\alpha(t)dt \ge 1. \end{cases}$
Indeed, based on the definition of the lower and upper bound of the fuzzy probability density function, we divide the interval $[\min \underline{x}_{\alpha,i} - \varepsilon, \max \overline{x}_{\alpha,i}]$ with small $\varepsilon > 0$, into k small non-overlapping intervals with distance $2h_n$.

$$I_1 = (\min \underline{x}_{\alpha,i} - \varepsilon, \min \underline{x}_{\alpha,i} + 2h_n],$$

$$I_2 = (\min \underline{x}_{\alpha,i} + 2h_n, \min \underline{x}_{\alpha,i} + 2 \times 2h_n],$$

$$\ldots$$

$$I_k = (\min \underline{x}_{\alpha,i} + (k-1) \times 2h_n, \max \overline{x}_{\alpha,i}].$$

Then, the intervals $(t - h_n, t + h_n]$ is denoted as $I_1, I_2, \ldots, I_k$. The integral of the lower bound is approximated by the sum:

$$\int_{[\min \underline{x}_{\alpha,i}, \max \overline{x}_{\alpha,i}]} \hat{f}_\alpha^L(t)dt = \lim_{h_n \to 0} \sum_{j=\overline{1:k}}^{t_j = \frac{\min I_j + \max I_j}{2}} 2h_n \hat{f}_\alpha^L(t_j)$$
$$= \sum_{j=\overline{1:k}}^{t_j = \frac{\min I_j + \max I_j}{2}} 2h_n \frac{1}{2nh_n} \# \left\{ x_i^* : [\underline{x}_{\alpha,i}, \overline{x}_{\alpha,i}] \subseteq (t_j - h_n, t_j + h_n] \right\}$$
$$= \frac{1}{n} \sum_{j=\overline{1:k}}^{t_j = \frac{\min I_j + \max I_j}{2}} \# \left\{ x_i^* : [\underline{x}_{\alpha,i}, \overline{x}_{\alpha,i}] \subseteq (t_j - h_n, t_j + h_n] \right\}$$

Because I_j is approximately the disjoint intervals, once $x^*_{\alpha,i} \subseteq I_{j'}$ is improbable case $x^*_{\alpha,i} \subseteq I_{j''}$ with $j' \neq j''$.

Therefore, there are at most n observations $x^*_{\alpha,i} \subseteq I_j, i = \overline{1,n}, \forall j = \overline{1,k}$, which means that:

$$\int_{[\min \underline{x}_{\alpha,i}, \max \overline{x}_{\alpha,i}]} \hat{f}_\alpha^L(t)dt \leq \frac{1}{n} \times n = 1. \tag{5}$$

Similarly, the integral of the upper bound is determined by the sum:

$$\int_{[\min \underline{x}_{\alpha,i}, \max \overline{x}_{\alpha,i}]} \hat{f}_\alpha^U(t)dt \tag{6}$$
$$= \frac{1}{n} \sum_{j=\overline{1:k}}^{t_j = \frac{\min I_j + \max I_j}{2}} \# \left\{ x_i^* : [\underline{x}_{\alpha,i}, \overline{x}_{\alpha,i}] \cap (t_j - h_n, t_j + h_n] \neq \emptyset \right\}.$$

Because I_j is approximately the disjoint intervals, once $x^*_{\alpha,i} \cap I_{j'} \neq \emptyset$ is still probable cases $x^*_{\alpha,i} \cap I_{j''} \neq \emptyset$ with $j' \neq j''$.
Therefore, there are at most n observations $x^*_{\alpha,i} \cap I_j \neq \emptyset, i = \overline{1,n}, \forall j = \overline{1,k}$, which means that

$$\int_{[\min \underline{x}_{\alpha,i}, \max \overline{x}_{\alpha,i}]} \hat{f}_\alpha^U(t)dt \geq \frac{1}{n} \times n = 1. \tag{7}$$

ii. Next, we point out the existence and the way to find a function $g(t)$ satisfying the theorem.
From

$$\int_{[\min \underline{x}_{\alpha,i}, \max \overline{x}_{\alpha,i}]} \hat{f}_\alpha^L(t)dt \leq 1 \leq \int_{[\min \underline{x}_{\alpha,i}, \max \overline{x}_{\alpha,i}]} \hat{f}_\alpha^U(t)dt. \tag{8}$$

Thus, there exists a function $g(.)$ such that $\int_{[\min \underline{x}_{\alpha,i}, \max \overline{x}_{\alpha,i}]} g(t)dt = 1$.

iii. Finally, the selection of function $g(.)$ can be as follows:

$$g(t) = \frac{\hat{f}_{\alpha}^{U}(t)}{\int_{[\min \underline{x}_{\alpha,i}, \max \overline{x}_{\alpha,i}]} \hat{f}_{\alpha}^{U}(\tilde{t})d\tilde{t}}, \tag{9}$$

with $t \in [\min \underline{x}_{\alpha,i}, \max \overline{x}_{\alpha,i}]$ and 0 elsewhere.
Clearly, $g(t) \geq 0, \forall t \in [\min \underline{x}_{\alpha,i}, \max \overline{x}_{\alpha,i}]$ and $\int_{[\min \underline{x}_{\alpha,i}, \max \overline{x}_{\alpha,i}]} g(t)dt = 1$.
Therefore, $g(t)$ is an estimation for the density function between two $\alpha-$level functions.

The above theorem suggest an algorithm that construct the fuzzy probability density function. We will show this algorithm in Sect. 3.2.

3 Application with Simulated Data and Empirical Data

3.1 *Data Description*

If data are presented as precise numbers, the probability density function is estimated by the standard method such as histogram, empirical density function, or kernel density.

If data are treated as the fuzzy numbers, including the upper and lower bounds of $1-$cut ($\alpha = 1$) of overlapping observations and the upper and lower bounds of $1-$cut of non-overlapping observations. In this case, the probability density function is inapplicable for $f^U(x) \neq f^L(x)$ at each non-overlapping observation. Therefore, the estimation of probability density function should not solely based on $f^U(x)$ or $f^L(x)$ at each x. The fuzzy probability density function is a promising estimating method for this problem.

Notably, the upper bound coincides with lower bound of $1-$cut of observations if the fuzzy probability density function coincides with the probability density function.

In this research, the simulated data are generated by Monte Carlo simulation, while the empirical data is the stock prices provided by Thomson Reuters with time series data.

For the Monte Carlo simulated data: Probability density function is estimated since the original data are considered as a set of precise numbers. If the original data are represented as fuzzy numbers, the fuzzy data are generated by the following Algorithm 3.1:

Algorithm 3.1

Input	The original data consist of $(n + 1)$ observations, which are independent And follow a probability distribution
Step 1	Split the original data into two subsets: —The first subset includes the values from the first observation to the nth observation —The second subset includes values from the second observation to the $(n + 1)^{th}$ observation
Step 2	For each position from 1 to n of two subsets, we find min and max Respectively, i.e. we have n values min and n values max
Output	The fuzzy data includes n fuzzy numbers, such that the n lower Boundaries be n value min, and the n upper boundaries be n value max

Obviously, the fuzzy data containing n fuzzy numbers, in which the n lower boundaries of the fuzzy data are independent and follows a probability distribution, and n the upper boundaries of the fuzzy data set also are independent and follows a probability distribution. This means that the lower boundaries and the upper boundaries of the fuzzy data follow the initial probability distribution.

For the generated fuzzy data, $\alpha-$cut observations is considered as $\alpha = 0$ or $\alpha = 1$ or $\alpha = c$, in which c is a constant in (0, 1).

*Case 1. If $\alpha = 1$, the lower boundary and the upper boundary are unequal suggesting the use the fuzzy probability density function to estimate the distribution of the fuzzy data.

*Case 2. If $\alpha = 0$, the appropriate membership functions is built for fuzzy data.

**Case 2i. If the membership function has the shape of a trapezoidal fuzzy number, it means the lower boundary and the upper boundary $1-$cut observations are unequal, the method introduced in Case 1 is used to estimate the fuzzy probability density function.

**Case 2ii. If the membership function has the shape of a triangular, exponential, Gaussian fuzzy numbers, it means the lower boundary and the upper boundary $1-$cut observations are equal, the fuzzy probability density function is estimated as a probability density function.

*Case 3. If $\alpha = c$ with c is a constant in (0, 1), the membership functions is built for the fuzzy data. If the lower boundary and the upper boundary $1-$cut observations are unequal, the fuzzy probability density function are estimated by the same method as Case 2i. Otherwise, the method in case 2ii is applied to estimate the fuzzy probability density function.

For real data: The data that we use is the stock prices provided by Thomson Reuters, at the Economic—Finance Research Center, University of Economics and Law, Vietnam National University, Hochiminh City, Vietnam. We select stock prices of SAM Holdings Corporation, which is one of the first two companies listed on the Stock market of Vietnam in 2000. Figure 1 illustrates SAM code as time-series price fluctuations in 2017 with the highest price and the lowest price at each trading session.

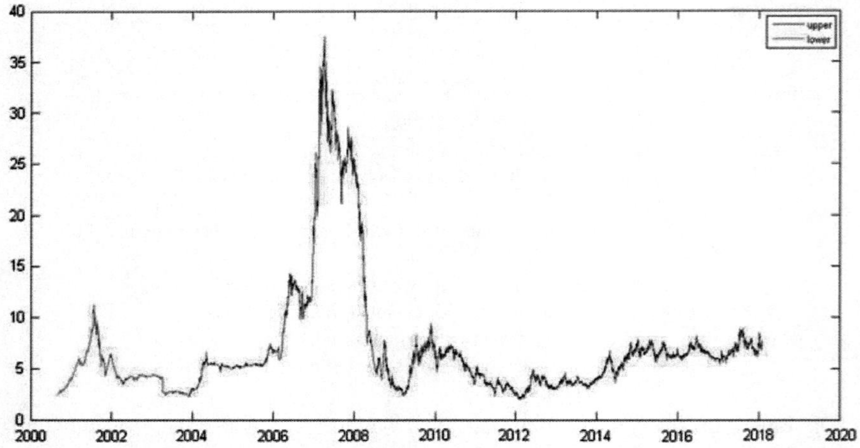

Fig. 1 The prices of all trading sessions of SAM code. *Source* Research results

3.2 *Estimating the Fuzzy Probability Density Function Based on Monte Carlo Simulation*

A set of data with known probability distribution is simulated for evaluating the aforementioned theory. Particularly, the first evaluation is carried on symmetric data from the standard normal distribution, the second one is performed on skewed gamma distribution. Then, the mixture of the symmetric and skewed data is considered for the last evaluation.

Details of the fuzzy probability density function estimation are described in the Algorithm 3.2:

Algorithm 3.2

Input	• Data includes the fuzzy numbers $x_i^*, i = 1, 2, \cdots, n$ with $\alpha-$cut $x_{\alpha,i}^* = [\underline{x}_{\alpha,i}, \overline{x}_{\alpha,i}]$, corresponding • And the arbitrary positive integer k
Step 1.	• Determine the minimum value: $\min \underline{x}_{\alpha,i} = \min\{\underline{x}_{\alpha,i}, i = 1, 2, \cdots, n\}$ • Determine the maximum value: $\max \overline{x}_{\alpha,i} = \max\{\overline{x}_{\alpha,i}, i = 1, 2, \cdots, n\}$ • Divide the range from the smallest value to the maximum value Just determined by k small non-overlapping intervals $I_1, I_2, \cdots, I_k$ With distance $2h_n$, according to theorem 2
Step 2.	Determine the number of corresponding intervals: • $fre_j^L = \#\left\{x_i^* : [\underline{x}_{\alpha,i}, \overline{x}_{\alpha,i}] \subseteq I_j, i = 1, 2, \cdots, n\right\}, j = 1, 2, \cdots, k$ • $fre_j^U = \#\left\{x_i^* : [\underline{x}_{\alpha,i}, \overline{x}_{\alpha,i}] \cap I_j \neq \emptyset, i = 1, 2, \cdots, n\right\}, j = 1, 2, \cdots, k$
Step 3	Determine the ratio of the corresponding intervals: $p_j^L = \frac{fre_j^L}{n}, p_j^U = \frac{fre_j^U}{n}, j = 1, 2, \cdots, k$
Step 4	Determine the lower and upper boundary probabilities: $p^L = \sum_{j=1}^k p_j^L, p^U = \sum_{j=1}^k p_j^U.$
Output	Determine the fuzzy probability density function $g(t), t \in [\min \underline{x}_{\alpha,i}, \max \overline{x}_{\alpha,i}]$: $g(t) = \frac{p_j^U}{2h_n p^U}, t \in I_j, j = 1, 2, \cdots, k$

3.2.1 Fuzzy Probability Density Function in Case of Uni-Modal and Symmetrical Shape

A sample of 10,001 observations generated from the standard normal distribution will create a fuzzy data set with 10,000 observations. Then, the range from the minimum value to the maximum values is divided into k equal length intervals.

Figure 2 illustrates the empirical fuzzy data with the lower bound f^L, the upper bound f^U, the estimating function g and the underlying probability density function.

Estimation of the lower and upper boundary probabilities are reported in the Table 1.

According to the Table 1, as k increases, the length of the segments decreases, the gap between the upper and lower boundary probabilities increases, the lower boundary probability decreases to 0. The fuzzy probability density function is estimated as an approximation to the underlying probability density function.

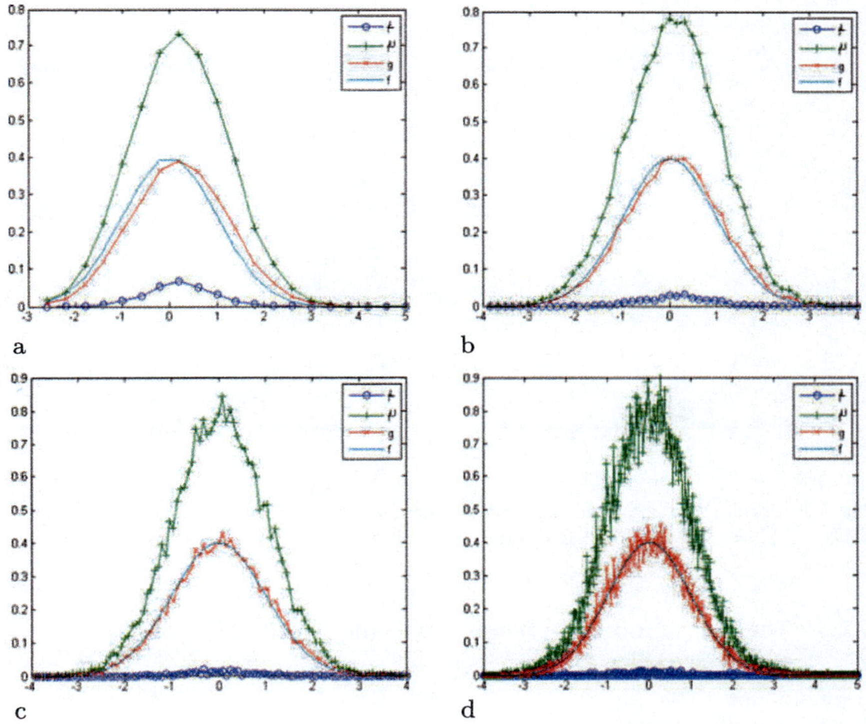

Fig. 2 Estimating the fuzzy probability density function for normal distribution with a to d, respectively, $k = 20$, 50, 100 and 500. *Source* Research results

Table 1 Lower and upper boundary probabilities of the standard normal distribution

k	Lower boundary probability	Upper boundary probability
20	0.1009	1.8958
50	0.0478	1.9521
100	0.0202	1.9795
500	0.0051	1.9948

Source Research results

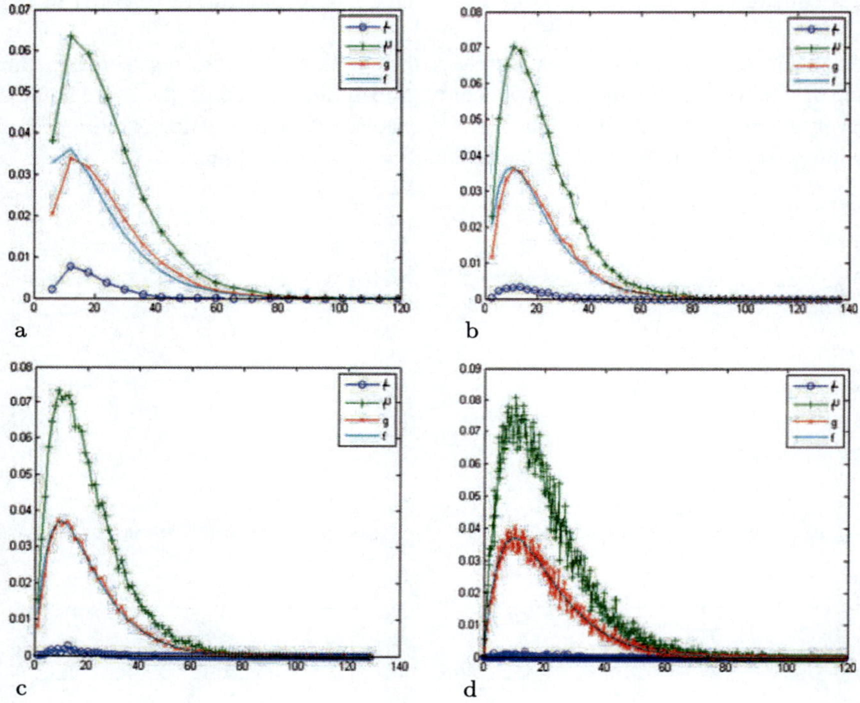

Fig. 3 Estimating the fuzzy probability density function for gamma distribution with a to d, respectively, $k = 20$, 50, 100 and 500. *Source* Research results

3.2.2 The Fuzzy Probability Density Function in Case of Uni-Modal and Skewed Shape

The same method is applied to the gamma distribution with shape parameter $a = 2$ and scale parameter $b = 10$. Result presented in Fig. 3 is based on the empirical fuzzy probability density function with the lower bound f^L, the upper bound f^U, the estimating function g and the initial probability density function f.

Table 2 shows the estimation results of lower and upper boundary probabilities.

Table 2 Lower and upper boundary probabilities of the gamma distribution

k	Lower boundary probability	Upper boundary probability
20	0.1406	1.8594
50	0.0640	1.9360
100	0.0304	1.9696
500	0.0068	1.9932

Source Research results

Table 3 Lower and upper boundary probabilities of a mixture of the normal and gamma distribution

k	Lower boundary probability	Upper boundary probability
20	0.4125	1.5855
50	0.2026	1.7962
100	0.1126	1.8866
500	0.0217	1.9773

Source Research results

The result in Table 2 with the case of asymmetrical distribution is the same as for the case of symmetry, the gap between upper and lower boundary probabilities increases, the lower boundary probability decreases to 0.

3.2.3 The Multi-modal Fuzzy Density Probability Function Through a Mixture of Distributions

The data is generated as mixtures of standard normal distribution $N(0, 1)$ and gamma distribution with two parameters $a = 2, b = 10$, with rates of 0.3 and 0.7, respectively.

Figure 4 reveals the results of the empirical fuzzy probability density function with the lower bound f^L, the upper bound f^U, the estimating function g and the initial probability density function f.

Table 3 report the estimated results of the lower and upper boundary probabilities, in which the gap between the upper and lower boundary probabilities increases, the lower boundary probability decreases to 0.

3.3 Estimating the Fuzzy Probability Density Function Based on Empirical Data

For stock-price data, we are interested in two approaches:

i. The first approach is the consideration of precise data like closing prices that have been solved standard method.

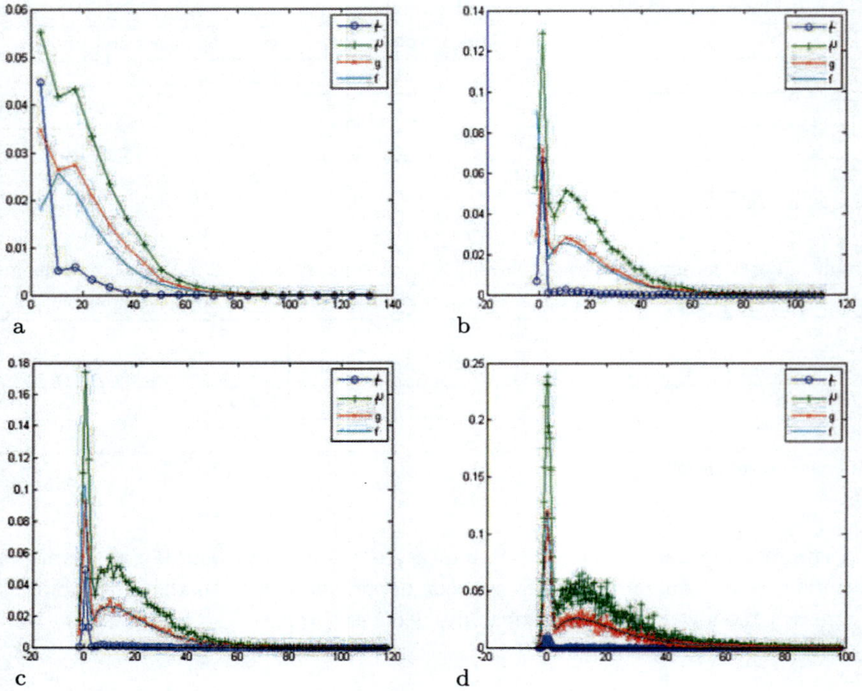

Fig. 4 Estimating the fuzzy probability density function for a mixture of normal and gamma dis. with a to d, respectively, $k = 20, 50, 100$ and 500. *Source* Research results

ii. The second approach is covering values that are actually traded at each session. Moreover, stock prices should be considered on its range from the smallest value to the highest value as well as the price at high volume. Real stock prices occur at each trading session as $\alpha-$cuts with $\alpha = 1$. At that time, stock price data at each trading session consists of ranges $[\underline{x}_{\alpha=1,i}, \overline{x}_{\alpha=1,i}]$, $i = 1, 2, \ldots, n$.

We calculate the fuzzy probability density function of the stock price $[\underline{x}_{\alpha=1,i}, \overline{x}_{\alpha=1,i}]$, $i = 1, 2, \ldots, n$. The fuzzy probability density represents the stock price data with common characteristics in multi-modal and more volatile requiring a very complex calculation.

For the real stock prices data, Table 4 provides the estimation of the lower and upper boundary probabilities.

Results of the empirical fuzzy probability density function with the lower bound f^L, the upper bound f^U, and the estimating function g are shown in Fig. 5.

Results showed that stock prices are mainly concentrated in the range of 2000 VND to 8000 VND with the prices range from 8000 VND to 38,000 VND has low density.

Table 4 Lower and upper boundary probabilities of the empirical distribution

k	Lower boundary probability	Upper boundary probability
20	0.9313	1.0687
50	0.8129	1.1871
100	0.6632	1.3368
500	0.3124	1.6876

Source Research results

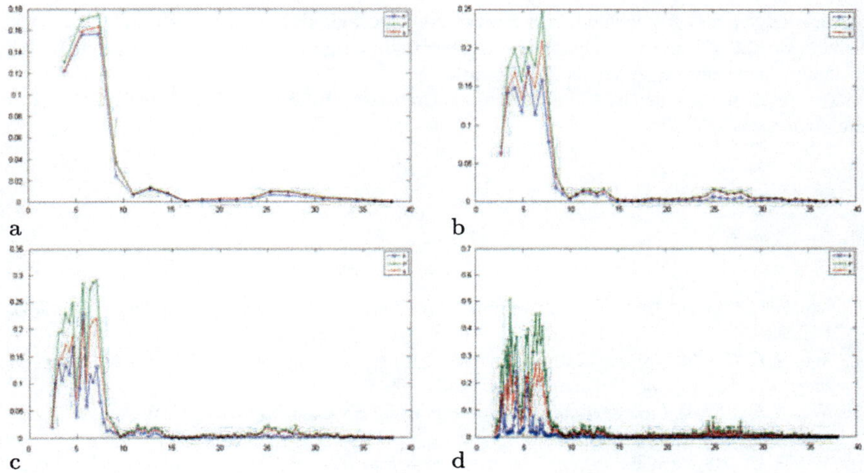

Fig. 5 Estimating the fuzzy probability density function for the empirical data with a to d, respectively, $k = 20$, 50, 100 and 500. *Source* Research results

Based on the upper and lower bound of the empirical probability density function f^L and f^U, we estimate the fuzzy density function g as shown in the Fig. 5. It should be noticed that the g is a mixture of the probability density functions.

4 Conclusion

The probability density function is critical both in frequency statistics and Bayesian statistics. When data appear as fuzzy numbers, estimating the probability density function in the way is inappropriate. Hence, it is required a new method to estimate the probability density function with fuzzy data characterized for the corresponding fuzzy probability density function.

By applying the empirical density function on fuzzy data, we verified that there exists a probability density function or a mixture of some probability density functions between the upper and lower limits. Estimations based on Monte Carlo simulated data show that our results are sufficient in comparison to the known underlying probability density function.

However, the estimating function g depends on dividing the interval $[\min \underline{x}_{\alpha,i} - \varepsilon, \max \overline{x}_{\alpha,i}]$, with $\varepsilon > 0$ small enough, into k small intervals. Therefore, for each problem, we need to select the appropriate value k so that the estimating function g is the function closest to the true function. This will be a research problem in our future focus.

Acknowledgements We would like to express our deep gratitude to professor Hung T. Nguyen of New Mexico State University/Chiang Mai University for his generous help in our research, for his encouragements and for numerous discussions.

This research is funded by Vietnam National University HoChiMinh City (VNU-HCM) under grant number B2021-34-03.

References

1. Wu, H. C. (2004). Fuzzy Bayesian estimation on lifetime data. *Computational Statistics, 19*(4), 613.
2. Wu, H. C. (2006). Applied Mathematical Modelling. *Fuzzy Bayesian system reliability assessment based on exponential distribution, 30*(6), 509–530.
3. Wu, H. C. (2004). Fuzzy reliability estimation using Bayesian approach. *Computers & Industrial Engineering, 46*(3), 467–493.
4. Wu, H. C. (2009). Statistical confidence intervals for fuzzy data. *Expert Systems with Applications, 36*(2), 2670–2676.
5. Gelman, A., Carlin, J. B., Stern, H. S., Dunson, D. B., Vehtari, A., & Rubin, D. B. (2014). *Bayesian data analysis* (Vol. 2). Boca Raton, FL: CRC Press.
6. Bolstad, W. M., & Curran, J. M. (2016). *Introduction to Bayesian statistics*. John Wiley & Sons.
7. Pak, A., Parham, G. A., & Saraj, M. (2014). Reliability estimation in Rayleigh distribution based on fuzzy lifetime data. *International Journal of System Assurance Engineering and Management, 5*(4), 487–494.
8. Viertl, R. (2011). *Statistical methods for fuzzy data.* John Wiley & Sons.
9. Viertl, R., & Hareter, D. (2004). Fuzzy information and imprecise probability. *ZAMM—Journal of Applied Mathematics and Mechanics/Zeitschrift fur Angewandte Mathematik und Mechanik, 84*(10–11), 731–739.
10. Viertl, R., & Hareter, D. (2004). Fuzzy information and stochastics. *Iranian Journal of Fuzzy Systems, 1*(1), 43–56.
11. Huang, H. Z., Zuo, M. J., & Sun, Z. Q. (2006). Bayesian reliability analysis for fuzzy lifetime data. *Fuzzy Sets and Systems, 157*(12), 1674–1686.
12. Tanaka, H., Hayashi, I., & Watada, J. (1989). Possibilistic linear regression analysis for fuzzy data. *European Journal of Operational Research, 40*(3), 389–396.
13. Shafiq, M., & Viertl, R. (2015). Empirical reliability functions based on fuzzy life time data. *Journal of Intelligent & Fuzzy Systems, 28*(2), 707–711.
14. Viertl, R. (2009). On reliability estimation based on fuzzy lifetime data. *Journal of Statistical Planning and Inference, 139*(5), 1750–1755.
15. Arefi, M., Viertl, R., & Taheri, S. M. (2012). Fuzzy density estimation. *Metrika, 75*(1), 5–22.

An Application of Quantum Optimization with Fuzzy Inference System for Stock Index Futures Forecasting

Namchok Chimprang and Roengchai Tansuchat

Abstract Futures contracts of stock index futures had been trading remarkably worldwide in 2020. To mitigate the market risks and make the best of investments, predicting the movement of stock index futures is considered significant to ensure the potential gain in futures market trading. However, the accurate and precise prediction is not easy to obtain in the financial data which normally have high fluctuation. While the conventional econometric methods for forecasting require many assumptions, the machine learning models generally involve complex and unintelligible rules as well as a complicated network structure. In addition, the machine learning model itself did not guarantee a global optimum solution. It easily falls to the local optimum answer that directly affects the model's predicted value accuracy. To handle these drawbacks, in this study, we propose using a novel hybrid Wavelet Transformation-Quantum-behaved Particle Swarm Optimization-Adaptive Neuro-Fuzzy Inference System (WT-QPSO-ANFIS) model to forecast stock index futures. Generally speaking, three approaches, Wavelet Transformation (WT), Quantum Particle Swarm Optimization (QPSO), and Adaptive Neuro-Fuzzy Inference System (ANFIS), are combined in our forecasting model. The result reveals that the hybrid WT-QPSO-ANFIS model provides higher efficiency and accuracy in predicting all 11 stock index futures considered in this study compared to the conventional Sugeno-type ANFIS model, ANN model and ARIMA model in terms of Root Mean Square Error (RMSE), Mean Absolute Error (MAE), Mean Absolute Percentage Error (MAPE), and Standard Error of Mean (SEM).

N. Chimprang · R. Tansuchat (✉)
Faculty of Economics, Chiang Mai University, Chiang Mai, Thailand
e-mail: roengchai.tan@cmu.ac.th

N. Chimprang
e-mail: namchok_c@cmu.ac.th

Center of Excellence in Econometrics, Chiang Mai University, Chiang Mai, Thailand

S. Sriboonchitta et al. (eds.), *Credible Asset Allocation, Optimal Transport Methods, and Related Topics*, Studies in Systems, Decision and Control 429,
https://doi.org/10.1007/978-3-030-97273-8_27

1 Introduction

Stock index futures are cash-settled contracts in the future which are based on a recent stock index price. A trader could open today's short or long position in a stock index to be settled in the coming times. In 2019, there are 19.24 billion futures contracts were traded worldwide, a substantial increase from 12.13 billion in 2013 [28]. There are two major types of people who trade on the stock index futures market. The first group is hedgers who use futures contracts to hedge against a portfolio of shares or the equity index options for reducing the risk on the investment. The second group is the speculators who trade for making a profit from the fluctuation of the stock index futures price. Traders in the Stock index futures market can be ranged from retail investors to proprietary traders or large banks [2].

To achieve the profit from stock index futures trading, skills, and abilities of correctly and accurately predicting stock index futures price are required. If the investor can buy at a low price, and can sell at a high price, he/she will make a high profit. Nonetheless, there are such a wide range and high fluctuation in stock index futures that is challenging for investors to make the right decisions. Accordingly, forecasting stock index futures is an attractive topic for numerous investors.

The conventional forecasting methods are usually classified into two classes, namely econometric and machine learning models. The econometric models such as Auto-Regressive model (AR), Moving-Average model (MA), and Auto-Regressive Integrated Moving Average model (ARIMA), are normally used to predict the value of a particular market index or the index's price at $t + 1$. However, these methods require some basic assumptions of variables to be satisfied for modeling and are not able to solve the problem with the non-linear system and the non-stationary financial signal. Recently, many studies have suggested using machine learning methods to forecast stock index futures value and indicating a complex chart price pattern [19, 32, 34, 21]. This new model requires fewer assumptions than the conventional econometric methods and can solve a complex non-linear problem, but this model produces complex and unintelligible rules and uses many variables [31]. Both methods as mentioned previously normally use the traditional gradient descent optimization algorithms to find an optimal solution with normally slow convergence and easy fall] to the local minimum which directly affects the accuracy of the forecasted value and the computational timing of the model [14]. The objective of this paper is to improve the optimization technique by combining quantum theory with particle swarm optimization (PSO) which ensures finding the global best solution and speeds up convergence [25].

As with other high-volume trading data, stock index futures price data has a high fluctuation and its own characteristic [13]. It is difficult to adjust and do data pre-processing to satisfy any assumptions of the model and some attributes of the data might disappear as a result of the pre-processing method. Therefore, in this study, we use the Adaptive Neuro-Fuzzy Inference System (ANFIS) model which does not require any restrictive assumptions [24]. This study uses full decomposition by Wavelet Transformation to denoise the original stock index futures data. The model uses the approximation and detail coefficients obtained from decomposing the original data as input variables of the ANFIS model to forecast stock index futures. Approximation coefficients represent the main structure to recognize the trends of the data and detail coefficients capture ruptures and singularities in the original data [1]. Furthermore, this study uses the particle swarm optimization (PSO) technique to modify membership function parameters of the ANFIS model to acquire the best solution for predicting the stock index futures. Lastly, we applied quantum theory with particle swarm optimization to assure the global optimal answer. The contribution of the present effort is a hybrid WT-QPSO-ANFIS model for forecasting the stock index futures. Furthermore, this novel method uses quantum theory for tuning the membership functions of the ANFIS model. The empirical results reveal that this model is highly efficient to predict future prices in stock markets.

Most previous studies have proposed methods for forecasting the stock index futures. Obviously, with the limitation of hardware and software and human knowledge at present, it is very difficult to make a perfect prediction. Therefore, many researchers have tried to apply any knowledge with complicated methods for advancing a forecasting model capable of minimizing the prediction error. In stock index futures market trading, placing importance on predicting future market trends is more reasonable than trying to indicate an exact stock index value. In this paper, the proposed model can offer a trading signal which is derived from the predicted stock index on the next day.

The paper organization is constructed as follows: Sect. 2 of this article provides literature reviews, Sect. 3 describes the hybrid WT-QPSO-ANFIS model, Sect. 4 provides the data collection and training a WT-QPSO-ANFIS model, Sect. 5 shows the results and makes comparisons, and the study's conclusions are presented in Sect. 6.

2 Literature Review

In this study, the hybrid model is constructed form three approaches, namely Wavelet Transformation (WT), Adaptive Neuro-Fuzzy Inference System (ANFIS), and Quantum Particle Swarm Optimization (QPSO). These three methods have been proposed in recent decades to improve forecasting performance.

For Wavelet Transform (WT) approach, Li and Kuo [21] proposed the Discrete Wavelet Transform (DWT) to extract features from the Taiwan Weighted Stock Index (TAIEX) and used it in a self-organizing map network. This visual map network helps to reveal the relationship of movements between primary bull markets and bear markets. Caetano and Yoneyama [4] discovered the suddenly abrupt changes in stock data such as Hang Seng and BOVESPA data by using Wavelet Transform. Wavelet coefficients' behavior can present the possibility of the stock market's short-term crashes. Zhao et al. [36] used wavelet transformation with a neural network in their study which used adjusted weight and learning of networks by Levenberg–Marquardt algorithm to predict Shanghai stock market returns. The results show that the improved wavelet neural network provides a more accurate prediction than the normal BP network. Hsieh et al. [17] proposed the Wavelet Transformation (WT) with the recurrent neural network system for optimization by the Artificial Bee Colony (ABC) algorithm to forecast the stock market price. Catalao et al. [5] proposed a combination of Wavelet Transform, Particle Swarm Optimization, and Adaptive Neuro-Fuzzy Inference System (Wavelet-PSO-ANFIS) to forecast electricity price in the Spanish market. The proposed method presents better accuracy in forecasting with more proper computation time than NN and ARIMA. Sun and Meinl [26] used wavelet in a denoising method that has flexible settings and is suitable for high-frequency German equity market data mining. Yang [34] used the wavelet transform as a pre-processing method on the stock index futures data by denoising the data for better results in BP neural network training. Kao [19] combined wavelet decomposition with Multivariate Adaptive Regression Splines (MARS) and Support Vector Regression (SVR) to select the sub-series of wavelet and improve the forecast efficiency on stock index price data.

In the case of ANFIS, Atsalakis and Valavanis [3] used several combinations of stock in Athens and the New York Stock Exchange (NYSE) to determine the best inputs of the ANFIS model for the prediction of short-term trends in terms of minimum Root Mean Square Error (RMSE). Chang et al. [6] presented an ANFIS model with volatility autoregression to solve problems from the use of only one variable input of the conventional time series model and the rules that are not easy to understand of ANN. This novel model was used to forecast Taiwan's stock exchange index prices. Tan et al. [29] proposed the use of reinforcement learning (RL) algorithm to determine the momentum and moving averages for the ANFIS model] to indicate the delay in the predicted cycle. This novel trading system can identify a trend of price movement for traders to make the decisions. Wei et al. [33] used an ANFIS model with multi-technical indicators to forecast stock price trends using the following steps: first, select popular technical indicators by a correlation matrix, second, clustering the technical indicators into linguistic values by using a data discretization method models on FIS to extract rules of linguistic terms; and then optimize the FIS parameters by an adaptive network for prediction. Escobar [15] proposed a novel technical indicator based on fuzzy logic by integrating a risk tendency of traders, instead of using mathematical models. They used the principle of fuzzy logic to suggest the investor for investment. Cheng et al. [9] used several past periods' data of Taiwanese stock markets as inputs of the ANFIS forecasting model and found that it is better than using a single previous-period data. Moreover, they used the ordered weighted averaging method (OWA) to integrate high-order data into a single attribute for indicating the best weight for each period. Wei [32] proposed a genetic algorithm (GA) to weigh an ANFIS model to forecast the Taiwan stock index because of the strong relationships between international trade and Taiwan's economy. Furthermore, the fluctuations in other stock markets were used as input parameters in the proposed model. Penawar et al. [23] used fuzzy logic model to forecast stock market momentum in Indonesia Stock Exchange (IDX). Janková and Dostál [18] proposed a supporting tool which modifies by neuro-fuzzy model to predict the selected stock indexes on the European stock markets. The result shows a strong predictive capacity of both efficient and less efficient stock markets. Lah et al. [20] used the ARIMA model with Triangular Fuzzy number to forecast the Nokia stock index and Zenith bank stock index. The result revealed that the proposed method to construct a fuzzy number for the ARIMA has a smaller error in the predictions.

Finally, the QPSO approach also has been used in predicting the financial data and there is evidence that this method could deliver high prediction accuracy. Van den Bergh [16] proved that particle swarm optimization is not a global optimization algorithm. Sun et al. [27] improved the particle swarm optimization (PSO) algorithm with quantum theory. This method guarantees to find the global optimal solution. By testing on various benchmark functions, the QPSO algorithm outperforms the standard PSO algorithm. Unfortunately, a QPSO ensures the global optimal solution in the case of infinite searching iterations. However, in practical optimization problems, the number of iterations must be a limited number. When solving complicated problems, QPSO tends to fall into local optimum. Liu et al. [22] used Teamwork Evolutionary (TE) strategy with the Quantum Particle Swarm Optimization (QPSO) to balance the global and the local search ability. The novel method can define the optimal point between global searching and convergence speed. Chen [7] used PSO to tune subtractive clustering parameters of the ANFIS model for forecasting business failures. Chiang [10] proposed a quantum version of ANFIS model with quantum membership functions (qMF) and used the genetic algorithm to select the quantum levels in qMF then update the qMF parameters by GD and QPSO method. The simulation results reveal that this method has higher performance and smaller error compared to ANFIS. Chun-tian [8] proposed an ANN-QPSO for solving problems of statistical techniques that rely on a number of assumptions. This novel method employed QPSO to improve learning speed in selecting the weights and thresholds of ANN for Daily Reservoir Runoff's prediction. The result shows that the proposed method can enhance the generalization performance of the forecasting model in terms of RMSE and performance indicators. Wang and Nie [30] proposed a novel model to predict stock index futures by applying wavelet analysis with the improved PSO-based to optimize the neural network. By using Shanghai and Shenzhen 300 stock index futures, the result shows that this novel method has the smallest RMSE compared to backpropagation (BP) neural network and wavelet-backpropagation (W-BP) neural network.

3 The Hybrid WT-QPSO-ANFIS Model

Wavelet Transformation (WT), Quantum Particle Swarm Optimization (QPSO), and Adaptive Neuro- Fuzzy Inference System (ANFIS) are combined to establish the hybrid WT-QPSO-ANFIS model. The WT is used to decompose or denoise the original poor-behaved price series which has high fluctuation into the better-behaved price series then a constitutive frequency of the series is used as input parameters of the ANFIS model [5]. Then, the ANFIS model is used to forecast the next day price of the data series. To improve the performance of ANFIS, the QPSO algorithm is used for tuning the membership functions to achieve a smaller error.

3.1 Wavelet Transform (WT)

Wavelet Transform is widely used to extract some features from time-series data, and it is a type of signal processing technique [12]. The mathematical functions are used in Wavelet Transform to decompose the original data or signals into several different frequencies. We can use the mathematical function to describe any data and decompose the signals into wavelets based on the wavelet principle, as described in Eq. (1)

$$F(fut) = \sum_k S_{j,k}\phi_{j,k}(fut) + \sum_k d_{j,k}\Psi_{j,k}(fut) + \sum_k d_{j-1,k}\Psi_{j-1,k}(fut) + \cdots + \sum_k d_{1,k}\Psi_{1,k}(fut) \tag{1}$$

where fut is the stock index futures series, ϕ is the father wavelets, and Ψ is the mother wavelets which are described as in Eqs. (2) and (3)

$$\phi_{j,k}(f) = 2^{-\frac{j}{2}}\phi(2^{-j}t - k) = 2^{-\frac{j}{2}}\phi\left(\frac{t - 2^j k}{2^j}\right) \tag{2}$$

$$\Psi_{j,k}(f) = 2^{-\frac{j}{2}}\Psi(2^{-j}t - k) = 2^{-\frac{j}{2}}\Psi\left(\frac{t - 2^j k}{2^j}\right) \tag{3}$$

$S_{j,k}$ are the smooth coefficients and $d_{j,k}$ are the detailed coefficients which are shown in Eqs. (4) and (5)

$$S_J(fut) = \sum_k S_{j,k}\phi_{j,k}(fut) \tag{4}$$

$$D_J(fut) = \sum_k d_{j,k}\Psi_{j,k}(fut) \tag{5}$$

Then the sum of the above signals can define as a signal $F(fut)$ as described in Eq. (6)

$$F(fut) = S_J(fut) + D_J(fut) + D_{J-1}(fut) + ... + D_1(fut) \tag{6}$$

where J is the wavelet decomposition level.

This study extracts the stock index futures data by using the wavelet full decomposition. The approximation and detail coefficients that are obtained from decomposing the wavelet of the original data are used as input variables of the model to forecast stock index futures.

3.2 Adaptive Neuro-Fuzzy Inference System (ANFIS)

In the Fuzzy Inference Systems (FIS), data are described as a member of more than one set. This fuzzy model is processed by using if–then rules and linguistic variables. There are three parts of the ANFIS model: fuzzy rules, membership functions, and a reasoning mechanism. In this study, we use the Takagi–Sugeno system which gives a real number as output to create an ANFIS model. For example, ANFIS consists of two rules as shown in Eqs. (7)–(8) [35].

$$\text{Rule 1: If } x \text{ is } A_1 \text{ and } y \text{ is } B_1 \text{ then } f_1 = p_1x + q_1y + r_1 \tag{7}$$

$$\text{Rule 2: If } x \text{ is } A_2 \text{ and } y \text{ is } B_2 \text{ then } f_2 = p_2x + q_2y + r_2 \tag{8}$$

The details of the functioning of each layer of the ANFIS are as follows:

Layer 1. This layer is the input layer which contains adaptive nodes with node functions. The number of nodes depends on input variables and can be set up into several sub-layers. Each node in this layer is defined as a linguistic label such as high, medium, or low. The output of this layer is the value of the membership function which is described in Eqs. (9)–(12).

$$O_{1,1} = \mu A_j(S_3) \tag{9}$$

$$O_{1,2} = \mu B_j(D_1) \tag{10}$$

$$O_{1,3} = \mu C_j(D_2) \tag{11}$$

$$O_{1,4} = \mu E_j(D_3) \tag{12}$$

where $O_{k,i}$ is the node in the position i of the layer k.

$\mu\left[A_j(S_3), B_j(D_1), C_j(D_2), E_j(D_3)\right]$ is the membership function j of a clustering group's data set $(A, B, C, and\ E)$ of input time series data S_3, D_1, D_2, and D_3 where $j = 1, 2, 3$.

S_i are approximation coefficients of wavelet decomposition at level i.

D_i are detail coefficients of wavelet decomposition at level i.

The membership function has various types. The bell-shaped membership function is the popular one that is usually used. It can be described as in Eq. (13):

$$\mu N(x) = \frac{1}{1 + \left|\frac{x - c_i}{a_i}\right|^{2b_i}} \tag{13}$$

where a_i, b_i, and c_i are known as premise parameters.

$N(x)$ is clustering data group's data set N of input time series data x where $N = A, B, C,\ and\ E,\ x = S_3, D_1, D_2,$ and D_3 respectively.

Layer 2. Every node in this layer multiplies the input signals from Layer 1 which can be written in Eq. (14)

$$O_{2,i} = w_i = \mu A_i(S)\mu B_i(D_1)\mu C_i(D_2)\mu E_i(D_3) \quad \text{where } i = 1, 2, 3 \tag{14}$$

Layer 3. The node i of this layer calculates the ratio of the firing strength of each rule to the sum of all the firing strengths as showm in Eq. (15)

$$O_{3,i} = \overline{w}_i = \frac{w_i}{w_1 + w_2 + w_3 + w_4} \tag{15}$$

Layer 4. The node i in this layer has the following node function which is described in Eq. (16)

$$O_{4,i} = \overline{w}_i f = \overline{w}_i(p_i S + q_i D_1 + m_i D_1 + n_i D_1 + r_i) \tag{16}$$

Layer 5. The node in this layer will compute the overall output as the sum of all input signals which can be described in Eq. (17):

$$O_{5,i} = \sum_i \overline{w}_i f = \frac{\sum_i \overline{w}_i f}{\sum_i \overline{w}_i} \tag{17}$$

The ANFIS's architecture of this study is shown in Fig. 2

3.3 Particle Swarm Optimization (PSO)

The Particle Swarm Optimization algorithm is a random search technique for the best solution by iteratively improving the best answer via a measure of quality. The concept of PSO imitates from the social behavior of a flock of birds, a school of fish or a swarm of flying ants which represents by a population of particles. In PSO each particle moves in a D-dimensional space according to a simple mathematical formula over the particle's position and velocity. The direction and step of particle moving is based on its own experience and the information of other particles. Each particle has a position and a velocity that are represented by the vectors $X_i = (x_{i1}, x_{i2}, \ldots, x_{iD})$ and $V_i = (v_{i1}, v_{i2}, \ldots, v_{iD})$ for the particle i. At each iteration, several data which are recorded in a form of position and dimension of particles are compared with another particle to find the best particle that gives a small error. The best position of each particle will be recorded as $P_i = (p_{i1}, p_{i2}, \ldots, p_{iD})$ which is called the personal best ($pbest$). The best position in the swarm of particles is called the global best ($gbest$) which is represented by $G = (G_1, G_2, \ldots, G_D)$. The velocity of each particle is described by Eq. (18).

$$V_{id} = w_1 v_{id} + c_1 \cdot rand_1() \cdot (pbest_{id} - x_{id}) + c_2 \cdot rand_2() \cdot (gbest_{id} - x_{id}) \tag{18}$$

where $i = 1, 2, \ldots, M$, $d = 1, 2, \ldots, D$, c_1 and c_2 are acceleration constants, $rand_1()$ and $rand_2()$ are random numbers between [1, 1], and w is the inertia weight [11]. The new position of the particle is determined by Eq. (19):

$$x_{id} = x_{id} + v_{id} \tag{19}$$

3.4 *Quantum Particle Swarm Optimization (QPSO)*

The main idea of the Quantum Particle Swarm Optimization was developed from quantum theory. In QPSO, each particle has quantum behavior which is treated as a spinless moving in the quantum space around the center of their potential field. The position of each particle is presented by the probability of the particle's appearing at that position instead of directly indicating a position and velocity to the particle. The wave function $w(x, t)$ is assigned to each particle in QPSO. The behavior of particles in QPSO is quite different from that in PSO. The probability density function $w(x, t)$ is used to calculate the probability of the particle i's appearance at position x. Each particle's movement is described according to Eqs. (20) and (21)

$$P_{id} = \varphi \cdot P_{id} + (1 - \varphi) \cdot P_{gd}, \quad \varphi = rand() \tag{20}$$

$$X_{id} = P_{id} \pm \alpha \cdot |mbest_d - X_{id}| \cdot \ln(1/u), \quad u \sim U(0, 1) \tag{21}$$

where, $mbest$ is the mean best position of the particles as shown in Eq. (22)

$$mbest = \frac{1}{M}\sum_{i=1}^{M} P_i = \left(\frac{1}{M}\sum_{i=1}^{M} P_{i1}, \frac{1}{M}\sum_{i=1}^{M} P_{i2}, \ldots \frac{1}{M}\sum_{i=1}^{M} P_{in}\right) \tag{22}$$

P_{id} is a random point between P_{id} and P_{gd} and a local attractor for the particle i on the dimension d. φ and u are random numbers in [0, 1], and α is the contraction–expansion coefficient.

3.5 Forecasting Accuracy Evaluation

The loss functions are used to evaluate the accuracy of a forecasting model. The loss functions such as Root Mean Square Error (RMSE), Standard Error of Mean (SEM), Mean Absolute Percentage Error (MAPE), and Mean Absolute Error (MAE) are calculated as in Eqs. (23)–(26)

$$RMSE = \sqrt{\frac{1}{N}\sum_{i=1}^{N}(Actual_i - Forecast_i)^2} \tag{23}$$

Root Mean Square Error (RMSE) is used to measure the difference between the predicted values and the actual values. It can measure accuracy and is often used to evaluate a forecasting model with the same type of data. The RMSE cannot be used for comparison between models with different types and different scales of data.

$$SEM = \frac{\sigma}{\sqrt{N}} = \sqrt{\frac{1}{N(N-1)}\sum_{i=1}^{N}(x_i - \overline{x})^2} \tag{24}$$

Standard Error of Mean (SEM) is used for estimating sampling distribution. The small SEM is expected because it represents many data points (samples) for calculation or the error term of the predicted value and the actual value are small compared to the number of samples.

$$MAPE = \left(\frac{1}{N}\sum_{i=1}^{N}\left|\frac{Actual_i - Forecast_i}{Actual_i}\right|\right) * 100 \tag{25}$$

Mean Absolute Percentage Error (MAPE) measures the accuracy of a forecasting model in terms of percentage. So, it can be used to measure across different datasets. The low MAPE means that the forecasting model is highly accurate for prediction because the difference between the predicted value and the actual value is low.

$$MAE = \frac{1}{N}\sum_{i=1}^{N}|Actual_i - Forecast_i| \tag{26}$$

Mean Absolute Error (MAE) is a scale-dependent accuracy measure which is used to measure the errors between paired observations expressing the same situation and with the same scaled data being measured. So, MAE cannot be used to compare the series with different scales.

4 Data Collection and Training a WT-QPSO-ANFIS Model

In this section, to demonstrate the Wavelet Transformation-Quantum Particle Swarm Optimization- Adaptive Network-based Fuzzy Inference System (WT-QPSO-ANFIS) model application step-by-step, the daily stock index futures of 10 major indexes and one minor index from the 1st January 2009 to 14th May 2020 are used as the empirical data. The 10 major and one minor stock index futures are shown in Table 1. The experimental procedures of the WT-QPSO-ANFIS model in this study are presented as follows.

4.1 Data Collection and Pre-processing Data

The daily stock index futures of 10 major indexes and one minor index indicated in Table 1 were collected from the Thomson Reuters database covering 2964 trading days from January 1st, 2009 to May14th, 2020. We used data of the first 2075 trading days, from January 1st, 2009 to December 14th, 2016 as training data, and that of the remaining 889 trading days, from December 15th, 2016 to February 14th, 2020 for testing the model. In this part, we decompose the data by using Daubechies 4 (db4) with three decomposition levels. The sample of full decomposition of Dow Jones stock futures is shown in Fig. 1. The wavelet approximation and detail coefficients received from decomposition of the stock index futures data are used as inputs of the ANFIS model. The sample of inputs data is shown in Table 2. Then, we generate the architecture of the ANFIS model which is illustrated in Fig. 2.

Table 1 11 stock index futures data

Code	Stock index futures	Code	Stock Index futures
DM1	Dow Jones	NK1	Nikkei 225
ES1	S&P 500	NZ1	Nifty 50
VG1	Euro STOXX 50	GX1	DAX 30
Z 1	FTSE 100	EO1	AEX
CF1	CAC 40	HI1	Hang Seng
BC1	SET 50		

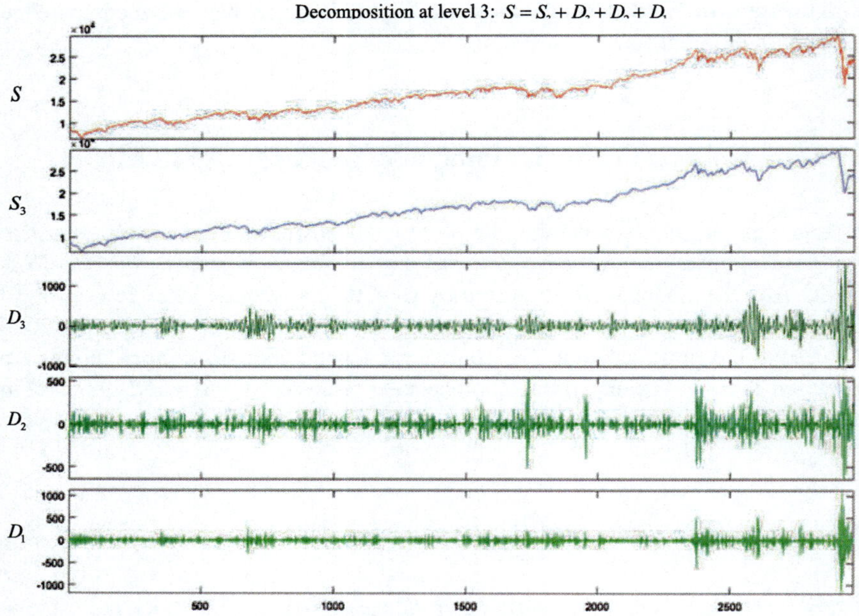

Fig. 1 The full decomposition of the Dow Jones stock futures. S is the Dow Jones stock index futures (red), S_3 is the wavelet approximation coefficient at level three (blue), and D_1, D_2, and D_3 are the wavelet details coefficient at levels one, two, and three (green)

4.2 Training the WT-QPSO-ANFIS Model

In this step, the membership function parameters of the ANFIS model are tuned for improving the model's performance. We use the Quantum Particle Swarm Optimization (QPSO) method to tune the membership function parameters to get the best parameters. We note that the architecture of our hybrid model can be illustrated as in Fig. 2. Then, we train the WT-QPSO-ANFIS model by using 2075 training data with 10,000 iterations to find the best value of membership function parameters in terms of the lowest RMSE. Then, we test the global best solution of membership function in the trained WT-QPSO-ANFIS model with the remaining 889 trading days' data. For instance, the results of using the WT-QPSO-ANFIS model on the training data and testing the model on the test set are shown in Fig. 3. Lastly, Root Mean Square Error (RMSE), Mean Absolute Percentage Error (MAPE), Mean Absolute Error (MAE), and the Standard Error of Mean (SEM) were calculated to evaluate the forecasting model. The results of evaluating the proposed model are shown in Table 3.

Table 2 The ANFIS input data set

Date	App.S_3	Detail D_1	Detail D_2	Detail D_3	App.S_3	Detail D_1	Detail D_2	Detail D_3
	AEX				FTSE 100			
2009.01.01	258.7708	−0.32268	−8.71518	−3.38289	4489.745	−2.66989	−94.036	−19.3812
2009.01.02	259.2679	0.73572	−1.13218	−0.67146	4472.11	0.182851	3.792227	21.28037
2009.01.03	259.7623	−1.55181	2.10024	1.089241	4453.353	−14.7712	34.79889	48.36251
⋮	⋮	⋮	⋮	⋮	⋮	⋮	⋮	⋮
2016.12.12	469.041	−0.89609	0.735205	1.469927	6933.19	−47.3904	−23.5458	23.29819
2016.12.13	470.7771	1.683072	0.033248	1.106606	6943.146	25.30044	−8.96522	27.97436
2016.12.14	472.7742	−1.42593	−0.23898	1.040751	6955.257	−8.99763	0.982068	28.86993
	Dow Jones mini				Hang Seng			
2009.01.01	8832.62	−13.0604	−148.517	7.96783	15,031.33	45.2443	−721.23	−58.6308
2009.01.02	8781.638	27.7876	12.19311	85.39964	14,916.17	−177.713	−80.5737	265.9588
2009.01.03	8728.303	−43.412	55.59606	124.1784	14,795.51	150.1454	169.6877	407.9991
⋮	⋮	⋮	⋮	⋮	⋮	⋮	⋮	⋮
2016.12.12	19,758.26	−38.9664	45.68292	36.82792	22,445.33	−129.752	−50.7021	28.20456
2016.12.13	19,788.35	63.62545	14.24817	75.86486	22,325.08	−58.8864	85.671	−14.1136
2016.12.14	19,828.83	−45.8565	−7.43619	88.93852	22,178.58	112.0325	74.1945	−24.3036

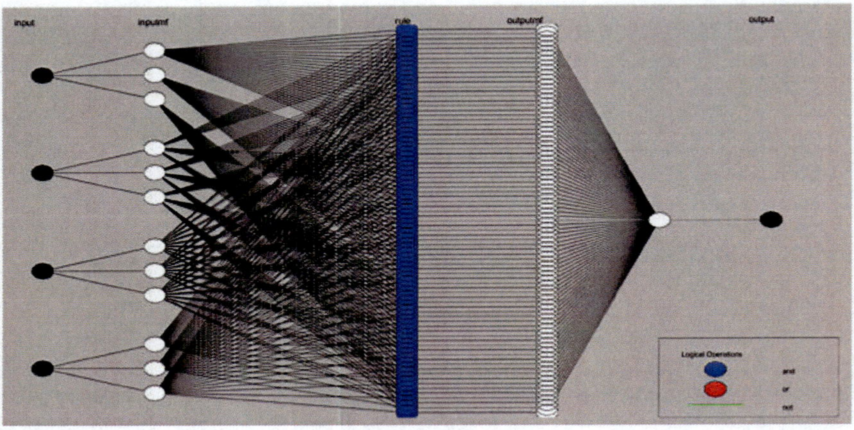

Fig. 2 The architecture of the ANFIS model

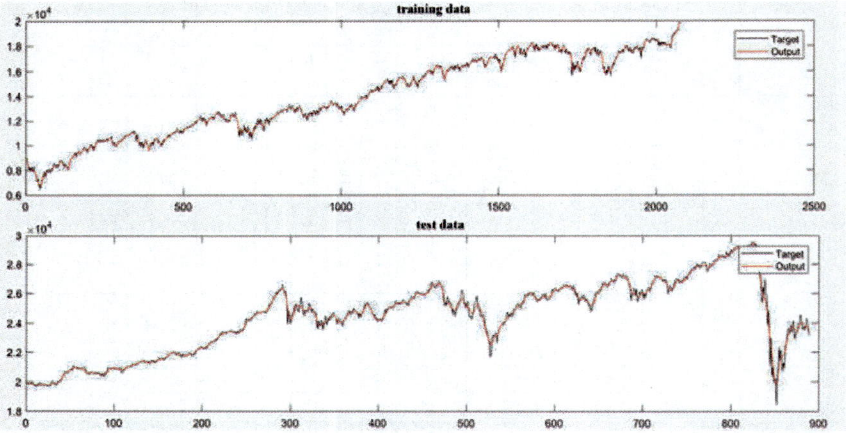

Fig. 3 The prediction performance of Dow Jones stock index futures on training data (from January 1st, 2009 to December 14th, 2016) and the result of model evaluation on test data (from December 15th, 2016 to February 14th, 2020). The red line shows the forecasted values, and the blue line shows the real stock index futures data

5 Results and Comparisons

As reported in Table 3, four loss functions were used to evaluate the WT-QPSO-ANFIS model and the conventional Sugeno-type ANFIS(S-ANFIS) model and the ARIMA model and the Artificial Neural Network model (ANN) for comparison. For the S-ANFIS model, we used the Neuro-Fuzzy designer for creating the model structure and training the parameters to forecast stock index futures and compare with real data. For the Artificial Neural Network model (ANN), we used Validann

Table 3 Evaluating the WT-QPSO-ANFIS model and comparison with other methods

Index	Models	RMSE	MAPE	MAE	SEM
Dow Jones	WT-QPSO-ANFIS	215.6224	0.540442	131.3468	7.235816
	Sugeno-ANFIS	6882.927	21.38271	5378.468	230.9759
	ARIMA (5,1,1)	2718.245	9.332601	2373.207	91.21834
	ANN	961.2103	3.392779	859.0531	15.65245
S&P 500	WT-QPSO-ANFIS	34.85533	0.985437	26.7501	1.169668
	Sugeno-ANFIS	611.5278	18.42421	520.663	20.52153
	ARIMA (5,1,1)	228.4419	6.691875	190.0691	7.666008
	ANN	114.6241	3.250861	93.44952	2.320706
FTSE 100	WT-QPSO-ANFIS	52.4224	0.48207	33.24919	1.759181
	Sugeno-ANFIS	325.8065	3.639981	268.1964	10.93335
	ARIMA (3,1,1)	729.0589	7.279075	485.7395	24.46561
	ANN	72.411	0.746868	52.34324	2.216399
STOXX 50	WT-QPSO-ANFIS	34.90934	0.715651	23.33672	1.171481
	Sugeno-ANFIS	64.4956	1.567383	52.47707	2.164331
	ARIMA (0,1,0)	271.2183	5.89619	188.1754	9.101491
	ANN	36.68222	0.762958	24.94383	1.230793
DAX 30	WT-QPSO-ANFIS	209.2875	1.196199	141.047	7.023229
	Sugeno-ANFIS	591.4747	3.623299	454.6668	19.84859
	ARIMA (0,1,0)	1418.055	9.113842	1049.328	47.5868
	ANN	301.2017	2.110909	260.9404	6.345075
CAC 40	WT-QPSO-ANFIS	49.79036	0.63178	32.22322	1.670855
	Sugeno-ANFIS	282.7998	3.670936	200.8132	9.490141
	ARIMA (1,1,0)	399.2711	5.982163	308.2909	13.39866
	ANN	361.0893	5.042877	326.9678	5.731521
AEX	WT-QPSO-ANFIS	4.929204	0.608332	3.216044	0.165413
	Sugeno-ANFIS	50.84157	7.593812	42.41298	1.706131
	ARIMA (3,1,0)	36.3623	5.565856	29.87099	1.220239
	ANN	7.736202	1.099421	5.911987	0.201811
HANG SENG	WT-QPSO-ANFIS	225.1228	0.622586	168.964	7.554627
	Sugeno-ANFIS	1202.819	2.840016	815.9319	40.36396
	ARIMA (2,1,2)	4359.89	13.31674	3773.037	146.3083
	ANN	1210.765	3.239618	933.6583	27.90088
NIFFY 50	WT-QPSO-ANFIS	96.14961	0.613062	63.57435	3.22657
	Sugeno-ANFIS	2013.289	16.15941	1775.888	67.56156
	ARIMA (0,1,2)	1555.342	13.67377	1470.775	52.19388
	ANN	219.0124	1.673288	181.8553	4.924452
NIKEI 225	WT-QPSO-ANFIS	279.6085	0.88161	184.7386	9.383046
	Sugeno-ANFIS	1724.841	5.934146	1319.904	57.88188
	ARIMA (0,1,2)	1658.186	5.619396	1171.164	55.64507
	ANN	337.4991	1.211879	260.3157	9.439921
SET 50	WT-QPSO-ANFIS	8.127255	0.521278	5.302642	0.272733
	Sugeno-ANFIS	53.67422	3.797767	41.71851	1.801189
	ARIMA (2,1,0)	122.7504	8.021126	78.67028	4.119234
	ANN	12.41941	0.851020	8.828713	0.379594

with 12 hidden layers to forecast stock index futures. In the ARIMA model, we used the best fit order of the ARIMA model in terms of the lowest BIC and AIC values for each stock index futures series. Consequently, the WT-QPSO-ANFIS model was found to give the lowest values for all the four loss functions of all of the 11 stock index futures data compared with the S-ANFIS model, the ANN model and the ARIMA model. This means the WT-QPSO-ANFIS model has higher accuracy and higher efficiency in forecasting stock index futures than the S-ANFIS model, the ANN model and the ARIMA model considering its lowest RMSE, MAPE, MAE, and SEM. The WT-QPSO-ANFIS model is the most suitable for prediction on FTSE 100 stock index futures because its MAPE is 0.48207 which is the lowest value in all of the 11 stock index futures. Furthermore, the ARIMA model gives all the four loss functions' values lower than those of the S-ANFIS in Dow Jones, S&P 500, AEX, NIFFY 50, and NIKEI 225 stock index futures. Apart from these five stock index futures, the S-ANFIS model outperforms the ARIMA model. In addition, the ANN model provides the lowest values for all the four loss functions of all the 11 stock index futures data, except for CAC 40 and HANG SENG, compared with the ARIMA model and the S-ANFIS model.

By the lowest RMSE criterion, the best forecasting model for Dow Jones, S&P 500, FTSE 100, STOXX 50, DAX 30, CAC 40, AEX, HANG SENG, NIFFY 50, NIKEI 225 and, SET 50 stock index futures is the WT-QPSO-ANFIS model with minimum RMSE at 215.6224, 34.85533, 52.4224, 34.90934, 209.2875, 49.79036, 4.929204, 225.1228, 96.14961, 279.6085, 8.127255 respectively.

6 Conclusion

The objective of this study was to improve the optimization technique by using the particle swarm optimization (PSO) method which has high potential in finding the best solution in the search space because of such particle's property as the ability to store the past data of the personal best position, and the ability to communicate with other particles for finding the global best position of swarms. Quantum theory was used to enhance a quantum behavior to particle's property such as the probability of appearance in the search space which ensures finding the global best position and speeds up the convergence. Furthermore, wavelet decomposition was used to characterize the main structure of the original data via the approximation coefficients along with capture ruptures, and identities of the original data by detail coefficients. Both coefficients were used as the system input parameters. The WT-QPSO-ANFIS model was used to solve the limitation of the econometric model which requires many assumptions. The fuzzy logic of the ANFIS model which allows data to be in more than one set can be described by probability that relates to quantum theory. It makes the WT-QPSO-ANFIS model have more powerful forecasting and higher efficiency in predicting stock index future than the conventional Sugeno-ANFIS model which has over-fitting and upper-lower bounded limit problems and the ANN model which did not guarantee the global solution and the ARIMA model which are usually used in linear problems and stationary series.

The WT-QPSO-ANFIS model can be used as a trader advisory system for traders. It can predict the next closing value day of stock index futures representing the short-term of movement of the stock index futures. If the expected value exceeds the current value, traders should open a short position and close in the next day for reducing risk on investment or making a higher profit in trading. But if the predicted value is below the current value, traders should open a long position and close in the next day for making a higher profit. Moreover, the WT-QPSO-ANFIS model can be used instead of the ANN model and the ARIMA model to predict the price of the asset at $t + 1$ that is used as the parameter for portfolio optimization.

Acknowledgements This research work was partially supported by Chiang Mai University.

References

1. Aggarwal Sanjeev Kumar. (2008). Price forecasting using wavelet transform and LSE based mixed model in Australian electricity market. *International Journal of Energy Sector Management, 2*(4), 521–546.
2. Archer, M. D. (2010). *Getting started in currency trading*. Wiley & Sons Inc.
3. Atsalakis, G., & Valavanis, K. (2009). Forecasting stock market short-term trends using a neuro-fuzzy based methodology. *Expert Systems with Applications, 36*, 10696–10707.
4. Caetano, M., & Yoneyama, T. (2009). A new indicator of imminent occurrence of drawdown in the stock market. *Physica A: Statistical Mechanics and its Applications, 388*, 3563–3571.
5. Catalao, J., Pousinho, H., & Mendes, V. (2011). Hybrid Wavelet-PSO-ANFIS approach for short-term electricity prices forecasting. *IEEE Transactions on Power Systems, 26*(1).
6. Chang, J.-R., Wei, L.-Y., & Cheng, C.-H. (2011). A hybrid ANFIS model based on AR and volatility for TAIEX forecasting. *Applied Soft Computing, 11*, 1388–1395.
7. Chen, M.-Y. (2013). A hybrid ANFIS model for business failure prediction utilizing particle swarm optimization and subtractive clustering. *Information Sciences, 220*, 180–195.
8. Cheng, C.-T. (2015). Daily reservoir runoff forecasting method using artificial neural network based on quantum- behaved particle swarm optimization. *Water, 7*, 4232–4246.
9. Cheng, C.-H., Wei, L.-Y., Liu, J.-W., & Chen, T.-L. (2013). OWA-based ANFIS model for TAIEX forecasting. *Economic Modelling, 30*, 442–448.
10. Chiang, C.-H. (2013). Quantum-membership-function-based adaptive neural fuzzy inference system. *Intelligent Technologies and Engineering Systems*, 227–233.
11. Clerc, M., & Kennedy, J. (2002). The particle swarm: Explosion, stability, and convergence in a multi-dimensional complex space. *IEEE Transactions on Evolutionary Computation, 6*(1), 58–73.
12. Cohen, A., Daubechies, I., & Vial, P. (1993). Wavelets on the interval and fast wavelet transform. *Applied and Computational Harmonic*, 54–81.
13. Ding, H., Zeng, S. H., & Guo, X. J. (2011). Influence of stock index futures on stock market price: Theoretical analysis and experiences of the Chinese and oversea markets. *International Journal of Economics and Finance, 3*(4), 113–118.
14. Divband, M. (2010). A comparison of particle swarm optimization and gradient descent in training wavelet neural network to predict dgps corrections. In *Proceedings of the World Congress on Engineering and Computer Science (WCECS), San Francisco, USA* (Vol. 1).

15. Escobar, A., Moreno, J., & Munera, S. (2013). A technical analysis indicator based on fuzzy logic. *Electronic Notes in Theoretical Computer Science, 292*, 27–37.
16. Frans van den Bergh. (2006). An Analysis of Particle Swarm Optimizers. *Information Sciences, 176*(8), 937–971.
17. Hsieh, T.-J., Hsiao, H.-F., & Yeha, W.-C. (2011). Forecasting stock markets using wavelet transforms and recurrent neural networks: An integrated system based on artificial bee colony algorithm. *Applied Soft Computing, 11*, 2510–2525.
18. Janková, Z., & Dostál, P. (2020). Prediction of European stock indexes using neuro-fuzzy technique. *Trends Economics and Management, 14*(35), 45–58.
19. Kao, L.-J., Chiu, C.-C., Lu, C.-J., & Chang, C.-H. (2013). A hybrid approach by integrating wavelet-based feature extraction with MARS and SVR for stock index forecasting. *Decision Support Systems, 54*, 1228–1244.
20. Lah, M. S. C., Arbaiy, N., & Lin, P. C. (2020). Stock index modelling using arima with standard deviation based triangular fuzzy numbers. *Journal of Critical Reviews, 7*(8), 1264–1268.
21. Li, S. T., & Kuo, S. C. (2008). Knowledge discovery in financial investment for forecasting and trading strategy through wavelet-based SOM networks. *Expert Systems with Applications, 34*(2), 935–951.
22. Liu, et al. (2019). A quantum particle swarm optimization algorithm with teamwork evolutionary strategy. *Mathematical Problems in Engineering, 19*, 12.
23. Penawar, H. K., & Rustam, Z. (2017). A fuzzy logic model to forecast stock market momentum in Indonesia's property and real estate sector. In *AIP Conference Proceedings* (Vol. 1862, No. 1, p. 030125). AIP Publishing LLC.
24. Şahin, M., & Erol, R. (2017). A comparative study of neural networks and ANFIS for forecasting attendance rate of soccer games. *Mathematical and Computational Applications, 22*(4), 43.
25. Sun, J., Fang, W., Palade, V., Wu, X., & Xu, W. (2011). Quantum-behaved particle swarm optimization with Gaussian distributed local attractor point. *Applied Mathematics and Computation, 218*, 3763–3775.
26. Sun, E., & Meinl, T. (2012). A new wavelet-based denoising algorithm for high frequency financial data mining. *European Journal of Operational Research, 217*, 589–599.
27. Sun, J., at el. (2007) A modified quantum-behaved particle swarm optimization. *In international conference on computational science* (pp. 294–301). Springer.
28. Szmigiera, M. (2020). Number of futures and options contracts traded globally 2013–2019. Available at: https://www.statista.com/global-futures-and-options-volume.html. Accessed 12 May, 2010
29. Tan, Z., Quek, C., & Cheng, P. Y. (2011). Stock trading with cycles: A financial application of ANFIS and reinforcement learning. *Expert Systems with Applications, 38*, 4741–4755.
30. Wang, W., & Nie, H. (2018). Comparative study on forecasting model for stock index future price. *Modern Economy, 9*, 4.
31. Wei, L.-Y. (2011). A hybrid model based on adaptive-network-based fuzzy inference system to forecast Taiwan stock market. *Expert Systems with Applications, 38*(11), 13625–13631.
32. Wei, L.-Y. (2013). A hybrid model based on ANFIS and adaptive expectation genetic algorithm to forecast TAIEX. *Economic Modelling, 33*, 893–899.
33. Wei, L.-Y., Chen, T.-L., & Ho, T.-H. (2011). A hybrid model based on adaptive network-based fuzzy inference system to forecast Taiwan stock market. *Expert Systems with Applications, 38*, 13625–13631.
34. Yang, C., Liu, X. H., Economics, C. O., et al. (2014). Forecasting model for stock index future price based on wavelet analysis and BP neural network. *Journal of Qingdao University, 1*, 101–105.
35. Zadeh, L. A. (1965). Fuzzy sets. *Information and Control, 8*, 338–353.
36. Zhao, Y., Zhang, Y., & Qi, C. (2008). Prediction model of stock market returns based on wavelet neural network. In *IEEE Pacific-Asia workshop on computational intelligence and industrial application.*

A Generalized Maximum Renyi Entropy Approach in Kink Regression Model

Payap Tarkhamtham and Woraphon Yamaka

Abstract This study aims to estimate the kink regression model with the small sample size since the small sample size will lead to an undetermined or ill-posed problem. Thus, to solve these problems, the generalized maximum entropy (GME), based Renyi measure is proposed. Specifically, we replace the Shannon entropy with Renyi entropy to GME estimator. Monte Carlo simulation and the real data are used to evaluate the performance of Renyi GME estimator in Kink regression model. The results demonstrate that Renyi measure does not perform better than Shannon measure when the error is assumed to be normal. However, it is comparable to the Shannon entropy when the errors are assumed to be uniform, chi-square, and Student-t distributions. We then conduct two application studies to validate the performance of the Renyi GME and similar performance is obtained. Thus, we increase the number of order α to be larger than 2 and the result indicates that our high-order Renyi GME is clearly better than the traditional Shannon GME.

Keywords Maximum entropy · Renyi entropy · Kink regression

1 Introduction

From the literature, one of the most common issues in much research is the limitation of data, partial, and incomplete data, which can bring the ill-posed problem or underdetermined [2, 10].

In general, the Maximum Likelihood (ML) and Least-square (LS) estimations have been employed as the applicable estimator in various econometric models. However, there have some drawbacks as the error assumption may not be normal

P. Tarkhamtham · W. Yamaka (✉)
Faculty of Economics, Chiang Mai University, Chiang Mai, Thailand
e-mail: woraphon.econ@gmail.com

W. Yamaka
Faculty of Economics, Center of Excellence in Econometrics, Chiang Mai University, Chiang Mai, Thailand

S. Sriboonchitta et al. (eds.), *Credible Asset Allocation, Optimal Transport Methods, and Related Topics*, Studies in Systems, Decision and Control 429,
https://doi.org/10.1007/978-3-030-97273-8_28

distribution and generally unknown. In particular, ML estimation, an appropriate likelihood function is not easily specified and if the likelihood function is miss-specified, the parameter estimates may be biased. Thus, to solve these problems, the Maximum Entropy (ME) approach is one of the estimations that is generally considered as an alternative estimator of the parametric estimation.

The ME principle based on Shannon entropy measure was firstly introduced by Jaynes [11] to estimate the unknown probability distribution of the undetermined problem. This estimator has been later widely used in many research problems. The maximum entropy problem has been generalized by Golan [5], where the individual noisy observations are considered in the entropy function. The ability of GME to estimate economic relationships has already been explored by several studies (see, [4, 5, 12]). Recently, Sriboochitta et al. [15] introduced GME estimator for estimating a nonlinear kink regression model of Hansen [9] and found that GME estimator is slightly better than conventional estimator namely; LS and ML estimates. Then, Tarkhamtham et al. [16] was extended their work by replacing the Shannon measure with Tsallis measure [17] and the result showed that Tsallis GME estimator is also used as an alternative potential estimator.

So, in this study, it is interesting to examine that if we replace the Shannon entropy measure with another entropy, namely Renyi measure [14], the performance of GME estimator could be improved or not. We note that Rényi entropy is a generalization of the Shannon entropy and Shannon's entropy is a particular case of Renyi's entropy when α approaches to one [1]. Therefore, in this study, two entropy measures, namely Shannon and Renyi, are also compared under the nonlinear kink regression context.

The remainder of the paper is organized as follows. Section 2 reviews the background of Kink regression model and Renyi GME estimator. Section 3 shows the Monte Carlo experiment and the results. Section 4 employs the proposed approach to two example data sets. Finally, Sect. 5 provides a conclusion.

2 Methodology

2.1 Kink Regression Model

According to Hansen [9], the two-regimes regression kink model can be given by

$$\begin{aligned} y_t = {} & \beta_1^-(x_{1,t}-\gamma_1)_- + \beta_1^+(x_{1,t}-\gamma_1)_+ \\ & + \cdots + \beta_K^-(x_{K,t}-\gamma_K)_- + \beta_K^+(x_{K,t}-\gamma_K)_+ + \phi z_t + \varepsilon_t \end{aligned} \tag{1}$$

The variables $(y_t, x_{1,t}, \ldots, x_{k,t}, z_t)$ are observed variables for $t = 1, \ldots, n$, whereas y_t is a dependent variable at time t. $x_{k,t}$ is independent variable i at time t for $k = 1, \ldots, K$. z_t is the regime independent exogenous variables. The parameters $(\beta_1^-, \ldots, \beta_K^-)$ are the coefficient with respect to variable $x_{1,t}, \ldots, x_{k,t}$ for value of $x_{1,t}, \ldots, x_{k,t} < \gamma_k$ and $(\beta_1^+, \ldots, \beta_K^+)$ are the coefficient with respect to variable

$x_{1,t}, \ldots, x_{k,t}$ for value of $x_{1,t}, \ldots, x_{k,t} > \gamma_k$. The relationship between y_t and z_t is linear while there is a nonlinear relationship between y_t and x_t. Therefore, the relationship between y_t and x_t changes at the unknown location called threshold or kink point γ_k. So, this model can be separated into two regimes. γ_k is the kink parameter which is compact and strictly in the interior of the support of $x_{k,t}$. In additional, we assume that the error term ε_t is free from the normal or any other distributions, but $E(\varepsilon_t) = 0$ is assumed here.

2.2 *Renyi Entropy Measure*

In this study, we apply the GME estimator of Golan et al. [5] by using Renyi entropy measure as an objective entropy function to estimate unknown parameters in the kink regression model of Hansen [9]. Renyi measure can be defined as follows

$$H^R(p; \alpha) = \frac{1}{1-\alpha} \ln \sum_{k=1}^{n} p_k^{\alpha}, \tag{2}$$

where p_k is the probability of entropy function and $\sum_{k=1}^{K} p_k = 1$, α is a positive value and $\sum_{k=1}^{K} p_k^{\alpha} = 1$. We can see that Renyi measure is indexed by a strictly positive order $\alpha > 0$. However, the Renyi measure becomes a Shannon measure when $\alpha = 1$. Following Sriboonchitta et al. [15], the parameters β_k^- and β_k^+ can be express as

$$\begin{aligned} \beta_k^- &= \sum_m p_{km}^- z_{km}^- \\ \beta_k^+ &= \sum_m p_{km}^+ z_{km}^+, \end{aligned} \tag{3}$$

where p_{km}^- and p_{km}^+ are the probability distribution of estimated parameters defined on the set of support z_{km}^- and z_{km}^+. We note that $z = \left[\underline{z}_{k1}, \ldots, \bar{z}_{km}\right]$ is the lower and upper bounds of support vector. For the kink value of γ_k, it can be computed by

$$\gamma_k = \sum_m h_{km} q_{km} \tag{4}$$

where γ_k is a discrete random variable with the set of $q_k = \left[\underline{q}_{k1}, \ldots, \bar{q}_{km}\right]$ in which $\underline{q}_{k1}$ is the lower bound and $\bar{q}_{km}$ is the upper bound. The error term is also estimated by

$$\varepsilon_t = \sum_m w_{tm} v_{tm}, \tag{5}$$

where ε_t is the set of sum value and w_t is the probability distribution of error defined on the set $v_t = [v_{t1}, \ldots, v_{tm}]$.

As we mentioned earlier, the unknowns $(\beta_1^-, \ldots, \beta_K^-)$, $(\beta_1^+, \ldots, \beta_K^+)$, $(\gamma_1, \ldots, \gamma_k)$, and ε can parameterized thereby rewrite Eq. (1) as

$$\begin{aligned} Y_t = & \sum_m p_{1m}^- z_{1m}^- \left(x_{1,t} - \sum_m h_{1m} q_{1m} \right)_- + \sum_m p_{1m}^+ z_{1m}^+ \left(x_{1,t} - \sum_m h_{1m} q_{1m} \right)_+ \\ & + \cdots + \sum_m p_{km}^- z_{km}^- \left(x_{k,t} - \sum_m h_{km} q_{km} \right)_- \\ & + \sum_m p_{km}^+ z_{km}^+ \left(x_{k,t} - \sum_m h_{km} q_{km} \right)_+ + \sum_m w_{tm} v_{tm} \end{aligned} \tag{6}$$

Then, the optimization problem of Renyi GME estimator in kink regression is

$$\begin{aligned} H(p, h, w) = & \underset{p,h,w}{\arg\max} \ \{H(p) + H(h) + H(w)\} \\ \equiv & -\frac{1}{1-\alpha} \left(\sum_m \sum_k (p_{km}^-)^\alpha \right) - \frac{1}{1-\alpha} \left(\sum_m \sum_k (p_{km}^+)^\alpha \right) \\ & - \frac{1}{1-\alpha} \left(\sum_m \sum_k (h_{km}^-)^\alpha \right) - \frac{1}{1-q} \left(\sum_t \sum_w (w_{tm})^\alpha \right) \end{aligned} \tag{7}$$

Subject to

$$\begin{aligned} Y_t = & \sum_m p_{1m}^- z_{1m}^- \left(x_{1,t} - \sum_m h_{1m} q_{1m} \right)_- + \sum_m p_{1m}^+ z_{1m}^+ \left(x_{1,t} - \sum_m h_{1m} q_{1m} \right)_+ \\ & + \cdots + \sum_m p_{Km}^- z_{Km}^- \left(x_{k,t} - \sum_m h_{Km} q_{Km} \right)_- \\ & + \sum_m p_{Km}^+ z_{Km}^+ \left(x_{k,t} - \sum_m h_{Km} q_{Km} \right)_+ + \sum_m w_{tm} v_{tm}, \end{aligned} \tag{8}$$

$$\begin{aligned} & \sum_m p_{km}^+ = 1, \sum_m p_{km}^- = 1, \sum_m h_{km} = 1, \sum_m w_{tm} = 1, \\ & k = 1, \ldots, K \ \text{and} \ t = 1, \ldots, T \end{aligned} \tag{9}$$

where $p_{1m}^-, \ldots, p_{Km}^-, p_{1m}^+, \ldots, p_{Km}^+, h_{1m}, \ldots, h_{Km}$, and $w_{1m}, \ldots, w_{tm}$ are the probabilities in the interval [0, 1] which require that the probabilities sum to one. To have a simple derivation, we consider only the kink regression with $K = 1$, then, the optimization problem can be solved as

$$
\begin{aligned}
L = H(p,h,w) &+ \lambda_1 \left(y_t - \sum_m p_{1m}^- z_{1m}^- \left(x_{1,t} - \sum_m h_{1m} q_{1m} \right)_- \right. \\
&\left. - \sum_m p_{1m}^+ z_{1m}^+ \left(x_{1,t} - \sum_m h_{1m} q_{1m} \right)_+ - \sum_m w_{tm} v_{tm} \right) \\
&+ \lambda_2 \left(1 - \sum_m p_{1m}^- \right) + \lambda_3 \left(1 - \sum_m p_{1m}^+ \right) + \lambda_4 \left(1 - \sum_m h_{1m} \right) \\
&+ \lambda_5 \left(1 - \sum_m w_{tm} \right),
\end{aligned} \tag{10}
$$

where λ_i, $i = 1, \ldots, 5$ are the vectors of Lagrange multipliers for the data constraint. Thus, first-order conditions for the Renyi GME problem are

$$
\frac{\partial L}{\partial p_{1m}^-} = -\frac{\alpha}{1-\alpha} \frac{(p_{1m}^-)^{\alpha-1}}{(p_{1m}^-)^{\alpha}} - \sum_m \lambda_{1m} z_{1m}^- \left(x_{1,t} - \sum_m h_{1m} q_{1m} \right)_- - \lambda_{2m} = 0, \tag{11}
$$

$$
\frac{\partial L}{\partial p_{1m}^+} = -\frac{\alpha}{1-\alpha} \frac{(p_{1m}^+)^{\alpha-1}}{(p_{1m}^+)^{\alpha}} - \sum_m \lambda_{1m} z_{1m}^+ \left(x_{1,t} - \sum_m h_{1m} q_{1m} \right)_+ - \lambda_{3m} = 0, \tag{12}
$$

$$
\frac{\partial L}{\partial w_{tm}} = -\frac{\alpha}{1-\alpha} \frac{(w_{tm})^{\alpha-1}}{(w_{tm})^{\alpha}} - \sum_m \lambda_{tm} v_{tm} - \lambda_{5t} = 0, \tag{13}
$$

$$
\begin{aligned}
\frac{\partial L}{\partial h_{1m}} = &-\frac{\alpha}{1-\alpha} \frac{(h_{1m})^{\alpha-1}}{(h_{1m})^{\alpha}} - \sum_m \lambda_{1m} p_{1m}^- z_{1m}^- \left(x_{1,t} - \sum_m q_{1m} \right)_- \\
&- \sum_m \lambda_{1m} p_{1m}^+ z_{1m}^+ \left(x_{1,t} - \sum_m q_{1m} \right)_+ - \lambda_{4m} = 0,
\end{aligned} \tag{14}
$$

$$
\begin{aligned}
\frac{\partial L}{\partial \lambda_{1m}} = &-(y_t - \sum_m p_{1m}^- z_{1m}^- \left(x_{1,t} - \sum_m h_{1m} q_{1m} \right)_- \\
&- \sum_m p_{1m}^+ z_{1m}^+ \left(x_{1,t} - \sum_m h_{1m} q_{1m} \right)_+ - \sum_m w_{tm} v_{tm}) = 0,
\end{aligned} \tag{15}
$$

$$
\frac{\partial L}{\partial \lambda_{2m}} = 1 - \sum_m p_{1m}^- = 0, \tag{16}
$$

$$
\frac{\partial L}{\partial \lambda_{3m}} = 1 - \sum_m p_{1m}^+ = 0, \tag{17}
$$

$$\frac{\partial L}{\partial \lambda_{4m}} = 1 - \sum_m h_{1m} = 0, \tag{18}$$

$$\frac{\partial L}{\partial \lambda_{5t}} = 1 - \sum_m w_{tm} = 0. \tag{19}$$

Finally, we have

$$\widehat{p}_{1m}^{-} = \left\{ \left(\frac{1-\alpha}{\alpha}\right) \frac{\exp\left(-z_{1m}^{-} \sum_t \widehat{\lambda}_{1t}\left(x'_{1,t} - \sum_m h_{1m} q_{1m}\right)_{-}\right)}{\sum_m \exp\left(-z_{1m}^{-} \sum_t \widehat{\lambda}_{1t}\left(x'_{1,t} - \sum_m h_{1m} q_{1m}\right)_{-}\right)} \right\}^{1/(\alpha-1)}, \tag{20}$$

$$\widehat{p}_{1m}^{+} = \left\{ \left(\frac{1-\alpha}{\alpha}\right) \frac{\exp\left(-z_{1m}^{+} \sum_t \widehat{\lambda}_{1t}\left(x'_{1,t} - \sum_m h_{1m} q_{1m}\right)_{+}\right)}{\sum_m \exp\left(-z_{1m}^{+} \sum_t \widehat{\lambda}_{1t}\left(x'_{1,t} - \sum_m h_{1m} q_{1m}\right)_{+}\right)} \right\}^{1/(\alpha-1)}, \tag{21}$$

$$\widehat{w}_{tm} = \left\{ \left(\frac{1-\alpha}{\alpha}\right) \frac{\exp\left(-\widehat{\lambda}_{1t} v_{1m}\right)}{\sum_m \exp\left(-\widehat{\lambda}_{1t} v_{1m}\right)} \right\}^{1/(\alpha-1)}, \tag{22}$$

$$\widehat{h}_{1m} = \left\{ \left(\frac{1-\alpha}{\alpha}\right) \frac{\exp\left(-\left(\begin{array}{l}\sum_t \widehat{\lambda}_{1t} p_{1m}^{-} z_{1m}^{-} \left(x'_{1,t} - \sum_m q_{1m}\right)_{-} \\ - \sum_t \widehat{\lambda}_{1t} p_{1m}^{+} z_{1m}^{+} \left(x'_{1,t} - \sum_m q_{1m}\right)_{+}\end{array}\right)\right)}{\sum_m \exp\left(-\left(\begin{array}{l}\sum_t \widehat{\lambda}_{1t} p_{1m}^{-} z_{1m}^{-} \left(x'_{1,t} - \sum_m q_{1m}\right)_{-} \\ - \sum_t \widehat{\lambda}_{1t} p_{1m}^{+} z_{1m}^{+} \left(x'_{1,t} - \sum_m q_{1m}\right)_{+}\end{array}\right)\right)} \right\}^{1/(1-\alpha)} \tag{23}$$

2.3 Testing for a Threshold Effect and Goodness of Fit Test

In this section, the study employs an entropy-ratio test which is introduced in Golan and Dose [2] and modified by Sriboochitta et al. [15] to examine the existence of Kink effect in the model. The null hypothesis of the entropy test set the restriction as $\beta_1^{-} = \beta_1^{+}$. Say that the kink or threshold point, γ_1, does not exist in the model structure. In the case of the alternative hypothesis, it is an unconstraint or Kink regression model. Therefore, the entropy ratio statistic can be given by

$$ER = 2\left|H_U(\beta_1{}^- \neq \beta_1{}^+) - H_R(\beta_1 = \beta_1{}^- = \beta_1{}^+)\right|, \tag{24}$$

where H_U is the unrestricted hypothesis (linear regression) and H_R is the restricted hypothesis (Kink regression). Under certain regularity assumptions (see [3, 8]), ER converges in distribution to χ_F^2 as $T \to \infty$, when the restriction is true and F is a degree of freedom. The approximate confidence level (κ) for this test is computed by setting $ER(\cdot) < C_\kappa$, where C_κ is the critical value of the χ_F^2 at a significance level at κ and it is chosen so that $\Pr(\chi_F^2 < C_\kappa) = \kappa$.

3 Monte Carlo Simulation Study

In this section, the Monte Carlo simulation study is conducted to examine the finite sample performance of the GME. We follow the simulation scheme of Sriboonchitta et al. [15] and separate this experiment into two part. In the first experiment, we compare the performance of Renyi GME estimator with varied the number of support (M) as $M = 3, 5, 7$ which $M = 3,\ z_{1m} = [-z, 0, z]$, $M = 5,\ z_{1m} = [-z, -z/2, 0, z/2, z]$, and $M = 7,\ z_{1m} = [-z, -z/2, -z/3, 0, z/3, z/2, z]$. The support space of the β, γ and ε are determined to be uniformly symmetric around zero. The support space for β are specified in the range of $[-z, 0, z]$ for $z = 2, 3, 4$. The supports space for γ are specified in the range of $[-h, 0, h]$ for $h = 3, 5, 6$. The support space for ε are specified in the range of $[-v, 0, v]$ for $v = 1$.

In the second experiment, we conduct the Monte Carlo simulation to compare the performance of Shannon GME and Renyi GME estimators under the kink regression model. To compare this, we use the Bias and Mean Square Error (MSE) which can express as follow

$$Bias = B^{-1}\sum_{b=1}^{B}\left(\tilde{\phi}_b - \phi_b\right), \tag{25}$$

$$MSE = B^{-1}\sum_{b=1}^{B}\left(\tilde{\phi}_b - \phi_b\right)^2. \tag{26}$$

where B is equal 100 which is the number of simulated datasets and $\tilde{\phi}_r$ and ϕ_r are the estimated value and the true value, respectively.

Our experiment based on kink regression model as follows

$$y_t = \alpha + \beta_1^-(x_{1,t} - \gamma_1)_- + \beta_1^+(x_{1,t} - \gamma_1)_+ + \varepsilon_t \tag{27}$$

In this experiment, we draw a random sample from Eq. (27) by simulating $x_{1,t} \sim Unif[-3, 5]$ and kink value $\gamma = 3$. The true value for coefficient α, β^-, β^+ are set to be 1, 2 and -1, respectively. To make an impartial comparison, we generated the error terms from various distribution $\chi^2(2)$, $N(0, 1)$, $t(0, 1, 4)$, and $Unif(-2, 2)$ with the sample size $n = 20$ and $n = 40$.

3.1 Monte Carlo Simulation Results

According to the works of Heckelei et al. [10], there are complicate and difficult to specify the range of support space and the number of support points. Thus, we propose the various number and range of supports and to make simple examine, we generated the error term from normal distribution $N(0, 1)$ for all case.

Table 1 shows the estimated parameters form Renyi GME with various specifications of support space and the number of support points. We can observe that if the support space is widened, bias and MSE are the increase in almost case. While the number of support points (M) is increasing, Bias and MSE are not increase in almost case. Thus, this result shows that GME estimator is responsive to the specific number of support points and the range of support. We can conclude that when the number of support points increases the Renyi GME estimator will provide a better result.

Table 1 Choice of number of supports and value of support space

Value of support space				Bias of parameters				MSE of parameters			
z	h	v	m	α	β_1^-	β_1^+	γ_1	α	β_1^-	β_1^+	γ_1
$N = 20$											
[−2, 2]	[−3, 3]	[−1, 1]	3	0.4494	0.0477	0.1320	0.2846	0.4345	0.0419	0.1398	0.1770
[−2, 2]	[−3, 3]	[−1, 1]	5	0.2416	0.0197	0.0583	0.1632	0.3618	0.0369	0.1004	0.1400
[−2, 2]	[−3, 3]	[−1, 1]	7	0.1277	0.0042	0.0059	0.0354	0.2779	0.0162	0.0613	0.0711
[−3, 3]	[−5, 5]	[−1, 1]	3	0.5741	0.4499	0.1731	0.6896	0.7201	1.2139	0.1349	0.8924
[−3, 3]	[−5, 5]	[−1, 1]	5	0.3849	0.2693	0.1170	0.6763	0.2348	0.3233	0.0343	0.8785
[−3, 3]	[−5, 5]	[−1, 1]	7	0.3374	0.2780	0.1100	0.6563	0.1859	0.2899	0.0288	0.8925
[−4, 4]	[−6, 6]	[−1, 1]	3	0.8829	1.1365	0.3496	1.3874	1.9331	4.1164	0.6601	3.3020
[−4, 4]	[−6, 6]	[−1, 1]	5	0.5291	0.6436	0.2557	1.3517	0.6999	1.4994	0.3512	3.7413
[−4, 4]	[−6, 6]	[−1, 1]	7	0.4197	0.4132	0.2235	1.2033	0.3119	0.4040	0.1655	2.7692
$N = 40$											
[−2, 2]	[−3, 3]	[−1, 1]	3	0.4902	0.2307	0.3170	0.0970	0.5238	0.3399	0.3412	0.1402
[−2, 2]	[−3, 3]	[−1, 1]	5	0.6413	1.0426	0.5658	1.4556	0.5944	1.9961	0.5369	4.3230
[−2, 2]	[−3, 3]	[−1, 1]	7	0.2042	0.1147	0.0786	0.0524	0.3229	0.1576	0.1437	0.0500
[−3, 3]	[−5, 5]	[−1, 1]	3	1.7348	2.3280	1.1527	2.0869	0.8601	0.8466	0.5772	1.0900
[−3, 3]	[−5, 5]	[−1, 1]	5	0.4744	0.7913	0.2254	1.0212	0.3959	0.4991	0.2451	0.6537
[−3, 3]	[−5, 5]	[−1, 1]	7	0.2594	0.4616	0.1712	0.7566	0.3359	0.4418	0.2002	0.5892
[−4, 4]	[−6, 6]	[−1, 1]	3	2.1977	3.9213	1.7202	4.4545	0.9553	1.4046	0.7195	1.7744
[−4, 4]	[−6, 6]	[−1, 1]	5	1.4652	2.6408	1.1340	3.7546	0.7152	1.0438	0.5810	1.3907
[−4, 4]	[−6, 6]	[−1, 1]	7	0.8416	1.2611	0.2234	3.8727	0.5067	0.7865	0.3184	1.4874

3.2 Comparing Shannon and Renyi GME Results

In the second experiment, we focus on the performance of GME estimator with the various distribution. The performance of Renyi GME estimator is compared with Shannon GME estimator. We specify the symmetric seven-points of support space $z = [-2, -1, -0.5, 0, 0.5, 1, 2]$, $v = [-1, -0.5, -0.25, 0, 0.25, 0.5, 1]$ and $h = [-3, -1.5, -0.75, 0, 0.75, 1.5, 3]$. We then generated the error from Chi-square distribution $(\chi^2(2))$, Normal distribution $(N(0, 1))$, Student-t distribution $(t(0, 1, 5))$, and Uniform distribution $(Unif(-2, 2))$.

Tables 2 and 3 shows that the performance of Shannon GEM is slightly better than Renyi GME, especially in case of error term was generated from normal distribution as the Bias and MSE values are obviously lower than Renyi GME in both sample size

Table 2 Bias parameter estimates

	Normal		Uniform		Chi-square		Student-t	
	Shannon	Renyi	Shannon	Renyi	Shannon	Renyi	Shannon	Renyi
$N = 20$								
α	0.2512	0.1277	0.9445	1.002	0.4471	0.5107	0.1719	0.2068
β^-	0.0354	0.0042	1.1258	1.3843	1.9347	2.0322	0.0788	0.0784
β^+	0.0469	0.0059	0.4416	0.5629	0.6927	0.7235	0.0359	0.0444
γ	0.0928	0.0354	0.0506	0.0436	0.1021	0.0009	0.0338	0.0628
$N = 40$								
α	0.1028	0.2042	1.292	1.2966	0.4696	0.484	0.3651	0.3737
β^-	0.008	0.1147	1.8693	2.1397	2.4812	2.4644	0.3641	0.3991
β^+	0.0051	0.0786	0.6514	0.7331	0.7979	0.8112	0.1347	0.1636
γ	0.0357	0.0524	0.0001	0.2702	0.0032	0.0001	0.0104	0.0195

Table 3 MSE of parameter estimates

	Normal		Uniform		Chi-square		Student-t	
	Shannon	Renyi	Shannon	Renyi	Shannon	Renyi	Shannon	Renyi
$N = 20$								
α	0.1028	0.2779	0.854	0.8584	0.3118	0.6515	0.595	0.3502
β^-	0.008	0.0162	1.0046	1.0645	0.1135	1.3982	1.3766	0.0879
β^+	0.0051	0.0613	0.6322	0.6973	0.0935	0.8417	0.8261	0.1141
γ	0.0357	0.0711	0.042	0.0437	0.0748	0.0033	0.0639	0.078
$N = 40$								
α	0.2512	0.3229	1.1047	1.0951	0.6679	0.6691	0.4213	0.4447
β^-	0.0354	0.1576	1.3513	1.4353	1.5734	1.5664	0.3506	0.3526
β^+	0.0469	0.1437	0.7987	0.8402	0.8928	0.8969	0.2217	0.2478
γ	0.0928	0.05	0.0004	0.0914	0.0104	0.0001	0.0363	0.0491

$N = 20$ and $N = 40$. For other error distribution cases, the performance of Shannon and Renyi are quite close as the Bias and MSE of estimated parameters seem to have a small difference. We note that the overall performance of the estimators in question remains quite stable when the sample size is increased from 20 to 40. It is well-known that this method normally requires a sizeable sample to attain its theoretical advantages [18]. Nonetheless, our results demonstrate that the Renyi GME shows a similar performance to Shannon GME when the error is not normal estimation. Interestingly, the value of Bias and MSE of kink parameters (γ) are lower for all case, it means that both GME estimators can capture the kink point very well, especially when the sample size is large ($N = 40$).

4 Empirical Illustration

4.1 Example 1

For empirical illustration, Renyi GME and Shannon GME are applied to the real data. We use GDP growth and Debt/GDP ratio as the example data set in this application study. Hansen [9] mentioned that the growth of the economy tends to slow when the government debt relative to GDP exceeds a threshold. Thus, we applied this model specification to Thai data and our model can be written as

$$GDP_t = \alpha + \beta_1^-(Debt/GDP_t \leq \gamma_1)_- + \beta_1^+(Debt/GDP_t > \gamma_1)_+ + \varepsilon_t \quad (28)$$

The data include GDP growth and the Debt/GDP ratio of Thailand, cover 20 observations from 1997 to 2016. The data are collected from the world bank database. Figure 1 presents the data plot.

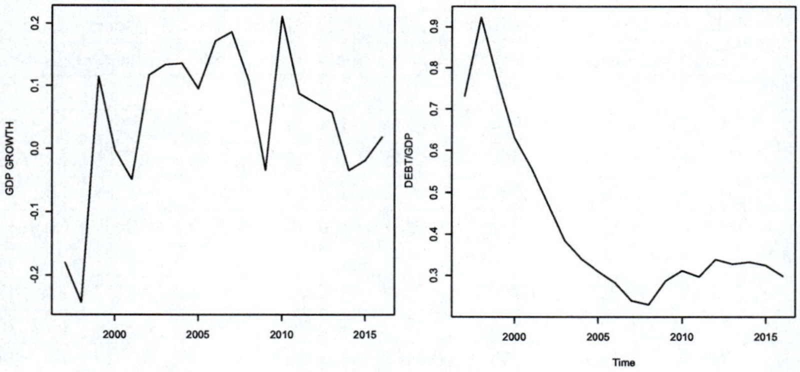

Fig. 1 GDP growth and debt/GDP ratio plot

Table 4 Kink regression application result

Regime	H_{linear}	H_{Kink}	Entropy ratio
1 versus 2	−1.6085	0.7607	4.7386 [0.000]

Note [] shows the Minimum Bayes factor computed by MBF = e^{plogp}, where p is p-value and the MBF can be interpreted as follows: MBF between 1 and 1/3 is considered weak evidence against H_0, from 1/3 to 1/10 considered moderate evidence, 1/10–1/30 considered substantial evidence, 1/30–1/100 considered strong evidence, 1/100–1/300 considered very strong evidence, and <1/300 considered decisive evidence (see, [6, 7])

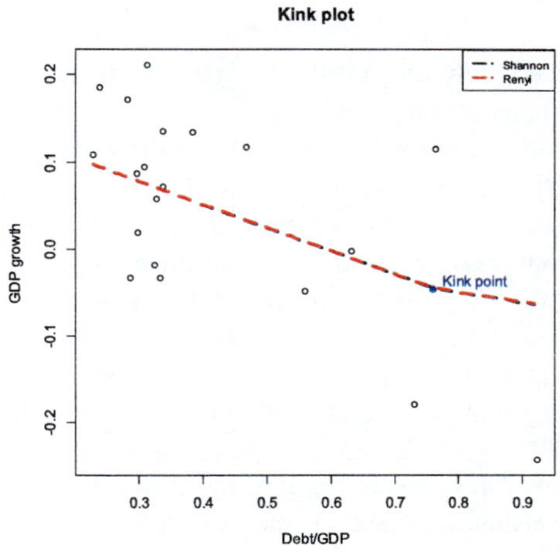

Fig. 2 Kink plot

In this application study, the study follows the suggestion from Golan et al. [3] in which suggested using the 3σ-rule of Pukelsheim [13] for setting the error bounds, where σ represents the standard deviation of error. Afterwards, the study conducted the entropy-ratio test presented in Sect. 3.2. The result is provided in Table 4 and we can see that the result shows decisive evidence against the null hypothesis of linear regression (H_{linear}) and the alternative hypothesis of Kink regression is therefore accepted in this application study. We can say that there exists a nonlinear relationship between GDP growth and Debt/GDP ratio, in the case of Thailand. In other words, the Kink regression model is preferred in this application and the kink specification should be incorporated into the model. In addition, the fitted kink regression line and there exists a significant kink point are shown in Fig. 2 and demonstrate a steep negative slope for low Debt/GDP ratio with a kink point (γ_1), and switching to low negative slope for Debt/GDP ratio.

Table 5 presents the parameter estimation for Kink regression model. The study fits two model using estimators described in Sect. 3.2. To compare then, the study considers the RMSE. According to Table 5, we can see that the RMSE of the kink regression model estimated by Shannon is equal to Renyi. According to these mea-

Table 5 Coefficients (standard errors) from kink regression (Example 1)

Parameter	Shannon GME	Renyi GME
α	−0.0442 (0.0592)	−0.0440 (0.0593)
β^-	−0.2659 (0.1476)	−0.2657 (0.1477)
β^+	−0.1158 (0.7195)	−0.1145 (0.7197)
γ	0.7607	0.7607
RMSE	0.0921	0.0921

sures, it can be concluded that the GME is a robust estimator because of the result of Renyi almost the same as Shannon.

Then, let investigate the effect of Debt/GDP on the economic growth, the result shows that Debt/GDP has a negative effect to the economic growth in both two regimes, namely regime 1 and 2. Surprisingly, the negative sign is obtained for both two regimes. Note that, regime 1 and 2 are interpreted as economic recession and expansion, respectively. This result does not correspond to the Keynesian microeconomics approaches which suggested that the Debt/GDP ratio should have a positive effect on economic growth during the economic recession (regime 1). This study expects that the Thai economy is quite stable along the last 20 years or since the financial crisis in 1997. Also, the Thai economic structure is different from those of advanced countries. Thus, it is difficult to boost the economy by expanding the Debt/GDP ratio. Nevertheless, the result shows a significant kink point, confirming the kink effect in the model.

4.2 *Example 2*

In this application study, we use GDP growth and Government expenditure on education of Thailand as the example data set. Our model specification can be written as

$$GDP_t = \alpha + \beta_1^-(EXP_t \leq \gamma_1)_- + \beta_1^+(EXP_t > \gamma_1)_+ + \varepsilon_t \tag{29}$$

The data covered 19 observations from 2001 to 2019 and it is collected from the Office of the national economic and social development council and Ministry of Education of Thailand. In this application, we try to investigate the performance of Renyi GME under a wide range of the real number of order α. Thus, the Akaike information criteria is used to find the best order α.

Figure 3 shows the AIC value for each α-order of Renyi GME. The lowest AIC is about -11.5163 located at $\alpha = 14$. From this figure, we can see that at the higher level of α-order, the lower AIC value is obtained and reach the lowest AIC at $\alpha = 14$. Then, the AIC value increase as the value of α-order increase. This result indicates that the higher-order of αdoes not guarantee the higher performance of the Renyi GME in estimating the kink regression model. Comparing o the Shannon GME, we consider the Renyi GME with $\alpha = 1$ and the result shows that the AIC value of Renyi GME with $\alpha = 14$ is lower than Renyi GME with $\alpha = 1$, indicating the

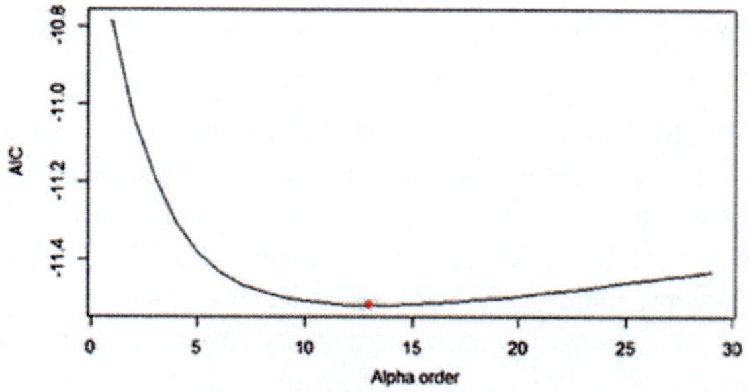

Fig. 3 AIC value with different values of order α

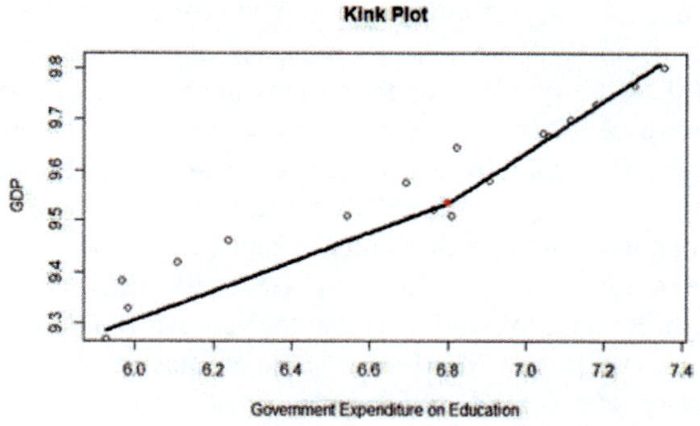

Fig. 4 Kink plot (order α)

Table 6 Coefficients (standard errors) from kink regression (Example 2)

Parameters	Renyi GME
α	9.5318 (0.0305)
β^-	0.2816 (0.0541)
β^+	0.5058 (0.1064)
γ	6.7979
AIC	−11.5164

higher performance of Renyi GME compared to Shannon GME. The slope of kink regression estimated from Renyi GME with $\alpha = 14$ is illustrated in Fig. 4.

Table 6 shows the result from Renyi GME at $\alpha = 14$. We can see that government expenditure on education has a positive effect on economic growth in both two regimes. It means that, if government expenditure on education increase 1%, GDP growth will increase 0.2816 in economic recession and GDP growth will increase 0.5058 in economic expansion.

5 Conclusion

In this study, we introduce a Renyi Generalized Entropy estimator to solve the ill-posed or the limited data problems in the kink regression model. Our work is similar to the studies of Sriboonchitta et al. [15] and Tarkhamtham et al. [16]. However, we differ those studies by replacing either Shannon and Tsallis entropy measure with the Renyi entropy measure. We attempt to fill the gap of applying Renyi GME to estimate the kink regression model. It proposes a more flexible as the order α is included to nest entropy measure. GME employs this high order entropy as the objective function in an estimation process of the kink regression model which have not been studied in earlier literature.

We then demonstrate the performance of Renyi GME through Monte Carlo simulation study. Two experiments are proposed in this simulation study. First, we examine the Bias and MSE of Renyi GME under various the number of support point and range of support space. Then, we compare the performance of Renyi entropy and Shannon entropy with the various distribution. The results show that if we widen the range of support space, the Bias and MSE are not lower. However, if we increase the number of supports points the Bias and MSE are lower. Then, when we compare the performance of Shannon GME and Renyi GME. We find that Shannon GME is slightly better than Renyi GME and finds that the value of Bias and MSE of the two models are not much different. The lowest value of Bias and MSE are obtained in the case of the error term is generated from a normal distribution. Finally, we compare Renyi GME and Shannon GME using the real data. The result shows the same performance as these two estimators. Therefore, we can conclude that the Renyi measure is an alternative entropy measure for GME estimation under the kink regression context. Besides, we also find evidence that the higher-order of α does not guarantee the higher performance of the Renyi-GME in estimating the kink regression model. Thus, the order of α is needed to be carefully selected.

Acknowledgements This research work was supported by Faculty of Economics, Chiang Mai University.

References

1. Ben-Bassat, M., & Raviv, J. (1978). Renyi's entropy and the probability of error. *IEEE Transactions on Information Theory, 24*(3), 324–331.
2. Golan, A., & Dose, V. (2001). A generalized information theoretical approach to tomographic reconstruction. *Journal of Physics A: Mathematical and General, 34*(7), 1271.
3. Golan, A., & Gzyl, H. (2006). *Ill conditioned linear estimation problems: An information theoretic approach with priors* (Working paper).
4. Golan, A., & Perloff, J. M. (2002). Comparison of maximum entropy and higher-order entropy estimators. *Journal of Econometrics, 107*(1–2), 195–211.
5. Golan, A., Judge, G., & Miller, D. (1996). *Maximum entropy econometrics: Robust estimation with limited data* (pp. 6–38). Wiley.

6. Goodman, S. N. (1999). Toward evidence-based medical statistics. 1: The P value fallacy. *Annals of Internal Medicine, 130*(12), 995–1004.
7. Goodman, S. N. (1999). Toward evidence-based medical statistics. 2: The Bayes factor. *Annals of Internal Medicine, 130*(12), 1005–1013.
8. Gzyl, H. (2003). Maximum entropy in the mean: A useful tool for constrained linear problems. *AIP Conference Proceedings, 659*(1), 361–385.
9. Hansen, B. E. (2017). Regression kink with an unknown threshold. *Journal of Business & Economic Statistics, 35*(2), 228–240.
10. Heckelei, T., Mittelhammer, R. C., & Jansson, T. (2008). A Bayesian alternative to generalized cross entropy solutions for underdetermined econometric models. *Food and Resource Economics*.
11. Jaynes, E. T. (1957). Information theory and statistical mechanics. *Physical Review, 106*(4), 620.
12. Pipitpojanakarn, V., Maneejuk, P., Yamaka, W., & Sriboonchitta, S. (2017). How does economic growth affect the well-being in Asia? *International Journal of Economic Research, 14*, 231–244.
13. Pukelsheim, F. (1994). The three sigma rule. *The American Statistician, 48*(2), 88–91.
14. Rényi, A. (1961). On measures of entropy and information. In *Proceedings of the Fourth Berkeley Symposium on Mathematical Statistics and Probability, Volume 1: Contributions to the Theory of Statistics*. The Regents of the University of California.
15. Sriboochitta, S., Yamaka, W., Maneejuk, P., & Pastpipatkul, P. (2017). A generalized information theoretical approach to nonlinear time series model. In *Robustness in econometrics* (pp. 333–348). Springer.
16. Tarkhamtham, P., Yamaka, W., & Sriboonchitta, S. (2018). The generalize maximum Tsallis entropy estimator in Kink regression model. *JPhCS, 1053*(1), 012103.
17. Tsallis, C. (1988). Possible generalization of Boltzmann-Gibbs statistics. *Journal of Statistical Physics, 52*(1–2), 479–487.
18. Wu, X. (2009). A weighted generalized maximum entropy estimator with a data-driven weight. *Entropy, 11*(4), 917–930.

How Does Economic Policy Uncertainty Affect Stock Market Returns: Evidence from a Markov-Switching Model with Mixture Distribution Regimes

Sukrit Thongkairat and Chatchai Khiewngamdee

Abstract In this study, the Global Economic Policy Uncertainty Index (GEPU) is considered the core factor affecting seven global stock markets (Japan, UK, USA, China, Singapore, Germany, and India). The Markov-switching(MS) model with mixed distribution regimes is employed to investigate the nonlinear effect of GEPU on these stock markets. To validate our results, 36 MS model specifications are considered for each stock market. The analysis can be illustrated into two parts. First, various model specifications are compared to find the best fit model to explain the EPU impacts. Second, we investigate the two-regime effects of EPU on stock market returns. The results show that each stock market return has characteristics of non-normal distribution regime. We also find that the GEPU negatively impacts the stock market in the bearish regime, but it turns to have a positive in the bullish regime.

Keywords Uncertainty index · Mixture distribution regimes · Stock markets

1 Introduction

The global economic interdependence has recently grown remarkably, resulting from the rapid and sophisticated advance in the communications technology and the economic policies and activities aiming at fostering economic growth through market liberalization like trade openness that facilitates each country to depend on others for necessary goods or services at a lower cost and allows for financial sector linkages. Many countries have exploited technological progress to make electronic finance easy, which becomes essential in all economic and financial sectors. The information and communication technology (ICT) evolution has helped support the increase

S. Thongkairat
Centre of Excellence in Econometrics, Chiang Mai University, Chiang Mai, Thailand

C. Khiewngamdee (✉)
Faculty of Economics, Chiang Mai University, Chiang Mai, Thailand
e-mail: chatchai.kh@cmu.ac.th

S. Sriboonchitta et al. (eds.), *Credible Asset Allocation, Optimal Transport Methods, and Related Topics*, Studies in Systems, Decision and Control 429,
https://doi.org/10.1007/978-3-030-97273-8_29

in economic productivity to attract investment from abroad and enabled people to access accurate and timely information. Due to the economic interdependence transmitted by identifiable channels, policymakers and investors are concerned about the economic uncertainty, e.g., sudden country-specific shocks, which can produce the spillover effect on another economy. For example, the outbreak of COVID-19, which started in China, caused the world's economic slowdown. The government responded with lockdowns and social distancing measures to prevent further spread of the virus. The measures have brought about significant uncertainty in economic activities and livelihoods. As a result of the virus outbreak, the oil price war of 2020 was triggered by Saudi-Arabia in response to Russia's rejection to reduce oil production to keep the price moderate, and this economic conflict led to a decrease in oil price. Kouam [1] reported that oil prices had dropped significantly since quarantine and social distancing measures have reduced manufacturing. The sudden change of policies and economic activities impacted not only the energy market but also the financial market. The pandemic has harmed the stock markets across all affected countries [2]. Furthermore, Baker et al. [3] and Onali [4] found that the number of COVID-19 related deaths affects stock market volatility.

A well-established global economic policy uncertainty index (GEPU) of Devis [5] reflects the global view's economic policy uncertainties. The Global EPU index is a GDP-weighted average of national EPU indices for 16 countries that account for two-thirds of the global output. Each country EPU index is constructed from own-country newspaper coverage frequency that incorporates a three-term condition, including economy, uncertainty, and policy-related matters. It is an appropriate index used to measure the global economic policy uncertainty, which is the concern of market participants. Many studies are interested in the inter-relation among the energy market, the stock market, and the economic policy uncertainty (EPU and Global EPU). For instance, Antonakakis et al. [6] and Degiannakis et al. [7] studied the relationship between oil prices and the uncertainty index's economic policy. Yu et al. [8] found Global EPU positively impacted the long-run correlation between oil price and stock return. Badshah et al. [9] investigated how EPU affects the relationship between stock and commodity and found a more substantial effect in the case of energy and industrial metals.

The U.S. economy has the largest economic size globally, and the U.S. dollar is the most widely used currency in international trade and financial transactions, so the change in U.S. policies and investor sentiment indicates a major role in driving global financial conditions. As shown in several studies, such as Kalu et al. [10], the U.S. monetary policy significantly affects African stock markets. Góes et al. [11] found the normalization had a statistically large impact on Brazil and Mexico's local bond yield. Yang and Hamori [12] confirmed that the U.S. interest rate harms the ASEAN stock markets during economic expansion periods. Colombo [13] and Kido [14] showed that the U.S. EPU index has a spillover effect on a currency market and industrial production in the Euro area, China, and Japan. Considering the U.S. energy market, the U.S. becomes the world's largest producer of natural gas and crude oil since the technological advances in the U.S. such as drilling and hydraulic techniques and the rise of the shale oil in the U.S. has enlarged its production capacity to surpass Russia and Saudi Arabia [15].

Meanwhile, the U.S. was the largest consumer of oil and natural gas globally [16] because its robust economic growth and industrial production tend to boost energy demand. As energy prices are influenced by the associated demand and supply, any change in the U.S. energy market will affect various kinds of energy throughout the country. Therefore, it is essential to investigate policy-related economic uncertainty changes in both energy and financial markets in the U.S.

Markov regime-switching models were commonly used in economics and finance, for example, Turner et al. [17], Gray [18], and Maheu and McCurdy [19]. Time-series data often contains a mixture of distributions as a result of unique data characteristics associated with different periods, for example, distinct consumer behavior during economic upturn and downturn, and several empirical studies of financial time series also show that most MS-type models indicate heavy tails in the data and skewness distribution. Many studies, particularly the GARCH-type studies, indicate a considerable and increasing number of candidate distributions. Haas et al., Haas [20, 21], and Ardia et al. [22] proposed that asymmetry was typically captured by authorizing specific means for the system. Nonetheless, data from the time series has various characteristics and includes a combination of distributions or populations. Maneejuk et al. [23] and Thongkairat et al. [24] thus proposed a Markov-Switching model of mixture distribution regimes that allows the distribution to take a different form in each regime, and they proved that this approach is superior to the traditional MS-AR model.

Therefore, this paper investigates the impact of economic policy uncertainty on the stock market return by implementing the basic Markov Switching model in the sense of nonlinear linear regression. This study uses the EPU index as a proxy of economic uncertainty with seven stock markets, namely Japan, UK, USA, China, Hong Kong, Singapore, Germany, and India. Although, this topic has already been in the literature [25–29], however, none of them considers the possible of nonlinear impact of global uncertainty on seven leading stock markets. Thus, we are the first attempt to investigate the two-regime impact of global uncertainty on seven leading stock markets.

The remainder of this paper, after the introduction, is divided into four sections. In Sect. 2, we briefly describe the approach used in this study. The descriptive statistics are provided in Sect. 3. The key findings of the empirical analysis are given and discussed in Sect. 4. Finally, Sect. 5 provides closing remarks and recommendations for future studies.

2 Methods and Procedures

2.1 A Markov-Switching (MS) Model

The two-regime MS model can be presented follows:

$$R_{it} = \alpha_{S_t} + \sum_{i=1}^{p} \beta_{p,S_t} Un_t + \varepsilon_{it,S_t}, \tag{1}$$

where R_{it} is the return on the stock market i at time t, Un_t is the index of uncertainty; α_{S_t} and β_{p,S_t} are a unknown regime dependent parameters. The vector of the error term ε_{it,S_t} is assumed to have a parametric distribution with mean zero and variance σ_{S_t}. The state variable is represented by S_t which follows a first-order Markov process, with transition probabilities

$$p_{ij} = P(S_t = i|S_{t-1} = j), \tag{2}$$

where p_{ij} is the probability of transition from regime i to regime j and $\sum_{j=1}^{2} p_{ij} = 1$. Thus, we can write the transition matrix as

$$Q = \begin{bmatrix} p_{11} & 1 - p_{22} \\ 1 - p_{11} & p_{22} \end{bmatrix}. \tag{3}$$

2.2 The Markov-Switching Model with Mixture Distribution Regimes

This study aims to explain the effect of economic policy uncertainty on seven global stock markets, assuming that different regimes in a Markov-switching model have different distributions [24], Maneejuk et al. [23]. The complete likelihood function is defined as

$$L(\theta_{S_t}|y_t, x_t) = \sum_{j=1}^{2} \left(\prod_{t=1}^{T} \left(f(\theta_{S_t=j}|y_t, x_t) \right) \left(\Pr(S_t = j|\theta_{S_t}) \right) \right), \tag{4}$$

where $f(\theta_{S_t=j}|y_t, x_t)$ is the density function, which can have different distributions across regimes. $\theta_{S_t=j} = \{\alpha_{S_t}, \beta_{S_t}, p_{11}, p_{22}, \sigma_{S_t}\}$ is a set of the state-dependent parameter of regime j, $j = 1, 2$. Note that six different distributions: normal, student-t, general error distribution (GED), skewed GED, skewed normal, and skewed student-t distributions are considered as the density function of our likelihood function Eq. (4).

2.3 *Hamilton's Filter*

Hamilton's filter, which Hamilton first proposed in 1989, applies to the estimate of the unobserved S_t variable. As it is unobserved, we never know which regime prevails [30]. According to Eq. (4), filter probability $\left(\Pr(S_t = j|\theta_{S_t})\right)$ is a necessary process. The calculated coefficient and variance are to be separated into two regimes. Hamilton's filter is calculated following Perlin [31], using the following algorithm.

1. Give an initial guess of transition probabilities Q in Eq. (3), say $p_{11} = p_{22} = 0.5$
2. Then update the transition probabilities of each regime with the past information, including the parameters in the system equation, θ_{S_t} for calculating the likelihood function in each state at time t, as shown in Eq. (4). Thus, the probability of each regime is updated by the following formula

$$\Pr(S_t = j|\theta_{S_t}) = \frac{f_j(y_t, x_t|S_t = j, \theta_{S_{t-1}})\Pr(S_t = j|\theta_{S_{t-1}})}{\sum_{j=1}^{h} f_j(y_t, x_t|S_t = j, \theta_{S_{t-1}})\Pr(S_t = j|\theta_{S_{t-1}})}, \quad (5)$$

3. Repeat steps 1 and 2 for $t = 1, \ldots, T$.

3 Data Description and Model Selection

This article collected data from the Thomson Reuters database and the Economic Policy Uncertainty website. The data are a monthly set of global economic policy uncertainty indices (Jan 2000 to May 2020) and stock market indexes representing the returns of seven global stock markets, including the NIKKEI 225 index, FTSE 100 index, S&P 500 index, SSE index, DAX index, SINGAPORE index and NIFTY 50 index. We convert all the data into a log-return. The descriptive statistics for the Global EPU (GEPU)index's log-return and seven stock market indexes are shown in Table 1. The mean value of all data series in the sample is very similar to null. Nevertheless, there are more differences between the EPU, SINGAPORE (SGP), and NIFTY than the others. NIKKEI, FTSE, S&P 500, SSE, DAX, and NIFTY series' log-returns indicate non-normality of negative skewness while EPU and SGP have positive skewness. Each had high values of kurtosis, though (>3). Additionally, the Jarque–Bera(J-B) test is considered to investigate the natural property of distribution.

According to Goodman [32] for the alternative hypothesis, Minimum Bayes Factor (MBF) values between 1 and 1/3 are considered weak evidence, 1/3–1/10 moderate evidence, 1/10–1/30 substantial evidence, 1/30–1/100 strong evidence, 1/100–1/300 very strong evidence, and <1/300 decisive evidence. For more detail, see Maneejuk

Table 1 Statistics of classification

	GEPU	NIKKEI	FTSE	S&P 500	SSE	SGP	DAX	NIFTY
Mean	0.01	0.00	0.00	0.00	0.00	0.01	0.00	0.01
Median	−0.02	0.01	0.00	0.01	0.01	0.00	0.01	0.01
Maximum	1.08	0.18	0.13	0.17	0.14	0.38	0.19	0.26
Minimum	−0.92	−0.26	−0.19	−0.20	−0.30	−0.31	−0.29	−0.30
Std. Dev	0.28	0.06	0.04	0.05	0.05	0.08	0.06	0.07
Skewness	0.52	−0.59	−0.80	−0.67	−0.80	0.57	−0.94	−0.45
Kurtosis	4.58	4.45	5.06	5.15	6.97	7.36	6.02	5.26
Jarque–Bera	36.42	35.99	69.81	65.79	186.99	206.99	129.56	60.52
Probability	0.00	0.00	0.00	0.00	0.00	0.00	0.00	0.00
ADF-test	−12.93[a]	−14.11[a]	−15.74[a]	−16.24[a]	−17.94[a]	−13.70[a]	−14.45[a]	−14.96[a]

Note [a] denotes the interpretation of MBF as decisive evidence for stationarity
Source Calculation

and Yamaka [33]. Consequently, the MBF values for the Jarque–Bera check are close to null. This means that the proof of normal distribution is decisive evidence. An Augmented Dickey-Fuller (ADF) test is also conducted to display a stationary data set. The results show that stationary effects are significant, as shown in Table 1.

4 Empirical Results

We analyzed whether economic policy uncertainty affects the return on the seven major stock markets. In doing so, we considered a two-regime MS model to investigate this impact and characterize a business cycle of the stock markets, namely the bearish and bullish regime. The findings can be broken down into two parts.

4.1 The Best-Fit Model for Each Market

In the first part, we consider a set of distributions to be an option for the distribution of errors. The most suitable distribution is then chosen according to the lowest Akaike Information Criterion (AIC). The six distributions are consisting of regular normal distribution (N), skew-normal distribution (SN), student-t distribution (T), skewed student-t distribution (ST), generalized error distribution (GED), and skewed general error distribution (SGED). The comparison of AIC values for all markets are reported in Tables 2, 3, 4, 5, 6, 7 and 8. The table's symbol means; the underlined is the minimum AIC value of the Markov-switching model with mixture distribution, and

Table 2 AIC of mixture MS models for Nikkei

Regime 2/	Regime 1					
Distribution	N	SN	T	ST	GED	SGED
N	65.34*	67.76	61.30	61.62	66.51	68.45
SN	67.84	70.64	75.21	73.94	70.95	77.34
T	58.75	74.02	82.39	84.43	61.79	75.64
ST	78.93	80.75	80.94	82.60	80.93	82.93
GED	66.51	70.94	60.27	64.56	69.42	71.37
SGED	68.45	71.05	77.12	74.47	71.82	72.28

Table 3 AIC of mixture MS models for FTSE

Regime 2/	Regime 1					
Distribution	N	SN	T	ST	GED	SGED
N	67.06*	68.29	78.43	71.97	69.43	67.68
SN	67.64	70.09	79.79	76.24	73.33	70.19
T	70.19	73.80	51.54	67.71	68.75	82.51
ST	74.90	74.55	83.58	76.52	72.39	80.53
GED	65.75	69.10	71.06	73.81	68.07	69.57
SGED	73.59	71.85	81.65	76.94	75.74	70.10

Table 4 AIC of mixture MS models for S&P 500

Regime 2/	Regime 1					
Distribution	N	SN	T	ST	GED	SGED
N	68.57*	74.53	70.64	86.32	86.10	83.37
SN	62.55	73.05	76.86	75.47	67.24	70.14
T	88.79	79.51	79.51	81.44	71.02	75.63
ST	75.08	74.02	67.75	73.97	77.08	74.35
GED	80.80	82.94	73.06	81.73	61.27	92.18
SGED	60.79	69.98	74.78	76.18	67.75	69.84

Table 5 AIC of mixture MS models for SSE

Regime 2/	Regime 1					
Distribution	N	SN	T	ST	GED	SGED
N	74.87*	75.09	72.23	76.98	73.96	75.12
SN	75.09	73.11	75.81	78.76	72.11	72.80
T	76.66	76.86	58.24	68.55	85.14	80.21
ST	86.99	77.69	83.93	81.74	66.74	77.45
GED	73.96	72.11	76.09	75.21	72.48	72.58
SGED	75.12	74.68	69.33	80.96	72.58	74.48

Table 6 AIC of mixture MS models for SINGAPORE

Regime 2/	Regime 1					
Distribution	N	SN	T	ST	GED	SGED
N	66.83*	69.15	73.23	58.35	68.01	69.58
SN	59.96	64.58	72.20	60.58	62.30	66.07
T	65.12	72.67	80.03	70.00	65.57	74.81
ST	86.58	75.37	79.65	68.16	69.01	70.37
GED	68.01	65.64	74.87	69.29	76.23	66.78
SGED	69.79	66.83	78.59	76.10	64.56	70.07

Table 7 AIC of mixture MS models for DAX

Regime 2/	Regime 1					
Distribution	N	SN	T	ST	GED	SGED
N	43.72*	28.39	47.87	44.86	46.56	45.51
SN	34.86	35.65	107.86	54.92	61.19	47.02
T	54.37	35.97	47.74	46.94	45.96	44.04
ST	36.98	53.32	55.13	55.30	40.03	61.11
GED	45.94	53.85	42.01	39.43	41.72	57.31
SGED	43.11	55.58	62.42	50.46	42.22	55.93

Table 8 AIC of mixture MS models for NIFTY

Regime 2/	Regime 1					
Distribution	N	SN	T	ST	GED	SGED
N	72.58*	73.05	75.22	72.64	72.45	74.25
SN	73.73	73.26	71.30	77.42	71.24	74.93
T	75.46	75.52	65.04	78.52	83.70	77.76
ST	79.27	76.71	65.01	78.94	68.23	80.29
GED	71.52	75.90	75.93	77.61	74.55	72.46
SGED	74.25	74.18	77.84	78.69	72.46	76.01

"*" is the AIC value based on the normal distribution and refers to the conventional MS model.

The results indicate that for Nikkei, the Markov-switching model of N-T distribution regimes is the optimal model, in which the errors in regime 1 are normally distributed, and those in regime 2 have Student-t distribution, with its AIC value (58.75) much lower than that of the original MS-AR model (65.34). For FTSE, the optimal model is the one that has student-t distribution in both regimes with the lowest AIC value of 51.54, which is lower than that 67.06 of the original MS-AR model. For S&P 500, the Markov-switching mixture distribution model with

N-SGED distribution regimes is the optimal model, where regime 1 has a normal distribution and regime 2 has the skew- generalized distribution of errors, with the corresponding AIC value (60.79) lower than that of the standard MS-AR model (68.57). For SSE, the optimal model also has student-t distribution in both regimes, with the AIC value of 58.24 much lower than that of the original MS-AR model (74.87). For SGP, the Markov-switching model with ST-N distribution regimes is the optimal model, in which regime 1 has skew student-t distribution and regime 2 has a normal distribution, which has the AIC value (58.35) lower than that of the original MS-AR model (66.83). For DAX, the Markov-switching model with N-SN distribution regimes is the optimal model, with regime 1 having normal distribution and regime 2 having skew-normal distribution and the corresponding AIC value (28.39) lower than that of the original MS-AR model (43.72). For NIFTY, the optimal model is the Markov-switching model with T-ST distribution regimes where regime 1 has student-t distribution, and regime 2 has skew student-t distribution. The results suggest that the AIC value of the optimal Markov-switching mixture distribution model (65.01) is lower than that of the original MS-AR model (72.58).

This indicates that the mixture regime model is more robust and better than the traditional model according to AIC values. The result implies that the data would continue to be distributed differently across various economic regimes. Moreover, the experiment shows that the return of the main stock-market index over regime 1 has more massive tails than regime 2, which means this method is more likely to yield values that fall far from its mean.

4.2 *Empirical Results*

In this section, the unknown parameters are estimated by the best fit model presented in Sect. 4.1. We define the regime 1 as a bearish market and regime 2 as a bullish market as the intercept term of regime 1is less than regime 2. Table 9 shows the estimated parameters obtained from the mixture MS model in each market. Our results show that the economic policy uncertainty has a weak effect on NIKKEI for both the bullish and bearish markets. However, the results present strong evidence in FTSE and DAX. This indicates that English and German stock market returns respond to global economic policy uncertainty during both the bull and bear market regimes. Additionally, we find strong evidence that the market return in S.P.500 and SINGAPORE responds to the change in economic policy uncertainty during the bull market but finds only weak evidence during the bear market regime. On the contrary, economic policy uncertainty tends to affect SSE and NIFTY only in the bear market regime according to the parameter estimation.

Considering the size of the impact of economic policy uncertainty on stock market returns, we found that global economic policy uncertainty has a higher impact on US and Europe stock market returns than Asian stock market returns. This implies that Asian stock market returns are less sensitive to economic policy uncertainty than in the US's and Europe's. Besides, the results also show that, in FTSE, SSE,

Table 9 Estimated parameters from the mixture MS-AR model

	NIKKEI	FTSE	S.P.500	SSE	SINGAPORE	DAX	NIFTY
	Regime 1 (Bearish Market)						
	N	T	SGED	T	ST	SN	ST
Intercept	−0.06[a]	−0.03[e]	−0.02[a]	1.06[e]	−0.59[b]	0.02[a]	−0.25 [e]
Coefficient	−0.12[b]	−2.99[e]	0.49[b]	−0.38[e]	−3.42[b]	1.36[e]	−1.87 [e]
Sigma	0.21[e]	0.01[a]	0.22[e]	0.00[e]	0.39[c]	0.25[e]	0.03 [e]
Degree of Freedom		2.51[a]	3.20[e]	2.12[e]	2.52[b]		2.47 [e]
Skew			1.00[e]		0.35[c]	1.31[e]	0.29[b]
Duration	2.7	3.23	5.26	1.01	1.01	100	2.27
	Regime 2 (Bullish Market)						
	T	T	N	T	N	N	T
Intercept	0.22[a]	0.00[b]	0.23[d]	2.00[c]	0.01[a]	0.77 [e]	−0.01[b]
Coefficient	3.53[a]	0.96[c]	8.26[d]	−0.22[a]	0.65[c]	2.98 [e]	0.52[b]
Sigma	0.71[c]	0.28[e]	0.34 [e]	0.28 [e]	0.28 [e]	0.03[b]	0.35 [e]
Degree of Freedom	2.10[b]	7.22[c]		17.19 [e]			3.07 [e]
Skew							
Duration	1.01	100	1.01	100	100	1.01	100
Transition matrix							
P1	0.63	0.69	0.81	0.01	0.01	0.99	0.56
P2	0.01	0.99	0.01	0.99	0.99	0.01	0.99

Note
[a] Denotes the interpretation of MBF as weak evidence
[b] Denotes the interpretation of MBF as moderate evidence
[c] Denotes the interpretation of MBF as substantial evidence
[d] Denotes the interpretation of MBF as strong evidence
[e] Denotes the interpretation of MBF as decisive evidence

SINGAPORE, and NIFTY, the impacts during the bear market are higher than during the bull market. On the other hand, in NIKKEI, S.P.500, and DAX, the impacts of economic policy uncertainty are high during the bull market. Unexpectedly, considering the direction of the impact, there is no obvious conclusion how economic policy uncertainty affects stock market returns; that is, the results show that, during the bull market regime, we found that economic policy uncertainty has a positive effect on most stock market returns except for SSE. Additionally, during the bear market regime, we found that economic policy uncertainty harms most stock market returns except for S.P.500 and DAX.

In addition, the transition matrix will allow tracking the asymmetry overtime of the business cycle. The consequence of the model proposed is much clearer. We find that in the case of NIKKEI, the bull market has a duration of approximately 2.7 months whilst the bear market has only 1.59 months, approximately. Although 63% is expected to stay in regime 1 (Bear Market), the chance of switching from regime 1 to regime 2 (Bull Market) is 37%. Similarly, the probability of moving from Regime 2 to Regime 1 is 1%, When in regime 2 the chance to stay is 99%. In other cases, such as FTSE, S&P 500, SSE, SINGAPORE, DAX, and NIFTY, the bear market has a duration of approximately 3.23, 5.26, 1.01, 1.01, 100, and 2.27 months, respectively, whilst the bull market has 100, 1.01, 100, 100, 1.01, and 100 months, approximately. The chance of remaining in regime 1 (Bear Market) is 69%, 81%, 1%, 1%, 99%, and 56% respectively, but the chance of switching from regime 1 to regime 2 (Bull Market) is 31%, 19%, 99%, 99%, 1%, and 44% respectively. Similarly, the chance of moving from regime 2 to regime 1 is 99%, 1%, 99%, 99%, 1%, and 99% respectively, while the chance of remaining in the regime 2 is 1%, 99%, 1%, 1%, 99%, and 1% respectively.

5 Conclusion

This paper aims to examine the effect of global economic policy uncertainty on several stock markets' returns. We apply a Markov-switching model with mixture distribution regimes to better capture the financial time series behavior and provide more reliable results. We focus on seven major stock markets worldwide, namely, NIKKEI 225 index, FTSE 100 index, S&P 500 index, SSE index, SINGAPORE index, DAX index, and NIFTY 50 index. To achieve our objective, we start with stationary testing of all variables. Then, we employ our proposed Markov Switching model, which allows us to mix a different distribution in each scheme and select the best fitting model among candidates using the Akaike information criterion (AIC). Finally, we estimate the parameters.

The empirical results present that, according to the best-fitting model of mixture distribution regimes, each stock market return has characteristics of non-normal distribution. This indicates that our proposed model is more suitable than the conventional model. Moreover, based on the minimum Bayes' factor, we found strong evidence that global economic policy uncertainty affects stock market returns during both bull and bear market regimes, except for Nikkei. We also found that the stock market return in China, Singapore, and India are less sensitive to economic policy uncertainty than the US's and Europe's. However, the results are mixed, considering the sign of the estimated parameters. In other words, the results present that economic policy uncertainty has a positive impact on stock market return in the US and German stock markets but shows a negative sign in Chinese and Indian stock markets.

Acknowledgements The authors are grateful to the Centre of Excellence in Econometrics, Chiang Mai University, for financial support.

References

1. Kouam, H. (2020). In the face of spillovers; Chinese ICT trade with the US, innovation and economic growth. *Chinese ICT Trade with the US, Innovation and Economic Growth* (June 29, 2020).
2. Liu, Y., Lee, J. M., & Lee, C. (2020). The challenges and opportunities of a global health crisis: the management and business implications of COVID-19 from an Asian perspective. *Asian Business & Management, 1*.
3. Baker, S. R., Bloom, N., Davis, S. J., Kost, K. J., Sammon, M. C., & Viratyosin, T. (2020). The unprecedented stock market impact of COVID-19 (No. w26945). *National Bureau of Economic Research.*
4. Onali, E. (2020). *Covid-19 and stock market volatility*. Available at SSRN 3571453.
5. Davis, S. J. (2016). *An index of global economic policy uncertainty (No. w22740)*. National Bureau of Economic Research.
6. Antonakakis, N., Chatziantoniou, I., & Filis, G. (2014). Dynamic spillovers of oil price shocks and economic policy uncertainty. *Energy Economics, 44*, 433–447.
7. Degiannakis, S., Filis, G., & Panagiotakopoulou, S. (2018). Oil Price Shocks and Uncertainty: How stable is their relationship over time? *Economic Modelling, 72*, 42–53.
8. Yu, H., Fang, L., & Sun, B. (2018). The role of global economic policy uncertainty in long-run volatilities and correlations of US industry-level stock returns and crude oil. *PLoS ONE, 13*(2), e0192305.
9. Badshah, I., Demirer, R., & Suleman, M. T. (2019). The effect of economic policy uncertainty on stock-commodity correlations and its implications on optimal hedging. *Energy Economics, 84*, 104553.
10. Kalu, E., Okoyeuzu, C., Ukemenam, A., & Ujunwa, A. (2020). Spillover effects of the US monetary policy normalization on African stock markets. *Journal of Economics and Development*.
11. Góes, C., Kamil, H., De Imus, P., Garcia-Escribano, M. M., Perrelli, M. R., Roache, M. S. K., & Zook, J. (2017). *Spillovers from US monetary policy normalization on Brazil and Mexico's Sovereign bond yields*. International Monetary Fund.
12. Yang, L., & Hamori, S. (2014). Spillover effect of US monetary policy to ASEAN stock markets: Evidence from Indonesia, Singapore, and Thailand. *Pacific-Basin Finance Journal, 26*, 145–155.
13. Colombo, V. (2013). Economic policy uncertainty in the US: Does it matter for the Euro area? *Economics Letters, 121*(1), 39–42.
14. Kido, Y. (2018). The transmission of US economic policy uncertainty shocks to Asian and global financial markets. *The North American Journal of Economics and Finance, 46*, 222–231.
15. U.S. Energy Information Administration. (2019). https://www.eia.gov/
16. CIA Central Intelligence Agency. (2017). *The world factbook 2017*. CIA.
17. Turner, C. M., Startz, R., & Nelson, C. R. (1989). A Markov model of heteroskedasticity, risk, and learning in the stock market. *Journal of Financial Economics, 25*(1), 3–22.
18. Gray, S. F. (1996). Modeling the conditional distribution of interest rates as a regime-switching process. *Journal of Financial Economics, 42*(1), 27–62.
19. Maheu, J. M., & McCurdy, T. H. (2000). Identifying bull and bear markets in stock returns. *Journal of Business & Economic Statistics, 18*(1), 100–112.
20. Haas, M., Mittnik, S., & Paolella, M. S. (2004). A new approach to Markov-switching GARCH models. *Journal of Financial Econometrics, 2*(4), 493–530.
21. Haas, M. (2010). Skew-normal mixture and Markov-switching GARCH processes. *Studies in Nonlinear Dynamics and Econometrics, 14*(4), 1–56.
22. Ardia, D., Bluteau, K., Boudt, K., Catania, L., & Trottier, D. A. (2016). Denis-Alexandre, Markov-switching GARCH models in R: The MSGARCH package.
23. Maneejuk, P., Yamaka, W., & Sriboonchitta, S. (2018). A Markov-switching model with mixture distribution regimes. In *International Symposium on Integrated Uncertainty in Knowledge Modelling and Decision Making* (pp. 312–323). Springer, Cham.

24. Thongkairat, S., Yamaka, W., & Sriboonchitta, S. (2019). Economic policy uncertainty effect on precious metal markets: A Markov-switching model with mixture distribution regimes. *Thai Journal of Mathematics*, 77–92.
25. Sum, V. (2012). Economic policy uncertainty and stock market performance: Evidence from the European Union, Croatia, Norway, Russia, Switzerland, Turkey and Ukraine. *Journal of Money, Investment and Banking, 25*, 99–104.
26. Pastor, L., & Veronesi, P. (2012). Uncertainty about government policy and stock prices. *The Journal of Finance, 67*(4), 1219–1264.
27. Bansal, R., & Yaron, A. (2004). Risks for the long run: A potential resolution of asset pricing puzzles. *The Journal of Finance, 59*(4), 1481–1509.
28. Bansal, R., Khatchatrian, V., & Yaron, A. (2005). Interpretable asset markets? *European Economic Review, 49*(3), 531–560.
29. Dzielinski, M. (2012). Measuring economic uncertainty and its impact on the stock market. *Finance Research Letters, 9*(3), 167–175.
30. Kole, E. (2010). Regime switching models: An example for a stock market index. *Unpublished manuscript, Econometric Institute, Erasmus School of Economics, Erasmus University Rotterdam April.*
31. Perlin, M. (2015). *MS_Regress-the Matlab package for markov regime switching models.* Available at SSRN 1714016.
32. Goodman, S. N. (1999). Toward evidence-based medical statistics. 2: The Bayes factor. *Annals of Internal Medicine, 130*(12), 1005–1013.
33. Maneejuk, P., & Yamaka, W. (2020). Significance test for linear regression: how to test without P-values? *Journal of Applied Statistics*, 1–19.

Analyzing the Influence of Transportation and Macroeconomic Determinants on Chinese Inbound Tourism: A Markov Switching Model Using Ridge and Lasso Estimation

Xuefeng Zhang and Wenbo Zhang

Abstract This empirical study examines the influence of transportation and Macroeconomic determinants on Chinese inbound tourism. The purposes of this study are: (1) To measure the cycle of Chinese inbound tourism and find the high-growth stage and low-growth stage; (2) To investigate the impact of transportation and Macroeconomic factors on Chinese inbound tourism. All data used in this paper are secondary annual data from 1995 to 2018. The tourism demand equations are estimated by the Markov switching model based on the Ridge and Lasso estimation. The innovations of this paper are: (1) introducing the transportation factors into tourism demand; (2) because the data samples used in this paper are small, it is impossible to use the traditional Markov switching model for regression. Therefore, innovation is made in the model, namely the Markov switching model based on the Ridge and Lasso estimation method. The results are as follows: (1) the nonlinear model is more suitable for analyzing the tourism business cycle than the linear model. (2) Feature selection screened out the main factors that affect China's inbound tourism. They are lnHWL (China highway length), lnIAPT (China international air passenger traffic), lnNLR (China navigable length of the river), lnADU (China aircraft daily utilization and UNPR (China's unemployment rate). (3) For China's inbound tourism business cycle, the possibility of being in a high-growth stage is large, with long persistence and stability. The possibility of being in a low-growth stage is small, with short persistence and variability.

Keywords Feature selection · Markov switching model · Ridge and Lasso estimation · Tourism demand · Transportation

1 Introduction

As one of the world's largest economic sectors, Travel and Tourism create jobs, drive exports, and generate prosperity across the world. According to the World Travel and

X. Zhang (✉) · W. Zhang
Faculty of Economics, Chiang Mai University, Chiang Mai 50200, Thailand
e-mail: xuefengzhang999@gmail.com

S. Sriboonchitta et al. (eds.), *Credible Asset Allocation, Optimal Transport Methods, and Related Topics*, Studies in Systems, Decision and Control 429,
https://doi.org/10.1007/978-3-030-97273-8_30

Tourism Council's (WTTC), the tourism sector accounted for 10.4% of global GDP and 319 million jobs or 10% of total employment in 2018. Domestic tourism, which represented 71.2% of all tourism spending in 2018 and had the strongest growth in developing nations, continues to support opportunities by spreading development and regional economic benefits and building national pride. In terms of the growth in the country-by-country share of global travel and tourism contribution in 2018, two countries dominated; 25% came from China, while the United States accounted for another 12%. Other major contributors include India, with a 6% share, while Turkey, France and Japan contributed 4% of the growth in this sector in 2018.

Since the implementation of the reform and opening policy in 1978, China's economy has continued to grow, and China's tourism industry has made great progress in recent years. Asia is China's most important source market for inbound tourists. In 2018, in the foreign tourist market, the top ten source countries for visiting China are Myanmar, Vietnam, South Korea, Japan, Russia, the United States, Mongolia, Malaysia, the Philippines, and India. Taken together, the structure of the inbound tourist market has shown an optimization trend, and the countries along the "Belt and Road" have continued to increase their activity in the inbound tourism market.

There are many factors affecting Chinese inbound tourism. The development of inbound tourism may be affected by factors such as international exchange rates, national income, and consumption levels. At the same time, some unexpected or major events may also promote or hinder inbound tourism. This paper focuses on analyzing the impact of transportation and macro factors on Chinese inbound tourism. Van and Shimizu have pointed out that to correctly evaluate the impact of transportation on tourism, the key is to incorporate transportation accessibility into the model [1]. Transportation infrastructure performs an extraordinary role in the development of the tourism industry. Khadaroo and Seetanah [2] found that transport infrastructure has an important impact on tourism demand, which is mainly reflected in the distance to the destination, the relative price of goods and the income of tourists. Guirao and Campa [3], taking Spain as an example, found that high-speed rail plays an indispensable role in Spain's tourism industry. Wang [4] discussed the position of the transport sector to recognize the dynamic development of tourism. Taking China as an example, Wang found that the growth of the macroeconomy and transportation sector has greatly promoted the development of China's tourism industry. Transportation directly connects supply (tourist destination) and demand (tourist). Transportation provides transportation convenience for tourism and increase the commercial activities of tourist destinations and improve the living standards of destinations. Research shows that if a region has a well-developed transportation infrastructure, it can attract tourists and tourism investment companies because it can reduce transportation costs. Button and Taylor's [5] research found that the development of metropolis is inseparable from air transportation. Air transport and tourism are interconnected. In some cases, tourism is the driving force to stimulate the change in air transport, and it is also an important driving force for the development of air transport. On the other hand, air transport has expanded new destinations and forms of tourism, such as inbound and outbound tourism. With the development of the times, there is a new business model, charter. Airlines in a region need to provide a variety of airline plans and clear corporate policies to boost the improvement of tourism in the region. In recent years, global tourism has developed rapidly, and the inbound and outbound tourism of destinations has also been positively affected, which is mainly due to the increasingly perfect global air transport network [6].

From the macroeconomic factor's perspective, tourism macroeconomic benefit is a part of social benefit. Besides the direct economic gains of tourism enterprises, the macroeconomic gains of tourism also include the indirect economic gains of the development of tourism to drive the development of other related industries. This paper uses commonly used macroeconomic indicators to measure macro factors, which are: China's GDP per capita growth rate, unemployment rate, inflation rate, and official exchange rate (USD / CNY).

To sum up, the growth rate of China's inbound tourism in 2018 was 1.2%, while China's outbound tourism is growing at 14.7%. Following this trend, the scissors gap between outbound and inbound tourists will continue to widen. From an economic point of view, the number of outbound tourists far exceeds the number of inbound tourists, which will lead to a continuous deficit in tourism trade and affect the balance of payments. Due to the relatively slow development of inbound tourism in China. In this regard, the China National Immigration Administration has issued a series of inbound tourism policies in recent years: suspension of port visas, Shanghai cruise exemption visa, Hainan entry visa exemption, 24/72/144-h transit visa exemption, and foreigners from Hong Kong and Macau to enter Guangdong Visa-free for 144 h, visa-free for ASEAN tourist groups to enter Guangxi, etc. The foremost contribution of this paper is to estimate the Markov switching model based on Ridge and Lasso estimation. On the other hand, most of the existing studies use the total number of arrivals or total inbound tourism revenue to measure the business cycle of the tourism industry of one country. This paper collects the top 10 countries' tourist arrivals to China and aims to find out the tourism business cycle of each source country.

This paper aims to portray the complex and dynamic process of China's inbound tourism economic cycle in multiple stages and to revitalize inbound tourism in China. Specifically, this study intends to measure the cycle of Chinese inbound tourism and find the high-growth stage and low-growth stage of inbound tourism.

2 Methodology

2.1 Markov Switching Model

Before estimating the Markov switching model, the data must be stationary, and the stationary data can be switched between M states. There are corresponding theoretical supports to distinguish the number of international tourist arrivals into high tourist arrivals and low tourist arrivals. The different states are described by the state variable S_t. Since we cannot directly observe the state variables, they are called hidden variables [7, 8]. The state variable S_t can take discrete values from 1 to M [9]. The observed variable x_t is assumed to have been drawn from a probability distribution whose parameters are conditional upon the state prevailing at time t. We also assume that given the state sequence, x_t is independent of previous observations. Although in principle there is no restriction on the type of distributions used to describe x_t, we will here assume that x_t is normally distributed with mean and variance that are specific to the state. Hence, given S_t, we could generate x_t as

$$x_t = \mu_{S_t} + \sigma_{S_t}\varepsilon_t, t = 1, \ldots, T \tag{1}$$

where $\varepsilon_t \sim N(0, 1)$ is a standard normal distribution variable.

Assume that the unobserved state variable S_t follows an M-state Markov chain with elements described by the transition probability matrix $\mathbf{P}$

$$p_{ij} = \Pr\{S_t = j|S_{t-1} = i\}, i, j = 1, \ldots, M \tag{2}$$

The persistence of x_t is completely inherited from the characteristics of this Markov chain. Therefore, the different characteristics of the sequence generated by the MS model are closely related to the transition probability of the Markov chain.

In summary, the MS model is described by a transition probability ($M \times M$) matrix P and a normal distribution of $2M$ parameters ($\mu_1, \sigma_1, \mu_2, \sigma_2, \ldots, \mu_M, \sigma_M$). The model considered here extends the two-state model considered by Thyer and Kuczera [10].

2.2 Model Properties

In this section, we examine the statistical and random properties of the above MS model. We are particularly interested in the mean, variance, and skewness coefficients of x_t, their statistical distributions over time, and their analytical expressions of autocorrelation functions. These are the attributes commonly used to judge whether a time series model is appropriate.

Many of the following expressions involve the stationary probability $\pi_i, i = 1, \ldots, M$, which is the probability that the Markov chain will be in state i at some time in the future when the initial state effect disappears. For an infinitely long sequence, π_i can be interpreted as a part of the time the chain is in state i. Given the transition probability matrix $\mathbf{P}$, the stationary probabilities $\boldsymbol{\pi} = (\pi_1, \pi_2, \ldots, \pi_M)'$ can be obtained by solving $\mathbf{P}'\pi = \boldsymbol{\pi}$ subject to $\sum_{i=1}^{M} \pi_i = 1$.

Given the stationary probability, the marginal distribution of x_t becomes a mixture of M normal distributions.

$$f(x_t) = \sum_{i=1}^{M} f(x_t|S_t = i)\boldsymbol{Pr}\{S_t = i\}$$

$$= \sum_{i=1}^{M} \pi_i f(x_t | S_t = i) \tag{3}$$

where

$$f(x_t | S_t = i) = \frac{1}{\sqrt{2\pi}\sigma_i} \exp\left\{-\frac{(x_t - \mu_i)^2}{2\sigma_i^2}\right\}. \tag{4}$$

2.3 Model Estimation

In recent years, there has been significant interest in MS models, and this has among other things led to the development of efficient algorithms for parameter estimation. Because of the relatively high number of parameters in MS models, the method of moments [11] is not likely to produce good estimates. The method of maximum likelihood holds more promise. Here, we briefly outline a possible implementation of the maximum likelihood method for MS models.

The following notation will be needed. Let $\mathbf{X}_T = (x_1, x_2, \ldots, x_T)'$ be the vector of observations and $\mathbf{S}_T = (s_1, s_2, \ldots, s_T)'$ the associated (unobserved) state variables. The MS model in **Eq. (1) is described by the parameters $\boldsymbol{\mu} = (\mu_1, \mu_2, \ldots, \mu_M)'$ and $\boldsymbol{\sigma} = (\sigma_1, \sigma_2, \ldots, \sigma_M)'$ of the normal densities associated with the M states and the transition probability matrix, $\mathbf{P}$. Since the rows in $\mathbf{P}$ must sum to 1, not all elements of $\mathbf{P}$ need to be considered. For example, a two-state Markov chain is entirely described by two probabilities.

The algorithm described below requires an assumption about the distribution of the first state variable, S_1. We will assume that the initial state is drawn from a discrete probability distribution $\boldsymbol{\varrho} = \varrho_1, \varrho_2, \ldots \varrho_M$, where $\varrho_i = = \mathbf{Pr}\{S_1 = i\}$. Hence the set of parameters to be estimated is $= \{\boldsymbol{\mu}, \boldsymbol{\sigma}, \mathbf{P}, \varrho\}$.

Given $\boldsymbol{\theta}$, the joint density of $(\mathbf{X}_T, \mathbf{S}_T)$ can be expressed as

$$p(\boldsymbol{X}_T, \boldsymbol{S}_T | \boldsymbol{\theta}) = f(\boldsymbol{X}_T | \boldsymbol{S}_T, \boldsymbol{\theta}) Pr\{\boldsymbol{S}_T | \boldsymbol{\theta}\} \tag{5}$$

where the two terms on the right-hand side are

$$f(\boldsymbol{X}_T | \boldsymbol{S}_T, \boldsymbol{\theta}) = \prod_{t=1}^{T} f\left(x_t | \mu_{s_t}, \sigma_{s_t}\right), \tag{6}$$

$$Pr\{\boldsymbol{S}_T | \boldsymbol{\theta}\} = \varrho_{s_1} \prod_{t=1}^{T-1} Pr\{(s_{t+1} | s_t) | \boldsymbol{P}\} \tag{7}$$

Note that we specifically have assumed that $f\left(x_t|\mu_{s_t}, \sigma_{s_t}\right)$ normal distributions. If both $\mathbf{X}_T$ and $\mathbf{S}_T$ had been observed, one could have computed the complete data likelihood function is:

$$L(\boldsymbol{\theta}|\boldsymbol{X}_T, \boldsymbol{S}_T) = p(\boldsymbol{X}_T, \boldsymbol{S}_T|\boldsymbol{\theta}) = \varrho_{s_1} \prod_{t=1}^{T-1} Pr\{(s_{t+1}|s_t)|\boldsymbol{P}\} \prod_{t=1}^{T} f\left(x_t|\mu_{s_t}, \sigma_{s_t}\right) \quad (8)$$

Since the likelihood function cannot be conditioned on the unobserved state variables, it is necessary to eliminate $\mathbf{S}_T$ from the joint density by marginalization. This requires summation over all possible state sequences:

$$\begin{aligned} L(\boldsymbol{\theta}|\boldsymbol{X}_T) &= p(\boldsymbol{X}_T|\boldsymbol{\theta}) = \sum_{All\, S_T} p(\boldsymbol{X}_T, \boldsymbol{S}_T|\boldsymbol{\theta}) \\ &= \sum_{s_1=1}^{M}\sum_{s_2=1}^{M} \cdots \sum_{s_T=1}^{M} \varrho_{s_1} \prod_{t=1}^{T-1} Pr\{(s_{t+1}|s_t)|\boldsymbol{P}\} \prod_{t=1}^{T} f\left(x_t|\mu_{s_t}, \sigma_{s_t}\right) \end{aligned} \quad (9)$$

In the above form, the likelihood function is intractable because of the need to sum over all M^T possible state sequences. However, noting that for each t, the state variable S_t appears only in a few factors, **Eq. (9) can be simplified as

$$L(\boldsymbol{\theta}|\boldsymbol{X}_T) = \sum_{s_1=1}^{M} \varrho_{s_1} f\left(x_1|\mu_{s_1}, \sigma_{s_1}\right) \sum_{s_2=1}^{M} p_{s_1 s_2} f\left(x_2|\mu_{s_2}, \sigma_{s_2}\right) \quad (10)$$

As we consider using Lasso [12] and Ridge estimation to obtain the unknown parameter set in the model, thus we can modify the density function ($f(x_t|S_t = i)$) by adding the penalization term of the parameter estimates, says μ_{s_t}. Thus, we can write the Lasso and Ridge density function as follows:

(1) Ridge density

$$f(x_t|S_t = i) = \frac{1}{\sqrt{2\pi}\sigma_{S_t}} \exp\left\{-\frac{\left(x_t - \mu_{s_t}\right)^2}{2\sigma_{S_t}^2}\right\} + \lambda \sum \mu_{s_t}^2 \quad (11)$$

(2) Lasso density

$$f(x_t|S_t = i) = \frac{1}{\sqrt{2\pi}\sigma_{S_t}} \exp\left\{-\frac{\left(x_t - \mu_{s_t}\right)^2}{2\sigma_{S_t}^2}\right\} + \lambda \sum |\mu_{s_t}| \quad (12)$$

3 Empirical Results

3.1 Data and Variables Description

As permitted by the data availability, the data required for this study was collected from many different data sources. In this study, the data for the top 10 source countries are secondary annual data collected from the CEIC database [9] from 1995 to 2018. They are South Korea, Japan, Russia, USA, Mongolia, Malaysia, Philippines, Singapore, India, and Canada. We use 11 variables to measure the transportation conditions in China. They are China railway length in operation, China highway length, China coastal major port length of quay, China river major port length of quay, China number of international air route, China length of the international air route, China number of airports, China international air passenger traffic, China fixed-asset investment on the railway, China navigable length of the river and China aircraft daily utilization. These 11 data are secondary annual data from 1995 to 2018, from China Railway Corporation [10], National Railway Administration [11], Ministry of Transport [12], Civil Aviation Administration of China [13]. We use China's GDP per capita growth rate, unemployment rate, inflation rate, and official exchange rate (USD / CNY) to measure China's macroeconomy. These macro data are secondary annual data from 1995 to 2018 from the World Bank [14]. The description of each variable is presented in Table 1.

As the objective of this paper is to analyze the impact of transportation and macro determinants on Chinese inbound tourism, only the transportation factors and macro factors are included in the tourism demand function. Due to the different national conditions of each source country, this paper uses the top 10 source countries of China's inbound tourism as the dependent variable, and 11 transportation variables and 4 macro variables as independent variables. Macro factors can be based on Real Business Cycle theory. The main assumption of this paper is that all transportation factors are positively correlated to Tourism Demand. For the tourism demand model, the majority of researchers have developed a linear regression model. According to Yamaka et al. [13], there may be structural changes in international tourism demand, which the nonlinear tourism demand model can capture. Therefore, in this study, the following demand function is introduced to model the demand for inbound tourism in China by tourists' number from each of the 10 source countries:

$$\begin{aligned} TD_{i,t} = f(&\mathrm{GDPR_t},\ \mathrm{UNPR_t}, \mathrm{INFR_t}, \mathrm{EXR_t},\\ &\mathrm{lnRLO_t}, \mathrm{lnHWL_t}, \mathrm{lnCPL_t},\\ &\mathrm{lnRPL_t}, \mathrm{lnNIAR_t}, \mathrm{lnLIAR_t}, \mathrm{lnAPT_t},\\ &\mathrm{lnIAPT_t}, \mathrm{lnRFAI_t}, \mathrm{lnNLR_t}, \mathrm{lnADU_t}) \end{aligned} \tag{13}$$

Note that in (13) we performed natural logarithms transformation for the 10 source countries' tourist arrival and 11 transportation variables to reduce the possible heteroscedasticity. Since the 4 macroeconomic variables are growth rates in the unit of

Table 1 Data definition

Variables	Description	Unit	Source
RLO	China railway length in operation	km	Ministry of Transport of the People's Republic of China
HWL	China highway length	km	Ministry of Transport of the People's Republic of China
CPL	China coastal major port length of quay	m	Ministry of Transport of the People's Republic of China
RPL	China river major port length of quay	m	Ministry of Transport of the People's Republic of China
NIAR	China number of international air route	Unit	Civil Aviation Administration of China
LIAR	China length of the international air route	km	Civil Aviation Administration of China
APT	China number of airports	Unit	Civil Aviation Administration of China
IAPT	China international air passenger traffic	Person mn	Civil Aviation Administration of China
RFAI	China fixed-asset investment on the railway	RMB mn	Ministry of Transport of the People's Republic of China
NLR	China navigable length of the river	km	Ministry of Transport of the People's Republic of China
ADU	China aircraft daily utilization	Hour	Civil Aviation Administration of China
GDPR	China's GDP per capita growth rate	%	World Bank
UNPR	China's unemployment rate	%	World Bank
INFR	China's inflation rate	%	World Bank
EXR	China's official exchange rate (USD / CNY)	%	World Bank
SK	South Korea tourist arrival	Person-Time th	CEIC database
JP	Japan tourist arrival	Person-Time th	CEIC database
RU	Russia tourist arrival	Person-Time th	CEIC database
USA	USA tourist arrival	Person-Time th	CEIC database
MO	Mongolia tourist arrival	Person-Time th	CEIC database
MA	Malaysia tourist arrival	Person-Time th	CEIC database
PH	Philippines tourist arrival	Person-Time th	CEIC database
SI	Singapore tourist arrival	Person-Time th	CEIC database
IN	India tourist arrival	Person-Time th	CEIC database
CA	Canada tourist arrival	Person-Time th	CEIC database

Table 2 Descriptive statistics of data

		Mean	Med	Max	Min	Std	Skew	Kurt	Jarque–Bera
Macro	GDPR	8.427	8.260	13.636	6.082	1.932	0.864	0.247	3.679
	UNPR	4.142	4.459	4.700	3.000	0.595	−0.889	−1.028	4.317
	INFR	2.810	1.961	16.791	−1.401	3.759	2.129	5.385	58.170
	EXR	7.441	7.790	8.351	6.143	0.882	−0.219	−1.824	3.173
Transportation	lnRLO	11.342	11.258	11.788	11.041	0.238	0.576	−1.141	2.461
	lnHWL	14.804	15.074	15.394	13.961	0.538	−0.377	−1.661	3.023
	lnCPL	12.929	12.987	13.684	11.978	0.607	−0.169	−1.684	2.587
	lnRPL	13.118	12.987	13.752	12.169	0.545	−0.079	−1.639	2.334
	lnNIAR	5.566	5.582	6.744	4.443	0.689	0.135	−1.187	1.137
	lnLIAR	13.799	13.757	15.096	12.760	0.670	0.363	−0.962	1.209
	lnAPT	5.084	4.980	5.451	4.836	0.188	0.635	−1.096	2.690
	lnIAPT	2.639	2.670	4.154	1.303	0.844	0.209	−1.179	1.233
	lnRFAI	12.390	12.355	13.644	10.969	1.069	−0.008	−1.880	3.171
	lnNLR	11.707	11.724	11.753	11.606	0.048	−1.137	−0.189	5.882
	lnADU	2.149	2.225	2.257	1.723	0.140	−1.401	1.266	11.633
Countries	lnJP	7.871	7.903	8.288	7.174	0.321	−0.635	−0.795	2.192
	lnSK	7.791	8.216	8.472	6.272	0.729	−0.871	−0.840	3.872
	lnMO	6.441	6.456	7.531	5.568	0.562	0.084	−1.161	1.027
	lnPH	6.403	6.588	7.094	5.392	0.551	−0.484	−1.270	2.312
	lnSI	6.454	6.747	6.969	5.566	0.493	−0.535	−1.487	3.133
	lnMA	6.598	6.881	7.164	5.529	0.559	−0.580	−1.350	2.979
	lnIN	5.669	6.042	6.760	3.807	0.970	−0.639	−1.116	2.754
	lnUSA	7.213	7.444	7.818	6.244	0.514	−0.549	−1.322	2.744
	lnRU	7.361	7.540	8.047	6.193	0.536	−0.734	−0.731	2.727
	lnCA	6.009	6.248	6.745	4.858	0.601	−0.422	−1.405	2.404

percentage, we are not taken logarithms transformation on them. Table 2 summarizes the basic descriptive statistics of the data in this study.

3.2 *Unit Root Test*

Before estimating the Markov Switching model for analyzing the business cycle of the Chinese tourism industry, we need to check whether the data is stationary through the unit root test. The calibrated p-value of the test statistic is obtained using the Minimum Bayes factor (MBF) method. The Goodman (1999) labelled intervals are used to interpret the MBF. MBF between 1 and 1/3 is considered weak evidence for H_1, from 1/3 to 1/10 considered moderate evidence for H_1, 1/10 to 1/30 is considered substantial evidence, from 1/30 to 1/100 strong, from 1/100 to 1/300 very strong, and < 1/300 decisive. Table 3 shows the unit root test results, which reveal that all of the data provide at least substantial evidence for the stationary at 1st difference. Therefore, it is appropriate to further investigate the relationship between Chinese inbound tourism and macro and transportation.

Table 3 Unit root test results

		Level		1st difference	
		t-statistic	MBF	t-statistic	MBF
Macro	GDPR	−0.888	0.6741	−4.39	0.0006
	UNPR	−2.1	0.1102	−2.98	0.0117
	INFR	−3.98	0.0003	−4.52	0.0003
	EXR	−2.17	0.0949	−2.28	0.0743
Transportation	lnRLO	−0.852	0.6956	−3.81	0.0007
	lnHWL	−1.06	0.5701	−4.86	0.0007
	lnCPL	−0.984	0.6162	−7.75	0.0000
	lnRPL	−1.67	0.2479	−3.17	0.0065
	lnNIAR	−2.33	0.0662	−5.16	0.0000
	lnLIAR	−1.29	0.4351	−4.44	0.0000
	lnAPT	−1.41	0.3700	−4.69	0.0001
	lnIAPT	−1.77	0.2087	−6.29	0.0000
	lnRFAI	−1.56	0.2961	−2.46	0.0485
	lnNLR	−3.06	0.0092	−2.95	0.0128
	lnADU	−1.196	0.4890	−4.96	0.0000
Countries	lnJP	−1.114	0.5376	−6.50	0.0000
	lnSK	−0.703	0.7810	−5.16	0.0000
	lnMO	0.319	0.9503	−5.54	0.0000
	lnPH	−0.874	0.6825	−5.46	0.0000
	lnSI	−0.643	0.8132	−6.32	0.0000
	lnMA	−0.475	0.8933	−4.55	0.0003
	lnIN	−1.076	0.5605	−4.24	0.0001
	lnUSA	−1.23	0.4693	−6.53	0.0000
	lnRU	−1.09	0.5520	−5.39	0.0000
	lnCA	−1.26	0.4521	−5.23	0.0000

Note: MBF is Minimum Bayes Factors $MBF_{01}(t) = \left(-t^2/2\right)$

3.3 Model Comparison

In this section, the linear models consisting of the Generalized linear regression model (GLM), Ridge regression and Lasso regression, and the Markov switching models, consisting MS-Ridge model and MS-Lasso model are considered and compared. Table 4 shows the model comparison in 5 different estimation methods from the lowest Akaike information criterion (AIC) [14, 15] and the highest log-likelihood (LL) [16]. Hence, the AIC values of the nonlinear models are lower than linear models, and the log-likelihood of the nonlinear models is higher than linear models for all source countries, it shows that the MS-Ridge model and MS-Lasso model are better than linear regression models, indicating that the nonlinear models are appropriate for analyzing the influence of transportation and Macroeconomic determinants on Chinese inbound tourism. In the nonlinear models, looking at these 10 source countries, the MS-Lasso model of the other 9 countries except Russia fits better than the MS-Ridge model, because the MS-Lasso model has the lowest AIC value and the highest log-likelihood value. In other words, the most suitable model for estimating Russia is the MS-Ridge model, and the most suitable model for estimating Japan, South Korea, Mongolia, the Philippines, Singapore, Malaysia, India, the United States and Canada is the MS-Lasso model.

3.4 Estimation Result

In Sect. 3.3, we have already obtained: Except for Russian tourists, the best-fitting model for tourists from 9 source countries is the MS-Lasso regression model. Therefore, in this section, we will only elaborate and explain the best-fit model applicable to each source country.

Table 5 presents the estimated results of the MS regression model for 10 source countries in regime 1, which is defined as a low-growth stage, and regime 2, which is defined as a high-growth stage. For the optimal lambda selection of the MS model, the value range of lambda in all 10 countries is between 0 and 1. In general, the estimated coefficients in regime 2 are all positive, but the estimated coefficients in regime 1 are positive or negative, which reveals that transportation and macro factors have an asymmetrical impact on the growth of Chinese inbound tourism. The estimated coefficients of regime 1 (low) are more complicated than that of regime 2 (high). A more reasonable explanation is that when China's inbound tourism economy is in a low-growth stage, it is susceptible to many factors. Moreover, the estimated coefficients of regime 1 (low) both have positive and negative values, indicating that China's inbound tourism is more volatile when it is in a low-growth stage, which reflects the instability when the economy is in a low-growth period.

For the arrival of Japanese tourists, the MS-Lasso model has the lowest AIC value, $AIC_{JP} = 0.8153$, so the best fitting model is the MS-Lasso regression model. The optimal value of the lambda for Japanese tourists selected by MS-Lasso model

Table 4 Information criterion for model selection

	Linear models						Nonlinear models			
	GLM		Ridge		Lasso		MS-Ridge		MS-Lasso	
Countries	AIC	LL	AIC	LL	AIC	LL	AIC	LL	AIC	LL
JP	33.398	0.3009	29.258	2.3706	37.556	−1.7783	19.309	26.345	**0.8153**	35.592
SK	29.938	2.0309	27.239	3.3805	29.810	2.0947	−11.785	41.892	**−16.185**	44.092
MO	38.650	−2.3253	44.060	−5.0302	38.092	−2.0464	27.520	22.239	**25.662**	23.168
PH	47.299	−6.6498	47.270	−6.6351	54.607	−10.304	2.4229	34.788	**−30.918**	51.459
SI	40.415	−3.2079	53.763	−9.8818	64.560	−15.280	1.4768	35.261	**−12.76**	42.379
MA	35.487	−0.7436	35.465	−0.7327	34.954	−0.4770	25.917	23.041	**0.3799**	35.810
IN	20.519	6.7403	18.207	7.8960	17.903	8.0480	8.1045	31.947	**−3.5109**	37.755
USA	68.355	−17.178	68.326	−17.163	68.300	−17.150	−1.5274	36.763	**−18.941**	45.470
RU	44.454	−5.2271	47.116	−6.5584	47.399	−6.6996	**−2.1898**	37.094	18.108	26.945
CA	24.446	4.7768	26.484	3.7577	25.817	4.0914	8.9701	31.514	**−13.276**	42.638

Note LL represents the log-likelihood

Table 5 The results of MS regression model for 10 source countries

		MS-Lasso	MS-Lasso	MS-Lasso	MS-Lasso	MS-Lasso
	Variables	JP	SK	MO	PH	SI
Regime 1 (low)	Intercept	2.3667	−4.7317	−0.0279	−6.9719	−2.052
	lnRLO	−2.5162	−0.4678	−2.1918	−1.5402	−1.678
	lnHWL	0.0460	0.3353	0.4528	0.3474	0.3160
	lnCPL	−0.1468	−0.7207	0.2303	−0.1344	0
	lnRPL	0.2966	0	0.2008	0	0.4974
	lnNIAR	0.3691	0	0	0.1331	−0.1798
	lnLIAR	−0.7169	−0.4922	−0.3092	−0.3544	0
	lnAPT	0	−1.3957	2.4714	1.5534	0
	lnIAPT	0.8419	1.2328	0.8945	0.7038	0.6896
	lnRFAI	0	0.1417	−0.4944	−0.1546	−0.1333
	lnNLR	2.7735	2.5552	1.3646	1.9920	1.4041
	lnADU	0.1986	0.0794	1.7347	1.0068	0.6104
	GDPR	0	0.0232	0	0.0204	0
	UNPR	−0.2285	−0.3635	−0.3877	−0.4170	−0.0343
	INFR	0	0	0	0	0
	EXR	0.1374	−0.1135	0	−0.0316	0
Regime 2 (high)	Intercept	6.2188	0.9847	0.7960	0.8477	0.1548
	lnRLO	0	0	0	0	0
	lnHWL	0	0.0296	0.0005	0.0851	0.0003
	lnCPL	0	0	0	0	0
	lnRPL	0	0	0	0	0.0003
	lnNIAR	0	0	0.0001	0	0
	lnLIAR	0	0	0	0	0
	lnAPT	0	0	0	0	0
	lnIAPT	0	0	0.0036	0.0898	0
	lnRFAI	0	0	0	0	0
	lnNLR	0	0.0001	0	0.0009	0
	lnADU	0.6790	0.0087	0	0.0856	0.0001
	GDPR	0.1637	0.0069	0	0	0.0001
	UNPR	0.0289	0.0370	0	0.0978	0.0006
	INFR	0	0	0	0	0
	EXR	0	0	0	0	0
AIC		0.8153	−16.184	25.662	−30.918	−12.759
Log-likelihood		35.592	44.092	23.168	51.459	42.379
Lambda		0.1494	0.1494	0.1494	0.4413	0.1494

(continued)

Table 5 (continued)

		MS-Lasso	MS-Lasso	MS-Lasso	MS-Ridge	MS-Lasso
	Variables	MA	IN	USA	RU	CA
Regime 1 (low)	Intercept	−2.6692	−0.1205	−2.0443	−0.0896 (1.0294)	−2.4930
	lnRLO	−2.1511	−1.0364	−1.3312	−1.3656[c] (0.5122)	−2.1939
	lnHWL	0.3282	0.2906	0.3817	0.8521[c] (0.1294)	0.3366
	lnCPL	0	−0.4016	−0.2699	−1.0930[c] (0.1821)	0
	lnRPL	0.5330	−0.0372	0.2603	0.0591 (0.1079)	0.3688
	lnNIAR	−0.5302	0.7749	0	−0.2045 (0.1615)	0.5374
	lnLIAR	0	−0.7648	−0.2125	0.4705[c] (0.1446)	−0.5195
	lnAPT	0.6818	0.6321	0	−0.7590[b] (0.3052)	1.1724
	lnIAPT	1.1024	0.7693	0.6320	0.5137[c] (0.1629)	0.6409
	lnRFAI	−0.1788	0	0	0.2529[b] (0.1018)	0
	lnNLR	1.7646	2.4845	1.4363	1.4235[c] (0.4024)	1.7964
	lnADU	0.2253	0.1202	0.7107	0.1758 (0.4131)	0.2245
	GDPR	0	0	0	−0.0118 (0.0130)	0
	UNPR	−0.2967	−0.5561	−0.0371	−0.3007[c] (0.0655)	−0.0835
	INFR	0	0	0	0.0573[c] (0.0095)	0
	EXR	0.0538	0	0	−0.0015 (0.0773)	0.0544
Regime 2 (high)	Intercept	0.1610	0.3718	0.1969	0.0312[c] (0.0006)	0.2571
	lnRLO	0	0	0	0.0433 (0.0277)	0
	lnHWL	0.0001	0.0034	0.0357	0.0639[c] (0.0020)	0.0013
	lnCPL	0	0	0	0.0489 (0.0485)	0
	lnRPL	0.0055	0	0	0.0434 (0.0374)	0.0051
	lnNIAR	0	0.0219	0	0.0337[c] (0.0071)	0
	lnLIAR	0	0	0	0.0619[c] (0.0037)	0
	lnAPT	0	0	0	0.0466[c] (0.0050)	0
	lnIAPT	0	0	0.0044	0.0564[c] (0.0016)	0.0311
	lnRFAI	0	0	0	0.0325[b] (0.0139)	0
	lnNLR	0	0.0006	0.0749	0.0498[a] (0.0292)	0
	lnADU	0.0006	0.0023	0.0155	0.0313[c] (0.0009)	0.0001
	GDPR	0.0062	0	0	0.0562[c] (0.0060)	0

(continued)

Table 5 (continued)

		MS-Lasso	MS-Lasso	MS-Lasso	MS-Ridge	MS-Lasso
	Variables	MA	IN	USA	RU	CA
	UNPR	0.0016	0.0006	0	0.0290[c] (0.0016)	0
	INFR	0	0	0	0.2322[c] (0.0014)	0
	EXR	0	0	0	0.1720[c] (0.0069)	0
AIC		0.3799	−3.5109	−18.941	−2.1898	−13.276
Log-likelihood		35.810	37.755	45.470	37.094	42.638
Lambda		0.1494	0.1049	0.4413	0.4413	0.1989

Note ***(a), **(b), *(c) are decisive (MBF < 0.001), strong (MBF < 0.01), and moderate (MBF < 0.1). The parentheses () are standard error

is $\lambda = 0.1494$. The model screened out the two most important factors affecting the arrival of Japanese tourists, lnADU and UNPR. For South Korean tourists, the selected value is $\lambda = 0.1494$. The variables affecting Korean tourists are lnHWL, lnNLR, lnADU, GDPR and UNPR. For Mongolian tourists, the optimal value of the lambda for Mongolian tourists selected by MS-Lasso model is $\lambda = 0.1494$. Variables affecting Mongolian tourists are lnHWL and lnIAPT. For Filipino tourists, the optimal value of the lambda for Filipino tourists selected by MS-Lasso model is $\lambda = 0.4413$. The variables affecting Filipino tourists are lnHWL, lnIAPT, lnNLR, lnADU and UNPR. For Singaporean tourists, the optimal value of the lambda for Singaporean tourists selected by MS-Lasso model is $\lambda = 0.1494$. The variables that affect Singaporean tourists are lnHWL, lnRPL, lnADU and UNPR. For Malaysian tourists, the optimal value of the lambda for Malaysian tourists selected by MS-Lasso model is $\lambda = 0.1494$. The variables affecting Malaysian tourists are lnHWL, lnRPL, lnADU and UNPR. For Indian tourists, the optimal value of the lambda for Indian tourists selected by MS-Lasso model is $\lambda = 0.1049$. The variables affecting Indian tourists are lnHWL, lnNIAR, lnNLR, lnADU and UNPR. For American tourists, the optimal value of the lambda for American tourists selected by MS-Lasso model is $\lambda = 0.4413$. The variables affecting American tourists are lnHWL, lnIAPT, lnNLR and lnADU. For Russian tourists, because the MS-Ridge model has the lowest AIC value, $AIC_{RU} = -2.1898$, so the best fitting model is the MS-Ridge regression model. The 8 variables lnHWL, lnLIAR, lnAPT, lnIAPT, lnRFAI, lnNLR, UNPR and INFR are significant in both regimes. For lnHWL, a 1% increase in lnHWL will lead to a 0.8521% and 0.0639% increase in the growth rate of Russian tourists for regime 1 and regime 2, respectively. For lnNLR, a 1% increase in lnNLR will lead to a 1.4235% and 0.0498% increase in the growth rate of Russian tourists for regime 1 and regime 2, respectively. The variables that have asymmetric effects on Russian tourists are lnRLO, lnCPL, lnNIAR, lnAPT, lnADU, GDPR and EXR. The optimal value of the lambda for Russian tourists selected by MS-Ridge model is $\lambda = 0.4413$. For Canadian tourists, the optimal value of the lambda for Canadian tourists selected by MS-Lasso model is $\lambda = 0.1989$. The variables affecting Canadian tourists are lnHWL, lnRPL, lnIAPT and lnADU.

Since the coefficient of lnHWL is positive in both regimes for all source countries, we can be sure that the extensive highway network provides international tourists with the opportunity to visit tourist attractions in China. This corresponds to the findings of Khadaroo et al. [17] used the gravity model to find that transportation infrastructure is an important determinant of destination tourist flow. lnIAPT indicates that China international air passenger traffic (IAPT) is also the main factor affecting China's inbound tourism. This is consistent with our expectations. International tourism is a tourism activity that crosses national borders. International air passenger traffic demonstrates the scale of air transportation in a country. The larger the scale, the greater the demand for tourism. It is worth noting that the logarithm coefficient of international air passenger traffic is greater than 1 in the two source countries, 1.2328 in regime 1 of Korean tourists, and 1.1024 in regime 1 of Malaysian tourists. This shows that when China's inbound tourism is at a low growth stage, the development of international air passenger traffic is most significant for attracting Korean and Malaysian tourists. lnNLR represents the navigable length of rivers in mainland China. Water operation is an important means of transportation, providing foreign tourists with the opportunity to view the scenery along the waterways by boat or raft. The logarithm coefficient of China navigable length of the river is positive for the 10 source countries in both regimes. The coefficient ranges from the lowest Mongolian tourist at 1.3646 to the highest Japanese tourist at 2.7735. For ADU, lnADU explained that the daily utilization of Chinese aircraft is likewise one of the essential determinants affecting foreign tourists' entry into China. The logarithm coefficient of China aircraft daily utilization is positive for 10 source countries in both regimes. It is worth noting that in regime 1, the coefficient of InADU is 1.7347 for Mongolian tourists, 1.0068 for Filipino tourists, and the coefficient of InADU is less than 1 for the other 8 source countries. For UNPR, the logarithm coefficient of China's unemployment rate in regime 1 is negative for all 10 source countries. International tourism is a service industry. In the tourism industry, the products are mainly services and experiences. Therefore, the development of the tourism industry requires more investment in human capital.

As mentioned earlier, because the AIC value of the Nonlinear model is smaller than that of the linear model for all cases, our test results largely support the existence of the nonlinear relationship. This means that the nonlinear model is more appropriate for analyzing the impact of transportation and macro factors on China's inbound tourism demand. Overall, the impact of transportation and macro factors considered in this paper on China's inbound tourism is mainly nonlinear. According to the empirical evidence of Feature Selection in the MS-Lasso model, lnHWL affects 8 source countries, lnRPL affects 3 source countries, lnNIAR affects 1 source country, lnIAPT affects 4 source countries, and lnNLR affects 4 source countries. lnADU affects 8 source countries, GDPR affects 1 source country, and UNPR affects 6 source countries. We can conclude that lnHWL, lnIAPT, lnNLR, lnADU and UNPR are the main factors affecting China's inbound tourism.

Table 6 shows the probabilities of being in regime 1 and regime 2 of MS model for the 10 source countries. P_{11} means the probability that Chinese inbound tourism economy is in a low-growth stage, and P_{22} means the probality that Chinese inbound

Table 6 The transition probabilities of MS model for 10 source countries

	Countries	JP	SK	MO	PH	SI
Transition probabilities	P_{11}	0.0100	0.0104	0.0100	0.0100	0.0298
	P_{22}	0.9900	0.9899	0.9900	0.9900	0.9900
	Countries	MA	IN	USA	RU	CA
Transition probabilities	P_{11}	0.0100	0.0100	0.0247	0.0100	0.0100
	P_{22}	0.9900	0.9900	0.9900	0.9899	0.9900

tourism economy is in a high-growth stage. We can conclude that China's inbound tourism cycle mainly occurs in the high-growth stage. We noticed that the probability of South Korea and Russia in the high-growth stage is slightly lower than that of the other eight countries, which is $P_{22} = 0.9899$. Because the value of P_{22} is very high, it means that when China's inbound tourism economy is in a high-growth stage, it is unlikely to switch from a high-growth stage to a low-growth stage, and it is more likely to remain in the high-growth stage. The P_{11} of the MS model is very low, indicating that the possibility of China's inbound tourism economy in a low-growth stage is very small.

This section lists the estimation results of the best fit MS model. The results show that the impact of transportation and macro factors considered in this paper on China's inbound tourism is mainly nonlinear. Moreover, China's inbound tourism cycle is more likely to be at a high growth stage. According to the estimation results of the best fit MS model, the MS-Lasso model has done feature selection and selected the main factors that affect China's inbound tourism demand: lnHWL, lnIAPT, lnNLR, lnADU and UNPR. Hence, we can conclude that lnHWL, lnIAPT, lnNLR, lnADU and UNPR are the main factors affecting China's inbound tourism. Four transportation factors confirm that areas with a well-developed transportation system provide tourists with opportunities to explore more tourist attractions. This result is consistent with the discovery of Khan et al. [6]. Khan et al. [6] pointed out, the gathering of different forms of transport such as air transportation, railway transportation, travel and transportation systems will help promote the development of international tourism. The theory of interdependence between transportation and tourism supports our and previous findings. In addition, by comparing the AIC values of the MS-Ridge model and the MS-Lasso model, we can find that, except for Russia, the AIC values of the MS-Lasso model of the other 9 source countries are lower than the MS-Ridge model. This shows that the MS-Lasso model is better than the MS-Ridge model in the goodness of model fitting in our empirical study.

3.5 *Filtered Probability Obtained from the Best Fit Model*

Figure 1 portrays the filtered probability of regime 2 (high tourism arrival regime) from the best fit model for the 10 source countries. In other words, except for Russia,

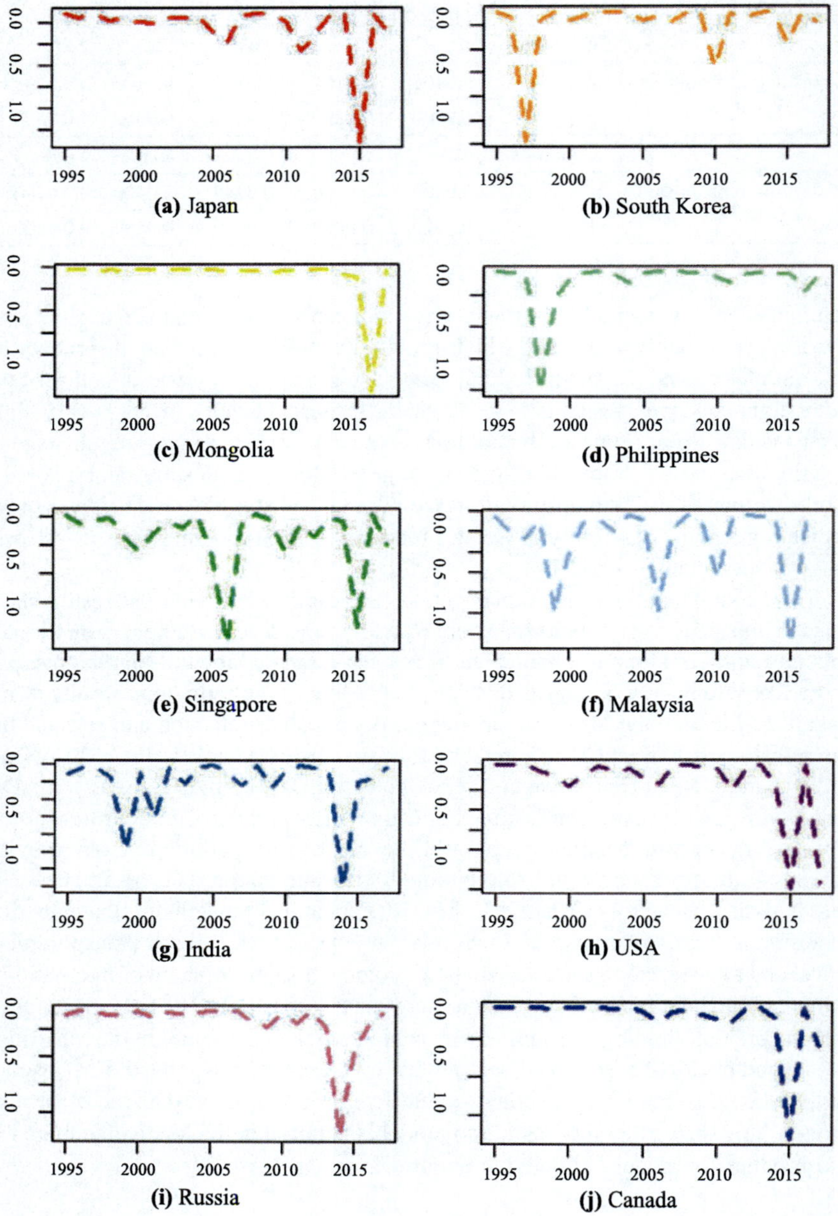

Fig. 1 Filtered Probability obtained from the best fit model

which is suitable for the MS-Ridge model, the other 9 countries have chosen the MS-Lasso model to portray the filtered probability. It can be noted that for China's inbound tourism business cycle, the possibility of being in a high-growth stage is large, with long persistence and stability. The possibility of being in a low-growth stage is small, with short persistence and variability. We also noticed that there is an asymmetric business cycle in Chinese inbound tourism economy because the duration of regime 2 is much longer than regime 1.

From the perspective of source countries, the filtered probability of regime 2 for Japan was at the trough in 2015, indicating that Japanese tourists have reached a period of low growth in 2015. It may be due to the political relationship between the two countries. The filtered probability of regime 2 for South Korea was at the bottom in 1997, indicating that South Korean tourists reached a period of low growth in 1997. It may be affected by the 1997 Asian financial crisis. The filtered probability of regime 2 for Mongolia was at the bottom in 2016, indicating that Mongolian tourists reached a period of low growth in 2016. It may be due to the 2016 national elections in the country. The filtered probability of regime 2 for the Philippines was at the bottom in 1997, and like South Korea, it may be affected by the Asian financial crisis. The filtered probability of regime 2 for Singapore was at the bottom in 2006 and 2015, probably because the country held national elections in 2006 and 2015. The filtered probability of regime 2 for Malaysia is the most volatile. It is certain that the country received the impact of the Asian financial crisis in 1998. The filtered probability of regime 2 for India was at the bottom in 2014, probably because the country held a national election in 2014. The filtered probability of regime 2 for the United States was at a bottom in 2015, probably due to the tense relationship between China and the United States in 2015. The filtered probability of regime 2 for Russia was at the bottom in 2014, which may be affected by the 2014 Russian financial crisis. The filtered probability of regime 2 for Canada was at its trough in 2015, which may be affected by the 2014 national election. It is worth noting that Korea, Singapore, Malaysia, and India have relatively large volatility in the filtered probability. We can conclude that the financial crisis, national elections, and political relations will affect international tourism.

3.6 Engle-Granger Cointegration Test

After completing the model estimation, the next step is to perform a cointegration test. We chose the Engle-Granger cointegration test [18]. It builds residuals based on static regression. The test uses the residuals to see if unit roots are present, using the Augmented Dickey-Fuller test [19]. The residuals will be practically stationary if the time series are cointegrated. Table 7 displays the results of the cointegration test. The statistic values from the best fit models reject the null hypothesis of nonstationary at 10% significance level, indicating that the residuals are stationary. So, the variables are cointegrated. Hence, it can be concluded that there is a significant long-run

Table 7 Cointegration tests among the best fit model

Model	Countries	Statistic
MS-Lasso	JP	−0.2705[a] (0.1345)
MS-Lasso	SK	−0.4120[a] (0.2106)
MS-Lasso	MO	−0.3418[b] (0.1615)
MS-Lasso	PH	−0.2675[b] (0.1135)
MS-Lasso	SI	−0.3437[a] (0.1853)
MS-Lasso	MA	−0.3323[b] (0.1459)
MS-Lasso	IN	−0.0823[b] (0.0351)
MS-Lasso	USA	−0.2994[b] (0.1295)
MS-Ridge	RU	−0.6188[b] (0.2401)
MS-Lasso	CA	−0.3223[c] (0.0924)

Note "a," "b," and "c" denote rejection of the null hypothesis at the 10%, 5%, and 1% significance level, respectively. The parentheses () is standard error

equilibrium relationship among China's inbound tourism demand, transportation, and macroeconomic variables of the top 10 source countries.

4 Conclusions and Suggestions

4.1 Conclusions

This empirical study examines the influence of transportation and Macroeconomic determinants on Chinese inbound tourism from 1995 through 2018 by using the Markov switching model based on Ridge and Lasso estimation. This study intends to measure the cycle of Chinese inbound tourism and find the high-growth stage and low-growth stage of inbound tourism. There are two innovations in this paper. Firstly, transportation factors are introduced into tourism demand; secondly, due to the small data samples used in this paper, the traditional Markov switching model cannot be used for regression. Therefore, the model is innovated, namely, Markov switching Model based on Ridge and Lasso estimation method. The empirical results show that the nonlinear model is more suitable for analyzing the tourism business cycle than the linear model. The estimation results of the two nonlinear models are roughly the same, but the MS-Lasso model is better than the MS-Ridge model in the goodness of model fitting. Feature selection screened out the main factors that affect China's inbound tourism. They are lnHWL (China highway length), lnIAPT (China international air passenger traffic), lnNLR (China navigable length of the river), lnADU (China aircraft daily utilization), and UNPR (China's unemployment rate). For China's inbound tourism business cycle, the possibility of being in a high-growth stage is large, with long persistence and stability. The possibility of being in a low-growth stage is small, with short persistence and variability.

4.2 Suggestions for Policy

According to the Feature selection results of the MS-Lasso model, the Ministry of Transport should increase highway construction, particularly in the middle and western provinces with backward transportation infrastructure in mainland China. Maintain or increase the navigable reach of inland rivers through remediation and dredging. The Civil Aviation Administration of China should use new technologies to implement operations outside the airport, simplify procedures, and improve efficiency. Chinese civil airlines should increase the daily utilization rate of their aircraft. Increasing aircraft utilization is of great significance for developing production, reducing costs, and improving airlines' economic benefits. The improvement of airlines' economic efficiency can moreover boost the development of China's inbound tourism. The Chinese government should focus on skills training in tourism-related industries to provide job opportunities for the unemployed. For example, foreign language skills training for workers can provide a more comfortable travel environment for foreign tourists.

Since international tourism is more susceptible to external factors, under the current circumstances, China can seize the "Belt and Road" construction opportunity to boost the establishment of tourism collaboration alliances between China and the top 10 source countries and establish a mutually beneficial tourism cooperation mechanism.

4.3 Limitations of This Study

One of such limitations is that we use annual data, while monthly or quarterly data may provide a better understanding of time series switch or transitions. Future research can collect monthly or quarterly data to better capture changes in the tourism business cycle. Secondly, the variables analyzed in this paper are transportation and macroeconomic factors. Future research can introduce other variables when investigating the business cycle of tourism. In addition, this paper adopts the number of arrivals of international tourists from the source country. Future research can start with market segmentation (age or gender) for microeconomic analysis.

References

1. Van Truong, N., & Shimizu, T. (2017). The effect of transportation on tourism promotion: Literature review on application of the Computable General Equilibrium (CGE) Model. *Transportation Research Procedia, 25*, 3096–3115.
2. Khadaroo, J., & Seetanah, B. (2007). Transport infrastructure and tourism development. *Annals of Tourism Research, 34*(4), 1021–1032.
3. Guirao, B., & Campa, J. L. (2015). The effects of tourism on HSR: Spanish empirical evidence derived from a multi-criteria corridor selection methodology. *Journal of Transport Geography, 47*, 37–46.
4. Wang, Z. (2010). Factors that influence the growth of Chinese domestic tourism arrivals (1985–2007)–An empirical research based on the VAR model. *Asia Pacific Journal of Tourism Research, 15*(4), 449–459.
5. Button, K., & Taylor, S. (2000). International air transportation and economic development. *Journal of air transport management, 6*(4), 209–222.
6. Khan, S. A. R., Qianli, D., SongBo, W., Zaman, K., & Zhang, Y. (2017). Travel and tourism competitiveness index: The impact of air transportation, railways transportation, travel and transport services on international inbound and outbound tourism. *Journal of Air Transport Management, 58*, 125–134.
7. MacDonald, I. L., & Zucchini, W. (1997). Hidden Markov and other models for discrete-valued time series (Vol. 110). CRC Press.
8. Rabiner, L. R., Wilpon, J. G., & Soong, F. K. (1989). High performance connected digit recognition using hidden Markov models. *IEEE Transactions on Acoustics, Speech, and Signal Processing, 37*(8), 1214–1225.
9. Akintug, B., & Rasmussen, P. F. (2005). A Markov switching model for annual hydrologic time series. *Water Resources Research, 41*(9).
10. Thyer, M., & Kuczera, G. (2000). Modeling long-term persistence in hydroclimatic time series using a hidden state Markov Model. *Water Resources Research, 36*(11), 3301–3310.
11. Timmermann, A. (2000). Moments of Markov switching models. *Journal of econometrics, 96*(1), 75–111.
12. Tibshirani, R. (1996). Regression shrinkage and selection via the lasso. Journal of the Royal Statistical Society: Series B (Methodological), *58*(1), 267–288.
13. Yamaka, W., Zhang, X., Maneejuk, P. (2021). Analyzing the Influence of Transportations on Chinese Inbound Tourism: Markov Switching Penalized Regression Approaches. *Mathematics, 9*(5), 515.
14. Sakamoto, Y., Ishiguro, M., & Kitagawa, G. (1986). Akaike information criterion statistics. Dordrecht, The Netherlands: D. Reidel, 81.
15. Akaike, H. (1974), A new look at the statistical model identification, *IEEE Trans. Autom. Control, 19*, 716–723.
16. Hansen, B. E. (1992). The likelihood ratio test under nonstandard conditions: Testing the Markov switching model of GNP. *Journal of applied Econometrics, 7*(S1), S61–S82.
17. Khadaroo, J., & Seetanah, B. (2008). The role of transport infrastructure in international tourism development: A gravity model approach. *Tourism Management, 29*(5), 831–840.
18. Engle, R. F., & Granger, C. W. (1987). Co-integration and error correction: Representation, estimation, and testing. Econometrica: journal of the Econometric Society, 251–276.
19. Mushtaq, R. (2011). Augmented dickey fuller test.

The ⟨Im|Possibility⟩ of Quantum Annealing for Maximum Likelihood Estimation

Yong Yoon

Abstract This paper provides a casual exploration of quantum ideas and quantum computing to solve maximum likelihood problems common in statistical and econometric estimations. A brief description of quantum ideas, and in particular, the quantum bit and its properties, is followed by how quantum annealing, specifically conducted by the D-Wave quantum computer, takes advantage of "quantum tunneling" to solve optimization problems that may typically be handled by simulated annealing. Although there is now established theoretical grounds in favor of quantum annealing over simulated annealing, there remain several limitations of its usefulness and useability in MLE, namely, improving its accuracy by formulating the problem more fully for quantum computers and/or reducing hardware noise which may mask the true solution. Currently, the process of embedding quantum maximum likelihood into the quantum machine reduces the scale of problems available and as of now non-artificial problems in which quantum annealing outperforms simulated annealing are hard to find.

Keywords Maximum likelihood estimation (MLE) · Quantum bit (qubit) · Quantum annealing · Quantum tunneling · D-wave

1 Introduction

Ideas in quantum theory present us with many exciting (and often controversial) new possibilities e.g., physical duality (wave-particle property), multiple universes associated with (quantum) entanglement, Q-biology as with plant photosynthesis, bird's consciousness, and so on, all of which have stimulated research not only in the hard sciences but also in philosophy (and everything in between)—the marvelous ideas of Staff that consciousness creates the universe, Rovelli that time does not

Y. Yoon (✉)
Faculty of Economics, Chulalongkorn University, Bangkok, Thailand
e-mail: yong.y@chula.ac.th

S. Sriboonchitta et al. (eds.), *Credible Asset Allocation, Optimal Transport Methods, and Related Topics*, Studies in Systems, Decision and Control 429,
https://doi.org/10.1007/978-3-030-97273-8_31

exist, Everett's quantum immortality, and so on.[1] In economics, Kreinovich et al. [21] acknowledges that the idea of using quantum techniques or to solve econometric problems has been successfully implemented in what is appropriately titled "quantum econometrics" [1].

Maximum likelihood is perhaps the most widely use estimation technique. However, many challenges remain especially when dealing with multi-parameter, highly complex, high-dimensional scenarios e.g. large network data, erratic log-likelihood functions, latent and mixture models, and so on.[2] Researchers therefore continue to explore ways to execute MLE more effectively and efficiently. Moreover, many authors have shown that simulated annealing may have advantages over other standard methods for maximum likelihood estimation, for example, when exploring the possible state space having multiple equilibria. The computational difficulty of statistical and econometric problems continues to increase as empirical researchers examine more complicated models and utilize more sophisticated estimation techniques. In this paper, we focus primarily on how quantum annealing may tremendously speed up maximum likelihood estimation. We however use a very simple econometric problem—a quadratic log-likelihood function—to demonstrate how quantum annealing in the D-wave system: This simple example is meant to allow us to get across the idea of quantum annealing to any interested reader with minimum technical background.

Section 2 gives a general description of quantum bits, focusing on its properties needed to understand quantum annealing, and a general description of quantum annealing through the adiabatic theorem. Section 3 describes quantum annealing under the adiabatic model as this directly relates to the D-Wave for maximum likelihood estimation.[3] Section 4 describes how to set up the maximum likelihood function for a simple two-parameter model in the Hamiltonian form which can then be handled by quantum machines. Section 5 provides a non-technical, intuitive comparison of simulated annealing and quantum annealing, highlighting the potential strengths of the latter for maximum likelihood estimation. Section 6 concludes.

[1] There is a lot of work from these pioneers, but perhaps see Stapp [35], Byrne [5], Rovelli [31], for a small glimpse into their ideas.

[2] See for example Jacquier et al. [17], Byshkin et al. [6], Crosson and Harrow [7], and others. Even apparently simple MLE problem for a 3-parameter Weibull distribution can pose serious challenges (see for example Yang et al. [40]).

[3] The D-Wave quantum computer is a rather controversial machine within the quantum computing community, with even some questioning how "quantum" the machine really is [34]. Johnson et al. [18] compared the results of quantum annealing using the D-Wave to classical techniques and showed that the D-Wave exhibited results that could only have occurred with quantum tunneling.

2 Some Properties of Quantum Bits (Qubits)

In the classical computer, the traditional bit takes on either 0 or 1, but never both simultaneously, while a quantum bit or qubit in short is a particle that exists in the form:

$$q = \alpha_0|0\rangle + \alpha_1|1\rangle \quad (1)$$

where $\alpha_0, \alpha_1 \in \mathbb{C}^d$ and $|0\rangle, |1\rangle$ are in standard braket notation of Dirac orthonormal basis vectors in a d-dimensional complex space, often referred to as the computational basis, for example, corresponding to the spin of a particle $|\uparrow\rangle$ and $|\downarrow\rangle$, respectively. As such, the qubit q can be in superposition, a blend of the spin up and spin down states simultaneously.

More formally, a superposition qubit may take the form as in Eq. (1) where α_0 and α_1 are complex coefficients satisfying $|\alpha_0|^2 + |\alpha_1|^2 = 1$, and with this constraint the set of values that q may take is often represented in as a 2-dimensional sphere in complex space $\mathbb{C}^2$. The qubit manages to exist as a linear combination of the vectors $|0\rangle$ and $|1\rangle$ which serve to connect the qubit to the traditional bit by associating either pure $|0\rangle$ and $|1\rangle$ states with numerical values, commonly $\{+1, -1\}$ or $\{0, 1\}$. We adopt the latter, defining 0 to be the numerical value associated with the superposition defined by $\alpha_0 = 1$ and $\alpha_1 = 0$, and likewise 1 as the numerical value associated with the superposition defined by $\alpha_0 = 0$ and $\alpha_1 = 1$. Unlike a classical bit which is either in states 0 or 1, for a qubit, we cannot determine its state and find values of α_0 and α_1 by examining it. Rather, upon observation the qubit q is said to "collapse" into either 0 or 1 with probability $Pr(0) = |\alpha_0|^2$ and $Pr(1) = |\alpha_1|^2$. We will see that this superposition property is fundamental for quantum annealing.[4]

As with the classical bit, the states of n_q qubits are unit vectors in $\mathbb{C}^d$, where $d = 2^{n_q}$, with computational basis states of the form $|x_1, x_2, \ldots, x_{n_q}\rangle$, where $x_i = 0$ or 1, and $j = 1, \ldots, n_q$. For example, the states of 2 qubits are unit vectors in $\mathbb{C}^4$, with the superposition states $\alpha_{00}|00\rangle + \alpha_{01}|01\rangle + \alpha_{10}|10\rangle + \alpha_{11}|11\rangle$, where α's are complex numbers satisfying $|\alpha_{00}|^2 + |\alpha_{01}|^2 + |\alpha_{10}|^2 + |\alpha_{11}|^2 = 1$. More interesting, any two qubits can be made to interact and stay in an entangled state so that they maintain persistent, instantly "influencing" each other no matter their distance apart and regardless of the nature of the intervening medium. As such, measuring only one of any two entangled qubits allow complete predictability about the state of the other entangled qubit. The quantum properties of superposition and entanglement of qubits continue to a fascinating area of research, and is the basis of quantum annealing in D-wave.[5]

---

[4] Nguyen and Dong [28] provides a more technical discussion on quantum properties and probability calculus.

[5] Scientists only recently photographed entangled particle in July 2019.

3 Quantum Annealing

A comprehensive overview of quantum annealing can be found in Kadowaki and Nishimori [19]. Among the various paradigms of quantum computing identified in the literature, quantum annealing [11] and universal gate-based paradigm (used by IBM, Google, Rigetti), both comprising discrete gate-based and continuous gate-based (Xanadu), are the most popular. The D-Wave machine, which is the focus of this paper, is essentially an optimizing machine which embraces the quantum annealing paradigm (D-Wave 2012) solving a very specific problem—to find a set of q_i that minimizes the energy landscape described by the Hamiltonian (for the classical Ising model)[6]:

$$H_0 = \sum_i \sum_j b_{ij} q_i q_j - \sum_i a_i q_i \tag{2}$$

where a_i and b_{ij} are any real numbers, while q_i are binary taking either $\{+1, -1\}$, and which essentially is just an unconstrained binary quadratic optimization problem.[7] Note that for a system of n_q qubits, there are 2^{n_q} possible solutions, and with more qubits to consider, complexity grows exponentially, and the computational challenge quickly exceed what classical computers may be able to handle [20, 27, 37, 38].

It is useful to think of each system as in Eq. (2) as a graphical network between the n_q qubits, with each qubit assigned a weight a_i to itself and any interactions between qubits assigned weight b_{ij}. When q_i follow the QUBO model, that is when q_i are binary taking either {0, 1}, the system can be represented by the upper-triangular matrix on the right, and the energy calculated by $x^\top Qx$, where x is vector of length n_q consisting of the q_i values and Q is an upper-triangular matrix consisting of the a_i along the diagonals and b_{ij} in the upper triangle (see Fig. 1).

The D-Wave performs quantum annealing using the adiabatic theorem, which, simply stated says that if a system begins in the minimum energy state of a Hamiltonian, where a Hamiltonian is a system of qubits, weights, and interactions as in Eq. (2) and Fig. 1, and transitions to a new Hamiltonian slowly enough, then the system will remain in the lowest-energy state the entire time [3, 13]. That is, from some initial state, H_0, say where every state is equally likely (i.e., a uniform superposition of all states of 0 and 1), the system is then allowed to transition to the final energy landscape H_T through the equation:

$$H(t) = (1 - t)H_0 + tH_1 \tag{3}$$

where t transitions from some normalized time 0–1. Note that as $t \rightarrow 1$ from 0 at the beginning of the anneal, we are left with H_1, the user-input Hamiltonian, which contain the biases and couplers or energy landscape (i.e. problem Hamiltonian) that

[6] The original paper discussing the Ising model for quantum annealing is Bian et al. [2]. For more on the Ising model see Irback et al. [16], Majewski et al. [22], Stauffer [36] and McGeoch [24].

[7] This is derived from the original quantum Hamiltonian, an operator in the Hilbert space.

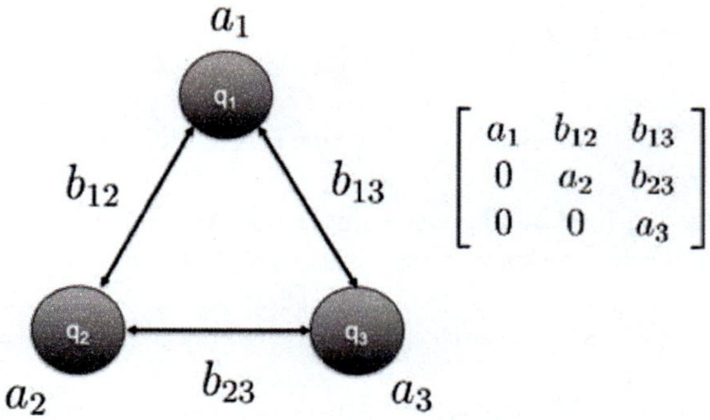

Fig. 1 A simple three-qubit QUBO system with each qubit having weight a_i and each pair of qubits connected with interaction b_{ij}

we want. If we do this slowly enough, the theorem states that we may reach our solution at the lowest-energy state, but in finite time there is no guarantee that the lowest-energy solution will be found. Moreover, the system is not guaranteed to stay in the lowest-energy state for a given anneal and hence it is common to perform many anneals and return the set of q_i and energy for each. In general, the necessary anneal time increases with the complexity of the energy surface, the larger the energy gap between the lowest-energy solution and the second lowest-energy solution at any point during the anneal process, at which the qubits are likely to tunnel from the optimal solution to a non-optimal solution thereby increasing instability.

4 Maximum Likelihood Estimation in Qubits

Suppose that independent and identically distributed data x_n for $n = 1, 2, \ldots, N$ is available from some two-parameter distribution, say, $f(x_n|\theta, \varphi)$. The maximum likelihood estimates are parameters that maximize the log-likelihood function, that is:

$$\hat{\theta}, \hat{\varphi} = \underset{(\theta\varphi)}{\text{argmax}} \sum_n l(\theta, \varphi|x_n) \tag{4}$$

D-Wave requires that the log-likelihood function be in the form of Eq. (2). Following O'Malley and Vesselnov [29], we can represent qubits as numbers by writing the parameters θ and φ in binary. Suppose that we have n_q qubits indexed by q_i for $i = 1, \ldots, n_q$. Let the subscript θ and φ on the index indicate that the qubit belongs to set of qubits which represent binary digits of θ and which represent powers of φ, respectively. The conversion back into decimal form is then:

$$\theta = \sum_i 2^{p_{i_\theta}} q_{i_\theta} \qquad \varphi = \sum_i 2^{p_{i_\varphi}} q_{i_\varphi} \tag{5}$$

where p_{i_θ} and p_{i_φ} are powers of 2 used for the calculation of each parameter. Also note that this forces $\theta, \varphi > 0$.[8]

Expanding the log-likelihood given a single datum x_i, $l(\theta, \varphi|x_i)$ around points θ_0 and φ_0 using Taylor series expansion gives the first two terms as:

$$\begin{aligned} l(\theta,\varphi|x_i) \approx\, & l(\theta_0,\varphi_0|x_i) + l_\theta(\theta_0,\varphi_0|x_i)(\theta-\theta_0) + l_\varphi(\theta_0,\varphi_0|x_i)(\varphi-\varphi_0) \\ & + \frac{1}{2}\Big[l_{\theta\theta}(\theta_0,\varphi_0|x_i)(\theta-\theta_0)^2 + 2l_{\theta\varphi}(\theta_0,\varphi_0|x_i)(\theta-\theta_0)(\varphi-\varphi_0) \\ & + l_{\varphi\varphi}(\theta_0,\varphi_0|x_i)(\varphi-\varphi_0)^2 \Big] \end{aligned} \tag{6}$$

where the subscripts on the log-likelihood l represent partial derivatives. A two-term Taylor expansion is taken since higher powers involve three-qubit interactions $c_{ijk}q_iq_jq_k$ or higher, which are not supported by the D-Wave machine.

Substituting Eq. (5) into Eq. (6), expanding polynomials and expanding terms give $l(\theta, \varphi|x_i)$ in the form of Eq. (2) which we want, with

$$a_i = \begin{cases} \sum_{i=1}^{n_q} \Big[2^{p_{i_\theta}} l_\theta(\theta_0,\varphi_0|x_i) - \frac{1}{2^{2p_{i_\theta}}} l_{\theta\theta}(\theta_0,\varphi_0|x_i) \\ \quad + 2^{p_{i_\theta}} \theta_0 l_{\theta\theta}(\theta_0,\varphi_0|x_i) + 2^{p_{i_\theta}} \varphi_0 l_{\theta\varphi}(\theta_0,\varphi_0|x_i) \Big], & i \text{ is a qubit of } \theta \\ \sum_{i=1}^{n_q} \Big[2^{p_{i_\varphi}} l_\varphi(\theta_0,\varphi_0|x_i) - \frac{1}{2^{2p_{i_\varphi}}} l_{\varphi\varphi}(\theta_0,\varphi_0|x_i) \\ \quad + 2^{p_{i_\varphi}} \varphi_0 l_{\varphi\varphi}(\theta_0,\varphi_0|x_i) + 2^{p_{i_\varphi}} \theta_0 l_{\theta\varphi}(\theta_0,\varphi_0|x_i) \Big], & i \text{ is a qubit of } \varphi \end{cases} \tag{7}$$

and

$$b_i = \begin{cases} \sum_{i=1}^{n_q} \Big[2^{p_{i_\theta}+p_{i_\theta}} l_{\theta\theta}(\theta_0,\varphi_0|x_i) \Big] & i, j \text{ are both qubits of } \theta \\ \sum_{i=1}^{n_q} \Big[2^{p_{i_\varphi}+p_{i_\varphi}} l_{\varphi\varphi}(\theta_0,\varphi_0|x_i) \Big] & i, j \text{ are both qubits of } \varphi \\ \sum_{i=1}^{n_q} \Big[2^{p_{i_\theta}+p_{i_\varphi}} l_{\theta\varphi}(\theta_0,\varphi_0|x_i) \Big] & i, j \text{ are qubits of different parameters} \end{cases} \tag{8}$$

[8] We can easily extend this to negative numbers by introducing a sign qubit [4].

Hence the set of qubits for the lowest-energy solution returned by the quantum annealer can now be converted into numerical estimates $\hat{\theta}$ and $\hat{\varphi}$ using Eq. (5) which is required by the D-Wave machine.

Note that because the numerical estimates $\hat{\theta}$ and $\hat{\varphi}$ maximize only the approximated likelihood constructed by the two-term Taylor series expansion in Eq. (6), and not the full likelihood function, the procedure should be iterated a number of times, similar to Newton and Raphson's original optimization method, by taking the $\hat{\theta}$ and $\hat{\varphi}$ returned by a quantum anneal, expanding the Taylor series around those values, and finding a new set of $\hat{\theta}$ and $\hat{\varphi}$ values, and so on, which maximize Eq. (6) using a new quantum anneal.

5 Simulated Annealing Versus Quantum Annealing

Systems of the form discussed in this paper are commonly optimized by techniques such as simulated annealing which can both climb and descend peaks in the energy surface in order to avoid becoming trapped in local optima.[9]

The basic algorithm for simulated annealing can be outlined in pseudo-code [10, 26].

Algorithm 1 Simulated annealing

procedure SET- UP
    $\omega \in \Omega \leftarrow$ *initial solution*
    $k = 0 \leftarrow$ *temp. change counter*
    $t_k \leftarrow$ *temp. cooling schedule*
    $T = t_0 > 0 \leftarrow$ *initial temp.*
    $M_k \leftarrow$ *repetition schedule* ▷ defines number of iterations for each temp. t_k
    while $m \leq M_k$ **do**
        $\omega' \in N(\omega)$ ▷ generate solution
        Calculate $\Delta_{\omega,\omega'} = f(\omega') - f(\omega)$
        if $\Delta_{\omega,\omega'} \leq 0$ **then**
            $\omega \leftarrow \omega'$
        else if $\Delta_{\omega,\omega'} > 0$ **then**
            $\omega \leftarrow \omega'$ with probability $exp(-\Delta_{\omega,\omega'}/t_k)$
            $m \leftarrow m + 1$
    $k \leftarrow k + 1$ ▷ until stopping criterion is met

Although easy to implement and use, and often provides optimal solutions to a wide range of problems, simulated annealing can take a long time to run especially if the annealing schedule is very long and often there can be a lot of tune-able parameters in this algorithm making it inefficient and time consuming.

Referring to Fig. 2, with simulated annealing indicated by the dashed line, the algorithm must climb over the hill in the energy landscape to escape a local minima

[9] The original simulated annealing paper is Metropolis et al. [25].

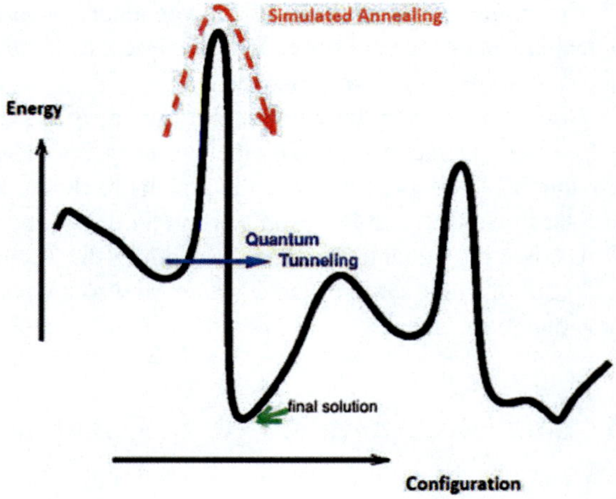

Fig. 2 Simulated annealing versus quantum annealing

and traverse to other lower-energy solutions. While simulated annealing relies on thermal fluctuations to make the system hop from state to state over intermediate energy barriers (indicated by the dotted line) in search for the desired lowest-energy state, for state transitions in quantum annealing, thermal hopping is replaced by quantum-mechanical fluctuations that relies merely on maintaining quantum coherence of the qubits, as in Eq. (1), allowing the annealing process to discover potentially lower-energy states by traveling directly through energy barriers i.e. through quantum tunneling (indicated by the solid line) rather than climbing over them [8]. Additionally, quantum annealing, at least in theory, allows for the direct estimation of the minimum-energy state without traversal of the entire energy surface. Hence, it should therefore theoretically outperform simulated annealing on wide energy surfaces which feature tall, thin peaks in the energy surface.[10]

Several issues need to be addressed when implementing quantum annealing in a quantum machine. The first challenge is how to encode and embed the problem into a format the quantum computer can understand. Translated into Eq. (2), the MLE model has so far been presented as a theoretical graphical network of qubits, but this must be embedded onto the physical hardware of the D-Wave for actual implementation. D-Wave's graph configuration is called a Chimera structure, in which qubits are arranged in sparsely connected groups of at most six other qubits, called unit cells. A complete K_{n_q} graph and the connected Chimera structure required for the maximum likelihood annealing step is shown in Fig. 3. On the left is a complete K_{16} graph representing all the connections between the qubits of θ and φ for the maximum likelihood problem in Eq. (6), while the right shows the embedding of the K_{16} graph onto the Chimera hardware of the D-Wave.[11]

[10] The tunneling process cannot be directly observed nor adequately explained by classical physics and is rather explained by the Heisenberg uncertainty principle and the wave-particle duality of matter in quantum physics [32, 33].

[11] D-Wave in 2019 announced their next generation of their quantum computing topology, Pegasus, a significant improvement to the Chimera technology.

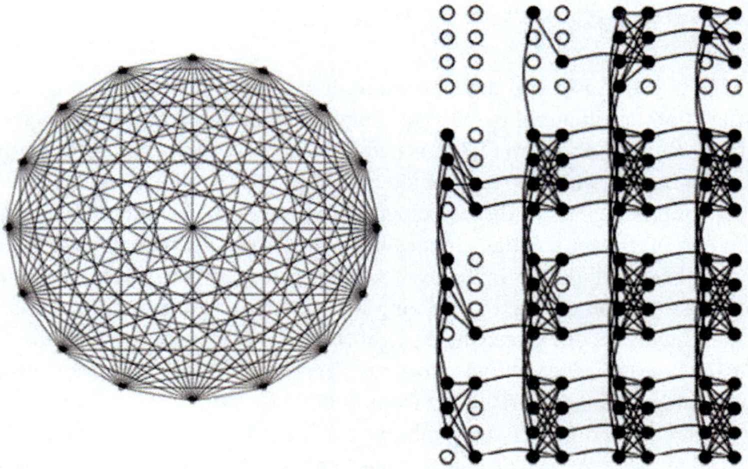

Fig. 3 Chimera graph for maximum likelihood annealing in D-wave. Adopted from Foster et al. [14]

The following steps are taken to run a program on the D-Wave:

1. Formulate the problem as a Ising model or QUBO model, as in Eq. (2).
2. Embed the graphical structure of the problem onto the Chimera graph in Fig. 3.
3. Specify any parameters of the anneal, such as the chain strength between qubits, anneal time, number of reads, etc.
4. Perform the quantum anneal and obtain the results.

At the end, after unembedding the results from the Chimera structure, the results are analyzed to obtain a solution and energy for each read of the anneal, which may involve simply choosing the solution returned with the lowest energy or performing some analysis over all returned solutions.

So far, the story seems to be good to be true. Indeed, as they say, "it is easier said than done". Step 1 for example is hardly non-trivial and there are other technical issues which we do not get into here regarding steps 2 and 3 (see Foster et al. [14] for detailed discussion). Furthermore, the machine itself suffers from hardware noise which may mask the true solution and the process of embedding reduces the scale of problems available. There is also the question of decoherence time, and so on (see DiVincenzo [9] criteria for implementation of quantum computation). There is also the question of decoherence time, and so on. Moreover, while the quantum machine uses quantum tunneling effects not available to classical computer, more research is still required to see whether these effects are indeed useful to more complex likelihood problems or scalable as the likelihood function grows.

6 Concluding Remarks

Simulated annealing is a popular algorithm that has been applied not only to maximum likelihood estimation, but also to many other complex problems such as the Travelling Salesman Problem (TSP), scheduling and task allocation, graph coloring and partitioning, as well as optimization problems for non-linear, multi-parameter functions. In this paper, we discuss quantum annealing as an alternative that may have more to offer over simulated annealing with respect estimating parameters for maximum likelihood estimation, at least when it comes to speed. However, with all types of quantum computing, maintaining the coherence of quantum particles as in Eq. (1), and with decoherance time typical of nanoseconds in quantum machines, lengthening time until decoherance remains a very difficult task. Moreover, the debate over quantum supremacy remains far from settled [23, 30].

For the simple two-parameter likelihood function discussed in this paper, Foster et al. [14] demonstrates that despite various hardware and technical limitations, the D-Wave algorithm performed well on a simulated data set closely converging to the given numerical resolution. However there much work needed to make quantum annealing useful for a broader class of maximum likelihood estimation. For example, inaccuracies creep in when maximizing the Taylor expansion likelihood rather than the full likelihood, thereby leading to the nearest local minima rather than the global minima. An obvious possible solution is to add more terms to the Taylor expansion of Eq. (6) but this would add complexity with the interactions of three or more qubits, which is not feasible with current hardware. There are other potential solutions, such as estimating the Ising or QUBO representation directly as the objective function, and so on. Furthermore, when is it worthwhile to write a MLE program to run in a quantum computer? Unfortunately, apart from the gain in speed, there is no universal answer. Further research on other possibilities with quantum annealing, quantum computing and indeed everything quantum is expected to grow exponentially as technical progress are made [15, 24, 39].

References

1. Baaquie, B. E. (2004). *Quantum finance: Path integrals and Hamiltonians for options and interest rates*. Cambridge University Press.
2. Bian, Z., Chudak, F. A., Macready, W. G., & Rose, G. (2010). *The Ising model: Teaching an old problem new tricks—D-wave systems* (Technical report).
3. Biswas, R., Jiang, Z., Kechezhi, K., Knysh, S., Mandrà, S., O'Gorman, B., Perdomo-Ortiz, A., Petukhov, A., Realpe-Gmez, J., Rieffel, E., Venturelli, D., Vasko, F., & Wang, Z. (2017). A NASA perspective on quantum computing. *Parallel Computing, 64*(C), 81–98.
4. Borle, A., & Lomonaco, J. (2019, February 27–March 2). Analyzing the quantum annealing approach for solving linear least squares problems. In *13th International Conference, WALCOM, Proceedings*, Guwahati, India.
5. Byrne, P. (2013). *The many worlds of Hugh Everett III: Multiple universes, mutual assured destruction, and the meltdown of a nuclear family*. Oxford University Press.

6. Byshkin, M., Stivala, A., Mira, A., Robins, G., & Lomi, A. (2018). Fast maximum likelihood estimation via equilibrium expectation for large network data. *Scientific Reports, 8*(1), 1–11.
7. Crosson, E., & Harrow, A. (2016). Simulated quantum annealing can be exponentially faster than classical simulated annealing. In *Proceedings of FOCS 2016* (pp. 714–723).
8. Denchev, V. S., Bloixo, S., Isakov, S. V., Ding, N., Babush, R., Smelyanskiy, V., Martinis, J., & Neven, J. (2016). What is the computational value of finite range tunneling? *Physical Review X, 6*(3). Available at arXiv:1512.02206v4
9. DiVincenzo, D. P. (2000). The physical implementation of quantum computation. *Protein Science, 48*, 771–783.
10. Eglese, R. W. (1990). Simulated annealing: A tool for operational research. *European Journal of Operational Research, 46*, 271–281.
11. Farhi, E., Goldstone, J., Gutmann, J., & Sipser, M. (2000). Quantum computation by adiabatic evolution. Available at arXiv:quant-ph/0001106v1
12. Farhi, E., Goldstone, J., Gutmann, S., Lapan, J., Lundgren, A., & Preda, D. (2001). A quantum adiabatic evolution algorithm applied to random instances of an NP-complete problem. *Science, 292*, 472–476.
13. Farhi, E., Goldstone, J., Gutmann, S., & Sipser, M. (2002). Quantum adiabatic evolution algorithms versus simulated annealing. Available at arXiv:quant-ph/0201031v1
14. Foster, R. C., Weaver, B., & Gattiker, J. (2019). Applications of quantum annealing in statistics. Available at https://arxiv.org/pdf/1904.06819.pdf
15. Ide, N., Asayama, T., Ueno, H., & Ohzeki, M. (2020). Maximum-likelihood channel decoding with quantum annealing machine. Available at https://arxiv.org/abs/2007.08689
16. Irback, A., Peterson, C., & Potthast, F. (1996). Evidence for nonrandom hydrophobicity structures in protein chains. *Proceedings of the National Academy of Sciences of the United States of America, 93*, 533–538.
17. Jacquier, E., Johannes, M., & Polson, N. (2007). MCMC maximum likelihood for latent state models. *Journal of Econometrics, 137*(2), 615–640.
18. Johnson, M. W., Amin, M. H. S., Gildert, S., Lanting, T., Hamze, F., Dickson, N., Harris, N., Berkley, A. J., Johansson, J., Bunyk, P., Chapple, E. M., Enderud, C., Hilton, J. P., Karimi, K., Ladizinsky, E., Ladizinsky, N., Oh, T., Perminov, I., Rich, C., ... Rose, G. (2011). Quantum annealing with manufactured spins. *Nature, 473*, 194–198.
19. Kadowaki, T., & Nishimori, H. (1998). Quantum annealing in the transverse Ising model. *Physical Review E, 58*(5), 5355.
20. Kaye, P., Laflamme, R., & Mosca, M. (2007). *An introduction to quantum computing*. Oxford University Press.
21. Kreinovich, V., Nguyen, H., & Sriboonchitta, S. (2018). Quantum ideas in economics beyond quantum econometrics. In L. H. Anh, L. S. Dong, V. Kreinovich, & N. N. Thach (Eds.), *Econometrics for financial applications*. Studies in Computational Intelligence (Vol. 760, pp. 146–151).
22. Majewski, J., Li, H., & Ott, J. (2001). The Ising model in physics and statistical genetics. *The American Journal of Human Genetics, 69*, 853–862.
23. Mandrà, S., Katzgraber, H. G., & Thomas, C. (2017). The pitfalls of planar spin-glass benchmarks: Raising the bar for quantum annealers (again). *Quantum Science and Technology, 2*(3), 38501.
24. McGeoch, C. C. (2014). Adiabatic quantum computation and quantum annealing. In *Synthesis lectures on quantum computing*. Morgan & Claypool Publisher.
25. Metropolis, N., Rosenbluth, A. E., Rosenbluth, M. N., Teller, A. H., & Teller, E. (1953). Equation of state calculations by fast computing machines. *The Journal of Chemical Physics, 21*, 1087–1092.
26. Nikolaev, A. G., & Jacobson, S. H. (2010). Simulated annealing, Chap. 1. In M. Gendreau & J.-Y. Potvin (Eds.), *Handbook of metaheuristics, 1*. International Series in Operations Research & Management Science (Vol. 146, pp. 1–39). Springer.
27. Nielsen, M. A., & Chuang, I. L. (2000). *Quantum computation and quantum information*. Cambridge University Press.

28. Nguyen, H., & Dong, L. S. (2018). An invitation to quantum econometrics. In L. H. Anh, L. S. Dong, V. Kreinovich, & N. N. Thach (Eds.), *Econometrics for financial applications*. Studies in Computational Intelligence (Vol. 760, pp. 44–62).
29. O'Malley, D., & Vesselinov, V. V. (2016). A high-level programming language for D-Wave machines based on Julia. In *2016 IEEE High Performance Extreme Computing Conference (HPEC)* (pp. 1–7).
30. Ronnow, T. F., Wang, Z., Job, J., Boixo, S., Isakov, S. V., Wecker, D., Martinis, J. M., Lidar, D. A., & Troyer, D. A. (2014). Quantum computing. Defining and detecting quantum speedup. *Science, 345*(6195), 420–424.
31. Rovelli, C. (2018). *The order of time*. Riverhead Books.
32. Sakurai, J. J., & Napolitano, J. (2010). *Modern quantum mechanics*, 2nd ed. Addison-Wesley.
33. Shankar, R. (1994). *Principles of quantum mechanics*, 2nd ed. Plenum Press.
34. Shin, S. W., Smith, G., Smolin, J. A., & Vazirani, U. (2014). How "quantum" is the D-wave machine? Available at arXiv:1401.7087 [quant-ph]
35. Stapp, H. P. (1996). The hard problem: A quantum approach. *Journal of Consciousness Studies, 3*(3), 194–210.
36. Stauffer, D. (2008). Social applications of two-dimensional Ising models. *American Journal of Physics, 76*, 470–473.
37. Wang, Y. (2011). Quantum Monte Carlo simulation. *Annals of Applied Statistics, 5*, 669–683.
38. Wang, Y. (2012). Quantum computation and quantum information. *Statistical Science, 27*, 373–394.
39. Wang, Y., Wu, S., & Zou, J. (2016). Quantum annealing with Markov chain Monte Carlo simulations and D-wave quantum computers. *Statistical Science, 31*(3), 362–398.
40. Yang, F., Ren, H., & Hu, Z. (2019). Maximum likelihood estimation for three-parameter Weibull distribution using evolutionary strategy. *Mathematical Problems in Engineering, 2019*(2019), 1–8.

Effects of Tourism Expenditure on Cambodia' Economy: A Computable General Equilibrium Model for Cambodia

Chantha Hor

Abstract The ten-year (2011–2020) tourism strategic development plan in Cambodia has diversified destinations into four strategic tourism destinations: Central (business tourism), Northwest (cultural tourism), Southwest (coastal tourism), and Northeast (eco-tourism). This paper analyzes the impact of a 10% increase in tourism expenditure for the tourism sector by employing a computable general equilibrium (CGE) model based on Macro-and Micro social accounting matrix year 2015. I find that 10% increased tourism expenditure stimulates the real GDP, private consumption, exports, and imports in the short run; it reduces real GDP in the long run. The estimation result suggests that a shock in tourism expenditure increases household income and improves household welfare at Cambodia's four strategic tourism destinations.

Keywords Tourism expenditure · CGE model · Social accounting matrices · Cambodia

JEL Classification E16 · E17 · O53 · Z32

1 Introduction

In many countries, especially developing countries, tourism has emerged as an important engine for economic growth. Cambodia's tourism sector has been re-emerging into the Cambodian economy since 1992 after nearly two decades of suffering from civil war and political turmoil (1975–1991) [27]. The government of Cambodia recognized its importance for economic growth and poverty reduction in 1995. Since then, Cambodia's government has made tourism a top priority in the country's growth strategy. Tourism was the first mentioned in the five-year Socio-Economic Development Plan (SEDP-I) (1996–2000) in 1995. The plan aimed to increase tourism

C. Hor (✉)
ACLEDA Institute of Business, #1397, Phnom Penh-Hanoi Friendship Blvd, Phum Anlong Kngan, Sangkat Khmuonh, Khan Saensokh, Phnom Penh 1149, Kingdom of Cambodia
e-mail: chatha.hor@gmail.com

S. Sriboonchitta et al. (eds.), *Credible Asset Allocation, Optimal Transport Methods, and Related Topics*, Studies in Systems, Decision and Control 429,
https://doi.org/10.1007/978-3-030-97273-8_32

attractiveness in Phnom Penh, Siem Reap, Sihanouk Ville, and Ratanakiri. Government also strengthened infrastructure, increased human capital, and attracted the private sector's investment in the tourism and related tourism sectors. In 2000, the government released the SEDP-II (2001–2005). It focused on diversifying tourism goods and services, ensuring effective tourism promotion and marketing, encouraging private sectors to develop innovative tourism products, encouraging local people to develop tourism products in the community, improving accessibility into and within the country, and upgrading infrastructure designated tourist sites. In 2005, Cambodia released the National Socio-Development Plan (NSDP) (2006–2010). The plan included important initiatives for Cambodia's tourism growth, including an open sky policy, visa facilitation on arrival, signing the ASEAN Tourism Agreement to enhance intra-regional tourism, tourism marketing, and tourism information offices establishment, and improvement of tourism products in prioritized areas (Phnom Penh, Siem Reap, Coastal Zone, and Northeast).

In addition, in 2009, the tourism industry received its first law. This law aimed to manage and sustain tourism growth, conserve and protect natural and cultural resources, promote tourism service quality, ensure security and safety, and encourage public and private sectors to work together to advance the tourism industry. Cambodia issued the national tourism strategic development plan (2012–2020) in 2011, which attempted to attract 7.5 million inbound tourists, generate US$5 billion in tourism receipts, and create one million employments in 2020. The government aimed to enhance tourism product quality, improve marketing and promotion, improve travel facilitation, transportation, regional and global connectivity, strengthen human resources, and diversify tourism destinations into four main strategic tourism destinations. In 2016, the Ministry of Tourism released the China-Ready White Paper Policy, which aimed to attract two million Chinese tourists per year. The policy supports Chinese signage and visa processing documents, promotes Chinese Yuan currency and language, employs Chinese speakers, and prepares tourism documents in Chinese. The government has accredited tourism sites as China-ready by ensuring that food and accommodation facilities satisfy Chinese tastes. Lately, the government released the national eco-tourism policy (2019–2030) in 2018. The policy has a long-term vision: "to transform Cambodia into a quality tourism destination and a new eco-tourism attraction with high competitiveness in the region and the world by creating eco-tourism products that give absorbing and high-quality experiences for tourists contributing to poverty reduction and promotion of natural resource conservation as well as an effective prevision of climate changes" [30].

Within the above series of tourism development plans and strategies implementation from 1995 to 2018 in Cambodia, tourism has accounted as the second-largest sector, earning over US$4.4 million in foreign exchange—equivalent to over 32% of GDP—and employing directly and indirectly over 3 million people in 2019 [29]. During the same year, inbound tourists visit Cambodia about 6.6 million, three times higher than ten years ago [25]. Meanwhile, Cambodia is moving toward a sector-specific approach focused on the high-value-added tourism sector. This study aims to measure what these programs have for Cambodia's economy. Hence, I attempt to examine changes in inbound tourists' expenditures on macroeconomic, sectorial effect, household income, and household welfare at the four main tourism strategic destinations. This paper simulates a 10% increase in tourism expenditures using the computable general equilibrium (CGE) model. The aim of a 10% shock on the tourism expenditure is to test the capability of the tourism sector has on Cambodia's economy. This study has two main contributions. First, I create the 2015 social accounting matrix (SAM) table for the tourism economic impact study in Cambodia, the latest SAM table at the time of the study. Second, it is the first study to employ the CGE modeling to measure the economic impact of tourism at the national and provincial levels in Cambodia.

The rest of this paper is structured as follows; the following section describes a literature review—section three highlights the research method and data description in section four. Section five discusses the findings. Section six is the conclusion and policy recommendation.

2 Literature Review

This section primarily focused on our literature review, methodology, and simulated scenarios. Some studies have used the IO model to assess economic impact in tourism studies. The multiplier studies (e.g., IO and SAM-based models) use disaggregated tourism-related sectors rather than a single sector to reflect the tourism sector. Some countries have used these models, including Ireland [14] and Turkey [12]. Most studies in Cambodia utilize these models in agriculture and fishery [16] and critical sectors specification [6]; however, none use them in tourism studies.

There are some drawbacks to the IO model. Certain assumptions, such as fixed rates and wages, are used in the model, which may or may not be appropriate for each economy [4]. The CGE model can overcome this constraint since it uses more realistic assumptions and more accurate findings than the IO model [8]. The CGE model has two advantages: (1) it allows prices and wages to adjust [7], and (2) it incorporates a wide variety of dynamic economies into a single model, making the findings more realistic. Tourism is a multi-sector activity by nature; therefore, it is likely to have economic consequences and connections. The CGE model is a policy analysis of the economic impact used in tourism studies. As a result, the CGE model fits the characteristics of the tourism industry. The CGE model has been widely used in empirical studies to address a broad range of tourism problems and simulations in a variety of economies: New Zealand [1], Brazil [3], Australia [15], Turkey [13], China [20, 21], Singapore [24], Indonesia [26], and Taiwan [31]. For example, a rise or decrease in tourism receipts [17], changes in tourism demand [24], shock or crisis in the tourism sector [1], tourism taxes [11], environment and globalization [28], poverty and income inequality [3, 23], productivity [5], and festival events.

For example, in the literature, [2] utilized the CGE model for Spain to study the impact of a 10% increase in tourism demand. He did not use a single tourism industry; instead, he used the Spanish tourism IO table from 1992, which included 49 sectors and six tourism industries (hotels, hostels, restaurants, other lodging establishments, and travel agencies), [13], on the other hand, looked at the effect of a 10% rise in international demand on Turkey's tourism industry.

In Cambodia, empirical research on policy effect assessments using the CGE model (see, e.g., [6, 9, 10, 16, 19]) excluded tourism. Dyna and Em [9] investigated the effect of tariff elimination on household welfare and labor market; [6] investigated the impact of promising export sectors on the domestic economy; [10] studied the impact of increased public education expenditure on the labor market. Furthermore, [16] investigated Cambodia's economic system and pro-poor development strategies, while [19] investigated the effect of graduation-based social production on Cambodia's extreme poor. These studies mainly focus on the Agriculture, agro-agriculture, taxes, and education sectors. The CGE model has not been used to test Cambodia's tourism sector or policies in any reports. The main reason for this seems to be data availability and the SAM table. This paper will first apply the CGE model to Cambodia's tourism sector to the best of the author's knowledge.

3 CGE Model

This study employed a single-country, multi-sector, static CGE model for the Cambodian economy. Labor and capital are the two types of production factors that each industry employs. The Cobb–Douglas production function is used, and each sector is assumed to have its capital stock. The model takes several tax instruments into account and simulates all conceivable agent transfers. The model and GAMs code of [22] are written in a comprehensive form based on conventional assumptions. These characteristics make it easier to apply to variously aggregated social accounting matrices. The model has structured in production, demand, international trade, income and savings, prices, equilibrium, closure, and exogenous variables follow the paper of [18, 22]. All equations for each box in the following sub-sections are heavily adapted from the [22].

More specifically, this study presents the specification for the 10% increased tourism expenditure on Cambodia's economy by shocking the export price parameter (δ_c^t) in the tourism sector as shown in the equation below:

$$\frac{QE_c}{QD_c} = \left(\frac{PE_c}{PDS_c} \cdot \frac{1-\delta_c^t}{\delta_c^t}\right)^{\frac{1}{\rho_c^t - 1}} \tag{1}$$

where δ_c^t is output transformation (CET) function share parameter; ρ_c^t is a CET function exponent, QE_c is quantity export; QD_c is the quantity of domestic sales, PE_c is the export price, PDS_c is the supply price for commodities produced and sold domestically.

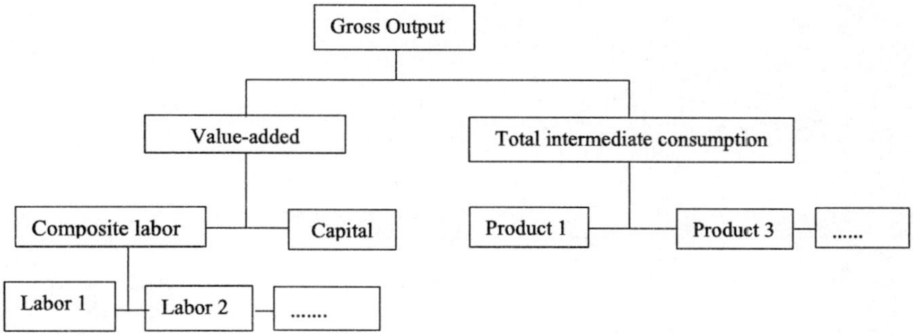

Fig. 1 Nesting structure of the production function in the CGE model

3.1 Production

Producer behavior defines by a series of equations that consider production technology, demand for production factors and intermediary goods, and factor substitution. Each firm attempts to operate in a perfectly competitive environment, in which case it optimizes profits by utilizing the existing manufacturing technology and engaging in price-taking conduct. Each production function is equated through a nested production structure (Fig. 1).

Each productive activity's sectorial output, at the top level, combines value-added and total intermediate consumption in fixed shares (e.g., aggregate outputs are considered to be strictly complementary with no possibility of substitution, following a Leontief production function).

Each sector value-added comprises a mix of labor and capital that follows a Cobb–Douglas formula at the second level. Profit maximization and cost minimization will drive firms to use each factor until the marginal product of each input matches its price. The goods and services that makeup aggregated intermediate consumption are diverse. Intermediate inputs are supposed to be wholly complimentary and combined by a Leontief production function.

At the bottom level, several categories of labor are merged using a constant elasticity of substitution type of technology, reflecting the imperfect substitutability of different forms of labor. Given the relative wage rate, the firm chooses its worker mix to minimize labor expenditures.

3.2 Demand and International Trade

Another set of equations utilizes to describe demand and trade relationships. We suppose that domestic sales and exports have imperfect transformability and that the aggregated output is divided between international and domestic markets. The

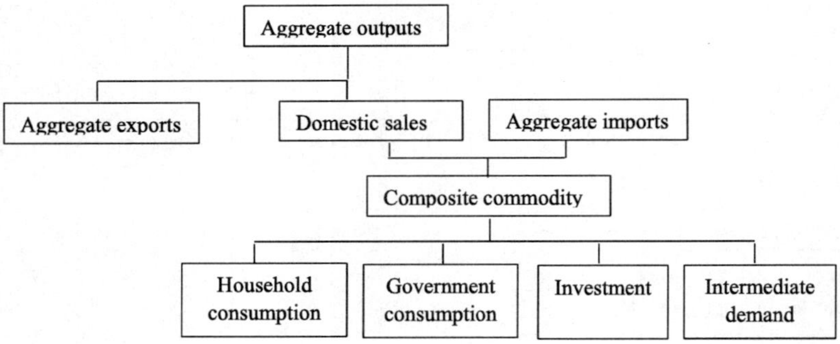

Fig. 2 Aggregate supply and demand

constant elasticity of the transformation function represents imperfect transformability between two goods. Composited goods are goods that combine domestically produced and imported commodities and are sold on the domestic market. Locally produced items and imported commodities, on the other hand, are assumed to be imperfect substitutes. The substitution function's constant elasticity is used to describe this imperfect substitutability. Household consumption, government consumption, investment and intermediate inputs are the four components of composite goods demand (Fig. 2).

3.3 *Income and Savings*

A set of equations describes the income and savings of various domestic institutions (households, firms, and the government) and international institutions (the rest of the world) in the economy. The income sources of households are depicted in Fig. 3. Lofgren et al. [22] provide a detailed explanation of the model. Labor, capital, and transfers from companies, the government, and the rest of the world are sources of income for households. With the profit they earn, they save and buy goods. Their disposable income is computed by deducting direct taxes and household transfers from total income. The remainder of the income is spent after it has been saved and

Fig. 3 Household income and savings

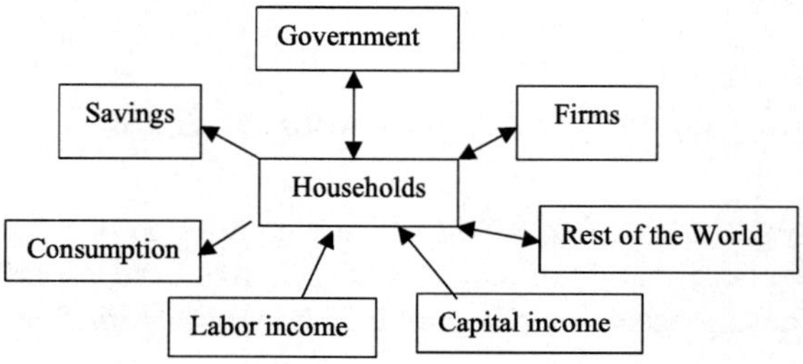

transferred to other agents. Income sources by labor and household are utilized to deal with the distributional issue in this model.

3.4 Prices

The prices in the model are specified by a set of equations based on assumptions and function forms. Producer pricing, export and import prices, and consumer prices are all included. The pricing mechanism sets up to match demand and supply. The weighted sum of the components determines the price of value-added and intermediate consumption. The price of value-added, for example, is made up of the price of composite labor and the price of capital.

3.5 Equilibrium

Equilibrium is one of the CGE model's assumptions, and it must be proven for all markets, including output markets as well as markets for factors of production. Demand and supply of domestically manufactured commodities must balance each export market, and total investment expenditure must equal total savings.

3.6 Simulation Designs and Macro Closure

As stated in the introduction, this paper uses the CGE model to examine the influence of tourism development strategies on Cambodia's economy. The simulation carries out by increasing tourism expenditure in the tourism sector by 10%. It is worth mentioning that CGE model simulations necessitate assumptions regarding macroeconomic closures and parameter behavior. Even though these closures represent some aspects of the economy, they can impact simulation results. I choose the closure based on the reality of the Cambodian economy. I conduct the simulation under the closure related to saving-driven investment, fixed foreign saving, government flexible, and flexible nominal exchange rate. The simulation and closure are set to assess the short- and long-term impact of tourism. In the short-run simulation, given the high unemployment rate across all labor categories, labor is unemployed and move across sectors. Wages are set for all job categories, and employment levels adjust to achieve labor market equilibrium. Capital is fully utilized and sector-specific, allowing returns to adjust to establish equilibrium in the capital market. Labour and capital are fully employed and mobile across sectors in the long-run simulation. In the top panel of each table, I present the simulation closure rules.

4 Model Databases

This study uses the SAM table to calibrate into the CGE model. I develop the 2015 SAM based on national accounts elaborated by the Cambodian National Institute of Statistics (NIS). To develop the table, I utilize various data sources to calculate the SAM and mathematical balancing approach to create a coherent SAM and align all data sources. The three crucial data sources used to construct the 2015 SAM: (1) macroeconomics data—the Ministry of Economic and Finance, National Bank of Cambodia, and IO table collected from the OECD database; (2) national socio-economic survey (also known as a household survey) in 2015; and (3) tourism domestic and inbound tourism expenditure survey in 2017. Since comprehensive national accounts and household survey data are only available for 2015, I choose it as the benchmark year. The IO table has 35 × 35 sectors and is then aggregated into 15 × 15 sectors in this study. Table 1 shows a list of these sectors along with their IO table codes.

As shown in Table 2, the SAM has 17 accounts: two production accounts, two production factors, four institutions, five taxes, the rest of the world, inbound tourism, changes in stock, and savings-investment accounts. The production accounts break down into 15 different industries. The activities account depicts production interactions, while the commodities account depicts supply and demand structure. Inbound tourism and domestic tourism are the two types of tourism accounts. Payments to the

Table 1 Sector classification in the macro SAM in 2015

Sector description	Codes in IO Table
(1) Agriculture	01
(2) Mining and quarrying	02–04
(3) Textiles	05
(1) Foods and beverages	06
(1) Other manufacturing	07–20
(1) Electricity	21
(1) Transportation and communication	22–25
(1) Construction	26
(1) Wholesale and retails	27
(1) Financial and insurance	28
(1) Real estates	29
(1) Public administration	30
(1) Education	31
(1) Human health and social work	32
(1) Tourism	33–35
(1) Domestic tourism	–
(2) Inbound tourism	–

Source IO table from IO-OECD database

Table 2 Structure of SAM for Cambodia in 2015

		Activities	Commodities	Labour	Capital	Households	Domestic tourism	Enterprises	Government	Rest of the world
		1	2	3	4	5	6	7	8	9
Activities	1		A_1							
Commodities	2	C_1				C_2	C_3		C_4	C_5
Labour	3	L_1								
Capital	4	K_1								
Households	5			H_1	H_2			H_3	H_4	H_5
Domestic tourism	6			D_1		D_2				
Enterprises	7				E_1					
Government	8				G_1			G_2		G_3
Rest of the world	9		R_1			R_2		R_3	R_4	
Inbound tourism	10		I_1							I_2
VAT-taxes	11		G_4							
Excise-taxes	12		G_5							
Import-taxes	13		G_6							
Export-taxes	14		G_7							
Direct-taxes	15					G_8				
Investments	16					S_1		S_2	S_3	S_4
Change in stocks	17									
	18	TPC	TAB	TVA		THE	TDE	TEE	TGE	TFE

(continued)

Table 2 (continued)

		Inbound tourism	VAT-taxes	Excise-taxes	Import-taxes	Export-taxes	Direct-taxes	Investments	Change in stocks	
		10	11	12	13	14	15	16	17	18
Activities	1									TDP
Commodities	2	C_6						C_7	C_8	TMS
Labour	3									TFI
Capital	4									
Households	5									THI
Domestic tourism	6									TDI
Enterprises	7									TEI
Government	8		G_4	G_5	G_6	G_7	G_8			TGI
Rest of the world	9									TFE
Inbound tourism	10									TR
VAT-taxes	11									TTAX
Excise-taxes	12									
Import-taxes	13									
Export-taxes	14									

(continued)

Table 2 (continued)

		Inbound tourism	VAT-taxes	Excise-taxes	Import-taxes	Export-taxes	Direct-taxes	Investments	Change in stocks	
		10	11	12	13	14	15	16	17	18
Direct-taxes	15									
Investments	16									TS
Change in stocks	17							C_8		TCHS
	18	TIBE	TTAX					TIV	TCHS	

Note A_1 marketed outputs, C_1 intermediate inputs, C_2 household consumption, C_3 domestic tourism expenditures, C_4 government consumption, C_5 Exports excluding tourism exports, C_6 inbound tourism expenditure (direct purchased by none-resident), C_7 saving-investment, C_8 change in stock, L_1 labour compensation, K_1 operating surplus, H_1 factor of labour income to households, H_2 factor of capital income to households, H_3 Distributed profits and other transfers, H_4 transfers to households, H_5 foreign transfer, D_1 domestic business tourism expenditure, D_2 transfer to domestic tourism consumption, E_1 factor income to enterprise, G_1 capital income to government, G_2 corporate taxes, G_3 foreign transfer, G_4 value added taxes, G_5 excise taxes, G_6 import taxes, G_7 export taxes, G_8 direct taxes, R_1:import exclude tourism imports, R_2 household transfer to ROW, R_3 surplus to *ROW*, I_1 outbound tourism expenditure (direct purchased abroad by resident), I_2 inbound tourism expenditure (direct purchased by non-resident), S_1 household saving, S_2 corporate saving, S_3 government saving, S_4 foreign savings, *TPC* total production costs, *TAB* total absorption, *TVA* total value added, *THE* total household expenditures, *TDE* total domestic tourism expenditures, *TEE* total enterprise expenditures, *TGE* total government expenditures, *TFE* total foreign exchange earnings, *TIBE* total inbound tourism expenditures, *TTAX* total tax expenditures, *TIV* total investments, *TCHS* total change in stock, *TDP* total domestic production, *TMS* total market supply, *THI* total household incomes, *TDI* total tourism domestic earnings, *TEI* total enterprise income, *TGI* total government income, *TFE* total foreign exchange earnings, *TR* total tourism receipts, *TTAX* total incomes from taxes, *TS* total savings, *TCHS* total stock changes tourism accounts from other sectors resulting from intermediary use of tourism activities and household tourism expenses are reported in the commodities to account for tourism. Payments from the tourism industry to other sectors for intermediate input purchases are recorded in the tourism activity column accounts. Export receipts are reported in the rest of the world column. Inbound tourism receipts are split from exports by separating inbound tourism from the rest of the world account. The domestic and inbound tourism expenditure survey for 2017 conducted by Cambodia's Ministry of Tourism was used to compile the tourism data used in the 2015 SAM.

The 2015 macro SAM and data sources are explained in detail for each cell. The IO table is based on data from the OECD database. Intermediate demand inputs, household consumption, government consumption, exports, imports, gross fixed capital information (investment), stock changes, labor compensation, gross operating surplus, and value-added taxes are all included in the table. The table uses the price paid by buyers and includes the cost of products and services at the point of sale to buyers. For the household account, I used the Cambodia National Institute of Statistics' 2015 socio-economic survey, also known as the households survey. I use data from the Ministry of Finance and Economics for taxable accounts. I used data from the Ministry of Tourism of Cambodia's 2017 tourism domestic expenditure survey and the IO table to calculate the domestic and inbound tourism accounts.

The structure of the 2015 SAM for Cambodia is depicted in Table 2. The row and column numbers are assigned to each cell in the macro SAM. For instance, (1, 2) corresponds to the first column (activities) and the second row (activities) (commodities). This cell displays the outputs that have been sold.

Marketed outputs (1, 2): Marketed outputs are determined by subtracting total exports from total outputs in the activity account's total column.

Intermediate inputs (2, 1): The data for intermediate inputs comes from the IO table.

Households consumption (2, 5): The private consumption of inhabitants collected from the IO table is household consumption expenditure.

Government consumption (2, 8): Government consumption refers to governments' purchases of goods and services, as obtained from the IO table.

Exports (2, 9): Non-residents purchase goods and services, which are referred to as exports. The purchases of products and services made by inbound tourists are excluded from the IO table's exports.

Investment (2, 16): Gross capital fixed formation is used as investment demand. It was spent on residential buildings, machinery, and equipment, and it was added to the stock of fixed assets. The investment demand is obtained from the IO table.

Change in stock (2, 17): Change in stock taken from the IO table.

Employee compensation (3, 1): Employees' wages and salaries are included in compensation. Employers contribute to pension funds and other social benefits. Employee compensation data is taken from the IO table.

Gross operating surplus (4, 1): Operating surplus is the amount paid to capital services after deducting employee salary and indirect taxes from gross output intermediate demand, except for commodities taxes.

Non-distributed profits (7, 4): Gross operational surplus, excluding production taxes, intermediate inputs, and employee compensation, is the non-distributed profit to enterprises. It contains the gross factor of capital income arising from capital employed, including capital consumption allowance and profit remittances from other countries. This data comes from the Cambodian National Bank.

Transfer to Households: Households receive income from various sources such as income, government transfer, and the rest of the world. Households in cells (5, 3) and (5, 4) receive compensation for their labor services and capital income from renting their properties or land. Data from the Cambodia socio-economic survey is used to calculate these figures. Households also receive incomes from enterprises that incorporate dividends shown in cell *(5, 7)* and from the rest of the world as remittances shown in cell *(5, 9)*. Data in cells *(5, 7)* and *(5, 9)* are measured from the balance of payment from the National Bank of Cambodia and socio-economic survey data.

Corporate income taxes on enterprises (8, 7) and direct income taxes on households (15, 5): The government received corporate and direct tax contributions from firms and households. These figures can be found in the government's tax accounts, which the Ministry of Economic and Finance records.

Value-added taxes (8, 11), excise taxes (8, 12), import taxes (8, 13), and export taxes (8, 14): Value-added taxes (VATs) are sales taxes on goods. Excise tax is the tax of specific commodities (e.g., foods and beverages and transportation). Import taxes are paid per unit of imported products, while export taxes are paid when commodities and services leave the economic territory. These figures come from Cambodia's Ministry of Finance and Economics.

Imports (9, 2): The imported goods and services are measured based on cost, insurances, and freight. This data is taken from the IO table.

Current transfer abroad: Enterprises transfer to the rest of the world in payment factors as in cell (9, 7). Households and government make a foreign transfer abroad shown in cells (9, 5) and (9, 8), respectively. These data are taken from the Balance of Payment, National Bank of Cambodia.

Household saving (16, 5) and corporate saving (16, 7): Household saving is the amount of money left over after all expenses have been paid. The residual balance value from the enterprise account is used to compute the corporate saving of enterprises.

Government saving (16, 8) and foreign saving (16, 9): The current surplus of the government budget, as calculated by the Ministry of Economics and Finance, is used to calculate government savings. The current account balance, or the difference between foreign receipts and expenditures, is referred to as foreign saving. It is from Cambodia's National Bank's Balance of Payments.

Domestic tourism account: Domestic tourism expenditures (6, 5) are calculated using the Ministry of Tourism of Cambodia's domestic tourism expenditure survey. Domestic tourism expenditure is divided into five categories: accommodation, food and beverages, transportation, shopping, and others. The accommodation and food and drinks are allocated to the sector of hotel and food services. The expenditure on transportation is allocated into the transportation and communication sector. Whereas the expenditure on shopping and others are allocated into the sectors of wholesale and retails, and other business service sectors, respectively. The domestic tourism expenditure includes holiday tourism expenditure paid by tourists (private tourism consumption) and business tourism expenditure paid by enterprises (business tourism consumption). The domestic tourism expenditure conducted by the Ministry of Tourism of Cambodia does not provide separate data; thus, the calculation should be done by separating them. When domestic tourism expenditures are matched into the 15-production sectors, it is found that total domestic tourism expenditure on hotel and food services, transportation, wholesale and retails, and other business services is smaller than the household consumption data in the IO table. Therefore, the difference is assumed to be the private tourism expenditure *(6, 5)*, and the remaining is considered business tourism consumption *(6, 3)*.

Inbound tourism account: Inbound tourism expenditure and outbound tourism expenditure consist of exports and imports in the IO table. Hence, inbound tourism expenditure is split from the rest of the world's account. The inbound tourism expenditure *(2, 10)* uses the data directly purchased by a non-resident in the IO table that matches the 15-production account. The outbound tourism expenditure *(10, 1)* uses the data directly purchased abroad by residents in the IO table that matches the 15-production account. Therefore, the numerical macro 2015 SAM is depicted as in Table 3.

5 Result and Discussion

5.1 Cambodian Economic Structure

Table 4 shows the industrial economic structure in 2015. Cambodia's economy is heavily reliant on agriculture, accounting for 20% of total production. Textile, tourism, transportation, and communication, wholesale and retail, and other manufacturing come after agriculture, with 17, 11.8, 9.6, 8.9, and 7.8%, respectively. These top seven industries also dominate the total value-added. Agriculture contributes 22.8% of the total value-added, followed by tourism (12.5%), transportation and communication (11.9%), textile (11.7%), wholesale and retail (9.7%), and construction (7.3%). Tourism contributes 21% of total labor income, followed by construction

Table 3 Macro-SAM for Cambodia in 2015 (in USD Million)

		Activities	Commodities	Labour	Capital	Households	Domestic tourism	Enterprises	Government	Rest of the world
		1	2	3	4	5	6	7	8	9
Activities	1		29,878							
Commodities	2	12,625				12,056	1168		1023	6764
Labour	3	7265								
Capital	4	9989								
Households	5			6795	5033			1200	148	435
Domestic tourism	6			470		697				
Enterprises	7				4939					
Government	8				17			348		387
Rest of the world	9		8622			49		37	114	
Inbound tourism	10		367							282
VAT-taxes	11		901							
Excise-taxes	12		599							
Import-taxes	13		450							
Export-taxes	14		74							
Direct-taxes	15					609				
Investments	16					198		3355	2,101	953
Change in stocks	17									
	18	29,878	40,891	7265	9989	13,610	1168	4939	3385	8822

(continued)

Table 3 (continued)

		Inbound tourism	VAT-taxes	Excise-taxes	Import-taxes	Export-taxes	Direct-taxes	Investments	Change in stocks	
		10	11	12	13	14	15	16	17	18
Activities	1									29,878
Commodities	2	649						6512	95	40,891
Labour	3									7265
Capital	4									9989
Households	5									13,610
Domestic tourism	6									1168
Enterprises	7									4939
Government	8		901	599	450	74	609			3385
Rest of the world	9									8822
Inbound tourism	10									649
VAT-taxes	11									901
Excise-taxes	12									599
Import-taxes	13									450
Export-taxes	14									74
Direct-taxes	15									609
Investments	16									6607
Change in stocks	17							95		95
	18	649	901	599	450	74	609	6607	95	

Source Author's calculation

Table 4 Structure of Cambodia's economy, 2015, *Source:* Author's calculation from the 2015 SAM

Sectors	Q (%)	VA (%)	EM (%)	X (%)	M (%)	X/XS	M/Q
Agriculture	20.5	22.8	11.1	6.3	1.7	6.6	4.6
Mining and quarrying	0.4	0.4	0.3	0.2	9.5	6.9	87.1
Textile	17.0	11.7	7.3	38.1	12.2	49.9	28.2
Food and beverage	5.9	5.1	3.8	12.2	1.2	46.5	10.4
Other manufacturing	7.8	5.7	6.2	5.5	46.0	15.6	65.0
Electricity	1.3	1.2	1.0	10.4	0.0	0.0	68.3
Transportation and communication	9.6	11.9	12.0	19.2	14.0	44.4	42.2
Construction	7.1	7.3	13.5	0.1	0.0	0.0	3.0
Wholesale and retail	8.9	9.7	9.5	4.8	2.3	12.0	9.0
Financial and insurance	1.2	1.2	0.7	1.2	0.3	18.1	9.7
Real estates	3.7	5.0	2.7	8.3	0.1	50.4	4.0
Public administration	2.0	2.2	5.7	0.0	0.0	0.0	1.1
Education	1.6	2.1	3.9	0.2	0.1	2.2	2.1
Human health and social work	1.2	1.3	0.7	0.1	0.0	2.4	1.4
Tourism	11.8	12.5	21.5	4.0	2.0	7.5	5.5
Total	100	100	100	100	100		

Note X/XS export share to sector output, *M/Q* import share to total domestic production, *Q* domestic production, *X* exports, *M* imports, *EM* employment, *VA* value-Added

13.5%, transportation and communication 12%, and agriculture 11.1%. It indicates that these industries are the most important sources of labor income and jobs.

Textile, transportation and communication, as well as food and beverage, are Cambodia's main exports. 69.5% of total exports come from these three industries, which also have the highest export intensity, with 50.4%, 49.9%, and 46.5% of their exported output, respectively. Other manufacturing, transportation and communication, and textile account for 72.2% of total imports. Surprisingly, mining is the sector with the most import competition, with foreign producers reaching 87.1% of total demand. Electricity (68.3%), other manufacturing (65%), and textile are the other industries that face major import competition (28.2%).

5.2 Simulation Results

5.2.1 Macroeconomic Impacts

Table 5 depicts the impact of a 10% increase in tourism spending on macroeconomic indicators in the short and long run. In the short run, a 10% increase in tourism spending boosts real GDP by 0.50%, but in the long run, it lowers real GDP by 0.06%.

Table 5 Macroeconomic impacts (%)

Saving-investment	Savings-driven	Savings-driven
Government savings	Flexible	Flexible
Foreign saving	Fixed	Fixed
Exchange rate	Flexible	Flexible
Short-run/Long-run	Short-run	Long-run
Macroeconomic variables (Percentage change compare to base)		
Absorption	0.66	0.17
Private consumption	0.64	0.39
Investment	0.82	−0.28
Exports	0.45	−0.90
Imports	0.86	−0.14
GDP at market price	0.51	−0.06
Indirect taxes	0.75	0.03
GDP at factor price	0.52	−0.002
Factor Return		
Labour	1.49	0.91
Capital	0.21	−0.39
Price changes		
Consumer price index	0.02	−0.21

Source Author's calculation from the model

The short-run expansion in real GDP can be attributed to increased aggregate labor income and capital income. In the long run, the real GDP contraction is explained by reducing the capital stock on the income side. The expenditure side can also be explained by referring to a drop in exports and investments. For example, exports and investments reduce by 0.90 and 0.28%, respectively. It is clearly seen that outputs in the textile, other manufacturing, and construction sectors decline in the long run. All these sectors generally have a higher rate of return on capital investment and tourism sector has low rate of return on capital investment because of labour-intensive nature. Hence, when allocation of labour and capital from these sectors to the tourism sector, causing a decrease of the economy's average rate of return on capital investment.

On the other hand, the significant expansion of imports, exports, and investments are driven by an increased tourism expenditure in the short run. It increases imports by 0.86%, outweighing the exports, leading to a trade deficit. There are two reasons that imports have increased. The first reason is because of higher total demand in the economy caused by tourism expenditure shock. Second, the consumer price index is increased, which makes imported goods and services cheaper for Cambodians.

5.2.2 Sectorial Impacts (%)

Table 6 shows the sectoral impacts of 10% increase tourism expenditure in the short and long run.

In the short and long run, the direction and size of the overall change in outputs reveal positive and negative impacts in different sectors. As expected, a 10% increase in tourism expenditure positively impacts tourism, construction, transportation and communication, electricity, and agriculture outputs by 4.7%, 0.78%, 0.52%, 0.16%, and 0.12%, respectively, in the short run. Outputs in these sectors also increase in the long run. These can be explained by an increased domestic price caused by tourism demand shock. On the contrary, most manufacturing industries (e.g., textile and other manufacturing) fall in outputs experiences from this demand shock. It increases exports in the tourism sector by nearly 61% in the short run and 84% in the long run. However, exports in other sectors decreased. This result indicates that an increased tourism expenditure leads to fewer goods and services exports because of a tourism demand shock. Industries sell goods to the domestic market rather than to the rest of the world, reflecting a shift in focus from export-oriented sectors to domestic-oriented sectors. As expected, Table 6 reveals that a 10% increase in tourism expenditure increases imported and domestic goods for all sectors in the short run. However, it shows positive and negative impacts in the long run. In particular, imports in the food and beverage sector increase by 2.98%, followed by tourism (2.46%), human health and social work (2.2%), and agriculture (1.88%) in the short run. In the long run, an increased tourism expenditure increases imports in most industries due to higher tourism demand. This result indicates that shock in tourism demand causes inadequate domestic goods to meet tourists' expansion.

5.2.3 Households Income

Figure 4 shows the impact of 10% increase tourism expenditure on households income by four tourism destinations in Cambodia for the short and long run. In the short-run, the 10% increase in tourism expenditure positive impact on households' income in Northwest (cultural tourism) and Northeast (eco-tourism) around 0.85%, followed by households in Central (business tourism) and Southwest (coastal tourism). In the long run, tourism expenditure positively impacts a household's income at all four main tourism destinations; however, it is lower than in the short run.

5.2.4 Households Welfares

Figure 5 shows the impact of 10% tourism expenditure on the household welfares at the four main tourism destinations for both the short and long run. The tourism expenditure shock has improved the household's welfare at all four main tourism destinations in the short run higher than in the long run. As seen in Fig. 5, the

Table 6 Sectorial impacts (%) in the short-and long run

Sectors	Domestic price	Composite price	Imports	Domestic goods	Output	Export
	Short-run					
Agriculture	0.23	0.23	1.88	0.47	0.12	−6.25
Mining and quarrying	−0.29	−0.78	0.93	0.18	−0.09	−3.46
Textile	−0.81	−0.78	0.17	0.02	−0.34	−0.69
Food and beverage	−0.40	−0.12	2.98	1.95	−0.26	−3.02
Other manufacturing	−0.46	−0.69	0.79	0.20	−0.28	−2.70
Electricity	0.03	−0.59	1.24	0.16	0.16	0.00
Transportation and communication	−0.69	−0.68	0.73	0.34	0.52	−1.58
Construction	0.03	0.00	1.87	0.78	0.78	0.00
Wholesale and retail	−0.25	−0.23	1.23	0.39	−0.10	−3.71
Financial and insurance	−0.34	−0.29	1.26	0.49	−0.16	−3.28
Real estates	−0.72	−0.56	1.14	0.76	−0.23	−1.12
Public administration	−0.09	−0.10	1.03	0.10	0.10	0.00
Education	0.06	0.06	1.54	0.40	0.28	−5.14
Human health and social work	0.61	0.62	2.26	0.44	0.27	−8.23
Tourism	1.52	0.72	2.44	0.39	4.69	60.79
	Long−run					
Agriculture	−0.12	−0.12	0.70	0.42	0.33	−0.94
Mining and quarrying	−0.05	−0.29	0.35	−0.02	−0.15	−1.85
Textile	0.05	0.20	−2.37	−3.25	−5.37	−7.50
Food and beverage	−0.04	0.15	4.12	3.44	1.90	0.13
Other manufacturing	0.00	−0.20	−0.06	−0.53	−0.89	−2.83
Electricity	−0.08	−0.25	0.47	0.16	0.16	0.00
Transportation and communication	0.03	0.04	0.19	−0.57	2.24	−4.33
Construction	0.26	0.24	0.44	−0.27	−0.27	0.00

(continued)

Table 6 (continued)

Sectors	Domestic price	Composite price	Imports	Domestic goods	Output	Export
Wholesale and retail	0.03	0.04	−0.27	−0.76	−1.06	−3.20
Financial and insurance	−0.15	−0.13	0.44	0.18	−0.06	−1.15
Real estates	−0.15	0.01	0.58	0.15	−0.93	−2.00
Public administration	0.52	0.51	1.02	−0.01	−0.01	0.00
Education	0.33	0.33	0.84	0.02	−0.06	−3.95
Human health and social work	−0.16	−0.16	0.70	0.49	0.46	−0.57
Tourism	0.25	−0.78	0.35	0.92	7.49	83.92

Source Author's calculation from the model

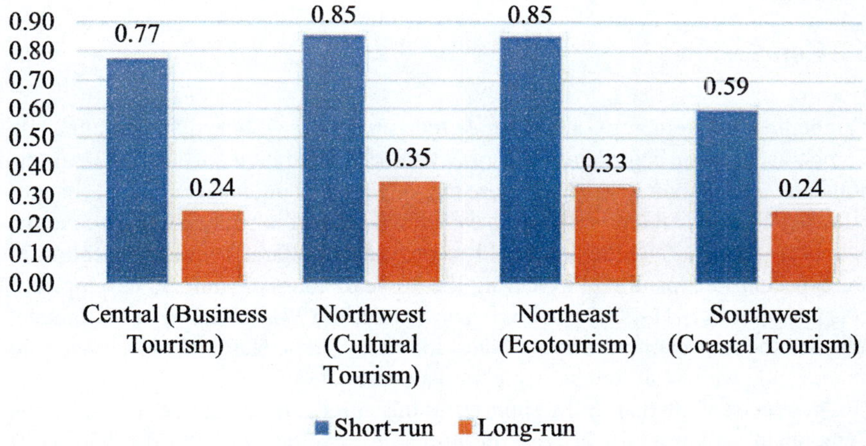

Fig. 4 Households income the four main tourism destinations. *Source* Author's calculation from model

improvement of household welfare in the central area is better than other tourism destinations in the short and long run. Notably, a 10% increase in tourism expenditure enhances household welfare by 0.006% in the short run and 0.004% in the long run in the central region, whereas it improves household welfare by 0.003% in the short run and 0.002% in the long run in the Northwest.

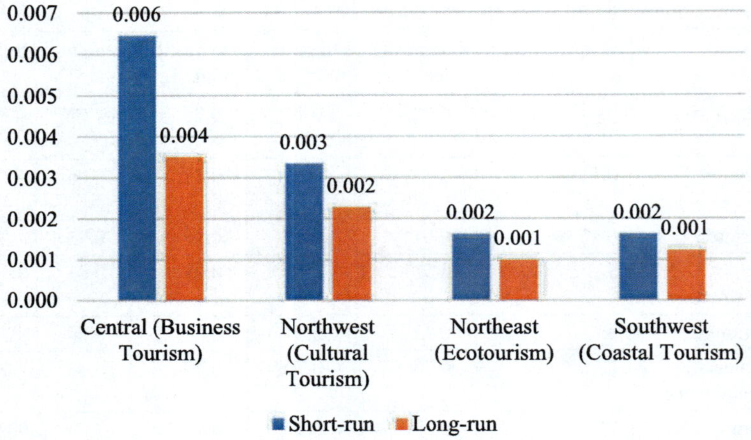

Fig. 5 Household welfare at the four main tourism destinations. *Source* Author's calculation from the model

6 Conclusion

Cambodia is a developing country whose tourism industry is heavily reliant. In 2018, Cambodian tourism contributed 33% of GDP, 8.1% of employment, and 30% exports, both directly and indirectly. Given these magnitudes, any internal or external changes affecting Cambodian tourism could have a wide-ranging impact on macroeconomic variables, sectorial effects, household income, and welfare. Tourism expenditures are expected to expand rapidly during the 10 years tourism strategic development plan and policy (2011–2020). In this paper, I utilize the 2015 SAM and a CGE model to assess the impact on macroeconomic variables, sectors, factor income, household income, and welfare at four strategic tourism destinations.

The results show that in the short term, this simulation would be helpful to the Cambodian economy. An increase in tourism expenditure by 10% can boost real GDP in the short term but not in the long term. It also suggests that an expansion in tourism increases household income and improves household welfare at the four main tourism strategic destinations in the short and long run. Consider the impacts on sectorial production, a tourism expenditure shock stimulates higher outputs in tourism, construction, electricity, and agriculture but decreases outputs in most industries in the short- and long run. A tourism expenditure shock distributes income unequally among the four main tourism strategic destinations. An increase in tourism expenditure stimulates development in services and domestic-oriented sectors more than manufacturing and export-oriented sectors. Policy shocks, as macroeconomic impacts, raise factor income, real private consumption, investment, and domestic prices at various levels, as expected.

To the best of the author's knowledge, the shock of tourism expenditure has never been examined in Cambodia using a CGE model. The findings can be used to inform economic policy in Cambodia. Simple intuition informs us that this study benefits tourism development planners and policy-makers for related tourism. More research is needed. Database improvements, modifications to the theoretical foundations that underpin CGE models, and simulations of various government policies influencing the tourism sector are just a few examples. For further research, additional databases and new simulations are all essential and interdependent tasks.

References

1. Becken, S., & Lennox, J. (2012). Implications of a long-term increase in oil prices for tourism. *Tourism Management, 33*(1), 133–142.
2. Blake, A. (2000). *The economic effects of tourism in Spain.* Christel DeHaan Tourism and Travel Research Institute.
3. Blake, A., Arbache, J. S., Sinclair, M. T., & Teles, V. (2008). Tourism and poverty relief. *Annals of Tourism Research, 35*(1), 107–126.
4. Blake, A., & Sinclair, M. T. (2003). Tourism crisis management US response to September 11. *Annals of Tourism Research, 30*(4), 813–832.
5. Blake, A., Sinclair, M. T., & Soria, J. A. C. (2006). Tourism productivity: Evidence from the United Kingdom. *Annals of Tourism Research, 33*(4), 1099–1120.
6. Chhuor, S. (2017). Potential roles of the export orientation of Cambodia's agriculture and agro-industry: An application of CGE analysis. *Journal of Economic Structures, 6*(1). https://doi.org/10.1186/s40008-017-0087-6
7. Dwyer, L., Forsyth, P., Madden, J., & Spurr, R. (2000). Economic impacts of inbound tourism under different assumptions regarding the macroeconomy. *Current Issues in Tourism, 3*(4), 325–363.
8. Dwyer, L., Forsyth, P., & Spurr, R. (2004). Evaluating tourism's economic effects: New and old approaches. *Tourism Management, 25*(3), 307–317.
9. Dyna, H., & Em, K. (2015). Impacts of Cambodia's tariff elimination on household welfare and labor market: a CGE approach. *Partnership for Economic Policy-PEP.*
10. Ear, S., Sim, S., & Khiev, P. (2016). Macroeconomic impacts of public consumption on education—A computable general equilibrium approach. *Partnership for Economic Policy-PEP.*
11. Forsyth, P., Dwyer, L., Spurr, R., & Pham, T. (2014). The impacts of Australia's departure tax: Tourism versus the economy? *Tourism Management, 40*, 126–136.
12. Gül, H. (2013). Economic impacts of an increase in the foreign tourism receipts: A SAM-based income multiplier analysis. *Advances in Hospitality and Tourism Research, 1*(1), 17–36.
13. Gül, H. (2016). Effects of foreign demand increase in the tourism industry: A CGE approach to Turkey. *An International Journal of Tourism and Hospitality Research, 26*(4), 598–611.
14. Henry, E. W., & Deane, B. (1997). The contribution of tourism to the economy of Ireland in 1990 and 1995. *Tourism Management, 18*(8), 535–553.
15. Ihalanayake, R. (2012). Economic effects of tourism tax changes in Australia: Empirical evidence from a computable general equilibrium model. *Tourism Economics, 18*(1), 181–202.
16. Kobayashi, S., Tanji, H., Saito, K., Huang, W., & Tada, M. (2009). The industrial structure of Cambodia and the role of agriculture and fishery in its development. *Japan Agricultural Research Quarterly, 43*(4), 309–316.
17. Narayan, P. K. (2004). The economic impact of tourism on Fiji's economy: Empirical evidence from the computable general equilibrium model. *Tourism Economics, 10*(4), 419–433.
18. Njoya, E. T., & Nikitas, A. (2020). The role of air transport in employment creation and inclusive growth in the Global South: The case of South Africa. *Journal of Transport Geography, 85*, 102738. https://doi.org/10.1016/j.jtrangeo.2020.102738

19. Levy, S. (2017). *Graduation-based social protection for Cambodia's extreme poor: a general equilibrium analysis of economic and poverty impacts.*
20. Li, S., Blake, A., & Cooper, C. (2011). Modeling the economic impact of international tourism on the Chinese economy: A CGE analysis of the Beijing 2008 Olympics. *Tourism Economics, 17*(2), 279–303.
21. Li, S., & Song, H. (2013). Economic impacts of visa restrictions on tourism: A case of two events in China. *Annals of Tourism Research, 43*, 251–271.
22. Lofgren, H., Harris, R. L., & Robinson, S. (2002). *A standard computable general equilibrium (CGE) model in GAMs.* Micro Computers in Policy Research No. 5. International Food Policy Research Institution, Washington, DC.
23. Mahadevan, R., Amir, H., & Nugroho, A. (2017). Regional impacts of tourism-led growth on poverty and income inequality. *Tourism Economics, 23*(3), 614–631.
24. Meng, S. (2014). The role of inbound tourism in the Singaporean economy: A computable general equilibrium (CGE) assessment. *Journal of Travel and Tourism Marketing, 31*(8), 1071–1089.
25. MOT. (2019). *Tourism annual statistics report*. Ministry of Tourism, Phnom Penh, Cambodia.
26. Pambudi, D., McCaughey, N., & Smyth, R. (2009). Computable general equilibrium estimates of the impact of the Bali bombing on the Indonesian economy. *Tourism Management, 30*(2), 232–239.
27. Soubert, S., & Lay, S. (1995). *Case study on the effects of tourism on culture and the environment.* UNESCO.
28. Sugiya, G., Blake, A., & Sinclair, M. T. (2003). Tourism and globalization: Economic impact in Indonesia. *Annals of Tourism Research, 30*(3), 683–701.
29. United Nation World Tourism Organization. (2020). *International tourism highlights.* https://doi.org/10.18111/9789284422456. Accessed July 25, 2021.
30. World Bank. (2020). *Enabling eco-tourism development in Cambodia.* Washington, DC.
31. Yang, H. Y., & Chen, K. H. (2009). A general equilibrium analysis of the economic impact of a tourism crisis: A case study of the SARS epidemic in Taiwan. *Journal of Policy Research in Tourism, Leisure and Events, 1*(1), 37–60.

The Nexus Between Regional Trade Integration and ASEAN Macroeconomic Indicators: Evidence from Bayesian Panel Regression Approach

Nathapornpan Piyaareekul Uttama and Popkarn Arwatchanakarn

Abstract As regionalization gains momentum globally, this paper aims to investigate the impact of regional trade integration on three macroeconomic indicators for ASEAN member countries: economic growth, inflation, and labor force participation. The regional trade integration is represented by three regional variables: trade intensity, export diversification, and value chain. This paper has deployed the Bayesian panel regression approach using a panel data set of 10 ASEAN countries with their major trading partners in Asia from 1995 to 2019. The empirical findings suggest that the regional export diversification and trade intensity promote economic growth for the overall ASEAN, whereas the regional value chain stimulates employment. Furthermore, the regional trade integration significantly reduces inflation. The key finding indicate that since 2015, ASEAN's deeper integration has enabled member countries with a greater regional trade intensity and regional value chain to enhance their economic growth and development. The ASEAN could also achieve the goal of the ASEAN economic community to uplift their people's standard living.

1 Introduction

As regionalization gains momentum globally, regional economic integration and cooperation policy catalyze intra-regional economic activities. The Association of Southeast Asian Nations (ASEAN) is a regional organization aiming to promote regional economic integration and cooperation among its member and other countries in Asia. However, the policy remains a challenge for the regional economy as a whole. In 2015, ASEAN accomplished the milestones of establishing the ASEAN Economic Community underlying the AEC blueprint 2015 with four pillars: "a single market and production base; a highly competitive economic region; a region of equitable

N. P. Uttama (✉) · P. Arwatchanakarn
Department of Economics and Office of Border Economy and Logistics Study (OBELS), School of Management, Mae Fah Luang University, Chiang Rai, Thailand
e-mail: nathapornpan@mfu.ac.th

P. Arwatchanakarn
e-mail: popkarn.arw@mfu.ac.th

S. Sriboonchitta et al. (eds.), *Credible Asset Allocation, Optimal Transport Methods, and Related Topics*, Studies in Systems, Decision and Control 429,
https://doi.org/10.1007/978-3-030-97273-8_33

economic development; and a region fully integrated into the global economy" [1]. Later on, the AEC blueprint 2025 was created and implemented concerning five achievements: "a highly integrated and cohesive economy; a competitive, innovative, and dynamic ASEAN; enhanced connectivity and sectoral cooperation; a resilient, inclusive, people-oriented, and people-centered ASEAN; and a global ASEAN" [2].

Enhancing the regionalization, expanding the regional trade integration, and strengthening the former regional integration have a significant influence on the future ASEAN. To be precise, the Regional Comprehensive Economic Partnership (RCEP) is an economic and collaborative architecture that has the capacity to assist ASEAN in achieving the AEC blueprint 2025. The RCEP is the mega-regional Asia-Pacific free trade agreement (FTA) among the ten members of ASEAN and five trading partners (China, Japan, South Korea, Australia, and New Zealand). It was formally negotiated in 2012, and dialogue partners have recently committed to officially signing the RCEP agreement by 2020. Many empirical studies on the RCEP agreement suggest that the formation of the RCEP with tariff removals and non-tariff measure reduction generates income gains, economic welfare, sectoral output adjustments, and an influx of foreign investment in ASEAN countries [31, 34, 40, 43, 44].

Similarly, there is evidence that expending and deepening trade integration in RCEP helps developing Asia countries such as the Philippines and Vietnam achieve greater welfare gain [6, 35]. In addition, [41] found that the ASEAN countries benefit from the ASEAN plus six RCEP memberships due to the increased export volume. These previous studies have reasserted that the RCEP and ASEAN Plus collaborative formations' deep regional trade integration would pose a challenge to ASEAN, not only in terms of economic performance but also in other development aspects. Regional trade integration can be measured in three constructive variables: regional trade intensity, regional export diversification, and regional value chain. These three variables have been increasingly important and should be considered in the empirical issue. Therefore, whether these constructive variables impact the ASEAN economy arises as an essential research question of this study.

With the importance of regional trade integration highlighted above, the main objective of this paper is to investigate whether the regional trade integration characteristics influence macroeconomic indicators in ASEAN countries. This paper specifies the regression model that includes regional trade intensity, regional export diversification, and regional value chain as the main determinants of macroeconomic indicators. Then, the specified models are estimated using the Bayesian panel regression model with a dataset between 10 ASEAN members and their major trading partners in Asia from 1995 to 2019. The overall results reveal that all the three aspects of regional trade integration can render either challenge or threat to economic growth, inflation, and employment in ASEAN countries. In a nutshell, an increase in regional export diversification and regional trade intensity supports economic growth for overall ASEAN countries, while an increase in the regional value chain fosters employment. Simultaneously, inflation is lowered by regional trade integration.

The remainder of the paper is organized as follows. The following section contains the literature review and regional economic outlook. In Sect. 3, the model and the estimation approach are presented. Section 4 reports the empirical results and robustness check and concludes in Sect. 5. This paper also has an appendix that provides diagnostics for the convergence of MCMC parameter estimates in the models.

2 Literature Review

Economic integration theory has been evolved in two phases: traditional and new economic integration [7]. The first phase (traditional economic integration) focuses on the gain from economic integration in terms of trade creation and trade diversion. It was known as Viner's customs unions theory [62]. Based on Viner's analysis, many scholars attempted to identify the related effects of economic integration such as trade expansion [49], production and consumption effects [45], and terms of trade effect [38]. The second phase (New economic integration) deals with the dynamic effects of economic integration such as large-scale economies, technological change, market structure and competition, productivity growth, risk and uncertainty, and investment activity [3], investment creation and diversion [5, 16], and an increase in foreign direct investment [19]. To date, economic integration theory has still developed according to the changing world. This section describes the current literature on the impacts of regional trade integration, including trade intensity, global value chain, and export diversification) on macroeconomic performance.

2.1 *Trade Intensity*

A wide range of literature empirically investigates whether trade affects economic growth and development. The Bayesian approach has been widely used to investigate the relationship between trade openness and macroeconomic indicators such as economic growth, inflation, and unemployment [18, 42, 51]. Their results indicated that trade openness had both positive and negative impacts on economic growth, negative impact on inflation, and positive impact on the labor force. However, relatively few studies highlight the role of trade intensity and barely employ the Bayesian analysis. In general, trade intensity measures one country's trade with other countries as a proportion of world trade. It is essential to consider geographical proximity to the market, availability of goods, and economic dependency. The effect of trade intensity on many economic aspects, namely economic growth and business cycle synchronization, has been investigated in many countries [11, 14, 33]. For example, trade intensity has a positive and significant effect on economic growth and business cycle synchronization [11, 33, 50, 54]. In addition, [14] found a negative relationship between exchange rate volatility and trade intensity in the long run. However, there is evidence that the direction and magnitude of trade intensity effect can vary with the country's nature (or region) and level of development [10, 54]. There remains a gap that needs to investigate in the ASEAN context. This paper, therefore, employs intra- regional trade intensity rather than a broad definition of trade intensity. The intra-regional trade intensity is typically defined as the strength of trade activities in the regional market and measured with the share of intra-regional trade to total trade [28].

2.2 Export Diversification

Export diversification has become increasingly important when considering trade sectoral concentration, trade structure vulnerability, and an economy's resilience to external shocks, including economic complexity [26]. It can also be used to assess the degree of trade integration. This paper believes that export diversification, as a component of trade integration, may have a wide-ranging impact on the economy. However, the effects of export diversification are explored as follows.

First, previous research found a positive relationship between export and export diversification, which led to economic growth [8, 47, 56, 58]. For example, Solano et al. [58] revealed that Chilean fruit exporters' product diversification strengthens their firm's export performance. Furthermore, Mania and Rieber [47] found that increasing export diversification leads to a marginal increase in export income elasticity. Their result reflected a structural improvement in competitiveness as well. However, Vannoorenberghe et al. [61] encountered that export diversification has a positive relationship with export volatility.

Second, export diversification can have an impact on economic stability, which includes output volatility and income inequality. According to Balavac and Pugh [4], the impact of export diversification on output volatility in transition economies depends on the degree of export diversification. Output volatility may be reduced in countries with a low degree of export diversification; however, it may not be reduced in countries with a medium or high degree of export diversification. Using a global sample of 90 countries, Le et al. [39] discovered an inverted-U-shaped relationship between export diversification and income equality. In addition to trade performance and economic stability, Massol and Banal-Estaol [48] discovered that resource-based industrialization (RBI) policies that diversify away from raw exports and other resource processing industries may increase the variability of the country's export earnings. Similarly, Ouedraogo and Sourouema [52] found that export concentration is positively associated with pro-cyclical fiscal policy in Sub-Saharan Africa; thus, increasing export diversification could pave the way for countercyclical fiscal policy.

According to the literature reviewed previously, there is a gap for the ASEAN context, as well as a question about the impact of export diversification on the ASEAN economy. To fill this gap, this paper employs the intra-regional export diversification index variable rather than a broad definition of export diversification. The intra-regional export diversification index is defined as the degree to which exports are distributed across different sectors within the region. Therefore, it is measured by comparing the average export diversification index over regional trading partners to the average export diversification index over total trading partners [28].

2.3 Global Value Chains

Global value chains (GVCs) are essential concepts for capturing fragmented and geographically dispersed supply chain production processes, leveraging supply chain capability and global sourcing, and specializing in the global value chain. Previous research has shown that the global value chain has an impact on the economy from both the demand and supply sides.

Initially, the global value chain stimulates demand-side economic growth and improves other macroeconomic indicators [9, 17]. For example, Tinta et al. [59] revealed that GVCs positively affect economic growth by increasing per capita income. In the case of China, Gangnes et al. [20] discovered that increased GVC participation results in an increase in income elasticity; however, Yu and Luo [64] discovered a real trade gain gap under GVCs. According to the ECB Working Group on Global Value Chains [17], participation in GVCs has significant implications for the eurozone economy. On the supply side, GVCs have a significant impact on technology transfer and business synchronization. Across Europe, there is evidence that GVC participation results in an increase in the host economy's total factor productivity (TFP) [17, 23, 53]. This implies that technological innovation and the process of technological catch-up are significantly correlated with the performance of host economies' firms in terms of TFP growth.

Furthermore, it is discovered that an increase in the foreign value-added component of exports is associated with a shift toward high - skilled workers. Similarly, Shen and Zheng [57] confirmed the negative relationship between GVC's low-end integration and China's skill-biased technological change. However, the effects of GVCs on the ASEAN economy remain indeterminate, and there are relatively few studies on the issue. This paper aims to close this gap for ASEAN by deploying an intra-regional value chain rather than GVCs. By definition, an intra-regional value chain is a degree of connectivity to an intra-regional value chain for foreign trade and a link in an intra-regional production chain that is both upstream and downstream. By definition, an intra-regional value chain is a degree of connectivity to an intra-regional value chain for foreign trade and a link in an intra-regional production chain that is both upstream and downstream. It is measured as the intra-regional value chain's contribution to the global value chain, where the global value chain is defined as the sum of foreign and indirect domestic value added in exports ([12].

The review of literature can measure regional trade integration by staring at trade intensity, export diversification, and the global value chain. This paper employs these three variables at the regional level to determine whether regional trade integration influences the ASEAN economy, as stated previously. The following section illustrates the methodology and dataset used in this paper.

3 Methodology

3.1 Empirical Model

The main objective of this paper is to investigate the impact of regional trade integration on key macroeconomic indicators for ASEAN countries. This study has incorporated the representative variables for regional trade integration to affect these economies, such as intra-regional export diversification, intra-regional value chain, and intra-regional trade intensity. Our empirical model is specified below for estimation with annual data for 10 ASEAN economies:

$$Y_{it} = \alpha + \beta_{1t} RED_{it} + \beta_{2t} RVC_{it} + \beta_{3t} RTI_{it} + \epsilon_{it} \tag{1}$$

where Y_{it} represents the macroeconomic indicators which are economic growth (GDP), inflation (INF), and labor force participation rate (LFP); RED_{it} is intra-regional export diversification; RVC_{it} is intra-regional value chain; RTI_{it} is intra-regional trade intensity; ϵ_{it} is a random disturbance, and the subscript *i* and *t* represent country and time, respectively.

3.1.1 Bayesian Panel Regression Approach

In order to examine the relationship among variables of interest, the Bayesian panel regression model is applied for this analysis. The Bayesian analysis using posterior distribution results has advantages over the frequentist (classical) regression analysis using p-value based statistical results [25, 27, 46]. Then, the Bayesian regression results provide a more robust estimation. In a nutshell, the Bayes Theorem formulated the Bayesian statistics that the probability distribution of all model parameters (posterior distribution) depends on the likelihood distribution of the observed data and the prior distribution of parameters [13, 21, 36, 37] as illustrated below:

$$p(\theta \mid y) \propto p(y \mid \theta) \times p(\theta). \tag{2}$$

where $p(\theta|y)$ is the posterior distribution of model parameters θ; $p(y|\theta)$ is the likelihood probability distribution of the observed data y given parameters θ; and $p(\theta)$ is the prior probability of parameters θ. It is assumed that both y and θ are random. Therefore, the Bayesian posterior distribution in Eq. (2) can be written as:

$$p(\beta, \sigma^2 \mid y, x) \propto p(y \mid x, \beta, \sigma^2) \times p(\sigma^2) \times p(\beta|\sigma^2). \tag{3}$$

where y is the macroeconomic indicators, Y_{it}; x is the regional trade integration such as intra-regional export diversification, intra-regional value chain, and intra-regional trade intensity, which are normally distributed with mean, μ, and variance σ^2, or $x \sim N(\mu, \sigma^2)$. It is assumed that the normal distribution of error for Bayesian regression is

$\epsilon \sim N(0, \sigma^2)$. In Bayesian inference, the Markov Chain Monte Carlo (MCMC) and adaptive random-walk Metropolis-Hasting (MH) algorithms are applied to estimate the posterior probability [22, 24, 55].

3.1.2 Testing of Unit Root and Normality

Prior to performing regression analysis, the unit root and normality must be tested. In this study, the Fisher-type is used to test the stationary of all variables in our model [15]. The null hypothesis is that each variable in the panel model has unit root properties. Moreover, the Jarque-Bera test is used to check the normality for all variables [32]. The null hypothesis is that the variable in the panel model is normally distributed.

3.2 Data Sources and Definitions

This study analyzed panel data pertaining to ten ASEAN member states and their major trading partners[1] with annual data from 1995–2019. The Bayesian panel regression has been employed as the primary approach of our analysis. The data used in this study are compiled and calculated based on various sources, including the International Monetary Fund's *International Financial Statistics*, the World Bank's *World Development Indicators*, the United Nations Conference on Trade and Development (UNCTAD), and International Labour Organization (ILO). The definition and descriptive statistics of all variables are shown in Table 1.

Prior to estimating the relationship between regional trade factors and macroeconomic indicators in ASEAN countries using the Bayesian panel regression approach, a descriptive analysis of all observed variables is conducted, see Table 1. It found that all observed variables in our model are not normally distributed, and almost all observed variables except for labor force participation are stationary at the level form.

4 Empirical Result

This section is divided into two subsections. Following that, the first subsection provides the empirical findings from the Bayesian panel regression approach. The second subsection includes a check for robustness.

[1] ASEAN (Association of South East Asia Nations) countries include Brunei, Cambodia, Indonesia, Malaysia, Myanmar, Philippines, Singapore, Thailand, and Vietnam. In addition, the major trading partners include ASEAN, Australia, China, India, Japan, South Korea, and New Zealand.

Table 1 Definition and descriptive statistics of all variables

Variables	Definition	Mean	Std. Dev.	Jarque Bera text	Fisher-type unit root text		Obs.
					Level	1st difference	
GDP growth	Growth rate of gross domestic product	5.496	3.659	1183[a]	1032.4[a]	2431.6[a]	1750
Inflation	Inflation rate	6.468	12.404	14000[a]	803.65[a]	2684.3[a]	1743
Labor force participation	Share of labor force to total working-age population	70.231	6.777	131.3[a]	140.51	691.84[a]	1750
Regional Export Diversification	Share of the average export diversification index over regional trading partners to the average export diversification index over total trading partners	1.130	0.193	86.59[a]	691.84[a]	2386.7[a]	1750
Regional Value Chain	Share of the intra-regional value chain to the global value chain, where the global value chain is the sum of foreign value added in exports and indirect domestic value-added in exports	1.408	1.748	899.7[a]	557.57[a]	2006.0[a]	1750
Regional Trade Intensity	Share of the intra-regional trade value to total trade value	0.078	0.108	6395[a]	244.52[a]	1705.9[a]	1750

Note: [a] indicates significance at the 1% level

4.1 Results of Bayesian Panel Estimation

The purpose of this study is to examine the effects of regional trade integration on three macroeconomic indicators, namely GDP growth, inflation, and employment. Three regression models (hereafter known as GDP, inflation, and employment models) are estimated. The Bayesian panel regression approach with MCMC sampling adapted by the random-walk MH-MCMC method is used to estimate the marginal posterior distribution of parameters in our three regression models. In the Bayesian estimation, the MCMC iterations are equal to 12,500 times, the burn-in is equal to 2500 times, and the MCMC sample size is equal to 10,000 simulation-based observations. The convergence diagnostics in three models are contained in Appendix. The estimated results of the three models are reported in Tables 2, 3 and 4. Overall, the acceptance rates obtained using the MH algorithm in the three models are greater than 10% of the MCMC sample size. It denotes the precision of random-walk MH-MCMC algorithm. The average simulation efficiency of the three models is greater than 10%, indicating that the MH-MCMC algorithm is valid.

The results of the GDP model are summarized in Table 2, which includes parameter estimates based on posterior means and the 95% credible intervals. The posterior means indicate that regional export diversification and regional trade intensity both benefit economic growth, whereas the regional value chain has a detrimental effect. The greater the diversification of ASEAN's exports, the more likely that ASEAN's economic growth will be favored, and the greater the trade complexity. Additionally, the increased trade intensity, which indicates a high degree of market penetration,

Table 2 The estimation results of GDP growth based on the MCMC simulation technique for Bayesian panel regression

	Posterier mean	Posterier standard deviation	Standard error of Posterior mean	95% credible Intervals
Regional export diversification	0.5416	0.6940	0.0312	(−0.8045, 1.8880)
Regional value chain	−0.2120	0.1322	0.0106	(−0.4777, 0.0568)
Regional trade intensity	0.5852	1.1675	0.0449	(−1.6489, 2.9174)
Constant	5.1398	0.8420	0.0417	(3.4947, 6.7580)
Acceptance rate	0.8083			
Average efficiency	0.2275			
Observations	1750			

Table 3 The estimation results of inflation based on the MCMC simulation technique for Bayesian panel regression

	Posterier mean	Posterier standard deviation	Standard error of Posterior mean	95% credible intervals
Regional export diversification	−8.8651	2.3195	0.0947	(−13.4323, −4.3138)
Regional value chain	−1.0495	0.3722	0.0217	(−1.7810, −0.3126)
Regional trade intensity	−32.0366	4.1456	0.1483	(−40.2775, −23.9795)
Constant	20.5344	2.8370	0.1271	(14.9497, 26.1009)
Acceptance rate	0.8080			
Average efficiency	0.1820			
Observations	1743			

tends to stimulate economic growth in ASEAN economies. Our findings are consistent with prior research [11, 33, 50, 54]. The regional value chain, on the other hand, has a detrimental effect on economic growth. The higher the regional value chain, the less likely ASEAN's economic growth is. Unfortunately, the findings contradict previous research [59]. It is worth noting that ASEAN's economies still lack mechanisms to transform regional value chain activities into economic growth.

As expected, the estimated inflation model results in Table 3 show that regional export diversification, regional trade intensity, and regional value chain all have a negative impact on inflation. These findings imply that greater regional trade integration benefits ASEAN member countries by lowering inflation. Our findings were in accordance those of Leon-Gonzalez and Vinayagathasan [42].

Furthermore, Table 4 contains the estimated results for the employment model. The Regional value chain benefits employment, whereas regional export diversification and trade intensity have a detrimental effect. This implies that regional value chain activities benefit ASEAN economies by creating employment. Regional export diversification and regional trade intensity, on the other hand, do not increase employment. These findings do not align with those of previous research [11, 33, 50, 54].

In summary, the empirical results suggest that the pursuit of higher trade intensity and export diversification in ASEAN countries are essential to expanding foreign trade activities and lowering the risk associated with doing business globally and other external shocks. In doing this, ASEAN countries would generate their economic growths and lower their inflation. In addition, an increase in the regional value chain is conducive to expanding trade and investment connectivity of ASEAN countries, improving specialization in the production process, and enhancing supply chain capacity. This impact leads to stimulating employment.

Table 4 The estimation results of employment based on the MCMC simulation technique for Bayesian panel regression

	Posterier mean	Posterier standard deviation	Standard error of Posterior mean	95% credible intervals
Regional export diversification	−0.0673	0.1074	0.0022	(−0.2770, 0.1420)
Regional value chain	0.0516	0.0119	0.0002	(0.0284, 0.0754)
Regional trade intensity	−0.1212	0.1844	0.0038	(−0.4765, 0.2412)
Constant	0.0404	0.1293	0.0027	(−0.2954, 0.2121)
Acceptance rate	0.8064			
Average efficiency	0.2913			
Observations	1680			

Note Dependent variable of employement based on the MCMC simulation technique for Bayesian panel regression

4.2 Robustness Check

As reported previously, the baseline results show a link between intra-regional trade integration and macroeconomic indicators in the Asian region. The baseline results, as shown previously, indicate the nexus between intra-regional trade integration and macroeconomic indicators in the Asian region. This section extends the analysis to focus on the nexus between intra-ASEAN and extra-ASEAN regional trade integration on GDP growth, inflation, and employment. Table 5 presents the estimated results for the three models obtained by using the Bayesian panel regression approach.

With regards to the GDP model, it is discovered that both intra-ASEAN regional trade intensity and extra-ASEAN regional export diversification have a positive effect on economic growth, which is consistent with the baseline results. It reveals that increased regional trade integration boosts economic growth in ASEAN member countries. In terms of inflation, the results reaffirm that integration of intra-ASEAN and extra-ASEAN trade significantly reduces inflation. Regional trade between ASEAN member countries and their trading partners has the ability to reduce inflation and inflationary pressure. Additionally, the results suggest that regional value chains within and outside ASEAN have a positive effect on employment. Overall, the results, which focused on intra-ASEAN and extra-ASEAN trade activities, support the findings from the baseline results.

Table 5 Robustness check for the estimation results

	GDP growth		Inflation		Employment	
	Posterior mean	Posterior Std. Dev	Posterior mean	Posterior Std. Dev	Posterior mean	Posterior Std. Dev
Intra-ASEAN						
Regional export diversification	−0.4569	2.5393	17.9780	8.4255	−0.2451	0.5306
Regional value chain	−0.0866	0.2757	−0.5468	0.8139	0.0368	0.0334
Regional trade intensity	3.0778	1.5078	−37.8588	5.1238	0.2644	0.3325
Constant	5.3505	2.5338	0.4200	8.5119	0.0216	0.5275
Acceptance rate	0.8280		0.8289		0.8247	
Average efficiency	0.2086		0.2304		0.3628	
Observations	250		249		240	
Extra-ASEAN						
Regional export diversification	0.4816	0.7435	−9.2671	2.5686	−0.1597	0.1322
Regional value chain	−0.2699	0.1573	−1.1549	0.4343	0.0635	0.0158
Regional trade intensity	−4.3705	2.0873	−41.7190	7.0948	−0.2969	0.3245
Constant	5.454	0.9204	20.8938	3.1216	0.0683	0.1545
Acceptance rate	0.8092		0.8085		0.8047	
Average efficiency	0.2094		0.2008		0.2953	
Observations	1500		1494		1440	

Note Dependent variable of employment is proxied by the first-difference of labor force participation rate

5 Conclusion

This paper has investigated whether regional trade integration impacts economic growth and stability, including inflation and employment, in the context of ASEAN countries. Using a panel data set of 10 ASEAN countries with their major trading partners in Asia over 1995–2019, this paper has deployed the Bayesian panel regression approach to address the issue. The empirical results suggest that the regional export diversification and regional trade intensity promote economic growth for overall ASEAN countries, while the influence of the regional value chain appears in employment. Furthermore, the regional export diversification, regional value chain, and regional trade intensity significantly reduce inflation. In addition, these findings are confirmatory by the robust results.

Based on the results of this paper, the potential policy implications for enhancing economic growth and stability in ASEAN could be expressed as follows. First, national policies for regional integration should be maintained, shifted, and innovated in both soft and hard policies. The negotiation on trade facilitation and connectivity, the protection of intra-regional economic activity, and the implementation of a national strategic plan are in place. Second, regional trade integration and cooperation in ASEAN and RCEP should be carefully established to attract regional economic activities and leverage regionalization. Moreover, economic cooperation with the second-tier counterpart should be considered. Third, policies toward intra-regional capacity building and economic connectivity should be set and monitored at the highest level so that a whole approach ensures tangibility, certainty, and stability. However, the collaboration between the public and private sectors of RCEP members should be vigorously performed to achieve an effective outcome. Last but not least, building resilience strategies and measures for sustaining regional economic growth and economic stability should be developed in response to the challenge or threat of ASEAN and RCEP membership [29, 30, 60, 63].

Appendix

This appendix contains diagnostics for the convergence of MCMC parameter estimates in GDP, inflation, and employment models. Figure 1 depicts the GDP growth (GDPGR) convergence diagnostics on regional export diversification (RED), regional value chain (RVC), and regional trade intensity (RTI). It found the well-mixing MCMC chains, high autocorrelation, and a unimodal histogram in the GDP model. Additionally, convergence diagnostics for the inflation (INF) and employment (LFP) models are depicted in Figs. 2 and 3, respectively, in accordance with the GDP model.

Declaration of Competing Interest

The authors have no competing interest to declare.

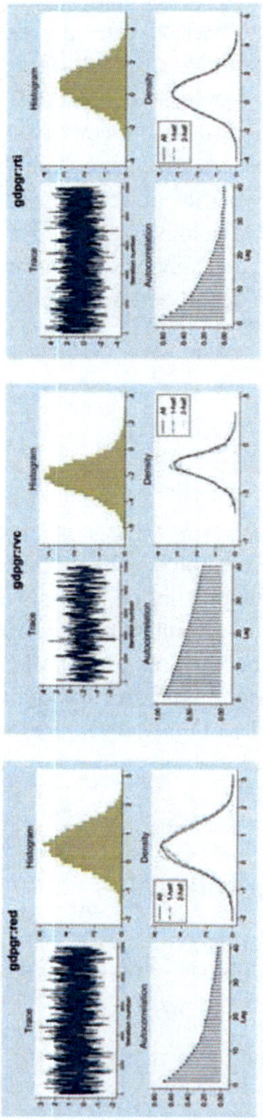

Fig. 1 Convergence diagnostics of GDP model

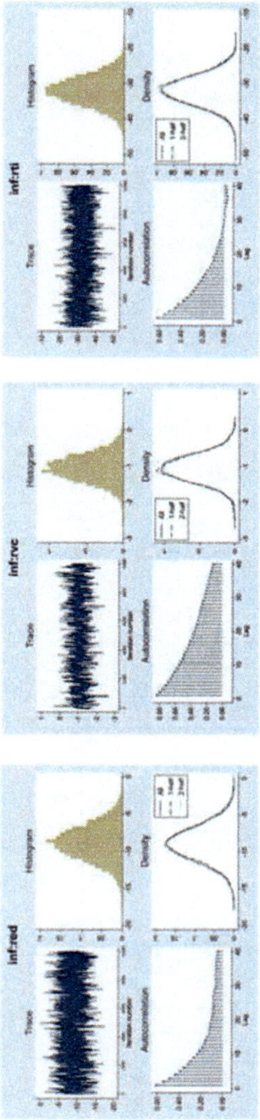

Fig. 2 Convergence diagnostics of inflation model

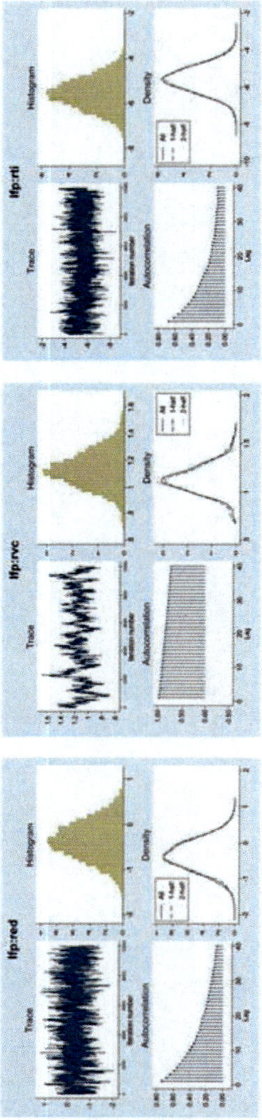

Fig. 3 Convergence diagnostics of employment model

References

1. ASEAN Secretariat. (2008). *ASEAN economic community blueprint*. ASEAN Secretariat, Jakarta, Indonesia.
2. ASEAN Secretariat. (2015). *ASEAN economic community blueprint 2025*. ASEAN Secretariat, Jakarta, Indonesia.
3. Balassa, B. (1965). *Economic development and integration*. Mexico: Centro de Estudios Monetario Latinoamericanos.
4. Balavac, M., & Pugh, G. (2016). The link between trade openness, export diversification, institutions and output volatility in transition countries. *Economic System, 40*(2), 273–287.
5. Baldwin, R., Forslid, R., & Haaland, J. (1995). *Investment creation and investment diversion: Simulation analysis of the single market programme*. NBER Working Paper no. 5364, Cambridge, Massachusettes
6. Balistreri, E. J., & Tarr, D. G. (2020). Comparison of deep integration in the Melitz, Krugman and Armington models: The case of The Philippines in RCEP. *Economic Modelling, 85,* 255–271.
7. Bhagwati, J., & Panagariya, A. (1996). The theory of preferential trade agreements: Historical evolution and current trends. *The American Economic Review, 86*(2), 82–87.
8. Boehe, D., & Jiménez, A. (1996). Does the sequencing of related and unrelated export diversification matter? Evidence from Colombian exporters. *International Business Review, 27*(6), 1141–1149.
9. Brumm, J., Georgiadis, G., & Trottner, F. (2019). Global value chain participation and current account imbalances. *Journal of International Money and Finance, 97,* 111–124.
10. Calderón, C., Chong, A., & Stein, E. (2007). Trade intensity and business cycle synchronization: Are developing countries any different? *Journal of International Economics, 7*(1), 2–21.
11. Caporale, G. M., Santis, R. D., & Giradis, A. (2007). Trade intensity and output synchronisation: On the endogeneity properties of EMU. *Journal of Financial Stability, 16,* 154–163.
12. Casella, B., Bolwijn, R., Moran, D., & Kanemoto, K. (2019). UNCTAD insights: Improving the analysis of global value chains: the UNCTAD-Eora Database. *Transnational Corporation, 26*(3), 115–142.
13. Chib, S. (2008). *Panel data modeling and inference: A Bayesian Primer*. Berlin Heidelberd: Springer-Verlag.
14. Cho, D., & Doblas-Madrid, A. (2014). Trade intensity and purchasing power parity. *Journal of International Economics, 93,* 194–209.
15. Choi, I. (2001). Unit root tests for panel data. *Journal of International Money and Finance, 20,* 249–272.
16. Dunning, J., & Robson, P. (1998). *Multinationals and the European community*. Oxford: Basil Blackwell.
17. ECB Working Group on Global Value Chain. (2019). *The impact of global value chains on the euro area economy*, ECB Occasional Paper 221, European Central Bank.
18. Eris, M. N., & Ulasan, B. (2013). Trade openness and economic growth: Bayesian model averaging estimate of cross-country growth regressions. *Economic Modelling, 33,* 867–883.
19. Ethier, W. (1998). The new regionalism. *The Economic Journal, 108*(449), 1149–1161.
20. Gangnes, B. S., Alyson, C. M., & Ari, V. A. (2014). Global value chains and trade elasticities. *Economic Letters, 124*(3), 482–486.
21. Gelman, A., Carlin, J. B., Stern, H. S., Dunson, D. B., Vehtari, A., & Rubin, D. B. (2013). *Bayesian data analysis*. Boca Raton, Florida: Chapman & Hall/CRC.
22. Giordani, P., & Kohn, R. (2010). Adaptive independent Metropolis-Hastings by fast estimation of mixtures of normals. *Journal of Computational and Graphical Statistics, 19*(2), 243–259.
23. Gunnella, V., Fidora, M., & Schmitz, M. (2017). The impact of global value chains on the macroeconomic analysis of the euro area. *Economic Bulletin Articles, 8,* 75–95.
24. Haario, H., Saksman, E., & Tamminen, J. (2001). An adaptive Metropolis algorithm. *Bernoullis, 7*(2), 223–242.

25. Halsey, L. G. (2019). The reign of the p-value is over: what alternative analyses could we employ to fill the power vacuum? *Biology Letters, 15,* 20190174.
26. Hausmann, R., Hidaldo, C. A., Bustos, S., Coscia, M., Simoes, A., & Yildirim, M. A. (2014). *The atlas of economic complexity: Mapping paths to prosperity*. Cambridge, MA: MIT Press.
27. Held, L., & Ott, M. (2018). On p-values and Bayes factors. *Annual Review of Statistics and Its Application, 5,* 393–419.
28. Huh, H. S., & Park, C. Y. (2018). Asia-Pacific regional integration index: Construction, interpretation, and comparison. *Journal of Asian Economics, 54,* 22–38.
29. International Labour Organization, *ILO statistics and databases*, Geneva, Switzerland, various years.
30. International Monetary Fund, *Internatinal Financial Statistics*, Washington, DC, various years.
31. Itakura, K. (2014). Impact of liberalization and improved connectivity and facilitation in ASEAN. *Journal of Asian Economics, 35,* 2–11.
32. Jarque, C. M., & Bera, A. K. (1987). A test for normality of observations and regression residuals. *International Statistical Review, 55,* 163–172.
33. Jiang, W., Li, Y., & Zhang, S. (2019). Business cycle synchronisation in East Asia: The role of value- added trade. *The World Economy, 42*(1), 226–241.
34. Kawasaki, K. (2015). The relative significance of EPAs in Asia-Pacific. *Journal of Asian Economics, 39,* 19–30.
35. Kikuchi, T., Yanagida, K., & Vo, H. (2018). The effects of mega-regional trade agreements on Vietnam. *Journal of Asian Economics, 55,* 4–19.
36. Koop, G., Osiewalski, J., & Steel, M. F. J. (1997). Bayesian efficiency analysis through individual effects: Hospital cost frontiers. *Journal of Econometrics, 75,* 77–105.
37. Koop, G., & Poirier, D. J. (2004). Bayesian variants of some classical semiparametric regression techniques. *Journal of Econometrics, 123,* 259–282.
38. Krauss, M. B. (1972). Recent developments in customs union theory: An interpretive surveys. *Journal of Economic Literature, 10*(2), 413–436.
39. Le, T. H., Nguyen, C. P., Su, T. D., & Tran-Nam, B. (2020). The Kuznets curve for export diversification and income inequality: Evidence from a global sample. *Economic Analysis and Policy, 65,* 21–39.
40. Lee, H., & Itakura, K. (2018). The welfare and sectoral adjustment effects of mega-regional trade agreements on ASEAN countries. *Journal of Asian Economics, 55,* 20–32.
41. Lee, J. W., & Oh, J. (2020). ASEAN or plus alpha? The effectiveness of regional economic cooperation. *Asia Pacific Management Review, 55*(1), 48–53.
42. Leon-Gonzalez, R., & Vinayagathasan, T. (2015). Robust determinants of growth in Asian developing economies: A Bayesian panel data model averaging approach. *Journal of Asian Economics, 36,* 34–46.
43. Li, C., Wang, J., & Whalley, J. (2016). Impact of mega trade deals on China: A computational general equilibrium analysis. *Economic Modelling, 57,* 13–25.
44. Li, Q., Scollay, R., & Gilbert, J. (2017). Analyzing the effects of the regional comprehensive economic partnership on FDI in a CGE framework with firm heterogeneity. *Economic Modelling, 67,* 409–420.
45. Lipsey, R. G. (1957). The theory of customs unions: Trade diversion and welfare. *Economica, 24,* 40–46.
46. Maneejuk, P., & Yamaka, W. (2020). Significance test for linear regression: how to test without P-values? *Journal of Applied Statistics, 45*(5), 827–845.
47. Mania, E., & Rieber, A. (2019). Product export diversification and sustainable economic growth in developing countries. *Structural Change and Economic Dynamics, 51,* 138–151.
48. Massol, O., & Banal-Estañol, A. (2014). Export diversification through resource-based industrialization: The case of natural gas. *European Journal of Operational Research, 237*(3), 1067–1082.
49. Meade, J. E. (1955). *The theory of customs unions*. Amsterdam: North Holland.
50. Mo, P. H. (2010). Trade intensity, net export, and economic growth. *Review of Development Economics, 14*(3), 563–576.

51. Moral-Benito, E. (2012). Determinants of economic growth: A Bayesian panel data approach. *The Review of Economics and Statistics, 94*(2), 566–579.
52. Ouedraogo, R., & Sourouema, W. S. (2018). Fiscal policy pro-cyclicality in Sub-Saharan African countries: The role of export concentration. *Economic Modelling, 74,* 219–229.
53. Qua, C., Shao, J., & Cheng, Z. (2020). Can embedding in global value chain drive green growth in China's manufacturing industry. *Journal of Cleaner Production, 268,* 121962.
54. Rana, P. B., Cheng, T., & Chia, W. M. (2012). Trade intensity and business cycle synchronization: East Asia versus Europe. *Journal of Asian Economics, 23*(6), 701–706.
55. Roberts, G. O., & Rosenthal, J. S. (2009). Examples of adaptive MCMC. *Journal of Computational and Graphical Statistics, 18*(2), 349–367.
56. Shahbaz, M., Gozgor, G., & Hammoudeh, S. (2019). Human capital and export diversification as new determinants of energy demand in the United States. *Energy Economics, 78,* 335–349.
57. Shen, C., & Zheng, J. (2018). Does global value chains participation really promote skill-biased technological change? Theory and evidence from China. *Economic Modelling, 86,* 10–18.
58. Solano, L. V. L., Brümmer, B., Engler, A., & Otter, V. (2018). Effects of intra- and inter-regional geographic diversification and product diversification on export performance: Evidence from the Chilean fresh fruit export sector. *Food Policy, 86,* 1–10.
59. Tinta, A. A., Sarpong, D. B., Ouedraogo, I. M., Hassan, R. A., Mensah-Bonsu, A., & Onumah, E. E. (2018). The effect of integration, global value chains and international trade on economic growth and food security in ECOWAS. *Cogent Food and Agriculture, 4*(1), 1–15.
60. United Nations Conference for Trade and Development (UNCTAD), *UNCTADl Statistics*, Geneva, Switzerland, various years.
61. Vannoorenberghe, G., Wang, Z., & Yu, Z. (2016). Volatility and diversification of exports: Firm-level theory and evidence. *European Economic Review, 89,* 216–247.
62. Viner, J. (1950). *The customs union issue*. New York: Carnegie Endowment for International Peace.
63. World Bank, *World Development Indicators*, Washingtod, DC, various years.
64. Yu, C., & Luo, Z. (2018). Volatility and diversification of exports: Firm-level theory and evidence. *China Economic Review, 47,* 263–273.

Impacts of Climate Variability on Rice Production in Thailand

Thanat Boonchai, Nachatchapong Kaewsompong, and Terdthiti Chitkasame

Abstract This study examines the role of climate variables and rice' price (THB) on rice production in Thailand. This research applied ridge, LASSO, and elastic net regression to specify which climate factors are affecting Thai rice production. The data set used in this study was monthly data starting from January 2010 to December 2020. The study shows that average temperature, rainfall, rice price, and lowest temperature have a positive impact on Thai rice production, while the highest temperature and humidity lead to a significant decrease on rice harvesting.

1 Introduction

The role of the agricultural sector in the development and economic growth process is important both in developed and developing countries. The world's agricultural products amounted to 232.2 million metric tons in 2020 [1].

Global agricultural trade had a total value of 1.5 trillion dollars in 2018, an increase of over 3.6% per year for the last two decades, with emerging and developing countries' exports on the rise and accounting for over one-third of the world's total agricultural trade [2]. The agricultural trade of developing countries grew quickly, led by growth in China's trade, while developed countries' agricultural trade grew much slower from 1995 to 2014 [3]. There is growing evidence that supports agricultural trade being an engine of economic growth in developing countries [4–8] and developed countries [9].

Thailand is an essential rice producer in the world, and it has a rice cultivation area of approximately 45% of the total agricultural area of the country area. In 2020, there was a total area of 10.30 million hectares, mainly in the lower North,

N. Kaewsompong (✉) · T. Chitkasame
Faculty of Economics, Chiang Mai University, Chiang Mai 50200, Thailand
e-mail: nachat_flysky@hotmail.com

T. Boonchai
Department of Mathematics and Statistics, Faculty of Science and Technology, Chiang Mai Rajabhat University, Chiang Mai 50300, Thailand

S. Sriboonchitta et al. (eds.), *Credible Asset Allocation, Optimal Transport Methods, and Related Topics*, Studies in Systems, Decision and Control 429,
https://doi.org/10.1007/978-3-030-97273-8_34

the Central, and the Northeast areas of Thailand. The growing of rice in Thailand focuses upon rainwater supply, an important planting time is in the rainy season from July to September of every year. Rice cultivation involves more than 17.5 million farmers or about 1 in 5 of the nation's population [1]. It is known that the amount of agricultural output, especially rice, depends on a number of physical factors, and climate change is an important physical factor that directly affects agricultural productivity. Some studies show that increasing temperature, decreasing humidity, and the changing pattern of rainfall have a substantial impact on food production [10–12]. As climate controls all performances for crop growth right from germination till harvesting, through its key factors like temperature, rainfall, and humidity and, others, a changing climate thus, is of critical importance for the country.

Climate change is a major global phenomenon in the twenty-first century, which has bought changes in rain patterns, temperature, and negative effects on water and land resources, floods, and droughts globally. Its impacts are widespread in developing countries due to their greater vulnerabilities and lesser ability to mitigate the effects of climate change. Since most developing countries, including Thailand, are agriculture-based economies, their agricultural sector is directly affected by climate change. Therefore, changes in rain patterns and temperature, floods, and droughts dramatically affect agricultural production.

In this study the aims are to define values on different variables through forecasts, it is necessary to handle the multicollinearity contexts of the data set. Ridge regression is a technique used when the data suffers from multicollinearity. In multicollinearity, even though the ordinary least squares estimate (OLS) are unbiased, their variances are large which deviates the observed value far from the true value. By adding a degree of bias to the regression estimates, ridge regression reduces the standard errors. Similar to ridge regression, LASSO (Least Absolute Shrinkage and Selection Operator) also penalizes the absolute size of the regression coefficients. In addition, it is capable of reducing the variability and improving the accuracy of linear regression models. Elastic net is a hybrid of LASSO and ridge regression, useful when multiple features are correlated [13].

This study aims to examine the role of climate variables on rice production in Thailand. This research applied ridge, LASSO, and elastic net regression to specify which direct and indirect factors are affecting the Thai rice production. The findings of this study would be beneficial to farmers and policy makers for the future management related to rice production.

The rest of the paper is set out as follows. Section 2 reviews the literature. The methodology used and the scope of the study are described in Sects. 3 and 4. Section 5 discusses the empirical results. The final section concludes the study.

2 Literature Review

Climate change influences precipitation and temperature directly, with rising temperatures causing future water shortages and floods, as well as modifying soil moisture status and pest and disease occurrence [14]. A number of studies have examined climatic variables on the yield of rice. Sarker et al. [15] examined the relationship between climate variables and rice yield, using the ordinary least squares and median (quantile) regression methods. The results revealed that climatic variables significantly affect the yield of rice but vary among different rice crops. Additionally, the effects of highest temperature and lowest temperature are prominent as compared to rainfall. Shakoor et al. [16] employing the Vector Auto Regression (VAR) model and using the feasible generalized least square (FGLS) and heteroscedasticity and autocorrelation (HAC) consistent standard error found that a rising highest mean temperature would increase rice production. In addition, some studies indicated that decreasing humidity and increasing rainfall would have a substantial impact on rice production [10–12].

In regard to Thailand, it may see drier spells in the middle of the rainy season, which could harm new plants, as well as floods at the end of the rainy season, which could disrupt the harvest. Increased spikelet sterility and lower grain yield can also be caused by rising temperatures or greater night temperatures in rice [17–19]. However, only a small amount of research has been done in Thailand to investigate the effects of climate conditions on rice production. According to Babel et al. [20], the production of Thai rice is anticipated to drop by 18% in the 2020s due to changes in the temperature and rainfall cycles, as well as changes in soil quality, pests, and diseases as a result of climate change. Related to these predicted results, rice growers in the basin areas face the risk of losing paddy fields to climate change-related floods and droughts [21].

3 Methodology

3.1 LASSO and Ridge Regression

Ridge regression is very similar to least squares, except that the coefficients are estimated by minimizing a slightly adjusted quantity. As with least squares, ridge regression seeks coefficients that fit the data, by making the residual sum of squares (RSS) as small as possible. However, the regression also seeks to minimize a second term in Eq. (1). This is also known as a shrinkage penalty, which is small when the regression coefficients are close to zero. This term will tend to shrink the coefficient estimates towards zero. The tuning parameter (λ) controls the relative impact of the penalty term. When the Ridge penalty does not affect ($\lambda = 0$), it will produce the least-squares estimates. On the other hand, the impact of the shrinkage penalty (as λ) gets larger, the coefficient estimates will not approach to zero. Unlike least squares, which generates only one set of estimates, ridge regression will produce a different set of coefficients for each value of λ. Therefore, selecting a good value for λ is critical, and will be addressed in the section on cross-validation as set out below [22].

$$\hat{\beta}_i = \arg\min_{\hat{\beta}_i} \left\{ \sum_{i=1}^{n} \left(Y - X\hat{\beta}_i\right)^2 + Penalty\left(\hat{\beta}_i\right) \right\}, \tag{1}$$

Ridge:

$$Penalty\left(\hat{\beta}_i\right) = \lambda \sum_{i=1}^{n} \left(\hat{\beta}_i\right)^2, \tag{2}$$

LASSO:

$$Penalty\left(\hat{\beta}_i\right) = \lambda \sum_{i=1}^{i} \left|\hat{\beta}_i\right|, \tag{3}$$

where n is the number of observations, and p is the number of candidate predictors.

LASSO regression (Least Absolute Shrinkage and Selection Operator) is similar to the Ridge regression, but it has a different penalty as shown in Eq. (3). As with ridge regression, the LASSO shrinks the coefficient estimates towards zero. However, in the case of LASSO, the penalty has the effect of forcing some of the coefficients to be exactly equal to zero when the tuning parameter (λ) is large enough. In contrast to the ridge regression which might shrink coefficients. So, they are close to zero, but it will never eliminate them. Hence, like some of the stepwise approaches outlined above, the LASSO includes an element of variable selection and will tend to produce a parsimonious model with fewer predictors [23, 24].

3.2 *The Elastic Net Regression*

The Elastic net regression contains a hybrid of the Ridge and LASSO penalties. The Ridge penalty will tend to shrink all coefficients proportionately; and for closely correlated variables, it will tend to move the coefficients toward one another, without choosing among them. The LASSO penalty, on the other hand, will produce a leaner model by focusing on a small subset of those variables and discarding the rest. Each approach has benefits, depending on the data, and there is no priori reason to prefer one over the other.

$$\sum_{i=1}^{n} \left(Y - X\hat{\beta}_i\right)^2 + \lambda \sum_{j=1}^{p} \left[\underbrace{(1-\alpha)\left(\hat{\beta}_j\right)^2}_{RIDGE} + \underbrace{\alpha \left|\hat{\beta}_j\right|}_{LASSO} \right]. \tag{4}$$

The Elastic net regression combines the strengths of both by selecting the best predictors to provide a parsimonious model, while still identifying groups as closely

Table 1 Descriptive statistics of returns

	Rice	Price	MaxT	MinT	AvgT	Rain	Humi	Wind	UVI
Mean	4.73	6.1	3.64	2.81	3.32	5	4.31	1.01	2.4
Median	4.88	6.05	3.65	2.91	3.33	4.9	4.32	1.03	2.3
Maximum	6.11	6.42	3.75	3.16	3.43	5.99	4.43	1.22	2.64
Minimum	2.6	5.87	3.55	1.06	2.98	3.87	4.11	0.58	2.2
Std. dev.	0.76	0.16	0.05	0.29	0.06	0.6	0.07	0.11	0.16
Skewness	−0.55	0.55	0.01	−2.15	−1.72	−0.02	−0.67	−1.24	0.23
Kurtosis	2.86	1.98	1.92	11.98	10.17	1.64	3.02	5.68	1.58
Jarque-Bera	6.13*	11.17*	5.79	495.91*	315.82*	9.32*	8.90*	66.86*	11.18*
MBF unit root test	0	0	0	0	0	0	0	0	0

Note "*" indicates the strong evidence support the alternative hypothesis by MBF

correlated predictors. The relative weights of the two penalties are determined by an additional tuning variable (α). And as with the Ridge and LASSO regressions, different values of the tuning parameters (α and λ) will produce different sets of coefficients. Therefore, selecting the right parameter value is important [24].

4 Data

Firstly, the monthly data was collected from January 2010 to December 2020, approximately 10 years, from the Meteorological Department of Thailand's database. The variables used in this study consist of the Thai rice production (Rice), the Thai rice spot price (Price), the highest temperature (MaxT), the lowest temperature (MinT), the average temperature (AvgT), the rainfall (Rain), the humidity (Humi), the wind speed (Wind), and the UV index (UV). All of the data was transformed into the logarithm. Table 1 shows the descriptive statistics of all variables. After transforming the p-value of the test to be the Minimum Bayes' Factor (MBF), the results found that the data was not normally distributed. Lastly, according to the MBF of the unit root test, the test was conducted to investigate the stationarity of the data and the results showed that all variables are stationary with decisive evidence. Subsequently, these variables can be used to estimate the ridge, LASSO, and Elastic net regression in a further step.

5 Result

According to the Elastic net penalty, it appeared to include the better properties of both the Ridge and the LASSO regression when estimating Ridge regression and its penalty. This tended to shrink all coefficients proportionately, and for closely

Table 2 Estimation result of regression models

	Ridge regression	LASSO regression	Elastic net regression
(Intercept term)	3.642	2.43	3.502
Price	8.3032	4.5859	4.9683
MaxT	−0.7554	–	−1.8574
MinT	0.0021	–	0.0025
AvgT	11.2811	8.6229	9.7271
Rain	7.0964	4.0638	7.1079
Humi	0.2787	–	−0.0005
Wind	0.1496	–	–
UVI	0.7865	–	–
MSE	4.104	3.695	3.521

correlated variables, it tended to move the coefficients toward one another without choosing among them. Secondly, the LASSO regression and its penalty produced a leaner model by focusing on a small subset of those variables and discarded the rest.

In addition to this, the mean squared error (MSE) of an estimator measures the average of the squares of the errors that is the average squared difference between the estimated values and the actual value [25]. Consequently, this research applied the MSE to check how close estimates or forecasts are to actual values. This is used as a model evaluation measure for regression models and the lower value indicates a better fit model.

Table 2 reported that the Elastic net regression was the suitable model for this data among all regression in terms of both prediction of accuracy and sparsity because it has the lowest value of MSE. Consequently, the ridge is also the worst reasonable method (compared the highest MSE's value). Therefore, the Elastic net estimator can solve both multicollinearity and variable selection simultaneously (from the Ridge and LASSO regression). Thus, this study will interpret the results from the elastic net regression. The result showed that only 2 variables do not have an impact on Thailand's rice production, wind speed (Wind) and the UV index. For variables that have a positive impact on Thai rice production these consisted of average temperature (AvgT), rainfall (Rain), rice spot price (Price), and lowest temperature (MinT) respectively. On the other hand, the highest temperature (MaxT) and humidity (Humi) revealed negative effects.

However, there are still concerns about the variable relationship problem (Multicollinearity), although Ridge, Lasso, and Elastic net regression can remove this issue. The VIF technique is then used to test the Multicollinearity problem (Table 3). The testing results show that the Centered VIF values are greater than 10. It confirms that less relationship exists between covariables [26, 27].

According to the results, it was observed that the main variable that had a majority impact on Thailand rice production was the average temperature (AvgT). This finding is consistent with TMD [28], where the optimum average temperature (between

Table 3 Multicollinearity test by variance inflation factors

Variable	Coefficient variance	Uncentered VIF	Centered VIF
C	77.21802	20057.1	NA
Price	0.171003	1654.46	1.150734
MaxT	7.09805	24478.83	4.761322
MinT	1.21095	2507.179	6.63244
AvgT	1.30384	61043.88	4.76953
Rain	0.011568	76.31201	1.073996
Humi	0.990586	4780.381	1.099098
Wind	0.406368	109.6745	1.217011
UVI	0.162974	245.825	1.044112

25 and 33 °C) will be suitable to affect suitable seed germination, leaf stretching, tillering, flower formation, and the growing process. However, if the temperature has a slight rise above the average temperature this will then be a negative impact on rice production, equally, a temperature drop will affect crop production in the same way [29], hence, supporting the coefficients of positive and negative effects in lowest and highest temperature variable changes, respectively. Due to the minority factor though, rice requires a much more consistent supply of water, which requires sufficient rain throughout the year, to grow and thrive [30]. Therefore, the results indicated that the average temperature had the majority effect on rice production. Furthermore, the highest and lowest temperature that has been rapidly changing will be a negative effect on rice growth and yield. In addition, the amount of rain should not be excessive so that crop growth is hampered.

6 Conclusion

This study aimed to investigate the role of climate variables related to Thai rice production, the highest temperature, the lowest temperature, the average temperature, the rainfall, the humidity, the wind speed, and the UV index, also including Thai rice spot price on rice production in Thailand. Therefore, this study applied the Ridge, LASSO, and Elastic net regression to specify which climate factors are affecting Thai rice production. They can solve both multicollinearity and variable selection simultaneously. The results reported that the Elastic net regression is the most suitable model among all regression in terms of both prediction of accuracy and sparsity. The results from this regression indicated that average temperature, rainfall, Thai rice price, and lowest temperature have a positive impact on Thai rice production. On the other hand, the research identified that highest temperature and humidity are a negative effect on rice production in Thailand. Additionally, the temperature in

Thailand can fluctuate and in some years is also prone to drought, consideration of these fluctuations are important before planning the production process for rice plantation preparation. Therefore, Thai farmers ought to take into account the average temperature and the rainfall as well as predicted weather.

References

1. United States Department of Agriculture. (2020). *World agricultural production*. United States Department of Agriculture.
2. Food and Agriculture Organization of the United Nations. (2020). *The state of agricultural commodity markets 2020*. Food and Agriculture Organization of the United Nations.
3. Beckman, J., Dyck, J., & Heerman, K. (2017). *The global landscape of agricultural trade, 1995–2014* (No. 1476-2017-5674).
4. Sanjuan-Lopez, A. I., & Dawson, P. J. (2010). Agricultural exports and economic growth in developing countries: A panel cointegration approach. *Journal of Agricultural Economics, 61*(3), 565–583.
5. Awokuse, T. O., & Xie, R. (2015). Does agriculture really matter for economic growth in developing countries? *Canadian Journal of Agricultural Economics/Revue canadienne d'agroeconomie, 63*(1), 77–99.
6. Cooke, B., Nigatu, G., Heerman, K., Landes, M., & Seeley, R. (2016). *Global macroeconomic developments drive downturn in US agricultural exports*. Department of Agriculture, Economic Research Service.
7. Ahmad, D., & Ahmad, J. (2018). An empirical analysis of agricultural export on economic growth: A case study of Pakistan. *International Journal of Advanced and Applied Sciences, 5*(2), 25–32.
8. Bakari, S., & Mabrouki, M. (2018). The impact of agricultural trade on economic growth in North Africa: Econometric analysis by static gravity model.
9. Shahbandeh, M. (2019). Total value of U.S. agricultural exports from 2012 to 2019. Retrieved November 18, 2019, from https://www.statista.com/statistics/220767/total-value-of-us-agricultural-exports-since-2000
10. Janjua, P. Z., Samad, G., Khan, N. U., & Nasir, M. (2010). Impact of climate change on wheat production: A case study of Pakistan [with comments]. *The Pakistan Development Review*, 799–822.
11. Mahmood, N., Ahmad, B., Hassan, S., & Bakhsh, K. (2012). Impact of temperature ADN precipitation on rice productivity in rice-wheat cropping system of Punjab province. *Journal of Animal and Plant Sciences, 22*, 993–997.
12. Kirby, J. M., Mainuddin, M., Mpelasoka, F., Ahmad, M. D., Palash, W., Quadir, M. E., Shah-Newaz, S. M., & Hossain, M. M. (2016). The impact of climate change on regional water balances in Bangladesh. *Climatic Change*.
13. Ray, S. (2015). 7 Regression techniques you should know! *Analytics Vidhya*.
14. Chinvanno, S. (2010). *Climate change adaptation as a development strategy: A major challenge for Southeast Asian countries*. Southeast Asia START Regional Center, Chulalongkorn University.
15. Sarker, M. A. R., Alam, K., & Gow, J. (2012). Exploring the relationship between climate change and rice yield in Bangladesh: An analysis of time series data. *Agricultural Systems, 112*, 11–16.
16. Shakoor, U., Saboor, A., Baig, I., Afzal, A., & Rahman, A. (2015). Climate variability impacts on rice crop production in Pakistan. *Pakistan Journal of Agricultural Research, 28*(1).
17. Wassmann, R., & Dobermann, A. (2007). Climate change adaptation through rice production in regions with high poverty levels. *Journal of ICRISAT Agricultural Research, 4*(1), 1–24.

18. Felkner, J., Tazhibayeva, K., & Townsend, R. (2009). Impact of climate change on rice production in Thailand. *American Economic Review, 99*(2), 205–210.
19. Kawasaki, J., & Herath, S. (2011). Impact assessment of climate change on rice production in Khon Kaen province, Thailand. *Journal of ISSAAS, 17*, 14–28.
20. Babel, M. S., Agarwal, A., Swain, D. K., & Herath, S. (2011). Evaluation of climate change impacts and adaptation measures for rice cultivation in Northeast Thailand. *Climate Research, 46*(2), 137–146.
21. MWBP [Mekong Wetlands Biodiversity Conservation and Sustainable Use Programme]. (2005). *Vulnerability assessment of climate risks in the lower Songkhram River Basin, Thailand.* Programme Management Unit. Mekong Wetlands Biodiversity Conservation and Sustainable Use Programme.
22. Hoerl, A. E., & Kennard, R. W. (1970). Ridge regression: Biased estimation for nonorthogonal problems. *Technometrics, 12*(1), 55–67.
23. Tibshirani, R. (1996). Regression shrinkage and selection via the lasso. *Journal of the Royal Statistical Society: Series B (Methodological), 58*(1), 267–288.
24. Tiffin, M. A. (2016). *Seeing in the dark: A machine-learning approach to nowcasting in Lebanon.* International Monetary Fund.
25. Lehmann, E. L., & Casella, G. (2006). *Theory of point estimation.* Springer Science & Business Media.
26. Myers, R. H., & Myers, R. H. (1990). *Classical and modern regression with applications* (Vol. 2, p. 488). Duxbury Press.
27. Bowerman, B. L., & O'connell, R. T. (1990). *Linear statistical models: An applied approach.* Brooks/Cole.
28. Ali, S., Liu, Y., Ishaq, M., Shah, T., Ilyas, A., & Din, I. U. (2017). Climate change and its impact on the yield of major food crops: Evidence from Pakistan. *Foods, 6*(6), 39.
29. Madan, P., Jagadish, S. V. K., Craufurd, P. Q., Fitzgerald, M., Lafarge, T., & Wheeler, T. R. (2012). Effect of elevated CO_2 and high temperature on seed-set and grain quality of rice. *Journal of Experimental Botany, 63*(10), 3843–3852.
30. Thanawong, K., Perret, S. R., & Basset-Mens, C. (2014). Eco-efficiency of paddy rice production in Northeastern Thailand: A comparison of rain-fed and irrigated cropping systems. *Journal of Cleaner Production, 73*, 204–217.

Herding Behavior During the COVID-19 Pandemic and the Disposition Effect Situation in the Stock Exchange of Thailand

Kunsuda Nimanussornkul and Chaiwat Nimanussornkul

Abstract The disposition effect is the tendency of investors to sell winning stocks and hold losing stocks and the herding behavior is an investment behavior in a group of investors who trade in the same direction. Both are regarded as irrational investment behaviors. This research examines the herding behavior in the Stock Exchange of Thailand during the COVID-19 pandemic and the disposition effect situation. We used daily data from January 2011 to December 2020 of complete observed stocks in the Stock Exchange of Thailand (SET). We modified the average holding period following Atkins and Dyl (Journal of Finance 52(1):309–325, 1997 [5]) and adapted the method for investigating the disposition effect after Visaltanachoti et al. (Applied Financial Economics 17(15):1265–1274, 2007 [45]). Then we constructed the dummy variable to represent the month in which the disposition effect took place in the market. After that, we tested the existence of the herding behavior during the COVID-19 pandemic and when the disposition effect was occurring. The herding behavior was tested following Yao et al. (International Review of Economics & Finance 29:12–29, 2014 [48]) and Filip et al. (Procedia Economics and Finance 32:307–315, 2015 [19]) using the cross-sectional absolute deviation (CSAD). The results reveal the herding behavior to occur during the disposition effect situation but not during the COVID-19 pandemic.

Keywords Herding behavior · Disposition effect · COVID-19 · Stock Exchange of Thailand

1 Introduction

In 2020, the rapid spread of coronavirus disease 2019 (COVID-19) had put the world at hazard. This novel coronavirus emerged in December 2019 in Wuhan, the capital city of Hubei Province in the People's Republic of China has subsequently caused outbreaks all over the globe leading to a massive number of infections and deaths in

K. Nimanussornkul (✉) · C. Nimanussornkul (✉)
Center of Excellence in Econometrics, Chiang Mai University, Chiang Mai 50200, Thailand
e-mail: kunsudan@gmail.com

S. Sriboonchitta et al. (eds.), *Credible Asset Allocation, Optimal Transport Methods, and Related Topics*, Studies in Systems, Decision and Control 429,
https://doi.org/10.1007/978-3-030-97273-8_35

more than 200 countries. The COVID-19 pandemic, thus, has not only affected global health but also brought about the global economic downturn. Markets and investors are facing uncertainty in both the physical and financial effects of the virus [8, 10, 32]. Thailand found its first patient, a Chinese visitor from Hubei, infected with COVID-19 on 12 January 2020 and the first locally infected patient on 31 January 2020. The COVID-19 outbreak has resulted in volatility in Thailand's financial markets. For example, the SET index decreased by 173.12 points or 11.47% from January 2020 to February 2020 [44]. Hence indicating the effect of the COVID-19 on the stock market.

A number of recent research works have focused on the impact of COVID-19 on stock market performance. Ashraf [3] showed a negative correlation between the growth in confirmed COVID-19 cases and stock market performance across 64 countries. Certain papers investigating the effect of COVID-19 on the stock market in the Asia region like that of Lee et al. [27] showed the adverse impact of this pandemic on the stock market's performance. However, Baek et al. [8], Onali [30], Yousef [49], and Panyagometh [33] showed the positive impact of COVID-19 on the volatility of the stock market while Papadamou et al. [32] found the COVID-19 have no direct effect on the volatility of the stock market but have an indirect effect via stock returns. However, a few pieces of the previous research investigating the effect of COVID-19 on the herding behavior like Espinosa-Méndez and Arias [18] found COVID-19 increased herding behavior in the stock market of Europe. Espinosa-Méndez and Arias [18] found that the herding behavior manifested itself during the COVID-19 crisis and intense periods in the financial market of Australia while Yarovaya et al. [47] found that the herding behavior did not get stronger during the COVID-19 pandemic in cryptocurrency markets—USD, EUR, JPY, and KRW. Chang et al. [15] investigated renewable and fossil fuel energy stock returns in the USA, Europe, and Asia and found herding during the COVID-19 pandemic as investors got so panicked that they might unintelligibly sell their assets. Only Goossens and Knoef [20] were found to study the disposition effect during the COVID-19 pandemic in the Netherlands.

The literature has concentrated on empirical studies of the COVID-19's effects on the stock market and irrational investment behaviors in some countries but not in Thailand. The present research thus fills the gap by examining the herding behavior in the Stock Exchange of Thailand during the COVID-19 pandemic and the disposition effect situation. These results are essential for investors and regulators to enhance their understanding of the irrational investment behaviors in the financial market brought about by the COVID-19 pandemic.

The structure of the rest of this paper is as follows. In the next section, provided are the disposition effect and the herding behavior. Section 3 presents data and methodology, Sect. 4 presents the empirical results. Finally, the conclusion is provided in Sect. 5.

2 The Disposition Effect and the Herding Behavior

2.1 The Disposition Effect

The efficient market hypothesis (EMH) assumes that a financial market is efficient if the prices reflect all relevant information. Investors have rational expectations about the evolution of future prices. Therefore, investors can adjust the portfolio optimization of stocks accordingly as suggested by some previous studies; for examples, Zhu et al. [50], Sirisrisakulchai et al. [37], and Autchariyapanitkul et al. [6]. However, irrational investment behaviors suffer from asymmetry in the information that has caused abnormal losses and returns in the financial market. Asymmetric information in financial markets can exist in any transaction, if one of the two parties has more information than the other like insider information and can make a more informed decision; and information asymmetry might lead to market failure. The disposition effect is the tendency of investors to sell winning stocks and hold losing stocks [4]. The disposition effect is due to avoiding regret because if investors sell losing stocks means recognizing a failure and investors tend to avoid the pain of failure by selling winning stocks instead; this is called loss aversion, acquired from the prospect theory by Kahneman and Tversky [24]. The prospect theory explains that the investors would become more risk-averse after facing gains (and thus will tend to sell winning stocks) and more risk-seeking after facing losses (and thus will hold losing stocks) [4].

Prior literature explains the disposition effect as arising from the prospect theory, mental accounting, regret aversion, and self-control [4, 12]. The first theory is the prospect theory [24] which explains that investors apply an S-shaped value function to compute and maximize their utility. The second theory is mental accounting, developed by Thaler [41, 42] and Tversky and Kahneman [43]. It describes investors' reference points (e.g., purchase price) for their accounts to determine losses and gains. Then, they track losses and gains in their mind on each stock they invested. The third theory is regret aversion that Shefrin and Statman [38] proposed. When a stock position is at a loss, investors hold on to their losing stocks to delay the feeling of regret. The fourth theory is self-control which explains that investors have lack self-control; therefore, they hold losing stocks rather than change their way of doing (e.g., stop-loss orders) [4, 12].

Many previous studies found the disposition effect in many countries, e.g., Visaltanachoti et al. [45] on Chinese stock markets. Ashraf et al. [4] on the Karachi Stock Exchange of Pakistan, and Bergsma et al. [9] on the U.S. equity options markets. Goossens and Knoef [20], in particular, found that the disposition effect increases the volatility of returns. In Thailand, previous studies used different methods to find the disposition effect including Sarajoti [36], Maneenil [28], Suppaudom [39], and Waiyasara and Padungsaksawasdi [46]. They found the disposition effect in the Stock Exchange of Thailand (SET) and particularly Sarajoti [36] found that the disposition effect is related to daily market return, volatility, and turnover rate.

2.2 The Herding Behavior

Herding behavior, one of the irrational behaviors, is an investment behavior in a group of investors who trade in the same direction [29]. The reasons behind the herding behavior are many. One reason is that the group of investors is better informed than individual investors. Therefore, the herding behavior may cause the financial market to become inefficient [11]. The methodologies for the investigation of the herding behavior can be classified into two main paths: the first path introduced by Lakonishok et al. [26] is the use of primary data, and the second path proposed by Christie and Huang [16] is the use of market data. Christie and Huang [16] used the cross-sectional standard deviation (CSSD) of stock returns to study the herding behavior. Then Chang et al. [14] tried to improve the method of Christie and Huang [16] by using the cross-sectional absolute deviation (CSAD) to study the herding behavior because the CSSD method has an outlier's problem. After that, Yao et al. [48] and Filip et al. [19] adopted the CSAD to reduce the multicollinearity problem. The results revealed herding behavior, as well as the impact of the crisis on herding behavior in some countries. For example, Putra et al. [34] found the herding behavior in Indonesia and Singapore and no evidence of herding behavior in Indonesia during the global financial crisis. The previous research works on the herding behavior in the Stock Exchange of Thailand such as Kulvanich and Boonvorachote [25], Rattanasri and Vichitthamaros [35] and Padungsaksawasdi and Pooissarakit [31]. Kulvanich and Boonvorachote [25] and Rattanasri and Vichitthamaros [35] found the herding behavior in some sectors, while Padungsaksawasdi and Pooissarakit [31] found no existence of the herding behavior in both the Stock Exchange of Thailand (SET) and the Market for Alternative Investment (MAI). However, they used the cross-sectional absolute deviation (CSAD) and estimated it by Ordinary Least Squares.

2.3 The Relationship Between the Disposition Effect and the Herding Behavior

Previous papers studying the disposition effect and herding behavior in the stock market and other markets include Awale et al. [7] and Haryanto et al. [21]. Awale et al. [7] acquired data by questionnaires from investors in Nepal and found the existence of the disposition effect, herding behavior and overconfidence bias in Nepal. Haryanto et al. [21] examined the disposition effect and herding behavior in the bitcoin markets and revealed that the disposition effect from investors' decisions to sell the winners stocks whose values have increased and keep the loser stocks whose values have decreased; might explain investors' herding behavior and the investors' profitability. There are very few articles dealing with the relationship between the disposition effect and the herding behavior. Chang [13] who did so measured the disposition effect from the ratio between the trading volume of stock whose value has increased and the trading volume of stock whose value has decreased and then investigated the herding behavior in the stock market by adapting the cross-sectional standard deviation (CSSD) of Christie and Huang [16]. The results showed the bi-directional relationship between the herding behavior and the disposition effect. However, the present research used the disposition effect to be the explanatory variable for the herding behavior because we think investors will delay their feelings of regret when stock prices fall. Therefore, they hold on to the losing stocks and sell the higher-priced stocks as a disposition effect. Then, if investors act in the same direction, it causes the herding behavior.

3 Data and Methodology

3.1 Data

We used daily data of complete observed stocks in the Stock Exchange of Thailand (SET), covering January 2011–December 2020, obtained from Datastream. Therefore, every month a total of 143 stocks are sampled.

3.2 Methodology for the Disposition Effect

Our methodology for investigating the disposition effect was adapted from Visaltanachoti et al. [45], in which estimates the relationship between the holding period and return of the stock was estimated. Therefore, we capture the disposition effect in the stock market by estimating the relationship between the monthly return of stock i; $i = 1, 2, \ldots, 143$ stocks and the investors' holding period using the Two-Stage Least Squares (2SLS) method with Newey-West heteroscedasticity and autocorrelation consistent standard errors as follows:

$$LNHP_{i,t} = \alpha_0 + \alpha_1 Ret_{i,t} + \alpha_2 LNILLIQ_{i,t} + \alpha_3 MV_{i,t} + \alpha_4 Volatility_{i,t} + \varepsilon_{i,t} \tag{1}$$

where $LNHP_{i,t}$ is the natural logarithm of the average holding period, which $HP_{i,t}$ is the average length of time that investors hold the stock of stock i during month t, $Ret_{i,t}$ is the monthly return of stock i, and $LNILLIQ_{i,t}$ is the predicted value from Eq. (2) which is the first-stage regression. $MV_{i,t}$ is the natural logarithm of the average market capitalization of stock i's shares during month t, and $Volatility_{i,t}$ is the natural logarithm of the variance of the stock i's daily stock returns, which represents the volatility of stock i's daily stock returns during month t in the natural logarithm term. The first-stage regression is as follows:

$$LNILLIQ_{i,t} = \beta_0 + \beta_1 LNILLIQ_{i,t-1} + \beta_2 MV_{i,t} + \beta_3 Volatility_{i,t} + \varepsilon_{i,t} \tag{2}$$

where $LNILLIQ_{i,t}$ is the natural logarithm of the average $ILLIQ$ on stock i's shares during month $t-1$.

We modified the average holding period ($HP_{i,t}$) following Atkins and Dyl [5]. The average holding period of stock i for each month is computed by dividing the number of outstanding shares in the stock by the stock's monthly trading volume. Therefore, for each month, the average holding period (in months) for each stock's common stock is measured as follows:

$$H_{i,t} = \frac{SharesOutstanding_{i,t}}{VOLD_{i,t}} \tag{3}$$

where $SharesOutstanding_{i,t}$ is the average daily shares outstanding on stock i in month t, which refers to a corporation's stock currently held by all its shareholders and $VOLD_{i,t}$ is the respective monthly volume of stock i, which is the number of shares traded within a month of stock i. The average holding period is used as a proxy for the average investor's investment horizon because investors are not likely to have the same length of time to hold the stock's shares.

The stock illiquidity ($ILLIQ$) following Amihud [1] is the average ratio between the absolute value of daily return and the trading volume on that day as in the following:

$$ILLIQ_{i,t} = \frac{\sum_{d=1}^{N} |R_{i,t,d}|/VOLD_{i,t,d}}{N} \tag{4}$$

where $R_{i,t,d}$ is the return on stock i on day d of month t, which is the change in the price of stock i, $VOLD_{i,t,d}$ is the respective daily volume in terms of Baht in month t and N is the total number of trading days for stock i during month t.

In Eq. (1), the market capitalization representing the stock's size is often related to liquidity because a more extensive stock has a more negligible price impact and a more minor bid-ask spread impact. The expected returns on stock are negatively related to market capitalization [45], so the market capitalization is a means of controlling the return size effect. The higher volatility implies the more significant asymmetry in information, which may induce more trading and shorten holding periods. The relationship between investors' holding periods and stock returns measures the disposition effect following the avoiding regret and seeking pride preferences. If $\alpha_1 < 0$ with the Minimum Bayes Factor (MBF) representing moderate and more substantial evidence, then the disposition effect holds. Then we set a dummy variable to represent the disposition effect as Dis_t. The month having the disposition effect will be assigned Dis_t equal to one.

3.3 Testing the Disposition Effect on the Herding Behavior

The methodology for testing the herding behavior following Chang et al. [14] is the use of the cross-sectional absolute deviation (CSAD).

$$CSAD_t = \frac{1}{N}\sum_{i=1}^{N} |r_{i,t} - r_{m,t}| \tag{5}$$

where $r_{i,t}$ is the stock return of stock i at time t, $r_{m,t}$ is the equal-weight average of N returns in the stock market at time t, and N is the number of stocks in the stock market.

Chang et al. [14] test the herding behavior following the regression model below:

$$CSAD_t = \gamma_0 + \gamma_1|r_{m,t}| + \gamma_2(r_{m,t})^2 + \varepsilon_t \tag{6}$$

The null hypothesis (H_0) is $\gamma_2 = 0$ implying this regression is a linearity model. Therefore the market does not have the herding behavior and the rational asset pricing model holds. On the other hand, the alternative hypothesis (H_1) is $\gamma_2 < 0$ implying a non-linear model that supports the existence of herding behavior in the market.

Yao et al. [48] and Filip et al. [19] reduced the multicollinearity problem by using the change between the variable $r_{m,t}$ and the arithmetic mean of the variable $r_{m,t}$ ($\bar{r}_{m,t}$) instead of $r_{m,t}$ and including a lag term of the dependent variable ($CSAD_{t-1}$) in the model to increase the power of the model. Therefore, this paper applies the model following Yao et al. [48] and Filip et al. [19], as shown below:

$$\begin{aligned} CSAD_t = {} & \gamma_0 + \gamma_1|r_{m,t}| + \gamma_2(r_{m,t} - \bar{r}_{m,t})^2 \\ & + \gamma_3 CSAD_{t-1} + \varepsilon_t \end{aligned} \tag{7}$$

where $\gamma_2 = 0$ or accepting the null hypothesis (H_0) means that no herding behavior exists in the market, and the rational pricing model holds, while the alternative hypothesis (H_1) is $\gamma_2 < 0$ implying the existence of herding behavior. To test the influence of disposition effect on the herding behavior, the procedure is as follows:

$$\begin{aligned} CSAD_t = {} & \gamma_0 + \gamma_1|r_{m,t}| + \gamma_2(r_{m,t} - \bar{r}_{m,t})^2 + \gamma_3 CSAD_{t-1} \\ & + \gamma_4 Dis_t(r_{m,t} - \bar{r}_{m,t})^2 + \varepsilon_t \end{aligned} \tag{8}$$

where Dis_t is equal to one when the month has the disposition effect. The alternative hypothesis (H_1) is $\gamma_4 < 0$ which implies that herding occurs mainly during the time the disposition effect is taking place in the market, while $\gamma_4 \geq 0$ indicates that there is no influence of herding during the disposition effect situation in the market.

The test of the COVID-19 pandemic effects on the herding behavior is conducted as follows:

Table 1 Interpretation from Minimum Bayes factor

Bayes factor (MBF)	Strength of evidence for H_1
1 to 1/3	Weak
1/3 to 1/10	Moderate
1/10 to 1/30	Substantial
1/30 to 1/100	Strong
1/100 to 1/300	Very strong
<1/300	Decisive

Remark Adapted from Held and Ott [22]

$$CSAD_t = \gamma_0 + \gamma_1|r_{m,t}| + \gamma_2(r_{m,t} - \bar{r}_{m,t})^2 + \gamma_3 CSAD_{t-1} + \gamma_4 COVID19_t(r_{m,t} - \bar{r}_{m,t})^2 + \varepsilon_t \quad (9)$$

where $COVID19_t$ is equal to one when the month has the COVID-19 pandemic. The alternative hypothesis (H_1) is $\gamma_4 < 0$ implying that herding occurs mainly during the COVID-19 pandemic, while $\gamma_4 \geq 0$ indicates that there is no influence of herding during the COVID-19 pandemic. However, the positive value of γ_4 with the Minimum Bayes Factor (MBF) representing moderate and higher evidence indicates that during the COVID-19 pandemic, the dispersion of returns has increased. This phenomenon is reflected in the change of the market return. We use the Ordinary Least Squares (OLS) with Newey-West heteroscedasticity and autocorrelation consistent standard errors on Eqs. (8) and (9).

After we estimate (1), (2), (8), and (9), we apply the Minimum Bayes Factor (MBF) approach using the $-eplogp$ calibration, where p is the p-$value$. The MBF is interpreted in Table 1.

4 Empirical Results

4.1 Summary Statistics

To show an overall view of the Stock Exchange of Thailand, the summary statistics for the common stocks are presented in Table 2. The complete availability of the data is required to compute the average month holding periods, market capitalization, daily return, volatility or daily return variances, and the average month $ILLIQ$.

A significant difference between the mean and median of the average month holding period suggests that the distribution of the holding period in the sample period is highly skewed. The most extended average holding period is 133.740 months in 2019, with a median holding period of 31.907 months in the same year. The shortest average holding period is 33.618 months in 2012, but the shortest median holding period is 13.049 in 2011. The holding periods exhibited an upward trend over the

Table 2 Descriptive statistics

	Holding period (months)			Market capitalization (MV) ($\times 10^4$ million Baht)			Return (%)			Volatility (% per month)			Illiquidity ($\times 10^{-4}$ per thousand Baht)		
	Mean	Median	Sd	Mean	Median	Sd	Mean	Median	Sd	Mean	Median	Sd	Mean	Median	Sd
2011	45.307	13.049	175.607	44.194	9.029	111.733	0.374	0.000	12.112	6.012	3.994	7.875	1.793	0.467	7.216
2012	33.618	13.296	97.274	52.037	10.563	124.381	3.788	2.406	9.740	4.215	2.799	5.778	1.112	0.336	2.614
2013	38.398	14.004	106.434	64.307	14.366	140.629	0.292	−0.893	14.717	7.753	5.085	9.177	2.059	0.447	11.311
2014	55.834	18.282	161.281	65.017	14.707	139.112	2.169	1.130	8.912	4.669	2.861	6.377	2.015	0.412	11.199
2015	68.397	23.070	193.734	65.969	16.180	136.372	−0.648	−1.015	9.794	5.181	3.305	9.974	1.987	0.481	5.737
2016	61.104	19.149	175.016	63.050	14.710	128.002	1.552	0.559	8.821	4.887	2.966	7.072	1.652	0.386	5.650
2017	48.903	19.374	125.036	72.195	17.615	151.054	0.880	0.000	7.987	3.388	2.055	4.664	0.941	0.306	2.373
2018	78.888	24.561	194.624	79.154	15.683	180.793	−1.973	−2.002	10.165	4.381	2.514	7.269	2.816	0.494	19.962
2019	133.740	31.907	435.283	74.362	12.923	172.581	−0.633	−0.746	8.163	3.737	2.127	6.161	3.884	0.597	32.285
2020	110.131	21.591	656.115	57.905	9.742	139.911	1.456	−0.957	17.709	10.976	6.301	14.423	6.149	0.572	47.229

sample period except the 2020 holding period with a decrease. The average market capitalization of the listed Thailand stocks increased from 44.194×10^4 million Baht in 2011 to 79.154×10^4 million Baht in 2018 and decreased to 57.905×10^4 million Baht in 2020.

The daily return and the variance of daily return are held from year to year and there is no clear trend over the period. However, the variance of daily return in 2020 is the highest. The mean and median $ILLIQ$ had no clear trend over the period, but the mean $ILLIQ$ throughout 2017–2020 exhibited an upward trend and the highest mean $ILLIQ$ was 6.149×10^{-4} per thousand Baht in 2020.

4.2 The Disposition Effect

The estimation results of Eq. (1) for all stocks in the Stock Exchange of Thailand are reported in Table 3. The Two-Stage Least Squares with Newey-West heteroscedasticity and autocorrelation consistent standard errors method for measuring the disposition effect of each month is employed for the period 2011–2020. The empirical results show a few negative coefficients and the Minimum Bayes factor (MBF) represents moderate and more substantial evidence on the monthly return. This indicates a weak disposition effect in the Stock Exchange of Thailand because almost every month has MBF representing weak evidence on the return and coefficients with the opposite sign. In summary, the results show that 49 out of 120 months or 40.83% of all observations are presented with the disposition effect.

The coefficient on the $LNILLIQ$ variable is positive, and MBF represents strong evidence in each month. This result strongly supports that the Thai investors' holding periods are related to transaction costs. These results are consistent with Amihud and Mandelson [2], Atkins and Dyl [5] and Visaltanachoti et al. [45] findings that when the transaction costs are higher, investors hold stocks with longer holding periods.

The regression coefficient on stock size (market capitalization) is the almost positive coefficient, and MBF represents moderate and more substantial evidence about 99 months of the total 120 months or 82.50% of all observations, except for a few discrete, especially for the period 2018–2020 the MBF represents weak evidence. These results indicate that the more extended holding periods are related to larger stocks in the Stock Exchange of Thailand mostly in 2011–2017, except in 2018–2020. The more extended holding periods are not associated with larger stocks in some months.

The regression coefficients on volatility (the return variance) are negative, and MBF represents strong evidence every month. The adjusted R^2 for the whole sample period is about 0.6. Then we create a new dummy variable to represent the disposition effect. The month with the presence of the disposition effect will have the dummy variable equal to one and otherwise the dummy variable equal zero.

Table 3 Disposition effect

Year		Jan	Feb	Mar	Apr	May	Jun	Jul	Aug	Sep	Oct	Nov	Dec
2011	Constant	4.597***	5.154***	5.005***	5.289***	5.022***	5.091***	5.054***	5.287***	6.030***	5.587***	5.501***	5.708***
	Return	−0.022***	−0.007	−0.004	−0.009	−0.010	−0.012***	−0.013*	−0.012*	−0.014*	−0.033***	0.012	−0.007
	Illiquidity	0.543***	0.583***	0.608***	0.654***	0.553***	0.680***	0.685***	0.617***	0.637***	0.631***	0.653***	0.723***
	Stock size	0.299***	0.300***	0.288***	0.327***	0.308***	0.351***	0.371***	0.284***	0.250***	0.308***	0.324***	0.321***
	Volatility	−0.635***	−0.659***	−0.479***	−0.500***	−0.799***	−0.527***	−0.384***	−0.582***	−0.609***	−0.501***	−0.737***	−0.303***
	Adj. R2	0.614	0.609	0.600	0.688	0.713	0.670	0.696	0.639	0.604	0.668	0.682	0.630
	F-stat	57.377	50.077	49.155	74.468	79.028	67.542	78.428	57.243	49.011	62.346	68.160	57.737
2012	Constant	5.597***	5.689***	5.322***	5.527***	5.574***	5.287***	4.928***	5.317***	4.988***	5.059***	5.588***	4.819***
	Return	−0.020*	−0.001	−0.002	0.011	−0.004	−0.019*	−0.015*	−0.028**	0.000	−0.010	−0.004	−0.005
	Illiquidity	0.593***	0.651***	0.623***	0.587***	0.556***	0.590***	0.582***	0.620***	0.667***	0.662***	0.614***	0.615***
	Stock size	0.256***	0.271***	0.280***	0.280***	0.239***	0.301***	0.307***	0.289***	0.339***	0.322***	0.249***	0.329***
	Volatility	−0.509***	−0.551***	−0.536***	−0.779***	−0.717***	−0.636***	−0.651***	−0.495***	−0.422***	−0.430***	−0.604***	−0.414***
	Adj. R2	0.682	0.651	0.616	0.655	0.624	0.639	0.656	0.651	0.699	0.680	0.675	0.646
	F-stat	68.588	64.277	53.522	63.811	54.589	55.735	59.961	65.053	74.098	77.299	70.782	66.476
2013	Constant	4.498***	4.892***	4.612***	4.827***	4.983***	5.847***	5.873***	6.593***	6.109***	5.412***	6.092***	6.447***
	Return	0.001	−0.005**	−0.009***	0.000	−0.010**	−0.010	0.004	0.002	0.002	−0.013	0.001	−0.016***
	Illiquidity	0.609***	0.558***	0.583***	0.570***	0.607***	0.571***	0.572***	0.575***	0.582***	0.591***	0.642***	0.562***
	Stock size	0.329***	0.286***	0.313***	0.315***	0.298***	0.215***	0.210***	0.173***	0.202***	0.256***	0.230***	0.158***
	Volatility	−0.481***	−0.641***	−0.528***	−0.627***	−0.526***	−0.568***	−0.507***	−0.626***	−0.552***	−0.435***	−0.341***	−0.551***
	Adj. R2	0.752	0.736	0.765	0.705	0.664	0.618	0.653	0.653	0.673	0.710	0.668	0.666
	F-stat	104.146	94.417	109.803	78.376	0.664	51.711	61.689	64.175	66.452	79.324	63.817	62.386

(continued)

Table 3 (continued)

Year		Jan	Feb	Mar	Apr	May	Jun	Jul	Aug	Sep	Oct	Nov	Dec
2014	Constant	6.297***	5.826***	6.073***	5.584***	6.070***	5.087***	5.120***	5.377***	5.082***	5.256***	5.228***	6.081***
	Return	−0.020**	−0.028***	−0.016*	−0.031***	−0.010	0.000	0.004	−0.010	−0.017**	−0.016***	−0.026***	−0.034***
	Illiquidity	0.525***	0.530***	0.629***	0.591***	0.526***	0.569***	0.610***	0.547***	0.600***	0.524***	0.533***	0.539***
	Stock size	0.160***	0.202***	0.234***	0.268***	0.179***	0.276***	0.297***	0.254***	0.289***	0.241***	0.259***	0.176***
	Volatility	−0.581***	−0.452***	−0.512***	−0.502***	−0.685***	−0.581***	−0.458***	−0.622***	−0.513***	−0.660***	−0.698***	−0.693***
	Adj. R2	0.671	0.658	0.676	0.704	0.694	0.733	0.760	0.767	0.778	0.785	0.817	0.690
	F-stat	63.556	60.794	69.201	87.746	73.216	91.932	111.105	110.527	123.292	122.085	152.179	67.896
2015	Constant	5.127***	5.318***	5.601***	5.980***	5.954***	5.449***	5.719***	6.448***	5.545***	5.499***	5.872***	6.387***
	Return	−0.015*	−0.002	−0.016*	−0.026***	−0.020***	−0.031***	−0.005	−0.017*	0.015	−0.027***	0.000	−0.021***
	Illiquidity	0.575***	0.567***	0.507***	0.500***	0.506***	0.516***	0.515***	0.551***	0.565***	0.524***	0.543***	0.465***
	Stock size	0.302***	0.277***	0.222***	0.186***	0.188***	0.219***	0.210***	0.181***	0.255***	0.231***	0.203***	0.123**
	Volatility	−0.547***	−0.701***	−0.816***	−0.636***	−0.613***	−0.545***	−0.679***	−0.577***	−0.613***	−0.547***	−0.524***	−0.612***
	Adj. R2	0.750	0.754	0.722	0.702	0.637	0.681	0.678	0.643	0.679	0.708	0.577	0.601
	F-stat	103.201	105.640	86.662	77.472	56.708	69.926	69.945	61.017	71.351	84.147	46.884	51.091
2016	Constant	6.037***	6.251***	6.072***	5.913***	6.169***	5.662***	5.860***	5.716***	6.259***	7.228***	5.909***	5.217***
	Return	0.000	0.006	0.003	−0.004	−0.001	−0.003	−0.015**	−0.003	−0.004	−0.019*	−0.008*	0.007
	Illiquidity	0.480***	0.505***	0.528***	0.502***	0.496***	0.504***	0.502***	0.451***	0.447***	0.480***	0.473***	0.484***
	Stock size	0.190***	0.156***	0.160***	0.199***	0.156***	0.190***	0.175***	0.139***	0.087*	0.040	0.138***	0.208***
	Volatility	−0.764***	−0.638***	−0.545***	−0.631***	−0.586***	−0.697***	−0.518***	−0.733***	−0.600***	−0.517***	−0.687***	−0.565***
	Adj. R2	0.691	0.673	0.636	0.627	0.652	0.702	0.677	0.704	0.625	0.625	0.693	0.627
	F-stat	75.220	68.721	59.587	56.648	60.487	82.052	69.335	82.122	54.301	55.491	75.447	53.570

(continued)

Table 3 (continued)

Year		Jan	Feb	Mar	Apr	May	Jun	Jul	Aug	Sep	Oct	Nov	Dec
2017	Constant	5.312***	5.147***	5.466***	5.159***	5.598***	5.128***	5.651***	5.639***	5.593***	5.452***	5.791***	6.049***
	Return	−0.012*	−0.004	−0.017***	−0.007	−0.045***	−0.002	0.000	−0.034***	0.000	−0.024**	−0.003	−0.019**
	Illiquidity	0.504***	0.450***	0.379***	0.379***	0.466***	0.436***	0.401***	0.442***	0.499***	0.436***	0.418***	0.431***
	Stock size	0.211***	0.194***	0.099*	0.149***	0.159***	0.170***	0.118***	0.146***	0.173***	0.151***	0.104**	0.100*
	Volatility	−0.498***	−0.586***	−0.624***	−0.503***	−0.513***	−0.603***	−0.736***	−0.632***	−0.651***	−0.555***	−0.641***	−0.502***
	Adj. R2	0.667	0.604	0.602	0.540	0.611	0.592	0.650	0.638	0.587	0.611	0.587	0.558
	F-stat	67.434	52.930	46.085	39.062	47.954	45.856	62.610	56.964	52.837	53.220	44.153	42.869
2018	Constant	5.857***	6.237***	7.023***	7.316***	6.888***	6.755***	7.217***	6.335***	6.099***	6.093***	6.746***	6.908***
	Return	−0.020*	−0.016**	−0.003	−0.008	−0.005	−0.006	−0.007	−0.004	0.002	0.001	0.016**	−0.017
	Illiquidity	0.448***	0.419***	0.442***	0.491***	0.451***	0.416***	0.476***	0.446***	0.491***	0.497***	0.518***	0.542***
	Stock size	0.114***	0.065	0.016	0.057	0.045	0.028	0.050	0.084*	0.139***	0.150***	0.114*	0.129**
	Volatility	−0.563***	−0.612***	−0.601***	−0.599***	−0.577***	−0.572***	−0.508***	−0.474***	−0.451***	−0.478***	−0.396***	−0.478***
	Adj. R2	0.605	0.593	0.587	0.654	0.598	0.575	0.637	0.609	0.613	0.633	0.633	0.629
	F-stat	55.051	47.608	45.247	63.681	50.018	42.012	58.378	46.921	54.440	55.912	58.308	57.140
2019	Constant	6.837***	6.912***	6.366***	7.046***	7.834***	7.174***	7.043***	6.882***	7.293***	7.124***	7.717***	7.770***
	Return	−0.009	0.000	−0.030***	−0.022***	−0.004	−0.044***	−0.027**	−0.020*	0.002	−0.007	0.006	−0.002
	Illiquidity	0.374***	0.484***	0.473***	0.511***	0.507***	0.500***	0.496***	0.481***	0.520***	0.577***	0.474***	0.463***
	Stock size	0.008	0.083*	0.108*	0.085*	−0.001	0.078	0.059	0.075	0.064	0.132**	0.000	−0.009
	Volatility	−0.560***	−0.590***	−0.439***	−0.446***	−0.565***	−0.440***	−0.576***	−0.564***	−0.496***	−0.456***	−0.500***	−0.377***
	Adj. R2	0.630	0.645	0.631	0.638	0.633	0.664	0.683	0.579	0.660	0.609	0.622	0.617
	F-stat	55.330	58.991	52.807	55.915	58.673	65.906	74.529	44.822	58.996	55.791	55.825	51.800

(continued)

Table 3 (continued)

Year		Jan	Feb	Mar	Apr	May	Jun	Jul	Aug	Sep	Oct	Nov	Dec
2020	Constant	7.407***	7.004***	7.746***	6.894***	7.513***	6.293***	6.891***	6.835***	6.914***	7.098***	6.839***	6.789***
	Return	0.007	−0.027***	0.011**	−0.005	−0.003	−0.008	−0.015***	−0.005	0.000	−0.010	−0.005	0.003
	Illiquidity	0.487***	0.512***	0.512***	0.525***	0.497***	0.537***	0.548***	0.466***	0.530***	0.506***	0.522***	0.502***
	Stock size	0.043	0.099	0.033	0.091	0.064	0.149***	0.130**	0.084	0.126**	0.101*	0.115*	0.104**
	Volatility	−0.514***	−0.527***	−0.363***	−0.301***	−0.616***	−0.394***	−0.479***	−0.582***	−0.624***	−0.666***	−0.510***	−0.516***
	Adj. R2	0.679	0.562	0.627	0.590	0.606	0.626	0.637	0.625	0.644	0.682	0.651	0.575
	F-stat	71.407	43.668	58.276	44.789	52.367	51.081	61.634	51.700	62.625	76.186	63.057	47.308

Note Calculated using Newey-West heteroscedasticity and autocorrelation consistent standard errors
***MBF represents strong evidence
**MBF represents substantial evidence
*MBF represents moderate evidence

Table 4 Testing the herding behavior during the disposition effect situation

Variable	Coefficient	t-Statistic	Prob.
Constant (γ_0)	4.491***	7.606	0.000
ABSRM (γ_1)	0.590***	3.817	0.001
RMINUSRBAR2 (γ_2)	0.007	0.725	0.769
CSAD(−1) (γ_3)	0.220***	3.929	0.000
DIS*RMINUSRBAR2 (γ_4)	−0.014**	−2.145	0.100
R-squared	0.560	Adjusted R-squared	0.545
F-statistic	36.335	Prob(F-statistic)	0.000

Note Calculated using Newey-West heteroscedasticity and autocorrelation consistent standard errors
***MBF represents strong evidence
**MBF represents substantial evidence
*MBF represents moderate evidence

4.3 *Testing the Herding Behavior During the Disposition Effect Situation*

Table 4 reports the estimation results of Eq. (8). The empirical results show the coefficient (γ_4) has a negative sign and the Minimum Bayes factor (MBF) represents substantial evidence, highlighting the herding behavior that occurs during the disposition effect situation in the market. This result is consistent with Chang [13], which found the bi-directional relationship between the herding behavior and the disposition effect in Taiwan's information technology stocks. When investors in the market are holding losing stocks for a long time and selling winning stocks with their decision in the same direction and magnitude, they are forming the herding behavior in the market.

4.4 *Testing the Herding Behavior During the COVID-19 Pandemic*

Table 5 reports the estimation results of Eq. (9). The results from testing the herding behavior during the COVID-19 pandemic reveal that the coefficient (γ_4) has a positive sign and the Minimum Bayes factor (MBF) represents strong evidence indicating that there is no influence of herding behavior during the COVID-19 pandemic and the dispersion of returns has increased. This phenomenon is reflected in the change of the market return. These results support the rational asset pricing models that periods of market stress from the announcement of the number of the infected with COVID-19 induced widened dispersions. All investors have been fully informed simultaneously, so they carefully consider it based on the fundamentals. This finding appears consistent with Demirer and Kutan [17] and Tan et al. [40], on concluding

Table 5 Testing the herding behavior during the COVID-19 pandemic

Variable	Coefficient	t-Statistic	Prob.
Constant (γ_0)	4.407***	6.983	0.000
ABSRM (γ_1)	0.965***	5.543	0.000
RMINUSRBAR2 (γ_2)	−0.041***	−3.357	0.004
CSAD(−1) (γ_3)	0.170**	2.601	0.034
COVIDDUMMY* RMINUSRBAR2 (γ_4)	0.031***	3.499	0.002
R-squared	0.598	Adjusted R-squared	0.584
F-statistic	42.368	Prob(F-statistic)	0.000

Note Calculated using Newey-West heteroscedasticity and autocorrelation consistent standard errors
***MBF represents strong evidence
**MBF represents substantial evidence
*MBF represents moderate evidence

that during the 1997 Asian financial crisis, investors did not exhibit herding behavior in the Chinese stock market. Moreover, the results are consistent with Yao et al. [48], that found investors did not have herd behavior during the financial crisis (October 2007–December 2008), and Filip et al. [19], which found the return dispersions are independent of the evolution of the global financial crisis of 2007–2009. However, our results are not in line with the findings by Espinosa-Méndez and Arias [18], Indars et al. [23]. Espinosa-Méndez and Arias [18] showing that the COVID-19 pandemic increased herding behavior in the Australian stock market. Meanwhile, Indars et al. [23] found that investors in the Moscow Stock Exchange exhibited the herding behavior during the subprime crisis.

5 Conclusion

In 2020, the rapid spread of the coronavirus disease 2019 (COVID-19) pandemic had put the world at hazard. The first patient was locally infected from the COVID-19 in Thailand on 31 January 2020. This research investigates the herding behavior in the Stock Exchange of Thailand during the COVID-19 pandemic and the disposition effect situation. We used daily data, with complete observations of stocks traded in the Stock Exchange of Thailand (SET), covering January 2011–December 2020, so the number of stocks in the sample is 143 stocks. We modified the average holding period following Atkins and Dyl [5]. Then we examined the disposition effect which was adapted from Visaltanachoti et al. [45] by using the Two-Stage Least Squares (2SLS). Then we created a new dummy variable to represent the disposition effect. The month having the disposition effect will be assigned the value of 1. After that, we tested the existence of the herding behavior during the COVID-19 pandemic and the time the disposition effect was occurring. The methodology for testing the herding behavior following Yao et al. [48] and Filip et al. [19] was the use of Ordinary Least Squares (OLS).

The empirical results show a weak disposition effect in the Stock Exchange of Thailand because as the disposition effect was detected in 44 out of the total 120 months or 36.67% of all observations. This result strongly supports that the Thai investors' holding periods are related to the level of transaction costs and stock size, except in the period 2018–2020 when the longer holding periods are irrelevant to larger stocks in some months. The empirical results show the herding behavior occurred during the disposition effect situation in the market. This result is consistent with Chang [13], which found the bi-directional relationship between the herding behavior and the disposition effect in Taiwan's information technology stocks. The results from testing the herding behavior during the COVID-19 pandemic show no influence of such behavior. These results support the rational asset pricing models that during the periods when a routine announcement was made of the number of the infected with COVID-19, all investors had complete information then carefully considered investment based on the fundamentals. This finding appears to be consistent with Demirer and Kutan [17], Tan et al. [40], and Yao et al. [48] which found the herding behavior did not occur during the financial crisis. The results reported in this paper have implications for investors in the Stock Exchange of Thailand. To decide on the allocation to various asset classes, they should consider that prices might deviate from the fundamental values when investors have herding behavior. The herding behavior is related to the other irrational behaviors of investors is the disposition effect of investors. On the other hand, the herding behavior occurs when the investors hold the losing stock and sell the winning stock. However, the herding behavior will not occur when the investors have complete information like the report of the number of confirmed COVID-19 cases. Therefore, the policymaker should clearly announce the news to the investors. Moreover, when investors get the news, they should recheck whether it is a fake or true one before considering trading.

Acknowledgements This paper has been partially supported by the Center of Excellence in Econometrics, Faculty of Economics, Chiang Mai University.

References

1. Amihud, Y. (2002). Illiquidity and stock returns: Cross-section and time-series effects. *Journal of Financial Markets, 5*(1), 31–56.
2. Amihud, Y., & Mendelson, H. (1986). Asset pricing and the bid-ask spread. *Journal of Financial Economics, 17*(2), 223–249.
3. Ashraf, B. N. (2020). Stock markets' reaction to COVID-19: Cases or fatalities? *Research in International Business and Finance, 54*, 7.
4. Ashraf, M., Waris, F., & Saeed, S. (2014). Disposition effect: Evidence from the Karachi stock exchange. *European Journal of Business and Management, 6*(11), 19–24.
5. Atkins, A. B., & Dyl, E. A. (1997). Transaction costs and holding periods for common stocks. *Journal of Finance, 52*(1), 309–325.

6. Autchariyapanitkul, K., Sriboonchitta, S., & Chanaim, S. (2014). Portfolio optimization of stock returns in high-dimensions: A copula-based approach. *Thai Journal of Mathematics,* 11–23.
7. Awale, S. L., Pandey, D. R., Sapkota, D. R., & Shrestha, K. C. (2018). A study of disposition effect, overconfidence bias and herding behavior in Nepalese investors. *Research Project Report,* 24.
8. Baek, S., Mohanty, S. K., & Glambosky, M. (2020). COVID-19 and stock market volatility: An industry level analysis. *Finance Research Letters, 37*, 1–10.
9. Bergsma, K., Fodor, A., & Tedford, E. (2020). A closer look at the disposition effect in U.S. equity option markets. *Journal of Behavioral Finance, 21*(1), 66–77.
10. Bora, D., & Basistha, D. (2021). The outbreak of COVID-19 pandemic and its impact on stock market volatility: Evidence from a worst-affected economy. *Journal of Public Affairs*. https://doi.org/10.1002/pa.2623
11. Bui, N. D., Nguyen, L., & Nguyen, N. (2015). Herd behavior in Southeast Asian stock markets—An empirical investigation. *Acta Oeconomica, 65*, 413–429.
12. Cao, X. (2010). *Disposition effect and momentum based on prospect theory/mental accounting in the Chinese stock markets* (Master of Business). Faculty of Business and Law, Auckland University of Technology, New Zealand.
13. Chang, C.-H. (2013). The relationship between the disposition effect and herding behavior: Evidence from Taiwan's information technology stocks. *World Academy of Science, Engineering and Technology International Journal of Economics and Management Engineering, 7*(7), 1971–1975.
14. Chang, E. C., Cheng, J. W., & Khorana, A. (2000). An examination of herd behavior in equity markets: An international perspective. *Journal of Banking & Finance, 24*, 1651–1679.
15. Chang, C. L., McAleer, M. J., & Wong, W. K. (2020). Risk and financial management of COVID-19 in business, economics and finance. *Journal of Risk and Financial Management, 13*(5), 1–7.
16. Christie, W. G., & Huang, R. D. (1995). Following the pied piper: Do individual returns herd around the market? *Financial Analysts Journal, 51*(4), 31–37.
17. Demirer, R., & Kutan, A. M. (2006). Does herding behavior exist in Chinese stock markets? *Journal of International Financial Markets, Institutions and Money, 16*(2), 123–142.
18. Espinosa-Méndez, C., & Arias, J. (2021). Herding behaviour in Australian stock market: Evidence on COVID-19 effect. *Applied Economics Letters, 28*(21), 1898–1901.
19. Filip, A., Pochea, M., & Pece, A. (2015). The herding behaviour of investors in the CEE stocks markets. *Procedia Economics and Finance, 32*, 307–315.
20. Goossens, J. T. G., & Knoef, M. (2020). Preferences, disposition effect and COVID-19. *Academic Final Report,* 40.
21. Haryanto, S., Subroto, A., & Ulpah, M. (2020). Disposition effect and herding behavior in the cryptocurrency market. *Journal of Industrial and Business Economics, 47*, 115–132.
22. Held, L., & Ott, M. (2018). On p-values and Bayes factors. *Annual Review of Statistics and Its Application, 5,* 393–419.
23. Indars, E. R., Savin, A., & Lublóy, A. (2019). *Herding behaviour in an emerging market: Evidence from the Moscow exchange* (Corvinus Economics Working Papers). Faculty of Economics, University of Budapest, 29 pages.
24. Kahneman, D., & Tversky, A. (1979). Prospect theory: An analysis of decision under risk. *Econometrica, 47,* 263–291.
25. Kulvanich, J., & Boonvorachote, T. (2013). Herding behavior analysis in the Stock Exchange of Thailand. *Kasetsart Journal, 34*, 43–59.
26. Lakonishok, L., Shleifer, A., & Vishny, R. W. (1992). The impact of institutional trading on stock prices. *Journal of Financial Economics, 32*(1), 23–43.
27. Lee, K. Y.-M., Jais, M., & Chan, C.-W. (2020). Impact of covid-19: Evidence from Malaysian stock market. *International Journal of Business and Society, 21*(2), 607–628.
28. Maneenil, P. (2012). *Disposition effect in securities trading: Evidence from Thai stock market* (Working Paper). Thammasat University.

29. Nimanussornkul, K., & Nimanussornkul, C. (2021). Herding behavior from loss aversion effect in the Stock Exchange of Thailand. In *Behavioral predictive modeling in economics*. Springer.
30. Onali, E. (2020). Covid-19 and stock market volatility, 25 pages.
31. Padungsaksawasdi, C., & Pooissarakit, I. (2015). Herd behavior in the Stock Exchanges in Thailand. *Chulalongkorn Business Review, 37,* 51–80.
32. Papadamou, S., Fassas, A. P., Kenourgios, D., & Dimitriou, D. (2020). *Direct and indirect effects of COVID-19 pandemic on implied stock market volatility: Evidence from panel data analysis* (MPRA Paper No. 100020).
33. Panyagometh, K. (2020). The effects of pandemic event on the Stock Exchange of Thailand. *Economies, 8*(90), 1–21.
34. Putra, A. A., Rizkianto, E., & Chalid, D. A. (2017). The analysis of herding behavior in Indonesia and Singapore Stock Market. *Advances in Economics, Business and Management Research, 36*, 197–206.
35. Rattanasri, W., & Vichitthamaros, P. (2018). The study of herding behavior in the Stock Exchange of Thailand during 2010–2015. *WMS Journal of Management Walaklak University, 7*, 9–20.
36. Sarajoti, P., & Puthpongsiriporn, T. (2009). *Evidences of the disposition effect across investor-types in Thailand.* SASIN Graduate Institute of Business Administration, Chulalongkorn University.
37. Sirisrisakulchai, J., Autchariyapanitkul, K., Harnpornchai, N., & Sriboonchitta, S. (2015). Portfolio optimization of financial returns using fuzzy approach with NSGA-II algorithm. *Journal of Advanced Computational Intelligence and Intelligent Informatics, 19*(5), 619–623.
38. Shefrin, H., & Statman, M. (1985). The disposition to sell winners too early and ride losers too long: Theory and evidence. *The Journal of Finance, 40*(3), 777–790.
39. Suppaudom, J. (2014). *Disposition effect and portfolio characteristics* (Working Paper). Thammasat University.
40. Tan, L., Chiang, T. C., Mason, J. R., & Nelling, E. (2008). Herding behaviour in Chinese stock markets: An examination of A and B shares. *Pacific-Basin Finance Journal, 16*(1), 61–77.
41. Thaler, R. (1980). Toward a positive theory of consumer choice. *Journal of Economic Behavior and Organization, 1,* 39–60.
42. Thaler, R. (1985). Mental accounting and consumer choice. *Marketing Science, 4*(3), 199–214.
43. Tversky, A., & Kahneman, D. (1981). The framing of decisions and the rationality of choice. *Science, 211*(4481), 453–458.
44. Urairoekkun, C. (2021). *The order of events of the corona virus outbreak.* Thai Health Literacy Promotion Association. http://doh.hpc.go.th/bs/wuhanDisplay.php
45. Visaltanachoti, N., Lu, L., & Luo, H. (2007). Holding periods, illiquidity and disposition effect in the Chinese stock markets. *Applied Financial Economics, 17*(15), 1265–1274.
46. Waiyasara, K., & Padungsaksawasdi, C. (2020). The V-shaped disposition effect in the Stock Exchange of Thailand. *Journal of Asian Finance, Economics and Business, 7*(11), 55–65.
47. Yarovaya, L., Matkovskyy, R., & Jalan, A. (2021). The effects of a "black swan" event (COVID-19) on herding behavior in cryptocurrency markets. *Journal of International Financial Markets, Institutions and Money, 75*, 16.
48. Yao, J., Ma, C., & He, W. P. (2014). Investor herding behaviour of Chinese stock market. *International Review of Economics & Finance, 29*, 12–29.
49. Yousef, I. (2020). Spillover of COVID-19: Impact on stock market volatility. *International Journal of Psychosocial Rehabilitation, 24*(6), 18069–18081.
50. Zhu, K., Yamaka, W., & Sriboonchitta, S. (2016). Multi-asset portfolio returns: A Markov switching copula-based approach. *Thai Journal of Mathematics, 14,* 183–200.

An Analysis of Market Cycle for Thai Cassava Chips

Kanthida Jaiin and Pathairat Pastpipatkul

Abstract This paper uses monthly data to examine factors affecting supply and demand, explore the time-varying magnitude effects of supply and demand, and clarify the market cycle for Thai cassava chips between January 2009 and December 2019, analyzed by using the Markov switching state-space model. The result shows that none of the variables affected the demand for cassava chips, while the amount of rainfall and yield of cassava roots in the previous month affected the supply of cassava chips. The price of substitution goods in regime 1 (downward market) and the GDP per capita of China in regime 2 (upward market) affected the time-varying of demand. The planting area, the amount of rainfall, the yield and the price of cassava root in the previous month affected the time-varying supply in regime 1. In regime 2, the planting area, the amount of rainfall and the yield of cassava roots in the previous month affected the time-varying supply. Moreover, the result is able to clarify the business cycle very well.

1 Introduction

Cassava is one of the important economic crops in Thailand. Thailand is the major cassava exporter in the world, exporting cassava products in the form of cassava chips, cassava pellets, native starch, modified starch, and ethanol. Cassava products are in high demand around the world due to the fact that they are cheaper than other starchy crops and have any industrial uses such as food, feed, and energy. In 2019, cassava chips accounted for 36.6% of Thailand's total exports of cassava products. The main export market is China [6].

The domestic purchasing and selling prices for cassava roots depend on the demand for cassava in the processing industries (chips, pellets, and starch). More-

K. Jaiin (✉) · P. Pastpipatkul
Faculty of Economics, Chiang Mai University, Chiang Mai, Thailand
e-mail: jkanthida@gmail.com

P. Pastpipatkul
e-mail: Pathairat.past@cmu.ac.th

S. Sriboonchitta et al. (eds.), *Credible Asset Allocation, Optimal Transport Methods, and Related Topics*, Studies in Systems, Decision and Control 429,
https://doi.org/10.1007/978-3-030-97273-8_36

over, the F.O.B. price will have an influence on determining the purchasing price of cassava roots and cassava products in the country because the production for export has a larger proportion than consumption within the country [1]. The price of cassava has been fluctuating in recent years. The reason for the volatility is the cassava production will be seasonal, especially from December to January. After March, the prices will continue to rise continuously, according to a report by Agri-Food International. There are two reasons why farmers rush to harvest cassava, namely concerns about the uncertainty of price and quality of starch [7].

While demand for cassava for transformation into cassava products for use in downstream sectors continues to rise, farmers continue to face the problem of declining prices during harvesting season and price volatility, which have an impact on their livelihood and quality of life. Therefore, the idea was to examine factors affecting the supply and demand for Thai cassava chips because the price and export volume of cassava products will reflect the price of cassava in the domestic. Another purpose is to explore the time-varying magnitude effects of supply and demand on Thai cassava chips and clarify the market cycle for Thai cassava chips by using the Markov-switching state-space model.

2 Methodology

The Markov switching model of Hamilton [4] involves multiple structures that can characterize time series behaviors in different regimes by permitting switching between various structures. The switching mechanism is controlled by an unobservable state variable that follows a first-order Markov chain. In particular, the Markovian property regulates that the present worth of the state variable depends on its immediate past worth. Therefore, the Markov switching model is suitable for describing correlated data that exhibits distinct dynamic patterns during different time periods [5].

Let y_t represent a p-variate vector with observations at occasion t. It is regressed on an unobserved r-variate vector α_t, and on a q-variate vector with observed predictors x_t, through the measurement equation

$$y_t = W_{s_t} a_t + B_{s_t} x_t + e_t, \quad e_t \sim N(0, R_{s_t}) \tag{1}$$

where W_{S_t} is a $p \times r$ matrix with regression coefficients for the regression of the observations on the latent variables, B_{S_t} is a pq matrix with regression coefficients for the regression of the endogenous observations y_t on the exogenous observations x_t, and S_t characterize the observed states, or regimes at time t. The residuals in the p-variate vector e_t are uncorrelated over time, and are uncorrelated with a_t and x_t, but may be correlated with other with each other at the same occasion (using S_t). To avoid confusion with the discrete state of the regime-switching process, we will refer to it as the latent variable vector. The latent variables evolve over time according to a first-order vector autoregressive model, as defined by the transition equation, that is

$$a_t = C_{s_t} + H_{s_t} a_{t-1} + G_{s_t} u_t, \quad u_t \sim N(0, Q_{s_t}) \tag{2}$$

where the $r \times r$ matrix H_{S_t} contains the autoregressive cross-lagged regression parameters by which the latent variables at occasion t are regressed on the latent variables at occasion $t-1$. The r-variate vector C_{S_t} contains the intercepts. The $r \times r$ matrix G_{S_t} is included because it allows for easier model specifications in some cases [3], but in many cases, it is restricted to be an identity matrix. The residuals in the r-variate vector u_t are uncorrelated over time and are also uncorrelated with the residuals from the measurement equation (i.e., e_t), and the latent variables at the previous occasion (i.e., a_{t-1}), although they may be correlated with each other at the same occasion (through Q_{S_t}).

$$p = \begin{pmatrix} p_{11} & p_{21} & \cdots & p_{M1} \\ p_{12} & p_{22} & \cdots & p_{M2} \\ \cdots & & & \\ p_{1M} & p_{2M} & \cdots & p_{MM} \end{pmatrix} \tag{3}$$

where $p_{ij} = P[S_t = j | S_{t-1} = i]$ denotes the probability for switching from state i to state j. Note that by definition $\sum_j p_{ij} = 1$. Together, the model defined in Eqs. (1)–(3) forms Kim's regime switching state-space model.

3 Data

The data for this study contains 9 variables, which are the monthly secondary data from the period January 2009 to December 2019, covering 132 observations. As with previous studies, GDP per capita, the price of substation goods, and crude oil prices affected demand for cassava products. The factors of all cassava plantations in Thailand affected the supply of cassava products [2]. The variables considered in the supply and demand for Thai cassava chips in this study were

$$Q_t^d = \alpha_0 + \alpha_1 Price_{t-1} + \alpha_2 Price_t^S + \alpha_3 GDP\_China_t \tag{4}$$

$$Q_t^S = \beta_0 + \beta_1 Price_{t-1} + \beta_2 A_t + \beta_3 R_t + \beta_4 Y_{t-1} + \beta_5 Price_{t-1}^c \tag{5}$$

where $Price_{t-1}$ is the F.O.B. price of cassava chips in the previous month, ($Price_t^S$) is the price of substitution goods (maize), (GDP_China_t) is the GDP per capita of China, (A_t) is the planting area, (R_t) is the amount of rainfall, (Y_{t-1}) and $Price_{t-1}^c$ are the yield and price of cassava root in the previous month, respectively. (Q_t^d) represents the total demand for cassava chips by assuming that the market is in equilibrium. (Q_t^s) represents Thailand's total cassava chip production, assuming that all of it is manufactured for export.

Table 1 Estimates parameter of demand

Variable	Coefficient	SD	t-statistic	t-probability
constant	17.973	3.319	5.414	0.000
$Price_{t-1}$	−0.322	0.397	−0.81	0.419
$Price_t^S$	−0.152	0.299	−0.508	0.611
GDP_China_t	−0.284	0.351	−0.81	0.419

Source Own calculation

Table 2 Estimated demand parameters of the Markov switching with time-varying

Variable	Coefficient	SD	t-statistic	t-probability
Regime 1				
$Price_{t-1}$	−8.580	11.867	−0.723	0.469
$Price_t^S$	10.321	3.098	3.330	0.000
GDP_China_t	0.997	7.678	0.129	0.896
Log(Sigma)	−0.791	0.324	−2.435	0.014
Regime 2				
$Price_{t-1}$	−0.407	0.301	−1.353	0.176
$Price_t^S$	0.288	0.192	1.502	0.133
GDP_China_{sst}	1.455	0.120	12.039	0.000
Log(Sigma)	−0.797	0.066	−12.025	0.000

Source Own calculation

4 Result

4.1 Demand

The demand of cassava chips was unaffected by any of the variables, as shown in Table 1.

The estimated demand parameters of the Markov switching with time-varying are shown in Table 2. This paper considers two regime models, consisting of the downward market denoted by Regime 1 and the upward market denoted by Regime 2. In Regime 1, only the parameters of the price of substitution goods (maize) is statistically significant at 99%. Namely, if the price of substitution goods increases by 1%, the demand for Thai cassava chips will increase by 10.321%. In Regime 2, only the parameters of the GDP per capita of China is statistically significant at 99%. In other words, if the GDP per capita of China increases by 1%, the demand for Thai cassava chips will increase by 1.455%.

The transition probability matrix of demand is presented in Table 3. The result shows that the probability of switching from regime 1 (downward market) to regime 2 (upward market) is 0.030, while the probability of remaining in regime 1 is 0.970. On the other hand, the probability of switching from regime 2 to regime 1 is 0.007, while the probability of remaining in regime 2 is 0.993.

Table 3 Transition probability matrix of demand

	Regime 1	Regime 2
Regime 1	0.970	0.030
Regime 2	0.007	0.993

Source Own calculation

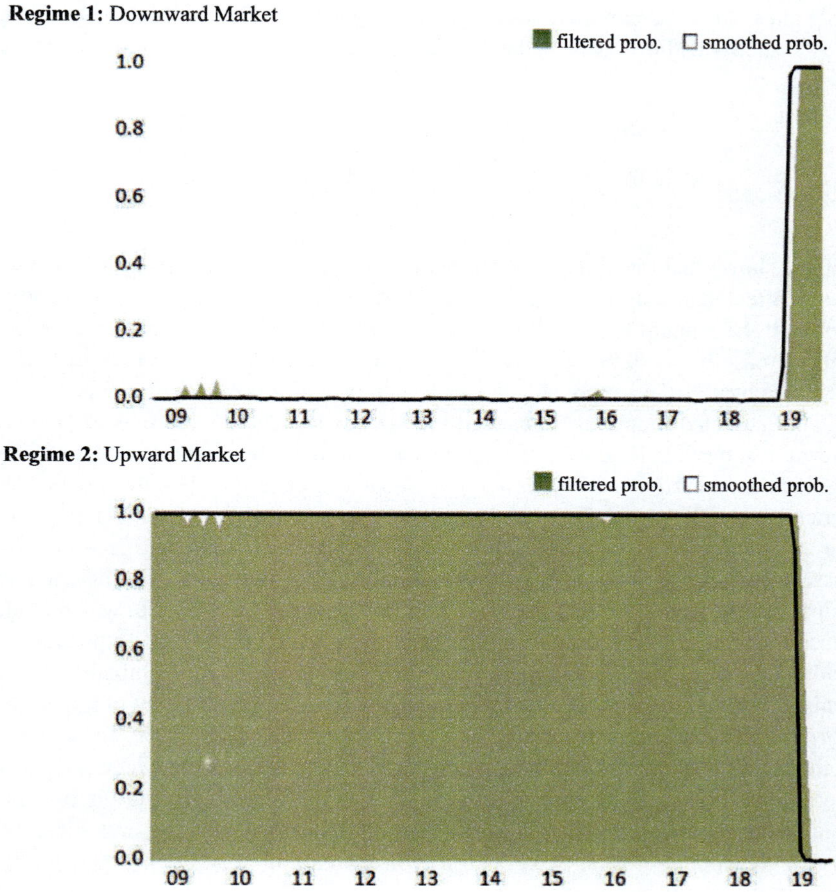

Fig. 1 Filtered and smooth probabilities plots for demand

Filtered probability refers to an estimate of the probability at time t based on data up to and including time t. Smoothed probability refers to an estimate of the probability at time t using all the data in the sample. When comparing Fig. 1 to the actual scenario, it can be seen that the demand for cassava chips has been declining since May of 2019 due to the global economic slowdown from the trade war between the United States and China. Moreover, China wants to deplete its low-cost maize stockpiles, so it has curtailed cassava chips imports. Another problem is that, due to African Swine Fever (ASF), the demand for cassava chips for feed industries has fallen.

4.2 Supply

Table 4 shows that the amount of rainfall and yield of cassava root in the previous month affected the supply of cassava chips statistically significantly at 99%. In other words, if the amount of rainfall decreases by 1%, the supply of Thai cassava chips will increase by 0.302%, while if the yield of cassava root in the previous month increases by 1%, the supply of Thai cassava chips will increase by 0.618%.

The estimated supply parameters of the Markov switching with time-varying are shown in Table 5. In Regime 1, the parameters of almost all variables are statistically significant except for the F.O.B. price of cassava chips in the previous month. The parameters of the price of cassava root in the previous month and the planting area are statistically significant at 90% and 95%, respectively. Namely, if the price of cassava root in the previous month increased by 10%, the supply of Thai cassava chips would increase by 0.525%. The planting area increased by 5%, which could cause a decrease in the supply of Thai cassava chips by 0.273%. The parameters of the amount of rainfall and yield of cassava root in the previous month are also statistically significant at 99%, meaning that an increase in the amount of rainfall by 1% can cause a decrease in the supply of Thai cassava chips by 0.273%. The increase in the yield of cassava root in the previous month increased by 1%, which caused an increase in the supply of Thai cassava chips by 0.597%. In Regime 2, the planting area is statistically significant at 95%, meaning that if the planting area increases by 5%, the supply of Thai cassava chips will increase by 0.305%. The parameters

Table 4 Estimates parameter of supply

Variable	Coefficient	SD	t-statistic	t-probability
Constant	8.383	11.665	0.718	0.473
$Price_{t-1}$	0.599	0.451	1.328	0.186
A_t	−0.377	0.765	−0.493	0.622
R_t	−0.302	0.054	−5.545	0.000
Y_{t-1}	0.618	0.059	10.388	0.000
$Price^c_{t-1}$	−0.51	0.348	−1.465	0.145

Source Own calculation

Table 5 Estimated supply parameters of the Markov switching with time-varying

Variable	Coefficient	SD	t-statistic	t-probability
Regime 1				
$Price_{t-1}$	−0.767	0.569	−1.346	0.178
A_t	0.525	0.179	2.927	0.003
R_t	−0.273	0.056	−4.873	0.000
Y_{t-1}	0.597	0.064	9.315	0.000
$Price^c_{t-1}$	0.836	0.451	1.854	0.063
Log(Sigma)	−1.160	0.133	−8.723	0.000
Regime 2				
$price_{t-1}$	−0.027	0.379	−0.071	0.942
A_t	0.305	0.150	2.028	0.042
R_t	−0.281	0.052	−5.343	0.000
Y_{t-1}	0.670	0.057	11.608	0.000
$Price^c_{t-1}$	−0.012	0.277	−0.043	0.965
Log(Sigma)	−1.369	0.099	−13.747	0.000

Source Own calculation

Table 6 Transition probability matrix of supply

	Regime 1	Regime 2
Regime 1	0.752	0.248
Regime 2	0.169	0.831

Source Own calculation

of the amount of rainfall and yield of cassava root in the previous month are also statistically significant at 99%, meaning that an increase in the amount of rainfall by 1% can cause a decrease in the supply of Thai cassava chips by 0.281%. The yield of cassava root in the previous month increased by 1%, which could cause an increase in the supply of Thai cassava chips by 0.670%.

The transition probability matrix of supply is presented in Table 6. The result shows that the probability of switching from regime 1 (downward market) to regime 2 (upward market) is 0.248, while the probability of remaining in regime 1 is 0.752. On the other hand, the probability of switching from regime 2 to regime 1 is 0.169, while the probability of remaining in regime 2 is 0.831.

When comparing Fig. 2 to the actual scenario, it can be seen that the supply of cassava chips will be in the upward market between June and March largely. Because it is the off-harvest season, the price is higher than it is during other times of the year, encouraging exporters to sell more cassava chips.

Regime 1: Downward Market

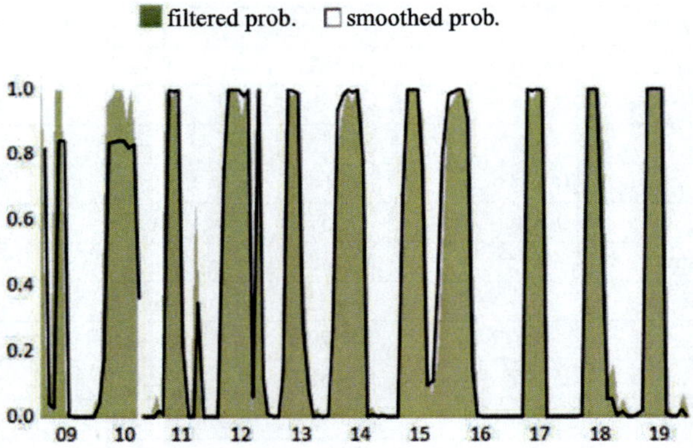

Regime 2: Upward Market

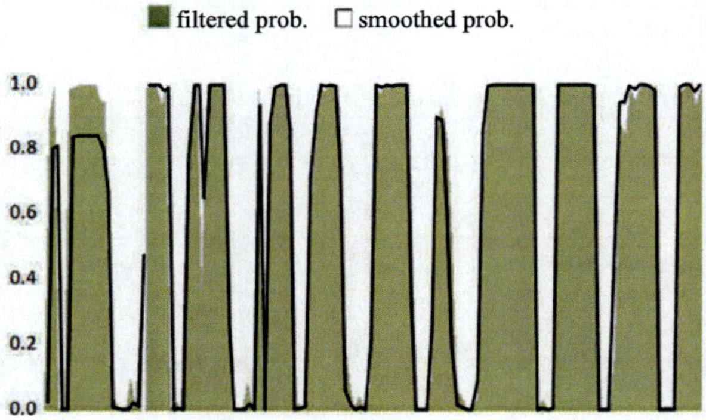

Fig. 2 Filtered and smooth probabilities plots for supply

5 Conclusion

This paper considers the factors affecting supply and demand for Thai cassava chips, and another purpose is to explore the time-varying magnitude effects of supply and demand on Thai cassava chips. We introduce the Markov switching state-space model as a contribution to the discussion on the factors affecting demand and supply for Thai cassava chips. In order to examine the market cycle for Thai cassava chips, we estimate the model using Kim's Filters to discover a regime probability.

The study found that none of the variables affected the demand for cassava chips. The time-varying of demand, the price of substitution goods, and the GDP per capita of China are statistically significant in regimes 1 (downward market) and 2 (upward market), respectively. Supply-side, the amount of rainfall and yield of cassava roots in the previous month affected the supply of cassava chips. The time-varying of supply, the planting area, the amount of rainfall, the yield and the price of cassava root in the previous month are statistically significant in regime 1. The planting area, the amount of rainfall and the yield of cassava roots in the previous month are statistically significant in regime 2. Moreover, the result is able to clarify the business cycle very well.

As a result of this research suggest the government should to findings a new market for cassava chips, to reduce dependence on the Chinese market in high proportion because the Thai cassava chips industry is sensitive to changes in Chinese policy. Nonetheless, this merits additional investigation. One of the things that should be improved in a future study is the use of different models.

References

1. AFTC. (2011). *Factors that determine the demand and supply of agricultural products*. Agricultural Futures Trading Commission. Retrieved from http://www.farmkaset.org/html15/contents.aspx?con_id=1797
2. Atipaeramanon, K. (2019). *Factors affecting the equilibrium of Thai cassava exported to China*. Economics NIDA.
3. Durbin, J., & Koopman, S. J. (2001). *Time series analysis by state space methods*. New York: Oxford University Press.
4. Hamilton, J. D. (1989). A new approach to the economic analysis of nonstationary time series and the business cycle. *Econometrica,57*, 357–384.
5. Kuan, C.-M. (2002). *The Markov switching model*. Institute of Economics.
6. LH BANK. (2014). Cassava industry. Retrieved from https://www.lhbank.co.th/Files/economic/economic_20161206154119.pdf
7. Tantaterdtham, et al. (2013). *Analysis of market demand and utilization of cassava in China*. National Research Council of Thailand.

Maximal Predictability Portfolio Optimization Model and Applications to Vietnam Stock Market

Bao Q. Ta, Vu T. Huynh, Khai Q. H. Nguyen, Phung N. Nguyen, and Binh H. Ho

Abstract The aim of this paper is to utilize the Maximal Predictability Portfolio optimization model proposed by Lo and MacKinlay to create an investment portfolio. The model can capture the market predictability and avoid forecasting errors which is a serious impediment to the Mean-Variance framework. The obtained results show that the Maximal predictability portfolio optimization model has a better performance than the traditional Markowitz mean-variance model, thereby partly showing its potential for further development in portfolio optimization progress.

Keywords Portfolio optimization · Maximal predictability · Mean-variance · Fama-French model

1 Introduction

Portfolio management presents the trade-off between the return and risk of investors, which is usually performed by maximizing expected returns at a specific level of risk [1]. However, in general, the assumption of Gaussian distribution for the asset returns is not fully sufficient in the Mean-Variance framework. Moreover, the major reason for the reluctance of academics and practitioners in using the framework of the Markowitz model is that the framework is extremely sensitive to the change of the input (the expected return of each asset and its corresponding covariance matrix), which leads to the implausible weight of the asset components in the portfolio.

Approximate two decades before, Lo and MacKinlay have devised a vigorous method in analyzing portfolios [2]. Maximal Predictability Portfolio (MPP, for short) seems to be a propitious method based on the idea of maximizing the predictive ability of the portfolio whose objective is constructed from underlying factor models' forecasting power, such as the CAPM and the Fama-French Three-Factor model

B. Q. Ta (✉) · V. T. Huynh · K. Q. H. Nguyen · P. N. Nguyen · B. H. Ho
Department of Mathematics, International University, Vietnam National University, Ho Chi Minh City, Vietnam
e-mail: baotq@hcmiu.edu.vn

S. Sriboonchitta et al. (eds.), *Credible Asset Allocation, Optimal Transport Methods, and Related Topics*, Studies in Systems, Decision and Control 429,
https://doi.org/10.1007/978-3-030-97273-8_37

(FF3). Despite its attractiveness to be applied in the financial industry, the MPP at that time failed to overcome the following primary hurdles. In specific, the problem of the MPP optimization is a non-convex fractional quadratic programming problem in which the objective function is a ratio of two convex quadratic functions, and there exists no efficient method to solve such thing. Fortunately, with the advancement in global optimization, the 0–1 integer programming approach proposed by Konno [3], and the Mean-Absolute-Deviation transformation approach proposed by Konno, Yamamoto and Morita [4] has facilitated the process of obtaining optimal weight. This study has adopted the latter approach to solving the maximal predictability portfolio optimization problem.

In this paper, we utilize the MPP optimization method in the Vietnam stock market and then compare it with the traditional Markowitz method. Moreover, there are two main important contributions to this empirical research literature. First, we provide evidence-based result on the performance of the MPP method and, in fact, even with a small amount, it did generate excess returns, while its counterpart failed to do so in a volatile, unstable Vietnam stock market that was in 'bearish' condition at that time. Second, the comparison between the traditional Markowitz and MPP models have partly proved the superior and the potential of the latter, and it also raised some questions regarding predictive power of the former.

The paper is organized as follows. In the next section, we give a short introduction to the Fama-French three-factor model. In Sects. 3 and 4, we will briefly explain the required structure to create the Markowitz mean-variance and MPP models, and also review the solution to the non-convex fractional problem. Section 5 will be devoted to showing the computational results on the Vietnam stock market. In Sect. 6, the final conclusion would be drawn from these empirical results.

2 The Fama-French Three-Factor Model

2.1 *Return of Asset and Portfolio*

Supposed that the interval $[0, T]$ is our historical time series and N is the number of equal sub-intervals between 0 and T. Denote P_{i,t_k}, where t_k is often measured in daily (or any frequency like weekly, monthly, yearly) and index i represents the ith asset in the portfolio, then the return of an asset is calculated as below

$$R_i(t_k) = \frac{P_{i,t_k} - P_{i,t_{k-1}}}{P_{i,t_k}}, \tag{1}$$

If there is a cash flow in the partition k of time t then return of the asset will be adjusted below

$$R_i(t_k) = \frac{P_{i,t_k} - P_{i,t_{k-1}} + C_{i,t_k}}{P_{i,t_k}},$$

where C_{i,t_k} is the cash flow at time t_k such as dividend. The cumulative return of asset i^{th} in N-periods between $[0, T]$ will be

$$R_i = \prod_{k=1}^{N}(1 + R_{i,t_k}) - 1. \tag{2}$$

Consider a portfolio consisting of n assets. Denote $W = (w_1, w_2, .., w_n)$ the vector of weights on assets of the portfolio. Then, the portfolio return $R_{p,k}$ at period k is

$$R_{p,k} = \sum_{i=1}^{n} w_i R_{i,t_k}, \tag{3}$$

and the cumulative return of the portfolio over $[0, T]$ is

$$R_p = \sum_{i=1}^{n} w_i R_i. \tag{4}$$

2.2 *Sharpe Ratio and Beta*

Modern Portfolio Theory (MPT) postulates that rational investors can maximize their expected returns using a diversified portfolio. A pioneer in the MPT theory is Harry Markowitz [5], who used the framework of choosing a portfolio that raises the optimum trade-off between variance and expected return. We will discuss the classical technique of Markowitz in the next section. William F. Sharpe developed a model called *Capital Asset Pricing Model* or CAPM in short. The model is followed by 7 assumptions [6]. Here, the first 5 assumptions deal with the decision-making of investors. The other two assumptions about the properties of the market will be introduced below for the sake of constituting the model

- There exists risk-free assets and there is no limitation in borrowing or lending at the same risk-free rate.
- The market is frictionless, meaning there appears no transaction cost and the market itself has to be competitive.

The model presents the linear relationship between the expected return of a portfolio and the combination of market portfolio (representing the choice of risky assets) and risk-free asset

$$\mathbb{E}[R_p] = R_f + \sigma_p \frac{\mathbb{E}[R_M] - R_f}{\sigma_M},$$

where R_f is the risk-free rate, R_p is the portfolio return and σ_p is the volatility of the portfolio.

Sharpe ratio:

$$\frac{\mathbb{E}[R_p] - R_f}{\sigma_p}. \tag{5}$$

This ratio is also called *Efficiency of investment* that measures the risk-adjusted return. The total risk may be interpreted as a mixture of systematic risks and non-systematic risks. Denote R_{bm} the market benchmark returns in the market then the systematic risk is defined as follows

$$\beta_p = \frac{Cov(R_p, R_{bm})}{Var(R_{bm})}.$$

2.3 The Fama-French Three-Factor Model

CAPM proposed by Sharpe, Lintner, Mossin, and Treynor is tremendously used by managers due to its simplicity. Albeit, many assumptions that are unfavorable in the real world. In 1992, Kenneth French and Eugene Fama proposed two new fundamental factors in their paper [7]. The equation of the Fama-French 3 factor model (FF3M in short) is given by

$$E[R_p] = \alpha + R_f + \beta_1(E[R_M - R_f]) + \beta_2 E[R_{SMB}] + \beta_3 E[R_{HML}], \tag{6}$$

where the *SMB* factor illustrates the observation that is concerned about the size of firms, which has negative effects on excess return. It is found that firms with small market capitalization outperform those with a big cap. The *HML* factor describes that high book-to-market ratio firm (usually it is referred to as value stocks) and Growth stocks have a low book-to-market ratio.

3 The Markowitz Mean-Variance and Mean Absolute Deviation Models

3.1 The Markowitz Mean-Variance Model

The framework of the Mean-Variance model has become the most influential and vital backbone for the quantitative management of portfolios. In this section, the scheme of the Mean-Variance model will shortly be introduced.

It is well-argued that the ingredients for Mean-Variance optimization are estimate of expected return and the portfolio variance together with many constraints. Markowitz stated that an adverse investor would choose a portfolio that raises the highest return at a favorable risk. Hence, solving the optimization problem will determine the efficient frontier which is the set of all possible portfolios and subsequently the optimal portfolio weights.

Suppose we have a set of n risky assets and a vector $\omega = (\omega_1, \omega_2, \omega_3, \ldots, \omega_n)$ where each ω_i represents the percentage of capital invested in the ith asset in the portfolio. For simplicity and practical issue, no short selling is introduced in this model because some kinds of assets are difficult to sell short. Additionally, in each allocation, we restrict the amount by imposing a number α_i.

Now consider the vector of asset returns $\mathbf{R} = (R_1, R_2, R_3, \ldots, R_n)$. Denote $\boldsymbol{\mu} = (\mu_1, \mu_2, \ldots, \mu_n)$ the expected returns and $\sigma_{ij} = Cov(R_i, R_j)$ the covariance between the ith and jth return. Note that $\sigma_i^2 = Var(R_i)$ and $\sigma_{ij} = \rho_{ij}\sigma_i\sigma_j$, where ρ_{ij} is the correlation between asset ith and asset jth. Assume that asset returns in the portfolio follow the normal distribution, i.e., $R_i \sim N(\mu_i, \sigma_i)$. The formulation of minimizing portfolio risk subject to a targeted expected portfolio return is given as follows

$$\begin{aligned} \min_{\omega} \quad & \sum_{i=1}^{n}\sum_{j=1}^{n} \sigma_{ij}\omega_i\omega_j \\ \text{s.t.} \quad & \sum_{i=1}^{n} \omega_i = 1 \\ & \sum_{i=1}^{n} \omega_i\mu_i \geq \upsilon \\ & 0 \leq \omega_i \leq \alpha, \qquad i = 1, 2, \ldots, n \end{aligned} \tag{7}$$

Where υ is the minimal that the investors will receive based on the behavior of risk-averse investors. The compact and elegant way to formulate the constrained optimization problem is illustrated [8]

$$\begin{aligned} \min_{\omega} \quad & \omega'\Sigma\omega \\ \text{s.t.} \quad & \omega'\mu \geq \upsilon \\ & \omega'\vec{1} = 1, \qquad \vec{1} = [1, 1, \ldots, 1] \end{aligned} \tag{8}$$

Where Σ is the covariance matrix, in which σ_{ij} is the ijth element and the main diagonal is $\sigma_i^2 = Var(R_i)$. Since the problem generally does not have an analytical (closed-form) solution, especially when our constraints are all inequality. The solution to the Markowitz optimization problem can be addressed by numerical method.

3.2 *Mean Absolute Deviation Model*

The quadratic programming problem corresponding to Mean-Variance analysis becomes such a huge obstacle when it comes to a large scale of assets. Hence, many researchers have attempted to devise a linearized version of Markowitz model (e.g., Sharpe 1971 and Konno and Yamazaki 1991). To reduce significantly the computational time of searching the optimal portfolio, an alternative risk metric was developed in [4]. We will start this section by introducing the definition of the Mean-Absolute-Deviation (MAD) model, and presenting a property which connects the result of the optimal solution of this model to Markowitz model. Consider a portfolio consisting of n assets $S_i, i = 1, 2, \ldots, n$ with corresponding returns R_i. Recall that the return of the portfolio with a set of n asset weights $\mathbf{x} = (x_1, x_2, x_3, \ldots, x_n)$ is computed as follows

$$R_p(x) = \sum_{i=1}^{n} R_i x_i, \tag{9}$$

In the conventional analysis of portfolio management, practitioners often use standard deviation as a measure of risk

$$V(x) = E[(R_p(x) - E[R_p(x)])^2], \tag{10}$$

$$\Leftrightarrow \sigma(x) = \sqrt{V(x)}, \tag{11}$$

In [9], Konno and Yamazaki replaced the L_2-norm risk by L_1-norm risk. Since these two risk metrics measure the dispersion of a sample around their mean, and, hence, there is not significantly different

$$W(x) = E[|R_p(x) - E[R_p(x)]|]. \tag{12}$$

If the random variable R is normally distributed, then MAD is proportional to standard deviation, as the below theorem presents.

Theorem 1 *If $R_p(x)$ follows a normal distribution with mean μ and standard deviation σ, then*

$$W(x) = \sqrt{\frac{2}{\pi}}\sigma(x).$$

This theorem also holds when R is a multivariate normal random variable [9].

Theorem 2 *The MAD model generates the alike optimal portfolio to the corresponding MV model when $R = (R_1, R_2, \ldots, R_n)$ follows a multivariate normal distribution.*

From now on, the establishment of the problem in terms of the linear programming will be shown. The formulation plays a crucial role in constituting the problem in the MPP model that will be demonstrated in the next section.

Suppose that the vector returns $\mathbf{R} = (R_1, R_2, \ldots, R_n)$ is distributed over a finite set of points $(r_{1t}, r_{2t}, \ldots, r_{nt}), t \in 1, 2, \ldots, T$. Moreover, $p_t = P[(R_1, R_2, \ldots, R_n) = (r_{1t}, r_{2t}, \ldots, r_{nt})]$ is assumed to be available through historical data.

As denoted above, x_i is the percentage that asset ith contributes to the portfolio. The absolute deviation *W(x)* can be written as follows

$$\begin{aligned} W(x) =& E[|R_p(x) - E[R_p(x)]|] \\ =& E[|\sum_{i=1}^{n} R_i x_i - E[\sum_{i=1}^{n} R_i x_i]|] \\ =& \sum_{t=1}^{T} p_t |\sum_{i=1}^{n} (r_{it} - r_i) x_i|, \end{aligned} \tag{13}$$

where $r_i = \sum_{t=1}^{T} p_t r_{it}$ is expected value of R_i.
The Mean-Absolute-Deviation model is defined as follows:

$$\begin{aligned} \max \quad & \sum_{i=1}^{n} r_i x_i \\ \text{s.t.} \quad & \mathbf{W(x)} \le w, \\ & \mathbf{x} \in X, \end{aligned} \tag{14}$$

where w is an acceptable level of risk and $X \subset R_n$ is an investable set defined by

$$X = \left\{ \mathbf{x} \in R^n \mid \sum_{i=1}^{n} x_i = 1, 0 \le x_i \le \alpha, i = 1, 2, \ldots, n \right\}.$$

Note that α is the upper bound for the weight of the asset jth. Now let us introduce a set of non-negative variables ϕ_t, ψ_t satisfying the following conditions

$$\begin{aligned} & \phi_t - \psi_t = p_t \sum_{i=1}^{n} (r_{it} - r_i) x_i, \quad t = 1, 2, \ldots, T \\ & \phi_t \psi_t = 0, \quad t = 1, 2, \ldots, T \\ & \phi_t, \psi_t \ge 0, \quad t = 1, 2, \ldots, T \end{aligned} \tag{15}$$

It is necessary to replace all the occurrence of the term $p_t|\sum_{i=1}^{n}(r_{it} - r_i)x_i|$ with ϕ_t, ψ_t in such a manner that $p_t \ge 0$, for all t. We have either $\phi_t = 0$ or $\psi_t = 0$, otherwise both ϕ_t and ψ_t would be reduced by the same amount and preserve feasibility, while maximizing the benefit which is in contradiction of optimality [10]. Having guaranteed that either $\phi_t = 0$ or $\psi_t = 0$, the desired relation $p_t|\sum_{i=1}^{n}(r_{it} - r_i)x_i| = \phi_t + \psi_t$. So, the MAD model (13) turns into

$$
\begin{aligned}
maximize \quad & \sum_{i=1}^{n} r_i x_i \\
\text{s.t.} \quad & \sum_{t=1}^{T} (\phi_t + \psi_t) \le w \\
& \phi_t - \psi_t = p_t \sum_{i=1}^{n} (r_{it} - r_i) x_i, \quad t = 1, 2, \ldots, T \\
& \phi_t \psi_t = 0, \qquad t = 1, 2, \ldots, T \\
& \phi_t, \psi_t \ge 0, \qquad t = 1, 2, \ldots, T \\
& \mathbf{x} \in X.
\end{aligned}
\tag{16}
$$

The above problem could be reformulated by using the linear programming method from Chvatal [10], complementary conditions $\phi_t \psi_t = 0, t = 1, 2, \ldots, T$ can be removed so that the problem becomes a linear programming problem [10].

$$
\begin{aligned}
maximize \quad & \sum_{i=1}^{n} r_i x_i \\
\text{s.t.} \quad & \sum_{t=1}^{T} (\phi_t + \psi_t) \le w \\
& \phi_t - \psi_t = p_t \sum_{i=1}^{n} (r_{it} - r_i) x_i, \quad t = 1, 2, \ldots, T \\
& \phi_t, \psi_t \ge 0, \qquad t = 1, 2, \ldots, T \\
& \sum_{i=1}^{n} x_i = 1, \\
& 0 \le x_i \le \alpha, \qquad i = 1, 2, \ldots, n.
\end{aligned}
\tag{17}
$$

4 Maximal Predictability Portfolio

Maximal Predictability Portfolio is amongst one of the most potential methods to be exploited in portfolio analysis. The MPP optimization problem can be formulated from a non-convex quadratic programming problem into the 0–1 linear programming problem. This algorithm was proposed by Yamamoto and Konno [4]. Furthermore, the algorithm is also used to solve the MPP optimization problem with hundreds of assets and a maximum of 15 factors [3].

4.1 Some Concepts

Consider the set of n risky assets during time period t; $r_t = (r_{1t}, r_{2t}, r_{3t}, \ldots, r_{nt})$, $t = 1, 2, \ldots, T$. Assume that r_t is jointly stationary and follows an ergodic process with finite expectation $\mu = (\mu_1, \mu_2, \ldots, \mu_n)^{\mathsf{T}}$ and finite autocovariance matrix. The factor model that represents the rate of return of the ith asset

$$r_i = \beta_{i0} + \beta_{i1}F_1 + \beta_{i2}F_2 + \cdots + \beta_{iK}F_K + \epsilon_i, \qquad i = 1, 2, 3, \ldots, n, \tag{18}$$

where $F_1, F_2, \ldots, F_K$ are factors and ϵ_i is independent from each other with $E[\epsilon_i] = 0$, $Cov(F_k, \epsilon_i) = 0$, $k = 1, 2, \ldots, K$ and we obtain the coefficient β_{ik}'s, $k = 1, 2, \ldots, K$ from the method of least squares. The values of r_i and F_k during period t are

$$\begin{aligned} r_t &= (r_{1t}, r_{2t}, \ldots, r_{nt})^{\mathsf{T}} \in R^n, \\ F_t &= (F_{1t}, F_{2t}, \ldots, F_{kt})^{\mathsf{T}} \in R^K, \\ \epsilon_t &= (\epsilon_{1t}, \epsilon_{2t}, \ldots, \epsilon_{nt})^{\mathsf{T}} \in R^n. \end{aligned}$$

Suppose that ϵ_t is a conditionally homoskedastic process with $E[\epsilon_t] = 0$ and the information set $\mathscr{F}_t$ is well behaved such that r_t is a stationary and ergodic process. Moreover, $\mathscr{F}_t$ is earlier observable economic variables, such as dividend yields, interest rate spreads, or other leading economic indicators. The set of optimal portfolio weights, as denoted by $\mathbf{x}$, is the optimal solution of the following optimization problem.

$$\begin{aligned} \max_x \quad & \frac{Var(\tilde{r}_t{}^{\mathsf{T}}x)}{Var(r_t^{\mathsf{T}}x)} \\ \text{s.t.} \quad & \sum_{i=1}^{n} x_i = 1 \\ & \sum_{i=1}^{n} x_i \tilde{r}_i \geq \upsilon \\ & 0 \leq x_i \leq \alpha, \qquad i = 1, 2, \ldots, n \end{aligned} \tag{19}$$

where $\tilde{r}_t = E[r_t \mid \mathscr{F}_{t-1}]$ is the forecast return based on information that is available at time $t-1$. We see that $r_t = \tilde{r}_t + \epsilon_t$, and υ is a constant representing the minimal return.

4.2 Portfolio Optimization Problems

Now apply Theorem 1, if r_t and $\tilde{r}_t$ are normally distributed then

$$\text{maximize } \frac{Var(\tilde{r}_t)}{Var(r_t)} \text{ is equivalent to maximize } \frac{W(\tilde{r}_t)}{W(r_t)}.$$

If the variance is replaced by absolute deviation then (19) can be reformulated as follows

$$\begin{aligned} \min_x \quad & \frac{E[|r_t^\mathsf{T} x - E[r_t^\mathsf{T} x]|]}{E[|\tilde{r}_t{}^\mathsf{T} x - E[\tilde{r}_t{}^\mathsf{T} x]|]} \\ \text{s.t.} \quad & \sum_{i=1}^{n} x_i = 1 \\ & \sum_{i=1}^{n} x_i \tilde{r}_i \geq \upsilon \\ & 0 \leq x_i \leq \alpha, \qquad i = 1, 2, \ldots, n \end{aligned} \tag{20}$$

Note that $r_t^\mathsf{T} x = \sum_{i=1}^{n} x_i r_{i,t}$, this relationship will be worked on sooner. From the Fama-French three factors model, the weighted rate of return at period t could be represented with explanatory factors F_1, F_2, F_3.

$$\begin{aligned} x_i r_{i,t} &= x_i \beta_0 + x_i \beta_{i,1} F_{1,t-1} + x_i \beta_{i,2} F_{2,t-1} + x_i \beta_{i,3} F_{3,t-1} + x_i \epsilon_{i,t} \\ E[x_i r_{i,t}] &= x_i \beta_0 + x_i \beta_{i,1} E[F_{1,t-1}] + x_i \beta_{i,2} E[F_{2,t-1}] + x_i \beta_{i,3} E[F_{3,t-1}]. \end{aligned} \tag{21}$$

Consequently,

$$\begin{aligned} x_i r_{i,t}^\mathsf{T} - E[x_i r_{i,t}^\mathsf{T}] &= x_i \beta_{i,1}(F_{1,t-1} - \hat{F}_1) + x_i \beta_{i,2}(F_{2,t-1} - \hat{F}_2) + x_i \beta_{i,3}(F_{3,t-1} - \hat{F}_3) \\ &= \sum_{i=1}^{n} x_i \sum_{k=1}^{3} \beta_{i,k}(F_{k,t-1} - \hat{F}_k) + \sum_{i=1}^{n} x_i \epsilon_{i,t}. \end{aligned} \tag{22}$$

Hence, with the relation $r_t = \tilde{r}_t + \epsilon_t$, the denominator becomes

$$\begin{aligned} x_i \tilde{r_{i,t}}^\mathsf{T} - E[x_i \tilde{r_{i,t}}^\mathsf{T}] &= x_i \beta_{i,1}(F_{1,t-1} - \hat{F}_1) + x_i \beta_{i,2}(F_{2,t-1} - \hat{F}_2) \\ &+ x_i \beta_{i,3}(F_{3,t-1} - \hat{F}_3) = \sum_{i=1}^{n} x_i \sum_{k=1}^{3} \beta_{i,k}(F_{k,t-1} - \hat{F}_k). \end{aligned} \tag{23}$$

Assume that (F_1, F_2, F_3) is distributed over a finite set $(f_{1t}.f_{2t}, \ldots, f_{3t})$ with equal probability $p_t = \frac{1}{T}$. Then

$$E[|r_t^{\mathsf{T}}x - E[r_t^{\mathsf{T}}x]|] = \frac{1}{T}|\sum_{t=1}^{T}\sum_{k=1}^{3}\beta_k(x)(f_{k,t-1} - \hat{f}_k) + \epsilon_t(x)|, \tag{24}$$

$$E[|\tilde{r}_t{}^{\mathsf{T}}x - E[\tilde{r}_t{}^{\mathsf{T}}x]|] = \frac{1}{T}|\sum_{t=1}^{T}\sum_{k=1}^{3}\beta_k(x)(f_{k,t-1} - \hat{f}_k)|, \tag{25}$$

where

$$\beta_k(x) = \sum_{i=1}^{n}\beta_{i,k}x_i, \quad k = 1, 2, 3 \tag{26}$$

$$\epsilon_t(x) = \sum_{i=1}^{n}\epsilon_{i,t}x_i, \quad t = 1, 2, \ldots, T \tag{27}$$

$$\hat{f}_k = E[f_{k,t-1}] = \frac{1}{T}\sum_{t=1}^{T} f_{k,t-1}, \quad k = 1, 2, 3 \tag{28}$$

So the maximal predictability portfolio construction problem becomes

$$\begin{aligned}
\underset{x}{minimize} \quad & \frac{\sum_{t=1}^{T} p_t|\sum_{k=1}^{3}\beta_k(x)(f_{k,t-1} - \hat{f}_k) + \epsilon_t(x)|}{\sum_{t=1}^{T} p_t|\sum_{k=1}^{3}\beta_k(x)(f_{k,t-1} - \hat{f}_k)|} \\
\text{subject to} \quad & \sum_{i=1}^{n} x_i = 1, \\
& \beta_k(x) = \sum_{i=1}^{n}\beta_{i,k}x_i, \qquad k = 1, 2, 3 \\
& \epsilon_t(x) = \sum_{i=1}^{n}\epsilon_{i,t}x_i, \qquad t = 1, 2, \ldots, T \\
& 0 \le x_i \le \alpha, \qquad i = 1, 2, \ldots, n, \\
& \sum_{i=1}^{n}\tilde{r}_i x_i \ge v.
\end{aligned} \tag{29}$$

It is vital to convert the problem into the 0–1 mixed-integer programming problem, which can be solved faster by computational approaches. First of all, applying the Charnes-Cooper linear fractional functional transformation [11]. We could define

$$y_0 := \frac{1}{\sum_{t=1}^{T} p_t |\sum_{k=1}^{3} \beta_k(x)(f_{k,t-1} - \hat{f}_k)|}. \tag{30}$$

This new variable is introduced to escape from the quotient form and, hence the system (29) now becomes

$$\begin{aligned}
\underset{x}{minimize} \quad & \sum_{t=1}^{T} p_t |\sum_{k=1}^{3} \beta_k(x)(f_{k,t-1} - \hat{f}_k) + \epsilon_t(x)| \cdot y_0 \\
\text{s.t.} \quad & \sum_{t=1}^{T} p_t |\sum_{k=1}^{3} \beta_k(x)(f_{k,t-1} - \hat{f}_k)| \cdot y_0 = 1, \\
& \sum_{i=1}^{n} x_i = 1, \\
& \beta_k(x) = \sum_{i=1}^{n} \beta_{i,k} x_i, \qquad k = 1, 2, 3 \\
& \epsilon_t(x) = \sum_{i=1}^{n} \epsilon_{i,t} x_i, \qquad t = 1, 2, \ldots, T \\
& 0 \le x_i \le \alpha, \qquad i = 1, 2, \ldots, n \\
& \sum_{i=1}^{n} \tilde{r}_i x_i \ge \upsilon.
\end{aligned} \tag{31}$$

Let $y = y_0 \cdot x$. Then the equivalent problem of the problem (31) is

$$\begin{aligned}
\underset{x}{minimize} \quad & \sum_{t=1}^{T} p_t |\sum_{k=1}^{3} \beta_k(y)(f_{k,t-1} - \hat{f}_k) + \epsilon_t(y)| \\
\text{s.t.} \quad & \sum_{t=1}^{T} p_t |\sum_{k=1}^{3} \beta_k(y)(f_{k,t-1} - \hat{f}_k)| = 1, \\
& \sum_{i=1}^{n} y_i = y_0, \\
& \beta_k(y) = \sum_{i=1}^{n} \beta_{i,k} y_i, \qquad k = 1, 2, 3 \\
& \epsilon_t(y) = \sum_{i=1}^{n} \epsilon_{i,t} y_i, \qquad t = 1, 2, \ldots T \\
& 0 \le y_i \le \alpha \cdot y_0, \qquad i = 1, 2, \ldots, n, \\
& \sum_{i=1}^{n} \tilde{r}_i y_i \ge \upsilon \cdot y_0.
\end{aligned} \tag{32}$$

The optimal solution to (32) is a pair (y^*, y_0^*), thereby deriving optimal weights is $x^* = \frac{y^*}{y_0^*}$ for the (29). Let introduce 4 new variables

$$\nu_t = (\nu_1, \nu_2, \ldots, \nu_T), \qquad \psi_t = (\psi_1, \psi_2, \ldots, \psi_T),$$
$$\xi_t = (\xi_1, \xi_2, \ldots, \xi_T), \qquad \eta_t = (\eta_1, \eta_2, \ldots, \eta_T),$$

The constraint (32) could be reconstructed as follows

$$\begin{aligned}
\min \quad & \sum_{t=1}^{T} p_t(\nu_t + \psi_t) \\
\text{s.t.} \quad & \sum_{t=1}^{T} p_t(\xi_t + \eta_t) = 1, \\
& \nu_t - \psi_t = \sum_{k=1}^{3} \beta_k(y)(f_{k,t-1} - \hat{f}_k) + \epsilon_t(y), \qquad t = 1, 2, \ldots, T, \\
& \xi_t - \eta_t = \sum_{k=1}^{3} \beta_k(y)(f_{k,t-1} - \hat{f}_k), \qquad t = 1, 2, \ldots, T, \\
& \nu_t \psi_t = 0, \xi_t \eta_t = 0, \qquad t = 1, 2, \ldots, T, \\
& \nu_t, \psi_t, \xi_t, \eta_t \geq 0, \qquad t = 1, 2, \ldots, T, \\
& (y, y_0) \in Y.
\end{aligned} \tag{33}$$

where set Y contains all the following relations.

$$Y = \{(y, y_0) \mid \sum_{i=1}^{n} y_i = y_0, \quad \beta_k(y) = \sum_{i=1}^{n} \beta_{i,k} y_i,$$
$$\epsilon_t(y) = \sum_{i=1}^{n} \epsilon_{i,t} y_i, \quad 0 \leq y_i \leq \alpha \cdot y_0, \quad i = 1, 2, \ldots, n,$$
$$\sum_{i=1}^{n} \tilde{r}_i y_i \geq \upsilon \cdot y_0, \quad y_0 \geq 0\}$$

with the result of Chvatal [10] and Konno et al. [9], discarding the constraint $\nu_t \psi_t \geq 0$ is permitted. However, there is one set of complementary conditions which need manipulating. By introducing a variable $z_t = \{0, 1\}$ and considering these linear inequalities.

$$\xi_t \leq a_t z_t, \qquad t = 1, 2, .., T,$$
$$\eta_t \leq b_t(1 - z_t) \qquad t = 1, 2, .., T,$$

where

$$a_t = \max\{\max\{\sum_{k=1}^{3} \beta_k(y)(f_{k,t-1} - \hat{f}_k) \mid (y, y_0) \in Y\}, 0\},$$

$$b_t = -\min\{\min\{\sum_{k=1}^{3} \beta_k(y)(f_{k,t-1} - \hat{f}_k) \mid (y, y_0) \in Y\}, 0\}.$$

From the system we see that ξ_t and η_t are forced to be either 0 and the other one will be unconstrained, which satisfies the complementarity. Finally, the 0–1 mixed integer linear programming will be in the form of (32) [12].

$$\begin{aligned}
\min \quad & \sum_{t=1}^{T} p_t(\nu_t + \psi_t) \\
\text{s.t.} \quad & \sum_{t=1}^{T} p_t(\xi_t + \eta_t) = 1, \\
& \nu_t - \psi_t = \sum_{k=1}^{3} \beta_k(y)(f_{k,t-1} - \hat{f}_k) + \epsilon_t(y), \quad t = 1, 2, \ldots, T, \\
& \xi_t - \eta_t = \sum_{k=1}^{3} \beta_k(y)(f_{k,t-1} - \hat{f}_k) \quad, t = 1, 2, \ldots, T, \\
& 0 \le \xi_t \le a_t z_t, \quad 0 \le \eta_t \le b_t(1 - z_t), \quad t = 1, 2, \ldots, T, \\
& z_t \in \{0, 1\}, \quad , t = 1, 2, \ldots, T, \\
& \nu_t, \psi_t, \eta_t, \xi_t \ge 0, \quad t = 1, 2, \ldots, T, \\
& (y, y_0) \in Y.
\end{aligned} \tag{34}$$

5 Data and Empirical Results

5.1 Data Description

We use the following abbreviations for more convenience

- MPP: Maximal Predictability Portfolio
- MVP: Mean-Variance Portfolio
- FF3: The Fama-French three-factor model
- HOSE: HO CHI MINH stock exchanges.

To construct a large portfolio, the weekly data in the Vietnam stock market [13] was just collected from Jan/2015 to Dec/2019. Since Vietnam is an emerging market then some stocks have not been publicly offered on HOSE in 2014. Some Bluechips have enormous capitalization, and maintain high stability in HOSE. So they are very important to calculate the SMB and HML factors in the framework of the FF3. Moreover, such data was divided into 3 main parts: In the first part, data from Jan/2015 to April/2019, would be the training set. The training set was fitted on the FF3, thereby finding expected return and residuals, and beta coefficients. In the second part, data from May/2019 to August/2019, and those parameters, expected return, residuals and beta coefficients, would be used as inputs for optimization problem (MPP), with $T = 12$ and $n = 12$, and then yielding optimal weights of each stock. The last part involved performance testing those optimal weights for both MVP and MPP methods in the last 12-week period between Sep/2019 and Dec/2019.

To construct our portfolio, we select weekly prices of 12 companies in VN30 index listed on in Ho Chi Minh Stock Exchange (HOSE). They are the most stable and trustworthy companies in HOSE. Furthermore, these companies have important impacts on the stock market with the largest market capitalization.

- Phu Nhuan Jewelry Joint Stock Company: PNJ
- Refrigeration Electrical Engineering Corporation: REE
- SSI Securities Corporation: SSI
- Sai Gon Thuong Tin Commercial Joint Stock Bank: STB
- Bank for Foreign Trade of Vietnam: VCB
- Vingroup Joint Stock Company: VIC
- Viet Nam Dairy Products Joint Stock Company: VNM
- DHG Pharmaceutical Joint Stock Company: DHG
- Hoa Sen Group: HSG
- Phuoc Hoa Rubber Joint Stock Company: PHR
- FPT Corporation: FPT
- PetroVietnam Gas Joint Stock Corporation: GAS.

The returns of data insights are also provided as Figs. 1, 2 and 3.

According to the covariance matrix, all the stocks are somehow positively correlated with each other. Therefore, it is quite hard to utilize the potential of diversification effect. However, some stocks are nearly non-correlated with others. So the MVP method can be used to build such portfolio with minimal risks.

5.2 *Empirical Results*

To begin with, after rigorously finding the 3 main factors of the FF3 model, the weekly data return of each stock was fitted on this model then the yielded beta and residuals are stored and later used for the MPP optimization. All of which are shown as follows.

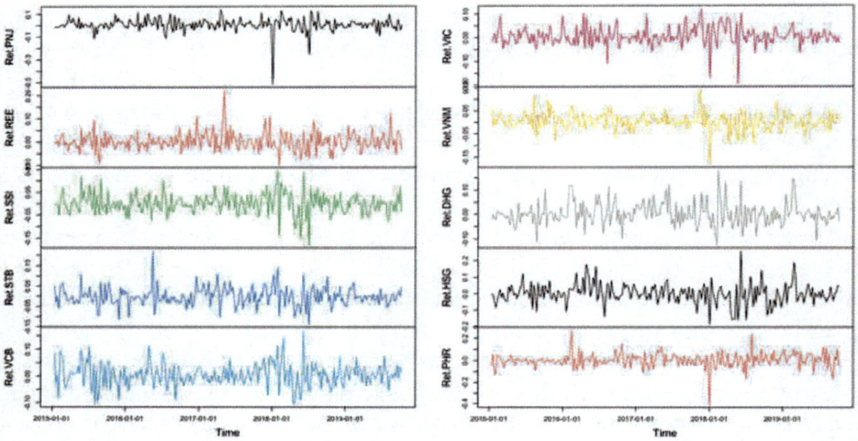

Fig. 1 Vietnam stock returns

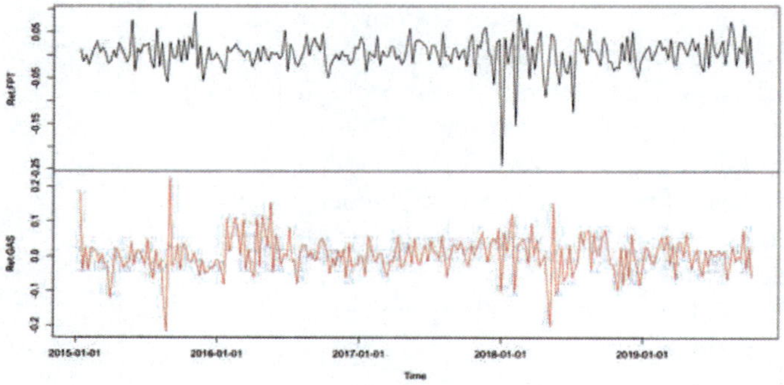

Fig. 2 Vietnam stock returns

Stock	PNJ	REE	SSI	STB	VCB	VIC	VNM	DHG	HSG	PHR	FPT	GAS
Beta 1	0.405	0.787	1.069	0.849	0.896	0.568	0.482	0.82	1.219	0.559	0.675	1.233
Beta 2	0.628	0.643	0.212	−0.019	−0.496	−0.270	−0.238	0.562	0.944	1.177	0.387	−0.160
Beta 3	−0.366	0.476	0.170	0.487	−0.119	−0.416	−0.422	−0.211	0.176	0.211	−0.017	−0.263

Using these inputs, the results of each method in terms of stock weights are yielded and given below

Stock	PNJ	REE	SSI	STB	VCB	VIC	VNM	DHG	HSG	PHR	FPT	GAS
MVP weight (%)	7.96	14.70	0.00	9.88	7.17	19.08	26.50	9.07	0.00	1.93	3.70	0.00
MPP weight (%)	15	14.02	0.00	6.631	15	15	8.353	15	15	0.00	7.818	0.793

We see that the amount of weight of each stock vary significantly between each portfolio. Now we use the backtesting method for period including 12 weeks to check

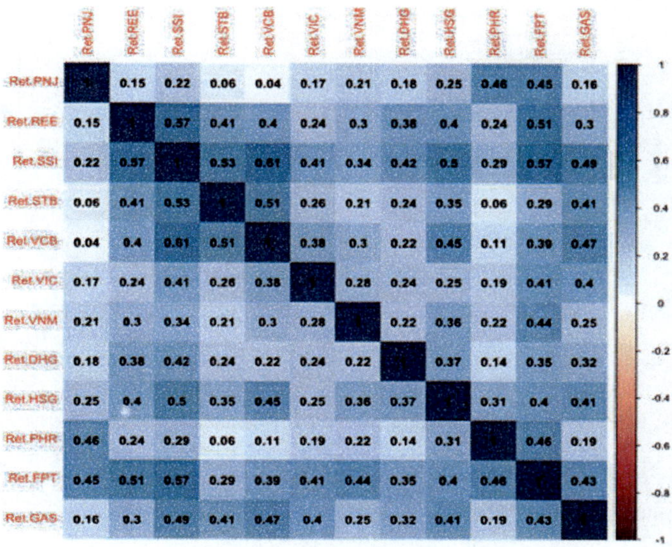

Fig. 3 Covariance matrix

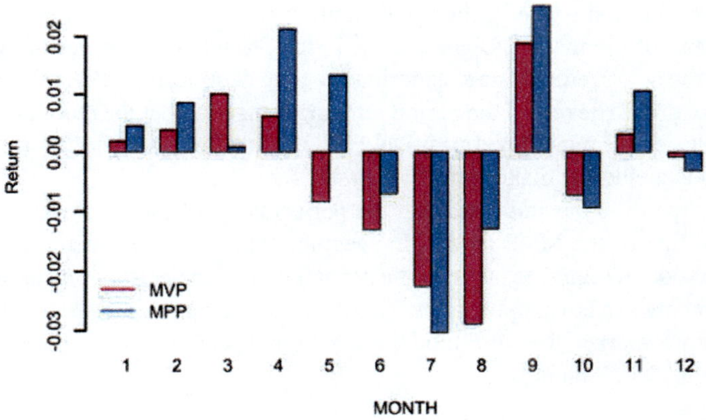

Fig. 4 The MVP and MPP single weekly return in Vietnam market

whether does the portfolio perform well or not. As the following bar chart presents the single weekly return of the MVP and MPP respectively (Fig. 4).

Looking at closer details of the single weekly return, over the first 4 weeks, Oct/2019, the MPP model shows good positive returns compared to the traditional method. However, there were drops in Nov/2019 of both portfolios, thus indicating that the market in the meantime is experiencing a downward trend. Fortunately, the MPP still demonstrates its efficiency by having a lower magnitude of loss in comparison with the MVP model. The overall single weekly performance is thus assessed with a better operation.

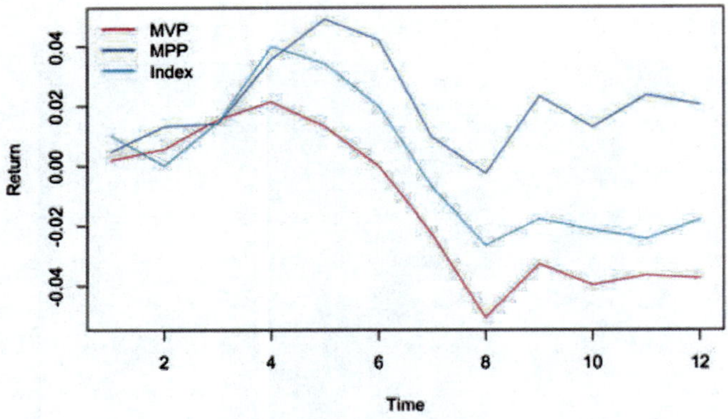

Fig. 5 MVP and MPP cumulative performance in Vietnam market

The cumulative performance of each in conjunction with Index performance would be used to achieve more divergent perspectives (Fig. 5).

Primarily, the cumulative performance of the MPP model is indeed outstanding compared to the traditional method as well as the index cumulative returns. In particular, over the first 3 weeks, the predictability of the MPP is strongly fit with an upward trend of the market. Significantly, in the 5th week of the period, while the index and the MVP return start to experience a long decline, the MPP still observes a rise in its returns. The rest of the period remained stagnant, but the cumulative return of the MPP is still maintained positively. The following *Sharpe ratio (investment efficiency)* is yielded to make further analysis (Table 1).

The Sharpe ratio has indicated that the performance of the MPP model is much better than that of the MVP model. In specific, while the investment efficiency of the MVP model remains negative in the second half of the period. Furthermore, the MPP model always hovers positively with only one exception. So we see that in the Vietnam stock market, the MPP model gives a good performance as compared with the traditional MVP model.

6 Conclusion

In this study, we utilize the maximal predictability portfolio optimization model (MPP) model for the optimization portfolio on the Vietnamese stock market. It shows that the model provides us with a useful quantitative model. The obtained results demonstrate the robustness of maximal predictability portfolio optimization compared to Mean-Variance portfolio optimization. Most of the promising features of the MPP model have been fairly exploited through the FF3 model, but further research should conduct on other models besides the FF3 to utilize its predictive power.

Table 1 Investment Efficiency

Week	1	2	3	4	5	6	7	8	9	10	11	12
MVP	0.0632	0.2592	0.7864	1.1052	0.6769	−0.0326	−1.1069	−2.494	−1.6611	−2.054	−1.8516	−1.9078
MPP	0.2008	0.6428	0.6931	1.8413	2.6413	2.2771	0.4237	−0.1603	1.1329	0.6083	1.1456	0.996

Acknowledgements The authors are grateful to the anonymous reviewers for their valuable comments and suggestions which lead to an improvement of the paper.

References

1. Kolm, P. N., Tütüncü, R., & Fabozzi, F. J. (2014). 60 years of portfolio optimization: Practical challenges and current trends. *European Journal of Operational Research, 234*(2), 356–371. ISSN 0377-2217.
2. Lo, A. W., & MacKinlay, A. C. (1997). Maximizing predictability in the stock and bond markets. *Macroeconomic Dynamics, 1*(1), 102–134.
3. Yamamoto, R., Ishii, D., & Konno, H. (2007). A maximal predictability portfolio model: Algorithm and performance evaluation. *International Journal of Theoretical and Applied Finance, 10*(06), 1095–1109.
4. Konno, H., Morita, Y., & Yamamoto, R. (2010). A maximal predictability portfolio using absolute deviation reformulation. *Computational Management Science, 7*, 47.
5. Markowitz, H. (1952). Portfolio selection. *The Journal of Finance, 7*(1), 77–91.
6. Sharpe, W. F. (1964). Capital asset prices: A theory of market equilibrium under conditions of risk. *Journal of Finance, American Finance Association, 19*(3), 425–442.
7. Fama, E. F., & French, K. R. (1992). The cross-section of expected stock returns. *The Journal of Finance, 47*(2), 427–465.
8. Fabozzi, F. J., Focardi, S. M., Kolm, P. N., & Pachamanova, D. A. (2012). Robust *Portfolio optimization and management* (Vol. 2). John Wiley & Sons, Inc.
9. Konno, H., & Koshizuka, T. (2005). Mean-absolute deviation model. *IIE Transactions, 37*(10), 893–900.
10. Chvatal, V. (1983). *Linear programming*. San Francisco: Freeman.
11. Charnes, A., & Cooper, W. W. (1962). Programming with linear fractional functionals. *Naval Research Logistics Quarterly 9*.
12. Zopounidis, C., Doumpos, M., & Pardalos, P. M. (2011). *Handbook of financial engineering*. Springer.
13. Vietnam stock data. https://www.cophieu68.vn/export.php

Driving Factors for Realizing the Fully Smart Transportation System: The Case of Individual-Use Autonomous Vehicle in Thailand

Tatcha Sudtasan

Abstract This research aims at investigating factors enhancing the transition to the fully smart transportation system and estimating the transition period to the fully Automated Driving System (ADS) in Thailand. Therefore, it is vital to investigate people's attitude toward ADS because they are the final evaluators on the adoption of new technologies. The contribution of this research is identifying the driving factors to attain the high level of trust in "emergency detection system", which is a key technology of autonomous vehicles (AVs). The ordered-probit model is applied to examine the marginal effects of driving factors to determine the different levels of confidence in "emergency detection system" in the next 1, 5, 10, and 20 years. The results illustrate that people living in Bangkok are likely to have a higher level of confidence in the ADS's safety system. Surprisingly, income seems to play a weak role in people's trust. Furthermore, the result of this study suggests that Thailand will not achieve the fully ADS in the next 10 years.

1 Introduction

Thailand 4.0 policy initiated by the Thai government aims at achieving a value-based and innovation driven economy through the promotion of technology utilization. It is expected that the use of new innovations and emerging technologies such as IoT, big data, and robotics will bring social and economic wealth to people's lives and lead to the sustainable development of the Thai economy. Within the scope of smart cities platform following many advanced countries like Japan and the USA, autonomous vehicle is expected to play an important role in the transformation to smart transportation system.

Autonomous vehicle (AV) is an important engine to reach sustainable smart cities. It represents a paradigm shift to a more flexible, seamless and efficient transport system [1]. Moreover, it creates network externalities in which the higher number

T. Sudtasan (✉)
Faculty of Economics, Chiang Mai University, Chiang Mai, Thailand
e-mail: tatcha.s@cmu.ac.th

S. Sriboonchitta et al. (eds.), *Credible Asset Allocation, Optimal Transport Methods, and Related Topics*, Studies in Systems, Decision and Control 429,
https://doi.org/10.1007/978-3-030-97273-8_38

of smart automated vehicles leads to the greater benefit in terms of safety, security, efficiency, and environmental friendliness [2]. The AV offers a new-way solution to several current issues in transportation such as real-time traffic information and warning through the Vehicle-to-everything (V2X) technology. The main ground of this technology are driving safety, traffic efficiency, and energy conservation.

For all of its advantages, however, AV is something of a double-edged sword. It is predicted that the excessive dependence on AVs may aggravate traffic congestion with extra empty cars. It can also be hacked, increase energy consumption due to the more technology utilization. The values and drawbacks of the ADS are coexisting. It is expected that the development process of AV will strengthen its advantages and weakened its disadvantages [3].

AV is representing a technological revolution that can create new perception of how individuals see mobility [4]. An unprecedented revolution experience is set to make the improvement of the way we experience personal mobility, such as boosting the mobility of unlicensed commuters, reducing accidents, enabling leisure and productive activities, and so forth [3, 5]. Many driving factors are hopeful for the new generation to adopt more connected devices and generate the growth and the diffusion. Some delaying factors that make the speed of the diffusion slower or even pause in some area of applications should also be emphasized at the policy level [2]. However, smart transportation mega projects in Thailand have currently concentrated in the utilization of telematics and GPS tracking equipment rather than ADS or AVs from the aspect of consumer behavior and preference. Concerning traffic situation in Thailand especially in big cities, in addition, heavy traffic congestion makes people exhausted in driving. Technological changes in the automotive industry are highly expected to bring a healthier road traffic. Therefore, although the road traffic with fully AVs in Thailand remains disbelieve for its realization, it is vital to identify the significant factors affecting people's perception towards AVs, especially at the beginning stage [3, 6].

With this background in mind, this research has the objectives to identify the factors capable of driving/enhancing the transition to the fully smart transportation system and estimate the transition period to attain a fully ADS in Thailand. It centers on public confidence on "Emergency Detection System" of the AVs. This is because sensing and machine communication technologies are core technologies enabling AV. According to the NHTSA, human error is the main reason behind 94% of all crashes in the USA [7]. The emergency detection system allows AVs to navigate safely without relying on human input to make decision on navigation. AVs can likely eliminate accidents caused by human errors [8]. Besides, Howard and Dai [4] and Fraedrich and Lenz [6] mention that public perception is vital to the demand for AVs which consequently links to future policies and investments in infrastructure. The contribution of this research is that it sheds some light on the policy direction in setting the right incentives for Thai people to support the transformation to the fully ADS as a part of smart transportation in Thailand 4.0 vision and shortening the critical transition period.

1.1 Autonomous Driving System (ADS)

For technological perspective, SAE International [9] defines a taxonomy classifying motor vehicle driving automation systems into six levels. Level 0 system is the level of no driving automation while level 5 represents full driving automation. In levels 0–2, the driver is expected to be responsible for the driving control. In levels 3–5, the driving is defined as "Automated Driving System (ADS)". Specifically, level 3 system is the conditional driving automation where the driver is expected to intervene when ADS requests as a result of system failure.

For microeconomic perspective, Mitomo and Sudtasan [2] review specific functions and characteristics of the Internet-of-Things (IoT) that affect people's perception toward the device. The concept of automatic communication among things without human intervention is at the heart of connected devices which is commonly referred to as an ability of identifying, sensing, and communicating through networks. Smart vehicle reveals a specific function of networked IoT, where number of current adopters affects the evaluation of IoT benefit made by potential adopters or overall adopters gain additional benefit when there are more people use the device or service, and the characteristic of visible-to-other.

1.2 Adoption and Diffusion of Innovation

According to Rogers [10], a social system consists of five groups of innovation adopters. Each group is determined by the degree of relative earliness of adoption compared to the other potential adopters. "Innovators" representing 2.5% of all adopters are the earliest relative to other members of a social system in adopting innovations. "Early adopters" representing 13.5% of potential adopters are considered to be the most important group in a social system because they critically consider an innovation before it is accepted and diffused throughout the social system, and they have the highest degree of opinion leadership. Innovators and early adopters represent consumers who are very quick in deciding to adopt new technologies. An innovation, then, spreads through mainstream publics, composed of 34% of "early majority" and another 34% of "late majority". The remaining 16% of potential adopters are "laggards" who later adopt the innovation and their decisions are mostly influenced by earlier adopters [10, 11].

2 Relevant Frameworks and Literature

There is a growing body of literature on ADS. Much of such literature focuses on the architecture and engineering aspects. Some research works center ADS within the smart cities platform. Perera et al. [12] illustrate smart cities scenarios as well as smart applications and services for realizing smart cities such as waste management, smart transportation, etc., while Jin et al. [13] focus more on network architecture for creating a smart city.

A look at the literature on intention to use smart or connected devices suggested that driving factors of consumer's intention to adopt connected devices include product specific factors, technology factors, personal innovativeness, factor of social context, and external factors.

For the aspect of user acceptance and confidence concerning the AV technology, Howard and Dai [4] reveal that cost, safety and control, convenience, amenities, and multitasking play important role in people's willingness to adopt AVs. Meanwhile, Payre et al. [14] address that attitude towards fully automated driving, considering its pleasantness, usefulness, and safety, is the main predictor of people's intention to adopt AVs. Later, Bansal and Kockelman [15] discover that elderly people are less willing to pay for AVs. It is not surprising that higher-income people are more likely to pay more for future vehicle and also more likely to purchase vehicles equipped with automated driving system (see [16]). In addition to these findings, the studies by Daziano et al. [17] and Schoettle and Sivak [18] discover that higher-income people are found to be wary more about the liability issues than the control issues, while lower-income people are more wary of safety and control issues.

3 Data

The data were retrieved from online survey questionnaires during August–September 2020. The questionnaires were sent to official social media of 16 universities throughout Thailand. Totally 73 volunteers were paid to send the questionnaires to at least 15 family members and colleagues. The questionnaires were sent to 1914 Thai people. The number of returned and completed questionnaires was 1334. Thus, the response rate was 69.7%.

The survey defines "autonomous car" as vehicle with capabilities to navigate and perform without human input, sense its surrounding environment, and respond appropriately to interventions. It mainly focuses on ADS in level 3–5, according to the definition given by SAE International [9] (see Sect. 1.1). Also, this research centers on Thai people's trust toward "emergency detection system" of autonomous cars. This is because safety is vital to promote public confidence in ADS [19].

Dependent variable is the level of confidence of the respondents in the emergency detection system of autonomous vehicles. Based on the respondents' experiences in technological development over time, the survey asked them to expect their level of confidence in the next 1, 5, 10, and 20 years.

For the explanatory variables, this study includes gender, age, monthly income, living region, and driving license type. The model also contains innovativeness and innovation adopters as proxies for technological acceptability. Furthermore, attitude towards driving is also included, as suggested by Payre et al. [14]. Table 1 provides the definition and description of all the variables.

Table 1 Description of variables

Variable	Description	Max	Min	Mean	SD
Dependent variables					
*y*1	Level of trust/confidence in "Emergency Detection System" of autonomous vehicles in the next year	5	1	2.442	1.255
*y*5	Level of trust/confidence in "Emergency Detection System" of autonomous vehicles in the next five years	5	1	2.981	1.349
*y*10	Level of trust/confidence in "Emergency Detection System" of autonomous vehicles in the next ten years	5	1	3.485	1.383
*y*20	Level of trust/confidence in "Emergency Detection System" of autonomous vehicles in the next twenty years	5	1	3.891	1.368
	y = 1, unconfident/untrusted. y = 2, somewhat unconfident y = 3, neutral or moderately confident y = 4, somewhat confident y = 5, confident				

(continued)

Table 1 (continued)

Variable	Description	Max	Min	Mean	SD
Explanatory variables					
Age	Age of the respondent	64	20	23.23	7.915
Gender	= 1, if the respondent is male; = 0, otherwise	1	0	0.316	0.465
MonthIncome	Monthly income of the respondent	150,000	2500	12,805.47	23,892.88
DriveLover	Willingness of respondent to drive by himself, ranges from 1 to 5	5	1	3.855	0.966
DriveFun	Level of enjoyment of driving, ranges from 1 to 5	5	1	3.56	1.058
RegionBKK	= 1, if current accommodation is in Bangkok and the vicinity; = 0, otherwise	1	0	0.09	0.286
RegionMiddle	= 1, if current accommodation is in the Central region of Thailand; = 0, otherwise	1	0	0.106	0.309
RegionNorthEast	= 1, if current accommodation is in the Northeastern region of Thailand; = 0, otherwise	1	0	0.103	0.305
RegionNorth	= 1, if current accommodation is in the Northern region of Thailand; = 0, otherwise	1	0	0.645	0.479
MotorbikeLicense	= 1, if the respondent has got a motorbike driving license; = 0, otherwise	1	0	0.561	0.496
CarLicense	= 1, if the respondent has got a car driving license; = 0, otherwise	1	0	0.328	0.470

(continued)

Table 1 (continued)

Variable	Description	Max	Min	Mean	SD
Innovativeness	The degree of new innovation acceptance, ranges from 1 (late acceptance) to 6 (early acceptance)	6	1	3.457	1.212
Innovator	= 1, if the respondent is eager to be pioneer to adopt the autonomous car; = 0, otherwise	1	0	0.061	0.239
EarlyAdopt	= 1, if the respondent wants to wait until 25% of adopters, then he or she will buy the technology; = 0, otherwise	1	0	0.103	0.305
EarlyMajority	= 1, if the respondent wants to wait until 50% of adopters, then he or she will buy the technology; = 0, otherwise	1	0	0.313	0.464
LateMajority	= 1, if the respondent wants to wait until 75% of adopters, then he or she will buy the technology; = 0, otherwise	1	0	0.356	0.479
Laggard	= 1, if the respondent feels not interested in adopting autonomous car; = 0, otherwise	1	0	0.088	0.283
Observations		1334			

4 Ordered Probit Model

To investigate the key characteristics of respondents that influence their attitudes (confidence or trust) concerning the emergency detection system of autonomous car to ensure safety, this research applies the ordered-probit regression model.

Let y_i^* denote the unobserved latent variable representing the confidence level of respondent i, for $i = 1, \ldots n$, and X_i is a $1 \times k$ vector of independent variables (factors) of respondents i, then the ordered probit model can be written as

$$y_i^* = X'_i\beta + \varepsilon_i \tag{1}$$

where β is a $k \times 1$ vector of coefficients, and ε_i is the error term of respondent i that has the standard normal distribution. In this study, we consider five confidence levels, therefore, four threshold parameters $\mu_1 < \mu_2 < \mu_3 < \mu_4$ are used to categorize five confidence levels as follows:

$$\begin{aligned} y_i &= 1 \ if \ y_i^* \le \mu_1 \\ y_i &= 2 \ if \ \mu_1 < y_i^* \le \mu_2 \\ y_i &= 3 \ if \ \mu_2 < y_i^* \le \mu_3 \\ y_i &= 4 \ if \ \mu_3 < y_i^* \le \mu_4 \\ y_i &= 5 \ if \ \mu_4 < y_i^* \end{aligned} \tag{2}$$

To estimate this model, we can construct the probability that respondent i chooses confidence level j as

$$\text{Prob}(y_i = j\,|X) = \Phi(\mu_j - X'_i\beta) - \Phi(\mu_{j-1} - X'_i\beta),\ j = 1, \ldots, 5 \tag{3}$$

where Φ is the cumulative distribution function of a standard normal distribution. Then, Maximum likelihood estimates of coefficients can be obtained by maximizing this log-likelihood function:

$$\text{LogL}(\mu, \beta\,|y, X) = \sum_{i=1}^{n}\sum_{j=1}^{5} z_{ij} \log(\text{Prob}(y_i = j\,|X)), \tag{4}$$

where z_{ij} is an indicator variable that equals 1 if $y_i = j$ and 0 otherwise.

In this study, the empirical ordered-probit model is defined as follows;

$$y_i = f\begin{pmatrix} \text{gender, age, monthincome, regionbkk} \\ \text{regionmiddle, regionnortheast, regionnorth,} \\ \text{motorbikelicense, carlicense, drivelover,} \\ \text{drivefun, innovativeness, innovator,} \\ \text{earlyadoptor, earlymajority, latemajority, laggard} \end{pmatrix} \tag{5}$$

where i = 1–5.

5 Estimation Results and Interpretations

The estimation was made based on the data from the 1334 respondents. The estimation results are shown in Table 2. The results suggest that monthly income, living region, and type of license affected respondents' confidence in the emergency detec-

Table 2 Ordered-probit model estimation results

Variable	y1	y5	y10	y20
Gender	−0.103 (0.078)	−0.130* (0.078)	−0.083 (0.082)	0.025 (0.093)
Age	0.003 (0.006)	0.002 (0.006)	0.003 (0.006)	0.009 (0.007)
MonthIncome	−2.63E−06 (0.000)	−2.37E−06 (0.000)	−6.31E−07 (0.000)	−6.32E−06*** (0.000)
RegionBKK	−0.210 (0.199)	−0.515*** (0.198)	−0.445** (0.215)	−0.348 (0.247)
RegionMiddle	−0.151 (0.189)	−0.254 (0.189)	−0.040 (0.207)	−0.246 (0.238)
RegionNorthEast	−0.035 (0.187)	−0.315* (0.188)	0.086 (0.207)	−0.361 (0.235)
RegionNorth	0.072 (0.158)	−0.167 (0.160)	0.097 (0.179)	−0.191 (0.207)
MotorbikeLicense	−0.145** (0.074)	−0.029 (0.074)	−0.057 (0.078)	0.053 (0.089)
CarLicense	−0.075 (0.079)	−0.011 (0.079)	−0.143* (0.083)	−0.015 (0.093)
DriveLover	0.003 (0.042)	−0.015 (0.042)	−0.035 (0.045)	−0.057 (0.052)
DriveFun	0.071* (0.039)	0.038 (0.039)	0.037 (0.041)	0.008 (0.047)
Innovativeness	−0.009 (0.037)	−0.004 (0.036)	0.021 (0.038)	−0.018 (0.041)
Innovator	0.041 (0.175)	0.103 (0.182)	−0.237 (0.199)	0.284 (0.242)
EarlyAdopt	−0.085 (0.149)	0.067 (0.156)	−0.138 (0.171)	−0.129 (0.211)
EarlyMajority	−0.236* (0.122)	−0.070 (0.122)	−0.063 (0.134)	0.030 (0.164)
LateMajority	−0.171 (0.118)	−0.098 (0.120)	−0.231* (0.134)	−0.048 (0.166)
Laggard	−0.293* (0.158)	−0.100 (0.158)	−0.316* (0.171)	−0.034 (0.215)

Note "*", "**", and "***" are the decisive, substantial, and strong evidence, respectively [20]. In the parentheses ( ) is standard error

tion system of autonomous car. Furthermore, the results also reveal that respondents who drive for pleasure, and the characteristics of innovation adopters as "late majority" and "laggard" play important role in determining the level of user confidence in the technology.

The interpretations of the results will be based on the marginal effects of the factors in contributing to the different levels of confidence in "Emergency Detection System" (Table 3—low level of confidence, Table 4—medium level of confidence, and Table 5—high level of confidence).

In sum, the results show that income significantly affects all—low, medium, and high—levels of trust in the emergency detection system. However, the effects are very weak since the degrees of marginal effect are close to zero. These results seem not to be in line with Kyriakidis [16] who finds that higher-income people are more likely to pay more for future vehicle and willing to purchase vehicles equipped with ADS. Nevertheless, Daziano et al. [17] and Schoettle and Sivak [18] suggest that higher-income people concern more about liability issues, thus the safety system and the passenger comfort features of autonomous vehicle may not induce high income people to buy it. Moreover, the result of this current study can be explained by the similarity in experience among high- and low-income people, and the trialability of AVs in Thailand. This technology is relatively new and sophisticate for overall

Table 3 Marginal effects of the ordered-probit model (low level of confidence $y_i = 1$ and $y_i = 2$)

Variable	Marginal effect ($y_i = 1$)				Marginal effect ($y_i = 2$)			
	y1	y5	y10	y20	y1	y5	y10	y20
Gender	0.015 (0.012)	0.001 (0.000)	1.74E−07 (0.000)	−2.35E−12 (0.000)	0.025 (0.018)	0.031 (0.019)	0.003 (0.003)	−2.58E−05 (0.000)
Age	0.000 (0.001)	−7.4E−06 (0.000)	−5.7E−09 (0.000)	−9.04E−13 (0.000)	−0.001 (0.001)	0.000 (0.001)	0.000 (0.000)	−9.77E−06 (0.000)
MonthIncome	0.000 (0.000)	8.79E−09 (0.000)	1.22E−12 (0.000)	6.12E−16*** (0.000)	6.45E−07 (0.000)	5.61E−07 (0.000)	2.05E−08 (0.000)	6.62E−09 (0.000)
RegionBKK	0.035 (0.037)	0.004 (0.003)	2.48E−06 (0.000)	1.05E−10 (0.000)	0.046 (0.038)	0.145** (0.064)	0.022 (0.016)	0.001 (0.001)
RegionMiddle	0.024 (0.033)	0.001 (0.001)	8.33E−08 (0.000)	5.00E−11 (0.000)	0.034 (0.039)	0.066 (0.053)	0.001 (0.007)	3.67E−04 (0.001)
RegionNorthEast	0.005 (0.028)	0.002 (0.002)	−1.4E−07 (0.000)	1.11E−10 (0.000)	0.008 (0.044)	0.083 (0.055)	−0.003 (0.006)	0.001 (0.001)
RegionNorth	−0.011 (0.024)	0.001 (0.001)	−2.0E−07 (0.000)	1.64E−11 (0.000)	−0.018 (0.038)	0.039 (0.036)	−0.003 (0.006)	1.85E−04 (0.000)
MotorbikeLicense	0.021** (0.011)	1.05E−04 (0.000)	1.09E−07 (0.000)	−5.28E−12 (0.000)	0.036* (0.019)	0.007 (0.018)	0.002 (0.003)	−5.63E−05 (0.000)
CarLicense	0.011 (0.012)	4.19E−05 (0.000)	3.17E−07 (0.000)	1.53E−12 (0.000)	0.018 (0.019)	0.003 (0.019)	0.005 (0.003)	1.63E−05 (0.000)
DriveLover	−0.001 (0.006)	5.59E−05 (0.000)	6.79E−08 (0.000)	5.55E−12 (0.000)	−0.001 (0.010)	0.004 (0.010)	0.001 (0.001)	6.01E−05 (0.000)
DriveFun	−0.010* (0.006)	−1.4E−04 (0.000)	−7.0E−08 (0.000)	−7.3E−13 (0.000)	−0.017* (0.010)	−0.009 (0.009)	−0.001 (0.001)	−7.94E−06 (0.000)
Innovativeness	0.001 (0.005)	1.42E−05 (0.000)	−3.9E−08 (0.000)	1.76E−12 (0.000)	0.002 (0.009)	0.001 (0.009)	−0.001 (0.001)	1.91E−05 (0.000)
Innovator	−0.006 (0.024)	−3.3E−04 (0.001)	8.0E−07 (0.000)	−1.38E−11 (0.000)	−0.010 (0.045)	−0.023 (0.039)	0.010 (0.010)	−2.00E−04 (0.000)
EarlyAdopt	0.013 (0.024)	−2.3E−04 (0.001)	3.56E−07 (0.000)	1.81E−11 (0.000)	0.020 (0.034)	−0.016 (0.035)	0.005 (0.007)	1.63E−04 (0.000)
EarlyMajority	0.037* (0.020)	2.72E−04 (0.001)	1.28E−07 (0.000)	−2.84E−12 (0.000)	0.055** (0.027)	0.017 (0.030)	0.002 (0.005)	−3.13E−05 (0.000)
LateMajority	0.026 (0.019)	3.80E−04 (0.001)	5.52E−07 (0.000)	4.90E−12 (0.000)	0.041 (0.027)	0.023 (0.029)	0.008 (0.005)	5.16E−05 (0.000)
Laggard	0.051 (0.032)	4.20E−04 (0.001)	1.26E−06 (0.000)	3.66E−12 (0.000)	0.060** (0.026)	0.025 (0.040)	0.014 (0.010)	3.78E−05 (0.000)

Note "*", "**" and "***" are the decisive, substantial, and strong evidence, respectively [20]. In the parentheses ( ) is standard error

Thai people, thus the expectation of advantages of AVs of high-income people is not much different from the low-income people. Therefore, income shows not to be a key factor driving level of confidence of people in the future.

For other factors, the results of this study show evidence that being a late majority and being a laggard are likely to affect the probability of respondents having, respectively, neutral confidence ($y_i = 3$) and high confidence ($y_i = 5$) in the emergency detection system of AVs in the next 10 years. These two types of potential adopters are more likely to have only moderate trust in the technology. However, these variables do not have any significant effect on the probability of being unconfident.

Table 4 Marginal effects of the ordered-probit model (medium level of confidence $y_i = 3$)

Variable	Marginal effect ($y_i = 3$)			
	y1	y5	y10	y20
Gender	−0.036 (0.027)	−0.006 (0.005)	0.030 (0.030)	−0.004 (0.016)
Age	0.001 (0.002)	6.92E−05 (0.000)	−0.001 (0.002)	−0.002 (0.001)
MonthIncome	−9.26E−07 (0.000)	−8.14E−08 (0.000)	2.31E−07 (0.000)	1.13E−06*** (0.000)
RegionBKK	−0.074 (0.069)	−0.067 (0.044)	0.151** (0.064)	0.074 (0.061)
RegionMiddle	−0.053 (0.0660	−0.021 (0.025)	0.015 (0.075)	0.049 (0.053)
RegionNorthEast	−0.012 (0.066)	−0.029 (0.029)	−0.032 (0.077)	0.077 (0.058)
RegionNorth	0.025 (0.056)	−0.004 (0.004)	−0.035 (0.065)	0.033 (0.035)
MotorbikeLicense	−0.051** (0.026)	−0.001 (0.003)	0.021 (0.029)	−0.010 (0.016)
CarLicense	−0.027 (0.028)	−3.96E−04 (0.003)	0.052* (0.030)	0.003 (0.017)
DriveLover	0.001 (0.015)	−0.001 (0.001)	0.013 (0.016)	0.010 (0.009)
DriveFun	0.025* (0.014)	0.001 (0.002)	−0.013 (0.015)	−0.001 (0.008)
Innovativeness	−0.003 (0.013)	−1.31E−04 (0.001)	−0.008 (0.014)	0.003 (0.007)
Innovator	0.015 (0.062)	0.001 (0.0030	0.084 (0.067)	−0.043 (0.031)
EarlyAdopt	−0.030 (0.053)	0.001 (0.002)	0.050 (0.061)	0.025 (0.043)
EarlyMajority	−0.083* (0.043)	−0.003 (0.006)	0.023 (0.049)	−0.005 (0.029)
LateMajority	−0.060 (0.041)	−0.004 (0.006)	0.084* (0.048)	0.009 (0.030)
Laggard	−0.102* (0.054)	−0.005 (0.012)	0.110** (0.056)	0.006 (0.040)

Note "*", "**" and "***" are the decisive, substantial, and strong evidence, respectively [20]. In the parentheses ( ) is standard error

Respondents who live in Bangkok have a weak probability (close to zero percent) of being unconfident ($y_i = 1$) in comparison with higher levels of confidence. In other words, respondents who live in Bangkok are likely to have a high level of trust in the emergency detection system in the next 5, 10, and 20 years. This can be explained by the fact that traffic congestion and road traffic safety problems in Bangkok are very severe. However, Bangkok people have more experience about monitoring and sensing technologies such as smart traffic system, and IoT devices available in the market. Therefore, Bangkok people can have higher probabilities of being confident in the emergency detection system of AVs.

Furthermore, being a driving-for-fun respondent corresponds to a higher probability of being unconfident and somewhat confident in the next year. This result can be explained by the respondents' familiarity with human driving and control ability. Since this group of people who perceive driving as for enjoyment is more likely to know driving system very well, the emergency detection sensor can be considered as a safety supporting tool. With their professional skill of driving, however, drivers for fun and adventure may not be interested in the emergency detection system thus leading to the higher probability of their low confidence in the system. Therefore, the emergency detection system cannot guarantee the adoption of the driving-for-fun people.

Interestingly, age is not a significant predictor of confidence level of AVs' emergency detection system while gender shows very mild effect. The results can be

Table 5 Marginal effects of the ordered-probit model (high level of confidence $y_i = 4$ and $y_i = 5$)

Variable	Marginal effect ($y_i = 4$)				Marginal effect ($y_i = 5$)			
	y1	y5	y10	y20	y1	y5	y10	y20
Gender	−0.004 (0.003)	−0.026* (0.015)	−0.031 (0.031)	0.001 (0.002)	−5.34E−06 (0.000)	−1.45E−04 (0.000)	−1.71E−03 (0.002)	0.004 (0.014)
Age	1.16E−04 (0.000)	4.13E−04 (0.001)	0.001 (0.002)	0.000 (0.000)	1.67E−07 (0.000)	2.43E−06 (0.000)	6.33E−05 (0.000)	0.001 (0.001)
MonthIncome	−1.02E−07 (0.000)	−4.85E−07 (0.000)	−2.38E−07 (0.000)	−1.66E−07 (0.000)	−1.47E−10 (0.000)	−2.86E−09 (0.000)	−1.35E−08 (0.000)	−9.71E−07*** (0.000)
RegionBKK	−0.007 (0.005)	−0.082*** (0.024)	−0.167** (0.077)	−0.031 (0.037)	−8.39E−06 (0.000)	−3.3E−04** (0.000)	−6.2E−03*** (0.002)	−0.044* (0.025)
RegionMiddle	−0.005 (0.006)	−0.046 (0.031)	−0.015 (0.079)	−0.017 (0.027)	−6.65E−06 (0.000)	−2.23E−04 (0.000)	−8.19E−04 (0.004)	−0.033 (0.028)
RegionNorthEast	−0.001 (0.007)	−0.056* (0.029)	0.032 (0.077)	−0.032 (0.036)	−1.83E−06 (0.000)	−2.5E−04* (0.000)	2.00E−03 (0.005)	−0.045* (0.024)
RegionNorth	0.003 (0.006)	−0.035 (0.035)	0.037 (0.068)	−0.003 (0.005)	3.88E−06 (0.000)	−2.22E−04 (0.000)	2.00E−03 (0.004)	−0.030 (0.035)
MotorbikeLicense	−0.006 (0.003)	−0.006 (0.015)	−0.022 (0.029)	0.001 (0.003)	−8.51E−06 (0.000)	−3.46E−05 (0.000)	−1.23E−03 (0.002)	0.008 (0.014)
CarLicense	−0.003 (0.003)	−0.002 (0.016)	−0.054* (0.031)	0.000 (0.003)	−4.00E−06 (0.000)	−1.35E−05 (0.000)	−2.88E−03* (0.002)	−0.002 (0.014)
DriveLover	0.000 (0.002)	−0.003 (0.009)	−0.013 (0.017)	−0.002 (0.002)	1.95E−07 (0.000)	−1.82E−05 (0.000)	−7.51E−04 (0.001)	−0.009 (0.008)
DriveFun	0.003* (0.002)	0.008 (0.008)	0.014 (0.015)	0.000 (0.001)	3.98E−06 (0.000)	4.64E−05 (0.000)	7.80E−04 (0.001)	0.001 (0.007)
Innovativeness	0.000 (0.001)	−0.001 (0.007)	0.008 (0.014)	0.000 (0.001)	−5.27E−07 (0.000)	−4.60E−06 (0.000)	4.38E−04 (0.001)	−0.003 (0.006)
Innovator	0.002 (0.007)	0.022 (0.0410	−0.090 (0.075)	−0.008 (0.022)	2.50E−06 (0.000)	1.45E−04 (0.000)	−3.95E−03 (0.003)	0.051 (0.051)
EarlyAdopt	−0.003 (0.005)	0.014 (0.034)	−0.052 (0.065)	−0.006 (0.015)	−4.14E−06 (0.000)	8.92E−05 (0.000)	−2.59E−03 (0.003)	−0.018 (0.028)
EarlyMajority	−0.008** (0.004)	−0.014 (0.024)	−0.024 (0.051)	0.001 (0.004)	0.000 (0.000)	−8.13E−05 (0.000)	−1.30E−03 (0.003)	0.005 (0.026)
LateMajority	−0.006 (0.004)	−0.020 (0.024)	−0.087* (0.050)	−0.001 (0.005)	−8.79E−06 (0.000)	−1.13E−04 (0.000)	−4.59E−03* (0.003)	−0.007 (0.0250
Laggard	−0.009** (0.004)	−0.019 (0.029)	−0.119* (0.064)	−0.001 (0.008)	0.000 (0.000)	−1.05E−04 (0.000)	−4.9E−03** (0.002)	0.005 (0.032)

Note "*", "**" and "***" are the decisive, substantial, and strong evidence, respectively [20]. In the parentheses ( ) is standard error explained by that safety is one of the basic needs of human. According to the World Health Organisation (WHO), Thailand's road traffic in 2020 is recognized as one of the worst in the world. The level of confidence of respondents toward the safety system remain moderate, even their expectation in the next 20 years. This means that, from Thai people's experiences, they've seen no sign of road safety correction despite a countless of the government's road traffic policies and acts. These results also in line with prior studies, such as Osswald et al. [21] and Zmud and Sener [22], that demographic variables are less predictive than psychological or personality variables especially on intention to use self-driving cars, attitudes towards perceived safety and anxiety while driving. Although demographic variables show strong predictive on willingness to pay for AVs in many previous studies, they are not significant predictors at this introduction stage of AVs.

Finally, the following determinants did not present significant effects either: (1) drive lover, (2) level of innovativeness, (3) being innovator, and (4) being early adopter. In summary, from the 17 variables included in the analysis, this research can conclude that six variables did not have any significant effect on the probability of being confident in the emergency detection system of AVs.

6 Implications and Conclusion

Among the personal mobility innovations nowadays, the Autonomous Vehicles (AVs) technology is popular across the world especially in developed and developing countries. The technology supports smoother driving cycles, shorter headway, and lighter vehicles.

The results of this study reveal that late majority and laggard are likely to trust in the emergency detection system of autonomous car in the next 10 years at the moderate level, and unlikely at the high level. This means that there are still questions among them on the safety-related function of ADS, which is in line with Howard and Dai [4] that found many people turn out to be resistant to use AVs because of safety, liabilities and control issues, although AVs have shown the potential to enhance safeness. Surprisingly, high income people and people living in the capital city of Thailand who are likely to be high potential adopters in most cases of new innovations, resist to have high confidence in ADS's safety supporting system. Besides, the effect of income is very weak.

Furthermore, the result of this study implies that Thailand will not achieve the fully ADS before the year 2030, since late majority and laggard are the final group of mainstream public in a society who would adopt AVs. Unlike the "fully smart transportation system" with 100% AVs, the "transition period" is a combination of no-automatic and ADS. This means that some vehicles can connect and sense surrounding environment and other vehicles while some cannot. The transition period will be very crucial for the government and policy makers to control and maintain safety of transportation system, especially in the early stage of the implementation of AVs where people will be shifting responsibility of driving to ADS. Therefore, ADS related bodies, such as Digital Government Development Agency (Public Organisation), the National Broadcasting and Telecommunication Commission (NBTC) of Thailand, need to focus a lot more on the regulation of the quality of sensing and communication technologies for secured emergency detection. AV firms should center the technological development on safety and emergency detection functions in order to build confidence among potential buyers and shorten the transition period to the fully ADS in Thailand.

References

1. Smart Cities Thailand Association. (2016). *Smart mobility and intelligent transportation system.*
2. Mitomo, H., & Sudtasan, T. (2017). The Internet of Things—A review of research trend and factors that drive the technology to consumer market.
3. Jing, P., Xu, G., Chen, Y., Shi, Y., & Zhan, F. (2020). The determinants behind the acceptance of autonomous vehicles: A systematic review. *Sustainability, 12*(5), 1719. https://doi.org/10.3390/su12051719
4. Howard, D., & Dai, D. (2014). Public perceptions of self-driving cars: The case of Berkeley, California. In *Transportation Research Board 93rd Annual Meeting* (Vol. 14, No. 4502, pp. 1–16).
5. Fagnant, D. J., & Kockelman, K. M. (2014). The travel and environmental implications of shared autonomous vehicles, using agent-based model scenarios. *Transportation Research Part C: Emerging Technologies, 40*, 1–13. https://doi.org/10.1016/j.trc.2013.12.001
6. Fraedrich, E., & Lenz, B. (2016). Societal and individual acceptance of autonomous driving. In *Autonomous driving* (pp. 621–640). Springer Berlin Heidelberg. https://doi.org/10.1007/978-3-662-48847-8_29
7. NHTSA. (n.d.). What are the safety benefits of automated vehicles? Retrieved November 20, 2020, from https://www.nhtsa.gov/technology-innovation/automated-vehicles-safety
8. NHTSA. (2019a). Automated driving systems. Retrieved November 20, 2020, from https://www.nhtsa.gov/vehicle-manufacturers/automated-driving-systems
9. SAE International. (2018). Taxonomy and definitions for terms related to driving automation systems for on-road motor vehicles.
10. Rogers, E. M. (1995). *Diffusion of innovations*. The Free Press.
11. Solomon, M., Bamossy, G., Askegaard, S., & Hogg, M. K. (2006). *Consumer behaviour: A European perspective* (3rd ed.). Pearson Education Limited.
12. Perera, C., Zaslavsky, A., Christen, P., & Georgakopoulos, D. (2014). Sensing as a service model for smart cities supported by Internet of Things. *Transactions on Emerging Telecommunications Technologies, 25*(1), 81–93. https://doi.org/10.1002/ett.2704
13. Jin, J., Gubbi, J., Marusic, S., & Palaniswami, M. (2014). An information framework for creating a smart city through Internet of Things. *IEEE Internet of Things Journal, 1*(2), 112–121. https://doi.org/10.1109/JIOT.2013.2296516
14. Payre, W., Cestac, J., & Delhomme, P. (2014). Intention to use a fully automated car: Attitudes and a priori acceptability. *Transportation Research Part F: Traffic Psychology and Behaviour, 27*, 252–263. https://doi.org/10.1016/j.trf.2014.04.009
15. Bansal, P., & Kockelman, K. M. (2017). Forecasting Americans' long-term adoption of connected and autonomous vehicle technologies. *Transportation Research Part A: Policy and Practice, 95*, 49–63. https://doi.org/10.1016/j.tra.2016.10.013

16. Kyriakidis, M., Happee, R., & de Winter, J. C. F. (2015). Public opinion on automated driving: Results of an international questionnaire among 5000 respondents. *Transportation Research Part F: Traffic Psychology and Behaviour, 32*, 127–140. https://doi.org/10.1016/j.trf.2015.04.014
17. Daziano, R. A., Sarrias, M., & Leard, B. (2017). Are consumers willing to pay to let cars drive for them? Analyzing response to autonomous vehicles. *Transportation Research Part C: Emerging Technologies, 78*, 150–164. https://doi.org/10.1016/j.trc.2017.03.003
18. Schoettle, B., & Sivak, M. (2014). *Public opinion about self-driving vehicles in China, India, Japan, the US, the UK, and Australia*. https://deepblue.lib.umich.edu/bitstream/handle/2027.42/109433/103139.pdf?sequence
19. NHTSA. (2019b). U.S. Department of Transportation recognizes companies for publicly releasing AV safety self-assessments. Retrieved November 20, 2020, from https://www.nhtsa.gov/press-releases/us-department-transportation-recognizes-companies-publicly-releasing-av-safety-self
20. Maneejuk, P., & Yamaka, W. (2021). Significance test for linear regression: How to test without P-values? *Journal of Applied Statistics, 48*(5), 827–845.
21. Osswald, S., Wurhofer, D., Trösterer, S., Beck, E., & Tscheligi, M. (2012). Predicting information technology usage in the car. In *Proceedings of the 4th International Conference on Automotive User Interfaces and Interactive Vehicular Applications—AutomotiveUI '12* (p. 51). ACM Press. https://doi.org/10.1145/2390256.2390264
22. Zmud, J. P., & Sener, I. N. (2017). Towards an understanding of the travel behavior impact of autonomous vehicles. *Transportation Research Procedia, 25*, 2500–2519. https://doi.org/10.1016/j.trpro.2017.05.281

TOUS: A New Technique for Imbalanced Data Classification

Bui T. T. My and Le T. T. An

Abstract Imbalanced data is an issue attracting much concern in classification. With imbalanced input, classification models often suffer from poor performance on the minority class. Moreover, the methods become inefficient if there are noise and overlapping samples in training sets. Although number of papers separately studied either imbalance, noise or overlapping issues, only few works addressed all of them. Besides, these method implementations were so far not accessible. These are the reasons of the new proposed algorithm called TOUS, which combines Tomek-link, random under and over sampling methods. TOUS generates a list of balanced training data sets with different numbers of samples. They are then the input data sets for parallel classifier ensembles. To clarify the performance of TOUS, we build two classifier ensembles, called LLE (Lasso-Logistic ensemble) and DTE (Decision tree ensemble), respectively based on Lasso-Logistic regression and Decision Tree. Our empirical results over six different data sets showed that TOUS helped LLE and DTE to outperform several popular methods for balancing data. Moreover, the execution time of LLE and DTE were relatively less than other tested methods.

Keywords Classification · Ensemble model · Imbalanced data · Re-sampling technique · Lasso- logistic regression · Decision tree

B. T. T. My (✉)
Department of Economic Mathematics, Ho Chi Minh University of Banking, Ho Chi Minh City, Vietnam
e-mail: mybtt@buh.edu.vn

Faculty of Mathematics and Statistics, University of Economics Ho Chi Minh City, Ho Chi Minh City 7000, Vietnam

L. T. T. An
Faculty of Economic Mathematics, University of Economics and Law, Ho Chi Minh City, Vietnam
e-mail: anltt@uel.edu.vn

Vietnam National University, Ho Chi Minh City, Vietnam

S. Sriboonchitta et al. (eds.), *Credible Asset Allocation, Optimal Transport Methods, and Related Topics*, Studies in Systems, Decision and Control 429,
https://doi.org/10.1007/978-3-030-97273-8_39

Abbreviation

AUC	Area under the curve
DT	Decision tree
DTE	Decision tree ensemble
LLE	Lasso-Logistic ensemble
LR	Logistic regression
LLR	Lasso-logistic regression
NCL	Neighborhood cleaning rule
ROS	Random over-sampling
RUS	Random under-sampling
RF	Random forest
SMOTE	Synthetic minority oversampling technique

1 Introduction

Nowadays, classification comes in many fields, such as medicine, finance, business administration, information retrieval, image identification, etc. Along with the development of computational tools, statistical and machine learning methods have been applied to increase performances of classifiers. This evolution is associated with the improvement of classification algorithms, utilizing ensemble models, and especially handling imbalanced data. The data for classification is divided into several classes according to the labels of the samples. The data is considered imbalanced if there is a class with much lower number of samples than other classes [1, 2]. For binary classification, identifying the label of samples in the minority class is often more important than the label of the other. On imbalanced input data, simple models often do not achieve high accuracy in the minority class. These models aim to maximize global accuracy, thus equating losses caused by error type I and error type II [3]. In the case of a rather high imbalanced ratio (the ratio of the number of samples of the majority and the minority class), these models often treat the minority as noise or outliers. Therefore, the classification results are often biased toward to the majority class [4, 5]. For example, in the set of loan customers, pointing out the bad is considered the main task. However, if the number of bad class is far less than the number of the good, popular classifiers often have high global accuracy but very low accuracy on the bad class. Experimental studies showed that if the imbalanced ratio increased, the overall model performance decreased [2]. Besides, some authors believed that imbalanced data was not only the main reason for the poor performance of classifiers, but the noise and the overlapping samples between the classes also degraded the performance of learning methods [1, 6]. The understanding of the nature of imbalanced data has prompted the development of new methods for solving this issue.

Inspired by ensemble models and data-level approach to tackle imbalanced data, we propose an algorithm called TOUS which consists of two steps. In the first step,

TOUS clears the noise and the overlapping samples of different classes by Tomek-Link; in the second step we combine random under-sampling and over-sampling techniques to create a list of C balanced training sets $\{B_i\}_{i=1}^{C}$, which have different numbers of samples. After creating C balanced data sets, we build two classifier ensembles called LLE and DTE by using Lasso-Logistic regression (LLR) and Decision tree (DT) as base learners, respectively. These two ensembles have the same protocols: On each balanced training set of the list $\{B_i\}_{i=1}^{C}$, base learner (LLR or DT) is applied to get C sub-classifiers. With a new data set, the classification result will be determined by the majority voting on C sub-classifiers.

The remainder of the paper consists of four sections. Section 2 reviews related works solving imbalanced data in details. Section 3 proposes the balanced data technique (called TOUS). Section 4 is the empirical study on six data sets with different characteristics to see the performance of the newly built classifier ensembles (LLE and DTE) compared to traditional approaches. Finally, Sect. 5 contains conclusions and discussions.

2 Imbalanced Data: Literature Review

This paper concentrates on binary classification. The labels of samples are denoted by "0" and "1". We call sample labeled "1" "positive sample", to indicate the label of interest. The sample of opposite nature is called "negative sample". When data is imbalanced, usually the positive class has far less samples than the negative class.

There are several approaches to solve imbalance problem, such as algorithm-level, data-level, and the hybrid of data or algorithm-level and classifier ensemble [6, 7]. Algorithm-level approach often imposes or adjusts the cost of loss in misclassification. This approach is usually an extension of existing learning methods to reduce the deviation toward the majority class [8, 9]. The assumption about the difference between the losses caused by the error type I and II is often constant, that reduces the actual meaning of this approach. For that reason, in this paper, we mainly focus on the data-level and the hybrid approaches.

2.1 The Data-Level Approach

The data-level approach involves re-sampling techniques to re-balance the distribution of classes in the original data. These techniques do not depend on learning methods training the classification model [1–3, 6, 10–12]. This section reviews the popular balancing techniques of data-level approach. It forms three main groups: under-sampling, over-sampling and the hybrid of these two techniques.

2.1.1 Under-Sampling

Under-sampling is a family of re-sampling techniques which removes the majority samples to re-balance or degrade the imbalance of the original data. The most common technique is random under-sampling (RUS). RUS creates a balanced subset of training set by randomly eliminating negative samples. This method is non-heuristic, easy to employ and shortens computational time. However, if the data is highly imbalanced, RUS may waste the majority class information because of removing too many negative samples.

Some authors developed heuristic methods to overcome the limitation of RUS, such as: Condensed Nearest Neighbor Rule (CNN) [13], Tomek-Link [14], One-side Selection (OSS) [15], Neighborhood Cleaning Rule (NCL) [16]. Among these methods, Tomek-Link and NCL are usually applied for the purposes of data cleaning and balancing.

Tomek-Link is a method developed by Tomek [14]. Let $d(e_i, e_j)$ be the distance between e_i and e_j. A pair of samples (e_i, e_j) is a Tomek-Link (or T-link), if for any sample e_k, we have $d(e_i, e_j) < d(e_i, e_k)$ and $d(e_i, e_j) < d(e_j, e_k)$. If two samples form a Tomek-Link, then one of them may be a noise or both of them are at borderline. Therefore, Tomek-Link can be applied as a cleaning or balancing method. With the purpose of data cleaning, two samples in Tomek-Link are removed. With the aim of reducing data imbalance, only the negative sample of Tomek-Link is eliminated.

NCL is the method to clean overlapping samples. For binary classification, NCL can be presented as follows: for each sample e_k in the training set, 3-Nearest Neighbors (3-NN) algorithm is applied to classify e_k. If e_k belongs to the majority class but the 3-NN method classifies it into the minority class, then e_k is eliminated. If e_k belongs to the minority class and the 3-NN method mis-classifies it, then the nearest neighbors belonging to the majority class are taken away (see [16]).

2.1.2 Over-Sampling

The idea of over-sampling techniques is opposite of under-sampling. Random over-sampling (ROS) is the baseline technique in this family. ROS randomly repeats positive samples until the data is balanced. Similar to RUS, ROS is non-heuristic and easy to apply. However, ROS leads to longer computational time and may duplicate the noise, the overlapping samples, therefore ROS may be the reason of an over-fitting classification model [1].

To avoid over-fitting, SMOTE (Synthetic Minority Oversampling Technique) was proposed in [17]. Instead of merely randomly repeating minority samples, SMOTE generates synthetic samples and adds them into the minority class. Therefore, SMOTE also identifies some specific region in the feature space for the minority class. SMOTE is believed to make a larger decision boundary for the minority class and increase the performance of learning methods [6, 18]. However, in some cases, the generated samples are inside the majority class, that pushes the decision boundary into the majority class space [19]. This may lead to an increased probability of error type II of learning methods.

2.1.3 The Combining Under and Over-Sampling Techniques

The effects of under and over-sampling on classifiers are not the same in experimental studies. Some authors concluded that over-sampling was better than under-sampling [1, 2, 6, 18] while others thought otherwise [10, 20]. It can be said that none of these techniques completely outperforms the others. Depending on the nature of the data, a specific technique of re-sampling can be more effective than others. As a result, the combination of under and over-sampling is expected to be better than the single ones. In particular, Tomek-Link or NCL were used to clean the noise or overlapping samples before applying re-sampling techniques to balance the data. Some typical works can be presented such as Tomek-Link+RUS+SMOTE [21]; NCL+SMOTE [22], SMOTE+RUS [23], SMOTE+ENN, SMOTE+Tomek-Link [1, 11, 24]. These methods use SMOTE as the main technique to balance data, therefore, can suffer the drawbacks of SMOTE which are pointed in the Sect. 2.1.2. Moreover, the implementation of these hybrid methods are not easily accessible.

2.2 The Hybrid of Data-Level or Algorithm Approach and Classifier Ensemble

2.2.1 Classifier Ensemble

An ensemble models is a collection of similar classifiers [4]. Ensemble models are inspired by the fact that the decision of a committee usually seems better than a single individual. From this idea, ensemble models use simple popular classifiers as the base learners. Classifier ensemble can operate in a parallel or sequential way. The first one is based on the majority votes of sub-classifiers, such as Bagging and Random forest. The second is based on innovation, the following sub-classifier is grown using the modification of the previous, such as Boosting [25].

2.2.2 The Hybrid of Data-Level Approach and Classifier Ensemble

The third way to solve imbalanced data is the hybrid of classifier ensemble and the methods of data-level or algorithm-level approach. It is thought that the hybrid approach can promote the strengths of each method to get the higher classification performance [3, 7, 17, 26, 27]. The hybrids of algorithm- level approach and classifier ensembles, which aim at the minimal overall cost of the classification process, still suffer from the subjective opinions of the researchers about the definitions of

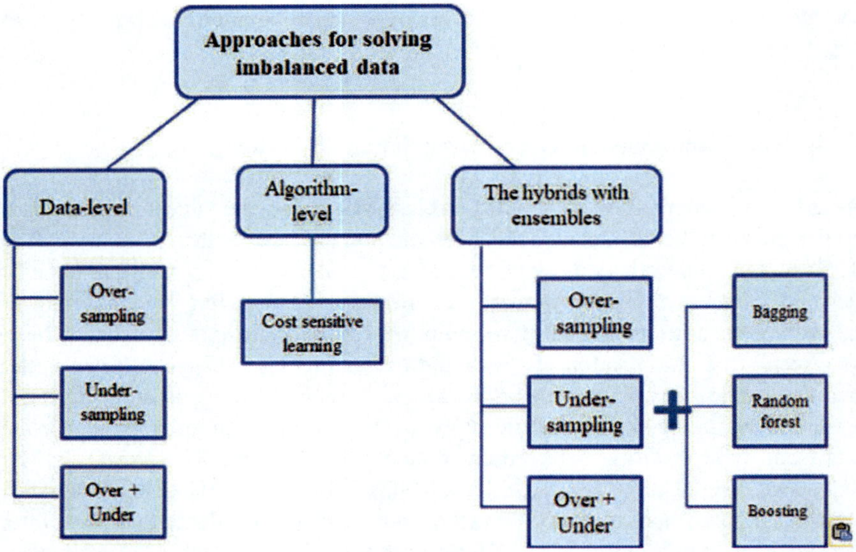

Fig. 1 The approaches for solving imbalanced data

the costs. In the case of combining a specific re-sampling technique and an ensemble learning algorithm, the new hybrid model usually balances the data before training sub-classifiers. The popular ensemble algorithms are bagging and boosting [4, 7]. From the previous study, the hybrid of a data-level approach and an ensemble algorithm improved the classification performance when compared to this ensemble. They also stress the better performance of RUSBoost than multiple complex algorithms (see [4]).

The approaches for balancing data in classification are summarized in Fig. 1.

3 The Proposed Algorithm

In this section, we propose two algorithms: TOUS(C), which is stated in Table 1, is for balancing data and TOUS-F(C), which is stated in Table 2, is for an ensemble model based classifier F. F may be LLR or DT.

TOUS(C) can solve the issue of cleaning and balancing data. According to previous conclusions, the imbalanced status and the noise or the overlapping samples of the training set are factors reducing the performance of learning methods [1, 2, 6]. Therefore, TOUS(C) has two steps corresponding to cleaning data and balancing data. In the first step, we employ Tomek-Link method to clear the noise and overlapping samples. The second step is the combination RUS and ROS to create a list of C balanced training sets which have different numbers of samples $\{B_i\}_{i=1}^{C}$.

Table 1 TOUS(C) algorithm

Input:	Training data set T, the number of balanced dataset C.
1.	Apply Tomek-Link method to find all pairs $\{(e_i,\ e_j)\}$ which are Tomek-Link.
2.	Remove all Tomek-Link $\{(e_i,\ e_j)\}$ from the original training set T.
3.	Determine MI, MA of the training set T': $T' = T \setminus \{(e_i, e_j)\}, T' = MI \cup MA,\ \lvert MI\rvert < \lvert MA\rvert$.
4.	Calculate $D = \lvert MA\rvert - \lvert MI\rvert$ and $S_c^{Add} = \text{round}\left(D \times \frac{c}{C}\right), c \in \{1, 2, \ldots, C\}$
5.	**For** $c \in \{1, 2, \ldots, C\}$, do following:
6.	$T_c' \leftarrow \text{ROS}(T', S_c^{Add})$. The new positive subset of T_c' called MI_c, where $\lvert MI_c\rvert = \lvert MI\rvert + S_c^{Add}$.
7.	$B_c \leftarrow \text{RUS}\left(T_c',\ (\lvert MA\rvert - \lvert MI_c\rvert)\right)$
8.	End **for**
Output:	The list of balanced data sets $\{B_c\}_{c=1}^{C}$.

Table 2 TOUS-F(C) algorithm

Input:	$\{B_c\}_{c=1}^{C}$: The list of balanced data set with the same number of features; F: Classifier
1.	**For** $c \in \{1, 2, \ldots, C\}$
2.	$\text{F}_c \leftarrow F(B_c)$
3.	End **for**
Output:	Classifier ensemble H: $\text{H}(x) = \begin{cases} 1 \text{ if } \sum_{c=1}^{C} F_c(x) \geq 0.5 \times C \\ 0 \text{ if } \sum_{c=1}^{C} F_c(x) < 0.5 \times C \end{cases} \quad \forall x$

In this section we use some notations:

- MA, MI: The majority and the minority class, respectively.
- T: The training set which consists of the majority and minority class $T = MA \cup MI$.
- $D = |MA| - |MI|$: The difference of sample quantities in two classes.
- C: The number of balanced training sets created by TOUS(C) algorithm.
- $S_c^{Add} = \text{round}\left(D \times \frac{c}{C}\right)$: The number of samples would be added to the original minority class by ROS, where function round(x) generates an integer closet to argument x.

The idea of the second step of TOUS algorithm is using ROS to duplicate S_c^{Add} samples in the original minority class to create the new positive class, called MI_c, where $|MI_c| = |MI| + S_c^{Add}$. Then, applying RUS in the original majority set to get the subset MA_c, where $|MA_c| = |MI_c|$. The training set $B_c = MA_c \cup MI_c$ is a balanced data. When c varies from 1 to C, a list of balanced data sets $\{B_c\}_{c=1}^{C}$ is generated. If $c = C$, then $S_c^{Add} = D$, B_C is born by ROS. With $c < C$, B_C is the result of the integration of ROS and RUS.

A note for TOUS(C) is that the list $\{B_c\}_{c=1}^{C}$ consist of C balanced data sets with different numbers of samples. Therefore, TOUS(C) can limit the bad effects of noise and overlapping samples while diversifying the input data of classification.

TOUS-F(C) generates an ensemble model based classifier F. The input of TOUS-F(C) are the output of TOUS(C), which is the list of C balanced training data sets $\{B_c\}_{c=1}^{C}$. On every B_c, F is applied to train a sub-classifier called F_c, $(c = 1, \ldots, C)$. The overall classification result of ensemble base TOUS-F(C) is the major voted by C sub-classifiers F_c. The proposed classifier ensemble is expected to reduce the over-fitting and exploit potential information of the training set.

4 Empirical Study

We have just mentioned the TOUS(C) and TOUS-F(C) algorithm. Here we present the classifier ensembles based on TOUS(C) and TOUS-F(C) and compare them to popular re-sampling techniques to solve imbalanced data. LLR and DT are chosen as the base learners in the proposed classifier ensembles which are called LLE(C) and DTE(C), respectively.

Although there are several effective learning methods in classification, LR is still popular because of some reasons. Firstly, the efficiency of LR is not inferior to other intelligent methods on many data sets [2, 18]. Secondly, LR even provides an easy model to explain the effect of the predictors on the response. In the proposed ensemble model, LLR is employed to take advantages of Lasso and Logistic. The performance of this combination is higher than LR applied individually [28]. The performance of LLE(C) will be compared with the model Lasso-Logistic after applying ROS, SMOTE, RUS, Tomek-link, and NCL.

Similar to LR, DT is a popular standard classification. DT is also considered a weak classifier which is suitable for classification ensembles [4]. The ensemble DTE(C) employs DT as based learner. The performance of DTE(C) will be compared with other classifier ensembles based tree, such as Bagging tree, Random forest (RF), AdaBoost with and without ROS, SMOTE, RUS, Tomek-link, and NCL.

In Sect. 4.1, we summarize some main ideas of LLR and DT, the description of data for empirical study, and the empirical results.

4.1 Logistic and Lasso-Logistic Regression

4.1.1 Logistic Regression (LR)

The LR model is commonly used in a wide variety of fields, especially in credit scoring. We refer to, e.g., Gareth et al. [29], Kuhn and Johnson [30] for a detailed discussions and treatments. The LR model presents the relationship between binary target or response variable $Y \in \{0, 1\}$ and independent or predictor variables $X :=$

$(X_1, X_2, \ldots, X_p)$. More precisely, the LR model describes the change of the probability $P(Y = 1|X)$ in the predictor variables X as following

$$P(Y = 1|X = x) = \frac{e^{\beta_0+\beta x^T}}{1 + e^{\beta_0+\beta x^T}} = f(\beta_0 + \beta x^T). \tag{1}$$

where $\beta = (\beta_1, \ldots, \beta_p)$ and β_0 are the parameters showing the effects of the features on the conditional probability $P(Y = 1|X = x)$, x^T denotes the transpose vector of x, and $f(x) := \frac{1}{1+e^{-x}}$ is sigmoid function with the range [0, 1].

Consider a independent sample $\{(x_{(i)}, y_i), i = 1, 2, \ldots, n\}$ of size n, where $x_{(i)} := (x_{i1}, \ldots, x_{ip}) \in \mathbb{R}^p$ is the vector expressing p features of sample i^{th} and y_i are class labels (binary responses). Then the coefficients $\beta = (\beta_1, \ldots, \beta_p)$ and β_0 in (1) can be estimated by Maximum Log-likelihood approach with the objective function:

$$\log(P(Y|X, \beta)) = \sum_{i=1}^{n} \left(-y_i \log\left(1 + e^{-(\beta_0+\beta x_{(i)}^T)}\right) - (1 - y_i) \log\left(1 + e^{\beta_0+\beta x_{(i)}^T}\right)\right) \tag{2}$$

A new sample x^* will be classified into the positive class if the conditional probability $P(Y = 1|X = x^*)$ is greater than a given threshold. Conversely, x^* will be treated as a negative sample if the conditional probability $P(Y = 1|X = x^*)$ is smaller than the given threshold.

4.1.2 Lasso-Logistic Regression (LLR)

The LLR model is similar to LR model in the assumption (1), however, Lasso imposes condition on the parameters in (2). The objective function of LLR is:

$$\begin{cases} \min_\beta \left\{\sum_{i=1}^{n} \left(y_i \log\left(1 + e^{-(\beta_0+\beta x_{(i)}^T)}\right) + (1 - y_i) \log\left(1 + e^{\beta_0+\beta x_{(i)}^T}\right)\right)\right\} \\ \sum_{j=1}^{p} |\beta_j| \le t \end{cases} \tag{3}$$

where t is called tuning parameter. According to knowledge of convex optimization, (3) is rewritten in the equivalent problem:

$$\min_{\beta} \left(\sum_{i=1}^{n} \left(\log\left(1 + e^{\beta_0+\beta x_{(i)}^T}\right) - y_i \left(\beta_0 + \beta x_{(i)}^T\right)\right) + \lambda \sum_{j=1}^{p} |\beta_j|\right) \tag{4}$$

where, λ is called penalty level, corresponding one - one to t in (3). When λ is zero, the solution of (4) is the same as the solution of (2). When λ is sufficiently large, the solution of (4) is zero. For some λ between the two extreme values, Lasso gives a solution with some of $\hat{\beta}_j = 0$. Therefore, some predictors are excluded from the model. Thanks to the ability of variable selection, Lasso is often employed to determine the important features impacting significantly on the response. The solution of (4) is solved by Coordinate descent algorithm or proximal—Newton interaction [31].

4.2 Decision Tree (DT)

DT is a class of non-parametric algorithms that partition the feature set into a number of non-overlapping sub-sets (called nodes) with similar response values. The most well-known algorithm for constructing a DT is the classification and regression tree (CART) [32]. Beginning with the best feature, CART splits the feature space into two sub-groups. Then, on each subgroup, the splitting process is repeated based on the best feature in the remaining ones. This process continues until a stopping condition is satisfied (for example, a maximum depth of the tree is reached). According to classification error rate or the measure of the impurity of a node (Gini index, Cross-entropy), there are some criteria for determining the best feature and its splitting point. With a new sample, following the series of statements "If-then" about the features, it can go along with the branches until reaching the terminal node. The predictive class is the class accounting for the largest proportion in this terminal node. The detailed knowledge of DT can be found in the machine learning documents [29, 32].

4.3 Data for Empirical Study

We will run the ensemble classification model based on TOUS(C) with six data sets. They are Bank personal loan (Bank), German credit (GER), Hepatitis C (Hepa), US credit (US), Vietnamese 1 (VN 1), and Vietnamese 2 (VN 2) credit. These data sets are chosen because of the diversity in the number of samples, the number of attributes, the imbalanced ratio, and the presence of the overlapping samples (which will be found by Tomek-Link and NCL method). The three heading data sets are obtained from a UCI machine learning repository. The Vietnamese 1 and 2 credit data sets are from the two commercial banks in Vietnam. Table 3 summarizes some characteristics of aforementioned data sets.

Some changes on the original data sets are made. For instance, with Hepa, observations with missing values are removed; the levels of variable Category are grouped into 2 labels which are denoted "0" for "Blood donor" and "1" for the remaining ones.

Table 3 The description of the experimental data-sets

Data sets	Sample size	Positive class size	Imbalanced ratio	No. features	No. numerical features	No. nominal attributes	Source
Bank	5000	480	9.42	11	6	5	https://archive.ics.uci.edu/ml/datasets
GER	1000	300	2.33	19	7	12	
Hepa	589	63	8.35	12	10	1	
US	8505	167	49.93	40	19	21	United State
VN 1	3232	454	6.12	10	0	10	Vietnam
VN2	11,124	837	12.29	12	0	12	

4.4 Implementation Protocol

Each data set is randomly split into the training and the testing sets at the proportion of 70–30%. For every value of C belonging to the set {3, 5, 7, 9}, TOUS(C) and then TOUS-LLR(C) or TOUS-DT(C) algorithm are applied on the training set to build the ensembles called LLE(C) or DTE(C), respectively. The values of AUC on the testing set are recorded to evaluate the classification performance of LLE(C) and DTE(C).

Figure 2 illustrates the implementation protocol of the proposed classifier ensembles. On each data set, this process is repeated 50 times for every value of C. The optimal proposed classifier ensembles on each data set are the ones corresponding to the highest average testing AUC.

4.4.1 The Implementation Protocol of LLE(C)

The process of classification is conducted following the steps showed in Fig. 2 with the base learner LLR. We also train single models based LLR and popular re-sampling techniques, such as ROS, SMOTE, RUS, Tomek-link, and NCL on the same training sets of the optimal LLE(C*). The performance of LLE(C*) and LLRs are compared based on the average testing AUC of 50 times runs.

Some computations on the data sets are made after conducting re-sampling. Firstly, for each nominal attribute, binary variables (dummy variables) are created to express all the levels. The numerical variables are scaled as formula:

$$X_{scale} = \frac{X - \overline{X}}{dev(X)}$$

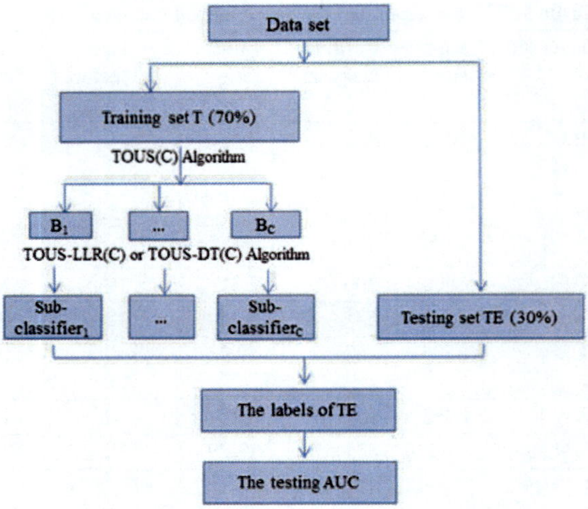

Fig. 2 The implementation protocol of LLE(C) and DTE(C)

where, $\overline{X}$ and $dev(X)$ are mean and deviation of variable X.

Secondly, on the step of training models by LLR, we design a grid of 500 values of penalty level λ on each data set. Coordinate Descent algorithm and 5-fold cross-validation procedure are applied to choose the best value of λ in the grid and estimate the parameters of the models (see [31]).

4.4.2 The Implementation Protocol of DTE(C)

All steps to construct the optimal classifier ensemble DTE(C*) is similar to the implementation protocol of LLE(C). However, DT is base learner instead of LLR. The performance of DTE(C*) is compared to the popular classifier ensembles based tree (Bagging, Random forest, and AdaBoost) combining with one of common re-sampling techniques (RUS, ROS, SMOTE, Tomek-link, and NCL).

RPART algorithm, which is the complementary of CART algorithm, is applied to construct DT classifier (see [33]). The parameters of RPART are assigned as following. The minimum number of observations in any terminal node is 10. The pruning process of each tree is determined by 5-fold cross-validation with the complexity parameter 0.001. For Bagging and AdaBoost, the number of nodes along the longest path from root node to the farthest terminal node is 10 (maxdept = 10). The number of trees in classifier ensemble is 30 (mfinal = 30). For Random forest, the number of trees is 300. The number of predictors of each tree is the squared root of the total predictors of the data set. The classification performance of DTE(C) and the popular ensemble are evaluated by comparing the average testing AUC of 50 running times.

4.5 *The Result of LLE(C) and DTE(C)*

AUC (Area under the receiver operating characteristic curve) is considered to be the most popular criteria for measurement classification performance, especially in case imbalanced data [1, 2].

To compute the value of AUC of LLE(C) and DTE(C), we fix a sequence of 300 threshold values α_i from 0.1 to 1 $\left(\{\alpha_i\}_{i=1}^{300}\right)$ which are the cut-off points to distinguish positive from negative samples. Therefore, on testing set, it can be determined the *true positive rate* $(TPR(\alpha_i))$ and the *false positive rate* $(FPR(\alpha_i))$ corresponding to each value of the threshold α_i. The value of AUC is calculated by the formula:

$$AUC = 0.5 * \sum_{i=2}^{300} |FPR(\alpha_i) - FPR(\alpha_{i-1})| \, (TPR(\alpha_i) + TPR(\alpha_{i-1}))$$

Specific details about the criteria for evaluation; the performance of classifiers can be found in related studies (see [34]).

Table 4 introduces the average testing AUC of LLE(C) and DTE(C) on six data sets, where C belongs to $\{3, 5, 7, 9\}$. For each data set, optimal classifier ensemble is the model with the greatest AUC. If two ensembles have the same greatest AUC, the better is one with smaller C. In Table 4, the underlined values of AUC are corresponding to optimal classifier ensembles on each data set. Because of the burden of the computational process, we do not check on the higher values of C. The upper bound of C is the difference of the numbers of samples in the majority and minority class. However, when C goes toward to the upper bound, the balanced data sets B_c, $(c = 1, \ldots, C)$ will be rather similar, therefore the sub-classifiers in the proposed ensemble will show the same classification results. Another result is that on each data set, DTE has higher AUC than LLE does.

4.6 *Evaluation Performance*

4.6.1 Evaluation Performance of LLE(C)

Table 5 shows the average testing AUC of the optimal LLE(C*) and the models based LLR post-applied the popular re-sampling techniques, such as ROS, SMOTE, RUS, Tomek-Link, and NCL. We also apply LLR without any re-sampling technique to rebalance the data and its AUC shown on the "Without re-sampling" column.

According to Table 5, LLE(C*) outperforms other popular models in AUC criteria. Therefore, it can be concluded that TOUS improves the performance of LLR even combining with popular re-sampling techniques. Moreover, on GER, Hepa, and US data set, the values of AUC of LLR without any re-sampling technique and LLR

Table 4 The values of average testing AUC of the proposed ensembles

Data sets	Ensembles	C = 3	C = 5	C = 7	C = 9
Bank	LLE(C)	0.9780	0.9782	0.9779	0.9780
	DTE(C)	0.9988	0.9989	0.9988	0.9987
GER	LLE(C)	0.7902	0.7893	0.7903	0.7900
	DTE(C)	0.8283	0.8151	0.8109	0.8077
Hepa	LLE(C)	0.9847	0.9849	0.9858	0.9839
	DTE(C)	0.9987	0.9986	0.9991	0.9984
US	LLE(C)	0.8592	0.8598	0.8588	0.8595
	DTE(C)	0.8898	0.8898	0.8897	0.8896
VN 1	LLE(C)	0.7639	0.7647	0.7639	0.7646
	DTE(C)	0.8764	0.8713	0.8685	0.8674
VN 2	LLE(C)	0.8148	0.8144	0.8146	0.8148
	DTE(C)	0.9442	0.9417	0.9422	0.9406

Table 5 The average testing AUC of the models based LLR

Data sets	Without re-sampling	ROS	SMOTE	RUS	Tomek-Link	NCL	LLE(C*)
Bank	0.9442	0.9487	0.9440	0.9475	0.9448	0.9445	0.9782
GER	0.7722	0.7634	0.7623	0.7646	0.7722	0.7722	0.7903
Hepa	0.9239	0.9243	0.9328	0.9256	0.9239	0.9239	0.9858
US	0.7574	0.8467	0.8260	0.8377	0.7574	0.7574	0.8598
VN 1	0.7515	0.7545	0.7523	0.7547	0.7528	0.7521	0.7647
VN 2	0.8018	0.8116	0.8088	0.8108	0.8026	0.8028	0.8148

with Tomek-Link or NCL are the same because these data sets do not have any noise and overlapping samples. On three remaining data sets, Tomek-Link can improve the values of AUC of LLR.

4.6.2 Evaluation Performance of DTE(C)

Table 6 shows the average testing AUC of the optimal ensemble DTE(C*) and the popular classifier ensembles based tree, such as Bagging, Random forest (RF), and AdaBoost with and without one of re-sampling techniques ROS, SMOTE, RUS, Tomek-Link, and NCL.

Similar to the case of LLE(C*), DTE(C*) has a great improvement in AUC compared to all methods considered. For some data sets which are not applied any balancing method, popular classifiers give very good classification performance (e.g. Bank and Hepa data set), DTE still pushes the AUC higher. For data sets with noisy and overlapping samples (e.g. VN 1 and VN 2 data set), DTE also demonstrates the

Table 6 The average testing AUC of the classifier ensembles based tree

Data sets	Ensemble algorithms	Without re-sampling	Re-sampling techniques					
			ROS	SMOTE	RUS	Tomek-link	NCL	DTE(C*)
Bank	Bagging	0.9472	0.9488	0.9510	0.9636	0.9527	0.9556	0.9989
	RF	0.9736	0.9724	0.9728	0.9715	0.9736	0.9739	
	AdaBoost	0.9744	0.9749	0.9748	0.9721	0.9746	0.9748	
GER	Bagging	0.7640	0.7605	0.7577	0.7551	0.7640	0.7640	0.8283
	RF	0.7783	0.7737	0.7716	0.7722	0.7783	0.7783	
	AdaBoost	0.7497	0.7529	0.7523	0.7432	0.7497	0.7497	
Hepa	Bagging	0.9810	0.9766	0.9784	0.9658	0.9810	0.9810	0.9991
	RF	0.9973	0.9968	0.9965	0.9941	0.9973	0.9973	
	AdaBoost	0.9977	0.9966	0.9967	0.9916	0.9977	0.9977	
US	Bagging	0.7308	0.8117	0.7649	0.8116	0.7308	0.7308	0.8898
	RF	0.8405	0.8407	0.8385	0.8210	0.8405	0.8405	
	AdaBoost	0.8390	0.8444	0.8333	0.8119	0.8390	0.8390	
VN 1	Bagging	0.7224	0.7490	0.7400	0.7597	0.7224	0.7223	0.8764
	RF	0.8194	0.8162	0.8165	0.7937	0.8194	0.8209	
	AdaBoost	0.8025	0.8000	0.7949	0.7831	0.8026	0.7993	
VN 2	Bagging	0.6203	0.7692	0.7681	0.7975	0.6313	0.6474	0.9442
	RF	0.7984	0.8319	0.8296	0.8262	0.7989	0.7984	
	AdaBoost	0.8212	0.8234	0.8223	0.8193	0.8225	0.8210	

ability to improve AUC. Moreover, since the numbers of trees in DTE(C*) are much lower than those in Bagging, RF, and AdaBoost, the computational time of DTE(C*) is shorter. This is also an important improvement of the proposed classifier ensemble.

Besides, some minor results are drawn. RUS+AdaBoost is not as well as RF and DTE(C*). This is in contrast to the conclusion of the previous study [4]. Bagging has much lower performance than the other considered classifier ensembles.

5 Conclusions

The paper proposes an algorithm called TOUS for dealing overlapping samples and imbalanced data in classification. TOUS algorithm combines Tomek-Link, random under and over-sampling technique to create a list of balanced data sets. The paper also proposes LLE and DTE which are classifier ensembles applied TOUS as the balancing method and LLR or DT as the base learners. An exhausted study was carried out on six data sets and they showed the efficiency of TOUS. Some important results are following. TOUS has a significant innovation in AUC compared with popular balancing techniques. On each data set, DTE has higher performance than LLE. With a small number of trees in ensemble, not only does DTE outperform Bagging, RF, and AdaBoost, but the computational time of DTE is also shorter. The results suggest empirical studies for another classifier as base learner of TOUS-F(C) algorithm to conclude the potential of the proposed method for solving imbalanced data.

Acknowledgements This research is funded by Vietnam National University Ho Chi Minh City (VNU-HCM) under grant number C2021-34-02. The authors thank Dr. Ta Quoc Bao's team at International University for the helpful comments and for supplying the two Vietnamese credit data sets.

References

1. Batista, G. E., Prati, R. C., & Monard, M. C. (2004). A study of the behavior of several methods for balancing machine learning training data. *ACM SIGKDD Explorations Newsletter, 6*(1), 20–29.
2. Brown, I., & Mues, C. (2012). An experimental comparison of classification algorithms for imbalanced credit scoring data sets. *Expert Systems with Applications, 39*(3), 3446–3453.
3. Shen, F., Zhao, X., Li, Z., Li, K., & Meng, Z. (2019). A novel ensemble classification model based on neural networks and a classifier optimisation technique for imbalanced credit risk evaluation. *Physica A: Statistical Mechanics and its Applications, 526,* 121073.
4. Galar, M., Fernandez, A., Barrenechea, E., Bustince, H., & Herrera, F. (2011). "A review on ensembles for the class imbalance problem: bagging-, boosting-, and hybrid-based approaches," IEEE Transactions on Systems, Man, and Cybernetics. *Part C (Applications and Reviews), 42*(4), 463–484.

5. Haixiang, G., Yijing, L., Shang, J., Mingyun, G., Yuanyue, H., & Bing, G. (2017). Learning from class-imbalanced data: Review of methods and applications. *Expert Systems with Applications, 73,* 220–239.
6. Kaur, P., & Gosain, A. (2018). Comparing the behavior of oversampling and undersampling approach of class imbalance learning by combining class imbalance problem with noise. In *ICT based innovations* (pp. 23–30). Springer.
7. López, V., Fernández, A., García, S., Palade, V., & Herrera, F. (2013). An insight into classification with imbalanced data: Empirical results and current trends on using data intrinsic characteristics. *Information Sciences, 250,* 113–141.
8. Xiao, J., Xie, L., He, C., & Jiang, X. (2012). Dynamic classifier ensemble model for customer classification with imbalanced class distribution. *Expert Systems with Applications, 39*(3), 3668–3675.
9. Xiao, J., Zhou, X., Zhong, Y., Xie, L., Gu, X., & Liu, D. (2020). Cost-sensitive semi-supervised selective ensemble model for customer credit scoring. *Knowledge-Based Systems, 189,* 105118.
10. Drummond, C., Holte,, R. C., et al. (2003) C4. 5, class imbalance, and cost sensitivity: why under-sampling beats over-sampling. In *Workshop on Learning from Imbalanced Datasets II,* (Vol. 11, pp. 1–8). Citeseer.
11. Koziarski, M. (2020). Csmoute: Combined synthetic oversampling and undersampling technique for imbalanced data classification. *arXiv preprint* arXiv:2004.03409.
12. Mansourifar, H., & Shi, W. (2020). Deep synthetic minority over-sampling technique. *arXiv preprint* arXiv:2003.09788.
13. Hart, P. (1968). The condensed nearest neighbor rule (corresp.). *IEEE Transactions on Information Theory, 14*(3), 515–516.
14. Tomek, I., et al. (1976). Two modifications of cnn. *IEEE Transactions on Systems, Man, and Cybernetics, SMC-6*(11), 769–772.
15. Kubat, M., & Matwin, S., et al. (1997). Addressing the curse of imbalanced training sets: one-sided selection. *ICML, 97,* 179–186.
16. Laurikkala, J. (2001) Improving identification of difficult small classes by balancing class distribution. In *Conference on Artificial Intelligence in Medicine in Europe* (pp. 63–66). Springer.
17. Chawla, N. V., Bowyer, K. W., Hall, L. O., & Kegelmeyer, W. P. (2002). Smote: Synthetic minority over-sampling technique. *Journal of Artificial Intelligence Research, 16,* 321–357.
18. Marqués, A. I., García, V., & Sánchez, J. S. (2013). On the suitability of resampling techniques for the class imbalance problem in credit scoring. *Journal of the Operational Research Society, 64*(7), 1060–1070.
19. Bellinger, C. (2016). *Beyond the boundaries of SMOTE: A framework for manifold-based synthetic oversampling*. Ph.D. thesis, Université d'Ottawa/University of Ottawa, 2016.
20. Huang, Y.-M., Hung, C.-M., & Jiau, H. C. (2006). Evaluation of neural networks and data mining methods on a credit assessment task for class imbalance problem. *Nonlinear Analysis: Real World Applications, 7*(4), 720–747.
21. Elhassan, T., & Aljurf, M. (2016). Classification of imbalance data using tomek link (t-link) combined with random under-sampling (rus) as a data reduction method," *Global J Technol Optim S, 1,* 2016.
22. Junsomboon, N., & Phienthrakul, T. (2017) Combining over-sampling and under-sampling techniques for imbalance dataset. In *Proceedings of the 9th International Conference on Machine Learning and Computing* (pp. 243–247).
23. Shamsudin, H., Yusof, U. K., Jayalakshmi, A., & Khalid, M. N. A. (2020). Combining oversampling and undersampling techniques for imbalanced classification: A comparative study using credit card fraudulent transaction dataset. In *2020 IEEE 16th International Conference on Control & Automation (ICCA)* (pp. 803–808). IEEE.
24. Ramentol, E., Caballero, Y., Bello, R., & Herrera, F. (2012). Smote-rs b*: a hybrid preprocessing approach based on oversampling and undersampling for high imbalanced data-sets using smote and rough sets theory. *Knowledge and Information Systems, 33*(2), 245–265.
25. Roncalli, T. (2020). *Handbook of financial risk management*. CRC Press.

26. Sun, J., Lang, J., Fujita, H., & Li, H. (2018). Imbalanced enterprise credit evaluation with dte-sbd: Decision tree ensemble based on smote and bagging with differentiated sampling rates. *Information Sciences, 425,* 76–91.
27. Seiffert, C., Khoshgoftaar, T. M., Van Hulse, J., & Napolitano, A. (2009). Rusboost: A hybrid approach to alleviating class imbalance. *IEEE Transactions on Systems, Man, and Cybernetics-Part A: Systems and Humans, 40*(1), 185–197.
28. Wang, H., Xu, Q., & Zhou, L. (2015). Large unbalanced credit scoring using lasso-logistic regression ensemble. *PloS One, 10*(2), e0117844.
29. James, G., Witten, D., Hastie, T., & Tibshirani, R. (2013) *An introduction to statistical learning* (vol. 112). Springer.
30. Kuhn, M., & Johnson, K. (2013). *Applied predictive modeling*. New York: Springer.
31. Hastie, T., Tibshirani, R., & Wainwright, M. (2019). *Statistical learning with sparsity: The lasso and generalizations*. Chapman and Hall/CRC.
32. Brieman, L., Friedman, J. H., Olshen, R. A., & Stone, C. J. (1984). *Classification and regression trees* (Vol. 67). Wadsworth Inc.
33. Therneau, T. M., Atkinson, E. J., et al. (1997). *An introduction to recursive partitioning using the RPART routines*. Technical report, Technical Report Mayo Foundation.
34. Řezáč, M., & Řezáč, F. (2011). How to measure the quality of credit scoring models. *Finance a úvěr: Czech Journal of Economics and Finance, 61*(5), 486–507.

Value at Risk Analysis and Investment Portfolio Optimization of Asian Stocks

Nuntawut Habkhonglek and Paravee Maneejuk

Abstract This study investigated the Value at Risk (VaR) and the Component Expected Shortfall (CES) to find an optimal investment portfolio considering the 15 Asian stock markets. The data is collected from 2006 to 2019, and then divided into three phases corresponding to the pre-crisis, during-crisis, and post-crisis periods of the 2008 Global Financial Crisis. Empirical results showed that the VaR increased from 0.0034 to 0.0062 (82.35%) during the crisis period but decreased from 0.0062 to 0.0035 (43.55%) after the crisis. Obviously, the crisis influenced the risk of stock markets in Asia. The results of portfolio analysis suggest that Malaysia stocks should have received the highest investment proportion during the crisis period with no investment proportion for Singapore, India, and Pakistan stocks. After the crisis, Indonesia stocks should constitute the highest investment proportion in the portfolio, while the investment in stock markets of Singapore, Turkey, and Russia is not suggested.

Keywords Crisis · Copula · Portfolios · Risk

1 Introduction

The world is now still under the globalization process with the interdependence among regions worldwide growing in scale and intensity in innumerable aspects. Without a doubt, nowadays, international financial markets are extremely interrelated [3]. However, this high degree of connectedness not only can generate investment and speculative prospects, but sometimes also bring about contagion and spillover effects across markets like what the world has recently witnessed in the time of the 2007–2008 financial crisis particularly in the U.S.

This crisis was mainly triggered by deregulation in the financial industry. U.S. banks utilized various programs for trading with derivatives. Then their demands for mortgages increased to facilitate the profitable sale of these derivatives. However,

N. Habkhonglek · P. Maneejuk (✉)
Faculty of Economics, Chiang Mai University, Chiang Mai, Thailand
e-mail: mparavee@gmail.com

S. Sriboonchitta et al. (eds.), *Credible Asset Allocation, Optimal Transport Methods, and Related Topics*, Studies in Systems, Decision and Control 429,
https://doi.org/10.1007/978-3-030-97273-8_40

many bad debtors could not pay the higher interest rates of these derivatives and thereby lead to a shortage of liquidity among financial institutions [3]. During this crisis, there was substantial evidence of high fluctuation in the financial and capital markets, in particular stock markets around the world [18].

With the interconnection of the financial markets worldwide, investors may be able to profit from investing in a diversity of stock markets, and one of the appealing markets is that of the Asian stocks. As supported by the FINNOMENA Investment Team [6], Asian countries are interesting market because, being the emerging countries, their economies are growing faster than that of developed countries. Asian stock markets have connected to each other as exhibited by their co-movements (Bloomberg 2018). More specifically, these stock markets had been on the rise before the 2008 global financial crisis until they declined in the same direction during the crisis. Afterward, they regained well the trading activities and volumes during the economic recovery stage. However, they always experienced violent fluctuations which posed capital investment risks throughout the pre-crisis, crisis, and post-crisis periods.

Markowitz categorized the financial investment risks into two types which are (1) systematic risk and (2) unsystematic risk. The systematic risk arises from macroeconomic factors, and investors could not eliminate or decrease it from diversification. The systematic risk will influence the whole stock market. For unsystematic risk, it is caused by the risks of money, administration, and industry [8]. This unsystematic risk could be avoided by making diversification [14]. Thus, to achieve risk reduction, investors considered the diversification of risk during pre-crisis, crisis, and post-crisis periods of the U.S. financial crisis.

Several studies have examined linkages among financial assets and investigated the optimal asset allocation and hedge ratios for the stock portfolio during the U.S. financial crisis. The risk of assets in five channels: industry, commerce, real estate, utility, and conglomerates group to manage portfolio optimization using the Bivariate copula-generalized autoregressive conditional heteroskedasticity (GARCH) [5]. They also quantified the VaR and Expected Shortfall (ES) to learn the portfolios' risk. Wang evaluated the allocation of investment portfolios among stocks in the Shanghai Stock Exchange using the Vine Copula-GARCH Model [17]. Nguyen and Nguyen [12] used the exponential generalized autoregressive conditional heteroskedasticity (EGARCH) model to investigate the risk in the Vietnamese stock market and reported that stock returns have intense volatility, especially between 2006 and 2010. They revealed that the market's volatility could be attributable to the effects of the 2008 global financial crisis [12]. Huong et al. [7], they confirmed that the global financial crisis in the 2008 period could decrease the financial inflows into Vietnam. Pastpipatkul et al. [13] Analyzed the co-movement and provided advice on the optimal portfolio investment for investors and risk managers in financial markets of the G7 countries. This is because the G7 countries have been impacted by the global financial system from the 2007–2008 crisis, and any catastrophic event hitting this group could easily trigger and spread to the global financial market. Sahadudheen [16] employed the GARCH and EGARCH models and found a negative relationship between exchange rate and stock prices in India.

Accordingly, the researcher realizes the importance of the fluctuation and risk in affecting the stock returns. Many studies have suggested that stock market returns' volatility is highly related to uncertainty in the world financial events [10, 13, 18, 20]. They suggested that investors could manage or avoid investment risk by understanding the risk behavior and sources of risk. Hence, this research aims to reinvestigate the VaR of Asian stock exchanges in the three sub-periods of the U.S. financial crisis: pre-crisis, crisis, and post-crisis. This study utilizes the Vine copula-EGARCH model. Driven by the fear generated by the threat of the U.S. financial crisis, this paper represents one of the first efforts to analyze the impact of the U.S. financial crisis on the return and volatility of the Asian stock market indices in the top 15 countries based on GDP. This study aims to fill the literature gap by decomposing the sources of Asian financial market risk in the three sub-periods of the U.S. financial crisis using the CES. The rest of the paper is organized as follows: Sect. 2 provides the data and research methods. Section 3 reports results and discussion. Section 4 concludes the whole discussion.

2 Data and Research Methods

This part presents the methods used for quantifying the Value at Risk (VaR) and the Component Expected Shortfalls (CES) to obtain an optimal portfolio investment in Asian stock exchanges in three sub-periods: pre-crisis, crisis, and post-crisis.

2.1 Data Used in the Study

In this study, weekly data of 15 Asian stock markets are used. During the studied period, there were 15 major stock markets in Asia which consist of Stock Exchange of Thailand (SET), Straits Times (ST) of Singapore, Philippine Stock Exchange Index (PSEI), Ho Chi Minh City Stock Exchange (HCMC) of Vietnam, The FTSE Bursa Malaysia Index Series (FTSE), Jakarta Stock Exchange Composite (JKT), Bombay Stock Exchange (BSE) of India, Karachi Stock Exchange (KSE) of Pakistan, China AMC CSI 300 Index ETF (CSI), Nikkei Stock Average (NIKKIE) of Japan, Korea Stock Exchange KOSPI Index (KOSPI), Taiwan Stock Exchange (TWSE), Borsa İstanbul Stock Exchange (BIST) of Turkey, Kazakhstan Stock Exchange (KASE), and the Russian Trading System (RTS). The data are collected from January 4, 2006, to December 30, 2019, and divided into three periods: pre-crisis period covering January 4, 2006, to July 31, 2007, crisis period covering August 1, 2007, to December 31, 2012, and post-crisis period from January 4, 2013, to December 30, 2019. If there was no data due to holidays or other reasons, the stock index price is given as the previous day's trading price [19]. All stock index prices are collected from the Bloomberg database; then each price is transformed as the logarithm return

$$\Delta r_{i,t} = \ln\left(\frac{P_{i,t}}{P_{i,t-1}}\right) \tag{1}$$

where $P_{i,t}$ is stock price i at time t; and $P_{i,t-1}$ is stock price i at time $t-1$.

2.2 *Research Methods*

2.2.1 The Exponential Generalized Autoregressive Conditional Heteroscedasticity (EGARCH) Model

The exponential GARCH model is an asymmetric volatility model. In this study, $p = 1$ and $q = 1$ are assumed to simplify the estimation of the conditional variance of the random variable [11]. The model can be estimated by

$$\log(h_t) = \omega + \sum_{i=1}^{p} \alpha_i \varepsilon_{t-i}^2 + \sum_{i=1}^{q} \beta_i \log(h_{t-i}) + \sum_{i=1}^{k} \gamma_i \left(\frac{|\varepsilon_{t-i}|}{h_{t-i}} - \sqrt{\frac{2}{\pi}}\right) \tag{2}$$

where h_t is the conditional variance at time t, ω, α, β and γ are the estimated parameters, $\Delta r_t = \varepsilon_t$. This model was used to obtain the standardized residuals of the Asian stock returns (marginals).

2.2.2 The Vine Copula Model

The first introduced the copula, which was used to investigate the dependence between the marginal distributions of the random variables [15]. Later the Vine copula of [2] introduced the Vine copula to deal with the complex dependence measures among the multivariate marginals. The Vine Copula could be written as

$$f_{j|i}\left(x_j \,|x_i\right) = \frac{f\left(x_i, x_j\right)}{f_i\left(x_i\right)} = c_{ij}\left(F_i\left(x_i\right), F_j\left(x_j\right)\right) \cdot f_j\left(x_j\right) \tag{3}$$

where x is the standardized residual obtained from EGARCH and $f(x_i, x_j)$ is the density function. $F(x_i, x_j)$ is marginal distribution which is assumed to have the skewed student-t form. There are two famous structures of Vine copula, namely C-vine and D-vine copulas.

For C-vine, the last x_i was selected to be the conditional variable, as in the following equation

$$F(x) = \prod_{k=1}^{n} f_k\left(x_k\right) \times \prod_{i=1}^{n-1}\prod_{j=1}^{n-i} c_{i,i+j|1:(i-1)}\left(F\left(x_i \,|x_1, \ldots, x_{i-1}\right), F\left(x_{i+j} \,|x_1, \ldots, x_{i-1}\right) \Big|\theta_{i,i+j|1:(i-1)}\right) \tag{4}$$

For D-vine, the first x_j was selected to be the conditional variable, as in the following equation

$$F(x) = \prod_{k=1}^{n} f_k(x_k) \times c_{j,j+i|(j+1):(j+i-1)}$$
$$\left(F\left(x_j \left| x_{j+1}, \ldots, x_{j+i-1}\right.\right), F\left(x_{j+i} \left| x_{j+i}, \ldots, x_{j+i-1}\right.\right) \left| \theta_{j,j+i|(j+1):(j+i-1)}\right.\right) \quad (5)$$

where *n* is the number of Asian stock returns. This model is used to measure the interdependence among the marginal distribution of Asian stock returns.This study consider 20 copula functions, Gaussian copula, Student t copula, Gumbel copula, Frank copula, Joe copula, BB1 copula, BB7 copula, Survival Clayton copula, Survival Gumbel copula, Survival Joe copula, Survival BB1 copula, Survival BB7 copula, Clayton copula (270 °C), Tawn type 1 copula, rotated Tawn type 1 copula (180 °C), rotated Tawn type 1 copula (90 °C), rotated Tawn type 1 copula, Tawn type 2 copula, rotated Tawn type 2 copula (180 °C) and rotated Tawn type 2 copula (90 °C) to construct a joint density between each pair of assets.

2.2.3 Value at Risk

They stated that "Value at Risk was the maximum loss of return (R_L) corresponding to probability (α)" [10]. Accordingly, the VaR can be measured by

$$VaR_i = \mu_{i,R} - z_{i,\alpha}\sigma_{i,R} \quad (6)$$

where $\mu_{i,R}$ is the mean of stock market return i ; $\sigma_{i,R}$ is the standard deviation of stock market return i. Z_α is the standard score (Z-score) with the probability corresponding to the probability of loss α. Note that we assume $\alpha = 0.05$. This means that there is a 0.05 probability that the portfolio will loss more than the value of VaR_i in each sub-period.

2.2.4 Component Expected Shortfall: CES

The expected shortfall was the expected tail loss from the asset investment [1], and this loss can be measured by

$$ES_{i,t+1} = E_{i,t+1}\left[r_{i,t+1} \left| r_{i,t+1} < -VaR_{i,t+1}\right.\right] \quad (7)$$

where $r_{i,t+1}$ is the return of the investment in the stock market at the period $t + 1$.later proposed CES to measure risk sources in the investment portfolio [4]. This risk measure can be obtained by

$$CES_{it} = w_{it}\frac{\partial ES_{i,t+1}}{\partial w_{it}} = w_{i,t+1}E_{t+1}\left[r_{i,t+1}\left|r_{i,t+1} < -VaR_{i,t+1}\right.\right] \tag{8}$$

2.2.5 The Portfolio Optimization Based on the Component Expected Shortfall (CES)

Min

$$CES_{it} = w_{i,t+1}E_{i,t+1}\left[r_{i,t+1}\left|r_{i,t+1} < -VaR_{i,t+1}\right.\right] \tag{9}$$

Subject to

$$r = \left[\left(w_1 \times r_{1,t+1}\right) + \cdots + \left(w_{15} \times r_{15,t+1}\right)\right] \tag{10}$$

$$w_1 + \cdots + w_{15} = 1 \tag{11}$$

$$0 \le w_i \le 1, \tag{12}$$

where $i = 1$ (SET), $i = 2$ (ST), $i = 3$ (PSEI), $i = 4$ (HCMC), $i = 5$ (FTSE), $i = 6$ (JKT), $i = 7$ (BSE), $i = 8$ (KSE), $i = 9$ (CSI), $i = 10$ (NIKKIE), $i = 11$ (KOSPI), $i = 12$ (TWSE), $i = 13$ (BIST), $i = 14$ (KASE), and $i = 15$ (RTS).

To sum up, the portfolio optimization of the Asian stock returns is based on the loss probability through the CES, and the optimal portfolio investment is achieved by minimizing the risk of portfolios.

3 Results and Discussion

3.1 The Results of Estimation from the ARMA-EGARCH Model

This study employs the EGARCH (1, 1)-skew-t model to model the marginal distributions for three sub-periods considered in this study. The estimation results for the margins are summarized in Table 1. It shows that the estimated ARCH coefficient (α) and GARCH coefficient (β) are most significant at the 1% level for all stock returns, signifying the presence of clustering volatility phenomenon in all stock returns during pre-crisis, crisis, and post-crisis periods. Moreover, the conditional variance equations ($\alpha + \beta$) are less than one for all cases, indicating that our market returns series satisfies the stationary condition. It is also observed that ($\alpha + \beta$) is very close to 1. This implies the high volatility persistence of our series. Moreover, the skew parameter deviates from one suggesting that the residual distribution cannot be normally distributed and that applying the Skew-t distribution is suitable for this EGARCH (1, 1) model.

Table 1 The results of the estimation from the ARMA-EGARCH model for pre-crisis, during crisis, and post-crisis periods of the U.S. financial crisis in 2008

Period	Pre-crisis			During-crisis			Post-crisis		
Parameter	α	β	γ	α	β	γ	α	β	γ
SET	−0.0788	0.9881	0.1618	−0.0931	0.9324	0.3618	−0.089	0.9991	0.172
	(0.0540)	(0.0045)	(0.1618)	(0.0898)	(0.0537)	(0.1043)	(0.0500)	(0.0050)	(0.0700)
ST	0.0489	0.8806	−0.5655	−0.2848	0.9894	0.2804	−0.114	0.9927	0.1646
	(0.0000)	(0.0001)	(0.0001)	(0.0691)	(0.0076)	(0.0423)	(0.0412)	(0.0007)	(0.0176)
PSEI	0.0432	0.7688	0.1512	−0.1226	1.000	0.2163	−0.1341	0.9933	0.1308
	(0.5676)	(0.9104)	(0.5698)	(0.0460)	(0.0007)	(0.0303)	(0.0308)	(0.0007)	(0.0269)
HCMC	0.3574	0.8902	−0.5857	−0.1351	0.8125	0.5621	−0.185	0.892	0.3793
	(0.0001)	(0.0001)	(0.0001)	(0.1060)	(0.1114)	(0.2029)	(0.0771)	(0.0506)	(0.0991)
FTSE	−0.1698	0.8300	0.5843	−0.2486	0.9931	0.1873	−0.1631	1.0000	0.0053
	(0.1692)	(0.0755)	(0.2112)	(0.0424)	(0.0000)	(0.0340)	(0.0329)	(0.0000)	(0.0195)
JKT	0.1338	0.4252	0.4395	−0.3173	0.9375	0.4881	−0.0921	0.9261	0.3086
	(0.2804)	(0.5026)	(0.3322)	(0.0880)	(0.0437)	(0.1408)	(0.0654)	(0.0327)	(0.0910)
BSE	0.7162	0.9152	−0.9316	−0.1307	0.9707	0.3402	−0.1140	0.9851	0.0791
	(0.0018)	(0.0002)	(0.0004)	(0.0557)	(0.0250)	(0.0977)	(0.0387)	(0.0000)	(0.0115)
KSE	−1.3781	0.9002	1.0601	−0.0566	0.8247	0.7019	−0.1606	0.7070	0.2824
	(0.4025)	(0.0092)	(0.3254)	(0.1120)	(0.0651)	(0.1640)	(0.0789)	(0.1307)	(0.1006)
CSI	0.2591	0.8867	−0.6340	−0.0933	0.9999	0.0736	0.0713	0.9776	0.2001
	(0.0000)	(0.0001)	(0.0002)	(0.0386)	(0.0001)	(0.0188)	(0.0402)	(0.0051)	(0.0573)
NIKKIE	0.0412	0.9278	0.5407	−0.0869	0.9989	0.0896	−0.2209	0.9276	0.2714
	(0.1412)	(0.0753)	(0.2432)	(0.0227)	(0.0016)	(0.0463)	(0.0696)	(0.0743)	(0.0743)
KOSPI	−0.0948	0.6312	0.3574	−0.3255	0.9899	0.20026	−0.1851	0.6978	0.1600
	(0.1712)	(0.3296)	(0.4105)	(0.0379)	(0.0048)	(−0.0618)	(0.0651)	(0.1449)	(0.1002)
TWSE	−0.1005	0.8242	0.3757	−0.1108	1.0000	0.0462	−0.1653	0.9999	0.1470
	(0.2477)	(0.2718)	(0.3134)	(0.2007)	(0.0000)	(0.0190)	(0.0452)	(0.0002)	(0.0297)
BIST	−0.1893	0.9172	0.2893	−0.1038	0.9736	0.2523	−0.0734	0.9999	0.0428
	(0.1625)	(0.0719)	(0.1871)	(0.0497)	(0.0143)	(0.0783)	(0.0204)	(0.0000)	(0.0111)
KASE	0.2741	0.9355	−0.3748	−0.3739	0.9823	−0.1414	−0.0132	0.7877	0.6594
	(0.0000)	(0.0001)	(0.0000)	(0.0000)	(0.0003)	(0.0000)	(0.0878)	(0.1111)	(0.1842)
RTS	0.7348	0.9746	1.3406	−0.1568	0.9695	0.3322	−0.1738	0.9907	0.2371
	(0.7423)	(0.0361)	(1.1914)	(0.0663)	(0.0220)	(0.1246)	(0.0466)	(0.0087)	(0.039)

Note The number in brackets is standard error

The results also show that BSE of India, HCMC of Vietnam, CSI of China, and KASE of Kazakhstan have the highest volatility in the pre-crisis period. During crisis, CSI of China and NIKKIE of Japan have the highest volatility. Besides, there is the highest volatility in the CSI of China, and BIST of Turkey after the crisis.

After the estimation of the EGARCH models, the standardized residuals for 15 market returns can be obtained. These standardized residuals will be transformed as marginal distributions in the C-, D-vine Copula models in three sub-periods. In order to guarantee that the marginal is uniformly distributed, we use the Kolmogorov-Smirnov test (KS-test) and Ljung-Box test to confirm whether the residuals are independent and identically distributed random variable. Hence, the standardized residuals estimation of ARMA-EGARCH (1, 1) is reliable and it could continuously be used in the Vine Copula model.

Table 2 The result of an optimal model

Period	Vine copula	Statistic for optimal model		
		AIC	BIC	LL
Pre-crisis	C-vine	**−2473.37**	**−2374.69**	**1277.68**
	D-vine	−2353.70	−2274.28	1209.85
During crisis	C-vine	**−1085.15**	**−968.49**	**574.57**
	D-vine	−1047.95	−909.42	561.97
Post-crisis	C-vine	**−1034.22**	**−874.32**	**558.11**
	D-vine	−967.35	−826.95	519.67

Note Bold number indicates the better Vine structure

Accordingly, the best fit Vine Copula based E-GARCH model is further employed to investigate the Value at Risk (VaR) and the Component Expected Shortfall (CES). We noted that VaR and CES allow us to find an optimal investment portfolio as well as the sources of risk. Moreover, this study also evaluated the effectiveness of the models in terms of their reliability. To clarify, the C-vine Copula and D-vine Copula models are compared based on the lowest Akaike Information Criterion (AIC) and Bayesian Information Criterion (BIC), and the highest Log-Likelihood (LL) to obtain the most optimal model, as shown in Table 2.

3.2 Model Selection

This section presents the comparison between C-vine copula and D-vine copula using AIC, BIC, and LL.

From Table 2, the AICs of C-vine and D-vine are −2473.37 and −2353.70, while BICs are −2374.69 and −2274.28, respectively, in the pre-crisis period. During the crisis, AICs are −1085.15 and −1047.95, and BICs are −968.49 and −909.42, respectively. Additionally, the AICs are −1034.22 and −967.35, whereas BICs are −874.32 and −826.95, respectively in the post-crisis period. In this comparison, the C-vine copula model has a lower AIC and BIC and a higher Log-Likelihood than the D-vine copula in all periods. Thus, the C-vine copula model is the best fit copula structure for constructing the joint distribution of 15 Asian stock markets' returns.

3.3 The Result of Parameter Estimates in the C-vine Copula Model in Each Period

This part presents the parameters estimated from the C-vine Copula model in Table 3. The results reveal that each pair in the C-vine structure shows a different optimal copula pair, which selected at each hierarchical level based on the lowest AIC and BIC. The C-vine copula parameters cannot be interpreted directly. Thus this study

Table 3 The result of parameter estimates from the C-vine copula model

Pre-crisis					During-crisis					Post-crisis				
Edge	Copula	Par1	Par2	Tau	Edge	Copula	Par1	Par2	Tau	Edge	Copula	Par1	Par2	Tau
2,1	G	1.26	–	0.21	2,8	F	1.66	–	0.18	2,11	J	1.42	–	0.19
2,8	SC	0.64	–	0.24	2,13	J	1.85	–	0.32	2,13	BB7[a]	1.28	0.21	0.21
2,7	SC	0.53	–	0.21	2,14	SC	1.16	–	0.37	2,3	SC	0.89	–	0.31
2,6	G	1.69	–	0.41	2,1	G	1.63	–	0.39	2,9	SJ	1.05	–	0.03
2,14	G	1.43	–	0.3	2,3	Tawn2[a]	1.76	0.48	0.26	2,8	BB7[a]	1.22	0.06	0.13
2,3	SC	69.01	–	0.97	2,7	G	1.75	–	0.43	2,5	G	1.47	–	0.32
					2,5	BB7[a]	1.99	0.22	0.39	2,12	J	1.14	–	0.07
										2,1	SBB1[a]	0.42	1.18	0.3
										2,14	G	1.27	–	0.21
										2,7	SBB7[a]	1.11	0.48	0.23
										2,15	N	0.42	–	0.28
11,12;2	G	1.54	–	0.35	10,8;2	G	1.12	–	0.1	7,13;2	N	0.15	–	0.1
11,4;2	J	1.68	–	0.27	10,14;2	F	1.11	–	0.12	7,3;2	SC	0.42	–	0.17
					10,11;2	BB1[a]	0.11	1.77	0.47	7,8;2	SJ	1.04	–	0.02
					10,3;2	G	1.15	–	0.13	7,6;2	C270	−0.15	–	−0.07
					15,10;2	G	1.16	–	0.14	7,5;2	J	1.33		0.16
										7,1;2	G	1.16	–	0.14
										15,7;2	F	1.19	–	0.13
3,8;11,2	Tawn270[a]	−3.25	0.39	−0.32	5,13;10,2	N	0.37	–	0.24	15,4;7,2	N	−0.12	–	−0.07
3,7;11,2	BBB1_90[a]	−1.07	−2.16	−0.7	5,14;10,2	F	1.35	–	0.14	15,13;7,2	F	1.6	–	0.17
3,9;11,2	Tawn180[a]	3.47	0.31	0.26	5,1;10,2	N	0.21	–	0.14	15,3;7,2	N	0.17	–	0.11
3,5;11,2	Tawn[a]	3.19	0.18	0.16	5,7;10,2	N	0.25	–	0.16	15,5;7,2	S.G.	1.15	–	0.13
3,14;11,2	Tawn2_270[a]	−13.38	0.06	−0.06						15,1;7,2	F	1.45	–	0.15
15,3;11,2	SG	4.4	–	0.77						15,14;7,2	N	0.16	–	0.1

(continued)

Table 3 (continued)

Pre-crisis					During-crisis					Post-crisis				
Edge	Copula	Par1	Par2	Tau	Edge	Copula	Par1	Par2	Tau	Edge	Copula	Par1	Par2	Tau
14,1;3,11,2	J	1.46	–	0.21	7,4;5,10,2	Tawn270[a]	−2.43	0.06	−0.06	10,4;15,7,2	J	1.17	–	0.09
14,4;3,11,2	N	−0.36	–	−0.24	7,13;5,10,2	G	1.09	–	0.08	10,11;15,7,2	T[a]	0.28	6.47	0.18
14,7;3,11,2	Tawn2_180[a]	13.97	0.03	0.03	7,1;5,10,2	G	1.18	–	0.15	10,9;15,7,2	J	1.12	–	0.07
					7,3;5,10,2	N	0.13	–	0.08	10,5;15,7,2	Tawn270[a]	−6.61	0.02	−0.02
										10,12;15,7,2	G	1.18	–	0.15
6,12;14,3,11,2	N	−0.28	–	−0.18	9,8;7,5,10,2	SG	1.06	–	0.05	1,3;10,15,7,2	G	1.12	–	0.1
6,4;14,3,11,2	N	−0.26	–	−0.17	9,11;7,5,10,2	SC	0.22	–	0.1	1,5;10,15,7,2	J	1.14	–	0.07
6,8;14,3,11,2	C	0.29	–	0.13										
15,8;6,14,3,11,2	C	0.39	–	0.16	12,4;9,7,5,10,2	G	1.18	–	0.15	12,4;1,10,15,7,2	BB7[a]	1.21	0.06	0.13
15,7;6,14,3,11,2	C90	−0.29	–	−0.13						12,11;1,10,15,7,2	SG	1.07	–	0.06
5,1;15,6,14,3,11,2	F	−1.27	–	−0.14	3,6;12,9,7,5,10,2	T[a]	0.02	5.05	0.01	5,13;12,1,10,15,7,2	SG	1.06	–	0.05
5,7;15,6,14,3,11,2	Tawn90[a]	−5.91	0.02	−0.02										
10,8;5,15,6,14,3,11,2	C	0.29	–	0.13	11,13;3,12,9,7,5,10,2	G	1.07	–	0.07					
9,1;10,5,15,6,14,3,11,2	Tawn[a]	9.57	0.01	0.01	15,6;1,11,3,12,9,7,5,10,2	Tawn2[a]	20	0.02	0.02					
9,4;10,5,15,6,14,3,11,2	C	0.48	–	0.19										
8,1;7,9,10,5,15,6,14,3,11,2	Tawn[a]	1.94	0.16	0.12										
13,4;8,7,9,10,5,15,6,14,3,11,2	Tawn2_90[a]	−20	0.04	−0.14										

Note (1). I = Independence copula, N = Gaussian copula, T = Student t copula (t-copula), G= Gumbel copula, F= Frank copula, J = Joe copula ,BB1 = BB1 copula, BB7 = BB7 copula, SC = Survival Clayton copula, SG = Survival Gumbel copula, SJ = Survival Joe copula, SBB1 = Survival BB1 copula, SBB7 = Survival BB7 copula, C270 = Clayton copula (270 °C), Tawn = Tawn type 1 copula, Tawn180 = rotated Tawn type 1 copula (180 °C), Tawn90 = rotated Tawn type 1 copula (90 °C), Tawn270 = rotated Tawn type 1 copula, Tawn2 = Tawn type 2 copula, Tawn2_180 = rotated Tawn type 2 copula (180 °C), Tawn2_90 = rotated Tawn type 2 copula (90 °C) and (2). "*a*" denotes the copula that contains two parameters, namely dependence parameter (Par1) and degree of freedom (Par 2) transforms them to be Kendall tau value. It can be observed that the Kendall tau values exhibit positive and negative values during the pre-crisis and post-crisis periods, while they exhibit only positive values during the crisis period. It shows that the C-vine structures in the three periods are different, indicating that the dependence structure among the stock markets is changing. To sum up, the overall correlations among stock markets are more likely to be positive during these three periods. This means that Asian stock markets are close to each other, and they move in the same direction.

3.4 The Result of Risk and Investment Portfolio

From the Value at Risk analysis in Table 4, the investment portfolio obtained from the equal weight method provides the lower return and the higher risk than the optimal weight method in all periods. In particular, the Value at Risk increases from 0.0034 to 0.0062 (82.35%) (pre-crisis to during crisis). In post-crisis, the Value at Risk decreases from 0.0062 to 0.0035 (−43.55%). Apparently, the 2008 global financial crisis affected the returns of the Asian stock markets and led to a relatively high risk in portfolio investment. This study then uses the CES to investigate the sources of risk in the Asian stock markets.

According to the CES discussion in Sect. 2.2.4, the C-vine copula-based ARMA-EGARCH model is used to simulate the return of the Asian stock market at time $t + 1$. Therefore, 10,000 simulated returns are obtained. From Table 5, the CES presents the sources of Asian stock markets' risk in three sub-periods. Figure 1 shows the top six stock markets which are contributing the risk to the Asian stock markets. Note that the higher percentage indicates a higher potential systemic risk. As shown in Fig. 1, the TWSE of Taiwan and JKT of Indonesia are among the top six stocks' returns contributing a high risk to Asian stock markets in all three periods, indicating that they have a strong influence on the Asian stock markets. It is also observed that the TWSE contributes the highest CES% in pre-crisis, while JKT contributes the highest CES% during crisis and post-crisis periods. These findings suggest that investors should be cautious in investing in TWSE and JKT markets.

After quantifying a systemic risk by the CES method, a portfolio investment analysis is conducted in each period. The trading decision can be made by finding

Table 4 The result of value at risk analysis and the return of investment portfolio

Period	Portfolio method	VaR	Expected return
Pre-crisis	Equal weight	0.0049	−0.0023
	Optimal weight	0.0034	−0.0029
During crisis	Equal weight	0.0072	0.0002
	Optimal weight	0.0062	0.0001
Post-crisis	Equal weight	0.0043	−0.0025
	Optimal weight	0.0035	−0.0029

Table 5 The result of Component Expected Shortfall (CES)

Period	Pre-crisis		During crisis		Post-crisis	
	Portfolio optimiza-tion : W	Source of risk : CES	Portfolio optimiza-tion : W	Source of risk : CES	Portfolio optimiza-tion : W	Source of risk : CES
SET	0.0416	0.0413	0.0135	0.0001	0.0119	0.0115
ST	0.0000	0.0000	0.0374	0.0262	0.0000	0.0000
PSEI	0.1461	0.1816	0.0495	0.0581	0.0574	0.0530
HCMC	0.0329	0.0660	0.0841	0.1892	0.0785	0.0864
FTSE	0.2531	0.1915	0.1793	0.0330	0.1098	0.0984
JKT	0.0600	0.0669	0.1153	0.2102	0.1613	0.2014
BSE	0.0000	0.0000	0.0000	0.0000	0.0961	0.0918
KSE	0.0000	0.0000	0.1070	0.1397	0.0725	0.0754
CSI	0.0609	0.097	0.0787	0.0689	0.0591	0.0542
NIKKIE	0.0709	0.0555	0.1446	0.0822	0.0190	0.0187
KOSPI	0.0424	0.045	0.0528	0.0365	0.1487	0.1265
TWSE	0.2348	0.1966	0.1208	0.1492	0.1439	0.1197
BIST	0.0000	0.0000	0.0005	0.0004	0.0000	0.0000
KASE	0.0146	0.0185	0.0000	0.0000	0.0418	0.0624
RTS	0.0426	0.0405	0.0163	0.0068	0.0000	0.0000

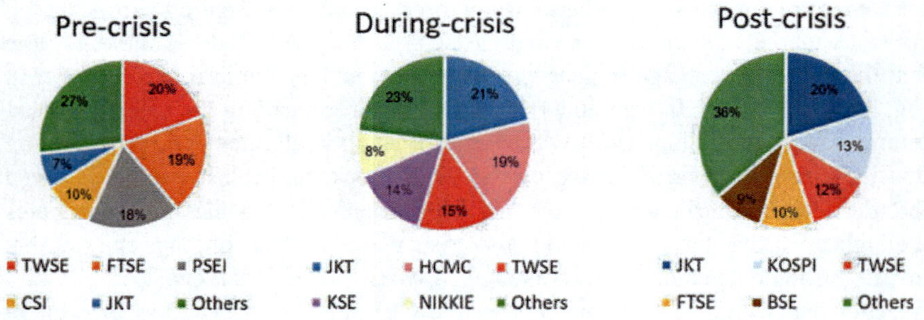

Fig. 1 Percentages of risk (CES%) from the top six stock markets during pre-crisis, during-crisis, and post-crisis (from left to right)

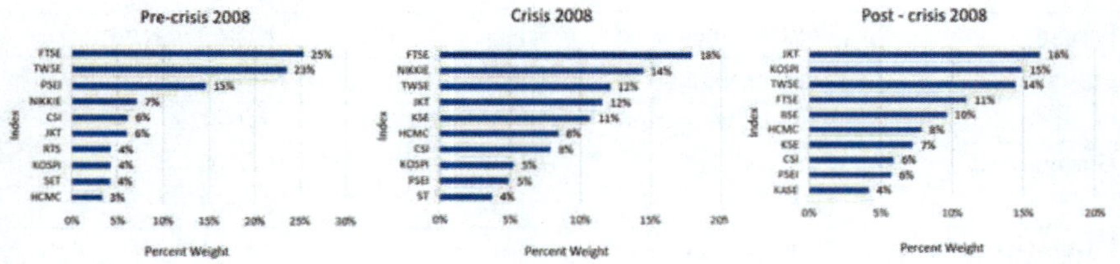

Fig. 2 The top ten average weight percentages for investment portfolio in pre-crisis, crisis, and post-crisis (from left to right)

optimal weights for each stock exchange in the Asian market in all crisis periods. The results are reported in Fig. 2. Each panel describes the top ten weights for portfolio investment in all periods. This study finds that the portfolio investment during the crisis period is more scattered as there is a bit gap between different stock markets. The results of the analysis suggest that Malaysia's FTSE index is the most suitable stock for pre-and during-crisis periods, while the JKT of Indonesia is an appropriate stock in the post-crisis.

4 Conclusions

This paper investigates the Value at Risk (VaR) and the Component Expected Shortfall (CES) using the C and D-vine copula-based EGARCH models to obtain an optimal investment portfolio of 15 Asian stocks as well as identify sources of risk. The results show that the ARMA-EGARCH (1, 1) model with skewed student-t provides reliable ARCH and GARCH effects in all three periods: pre-crisis, crisis, and post-crisis. It is observed that Asian stock markets exhibit high fluctuation in all subperiods, particularly during the financial crisis. The next step is the investigation on the linkage among the Asian financial markets using the C and D-vine copulas, and the result shows that the C-vine Copula has the lower AIC and BIC, and the higher log-likelihood, so the C-vine Copula model is the most suitable model to construct the linkage among the Asian stocks. Then, the C-vine copula-based EGARCH model is further used to measure the VaR, and CES. The results show that the VaR increased from 0.0034 to 0.0062 (82.35%) during the crisis period but decreased from 0.0062 to 0.0035 (43.55%) after the crisis. Obviously, the crisis intensified the risk of Asia stock markets. Therefore, the pre-crisis optimal investment is FTSE of Malaysia, whereas S.T. of Singapore, BSE of India, KSE of Pakistan, and BIST of Turkey are not suggested to invest. During the crisis, Malaysia's FTSE is suggested to invest with the highest proportion, while there is zero proportion in BSE of India and KASE of Kazakhstan. In the post-crisis, the JKT of Indonesia is found to have the highest proportion, while S.T. of Singapore, BIST of Turkey, and Russia's RTS are not suggested.

The investor could utilize these results to decide on the investment in the Asian stock markets. Although this study can help investors in their decision-making, investors ought to consider the situations and current news that affect the stock markets to reduce the risk.

References

1. Artzner, P., Delbaen, F., Eber, J.-M., & Heath, D. (1997). Thinking Coherently. *Risk, 10*, 68–71.
2. Aas, K., Czado, C., Frigessi, A., & Bakken, H. (2009). Pair-copula constructions of multiple dependence. *Insurance: Mathematics and Economics, 44*(2), 182–198. https://doi.org/10.1016/j.insmatheco.2007.02.001

3. Ayuba, S. (2011). *The impact of the global financial market on the Nigerian stock market.* Project, Department of Economics, Ahmadu Bello University, Zaria: An Unpublished M.Sc.
4. Banulescu, G.-D., & Dumitrescu, E.-I. (2015). Which are the SIFIs? A component Expected Shortfall approach to systemic risk. *The Journal of Banking and Finance, 50,* 575–588. https://doi.org/10.1016/j.jbankfin.2014.01.037.
5. Fan, G., Zeng, Y., & Wong, W. K. (2012). Decomposition of portfolio VaR and expected shortfall based on multivariate Copula simulation. *International Journal of Management Science and Engineering Management, 7*(2), 153–160. https://doi.org/10.1080/17509653.2012.10671219.
6. FINNOMENA Investment Team. (2019). The reason why "Asia" is such an interesting region? Finnomena Investment Team. Retrieved January 20, 2020, from: https://www.finnomena.com/finnomena-ic/interesting-asia/
7. Huong, T. T. X., Nguyen, M. L. T., & Lien, N. T. K. (2021). Exchange rate volatility and FDI response during the financial crisis: Empirical evidence from Vietnam. *Journal of Asian Finance, Economics and Business, 8*(3), 119–126. https://doi.org/10.13106/jafeb.2021.vol8.no3.0119
8. Markowitz, H. (1952). Portfolio selection. *The Journal of Finance, 7*(1), 77–91. https://doi.org/10.1111/j.1540-6261.1952.tb01525.x.
9. McNeil, A. J., Frey, R., & Embrechts, P. (2015). *Quantitative risk management* (Revised edition, Vol. 2015). Princeton University Press.
10. Maneejuk, P., & Yamaka, W. (2019). Predicting contagion from the U.S. Financial crisis to international stock markets using dynamic copula with google trends. *Mathematics, 7*(11), 1032. https://doi.org/10.3390/math7111032
11. Nelson, D. B. (1991). Conditional heteroskedasticity in asset returns: A new approach. *Econometrica, 59*(2), 347–370. https://www.jstor.org/stable/2938260 https://doi.org/10.2307/2938260
12. Nguyen, C. T., & Nguyen, M. H. (2019). Modeling stock price volatility: Umpirical evidence from the Ho Chi Minh City stock exchange in Vietnam. *Journal of Asian Finance, Economics and Business, 6*(3), 19–26. https://doi.org/10.13106/jafeb.2019.vol6.no3.19
13. Pastpipatkul, P., Yamaka, W., & Sriboonchitta, S. (2016). Dependence structure of and co-movement between Thai currency and international currencies after introduction of quantitative easing. In *Causal inference in econometrics* (pp. 545-564). Springer.
14. Reeb, D. M., Kwok, C. C. Y., & Baek, H. Y. (1998). Systematic risk of the multinational corporation. *Journal of International Business Studies, 29*(2), 263–279. https://doi.org/10.1057/palgrave.jibs.8490036
15. Sklar, A. (1959). Fonctions de Répartition à n Dimensions et Leurs Marges. *Publications de l'Institut Statistique de l'Université de Paris, 8,* 229–231.
16. Sahadudheen, I. (2015). An exponential garch approach to the effect of impulsiveness of euro on indian stock market. *Journal of Asian Finance, Economics and Business, 2*(3), 17–22. https://doi.org/10.13106/jafeb.2015.vol2.no3.17
17. Thongkairat, S., Yamaka, W., & Chakpitak, N. (2019). Portfolio optimization of stock, oil and gold returns: A mixed copula-based approach. In *International conference of the Thailand econometrics society* (pp. 474–487). Springer.
18. Yakubu, Z., & Akerele, A. O. (2012). An analysis of the impact of global financial crisis on the Nigeria stock exchange. *Current Research Journal of Social Sciences Journal of Social Sciences, 4*(6), 369–399.
19. Yamaka, W., & Maneejuk, P. (2020). Analyzing the causality and dependence between gold shocks and Asian emerging stock markets: A smooth transition copula approach. *Mathematics, 8*(1), 120.
20. Zhu, K., Yamaka, W., & Sriboonchitta, S. (2016). Multi-asset portfolio returns: a Markov switching copula-based approach. *Thai Journal of Mathematics*, 183–200.

Cash-Flow Volatility and Capital Structure Decisions

Trang Thi Van Do and Thuy Duong Phan

Abstract This article estimates the effect of cash flow volatility on the capital structure decisions of 531 companies listed on Vietnamese Securities Market the period from 2012 to 2020. The research utilizes a new and unique data set by employing Bayes estimation. The empirical results indicate that the determinants of capital structure decisions and firms with higher cash flow volatility have intended to reduce the financial leverage in order to control the financial risk. Furthermore, the finding suggests that it is necessary for internal managers to consider the cash flow volatility while making decisions of optimal capital structure.

Keywords Capital structure · Cash-flow · Vietnamese Security Market · Bayes methods

1 Introduction

Capital structure is one of the most important policies of a business. Research on capital structure was initially developed by the theories of Modigliani and Miller [12, 13]. After that, many researchers have conducted studies on this issue in different aspects, including the factors affecting capital structure. A numerous studies indicated that micro and macro the determinants have influence on the capital structure of enterprises in developed financial markets. Recently, this issue has been conducted in emerging countries by the researchers of Faris [5] and Vo [18]. Furthermore, cash flow risk is also considered an essential factor, but its influence on capital structure has not been studied widely and has not been concerned by managers. Bates et al. [1] pointed out companies with increasing volatility of cash flow could retire all their debt obligations. Meanwhile, research studies by Denis and McKeon [2]

T. T. Van Do
Postgraduate Faculty, Banking Academy, Hanoi, Vietnam
e-mail: trangdtv@hvnh.edu.vn

T. Duong Phan (✉)
Faculty of Transport Economics, University of Transport Technology, Hanoi, Vietnam
e-mail: duongpt@utt.edu.vn

S. Sriboonchitta et al. (eds.), *Credible Asset Allocation, Optimal Transport Methods, and Related Topics*, Studies in Systems, Decision and Control 429,
https://doi.org/10.1007/978-3-030-97273-8_41

and Huang and Ritter [8] showed that those with negative net cash flow would have insufficient amount of cash to meet the company's growth in the future. In addition, firms that generate negative operating cash flow intend to issue equity rather than debts. Moreover, Harris and Roark [7] has detected a relationship between cash flow risk and capital structure decisions. The results found that increasing cash flow volatility would raise financial leverage, especially for companies with poor financial performance. Therefore, the objective of this article is to examine whether cash-flow volatility can influence the capital structure of firms listed on Vietnam Stock Exchange during 2012–2020.

This study is structured as follows. After the introduction, this article has reviewed all economic theories related to the capital structure decisions. In part three, the paper has focused on data and methodology. In the fourth section, the empirical results are described. Finally, the conclusions and implications are discussed.

2 Literature Review

Since the theories of capital structure decision have been implemented by many scholars. Modigliani and Miller theory [12, 13] mentioned the capital structure issue without tax and with tax. This study indicated the relationship between financial leverage and return to shareholder, which also determined the optimal debt to assets ratio. Accordingly, the change in capital structure does not cause the change in the firm value. If the corporate income tax is added, firm value was an increasing function of financial leverage but ignoring the bankruptcy costs. Tax shield is also paid much attention by internal managers. Its advantages decrease if firms increase the proportion of debt. Furthermore, the existence of personal income taxes may decline the theoretical tax shield when considering corporate borrowing. The personal income tax changes as the difference between tax rates on capital gains and regular income. Nevertheless, Modigliani and Miller examined the capital structure in a perfect world that the capital market is incomplete, therefore the corporate financing may not be explained as limited. Trade-off theory by Jensen and Meckling [9] indicated that despite low cost of debts and benefit of tax shield, cost of capital increased and firm value decreased. Regarding the cost of debt, firms need to consider the corporate debt maturity policy in order to minimize the interest cost. The trade-off model is concentrated on one period only by Myers [14], whereas the dynamic trade-off model is highlighted in multiple periods by Fischer et al. [6], consisting of the expenses of transactions and adjustments toward optimal financial structure.

The pecking-order theory by Myers [14] identified the priority for different sources of financing since the managers considered issuing debt or equity. The usage of internal financing is taken into account initially before issuing the safe securities. If there is a requirement for external finance, issuing debt is prioritized before share issuance. For the debt financing, it is necessary to identify the proportion of short-term and long-term in order to match with the different kinds of assets. Furthermore, the signaling theory by Ross [17] observed that the increase in debt finance has a good signal for on-going business, and debt structure maturity had been taken into account.

The implications of capital structure theories have been demonstrated in many empirical research studies. Dudley and James [4] has examined the relationship between cash-flow volatility and capital structure policy of American firms by applying the GARCH model. The authors concluded that constrained firms would issue more debt than equity when the volatility is low and vice versa. Keefe and Yaghoubi [11] has detected that cash flow volatility played an important determinant, which significantly impacted the financial leverage and the usage of debt finance of different maturities. Douglas et al. [3] has measured the cash flow risk and the effects of cash flow volatility on the default risks. The result provides empirical findings to capital structural models and the effect of information uncertainty on bond issuance. Karimli [10] has shown that the negative volatility in cash flow influenced corporate debt decisions and its maturity structure. Besides, the proportion of longer debt maturity intends to be higher with shorter maturity. Harris and Roark [7] has revealed the influence of firm debt and cash flow risk is positive significantly. But for the firms with a high operating cash flow, the relationship between cash flow risk and financial leverage is not statistically significant.

In Vietnam, the determinants of capital structure have been studied in recent years by the researches of Nguyen and Dinh [16] and Vo [18]. However these authors have not indicated the effect of the cash-flow volatility yet. Nguyen [15] has investigated the relation between cash-flow volatility and capital structure by applying the quartiles regressions to measure the different impact levels. Therefore, this article measured the impact of cash-flow volatility on capital structure decisions of listed firms based on employing the Bayesian estimation. According to previous research studies, the article hypothesizes as follows:

Hypothesis 1: The volatility of cash flows positively influences the financial leverage of listed firms.
Hypothesis 2: The size of a company positively influences the financial leverage of listed firms.
Hypothesis 3: The profitability of a company positively influences the financial leverage of listed firms.
Hypothesis 4: Proportion of tangible assets positively influences the financial leverage of listed firms.
Hypothesis 5: The market to book value ratio positively influences the financial leverage of listed firms.
Hypothesis 6: GDP positively influences the financial leverage of listed firms.

3 Data and Methodology

Due to the rapid development of computer sciences and informative technology during the past decades, the Bayesian paradigm has been more and more widely used due to its superiorities. Bayesian estimates are an entire probability distribution

Table 1 List of dependent and independent variables

Variable	Notation	Definition
Financial leverage	LEV	LEV is calculated by total debts divided by total shareholders' equity
Cash flow volatility	CFV	CFV is measured based on the logarithm of standard deviation of cash flow for 3 years
Firm size	SZ	SZ is measured based on the logarithm of firm total assets
Firm profitability	PRO	The ratio is measured by earnings after tax divided by average total assets
Tangible fixed assets	TANG	TANG is measures by tangible fixed assets divided by total assets
Growth	MB	MB is measures by market value divided by book value of common equity
Gross domestic product	GDP	World Bank

of a particular coefficient, while frequentist results are point estimates. Besides, combining prior distributions with observed data results in a more balanced, accurate and valid Bayesian inference. Furthermore, while data-driven frequentist methods face the effect of small sample size, Bayesian methods do not. Unlike data-driven frequentist inference, the combination of data likelihood with prior distribution in Bayesian inference allows for overcoming the endogeneity problem attributed to the former method (Table 1).

Based on the empirical studies in Sect. 2, we propose the following research model:

$$\begin{aligned} \text{LEVit} = {} & \beta 1\text{CFVit} + \beta 2\text{SZit} + \beta 3\text{PROit} \\ & + \beta 4\text{TANGit} + \beta 5\text{MBit} + \beta 6\text{GDPit} + \text{uit} \end{aligned}$$

To estimate the relationship between cash flow volatility and the use of debt in Vietnamese listed firms, this study collects a sample of 531 listed firms on Vietnam Stock Exchange, from 2012 to 2020 by using panel data analysis. The data is obtained from the Fiinpro data, GDP is obtained from the World Bank data source.

4 Empirical Results

As can be seen from Table 2, the descriptive statistics indicate the mean, median, and standard deviation as well as the minimum and maximum values of all variables.

The result of analysis and statistics stated in Table 2 shows a difference in the level of debt utilization, cash flow fluctuation and other factors that affect the capital structure of companies listed on Vietnamese Securities Market. The debt ratio from

Table 2 Descriptive statistics

Variable	Obs	Mean	Std. Dev	Min	Max
LEV	4779	0.4941	0.2219	0.0019	1.2945
CFV	4779	23.52	1.7462	17.63	29.48
SZ	4779	27.26	1.5652	23.33	33.68
PRO	4779	0.5326	6.5636	−9.0948	252.77
TANG	4779	0.2179	0.1991	0	0.9654
MB	4779	0.6099	0.7704	0.0055	13.41
GDP	4779	5.9289	1.2371	2.91	7.08

listed companies is 0.4941 on average. The minimum and maximum debt ratio from companies is 0.0019 and 1.2945, respectively. It indicates a huge difference in the level of debt utilization from listed companies. A numerous companies almost use no debt while many loss-making companies have higher liabilities than their total assets.

The cash flow volatility is estimated in the period of 3 years. CFV is a variable used for examining the effect of cash flow fluctuation on the debt utilization, calculated by the logarithm of standard deviation of the cash flow within 3 years. The average level of the entire sample is 23.52. Meanwhile, the minimum and maximum values are 17.63 and 29.48, respectively.

In respect of control variables, the company size varies in listed companies. The return on assets is 0.5326 on average while its minimum and maximum values are −9.01 and 252.77, respectively. This proves that in addition to companies with very high profitability, there are still many loss-making ones. The fixed asset investment ratio is 0.2179 on average, while its minimum and maximum values are 0 and 0.9654 respectively. The ratio shows a huge difference in the level of fixed assets investment in companies due to different business lines. From 2012 to 2020, Vietnam's lowest GDP growth level is recorded in 2020 at 2.91 while the highest growth rate reported in 2018 is 7.08.

To analyze the impact of cash flow fluctuation and other factors on the level of debt utilization of listed companies, the authors implement regression on the basis of Bayes' rule with the Gibbs sampling algorithm. The diagnosis of convergence of MCMC is carried out in order to ensure that MCMC-based Bayesian Inference is reasonable. To test the convergence of MCMC, we can use Trace plot, Autocorrelation, Histogram and Kernel Density plot.

These tests indicate no sign of non-convergence of MCMC chains. For all the model parameters, trace plots traverse rapidly through the posterior domain without exhibiting any trend (Fig. 1).

The autocorrelation plots for all the parameters show that the plots die off after only a few lags. Even that the autocorrelation plot for GI dies off after about 28 lags can be considered acceptable (Fig. 2).

Graphical posterior summaries such as histograms and kernel density estimates provide useful additions to the various numerical statistics for summarizing MCMC

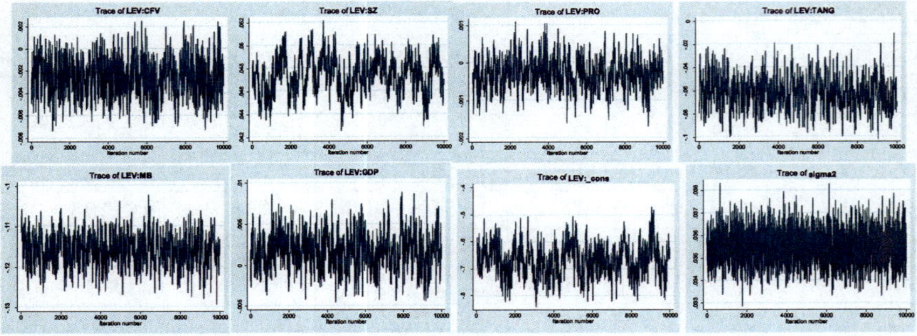

Fig. 1 Convergence test via trace plots

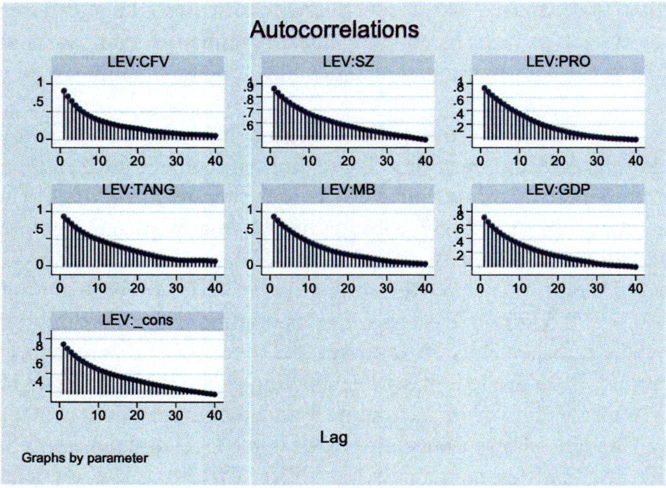

Fig. 2 Convergence test via autocorrelation plots

output. It is always a good practice to inspect the histogram and kernel density estimates of the marginal posterior distributions of parameters to ensure that these empirical distributions behave as expected. These plots can be used to compare the empirical posterior and the specified prior distributions to visualize the impact of the data.

A histogram depicts the general shape of the marginal posterior distribution of a model parameter (Fig. 3).

Kernel density plots provide alternative visualizations of the simulated marginal posterior distributions. They may be viewed as smoothed histograms. By default, the bayes graph kdensity command shows an overall density of the entire MCMC sample. To explore the convergence, the command provides the option, which additionally plots two density curves: the first-half density obtained using the first half of the

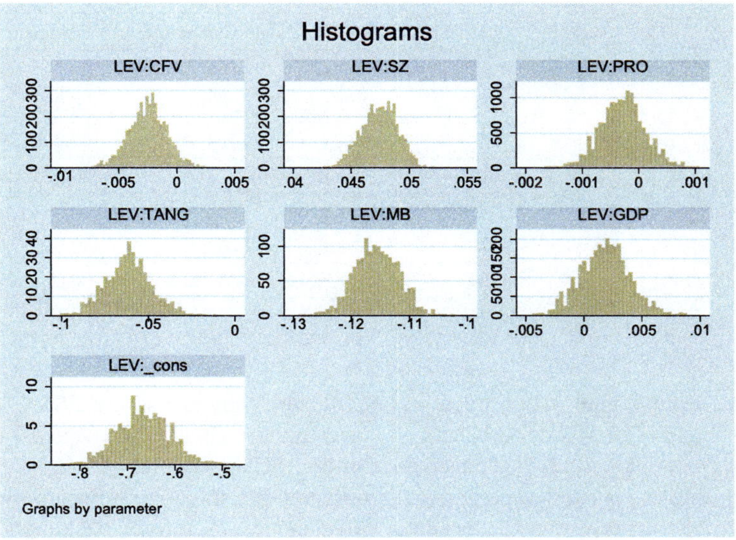

Fig. 3 Convergence test via histogram plots

MCMC sample and the second-half density obtained using the second half of the MCMC sample. The chain has converged and mixes well; we expect the three density curves to be close to each other (Fig. 4).

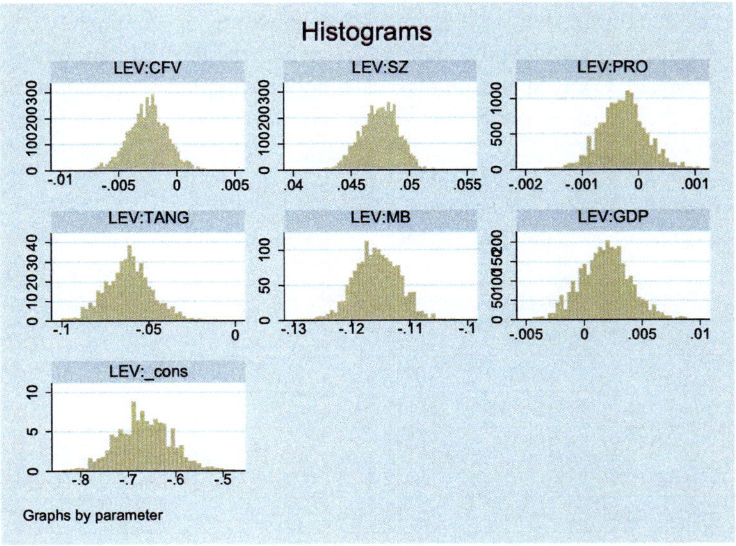

Fig. 4 Convergence test via densities plots

Table 3 ESS test

Variable	ESS	Corr. Time	Efficiency
LEV			
CFV	329.10	30.39	0.0329
SZ	60.86	164.31	0.0061
PRO	237.70	42.07	0.0238
TANG	345.03	28.98	0.0345
MB	374.98	26.67	0.0375
GDP	219.02	45.66	0.0219
Sigma2	1820.33	5.49	0.1820

Aside from diagnosis by images of MCMC, the authors also use ESS to assess the convergence of MCMC. MCMC's value is 10,000 which nearly equals MCMC sample's size. Therefore, the convergence of MCMC is very high (Table 3).

The sampling is large-scale with 4779 observations, therefore, prior information does not significantly affect the accuracy of the model. Therefore, the authors apply Normal priori distribution (1, 100) for the regression coefficient and Invagamma (2.5, 2.5) for variance of observational error.

Table 4 shows that on the basis of the regression model as per Bayes method, variables that are proportional to the level of debt utilization are company size and economic growth rate. Meanwhile, cash flow fluctuation, tangible asset profitability and growth rate are inversely proportional to the level of debt utilization.

For CFV, the cash flow fluctuation is inversely proportional to the level of debt utilization from companies. It proves that if the cash flow fluctuation increases by 1%, the level of debt utilization from companies decreases by approximately 0.0025%. This means higher cash flow fluctuation may reduce the level of debt utilization at listed companies. Companies with high cash flow fluctuation encounter difficulties in fulfilling their financial obligations under the contract with the creditors and debt

Table 4 Regression results

	Mean	Std. Dev	MCSE	Median	Equal-tailed [95% Cred. Interval]	
CFV	−0.0025	0.0016	0.0000	−0.0025	−0.0057	0.0007
SZ	0.0474	0.0015	0.0002	0.0475	0.0444	0.0503
PRO	−0.0003	0.0004	0.0000	−0.0003	−0.0010	0.0005
TANG	−0.0617	0.0134	0.0007	−0.0615	−0.0873	−0.0341
MB	−0.1157	0.0038	0.0002	−0.1158	−0.1233	−0.1085
GDP	0.0018	0.0022	0.0001	0.0018	−0.0023	0.0062
_cons	−0.6665	0.0570	0.0064	−0.6687	−0.7769	−0.5533
Sigma2	0.0355	0.0007	0.0000	0.0355	0.0341	0.0369

utilization is also disadvantageous, which increases the risk of financial distress. Therefore, such companies will reduce the loan for lower financial risk.

Moreover, the finding also shows that when companies have good profitability, hold a large number of tangible assets and display good growth rate, they tend to use less external finance via loans. Finally, due to the high economic growth rate, listed companies increase their demand for loans.

5 Conclusion

The research study shows empirical evidence regarding the level of effect of cash flow volatility and other factors on the level of debt utilization of companies listed on Vietnamese Securities Market. By the approach using Bayes method, the finding indicates that companys with higher cash flow volatility may reduce the level of debt utilization to keep the financial risk at a low level.

The result provides empirical evidence to listed companies as well as state management authorities, therefore the authorities can come up with appropriate business strategies and policies that help companies achieve sustainable development. The detail is as follows:

Companies need to take cash flow volatility into consideration before making their decisions on using loans as funding. It may help companies avoid financial risk because of cash flow deficiency during the production and business activities. Furthermore, in respect of direction and strategies for making appropriate financial decisions, especially the decision on funding, companies should consider the effect of other factors such as profitability, company size, tangible fixed assets and market value/book value ratio because such factors have an impact on the level of debt utilization in capital structure of companies.

This finding provides additional empirical evidence concerning the impact of cash flow volatility and other factors on the level of debt utilization of companies listed on Vietnamese Securities Market. In further study, the authors will propose tests for different business lines and measure the cash flow volatility in different lags.

Acknowledgements The authors are thankful to Banking Academy and University of Transport Technology for funding this research. We would like to thank the referees for their helpful comments and suggestions.

References

1. Bates, T. W., Kahle, K. M., & Stulz, R. M. (2009). Why do U.S. Firms hold so much more cash than they used to? *The Journal of Finance, 64*, 1985–2021.
2. Denis, D. J., & McKeon, S. B. (2018). *Persistent operating losses and corporate financial policies*. Available at SSRN 2881584.

3. Douglas, A. V., Huang, A. G., & Vetzal, K. R. (2016). Cash flow volatility and corporate bond yield spreads. *Review of Quantitative Finance and Accounting, 46*(2), 417–458.
4. Dudley, E. & James, C.M. (2015). *Cash flow volatility and capital structure choice.* Available at SSRN 2492152. https://doi.org/10.2139/ssrn.2492152
5. Faris, M. A. M. (2011). Determinants of capital structure of Palestine-listed companies. *The Journal of Risk Finance, 12*(3), 226–241.
6. Fischer, E. O., Heinkel, R., & Zechner, J. (1989). Dynamic capital structure choice: Theory and tests. *The Journal of Finance, 44*(1), 19–40.
7. Harris, C., & Roark, S. (2019). Cash flow risk and capital structure decisions. *Finance Research Letters, 29*, 393–397.
8. Huang, R., & Ritter, J. R. (2021). Corporate cash shortfalls and financing decisions. *The Review of Financial Studies, 34*(4), 1789–1833.
9. Jensen, M. C., & Meckling, W. H. (1976). Theory of the firm: Managerial behavior, agency costs and ownership structure. *Journal of financial economics, 3*(4), 305–360. https://doi.org/10.1016/0304-405X(76)90026-X
10. Karimli, T. (2018). *The impact of cash flow volatility on corporate debt decisions.* Central European University.
11. Keefe, M. O. C., & Yaghoubi, M. (2016). The influence of cash flow volatility on capital structure and the use of debt of different maturities. *Journal of Corporate Finance, 38*, 18–36.
12. Modigliani, F., & Miller, M. H. (1958). The cost of capital, corporation finance and the theory of investment. *The American Economic Review, 48*(3), 261–297. https://www.jstor.org/stable/1809766
13. Modigliani, F., & M. Miller. (1963). Corporate income taxes and the cost of capital: A correction. *American Economic Review*, 53, 433–4. https://www.jstor.org/stable/1809167?seq=1
14. Myers, S. C. (1977). Determinants of corporate borrowing. *Journal of Financial Economics, 5*(2), 147–175. https://doi.org/10.1016/0304-405X(77)90015-0
15. Nguyen, H. Y. (2021). Mối quan hệ giữa biến dòng tiền và việc sửdụng nợ của các doanh nghiệpniêm yết tại 79–88.
16. Nguyen, K. V., & Dinh, T. T. T. (2016). Factors effect on capital structure the caseof delisted companies on the Vietnam stock market. *VNU Journal of Science: Economics and Business, 32*(5E).
17. Ross, S. A. (1977). The Determination of Financial Structure: The Incentive-Signalling Approach. *The Journal of Economics*, 8(1), 23–40. Retrieved from https://www.jstor.org/stable/3003485?seq=1
18. Vo, X. V. (2017). Determinants of capital structure in emerging markets: Evidence from Vietnam. *Research in International Business and Finance, 40*, 105–113.

Consumer's Online Shopping in COVID-19 Pandemic: Evidence from Vietnam

Ngo Van Tuan and Bui Huy Khoi

Abstract The paper was conducted to understand the factors affecting the online shopping intention of Vietnamese consumers in the Covid-19 pandemic. The official study carried out an online survey through Google forms using a questionnaire with the participation of 193 samples. The BIC Algorithm shows that 04 factors are affecting the online shopping intention of Vietnamese consumers in the Covid-19 pandemic (COSI), which are: (1) COVID-19 risk perception (CPR), (2) Perceived usefulness (PU), (03) Subjective norm (SN), and (4) Attitude (ATT). From the results, we have proposed many implications to help increase the online shopping intention of Vietnamese consumers in the Covid-19 pandemic. This study uses the optimal choice of the BIC Algorithm for the online shopping intention of consumers.

Keywords BIC algorithm · COVID-19 risk perception · Perceived usefulness · Subjective norm · Attitude · Online shopping intention

1 Introduction

Online shopping is becoming more and more popular and receiving great attention from people, especially after the outbreak of the Covid-19 epidemic, consumers are changing their traditional shopping habits to online consumption. This will create opportunities for domestic e-commerce businesses to develop online business activities and overseas commercial sites that are integrating into Vietnam. E-commerce websites are working very well in our country Tiki, Lazada, Shoppee, Sendo, Alibaba, ... and websites such as Facebook, Zalo, Tiktok, Instagram, ... According to the report of the Department of E-commerce and Business The digital economy [34] with 53% of the population taking part in online buying and selling, has made the

N. Van Tuan
Banking University of Ho Chi Minh City, Ho Chi Minh City, Vietnam
e-mail: tuannv@buh.edu.vn

B. Huy Khoi (✉)
Industrial University of Ho Chi Minh City, Ho Chi Minh City, Vietnam
e-mail: buihuykhoi@iuh.edu.vn

S. Sriboonchitta et al. (eds.), *Credible Asset Allocation, Optimal Transport Methods, and Related Topics*, Studies in Systems, Decision and Control 429,
https://doi.org/10.1007/978-3-030-97273-8_42

e-commerce market in Vietnam grow by 18% in 2020 and reach US$11.8 billion estimated to account for 5.5% of the total retail sales of consumer goods and services nationwide. In particular, in Hanoi and Ho Chi Minh City, e-commerce activities grew strongly and accounted for 70% of the Total E-commerce transactions of the country. Vietnamese people increasingly prefer to use services on the Internet, from commuting and eating to entertainment. It is revealed when the penetration rate of the Internet in Vietnam is up to 66%, of which 50.6% of Vietnam's population has engaged in online shopping [19].

The occurrence of Covid-19 has changed the trend of shopping consumption and is a very reasonable excuse for the online shopping environment to explode. The victory will belong to businesses if they know how to change quickly and seize opportunities quickly. A recent Nielsen survey released at the Online Marketing Forum 2020 showed that the number of consumers shopping online increased by 25%, while traditional channels such as supermarkets, markets, and groceries only increased by 25%. Because of the epidemic, many businesses and business households have completely converted their sales model to online or promoted online business more than before. Google's latest report at the end of September showed that the number of people moving to shops and entertainment areas decreased by 19% after Covid-19. Especially in February and March 2020, the number of people moving to restaurants, cafes, cinemas... decreased by 52%. When the pandemic is complicated in the world, all of them change their daily living habits. 82% said they shopped online during the period of social isolation, of which 98% said they continued to shop online even after the quarantine. The emergence of e-commerce to the way traditional e-commerce business operates then brings about fundamental changes to consumer expectations.

Ha and Thuy [16] studied the factors of attitude and subjective standards that have a positive impact on online shopping intention during the Covid-19 pandemic [6] has shown that perceived usefulness, perceived ease of use, attitudes, and subjective standards have an impact on consumers' intention to adopt information technology. Rezaee [40] based on the TAM extension technology acceptance model and showed that factors such as perceived usefulness have an impact on online shopping intention. Kamtarin [21] has pointed out that the factors of online trust, word of mouth, and online shopping intention have an impact on online shopping intention. This study uses the optimal choice of the BIC Algorithm for the online shopping intention of consumers in the context of the Covid-19 pandemic.

2 Literature Review

2.1 Consumer's Online Shopping Intention (COSI)

According to research by [26], online shopping behavior (known as online buying behavior, Internet shopping behavior) is the process of purchasing products and services over the Internet. According to the definition in the study of [2], online shopping is the behavior of consumers in purchasing products and services through Internet stores and websites using online shopping transactions. The intention is a factor used to assess the likelihood of future behavior. According to [1], the intention is a motivating factor and it motivates an individual to willingly perform the behavior. Therefore, [3] say that "Online shopping intention is the certainty of a consumer's ability to purchase over the Internet". Online purchase intention is the willingness to make an online purchase [26]. Online purchase intention is the context in which a customer indicates a willingness to make an online transaction [28].

2.2 *Perceived Ease of Use (PEU)*

Perceived ease of use is "The degree to which a person believes that using a particular system is effortless" [6]. Previous studies have shown that ease of use increases perceived usefulness and influences the purchase intention of consumers [6]. They found that perceived ease of use has a positive effect on trust as it helps to motivate customers to use the online shopping service for the first time and further. According to [6], perceived ease of use has a positive influence [47]. Perceived ease of use on the electronic system is very easy to use, easy operations do not require much intelligence of the customer using electronic systems [7].

2.3 *Attitude (ATT)*

According to the TPB model, attitude is an individual's assessment [11] of how good or bad a behavior is. When customers shop online, attitudes come from their evaluations in the process of purchasing goods and services from retail websites on aspects such as incentives and benefits for consumers [10], convenience, or compatibility of online services. Consumers' attitudes in online shopping are assessed to have a positive influence on their purchase intention and buying behavior. At the time of the COVID-19 outbreak, the Ministry of Industry and Trade said that revenue at traditional markets decreased by 50–80%, while revenue from online shopping via e-commerce sites of some businesses increased by 20 percent. −30% [34]. In Vietnam, the impact of the COVID-19 epidemic has also caused people to change their daily shopping behavior to concentrated shopping through the websites of supermarkets or convenience stores. In addition, in that context, many businesses also offer many incentives for customers such as discounts, promotions, free delivery, or shipping subsidies for orders far away. In the face of increasingly complicated developments of the epidemic along with encouraging policies from the State and businesses, people's online consumption attitudes have also changed more or less. Attitudes have a positive influence on online shopping intentions, consumers' attitudes when shopping online can be economical, convenient, easy, assurance of trust due to the preferential price of the product and the return policy of the seller [16].

2.4 *Perceived Usefulness (PU)*

Usefulness is defined as "the degree to which a person believes that using a particular system will increase their efficiency or productivity" [6]. Many studies have shown the relationship between perceived usefulness and user acceptance of information systems [15]. Therefore, the researchers suggest that consumers will shop online if they are aware that online shopping will help them achieve the desired performance [5]. McCloskey [33] also mentioned that consumers find that manipulation through applications or online helps them save time, reduces effort, and can be done anytime that they want, thereby demonstrating the usefulness of use. When using an application or service to fulfill a purpose, users often consider its usefulness first. According to [40], perceived usefulness has a positive influence on online shopping intention. According to [6], the perceived usefulness of electronic systems can save time and costs, easily make work more efficient and help consumers solve their consumer needs.

2.5 *Online Word of Mouth (WOM)*

Word of mouth (WOM) is the act of providing one consumer to another. According to [23], word-of-mouth is the verbal, person-to-person, recipient-to-transmitter communication related to a brand, product/service, or information on the market. Currently, with the development of information technology, the traditional form of a word of mouth (WOM) has gradually been replaced by online word of mouth (E-WOM). E-WOM is "Any positive or negative statements made by existing or previous potential customers about an application or product-related feature over the internet" [17, 29]. Compared to traditional, online word-of-mouth is considered more useful because of its speed, convenience, scalability, and lack of face-to-face elements [9]. Online word of mouth created and transmitted by customers themselves is always a reliable source of information for them [42]. According to Wu and [46], customers trust word of mouth about products more than other sources of information when making purchasing decisions. A recent study by the Society for New Communications Research shows the extent of word of mouth in online media. According to this study: 59.1% of the respondents use online means to vent about the customer service experience of a brand, 72.2% of the respondents, research translation online customer care service before buying. According to [21], online word of mouth has a positive influence on online shopping intention. According to [14], online word-of-mouth affects online shopping intention because consumers often discuss and recommend the price, quality, and diversity of online shopping channels.

2.6 Subjective Norm (SN)

According to the theory of TRA [1] or TAM [6], social influence is a determinant of behavioral intention as the subjective norm. Due to the specificity of online purchases, social influence is even more important for online purchase intention. Social influence not only affects online shopping intention but also has a significant impact on customer attitude [4]. The influence of people around such as family, friends, colleagues, celebrities, etc. has a direct or indirect impact on the trust of customers [13, 27]. According to research by Ajzen [1], the subjective norm is the individual's perception of surrounding influences on the performance of certain behavior. Many studies have divided the types of impact on consumers according to two aspects: Impact from reference group and impact from society. The influence from the reference group includes the influence of calling from friends, colleagues, and family, while the social impact refers to the information coming from the media such as newspapers, radio, television, and the advice of the experts. In the field of e-commerce, subjective norms reflect consumers' perceptions of the influence of the surrounding reference group and media on the ability to purchase online. In the context of the COVID-19 epidemic, people seriously staying at home, self-isolating, and minimizing non-essential activities outside may make it difficult for them to live and shop daily, but this will help reduce the burden on the current much-stressed health system. Within the scope of this study, the subjective normative factor is understood as the impact from the surrounding reference group, the impact from government policies, and the propaganda of traditional agencies. Davis [6] and Ha and Thuy [16] found that subjective norm has a positive influence on online shopping intention. According to Ha and Thuy [16], consumers' subjective standards are affected by the epidemic, so customers actively shop online, due to the government's call and media messages on social media.

2.7 Covid-19 Risk Perception (CPR)

People use their emotions to assess risk, prompt action, and focus thoughts. These emotional influences are generally beneficial, but can also be harmful [45]. The first particular responses to risky situations often differ from perceived assessments and eventually lead to perceived risk [43, 45]. People rely on their feelings as a substitute for other information, such as actual numerical risk. In this case, someone with a more negative perception during the pandemic would perceive a greater risk if they experienced less negative emotions [20]. Emotions when faced with a risky situation also influence the judgment in two stages [37]. First, the quality of emotions (Example: Positive and Negative) focuses decision-makers on congruent information. That information, rather than the sensation itself, is then used to guide judgment. For example, smokers exposed to more emotional health warnings have more negative feelings about the warnings and smoke, spend more time reviewing warnings, and recall more risks with subsequent effects on risk perception and intention to quit [45]. In the case of COVID-19, when negative emotions increase, people may seek or consider negative information about COVID-19 more than others. Emotions also act as a strong driver of behaviors [12], such as social distancing and frequent hand washing, as well as food hoarding and enforcement of government decisions. The function of emotion as a direct motivator also means that, with particularly strong responses, people often overlook important numerical information such as probabilities [41] of the extent of a problem topic [18] and the influence of time [36, 45]. Risk perception is understood as the degree to which the conscious mind is aware of the factors associated with potentially hazardous activities or technologies that define human risk judgments [35]. The importance of perceived risk to the economy has been studied very early hundreds of years ago. To date, there have been many studies that acknowledge the impact of risk perception on behavior in life, including consumer behavior. The measure of risk varies from condition to situation, and individuals' decision-making is determined by costs, benefits, harms, and risk preferences [8]. Previous studies have considered Risk perception as an important factor affecting purchase decisions [30] or purchase intention [31]. During the ongoing COVID-19 situation where negative emotions are on the rise, people may seek and consider negative information about COVID-19 more than others. Emotions also act as a strong driver of behaviors [12]. Applying this study, the author believes that risk perception with online shopping needs to be measured in a social dimension similar to [32], disease risk perception has an impact on shopping intention because customers are afraid of the possibility of an outbreak and are aware of the danger of disease spread.

3 Methodology

3.1 Sample Approach

The official quantitative study was carried out by surveying 193 customers who have ever online shopping in Ho Chi Minh City, Vietnam. Table 1 describes statistics of sample characteristics.

Collected data will be encrypted, entered (Table 1). Through the above statistical results, Table 1 shows that the number of women accounts for a higher proportion than men and accounts for 62.2% and while men only account for 37.8%. Table 1 shows that the group of consumers aged from 18 to 25 years old accounted for the highest proportion at 64.8%, followed by the age group from 26 to 30 years old accounting for 22.8% than from 31 to 35 years old accounts for 9.3% and the lowest age is over 35 years old accounts for 3%. Table 1 shows that the group of consumers with careers as students accounted for the highest percentage of 57%, followed by the group of consumers with occupations as officers account for 22.7%, then the consumer group with occupation as freelance accounts for 11.9% and finally the consumer group with

Table 1 Statistics of Sample

Characteristics		Amount	Percent (%)
Sex and age	Male	73	37.8
	Female	120	62.2
	18–25	125	64.8
	26–30	44	22.8
	31–35	18	9.3
	Above 35	6	3.1
Income/month	Below 5 million VND	65	33.7
	5–10 million VND	72	37.3
	11–15 million VND	38	19.7
	Over 15 million VND	18	9.3
Job	Student	110	57.0
	Officer	40	20.7
	Freelance	23	11.9
	Business executives	20	10.4
Shopping/month	Below 3 times/month	106	54.9
	3–5 times/month	52	26.9
	5–7 times/month	21	10.9
	Over 7 times/month	14	7.3

business executive occupation accounts for the lowest 10.4%. Table 1 shows that the frequency of consumers shopping less than 03 times/month accounts for the highest rate of 54.9%, followed by the frequency of shopping from 03 to 05 times/month accounting for a percentage of 26.9%, then the frequency of shopping from 05 to 07 times/month accounts for 10.9% and the frequency of shopping over 07 times/month accounts for the lowest rate of 7.3%. Factors and items are in Table 2.

For the duration of the study, all study staff and respondents were blinded. No one from the outside world had any contact with the study participants.

3.2 Bayesian Information Criteria

In Bayesian statistics, the prior information constitutes a theoretical basis, and the conclusions derived from prior information are combined with observed data [44].

Table 2 Factor and Item

Factor	Mean	Item
PEU	4.1645	I find online shopping very easy to use
		Online shopping doesn't require much of my mental effort
		I easily perform operations when shopping online
		Because it's easy to use, I'm good at shopping online
ATT	4.0352	I save money because online shopping channels have preferential rates
		Online shopping is convenient because there are many incentives for freight forwarding
		I feel easy because there are many choices of online shopping
		I believe in return policies on online shopping
		I feel safe because I limit contact with people during the COVID-19 epidemic season
PU	4.1839	Online shopping helps me save more time
		Online shopping makes it easier for me to choose products
		I can buy essentials through online shopping
		Online shopping helps me save a lot of moving costs
		My actions on the online shopping channel help me shop more efficiently
WOM	3.9959	I often recommend reputable online shopping channels to everyone
		I often discuss the price of online shopping channels with people
		I often discuss product quality on the online shopping channel
		I often discuss product variety on the online shopping
		Word-of-mouth influences on the online shopping intentions
SN	3.9767	Because of the severity of the epidemic, I actively used the form of the online shopping
		Because of the severity of the epidemic, I have to give advice
		Because of the influence of people around me, I switched to the online shopping
		Because of government propaganda, I chose the form of the online shopping
		Because of the anti-epidemic messages in the media, I chose the online shopping
CPR	4.1209	I fear the possibility of a resurgence of the Covid-19 pandemic
		I am terrified of being infected with Covid-19
		I am aware of the danger of an outbreak of Covid-19
COSI	4.0699	I am planning to shop online during the epidemic season
		I plan to shop more online in the future
		I will recommend friends and family to use online shopping
		I will continue to shop online regularly

The Bayesian perspective interprets probability as the information about uncertainty, i.e., probability quantifies the (un)certainty of information [25]. Therefore, the Bayesian approach is used more and more, especially in the field of social sciences. With the strong development of data science, big data as well as computational software, Bayesian statistics became a favorite tool [24]. The BIC is an important and useful metric for determining a full and straightforward model. A model with a lower BIC is chosen based on the BIC information standard. When the minimal BIC value is reached, the best model will come to an end [22].

4 Results

4.1 BIC Algorithm

BIC (Bayesian Information Criteria) was used to choose the best model for R software. BIC has been used in the theoretical context for model selection. As a regression model, BIC can be applied, estimating one or more dependent variables from one or more independent variables [39]. An essential and useful measurement for deciding a complete and straightforward model is the BIC. Based on the BIC information standard, a model with a lower BIC is selected. The best model will stop when the minimum BIC value [22, 38, 39]. R report shows every step of searching for the optimal model. BIC selects the most suitable 3 of 64 models in Table 3.

There are six independent and one dependent variable. Attitude (ATT), perceived usefulness (PU), subjective norm (SN), COVID-19 risk perception (CPR) influence Consumer's Online Shopping intention (COSI) with a probability of 100%. Perceived ease of use (PEU) and online word of mouth (WOM) have a probability of 12.5 and 14%.

Table 3 BIC model selection

COSI	Probability (%)	SD	model 1	model 2	model 3
Intercept	100.0	0.19462	−0.04886	−0.10065	−0.16874
PEU	12.5	0.02293			0.05348
ATT	100.0	0.05663	0.23608	0.22371	0.22042
PU	100.0	0.06786	0.29677	0.27138	0.27899
WOM	14.0	0.03451		0.07830	
SN	100.0	0.04900	0.19965	0.18034	0.20626
CPR	100.0	0.05406	0.27435	0.26751	0.27640

Table 4 Model test

Model	nVar	R^2	BIC	Post prob
model 1	4	0.748	−245.03514	0.735
model 2	5	0.751	−241.71286	0.140
model 3	5	0.750	−241.48856	0.125

BIC = -2 * LL + log(N) * k

4.2 Model Evaluation

According to the results from Table 4, BIC shows model 1 is the optimal selection because BIC (−245.03514) is minimum. Attitude (ATT), Perceived usefulness (PU), Subjective norm (SN), COVID-19 risk perception (CPR) impact Consumer's Online Shopping intention (COSI) is 74.8% in Table 4. BIC finds model 1 is the optimal choice and four variables have a Probability of 73.5%. The above analysis shows the regression equation below is statistically significant.

$$\begin{aligned} \mathbf{COSI} = -\mathbf{0.04886} + \mathbf{0.23608ATT} + \mathbf{0.29677PU} \\ + \mathbf{0.19965SN} + \mathbf{0.27435CPR} \end{aligned}$$

4.3 Discussion

4.3.1 Attitude (ATT)

Return policies when shopping online must bring trust to customers and bring benefits to customers such as product information being accurate, clear, transparent, and convenient for exchange. Return the product when the customer is not satisfied with the product. Offer reasonable delivery and return policies for consumers and attractive product replacement policies. Use the media to relieve the fear of consumers when shopping online. In order to attract more customers to shop online during the epidemic season, e-commerce sites need to improve and consider prices to give customers the feeling of saving more when shopping online: Offering diverse, rich, and special promotions and massive advertising on media pages create good psychological effects for customers so that customers make decisions to choose to shop online. To promote customers to make online shopping intentions, businesses need to offer many preferential policies on goods delivery and delivery in the form of Payment via Momo e-wallet, payment via bank card customers, reduce product prices with the corresponding value of products, and apply discount codes to benefit customers. In order for customers to make online shopping intentions, e-commerce sites need to have a variety of product designs, so that customers can search and choose products faster and easier. Provide a full range of products according to customer needs. In the context of the current pandemic, limiting human-to-human contact is quite important to limit the spread of the disease, so businesses need to let customers feel safe when switching from buying. From traditional shopping to online shopping, limit going out to limit contact. To influence customers to shop online, businesses need to promote messages during the epidemic season to create a strong viral effect to encourage customers to shop online.

4.3.2 Perceived Usefulness (PU)

Online shopping by smart electronic means on computers and smartphones makes customers aware that online shopping will limit travel compared to going out of the store to shop. Save more costs so businesses promote messages using smart electronic means such as smartphones, computers to change consumers' habits. At that time, customers will have easier access to online shopping. The use of electronic systems in the media will make it easier for customers to make online shopping, so businesses need to offer simple and fast shopping operations. Make it easy for customers to easily search, choose and make online shopping decisions. Customers feel that shopping online by smart means will save more time. Therefore, businesses need to promote programs that bring benefits to customers so that customers continue to shop online, which will save more time. The use of smart means for online shopping is being most satisfied and appreciated by customers, but administrators also need to learn more about customers' needs to be able to meet the needs of customers client.

4.3.3 Covid-19 Risk Perception (CPR)

The current epidemic situation is very complicated and spreads rapidly because customers are very afraid of the outbreak of the disease again, so it is necessary to grasp customer psychology and encourage customers to limit contact instead. Various forms of shopping to prevent the spread of disease. In order for people to be aware of the danger, it is expected that the government will always assess and update the disease risk level so that people can be aware of the danger level as the pandemic outbreak continues. Therefore, businesses need to use this leverage to encourage customers to limit contact so as not to be infected with the disease.

4.3.4 Subjective Norm (SN)

In the context of the current pandemic, in order for customers to actively switch to online shopping, businesses need to step up anti-epidemic campaigns so that customers can understand the danger of the disease. Businesses need to promote and promote anti-epidemic messages to protect the health of the community and online shopping intentions in the following forms: Outlining the severity of the epidemic in the mass media and social media. Promote the spirit of family and friends together to protect the health of the epidemic by shopping online at home. Offer more incentive programs to motivate consumers to shop more online.

5 Conclusions

The aim of the study shows the optimal choice by the BIC Algorithm for the online shopping intention of consumers. We have provided the theoretical bases to build the proposed research model for the topic. The study was carried out according to qualitative and quantitative research methods, including 7 scales and 32 observed variables. Regarding the online shopping intention of consumers in Ho Chi Minh City, Vietnam in the current Covid-19 pandemic, this study has shown that the online shopping intention in the Covid-19 pandemic is at a relatively dark level, above the average (mean value = 4.0699). Thus, administrators need solutions to improve the online shopping intention of consumers in the Covid-19 pandemic more shortly. On the factors affecting the online shopping intention of consumers in Ho Chi Minh City in the Covid-19 pandemic. BIC Algorithm shows that 4 components are affecting online shopping intention from the strongest to the weakest: The factor that has the strongest impact is perceived usefulness (PU) ($\beta = 0.29677$), the second strongest factor is the COVID-19 perceived risk (CPR) ($\beta = 0.27435$), the third strongest factor is the attitude (ATT) ($\beta = 0.23608$), and last the same is the subjective norm (SN) ($\beta = 0.19965$).

In previous studies, the variable perceived risk always negatively affects attitude and cognitive behavior, and intention. In contrast, in this study, the results of the regression analysis show that the independent variable "COVID-19 perceived risk" (CPR) has a positive impact on the dependent variable of consumers' intention to shop online during the Covid-19 pandemic.

6 Limitations

The paper "BIC Algorithm for Consumer's Online Shopping in COVID-19 Pandemic: Evidence from Vietnam" only stops at the factors affecting the online shopping intention of consumers in Ho Chi Minh City, Vietnam in the Covid-19 pandemic, represented as Attitude (ATT), Perceived usefulness (PU), Subjective norm (SN), COVID-19 risk perception (CPR), Perceived ease of use (PEU), and Online word of mouth (WOM).

Because of the situation of the COVID-19 epidemic, we limited the survey to many places. But we have completed a survey for 193 questionnaires to conduct an online survey. Future studies can be carried out with a larger sample, with a larger population, not only in Ho Chi Minh City but also in other big cities or provinces in Vietnam such as Ha Noi, Can Tho, Da Nang, Can Tho, and other places across the country from which it is possible to research and gather more insights and to recommend helping managers overcome the limitations.

Acknowledgements This research is funded by the Industrial University of Ho Chi Minh City, Vietnam.

References

1. Ajzen, I. (1991). The theory of planned behavior. *Organizational Behavior and Human Decision Processes, 50*, 179–211.
2. Al-Matarneh, N. S. (2016). Online Shopping Risks and Effect on Consumers Purchasing Behavior. *International Journal of Online Marketing Research, 2*, 62–69.
3. Aldousari, A. A., Delafrooz, N., Ab Yajid, M. S., & Ahmed, Z. U. (2016). Determinants of consumers' attitudes toward online shopping. *Journal of Transnational Management, 21*, 183–199.
4. Barkhi, R., Belanger, F., & Hicks, J. (2008). A model of the determinants of purchasing from virtual stores. *Journal of Organizational Computing and Electronic Commerce, 18*, 177–196.
5. Datta, A. & Acharjee, M. K. (2018). Consumers attitude towards online shopping: factors influencing young consumers to shop online in Dhaka, Bangladesh. *International Journal of Management Studies.*
6. Davis, F. D. (1989). Perceived usefulness, perceived ease of use, and user acceptance of information technology. *MIS quarterly*, 319–340.
7. Denaputri, A., & Usman, O. (2019). Effect of perceived trust, perceived security, perceived usefulness and perceived ease of use on customers' intention to use mobile payment. *Perceived Security, Perceived Usefulness and Perceived Ease of Use on Customers' Intention to Use Mobile Payment (December 16, 2019).*
8. Dowling, G. R., & Staelin, R. (1994). A model of perceived risk and intended risk-handling activity. *Journal of Consumer Research, 21*, 119–134.
9. Eisingerich, A. B., Chun, H. H., Liu, Y., Jia, H. M., & Bell, S. J. (2015). Why recommend a brand face-to-face but not on Facebook? How word-of-mouth on online social sites differs from traditional word-of-mouth. *Journal of Consumer Psychology, 25*, 120–128.
10. Ene, S., & Özkaya, B. (2013). A study regarding the attitudes that are effective on the price perception of consumers that shop from the retail stores. *Mediterranean Journal of Social Sciences, 4*, 451.
11. Fan, W., Haung, Y., Hsu, H., & Chen, C. (2013). An Analysis of the blog-user attitude employing structural equation modeling combine TAM and TPB Model. *Applied Mechanics and Materials. Trans Tech Publ*, 90–93.
12. Frijda, N. H. (1986). *The emotions*. Cambridge University Press.
13. Gefen, D., & Straub, D. (2000). The relative importance of perceived ease of use in IS adoption: A study of e-commerce adoption. *Journal of The Association for Information Systems, 1*, 1–30.
14. Goyette, I., Ricard, L., Bergeron, J., & Marticotte, F. (2010). e-WOM scale: Word-of-mouth measurement scale for e-services context. *Canadian Journal of Administrative Sciences/Revue Canadienne des Sciences de l'Administration, 27*, 5–23.
15. Grover, P., Kar, A. K., Janssen, M., & Ilavarasan, P. V. (2019). Perceived usefulness, ease of use and user acceptance of blockchain technology for digital transactions–insights from user-generated content on Twitter. *Enterprise Information Systems, 13*, 771–800.
16. Ha, N. T., & Thuy, H. D. L. (2020). Consumer behaviour towards Vietnamese online shopping websites in the Covid-19 pandemic. *VNU Journal of Science: Economics and Business,* 36.
17. Hennig-Thurau, T., Gwinner, K. P., Walsh, G., & Gremler, D. D. (2004). Electronic word-of-mouth via consumer-opinion platforms: What motivates consumers to articulate themselves on the internet? *Journal of Interactive Marketing, 18*, 38–52.

18. Hsee, C. K., & Rottenstreich, Y. (2004). Music, pandas, and muggers: On the affective psychology of value. *Journal of Experimental Psychology: General, 133*, 23.
19. InvestmentNewspaper. (2021). *E-commerce boom in 2021 (Vietnamese)* (Online). Available: https://baodautu.vn/thuong-mai-dien-tu-bung-no-trong-nam-2021-d137452.html
20. Johnson, E. J., & Tversky, A. (1983). Affect, generalization, and the perception of risk. *Journal of Personality and Social Psychology, 45*, 20.
21. Kamtarin, M. (2012). The effect of electronic word of mouth, trust and perceived value on behavioral intention from the perspective of consumers. *International Journal of Academic Research in Economics and Management Sciences, 1*, 56.
22. Kaplan, D. (2021). On the quantification of model uncertainty: A Bayesian perspective. *Psychometrika, 86*, 215–238.
23. Kirby, J., & Marsden, P. (2006). *Connected marketing: The viral, buzz and word of mouth revolution*. Elsevier.
24. Kreinovich, V., Thach, N. N., Trung, N. D., & Van Thanh, D. (2018). *Beyond traditional probabilistic methods in economics*. Springer.
25. Kubsch, M., Stamer, I., Steiner, M., Neumann, K., & Parchmann, I. (2021). Beyond p-values: Using Bayesian data analysis in science education research. *Practical Assessment, Research, and Evaluation, 26*, 4.
26. Li, N., & Zhang, P. (2002). Consumer online shopping attitudes and behavior: An assessment of research. *AMCIS 2002 proceedings*, 74.
27. Lin, H.-F. (2007). Predicting consumer intentions to shop online: An empirical test of competing theories. *Electronic Commerce Research and Applications, 6*, 433–442.
28. Ling, K. C., Chai, L. T., & Piew, T. H. (2010). The effects of shopping orientations, online trust and prior online purchase experience toward customers' online purchase intention. *International Business Research, 3*, 63.
29. Litvin, S. W., Goldsmith, R. E., & Pan, B. (2008). Electronic word-of-mouth in hospitality and tourism management. *Tourism Management, 29*, 458–468.
30. Lo, A. Y. (2013). The role of social norms in climate adaptation: Mediating risk perception and flood insurance purchase. *Global Environmental Change, 23*, 1249–1257.
31. Lobb, A. E., Mazzocchi, M., & Traill, W. (2007). Modelling risk perception and trust in food safety information within the theory of planned behaviour. *Food Quality and Preference, 18*, 384–395.
32. Long, N. N., & Khoi, B. H. (2020). An empirical study about the intention to hoard food during COVID-19 pandemic. *Eurasia Journal of Mathematics, Science and Technology Education,* 16, em1857.
33. McCloskey, D. W. (2006). The importance of ease of use, usefulness, and trust to online consumers: An examination of the technology acceptance model with older customers. *Journal of Organizational and End User Computing (JOEUC), 18*, 47–65.
34. MinistryofIndustryandTrade. (2020). *Vietnam e-commerce white book (Vietnamese)* (Online). Available: http://thongketmdt.dvctt.gov.vn/NewsDetail.aspx?CateAlias=tin-tuc&DocId=3
35. Oltedal, S., Moen, B.-E., Klempe, H., & Rundmo, T. (2004). Explaining risk perception: An evaluation of cultural theory. *Rotunde, 85*, 1–33.
36. Peters, E., Kunreuther, H., Sagara, N., Slovic, P., & Schley, D. R. (2012). Protective measures, personal experience, and the affective psychology of time. *Risk Analysis: An International Journal, 32*, 2084–2097.
37. Peters, E., Lipkus, I., & Diefenbach, M. A. (2006). The functions of affect in health communications and in the construction of health preferences. *Journal of Communication, 56*, S140–S162.
38. Raftery, A. E. (1995). Bayesian model selection in social research. *Sociological Methodology*, 111–163.
39. Raftery, A. E., Madigan, D., & Hoeting, J. A. (1997). Bayesian model averaging for linear regression models. *Journal of the American Statistical Association, 92*, 179–191.
40. Rezaee, H. S. (2011). *The flat analysis of properties and the unity of science*. Springer.

41. Rottenstreich, Y., & Hsee, C. K. (2001). Money, kisses, and electric shocks: On the affective psychology of risk. *Psychological Science, 12*, 185–190.
42. Siddiqui, M. S., Siddiqui, U. A., Khan, M. A., Alkandi, I. G., Saxena, A. K., & Siddiqui, J. H. (2021). Creating electronic word of mouth credibility through social networking sites and determining its impact on brand image and online purchase intentions in India. *Journal of Theoretical and Applied Electronic Commerce Research, 16*, 1008–1024.
43. Slovic, P., Finucane, M. L., Peters, E., & MacGregor, D. G. (2013). *Risk as analysis and risk as feelings: Some thoughts about affect, reason, risk and rationality*. Routledge.
44. Thach, N. N. (2020). How to explain when the ES is lower than one? A Bayesian nonlinear mixed-effects approach. *Journal of Risk and Financial Management, 13*, 21.
45. Van Bavel, J. J., Baicker, K., Boggio, P. S., Capraro, V., Cichocka, A., Cikara, M., Crockett, M. J., Crum, A. J., Douglas, K. M., & Druckman, J. N. (2020). Using social and behavioural science to support COVID-19 pandemic response. *Nature Human Behaviour, 4*, 460–471.
46. Wu, P. C., & Wang, Y. C. (2011). The influences of electronic word-of-mouth message appeal and message source credibility on brand attitude. *Asia Pacific Journal of Marketing and Logistics*.
47. Yuannitha, I. I. (2018). *Influence of trust, perceived usefulness, perceived ease of use, and attitude on purchase intention (case studies in Padang)*. Universitas Negeri Padang.

The Role of Bond Yield in Financial Asset Markets: Application of the Regression Kink Model

Chaiwat Klinlampu, Piangtawan Polard, and Woraphon Yamaka

Abstract This study aims at exploring the nonlinear impact of bond yield on six major financial markets, consisting of gold, crude oil, the stock price index which comprise the MSCI Europe Index and the MSCI Asia Index, the US$ Index, and bitcoin. In addition, this study also aims to reveal the turning point of the relationship between bond yield and the six financial markets. To explore these issues, this study utilizes the kink regression model. The results confirm the presence of the kink effect in the linkage between bond and other six financial markets and also show that the negative impacts of the bond market have a clearer effect than the positive ones on bitcoin, gold, MSCI Asia, and USD markets. This implies bond is a hedge against those markets. Finally, we also find that the point where there will be a new form of change (turning points) of bond against gold, crude oil, MSCI EU, MSCI Asia, and bitcoin are 0.436, 0.910, 0.555, 0.608, 0.878, and 0.496, respectively.

Keywords Bond yield · Kink regression · Stock

1 Introduction

The analysis of the correlation between bond yields and financial asset returns has been the focus of extensive literature in financial economics since the sign of this relationship plays a vital role in asset allocation, portfolio optimization, and risk management (see, e.g., [2, 4, 8, 13]). Pastpipatkul, Yamaka, and Sriboonchitta (see, e.g., [14]) revealed that bond yields are currently the most important influence on price dynamics and on the value of returns on other assets.

C. Klinlampu · W. Yamaka (✉)
Center of Excellence in Econometrics, Chiang Mai University, Chiang Mai 50200, Thailand
e-mail: woraphon.econ@gmail.com

P. Polard
Faculty of Management Sciences, Chiang Mai Rajabhat University, Chiang Mai 50200, Thailand
e-mail: piangtawan@yahoo.co.th

S. Sriboonchitta et al. (eds.), *Credible Asset Allocation, Optimal Transport Methods, and Related Topics*, Studies in Systems, Decision and Control 429,
https://doi.org/10.1007/978-3-030-97273-8_43

Although the relationship between bond and other financial assets has been discussed and explored in the literature, an inconclusive result has been found. Lin et al. (see, e.g.,[9]) found that the relationship between stock and bond is significantly negative in the long term; however, in periods of market turmoil or crisis, this relationship tends to reverse and present a positive correlation (see, e.g., [1]). Johansson (see, e.g., [7]) found the relationship between stock and bond markets in many countries and confirmed that the stock and bond nexus will become stronger during periods of instability as investors shift their capital from high-risk assets to low-risk assets. Baele et al. (see, e.g., [3]) also provided pieces of convincing evidence that such a stronger negative relationship reflects the “flight-to-quality" phenomenon, characterized by a movement of funds from risky assets to safer assets during crisis periods.

In addition to the stock market, the change of bond yield will also result in greater or worse values of major financial assets around the world, such as those in the currency, cryptocurrency, gold, and oil markets. According to Naeem, Adekoya, and Oliyide (see, e.g., [12]), their investigation on the asymmetric spillovers between bond and commodity revealed that gold and silver have the strongest connectedness with green bonds, but the magnitude of connectedness is varying over time. Crude oil was also observed to have a strong connection with bond in the long term (see, e.g., [6, 16]). However, the relationship between bond yield and gold may sometimes be inverse (see, e.g., [16]).

As mentioned above, the inconclusive results were found in the literature, and we can see that bond has played different roles in the financial markets. Although the literature has investigated bond's ability to act as a hedge, as a safe haven, and as a means of diversification in its association with other classes of financial assets, to the best of our knowledge, little research discusses the turning points of the relationship between bond and financial assets. As many studies indicate an either positive or negative relationship between bond and financial markets in some periods of time, it is worth identifying the turning point of the impact of bond on financial markets. This finding will guide investors in planning their investment and making investment decisions when bond yields fluctuate.

To achieve our goal, this study applies a kink regression model (see, e.g., [5, 15]) to test the nonlinear effect and explore the turning or kink point of the impact of bond on six major assets (gold, crude oil, the MSCI Europe(EU), the MSCI Asia, US$ Index (USD), and bitcoin). In addition, we use the estimated slope coefficients before and after the kink point to analyze whether or not bond is a hedging asset or a safe haven for other assets.

The rest of the paper is organized as follows. Kink regression is described in Sect. 2. Data description is provided in Sect. 3. Empirical results are described and analyzed in Sects. 4, and 5 presents conclusions.

2 The Regression Kink Model

The model can be written as

$$y_t = \beta_1 (x_t - \gamma)^- + \beta_2 (x_t - \gamma)^+ + e_t, \quad t = 1, ..., T \tag{1}$$

where y_t, x_t and e_t are dependent variable, independent variable, and the error term, respectively. β_1, β_2 are the estimated coefficients and γ is the turning or kink point of the relationship between y_t and x_t. The negative and positive parts of $(x_t - \gamma)^-$ and $(x_t - \gamma)^+$ are defined as $\min\left[(x_t - \gamma), 0\right]$ and $\max\left[(x_t - \gamma), 0\right]$, respectively. When the value of x_t is less than that of γ, the slope with respect to x_t equals β_1 and when the value of x_t greater than that of γ, the slope with respect to x_t equals β_2. To estimate this model, we can use least squares estimations as follows: Let

$$X_t(\gamma) = \begin{pmatrix} (x_t - \gamma)^- \\ (x_t - \gamma)^+ \end{pmatrix} \tag{2}$$

Eq. 1 can be rewritten as

$$y_t = \beta' X_t(\gamma) + e_t \tag{3}$$

and the loss function can be obtained by

$$S_T(\beta, \gamma) = \frac{1}{T} \sum_{t=1}^{T} (y_t - \beta' X_t(\gamma))^2 \tag{4}$$

Then, $(\hat{\beta}, \hat{\gamma})$ is estimated by minimizing the joint loss function

$$(\hat{\beta}, \hat{\gamma}) = \underset{\beta \in \mathbb{R}, \gamma \in \Gamma}{\arg\min} \; S_T(\beta, \gamma) \tag{5}$$

According to Hansen (see, e.g., [5]), the loss function $S_n(\beta, \gamma)$ is quadratic in β, but it is non-convex in γ. Thus, we use a combination of concentration and grid search as

$$\begin{aligned} \hat{\gamma} &= \underset{\gamma \in \Gamma}{\arg\min} \; S_T^*(\tilde{\beta}, \gamma) \\ \tilde{\beta} &= \underset{\beta \in \mathbb{R}}{\arg\min} \; S_T(\beta, \tilde{\gamma}), \end{aligned} \tag{6}$$

In this estimation, the optimal $\hat{\gamma}$ can be found numerically by a grid search on γ. After $\hat{\gamma}$ is found, the estimated $\tilde{\beta}$ by standard least squares. The estimator could be iteratively updated.

3 Data Description

This study uses bond yield, bitcoin, Crude oil, Gold, MSCI EU Index, MSCI Asia Index, and US dollar Index covering the period of January 2011-March 2021. All variables are transformed into log-difference (except for bond yield). The description of each variable is presented as follows:

Bond yield (%): It is the return on investment that an investor would expect from holding different ages of government bonds. In this study, we consider US 10-year government treasury bond yield as a proxy.

Bitcoin: It is a digital currency or cryptocurrency that was developed in January 2009. It is a type of monetary unit. There are no physical bitcoins; only balances are stored on a public ledger to which anyone can access.

Crude oil: This study considers Brent crude oil price as it is one of the vital oil benchmark prices in the world.

Gold: This study considers spot gold price or closing price of AU99 corresponding to gold with a purity of 99.95

MSCI Index: MSCI EU and MSCI Asia are considered representative indicators of the European and Asian stock markets, respectively. MSCI EU captures large and mid-cap representation across 15 developed markets in Europe, while MSCI Asia captures large and mid-cap representation across developed markets and emerging markets in Asia.

US dollar Index (USD): The US dollar index measures the value of the US dollar about a basket of currencies representing the majority of the US's major trading partners. This index is comparable to other trade-weighted indexes, which use the same major currency exchange rates.

4 Result

All statistical data descriptions of the variables are shown in Table 1. The unit root test (for examine the data's stationarity) is also conducted to investigate the stationarity of our return series. Table 1 shows average, minimum, maximum, standard deviation, skewness, and kurtosis of daily bond yield and other financial returns. We can observe that MSCI EU performs the highest mean return, while bond presents the lowest mean return. Bitcoin is more volatile than other assets, while USD exhibits the lowest standard deviation. Considering the skewness values, we can see that bond, bitcoin, crude oil, and USD show a negative skewness, indicating the frequent small gains and a few large losses of these returns. Conversely, Gold, MSCI EU, and MSCI Asia present a positive skewness, indicating a higher gain of these markets during January 2011-March 2021.

This study first explores the linear link between the bond and other financial assets. The results are reported in Table 2. β_0 and β_1 are the intercept term and the impact of bond on other financial assets. The regression coefficients of are 0.537,

Table 1 Descriptive statistics

Variable	Mean	Max	Min	Std. Dev.	Skewness	Kurtosis	Unit root test
Bond	0.687	1.244	−0.629	0.385	−1.489	5.215	−0.054[0.053]
Bitcoin	6.227	10.981	−1.217	2.837	−0.618	2.520	−0.024[0.023]
Crude oil	4.259	4.835	3.124	0.390	−0.284	2.317	−0.058[0.031]
Gold	7.239	7.588	6.967	0.153	0.493	2.145	−1.095[0.000]
MSCI EU	7.362	7.553	7.060	0.104	0.560	2.822	−0.095[0.020]
MSCI Asia	4.875	5.264	4.600	0.147	0.409	2.800	−0.964[0.000]
USD	4.492	4.627	4.289	0.094	−0.493	1.767	−1.073[0.000]

Note [ ] denotes the Maximum Bayes Factor (see, e.g.,[10])

Table 2 Linear regression results

Parameter	Crude oil	Gold	MSCI EU	MSCI Asia	USD index	USD
β_0	3.889***	7.409***	7.320***	4.924***	4.539***	7.690***
	(0.061)	(0.022)	(0.018)	(0.026)	(0.016)	(0.504)
β_1	0.537***	−0.246***	0.061***	−0.071***	−0.0673***	−2.127***
	(0.077)	(0.028)	(0.023)	(0.0341)	(0.021)	(0.639)
AIC	−265.518	−514.826	−555.117	−467.819	−582.832	252.821

Note ***, **, * are decisive ($MBF < 0.001$), strong ($MBF < 0.01$), and moderate ($MBF < 0.1$) evidence supporting the impact of bond yield, respectively

−0.246, 0.061, −0.0711, −0.0673, −2.127 for crude oil, gold, MSCI EU, MSCI Asia, USD, and Bitcoin, respectively. The results show that the impact of bond is negative for virtually all assets, except for crude oil and MSCI EU. This indicates that the higher bond yield could generate lower returns of gold, MSCI Asia, USD, and bitcoin. Interestingly, bond plays an important role in all six assets with decisive evidence. However, this study expects that it might have a nonlinear effect of bond on other assets. Thus, the kink regression model is utilized and compared with the linear regression result. The Akaike information criterion (AIC) is used to compare the performance between linear and kink regression models. Our comparison shows that the AIC of the kink regression model is lower than that of the linear regression for all assets, indicating strong evidence of the nonlinear impact of bond on the other six financial assets. Therefore, we report the estimation results of the kink regression model in Table 3.

According to Table 3, we find strong evidence of kink effects for all pairs (γ_1),implying the nonlinear impact of bond on six financial returns. Considering the regime-dependent coefficients, we can see a negative effect of bond on gold, MSCI Asia, and bitcoin before the turning point (γ_1), while there is weak evidence after the turning point. In contrast, the result shows a negative effect of bond on USD when the bond yield is greater than 0.878%, but there is no evidence supporting the impact of bond on USD when the bond yield is lower than 0.878%. These results imply that if bond yield increases, investors will move their investments away from

Table 3 Kink regression results

Parameter	Oil	Gold	MSCI EU	MSCI Asia	USD	Bitcoin
β_0	4.1825***	7.1503***	7.308***	4.808***	4.501***	5.7414***
	(0.134)	(0.102)	(0.143)	(0.110)	(0.097)	(0.120)
β_1^-	0.694**	−0.306*	−0.052	−0.239***	−0.0302	−4.423***
	(0.326)	(0.168)	(0.328)	(0.079)	(0.201)	(0.301)
β_1^+	0.392**	0.256	0.218***	0.212	−0.317***	0.452**
	(0.109)	(1.017)	(0.032)	(0.423)	(0.107)	(0.227)
γ_1	0.436***	0.910***	0.555***	0.608**	0.878****	0.496***
	(0.080)	(0.213)	(0.154)	(0.302)	(0.200)	(0.007)
AIC	−266.968	−526.903	−568.139	−485.069	−589.577	246.905

Note ***, **, * are decisive ($MBF < 0.001$), strong ($MBF < 0.01$), and moderate ($MBF < 0.1$) evidence supporting the impact of bond yield, respectively

bitcoin, gold, MSCI Asia, and USD. These results are consistent with the findings obtained from linear regression. However, the negative impact of bond is exhibited regardless whether the bond return is greater or lower than the level at the kink point

Considering the oil and bitcoin returns, we find a decisive impact of bond on oil and bitcoin returns for both regimes. The result shows a positive impact of bond on oil return in both two regimes, while the impact of bond on bitcoin is positive when the bond yield is less than 0.496%, but it will reverse to positive when bond yield is larger than 0.496%.

To have a better picture of the nonlinear effects, we illustrate the kink-fitted line in Fig. 1. The red line represents the linear regression line, while the black dashed line presents the kink-fitted line. Clearly, we can observe the structural change of the impact of bond yield on all six financial returns, confirming the robustness results from the kink regression model.

5 Conclusion

The study examines the nonlinear impact of bond yield on six financial market returns: crude oil, gold, MSCI EU Index, MSCI Asia Index, USD Index, and bitcoin. We utilize the kink regression to examine this nonlinear relationship and explore the turning point of the relationship. Our results show that the impact of bond yield on six financial market return is nonlinear as the AIC of the kink regression is lower than that of the linear regression model. We also find that the kink points are 0.436%, 0.910%, 0.555%, 0.608%, 0.878%, and 0.496%, respectively, for crude oil, gold, MSCI EU, MSCI Asia, USD and bitcoin.

Bond yield is noted to play an important role in oil and bitcoin markets as the coefficients of regimes 1 and 2 are statistically significant. Interestingly, the coefficients of these two regimes are quite different in terms of sign and magnitude of the impact

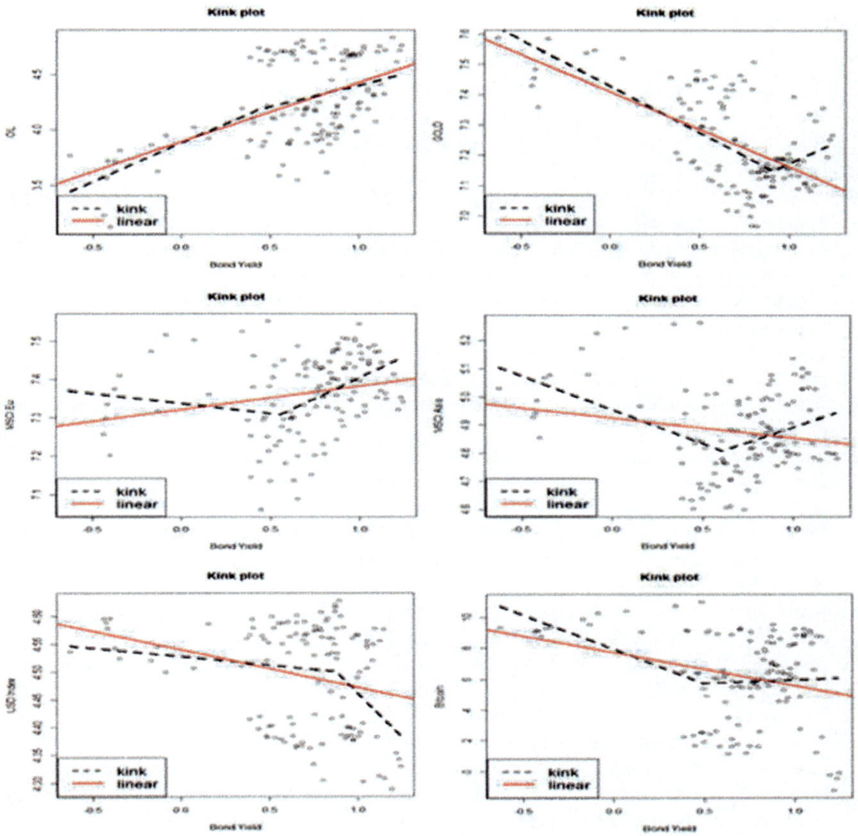

Fig. 1 The relationship between bond yield and other variables from the kink regression model

of bond across different financial markets. These results confirm the usefulness of the nonlinear analysis of this study. As the negative impact of bond yield is found in all markets (except for oil), this indicates that bond could act as a safe haven asset of the financial markets.

We suggest exploring the higher number of kink points in the relationship between bond and other financial assets in future studies. In addition, the smooth kink regression of Maneejuk, Yamaka, and Sriboonchitta (see, e.g.,[11]) is also suggested to reveal the smooth nonlinear impact of bond on other financial markets.

References

1. Andersson, M., Krylova, E., & Vähämaa, S. (2008). Why does the correlation between stock and bond returns vary over time? *Applied Financial Economics, 18*(2), 139–151.
2. Baur, D. G., & Lucey, B. M. (2009). Flights and contagion—An empirical analysis of stock-bond correlations. *Journal of Financial Stability, 4*, 339–352.
3. Baele, L., Bekaert, G., Inghelbrecht, K., & Wei, M. (2020). Flights to safety. *The Review of Financial Studies, 33*(2), 689–746.
4. Connolly, R., Stivers, C., & Sun, L. (2005). Stock market uncertainty and the stock-bond return relation. *Journal of Financial and Quantitative Analysis, 40*(1), 161–194.
5. Hansen, B. E. (2017). Regression kink with an unknown threshold. *Journal of Business & Economic Statistics, 35*(2), 228–240.
6. Jena, P. K., & Goyari, P. (2016). Empirical relationship between commodity, stock and bond prices in India: A DCC model analysis. *IUP Journal of Applied Finance, 22*(1), 37.
7. Johansson, A. C. (2010). *Stock and bond relationships in Asia*. Stockholm: Stockholm School of Economics.
8. Kim, S. -, J., Moshirian, F. & Wu, E. (2006). Evolution of International Stock and Bond Market Integration: Inference from the European Monetary Union. *Journal of Banking and Finance, 30*, 1507–1534.
9. Lin, F. L., Yang, S. Y., Marsh, T., & Chen, Y. F. (2018). Stock and bond return relations and stock market uncertainty: Evidence from wavelet analysis. *International Review of Economics & Finance, 55*, 285–294.
10. Maneejuk, P., & Yamaka, W. (2021). Significance test for linear regression: how to test without P-values? *Journal of Applied Statistics, 48*(5), 827–845.
11. Maneejuk, P., Yamaka, W., & Sriboonchitta, S. (2019). Does the Kuznets curve exist in Thailand? A two decades' perspective (1993-2015). *Annals of Operations Research*, 1-32.
12. Naeem, M. A., Adekoya, O. B., & Oliyide, J. A. (2021). Asymmetric spillovers between green bonds and commodities. *Journal of Cleaner Production*, 128100.
13. Pastpipatkul, P., Yamaka, W., & Sriboonchitta, S. (2015). Co-movement and dependency between New York stock exchange, London stock exchange, Tokyo stock exchange, oil price, and gold price. *In International symposium on integrated uncertainty in knowledge modelling and decision making* (pp. 362-373). Springer, Cham.
14. Pastpipatkul, P., Yamaka, W., & Sriboonchitta, S. (2018). Portfolio selection with stock, gold and bond in Thailand under vine Copulas functions. *In International Econometric Conference of Vietnam* (pp. 698-711). Springer, Cham.
15. Tibprasorn, P., Maneejuk, P., & Sriboochitta, S. (2017). Generalized information theoretical approach to panel regression kink model. *Thai Journal of Mathematics*, 133-145.
16. Ziaei, S. M. (2012). Effects of gold price on equity, bond and domestic credit: Evidence from ASEAN+ 3. *Procedia-Social and Behavioral Sciences, 40*, 341–346.

Can Cryptocurrency Be a New Safe-Haven?

Nathaphat Na Chiangmai, Nathee Naktnasukanjn, and Piyachat Udomwong

Abstract In the digital era, there are many investable assets and make profit from it. Because investments always have risk, the investors then decrease their risk by finding assets that are expected to retain or increase in their values during times of market turbulence. Herein the assets which can retain or increase in their values will be called as safe-haven assets. Gold has been known as the safe-haven asset, however cryptocurrency is a class of digital assets which is claimed as the new gold in the digital era. Consequently, this study aims to answer a question "Can cryptocurrency be a new safe-haven?".

1 Introduction

The definition of safe-haven assets is the assets provide hedging benefit during periods of market turbulence which is the basic concept of the safe-haven assets. Gold is one example of many safe-haven assets that can store value for centuries and be safe-haven for the investors in the several decades. However, Baur et al. (see, e.g., [1]) said that the duration of the safe-haven effect during the subprime crisis in 2008 was reduced significantly. It can be explained that there is a chance that the old safe-haven asset cannot retain its own value. This results in finding the digital asset such as cryptocurrency to be a new safe-haven. In 2008, there was a spur of creating a large number of cryptocurrencies known as altcoins by Nakamoto (see, e.g., [2]). One of the most popular altcoins is bitcoin which has the largest volume in the market. The cryptocurrencies are now becoming a focal point of the investors for several years (see, e.g., [3]). These cryptocurrencies are currently becoming easier to exchange with real money or to trade with other cryptocurrencies on digital assets exchange

N. N. Chiangmai (✉) · N. Naktnasukanjn · P. Udomwong
International College of Digital Innovation Chiangmai university, Chiangmai, Thailand
e-mail: nathaphat_na@cmu.ac.th

N. Naktnasukanjn
e-mail: nathee@cmuic.net

P. Udomwong
e-mail: piyachat.u@cmu.ac.th

S. Sriboonchitta et al. (eds.), *Credible Asset Allocation, Optimal Transport Methods, and Related Topics*, Studies in Systems, Decision and Control 429,
https://doi.org/10.1007/978-3-030-97273-8_44

platforms. For this reason, the cryptocurrency is probably considered to be a new safe-haven. The purpose of this study is to find a conclusion on the cryptocurrency can be the new safe-haven asset. In order to achieve the purpose, historical price data included price data of cryptocurrencies, gold, and top stock index around the world. The price data were analyzed to find insights from data such as cumulative return, risk, and correlation among assets.

2 Related Literature

Safe-Haven Asset

The safe-haven asset was a well-known investment that can retain or increase in value during the time of market turbulence. To qualify a safe-haven asset, an asset must be able to retain or increase in value during downturns of the market. Statistically, the returns from the safe-haven asset should be uncorrelated or negatively correlated with the returns of other assets during a crisis period (see, e.g., [4]). There is another concept of a safe asset by (see, e.g., Gorton et al. [5]) who defined the safe asset as an asset whose value is insensitive to information can be clearly distinguished as the "safe-haven" asset. A typical example of safe-haven asset is gold.

Safe-Haven Analysis

Under such a notion, gold has typically been considered as the safe-haven asset which has a long historical role as natural money or a store of value (see, e.g., [6]). There is a large volume of literature investigation whether the gold can act as the safe-haven asset. Results from the previous literatures show both positive and negative results. For example, Gözde Gürgün (see, e.g., [7]) concluded that gold performed as a hedge and the safe-haven in most of the financial markets. Baur (see, e.g., [8]) found that gold offered the investors distinctive forms of "safety". To find motivation factors of the investors on purchasing of safe-haven assets, Baur (see, e.g., [1]) performed the study showed that the duration of the safe -haven effect during the subprime crisis in 2008 was remarkably declined. Ghazali (see, e.g., [9]) pointed out that gold, at best, is a weak safe-haven during the financial stress for stockholders. Rather than the studys that focused on gold being the safe-haven assets, there is a study that studied on the cryptocurrencies whether they can be safe-haven. Conlon (see, e.g., [10]) discovered that Bitcoin and Ethereum are not, in general, found as the safe-haven asset for the international equity markets. However, Tether is found as the safe-haven asset over the most recent period including the COVID-19 crisis. The studys performed various methods, but the most popular methods are downside risk analysis and correlation analysis.

Cryptocurrencies

Cryptocurrencies have became well-known investable assets since 2008. There was a spur of creating a large number of cryptocurrencies known as altcoins by Satoshi Nakamoto (see, e.g., [2]). Yilmaz and Hazar (see, e.g., [11]) studied the importance of cryptocurrencies which has been increasing day by day and is taking their place in the market. The bright future of cryptocurrencies was mentioned by Fauzi (see, e.g., [12]) "Cryptocurrencies are here to stay." The future of trading lies well with new emerging technologies that benefit for mankind. In term of investing portfolios, including the cryptocurrencies to investors' portfolios probably offers attractive returns and hedging properties (see, e.g., [13]).

An authorization not only confronts with several legal issues which related to a payment institution's definition, but also introduces novel risks whose consequences are previously unknown. (see, e.g., [14]). Lastly the size effect and momentum effect are strong in cryptocurrency markets (see, e.g., [15]). Consequently, the investors need to understand their investment goals and tolerance for risk.

3 Methology

Samples, which were collected from trustable sources, were cleaned and tested. Later descriptive analysis and cumulative return calculated along with VaR, CVaR were processed in order to measure the risk of the sample assets. Bayesian correlation was applied to the sample assets to quantify the relationship among assets. At last, it can be concluded whether cryptocurrencies can be the novel safe-haven asset or not.

Cumulative Returns

The cumulative returns are the total change in the investment price over a set time. After the results of cumulative return of each asset the result were received and separated into 6 portfolios. Each portfolio contained one of cryptocurrencies to monitor its performance with the same weight of each asset. An equation of cumulative return was illustrated as Eq. 1 where cumulative return $[t]$ is the cumulative return of a particular day t, price[t] is the price of particular day and price [0] was a price of the beginning date.

$$\text{Cumulative return } [t] = \frac{\text{price}[t]}{\text{price}[0]} - 1 \quad (1)$$

Value of Risk and Conditional Value at Risk

The value at risk (VaR) was calculated by using historical method which is the simplest method for calculating Value at Risk. Market data for the last 250 d is taken to calculate the percentage change for each risk factor on each day. Each percentage change is then calculated with current market values to present 250 scenarios for future value. For each of the scenarios, the portfolio is valued using full, non-linear pricing models. The third worst day selected is assumed to be 99% VaR. The Var was computed by using Eq. 2 where v_i is the number of variables on day i and m is the number of days from which historical data is taken.

$$\text{Value at risk} = v_m * \frac{v_i}{(v_i - 1)} \tag{2}$$

Conditional value at risk (CVaR) was applied after completing VaR calculation. CVaR values were derived from the calculation of VaR itself. The computational assumptions of VaR were based on the shape of the distribution of returns and the cut-off level used. It is the average of the values that fall beyond the VaR. the result from Cvar can be use to quantifies the amount of tail risk an investment portfolio has. The equation of CVaR was in Eq. 3 where $p(x)dx$ is the probability density of getting a return with value"x". c is the cut-off point on the distribution where the analyst sets the Var breakpoint. Lastly VaR in the equation is the agreed-upon VaR level.

$$\text{CVaR} = \frac{1}{1-c} \int_{-1}^{\text{VaR}} xp(x)\mathrm{d}x \tag{3}$$

Standard Deviation

The standard deviation is use to calculated volatility of each asset which will be visualize in window of 30 day. The equation of Standard Deviation was in where Σ means "sum of", x is a value in the data set, μ is the mean of the data set, and N is the number of data points in the population. The equation of standard deviation was in Eq. 4

$$\text{SD} = \sqrt{\frac{\sum |x - \mu|^2}{N}} \tag{4}$$

Differencing

The Differencing method will be apply before using Bayesian Pearson Correlation in order to remove trend and seasonality from sample data. The equation of Differencing method is in Eq. 5.

$$Z_i = y_i - y_{i-1} \tag{5}$$

Bayesian Pearson Correlation

Bayesian Pearson correlation (see, e.g., [16]) was upgrade version of Pearson Correlation which change measure of the probability from p-value to more reliable measurement which is Bayesian probability. The correlation we got from Bayesian Pearson Correlation will contain strength of evidence for the hypothesis which can help determine that the result is trustworthy or not (see, e.g., [17]).

4 Empirical Results

Data

The range of our collected data is from 1 January 2018 to 25 November 2021. The sample period consists of covid-19 pandemic period as the early cluster of confirmed cases of COVID-19 were reported at the end of December 2019. Hence, our sample data was covering the entire period of covid-19 (see, e.g., [18]) .

Sample data we have are almost 48 months of daily price of 4 major world stock indexes referenced from Black Well Global website (see, e.g., [19]): which include, DowJones index (DJI), NASDAQ-100 (NDX) index, Hang Seng index (HSI), and German DAX index and other 1 of thai major index SET50. The cryptocurrencies are selected from top five market capitalization referenced from CoinMarketCap website (see, e.g., [20]): which include Bitcoin (BTC)—the first Cryptocurrencies introduced to the world, Ethereum (ETH)-coin of well-known open-source Blockchain-based platform, Tether (USDT)-known as digital dollar, Binance (BNB)-coin of the world's biggest cryptocurrencies exchange platform, Cardano (ADA)-coin of public Blockchain platform. Lastly, we also include gold daily price sample data. All samples were collected from investing.com website. After collected, the data was clean to make sure that there's no abnormal data.

Descriptive Analysis

We start from listing all the daily price raw data in Table 1 where we have 11 assets in total 5 stock indexes, 5 cryptocurrencies, and gold. Then we calculate the daily financial return and put in Table 2. The result of descriptive statistics in Table 3 show total sample data we have are 861 daily return data on 11 assets. Among them, USDT has the lowest mean and standard deviation. After that we calculate the cumulative returns and plot graph to compare them.

Figures 1, 2 and 3 show that the cumulative returns of most of the cryptocurrencies, except USDT, are far higher than all the stock market indexes, which is a good sign that these cryptocurrencies are doing well in term of cumulative return in Covid-19 pandemic period. Among cryptocurrencies, USDT has the lowest cumulative returns. Figure 4 is visualize of cumulative return of stock index and gold, which show that the best performer in cumulative returns among them was Gold in the first half of covid-19 incidence. However, in the second half it seems that NDX was doing better on cumulative returns follow by FBX. Other thing that we get from visualize the

Table 1 Price historical data

Date	NDX	SETO	HSI	DJI	DAX	GOLD	BTC	ETH	USDT	BNM	ADA
28-01-18	6811.38	1182.70	32121.94	26017.81	13281.43	1327.19	11045.0	991.01	1.0235	14.14	0.6500
28-01-19	6834.33	1183.60	32254.89	26017.72	13434.45	1331.80	11476.0	979.57	1.0174	14.93	0.6458
28-01-22	6906.28	1185.78	32393.41	26214.60	13463.69	1334.11	107710.0	999.37	1.0093	12.90	0.5675
28-01-23	6963.46	1192.74	32930.70	26210.81	13559.60	1341.39	10819.0	1042.47	1.0030	13.10	0.5598
28-01-24	6919.35	1195.60	32958.69	26252.12	13414.74	1358.15	11414.0	1051.95	1.0036	13.7	0.6290
...	...	...	...	...	...	...	...	...	...	...	...
28-01-18	16482.97	989.18	25319.72	35871.34	16221.73	1858.46	57014.0	4003.13	1.0005	531.90	1.7886
28-01-19	16573.34	984.39	25049.97	35602.18	16159.97	1844.60	58130.0	4295.39	1.0009	581.00	1.86637
28-01-22	16380.98	989.18	24951.34	35619.26	16115.69	1805.04	56299.0	4086.52	1.0007	559.70	1.7765
28-01-23	16306.72	987.26	24651.58	35813.74	15937.00	1789.73	57602.0	4340.45	1.0003	591.30	1.7508
28-01-24	16367.81	989.19	24685.50	35805.17	15878.39	1788.50	57209.0	4270.00	1.0003	590.60	1.6664

Table 2 Percent change historical data

Date	NDX	SETO	HSI	DJI	DAX	GOLD	BTC	ETH	USDT	BNM	ADA
2018-01-19	0.003369	0.000761	0.004139	0.002072	0.011521	0.003474	0.039022	−0.011544	−0.005960	0.055870	−0.006462
2018-01-22	0.010528	0.001842	0.004295	0.005480	0.002176	0.001734	−0.061433	0.020213	−0.007961	0.135968	−0.121245
2018-01-23	0.008279	0.005870	0.016586	−0.000145	0.007124	0.005457	0.004456	0.043127	−0.006242	0.015504	−0.013568
2018-01-24	0.006334	0.002398	0.000850	0.001576	0.010683	0.012495	0.054996	0.009094	0.000598	0.051145	0.0123616
2018-01-25	0.000441	−0.011743	−0.009231	0.0005358	−0.008676	−0.007437	−0.023480	−0.010238	−0.000797	0.042121	0.020668
...	...	...	...	...	...	...	...	...	...	...	...
2018-01-18	0.010725	0.002696	−0.012879	−0.001675	0.001809	−0.004553	−0.055842	−0.067085	0.000100	−0.080235	−0.047046
2018-01-19	0.005483	−0.004842	−0.010654	−0.007503	−0.003807	−0.007458	0.019574	−0.073008	0.000400	−0.092311	−0.041988
2018-01-22	0.0011607	−0.004866	−0.003937	0.000480	−0.002740	−0.021446	0.031498	−0.048627	−0.000200	−0.036661	−0.046789
2018-01-23	0.004533	−0.001941	−0.012014	0.005460	−0.0011088	−0.008482	0.023144	−0.062138	−0.000400	−0.056459	−0.014467
2018-01-24	0.003746	−0.001955	−0.001376	0.000239	−0.003678	−0.000687	0.006823	−0.016231	−0.000000	−0.001184	−0.048207

Table 3 Descriptive statistics result

	NDX	SET50	HSI	DJI	DAX	GOLD	BTC	ETH	USDT	BNB	ADA
count	860.000000	860.000000	860.000000	860.000000	860.000000	860.000000	860.000000	860.000000	860.000000	860.000000	860.000000
mean	0.001154	−0.000114	−0.000215	0.000482	0.000306	0.000391	0.003147	0.003871	−0.000023	0.007004	0.003927
std	0.016335	0.013584	0.013470	0.014818	0.013958	0.009328	0.049259	0.065509	−0.002868	0.073856	0.075762
min	−0.121932	−0.0116964	0.081822	−0.129265	0.122386	−0.057225	0.381776	0.0447032	−0.027517	0.0439614	0.415617
25%	−0.005267	−0.005678	−0.007683	−0.004384	−0.004589	−0.004083	−0.019629	−0.026172	−0.000600	−0.028596	−0.036023
50%	0.001802	−0.000095	0.000685	0.001108	0.000879	0.000793	0.002064	0.001765	0.000000	0.003641	−0.001062
75%	0.008848	0.005386	0.007517	0.006551	0.006591	0.005218	0.025386	0.035325	0.000600	0.038391	0.038869
max	0.100722	0.092660	0.050483	0.113650	0.109759	0.037005	0.236010	0.387285	0.021375	0.699683	0.337025

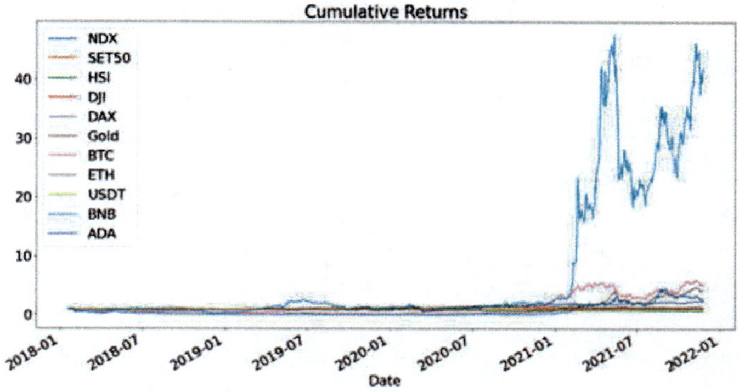

Fig. 1 Cumulative returns

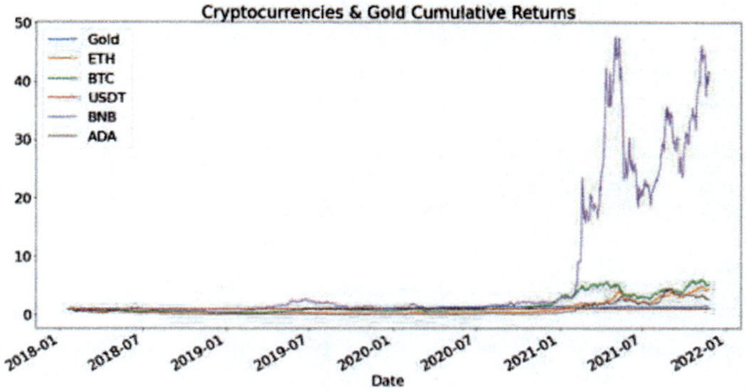

Fig. 2 Cumulative returns of cryptocurrencies and gold

Fig. 3 Cumulative returns of cryptocurrencies and gold without out BNB

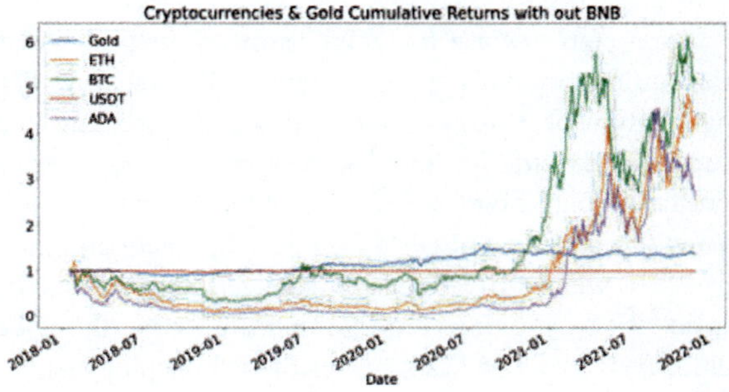

cumulative returns is when a cryptocurrencies return was starting to increase in the same way with stock index but with more value in the last half of the sample period. Lastly USDT seem to act difference from the other assets because it has steady cumulative returns along the sample period.

Fig. 4 Cumulative returns of stock index and gold

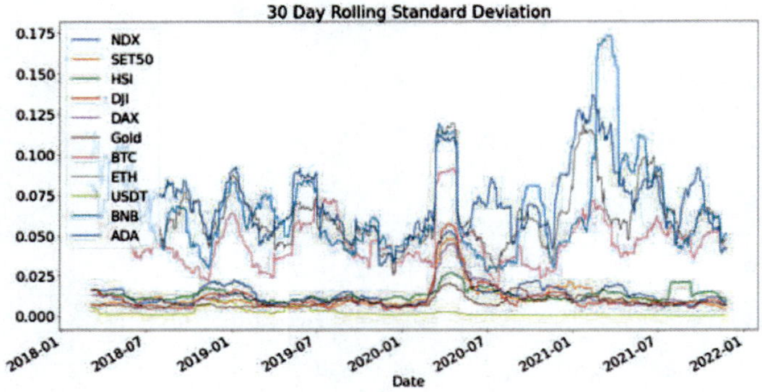

Fig. 5 30 day rolling standard deviation

Next step, we use financial return to measure volatility by calculating standard deviation from daily financial return. The result in Fig. 5 show the visualize of the calculation of 30-day standard deviation. It show that the volatility of cryptocurrencies are going in the same way with stock index, which seems to be fluctuated but cryptocurrencies are more fluctuated than stock index. On the other hand, the volatility of gold and USDT seem to be lower and more stable especially USDT .

Next, we use Box-Plot to visualize the daily financial return to observe the portfolio risk by looking into interquartile range, the wider the box the more risk that asset has. In Fig. 6 was the result of box plot visualize which show that most of cryptocurrencies, excepts USDT, clearly have more risk than the other asset as we can clearly see the interquartile range was wider than other assets. The assets that have shortest interquartile range is USDT, and Gold.

From overall descriptive analysis, we see that most of cryptocurrencies except USDT are better in performance of cumulative return during the second half of the sample period which increasing significantly on cumulative returns. However, it come with higher risk and volatility than stock indexes, gold and USDT.

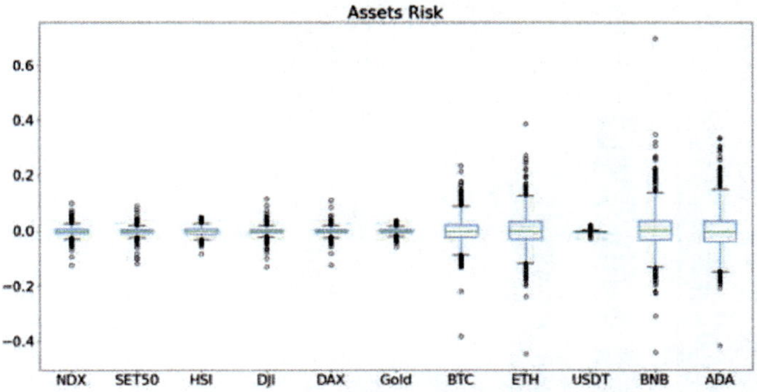

Fig. 6 Asset risk

Risk Analysis

The risk will be analyzed by using the Value At Risk (VAR) to quantify the level of financial risk and Conditional Value At Risk (CVAR) to quantify the amount of tail risk. The result of calculating VAR visualize in Fig. 7 show that the financial risk of most of cryptocurrencies, except USDT, was higher than all of stock indexes. From our sample data, the highest was ADA follow by ETH, BNB and BTC. Gold and USDT have lowest level of financial risk among all sample assets. The result of CVAR in Fig. 8 show that the asset with high tail risks are cryptocurrencies which are BNB follow by ADA, ETH and BTC. Gold and USDT have the lowest tail risk. From the result of risk analysis, most cryptocurrencies have high value of both VAR and CVAR, except for USDT which have the lowest value of VAR and CVAR among all sample data. Another interesting point to mention is that BTC has the lowest value in both VAR and CVAR among all the cryptocurrencies, excluding USDT.

Portfolio Analysis

We simulate portfolio that have all of stock indexes sample in it and add one of cryptocurrency for each portfolio to monitor the effect of cumulative return in each portfolio. The results in Fig. 9 show that most of cryptocurrencies, except USDT, are doing good on preventing loss of cumulative return. We can clearly see that when compare the visualize of portfolio with BTC, ETH, BNB, ADA with portfolio with USDT, Gold and portfolio without any cryptocurrencies, the return in downside period of the portfolios with BTC, ETH, BNB and ADA coins are better than portfolio without cryptocurrencies. However, as the result of portfolio risk calculation in Fig. 10, risk of portfolios with cryptocurrencies are higher than portfolios without cryptocurrencies.

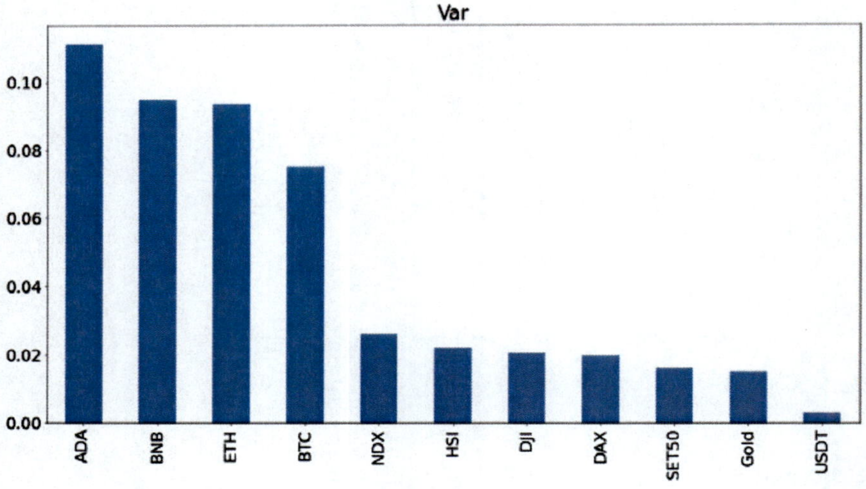

Fig. 7 Value of risk (VAR)

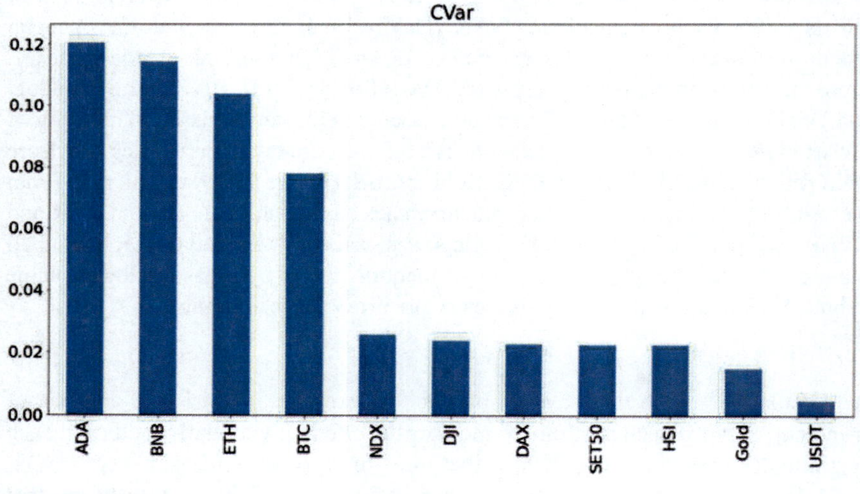

Fig. 8 Conditional value at risk (CVaR)

Correlation Analysis

Figure 11 show the result of Bayesian Correlation. The strong evidence from BF_{10} and Pearson's r show us that BTC and ETH was correlated with some of stock index which is SET50, NDX and DAX. However BNB and ADA was uncorrelated with any of stock index but seem to correlate with other cryptocurrencies. On the other hand USDT was the only cryptocurrencies that uncorrelated with all sample assets.

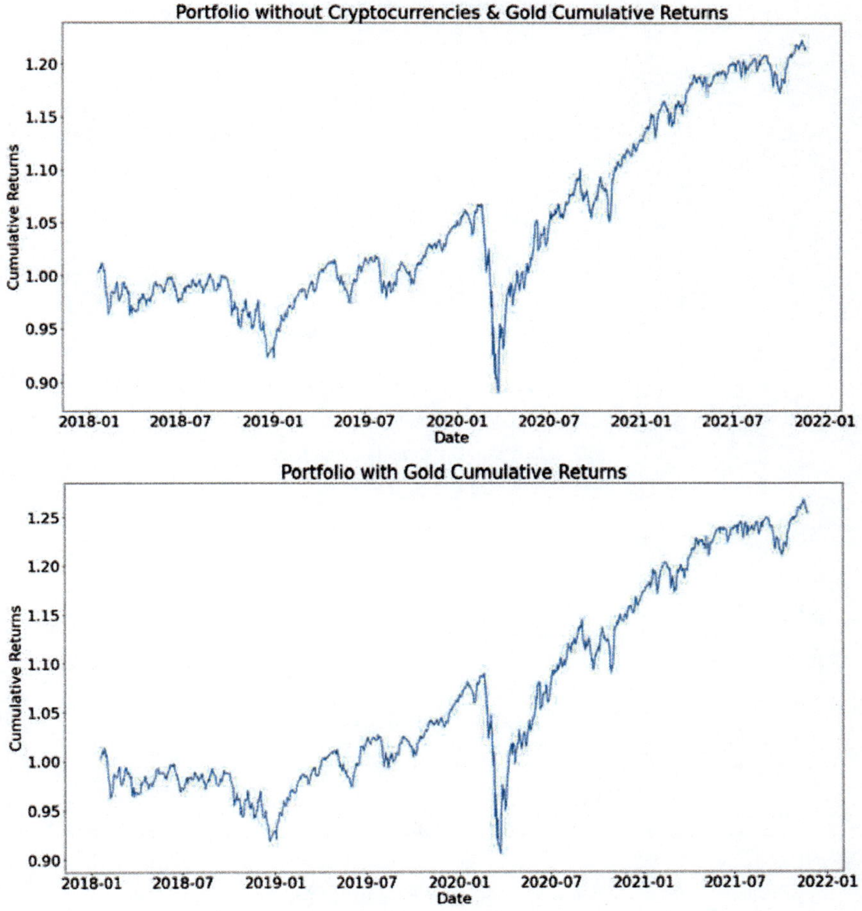

Fig. 9 All portfolio

5 Conclusion

From the empirical result, most of cryptocurrencies show good performance in cumulative returns with high financial risks during Covid-19 pandemic. Bayesian correlation show correlation signal between most stock indexes and most Cryptocurrencies. Portfolio Analysis show that most Cryptocurrencies might be good assets for speculating or to defense loss of cumulative return from stocks. But those Cryptocurrencies have only one feature of safe-haven asset, which is to retain or increase in value in case of market downturns. However, most Cryptocurrencies did not have the other

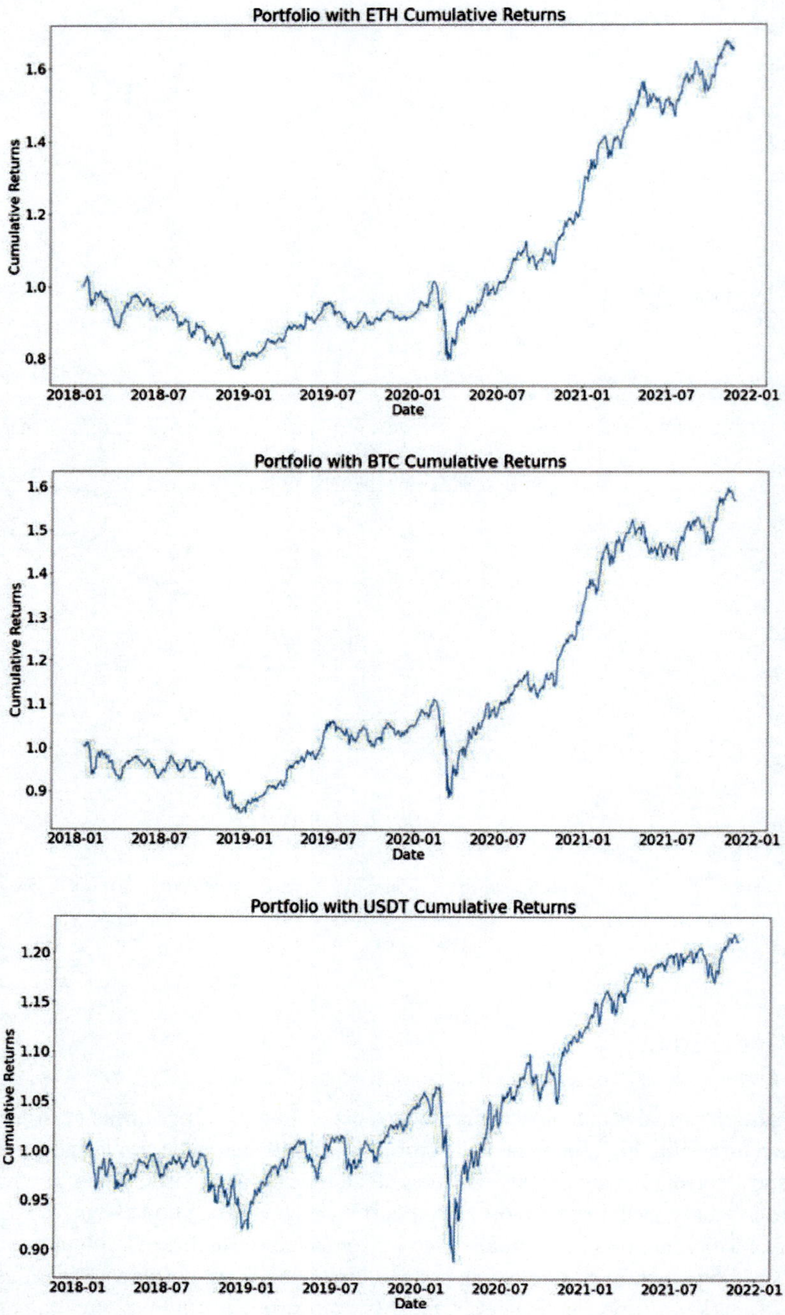

Fig. 9 (continued)

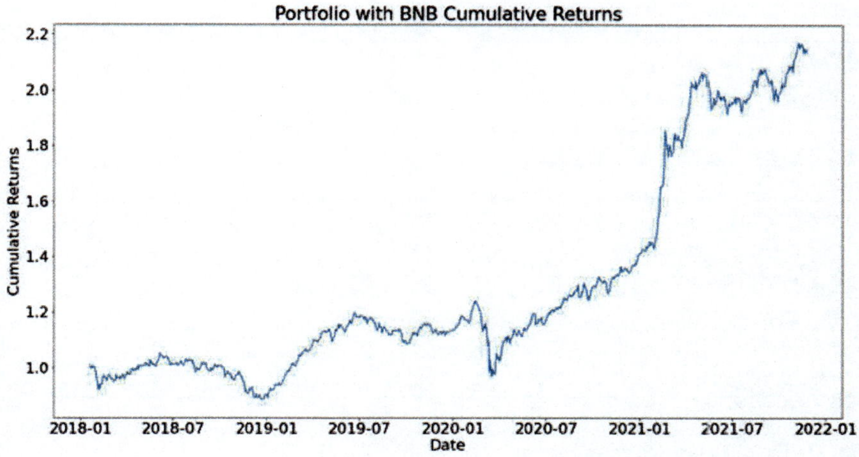

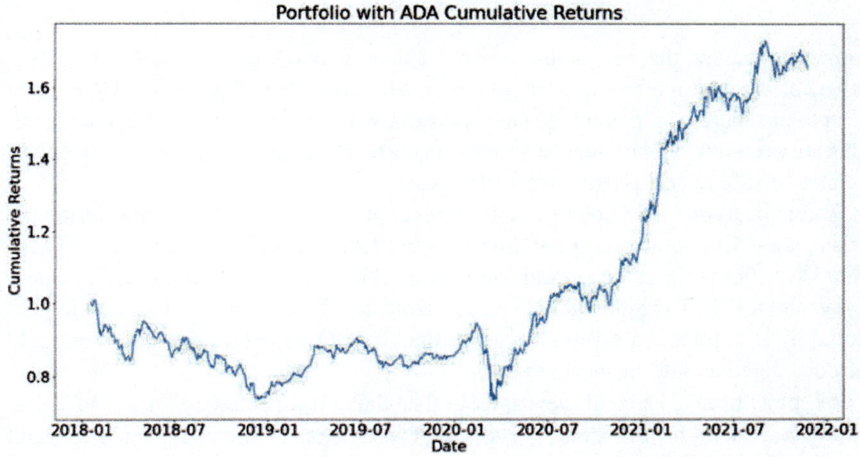

Fig. 9 (continued)

```
risk of portfolio with Gold : 8.962299608143194 %
risk of portfolio with ETH : 15.14049630158504 %
risk of portfolio with BTC : 12.793119742127306 %
risk of portfolio with USDT : 8.74050772874318 %
risk of portfolio with BNB : 15.97962030386866 %
risk of portfolio with ADA : 16.47133471328846 %
risk of portfolio with out cryptocurrencies :8.715385870884226 %
```

Fig. 10 Portfolio risk

Bayesian Pearson Correlations ▼

Variable		NDX	SET50	HSI	DJI	DAX	Gold	BTC	ETH	USDT	BNB	ADA
1. NDX	Pearson's r	—										
	BF_{10}	—										
2. SET50	Pearson's r	0.299***	—									
	BF_{10}	7.305e+8	—									
3. HSI	Pearson's r	0.298***	0.548***	—								
	BF_{10}	5.912e+8	1.538e+37	—								
4. DJI	Pearson's r	0.827***	0.407***	0.312***	—							
	BF_{10}	6.140e+122	1.890e+18	6.212e+9	—							
5. DAX	Pearson's r	0.540***	0.538***	0.483***	0.679***	—						
	BF_{10}	7.933e+35	3.589e+35	2.576e+27	2.112e+65	—						
6. Gold	Pearson's r	0.135	0.064	0.108	0.086	0.086	—					
	BF_{10}	5.342	0.155	1.042	0.360	0.347	—					
7. BTC	Pearson's r	0.143	0.148*	0.026	0.125	0.150*	−0.016	—				
	BF_{10}	9.574	13.260	0.067	2.782	15.697	0.060	—				
8. ETH	Pearson's r	0.100	0.142	0.056	0.115	0.152*	0.043	0.679***	—			
	BF_{10}	0.673	8.486	0.121	1.500	19.161	0.089	1.939e+65	—			
9. USDT	Pearson's r	0.085	−0.076	−0.032	0.099	−0.015	0.022	0.014	−0.017	—		
	BF_{10}	0.333	0.232	0.072	0.639	0.059	0.063	0.059	0.060	—		
10. BNB	Pearson's r	0.091	0.061	0.027	0.069	0.060	0.011	0.557***	0.601***	−0.023	—	
	BF_{10}	0.442	0.142	0.067	0.182	0.139	0.058	6.926e+38	8.313e+46	0.064	—	
11. ADA	Pearson's r	0.079	0.098	0.023	0.062	0.037	0.077	0.526***	0.560***	0.030	0.501***	—
	BF_{10}	0.261	0.611	0.064	0.145	0.078	0.249	3.795e+33	1.908e+39	0.070	8.227e+29	—

* $BF_{10} > 10$, ** $BF_{10} > 30$, *** $BF_{10} > 100$

Fig. 11 Bayesian pearson correlation

important feature, that says safe-haven asset should not correlate or should have negative correlate with other investment assets. Moreover, the VAR and CVAR of most Cryptocurrencies are higher than stock indexes, which mean that most Cryptocurrencies are even riskier. And thus, it show a strong evidence that most Cryptocurrencies cannot be safe-haven asset, at least for now.

But USDT coin is different from the other coins. It can sustain the value during the crisis, it has low volatility, and it does not correlate with other stock indexes. These characteristics is the same as gold, which is well known safe-haven asset. But again, the value of USDT is bind with US dollar. Holding USDT is the same as holding US dollar. If US dollar is a safe-haven asset, then Cryptocurrencies that peg to it could be considered as safe-haven asset too.

Cryptocurrencies are still new and our data that we collected might be too few to conclude with high confidence, we hope that when there are more available data and more available digital assets in the future, we could revisit this question and have a clearer picture.

References

1. Baur, D. G., & Glover, K. J. (2012). The destruction of a safe haven asset? *Applied Finance Letters, 1*, 8–15.
2. Nakamoto, S. (2008). https://bitcoin.org/en/.
3. Giudici, G., Milne, A., & Vinogradov, D. (2020). Cryptocurrencies: market analysis and perspectives. *Journal of Industrial and Business Economics, 47*, 1–18.
4. Tronzano, M. (2020). Safe-haven assets, financial crises, and macroeconomic variables: evidence from last two decades (2000–2018). *Risk and Financial Management, 71* (2020).
5. Gorton, G., Lewellen, S., & Metrick, A. (2012). The safe-asset share. *American Economic Review, 102*(3), 101–06.

6. Bouri, E., Hussan Shahzad, S. J., Roubaud, D., Kristoufek, L., & Lucey, B. (2020). Bitcoin, gold, and commodities as safe havens for stocks: new insight through wavelet analysis. *The Quarterly Review of Economics and Finance, 77*, 156–164.
7. Gürgün, G., & Ünalmıs, I. (2014). Is gold a safe haven against equity market investment in emerging and developing countries? *Finance Research Letters, 11*(4), 341–348.
8. Baur, D. G., & McDermott, T. K. (2016). Why is gold a safe haven? *Journal of Behavioral and Experimental Finance, 10*, 63–71.
9. Ghazali, M. F., Keab, H. H., & Bahari, Z. (2013). Is gold a hedge or a safe haven? An empirical evidence of gold and stocks in Malaysia. *International Journal of Business and Society, 14*, 428–443.
10. Conlon, T., Corbet, S., & McGee, R.J. (2020). Are cryptocurrencies a safe haven for equity markets? An international perspective from the COVID-19 pandemic. *Research in International Business and Finance, 54*.
11. Yilmaz, N. K., & Hazar, H. B. (2018). Predicting future cryptocurrency investment trends by conjoint analysis. *Journal of Economics Finance and Accounting, 5*(4), 321–330
12. Fauzi, M. A., Paiman, N., & Othman, Z. (2020). Bitcoin and cryptocurrency: challenges, opportunities and future works. *Journal of Asian Finance, Economics and Business, 7*(8), 695–704.
13. Borri, N. (2019). Conditional tail-risk in cryptocurrency markets. *Journal of Empirical Finance, 50*, 1–19.
14. Naibilou, H. (2019). The Dark side of licensing cryptocurrency exchanges as. *Law and Financial Markets Review, 14*(1), 39–47.
15. Liu, W., Liang, X., & Cui, G. (2020). Common risk factors in the returns on cryptocurrencies. *Economic Modelling, 86*, 299–305.
16. Nuzzo, R. L. (2017). An introduction to Bayesian data analysis for correlations. *PM&R, 9*(12), 1278–1282 (2017)
17. van Doorn, J., van den Bergh, D., Böhm, U., Dablander, F., Derks, K., Draws, T., Etz, A., Evans, N. J., Gronau, Q. F., Haaf, J. M., Hinne, M., Kucharský, Š., Ly, A., Marsman, M., Matzke, D., Komarlu, A. R., Gupta, N., Sarafoglou, A., Stefan, A. (2020). The JASP guidelines for conducting and reporting a Bayesian analysis. *Psychonomic Bulletin & Review, 28*, 813–826.
18. WHO. (2020). World Health Organization, 27 Apr 2020. https://www.who.int/news/item/27-04-2020-who-timeline-covid-19
19. "BlackWellGlobal", BlackWellGlobal, https://blackwellglobal.com/what-are-the-major-world-indices/
20. "CoinMarketCap", https://coinmarketcap.com/

Modelling the Relationship Among Telecommunication Infrastructure, Foreign Direct Investment and Economic Growth in ASEAN Countries

Nachatchapong Kaewsompong and Kasem Kunasri

Abstract The interactions between telecommunication infrastructure, foreign direct investment and economic growth were investigated for Indonesia, Cambodia, Malaysia, the Philippines, Thailand for the period 2000–2014. The study proceeded with the panel VAR model analysis, impulse response functions and forecast error variance decompositions. This panel VAR reveals that there exists a weak relationship among these three variables. However, when the shocks of these variables are examined, we found that shock of telecommunication infrastructure of ASEAN-5 countries has a greater impact on foreign direct investment and economic development.

1 Introduction

After the end of the Second World War, virtually all developing countries that used to be under colonial powers obtained independence from their former rulers and some of them started to enjoy the rising economic wealth. The ASEAN countries that were once under the western rules except Thailand only, in particular, have exhibited their rapid economic development making the ASEAN economy as a whole a dazzling star of the world economic stage. In 2015, the Gross Domestic Product of ASEAN was 2438.13 billion US dollars, with a 4.83% growth rate, about 300 percent higher than the year 2000 level. The rapid advance of the ASEAN economy to become an important regional economic power has been partly driven by an indispensable mechanism acknowledged as foreign direct investment. By attracting foreign-based firms or other entities to invest directly in their countries, the ASEAN nations can acquire not only enormous amount of funds quickly but also a practically instant transfer of

N. Kaewsompong
Faculty of Economics, Chiang Mai University, Chiang Mai, Thailand
e-mail: Nachat_flysky@hotmail.com

K. Kunasri (✉)
Department of Economics, Faculty of Management Sciences, Chiang Mai Rajabhat University, Chiang Mai 50300, Thailand
e-mail: kasem_kun@g.cmru.ac.th

S. Sriboonchitta et al. (eds.), *Credible Asset Allocation, Optimal Transport Methods, and Related Topics*, Studies in Systems, Decision and Control 429,
https://doi.org/10.1007/978-3-030-97273-8_45

advanced foreign technology and management capability which help speed up the process of domestic industrial structural improvement as well as economic recovery from previous or on-going perils.

In 2014, ASEAN became one of the main and attractive destinations of global foreign direct investment (FDI) again with FDI flows to the region rising about 5 times compared with 2001, at 135.03 billion US dollars in total that constituted about 6.20% of the regional GDP, making ASEAN the largest FDI recipient in the developing world. In December 2015, the ASEAN block announced its establishment as an economic community or the AEC, which greatly increased the confidence of investors in the future development of ASEAN. More and more foreign investors have already made the ASEAN Economic Community as one of the key reference factors for ASEAN's development strategy, and many investors are planning to expand their business to more ASEAN countries. Large multinational enterprises are the most important foreign investors in ASEAN, and they are expanding their investment in ASEAN in areas such as manufacturing, finance, infrastructure and other services.

The importance of infrastructure to ASEAN Member States cannot be ignored. In October 2010, the 17th ASEAN Leaders' Meeting adopted the "Master Plan on ASEAN Connectivity". ASEAN Connectivity, which contains the three pillars of infrastructure connectivity, institutional connectivity and people-to-people connectivity, is a key strategy for implementing the ASEAN community. A 'connected ASEAN' has important implications. It will increase further the competitiveness of the region, enhance production efficiency, reduce transaction costs and attract FDI. Member states of ASEAN have been investing in infrastructure with varying degrees of spending and development. The private sector, domestic and foreign, has been a significant player in the region's infrastructure development. In various infrastructure sectors, the presence of foreign multinational enterprises is highly visible. Investing in infrastructure to meet the rapid growth in demand has been a challenge for most ASEAN member states and some of them have already taken steps to liberalize their infrastructure industries into privatized infrastructure assets, such as those in telecommunication and power generation.

The telecommunication penetration rates in some ASEAN member states are among the highest in the developing countries, and the industry has attracted significant interest and participation from the private sector, which has contributed to the rapid growth of the industry in the region. The telecommunications sector in ASEAN grew rapidly in the last decade, particularly in access to mobile voice and data networks. The number of mobile subscriptions increased almost five times between 2005 and 2014. In 2020, mobile broadband became far more than fixed broadband in subscriptions as mobile broadband penetration in ASEAN reached more than 50% of the population compared with only 3% for fixed broadband (ASEAN Investment Report 2020). Telecommunication service revenues have increased in parallel with the growth in access. The ASEAN telecommunication service market was worth 63 billion US dollars in 2014, more than twice the figure of ten years ago. Investment in telecommunication infrastructure has enabled more efficient telephone services, internet service and increased the degree of penetration, with more people able to use telecommunication service even in remote areas. This development has contributed to national and global connectivity, enhancing communication and ICT connections, and bringing remote areas into the modern information age.

In this study, the relationship among ICT, FDI and Economic Growth of ASEAN countries is investigated. The interaction of these variables has been neglected or limited in the context of ASEAN countries. We overcome this deficiency and examine more broadly the causes and consequences of energy and transportation services. In particular, we examine relationship among ICT, FDI and Economic Growth of ASEAN, and two key indicators of a modern economy: (i) electric power consumption per capita, (ii) transportation services value. To achieve this goal, we utilize a panel vector autoregressive (VAR) model in order to identify the possible causal nexus between the variables. The major advantage of this method is that it exploits individual time series and cross-sectional variations in data and avoids biases associated with cross-sectional regressions by taking into account the country-specific fixed effect.

This paper proceeds as follows. Section 2 provides a brief review of the nexus among ICT, FDI and Economic Growth. Section 3 offers methodology and variables used in our analysis. Section 4, our results are presented. Finally, we offer pertinent conclusions and implications in Sect. 5.

2 Literature Review

In the recent literature, many researchers already worked on telecommunication infrastructure, foreign direct investment and economic growth relations but mostly with an interest in the bi-variate relationship. Few studies were conducted on the multivariate relationship among the three variables.

Some researchers studied the relationship between telecommunication infrastructure and foreign direct investment (FDI), like Zeb, Qiang and Shabbir [1] who explored the role of the telecommunication infrastructure in stimulating FDI in Pakistan and found the positive significant effect of the infrastructure in attracting FDI to Pakistan; and Pradhan [2] who examined the impact of changes in infrastructure on foreign direct investment (FDI) inflows in Asian-21 countries and concluded that state investment into larger and more reliable telephone networks could potentially encourage future foreign investment.

Some researchers studied the relationship between telecommunication infrastructure and economic growth. Roeller and Waverman [3] examined the impact of investment in telecommunication infrastructure on the GDP of 21 OECD countries and found evidence of a significant positive causal link. Cieslik and Kaniewsk [4] established a positive and statistically significant causal link from telecommunications infrastructure to income at the regional level in Poland. Lei and Kingsley [5] empirically investigated the role of telecommunication infrastructure on regional economic growth in China and their results show that telecommunication was both statistically significant and positively related to regional economic growth in real GDP per capita in China. Karner and Onyeji [6] examined the contribution to economic growth of private telecommunications investment in African countries and their regression results indicated that the contribution was positive but insignificant. Shiu and Lam [7] studied the relationship between telecommunication infrastructure development and economic growth in 105 countries across the world and found a bidirectional relationship between real gross domestic product (GDP) and telecommunications development for European and high-income countries.

Other researchers have studied the relationship between foreign direct investment and economic growth. Zhang et al. [8] examined the FDI and economic growth relationship for 11 developing countries in East Asia and Latin America and found a positive relationship between the two variables. Wang [9] found that total FDI inflows significantly affected economic growth in 12 Asian economies. Choe [10] investigated how FDI and economic growth were related in 80 countries and confirmed that FDI and economic growth are Granger causality for each other. Li and Liu [11] investigated the relationship between foreign direct investment and economic growth in both developed and developing countries and revealed a strong complementary connection between FDI and economic growth. Khaliq and Noy [13] performed an analysis covering 12 sectors in the economy of Indonesia on the effectiveness of FDI and concluded that FDI can have a positive effect on the aggregate economic growth. In contrast to those researchers who studied the relationship between the two variables, we determine to pursuit the investigation on all three vital macro variables. Recently, Maneejuk and Yamaka [12] studied the impact of telecommunication technology and FDI on the growth of developing and developed countries and revealed heterogenous effects of these variables on economic growth.

3 Data and Methodology

In this study, annual data for the five ASEAN developing countries: Indonesia, Cambodia, Malaysia, the Philippines, and Thailand, spanning over 2000–2014 were obtained from the World Development Indicators published by the World Bank. In this analysis, economic growth is represented by GDP per capita growth; foreign direct investment is represented by foreign direct investment net inflows growth rate; telecommunications infrastructure is represented by the two possible variables using the principal component analysis: investment in telecoms with private participation and information and communication technology service exports. There are some other infrastructures that will interact with economic growth and foreign direct investment, such as electricity infrastructure, transport infrastructure; thus, we use electric power consumption per capita growth rate and transportation services growth rate as the control variables. Table 1 presents the definition of these variables.

The variables INTC and ICTE are converted into the natural logarithm for this estimation. To reach the objective of the study on modelling the relationship among telecommunication infrastructure, foreign direct investment and economic growth, it is required to satisfy the five main stages of econometric approach.

Table 1 Data description

Variable	Definition	Unit
TEL	Growth rate of composite index of telecommunications infrastructure: Combining INTC and ICTE using principal component analysis	%
INTC	Investment in telecoms with private participation	US dollars
ICTE	Information and communication technology service exports	US dollars
GDP	Gross domestic product per capita growth	%
FDI	Foreign direct investment net inflows growth rate	%
ELEC	Electric power consumption per capita growth rate	%
TRAE	Transport services growth rate	%

The panel VAR model entails the following three steps: first, a panel unit root test is performed to identify the stationarity (order of integration) of time series variables; second, a panel cointegration test is conducted to determine the existence of a long-run relationship between the time series variables; and third, a VAR model is constructed to ascertain the direction of causality between the variables. These three tests are discussed in more detail below.

3.1 *Principal Component Analysis*

Principal component analysis is a statistical procedure that uses an orthogonal transformation to convert a set of observations of possibly correlated variables into a set of values of linearly uncorrelated variables called principal components (or sometimes, principal modes of variation). In this study, we consider $INTC$ and $ICTE$ as two components of telecommunication indicators.

3.2 *Panel Vector Autoregressive Model*

The Panel Vector Autoregressive Model (PVAR Model) is a preliminary Multivariable analysis of the economic cycle. This model resembles a structural connection between the variables from one variable that increases. All variables will be analyzed using this model to consider a variable that looks stagnant, called stationary; in the sense that all variables are assumed to be endogenous and interdependent. The panel VAR model of order p in this study is presented by

$$Y_{it} = Y_{it-1}A_1 + \ldots + Y_{it-p}A_p + X_{it}\beta + u_i + e_{it}, \quad (1)$$

where $i \in \{1, 2, ..., N\}$ and $t \in \{1, 2, ..., T\}$. Y_{it} is the vector of endogenous variables (GDP, TEL, FDI), X_{it} is the vector of exogenous covariate; $A_1, ..., A_p$ and β are the parameters to be estimated. u_i and e_{it} are vectors of dependent variable-specific panel fixed-effects and idiosyncratic errors, respectively.

4 Estimation Results

Panel unit root test is used to verify the property of the data and verify the integration order to avoid spurious estimation. There are two main processes of panel unit root test: Common unit root processes and Individual unit root processes. Two kinds of unit root tests are utilized in this paper, Levin, Lin and Chu (2002) (LLC) test, and Im, Pesaran and Shin (2003) (IPS) test. The LLC test assumes homogeneity in the dynamics of the autoregressive coefficient for all the time series across the cross section units. The IPS test is more flexible, accurate and powerful than the LLC test because it is based on the assumption of cross-sectional independence and allows heterogeneity of coefficients across groups.

Table 2 presents the results of the panel unit root tests for each variable, with the existence of intercept and trend in the equation. The results shown are from the test of time-series level. Both LLC and IPS tests results indicate all series are stationary at the 1% significant level. Hence, the series are stationary at I (0).

In modelling the PVAR model, the optimal lag order in panel VAR is selected based on log-likelihood(LogL); sequential modified LR test statistic (LR); Akaike information criterion(AIC); Schwarz information criterion(SC) and Hannan-Quinn information criterion (HQ). The lag selection result is reported in Table 3 and find that the first-order panel VAR is the preferred model because this has the smallest AIC, SC and HQ.

Table 2 Panel unit root test

Variable	LLC	IPS
TEL	−6.9466***	−5.2286***
	(0.0000)	(0.0000)
GDP	−5.1626***	−3.2146***
	(0.0000)	(0.0007)
FDI	−7.5173***	−4.2569***
	(0.0000)	(0.0000)
ELEC	−8.6104***	−5.6879***
	(0.0000)	(0.0000)
TRAE	−795.4860***	−313.5150***
	(0.0000)	(0.0000)

Note "***" indicates the decisive evidence of rejecting the null hypothesis of unit root. Theparantheses ( ) denotes the Maximum Bayes factor (MBF) (see, [14])

Table 3 PVAR: Lag length criteria

X	LogL	LR	AIC	SC	HQ
1	−735.2409	80.6365	33.8773*	34.9613*	34.2814*
2	−734.3629	1.2877	34.2383	35.6836	34.7771
3	−731.9452	3.2235	34.5309	36.3375	35.2044
4	−731.0714	1.0485	34.8920	37.0600	35.7002
5	−712.4240	19.8905*	34.4632	36.9926	35.4062

To proceed with the PVAR modelling, there is a need to first check the lag length order and the stationarity of the time series. Based on the results of lag order selection in Table 3, the results confirm lag order 1 is the optimal lag length for this modelling. Meanwhile, Fig. 1 shows that all AR roots are in the unit circle, indicating that the panel var is stationary and satisfies the conditions for establishing the impulse response function.

Consequential to the PVAR estimation, the results are shown in Table 4. TEL_{t-1} shows a positive effect on GDP_t while TEL_{t-1} contributes a positive effect on FDI_t. Considering the exogenous variables, we find that $ELEC_t$ and $TRAE_t$ show a strong evidence impact on telecommunications infrastructure and foreign direct investment of ASEAN countries. To complement the analysis, the impulse response function is used to trace the effect of a one-off shock in one of the variables on the current and future values of the endogenous variables. The impulse responses offer additional insight into how shocks to the variables TEL, GDP and FDI impacted one another.

To have a robust result, this study conducts a stability test. The Inverse AR root characteristic is illustrated in Fig. 1. We can see that the model is stable because the roots of the companion matrix are all inside the unit circle. Now that we have established that the panel VAR model is stable, we can calculate Impulse Response Functions (IRFs) and Forecast Error Variance Decompositions (FEVDs).

As clearly shown in Fig. 2, the responses of TEL to the one standard deviation shocks of GDP and FDI are weak along the 10 periods ahead. However, we observe that the responses of GDP to the one standard deviation shocks of TEL and FDI are negative and positively, respectively. Regarding the responses of FDI, the results show that the effects of one standard deviation shock on the TEL increases the FDI growth rate above its steady-state value for all periods of analysis. However, this response is negative according to the shock of GDP, but this effect is short lived.

In order to know more exactly the degree of mutual influence between TEL, GDP and FDI, and the contribution of various shocks to endogenous variables, variance decomposition will be used to further estimate the importance of various shocks.

The variance decompositions from this model are presented in Table 5. These results suggest that GDP explains about 0.88% and 8.20% of the variation in TEL and FDI, respectively, in the next ten year. In addition, the TEL accounts for about

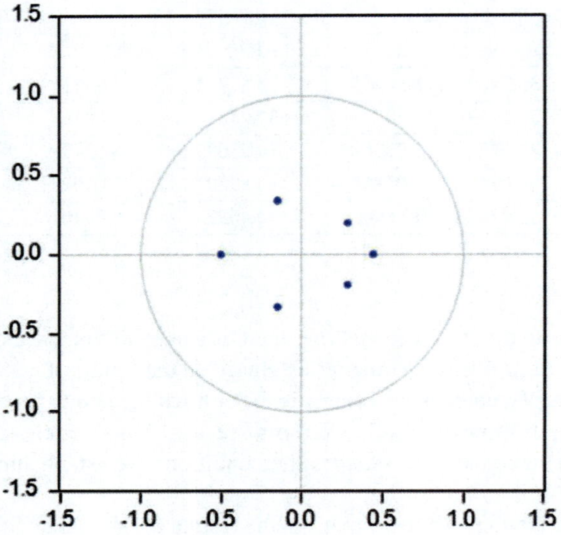

Fig. 1 Inverse AR roots

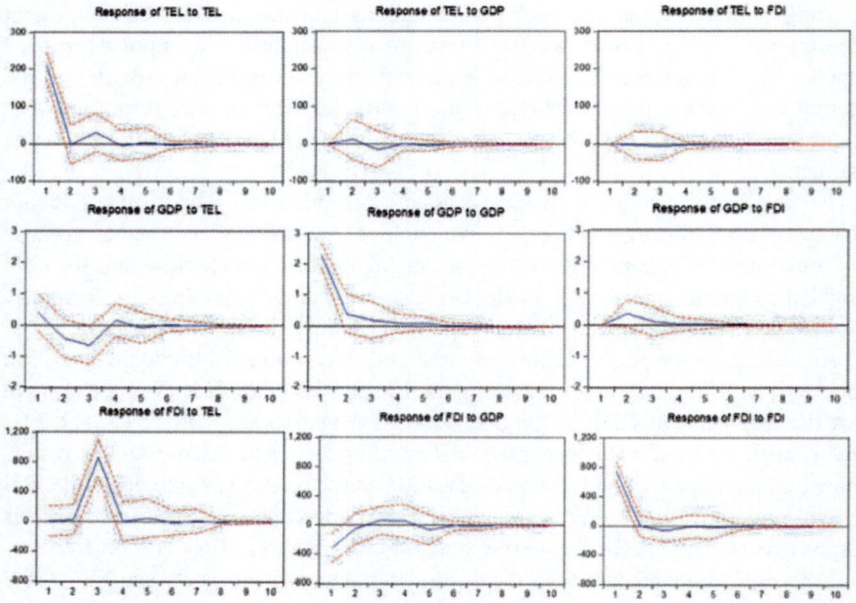

Fig. 2 Impulse response functions (Response to Cholesky One S.D. Innovations ± 2 S.E.)

Table 4 Estimation result

Dependent variable	TEL_t	GDP_t	FDI_t
Intercept	1.0823	1.3334***	0.6744***
	(0.2900)	(0.000)	(0.0000)
TEL_{t-1}	0.8340	0.5885	0.1777
	(1.0000)	(0.7451)	(0.9150)
GDP_{t-1}	6.7662**	3.0923	3.1047
	(0.0339)	(0.7839)	(0.2117)
FDI_{t-1}	6.4496***	0.2799	0.4367
	(0.0000)	(0.8694)	(0.7839)
$ELEC_t$	0.4324**	1.0824	1.3493***
	(0.0000)	(0.3849)	(0.0000)
$TRAE_t$	0.9993***	0.7388	0.7883**
	(0.0000)	(0.3553)	(0.0294)

Note "***" and "**" indicate the decisive and strong evidences of rejecting the null hypothesis of non-causal effect, respectively. The paratheses ( ) denotes the Maximum Bayes factor (MBF) (see, [14])

Table 5 Variance decomposition results in next ten years

		Impulse		
		TEL	GDP	FDI
Response	TEL	99.01%	0.88%	0.11%
	GDP	12.09%	85.53%	2.38%
	FDI	51.74%	8.20%	40.06%

12.09% and 51.74% of the change in GDP and FDI, respectively, after an equivalent period of analysis. Finally, for the case of FDI, it accounts for about 0.11% and 2.38% of the change in TEL and GDP, respectively.

5 Conclusion

The objective of this paper is to discover the relationship among telecommunication infrastructure, foreign direct investment, and economic growth in the ASEAN region. To achieve the objective, we employ the panel Vector Autoregressive model, Impulse Response Functions(IRFs), and Forecast Error Variance Decompositions(FEVDs), as the tools for analysis. Our results reveal that there exists a weak relationship among telecommunication infrastructure, foreign direct investment, and economic growth. When considering the IRFs, there is no difference in the telecommunication infrastructure if the shock stems from economic growth or foreign direct investment. However, the shock of telecommunication infrastructure produces a negative effect on economic growth. We also notice that the shock of this variable also contributes a positive shock on the foreign direct investment of ASEAN.

In order to shed more light on the analysis we applied the FEVD to our analysis. we find that telecommunication explains the greater variation in economic growth and foreign direct investment.

References

1. Nayyra, Z., Qiang, F., & Muhammad, S. (2014). Telecommunication infrastructure and foreign direct investment in Pakistan: An empirical study. *Global Journal of Management and Business Research, 14*(4), 1–6.
2. Pradhan, R. P., Arvin, M. B., Nair, M., Mittal, J., & Norman, N. R. (2017). Telecommunications infrastructure and usage and the FDI-growth nexus: evidence from Asian-21 countries. *Information Technology for Development, 23*(2), 235–260.
3. Roller, L. H., & Waverman, L. (2001). Telecommunications infrastructure and economic development: A simultaneous approach. *American economic review, 91*(4), 909–923.
4. Cieślik, A., & Kaniewska, M. (2004). Telecommunications infrastructure and regional economic development: The case of Poland. *Regional Studies, 38*(6), 713–725.
5. Ding, L., & Haynes, K. (2006). The role of telecommunications infrastructure in regional economic growth in China. *Australasian Journal of Regional Studies, The, 12*(3), 281.
6. Karner, J., & Onyeji, R. (2007). *Telecom private investment and economic growth: The case of African and Central & East European countries*. Jonkoping International Business School.
7. Lam, P. L., & Shiu, A. (2010). Economic growth, telecommunications development and productivity growth of the telecommunications sector: Evidence around the world. *Telecommunications Policy, 34*(4), 185–199.
8. Zhang, K. H. (2001). Does foreign direct investment promote economic growth? Evidence from East Asia and Latin America. *Contemporary economic policy, 19*(2), 175–185.
9. Wang, M. (2009). Manufacturing FDI and economic growth: evidence from Asian economies. *Applied Economics, 41*(8), 991–1002.
10. Choe, J. I. (2003). Do foreign direct investment and gross domestic investment promote economic growth? *Review of Development Economics, 7*(1), 44–57.
11. Li, X., & Liu, X. (2005). Foreign direct investment and economic growth: an increasingly endogenous relationship. *World development, 33*(3), 393–407.
12. Maneejuk, P., & Yamaka, W. (2020). An analysis of the impacts of telecommunications technology and innovation on economic growth. *Telecommunications Policy, 44*(10), 102038.
13. Khaliq, A., & Noy, I. (2007). Foreign direct investment and economic growth: Empirical evidence from sectoral data in Indonesia. *Journal of Economic Literature, 45*(1), 313–325.
14. Maneejuk, P., & Yamaka, W. (2021). Significance test for linear regression: how to test without P-values? *Journal of Applied Statistics, 48*(5), 827–845.
15. Holtz-Eakin, D., Newey, W., & Rosen, H. S. (1988). Estimating vector autoregressions with panel data. Econometrica: *Journal of the econometric society*, 1371–1395.

Industry Characteristics and Elders' Labor Demand in Thailand

Pacharaporn Arkornsakul, Pacharakamon Arkornsakul, Thitima Puttitanun, and Woradee Jongadsayakul

Abstract Current demographic shifts leading to aging workforce pose a threat to Thailand's economic growth. With this change, a research on elderly labor market is needed. This study aims to understand the overall labor demand opportunities for the elders in Thailand. Using the data from the National Statistical Office's Labor Force Survey Whole Kingdom from 2010 to 2018, the results show that the industrial sector that hire more of the elderly workers have high gross capital or large firm size, and companies with less technology advancement. This article suggests that the elderly should acquire technical skills in order to make themselves attractive which would lead to an increase in employability. Moreover, government units should provide technical training to the elderly who require it, as well as design specific policies to meet regional needs.

Keywords Elderly labor · Elderly worker · Labor demand · Labor market

1 Introduction

Aging society is a critical situation in which many countries are currently aware of the impact of demographic transition on their societies. Thailand has also been facing with this crisis. The number of Thai elderly people continue to climb that can cause a wide range of consequences, such as labor shortage, decrease in productivity, low savings, etc. The national development plan, such as the Eleventh National Economic and Social Development Plan (2012–2016) and the Twelfth National Economic

P. Arkornsakul (✉)
Department of Economics, Faculty of Management Sciences, Chiang Mai Rajabhat University, Chiang Mai, Thailand
e-mail: pacharaporn_ark@g.cmru.ac.th

P. Arkornsakul
Faculty of Arts, Thailand National Sports University Chiang Mai Campus, Chiang Mai, Thailand

T. Puttitanun · W. Jongadsayakul
Department of Economics, Faculty of Economics, Kasetsart University, Bangkok, Thailand

S. Sriboonchitta et al. (eds.), *Credible Asset Allocation, Optimal Transport Methods, and Related Topics*, Studies in Systems, Decision and Control 429,
https://doi.org/10.1007/978-3-030-97273-8_46

and Social Development Plan (2017–2021) highlights the aging society problem and guide the relevant entities in resolving this crisis. To cope with labor shortage problem, many European countries try to prolong a retirement of elderly workers. Therefore, we need to understand whether employers would like to hire elderly workers, and if so, what types of work they want to hire.

According to the definition of "Aging" given by the United Nations [34], aging is an individual who is over 60 years old. Similarly, the elderly act of Thailand defines "Elderly" as an individual who is over 60 years old (The Act on the Elderly, 2003 A.D.). The UN also defines a country with a population aged over 60 years old for more than 10%, or aged over 65 years old for more than 7% of the overall population as becoming an "Aging Society". The country will turn into an "Aged Society" when the number of populations aged over 60 years old rises to 20%, or 14% of the population aged over 65 years old. The country will eventually become a "Super-Aged Society" when the population aged over 60 years old reaches 28%, or 20% of the population aged over 65 years old. Therefore, in 2005, Thailand became an aging society by having 6,161,712 aging populations out of 60,549,342 populations, which accounted for 10.18% [24]. According to the prediction from the Office of the National Economic and Social Development Council, Thailand will become a complete aged society in 2028, and in 2031 it will become a super-aged society [6].

The tremendous increase of elderly populations was affected mainly by the advancement of technology and policy, such as a decrease in death rate and an increase in longevity due to the advancement in medical technology and public health service, a reduction in the birth rate due to the government birth control campaign in 1970, a tightening of economic status which preserves elderly in the labor market, and an independent woman trend [11].

Even though Thailand in the past decade had a high economic growth rate, the rising of the aging population could negatively affect the per capita income due to the reduction of employment rate when the age went up [7]. Changing in population structure results in economic and social structural changes, such as a decrease of labor force leading to a reduction of overall production, a loss of government's tax income, an increase of caretaking expenses for elderly.

Nevertheless, healthy elderlies still have a capability as a workforce, and they are likely to continue working to enhance their self-esteem, gain income, and reduce dependency. As shown in Fig. 1, the numbers of elderly workers were increasing from 2012 to 2019 which implies an increase of demand for work among elders.

Even though at the present elderly people's demands for work are increasing, the opportunity to enter the labor markets for the elderly is low [32]. Several institutions both public and private have prepared to support and promote elderlies to work. Recently, the government has created a project that aims to promote the elderly to work, to have income, and to utilize their knowledge and experience for benefiting organizations. The project has been announced as a Royal Decree issued under the Revenue Code regarding exemption of income tax No. 639 in 2017. The decree aims to promote and support private organizations on hiring elderlies by giving tax benefits to the organization that hires older people aged from 60 years old. Another benefit

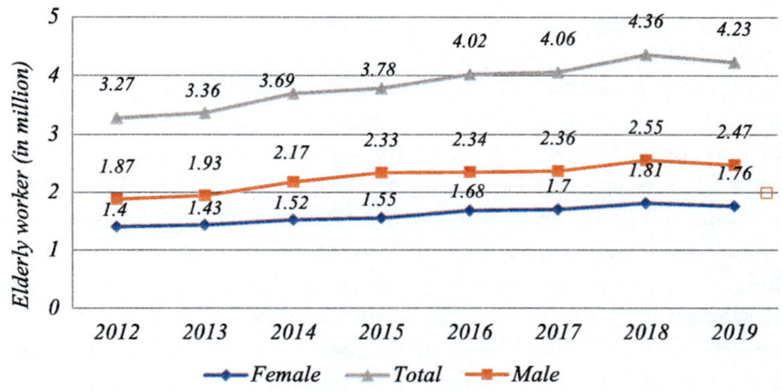

Fig. 1 Number of the elderly workers aged over 60 years old categorized by gender in 2012–2019. [25]. *Source* The surveys of the working population from 2012 to 2019. National Statistical Office

for the organizations that hiring the elderly is that they will be allowed to pay tax at less rate.

There are many studies focused on analyzing the guidelines, measures, and factors for promoting elderly employment especially in economic sectors, such as manufacturing, wholesale and retail trade, hotels and restaurants, etc. [2, 31]. However, there is a lack of study on factors determining industry employment of elderly workers in Thailand. Therefore, the aims of this study is to understand the current elderly employment and identify industry characteristics that lead to elders' labor demand in Thailand. The results of this study will help the government or policymakers in designing the policy on promoting elderly employment in Thailand, which will help to improve and enhance the elderly's quality of life and well-being in the future.

2 Literature Review

2.1 Theory of Demand for Labor

The labor market provides a means by which employers find the labor they need, while millions of individuals offer different types of labor services. The demand for any factors of production, including labor, is derived from demand for goods and services. This means that the demand for labor depends on, or is derived from consumers' demand for the products made using labor as an input. Thus, if consumers demand more goods or services, firms will require more labor to supply these products or services. Since the demand for labor is a "derived demand", the strength of the demand for any type of labor will depend on (1) how effective the employed labor is in producing a product or service, and (2) the market value of that item, including the number of competitors producing the same product, and the factors affecting the decision to produce it.

In this theory, the appropriate number of workers required by a firm is simply determined by profit maximization. Employers will hire workers so long as each successive worker adds more to the firm's total revenue than its total cost. The amount that each successive unit of labor adds to total revenue is measured by the marginal revenue product (MRP). The amount that a worker adds to total costs is measured by the marginal wage cost (MWC), which means that adding one more labor unit could affect the total wage cost. Therefore, in order to reach profit maximization, the firm will hire workers up to the point at which MRP = MWC [20].

Moreover, the theory concludes that the main factors affecting the demand for labor are the wage rate, demand for the products, labor productivity, number of employers, and prices of other resources. When the wage rate is getting higher, the demand for labor will be decreasing because it is more costly to hire extra employees. Firms will look to substitute capital for labor, or cheaper labor for relatively expensive labor. If demand for a firm's product increases, it will demand more labor and thus hire more workers. On the other hand, if demand for a firm's output of goods and services decreases, in turn, it will require less labor. The demand for labor also depends on the productivity, or the output per worker. If workers are more productive, they will be in greater demand. The productivity is influenced by skill levels, education and training, and the use of technology. Price changes in other inputs such as capital, land, and raw materials can also shift the demand curve for labor. Labor and capital are substitutes in production, meaning that a given quantity of output can be produced with significant capital and little labor, or significant labor and little capital. Thus, there is an impact on the demand for labor when the price of capital declines.

2.2 *Previous Studies on Elder Labor Demand*

Khemthong et al. [14] found that the size of the elderly population affects elderly labor demand. As the number of elderly increases, the demand to employ elderly workers will also increase. However, Zacher [36] stated that firms might perceive it to be a disadvantage to hire older workers because of their lower productivity or declining health, potentially reducing management flexibility and increasing personal expenses. Furthermore, the technology and innovation industry is unlikely to employ elderly workers due to the potential difficulties they may encounter with new technology. Firms may be reluctant due to the cost involved and the likelihood that such investment may not be worthwhile [28].

Regarding to firm characteristics, Suksiriserekul [32] revealed that organizations with a high level of employee expenditure, in other words, a high number of employees, were not likely to employ elderly workers when they want alleviate the shortage of skilled labor. Moreover, Raksasap et al. [28] found that industry sector type affected the decision to hire elderly workers. Soonthornchawakan and Cintakulchai [31] studied the number of Thai elderly workers in the economic sectors classified by the International Standard Industrial Classification (ISIC). These researchers pointed out that apart from agriculture and fisheries, the economic sectors such as retail trade, automotive repairs, and personal and household goods have the highest concentrations of male and female elderly workers. The industrial sector was ranked second highest for the proportion of elderly workers. The hotel and restaurant sector, which has an intensive labor force, has the third largest concentration of elderly workers and older adults. Besides agriculture and fisheries, elderly people are also employed as service staff, such as a shop salesperson, product presenter, or product demonstrator. Furthermore, firms planning for elderly employment were likely to hire elderly workers than those without any elderly employment plan. The employment plans are, such as, re-employment policy for retirees who perform well, creating family working environment, and providing work flexibility in terms of time and place. However, hazardous occupations or the exposure of the extreme environment conditions of the job (noise, chemicals, etc.) can also prevent elderly to continue to work both male and female [10].

Lee [16] studied the effects of industrial characteristics on the employment of older manufacturing workers in the USA during the early twentieth century. The findings reveal that the industrial characteristics favorably relate to the employment of older industrial workers, such as high labor productivity, less capital and material-intensive production, short workdays, low work intensity, high job flexibility, and formalized employment relationships. Furthermore, Raksasap et al. [28] indicated that the employer's decision to hire elderly workers depended on the economic situation. If a country's economic situation is good, employers are likely to hire more elderly workers. In contrast, if a country lacks economic growth, elderly workers are likely to be laid off.

According to the literature, most studies have studied labor demand gathered data by interviewing senior workers in specific business owner, specific organizations, or specific provinces. None had looked at the issue at the aggregate country level. Therefore, to full in the gap in the literature, the current study examines the demand for elderly employment, as well as individual and macroeconomic characteristics. In addition, this study reveals the need for elderly employment across all industrial sectors, allowing for the identification of the industries that best suit elderly employment.

3 Methodology

3.1 Data Collection

This study uses the secondary data on the number of elderly workers in each industry sector in Thailand surveyed by the National Statistical Office. The industry classification is classified based on the International Standard Industrial Classification (ISIC) Revision 3, 1990 of the United Nations. The data used in this research is a panel data of twelve industries between 2010 and 2018 (9 years). The twelve industries are (1) agriculture, forestry and fishing, (2) manufacturing, (3) construction, (4) wholesale and retail trade, (5) transportation storage, (6) accommodation and food services, (7) financial and insurance activities, (8) real estate activities, (9) public administration and defense, (10) education, (11) human health and social work, and (12) other service activities.

To proxy for the demand for elderly worker, this research will use the proportion of elderly workers to the total employment. Even though the proportion might not be the actual demand for elderly workers but at least it is employment of the elderly workers which is the demand for elderly workers that is met by the supply of elderly workers.

For the independent variables, the gross industrial production and the gross capital stock are used to represent firm size. The wage rate is used to indicate the elderly labor's abilities and experience. The number of computer using and internet using are used to represent skills and knowledge of the elderly labor. Work flexibility is represented by the number of working hours. The value of computer imports and the number of patent applications are used to indicate the technological advancement that each industry adopted. The number of workplace hazards is used to illustrate the industries that are particularly hazardous to senior workers. Table 1 shows the sources of all variables.

3.2 Data Analysis

As we used a panel data, a collection of cross-sectional data and time-series data, thus, we need to test a stationary of the data before putting it into the analyzing process in order to make it consistent and reliable as well as to prevent spurious regression. In this study, the researcher adopted Levin, Lin and Chu (LLC) method [17], one of panel unit root tests, to test the stationary of the data. Next, we estimate the fixed effect model and the random effect model by using the Hausman test to select the model for our analysis.

The empirical model used in this study is formed as follows:

$$\begin{aligned}\ln(d\_elder_{it}) = {} & \beta_0 + \beta_1 \ln(gip_{it}) + \beta_2 \ln(wage_{it}) \\ & + \beta_3 \ln(workhours_{it}) + \beta_4 \ln(capital_{it}) \\ & + \beta_5 \ln(computer_{it}) + \beta_6 \ln(internet_{it}) \\ & + \beta_7 \ln(hazard_{it}) + \beta_8 \ln(computer\_imp_t) \\ & + \beta_9 \ln(patent_t) + \alpha_i + \varepsilon_{it}\end{aligned}$$

where

$\ln(d\_elder_{it})$ is natural log of the proportion of elderly employment in the industry i at time t (the dependent variable).

Table 1 Data sources used to construct each variable

Variables	Sources
Dependent variable	
Proportion of elderly workers to total employment collected from the Labor Force Survey Whole Kingdom	National Statistical Office
Independent variable	
Gross industrial production, chain volume measures (reference year = 2002) collected from the Household Survey on the Use of Information and Communication Technology (millions of baht)	National Statistical Office
Average wage in each industry (baht per month)	Bank of Thailand
Average working hours in each industry (hours per week)	National Statistical Office
Gross Capital Stock in each industry (millions of baht)	Office of the National Economic and Social Development Council
The employed population aged 15 years and over that use computers in each industry collected from the Household Survey on the use of Information and Communication Technology (persons)	National Statistical Office
The employed population aged 15 years and over that use internet in each industry collected from the Household Survey on the use of Information and Communication Technology (persons)	National Statistical Office
The number of workplace hazard (persons)	National Statistical Office
Thai computer import as a percentage of commercial service imports by year	World Development Indicators
The number of patents application (numbers)	Department of Intellectual Property, Ministry of Commerce

$\ln(gip_{it})$ is natural log of gross industrial production by industry i at time t.

$\ln(wage_{it})$ is natural log of wage by industry i at time t.

$\ln(workhours_{it})$ is natural log of working hours per week by industry i at time t.

$\ln(capital_{it})$ is natural log of gross capital stock classified by industry i at time t.

$\ln(computer_{it})$ is natural log of the number of computer using by industry i at time t.

$\ln(internet_{it})$ is natural log of the number of internet using by industry i at time t.

$\ln(hazard_{it})$	is natural log of the number of workplace hazard by industry i at time t.
$\ln(computer\_imp_t)$	is natural log of the computer using at time t.
$\ln(patent_t)$	is natural log of the computer using at time t.
β	is coefficient parameters.
α_i	is an industry specific effect.
ε_{it}	is an error term.

All variables are in natural log for the ease of interpretation. Since taking natural log is a monotonic transformation, the direction of the effects of dependent variables on the demand for elderly is still preserved. Moreover, the interpretation of all variables will be in the same unit (percentage change).

4 Results and Discussion

4.1 Descriptive Statistics

A summary of descriptive statistics (mean values and standard deviations for relevant variables) is presented in Table 2. As can be inferred from the relatively large values of the standard deviation, the "other service activities" has the highest proportion of elderly employment among 12 industrial sectors (S.D. = 0.439, 0.232% of total workers), which the proportion of elderly employment in each of Thailand's industrial sectors is calculated from the number of elderly workers divided by the total number of workers in the same sector. Next, the agricultural sector becomes the second largest sector regarding the proportion of elderly workers (S.D. = 0.035, 0.164% of total workers). Thailand's development has been mostly based on agriculture, creating numerous career opportunities for Thai citizens. It is also a popular choice for many elderly individuals since it tends to suit their level of knowledge and capability. The result confirms the studies in Asian and European countries which found that agricultural works are mostly run by elderly people and the number of elderly farmers has been greater than before while the number of young farmers has been decreasing [16, 27, 37]. Furthermore, the sector with the lowest number of elderly workers is the financial sector (S.D. = 0.003, 0.012% of total workers). Senior people working in the financial sector mostly hold executive positions with a high degree of experience and pay. As a result, fewer old people can work in this field. Moreover, working in financial firms nowadays requires knowledge and skills related to technology such as computer skills. It obstructs elderly to work in this field and frequently be disqualified when comparing with the younger people [22].

According to Conen et al. [5], attempts to hire older employees vary depending on the economic climate. During the recession, older worker recruitment dropped dramatically. Therefore, the macroeconomics variables such as GIP, wage, and gross capital stock are also included in this study. The descriptive statistics showed that the manufacturing sector has the highest GIP with an average of 2,655.614 billion

Table 2 Descriptive statistics and summary of the key variables

Variable	Agriculture (S.D.)	Manufacturing (S.D.)	Construction (S.D.)	Trade (S.D.)	Transportation (S.D.)	Accommodation (S.D.)
Elderly workers (unit: per thousand persons)	2,252.840 (211.130)	292.306 (44.742)	99.040 (20.053)	544.940 (68.660)	62.754 (15.679)	187.944 (36.380)
All workers (unit: per thousand persons)	14,100 (1,934.627)	5,798.377 (573.208)	2,113.022 (135.990)	6,124.120 (127.787)	1,115.369 (143.587)	2,571.724 (214.965)
Proportion of elderly workers, d_elder (unit: % of total workers)	0.164 (0.035)	0.050 (0.004)	0.047 (0.010)	0.089 (0.010)	0.056 (0.009)	0.073 (0.009)
Gross industrial production, gip (unit: billion baht)	638.650 (25.405)	2,655.614 (144.581)	252.760 (28.734)	1,347.515 (141.410)	592.629 (80.390)	461.837 (118.120)
Wage, wage (unit: baht per month)	13,936.750 (902.923)	17,126.860 (1,907.428)	12,184.150 (1,228.338)	18,329.180 (1,472.478)	13,334.750 (1,446.138)	13,136.810 (1,030.846)
Working hours, workhours (unit: hours per week)	39.128 (1.321)	48.856 (0.678)	45.833 (2.570)	49.244 (3.066)	47.689 (4.631)	48.311 (3.216)
Gross capital stock, capital (unit: billion baht)	3,337.384 (376.988)	6.766 (622.112)	1,196.964 (152.208)	2,563.275 (114.706)	7,523.356 (672.823)	1,403.942 (146.230)
Number of computers used, computer (unit: per thousand persons)	677.295 (182.758)	1,487.350 (300.625)	304.293 (42.341)	1,946.661 (252.437)	338.811 (61.674)	558.082 (116.633)
Number of internet used, internet (unit: per thousand persons)	3,009.222 (5,050.204)	2,394.358 (1,368.495)	500.554 (310.029)	2,711.632 (1,222.061)	482.229 (240.938)	969.325 (544.743)
Workplace hazards, hazard (unit: persons)	697.333 (132.604)	69,931.440 (26,725.160)	10,176 (1,400.811)	14,433.780 (2,398.968)	3,935.667 (764.752)	5,152.889 (437.482)
Computer import, computer_imp (unit: % of commercial service imports)	36.123 (2.248)	36.123 (2.248)	36.123 (2.248)	36.123 (2.248)	36.123 (2.248)	36.123 (2.248)
Patents applications, patent (unit: number)	4,006.778 (624.679)	4,006.778 (624.679)	4,006.778 (624.679)	4,006.778 (624.679)	4,006.778 (624.679)	4,006.778 (624.679)

(continued)

Table 2 (continued)

Variable	Financial (S.D.)	Estate (S.D.)	Defense (S.D.)	Education (S.D.)	Health (S.D.)	Others (S.D.)
Elderly workers (unit: per thousand persons)	5.806 (1.761)	19.568 (6.755)	39.372 (8.520)	29.358 (8.764)	17.017 (3.101)	63.781 (18.061)
All workers (unit: per thousand persons)	478.436 (68.047)	226.007 (190.505)	1,596.650 (69.852)	1,205.246 (50.405)	680.148 (37.250)	662.639 (249.286)
Proportion of elderly workers, d_elder (unit: % of total workers)	0.012 (0.003)	0.103 (0.029)	0.025 (0.005)	(0.025) 0.008)	0.025 (0.005)	0.232 (0.439)
Gross industrial production, gip (unit: billion baht)	595.819 (131.299)	349.673 (35.368)	496.890 (20.682)	315.400 (14.107)	199.564 (21.924)	133.751 (17.796)
Wage, wage (unit: baht per month)	12,457.710 (1,343.739)	12,534.600 (1,176.546)	14,718.150 (1,306.546)	14,705.780 (1,171.772)	12,807.280 (1,278.111)	11,659.270 (1,127.830)
Working hours, workhours (unit: hours per week)	42.456 (0.361)	46.933 (1.576)	41.044 (3.828)	37.767 (1.448)	44.556 (1.849)	45.789 (1.995)
Gross capital stock, capital (unit: billion baht)	418.096 (20.902)	7,584.734 296.288	630.674 (52.862)	732.709 (72.846)	1,350.122 (162.732)	840.587 76.231
Number of computers used, computer (unit: per thousand persons)	417.090 (106.025)	114.581 (106.025)	1,035.482 (46.335)	1,027.980 (27.173)	472.003 (23.722)	177.060 (39.133)
Number of internet used, internet (unit: per thousand persons)	428.361 (97.120)	131.213 (89.414)	1,117.363 (161.475)	1,035.380 (99.587)	504.655 (110.728)	296.680 (162.976)
workplace hazards, hazard (unit: persons)	672 (306.560)	710.333 (65.374)	2,127 (869.202)	678.667 (293.641)	1,145.333 (194.257)	2,101.111 (1,173.701)
Computer import, computer_imp (unit: % of commercial service imports)	36.123 (2.248)	36.123 (2.248)	36.123 (2.248)	36.123 (2.248)	36.123 (2.248)	36.123 (2.248)

(continued)

Table 2 (continued)

Variable	Financial (S.D.)	Estate (S.D.)	Defense (S.D.)	Education (S.D.)	Health (S.D.)	Others (S.D.)
Patents applications, patent (unit: number)	4,006.778 (624.679)	4,006.778 (624.679)	4,006.778 (624.679)	4,006.778 (624.679)	4,006.778 (624.679)	4,006.778 (624.679)

Source Researcher's calculation

baht followed by the trade sector and agricultural sector with 1,347.515 and 638.650 billion baht, respectively. In terms of wage, the financial sector has the highest average wage of 24,928.78 baht per month, followed by the education sector at 21,781.50 baht per month. These results imply that sectors requiring highly skilled (white-collar) workers provide a higher wage rate than those requiring low-skilled (blue-collar) workers such as the agricultural sector. However, the proportion of elderly workers in high wage industries is relatively low (as can be seen from Table 2), implying that the elderly tend to be less qualified in knowledge-intensive sectors and less likely to earn a high wage. Regarding gross capital stock, the estate sector has the highest level at 7,584.734 billion baht, followed by transportation and manufacturing sectors with 7,523.356 and 6,766.014 billion baht, respectively. Industries with large capital stock tend to have a better opportunity to hire elderly workers. As indicated by Table 2, the proportion of elderly workers in the estate sector is relatively high. Even though elderly workers are often considered by employers to be more costly than younger workers, they are often recruited for their positive attributes of experience and wisdom (Recruitment and Consulting Services Association, 2015).

In terms of working hours, the trade sector has the highest average number of working hours with 49.244 h per week followed by the manufacturing and accommodation sectors where the average is 48.856 and 48.311 h per week, respectively. In the trade sector, the work is less physically demanding, and also has flexible working practices in terms of time and place. As a results, this sector has the second highest proportion of elderly workers. The result is consistent with Phijaisanit [26] which stated that elders are more likely to work in a firm that allows them to live the life of an elderly person.

4.2 Panel Data Analysis

Basically, time series data on macroeconomics variables varies according to time, known as non-stationary. Therefore, the panel unit root test is used prior to analyzing the data to test for stationarity and prevent spurious regression. In this study, the Levin-Lin-Chu test is applied to test the stationarity of the data. The results from the LLC test show that all nine variables potentially affecting elderly labor opportunities are stationary with decisive evidence, according to MBF. As shown in Table 3, the results are all stationary.

Then, to ensure robust results, the random effect and fixed effect performances need to be compared. Based on Hausman test result, the factors affecting elderly labor opportunities in Thailand can be estimated by the random effect model as shown in Table 4.

GIP (gip) was found to have an inverse relationship with opportunities for elderly workers. This implies that when other factors remain constant, a 1% increase in the country's GIP decreases the proportion of opportunities for elderly workers by 0.536% (as shown in Table 4). Generally, when GIP increases, firms tend to accelerate productivity to maximize income. Adopting technologies in the production process

Table 3 Panel unit root test at level

Variable	LLC (MBF.)	Stationary level	Result
ln(gip)	−15.5429[a] (0.0000)	I(0)	Stationary
ln(wage)	−6.6978[a] (0.0000)	I(0)	Stationary
ln(workhours)	−6.2459[a] (0.0000)	I(0)	Stationary
ln(capital)	−14.2113[a] (0.0000)	I(0)	Stationary
ln(computer)	−4.7576[a] (0.0000)	I(0)	Stationary
ln(internet)	−9.7702[a] (0.0000)	I(0)	Stationary
ln(hazard)	−4.0764[a] (0.0000)	I(0)	Stationary
ln(computer_imp)	−8.0796[a] (0.0000)	I(0)	Stationary
ln(patents)	−9.0612[a] (0.0000)	I(0)	Stationary

Note Values in parenthesis refer to Maximum Bayes factor (MBF)
[a] Indicates decisive evidence supporting the stationary [18]

seems to be one of the most effective ways to help increase productivity. Many firms tend to replace the human workforce by adopting new technological innovations to help solve the shortage of labor and boost productivity. As firms apply more technology in their production processes, many employees feel that young workers are significantly more productive than older workers, and can easily adapt to new technologies due to their cognitive abilities and other factors which generally decrease with age [15]. Allen [1] concluded that GDP per capita grows slowly when the aging population becomes larger because of lower growth in productivity.

Wage (ln(wage)) was found to have a positive relationship with opportunities for elderly workers. Wage is one of the most important factors affecting the employer's decision to re-employ older workers. Employers always consider the wage rate offered to older workers based on their experience and health [12, 13]. Retirees always expect to receive a high salary or, at least, the same rate as before, based on their experience, skills, and wisdom, with most occupying senior positions before mandatory retirement. Therefore, firms that can pay a high wage rate will be able to rehire valuable retirees. Employers probably decide to reduce the wages of the elderly depending on their productivity during the post-retirement contract [23]. Therefore, elderly workers need to maintain or even develop their skills and abilities to retain their existing wage rate and to show that they are worth paying.

The gross capital stock (ln(capital)) of each industrial sector was found to have a positive relationship with the proportion of opportunities for elderly workers. Gross

Table 4 Analysis of factors affecting the elderly labor demand using the panel model

Variables	(1) Fixed effect	(2) Random effect
ln(gip)	−0.700	***−0.536***[c]
	(0.426)	**(0.150)**
ln(wage)	1.033	***2.553***[b]
	(1.595)	**(1.293)**
ln(work hours)	0.259	**0.717**
	(0.691)	**(0.872)**
ln(capital)	0.0549	***0.441***[c]
	(0.568)	**(0.104)**
ln(computer)	−0.138	***−0.632***[c]
	(0.216)	**(0.237)**
ln(internet)	−0.047	***0.466***[c]
	(0.087)	**(0.139)**
ln(hazard)	0.067	**0.036**
	(0.085)	**(0.064)**
ln(computer_imp)	−0.156	**0.406**
	(0.715)	**(1.280)**
ln(patent)	0.874	***−1.700***[a]
	(0.925)	**(0.975)**
Constant	−10.240	**−14.98**[b]
	(10.390)	**(6.579)**
Observations	108	**108**
Number of id codes	12	**12**
R-squared within	0.1627	**0.0227**
R-squared between	0.0195	**0.6089**
R-squared overall	0.0279	**0.4592**

Note Values in parenthesis refer to t-statistics. Then this t-statistics is calibrated as Maximum Bayes factor (MBF)
[a,b,c] Indicate strong, substantial, and decisive evidence supporting the impact of this factor on elderly labor demand, respectively. Maneejuk and Yamaka [18]

capital stock is one of the factors indicating the size of a firm and its financial status. Therefore, larger firms are more likely to be able to pay high wages than smaller ones. Even though older workers are more expensive than younger workers, this does not cause any problems for large-scale firms. According to the Survey of Attitudes towards Older Workers [30], workers in their 50 s are rated as good regarding their positive attributes (e.g., leadership, experience, and wisdom) by recruiters. Moreover, older workers are less likely to leave their jobs than younger workers, indicating that older workers are more loyal [3, 8].

The number of patents (ln(patent)) has a negative impact on labor opportunities for the elderly. The number of patents is used to indicate the technology activities of the industry. Firms adopting high technology may feel reluctant to employ elderly workers since cognitive abilities generally decrease with age [15]. Moreover, not only do employers feel reluctant to hire elderly workers, but the elderly themselves may also feel uncomfortable working in a firm with high technology.

The computer usage (ln(computer)) in each industry has a negative relationship with the proportion of opportunities for elderly workers. Technological progress has made computer-based technologies essential for many businesses. Firms tend to focus more on recruiting workers with high technology skills. Many past studies have shown that employers see elderly people as possessing low technological skills, find it difficult to adapt to new computer technology, while also being disinterested in technological change [19, 21, 33]. Even though the number of older adults using computers, tablets, and other technology is growing [9], firms may be reluctant to invest in the training of older workers since: (1) they have only a short time in which to benefit the firm, (2) it is hard for them to learn something new due to decreasing cognitive skills, and (3) older workers are less motivated to spend their time in training sessions [15]. Therefore, our result supports the study done by Beckmann [4] which stated that the adoption of technological and organizational innovations by firms decreases the demand for older workers and increases the demand for younger workers.

The internet usage (ln(internet)) positively related to the proportion of opportunities for elderly workers. For elderly workers, changes in the work environment through the adoption of new technology may benefit older workers in terms of flexibility. According to Ratchagit [29], firms tend to hire elderly people to observe technological systems rather than inventing related technology. Since the internet has now become part of everyday life for people of all ages, the elderly need to acquire digital skills to improve their quality of life and meet the employer demand [35].

5 Conclusions and Implications

The random effect models are chosen based on the Hausman test to analyze the data. The results indicate that six variables were found to have a significant correlation with elderly labor demand. Those are gip, wage, gross capital stock, patents, computer use, and internet use. The results suggest that the sectors adopting high technology are less likely to hire elderly workers. Large firms have greater capability for hiring elderly workers since they can provide high salaries. Moreover, the industries that prefer hiring elderly workers are those with low technology. This might be because they are not as technology-able as those who are youngers. Combining these two sides of the labor market for the elderly, the elderly who would like to be hired need to increase and enhance their knowledge especially in terms of new technology to be attractive in the labor market.

This study suggests that organizations should provide work to match the knowledge and skills of the elderly. At the sometime, governmental units should also provide training to help the elderly develop the knowledge and skills necessary for working with other generations and adapt to the changes they face. Even though, the people who are near their retirement age have a better technological knowledge than the older generation, their ability to learn and acquire technical skills and knowledge are still going to be slower than the new generations such as Generation Z and Generation Alpha. Moreover, a campaign and policies should also be implemented to attract the elderly into the labor market and encourage employers to retain and hire older people, such as health insurance, tax reduction, and other benefits specifically for employers who hire elderly workers. Moreover, health conditions of the elderly are also one of the main reasons that affect employer decisions whether to reemploy the retirees. Therefore, the government should promote physical activities and exercise to all group of people in order to make them to be a healthy elderly in the future and to prolong their working life.

For the future study, to study the demand for elderly workers, more detailed data is needed. Currently, we use industrial level data, in the future, there should be firm survey data on what is needed for them to hire elderly workers.

References

1. Allen, S. G. (2019). *Demand for older workers: What do economists think? What are firms doing?* (Online). www.nber.org/papers/w26597. February 11, 2021.
2. Amonvatana, C., et al. (2008). *The study on guidelines and measures to promote employment in the elderly.* Developmemt Strategies During the Tenth National Economic and Social Development Plan into Practice Project. Report to Office of the National Economic and Social Development Council. Chula Unisearch
3. Bangali, L. (2004). *Are older workers genuine assets for the economy? Strategies and possibilities for the effective use of their human resource potential.* Working paper, Institute for Political Science in Cooperation with Research Institute for Work, Technology and Culture (F.A.T.K.1) Tuebingen, Germany: Eberhard Karls University Tuebingen.
4. Beckmann, M. (2007). *Age-biased technological and organizational change: Firm- level evidence and management implications* (No. 05/07). WWZ Discussion Paper. University of Basel.
5. Conen, W. S., Henkens, K., & Schippers, J. J. (2011). Are employers changing their behavior toward older workers? An analysis of employers' surveys 2000–2009. *Journal of Aging & Social Policy, 23*(2), 141–158.
6. Foundation of Thai Gerontology Research and Development Institute. (2016). *Situation of Thai Elderly 2013.* Bangkok.
7. Giles, J., & Huang, Y. (2015). Are the elderly left behind in a time of rapid demographic and economic change? A comparative study of the poverty and well-being of East Asia's elderly. *Journal of Public Economics, 90*(12), 2229–2255.
8. Hedge, J. W., Borman, W. C., & Lammlein, S. E. (2006). *The aging workforce: Realities, myths, and implications for organizations.* American Psychological Association.
9. Hewitt, A. (2015). A business case for workers age 50+: a look at the value of experience (Online). www.states.aarp.org/wp-content/uploads/2015/08/A-Business-Case-for-Older-Workers-Age-50-A-Look-at-the-Value-of-Experience.pdf. March 2, 2021.

10. Hirsch, B. T., Macpherson, D. A., & Hardy, M. A. (2000). Occupational age structure and access for older workers. *Internation Labour Review, 53*(3), 401–418.
11. Institute of Research and Development. (2015). *The research on mechanisms for promotion of employment of older persons who are informal laborers*. Khon Kaen University. (Mimeographed).
12. Karpinska, K., Henkens, K., & Schippers, J. (2011). The recruitment of early retirees: A vignette study of the factors that affect managers' decisions. *Ageing and Society, 31*, 570–589.
13. Karpinska, K., Henkens, K., & Schippers, J. (2013). Hiring retirees: Impact of age norms and stereotypes. *Journal of Managerial Psychology, 28*, 886–906.
14. Khemthong, S., et al. (2018). *Employment policy on elderly staffs in the Thai hotels chain*. Sukhothai Thammathirat Open University.
15. Lallemand, T., & Rycx, F. (2009). Are older workers harmful for firm productivity? *De Economist, 157*(3), 273–292.
16. Lee, C. (2015). Industrial characteristics and employment of older manufacturing workers in the early-twentieth-century United States. *Social Science History, 39*(4), 551–579.
17. Levin, A., Lin, C. F., & Chu, C. S. J. (2002). Unit root tests in panel data: Asymptotic and finite-sample properties. *Journal of Econometrics., 108*(1), 1–24.
18. Maneejuk, P., & Yamaka, W. (2021). Significance test for linear regression: How to test without P-values? *Journal of Applied Statistics, 48*(5), 827–845.
19. McCann, R., & Giles, H. (2002). Ageism in the workplace: A communication perspective. In T. D. Nelson (Ed.), *Ageism: Stereotyping and prejudice against older persons* (pp. 163–199). The MIT Press.
20. McConnell, C. R., Brue, S. L., & Macpherson, D. A. (2003). *Contemporary labor economics* (6th ed.). McGraw-Hill, Irwin.
21. McGregor, J., & Gray, L. (2002). Stereotypes and older workers: The New Zealand experience. *Social Policy Journal of New Zealand,* 163–177.
22. Meyer, J. (2011). Workforce age and technology adoption in small and medium-sized service firms. *Small Business Economics, 37*(3), 305–324.
23. Mulders, J. O., van Dalen, H. P., Henkens, K., & Schippers, J. (2014). How likely are employers to rehire older workers after mandatory retirement? A vignette study among managers. *De Economist, 162*(4), 415–431.
24. National Statistical Office. (2017). *The labor force survey whole Kingdom*. National Statistical Office.
25. National Statistical Office. (2020). *The labor force survey whole kingdom*. Bangkok, National Statistical Office.
26. Phijaisanit, E. (2016). How can promoting "desirable" elderly employment opportunities alleviate the shortfalls of Thailand's ageing society? *Thammasat Review of Economic and Social Policy, 2*(1), 124–171.
27. Puey Ungphakorn Institute for Economic Research. (2019). Situation of aging and productivity and agriculture in Thai agricultural households. *aBRIGDEd Making Research Accessible, 13*,1–11.
28. Raksasap, S., Ninbai, T., & Intrawisaes, N. (2008). *The pilot study on appropriate work for older persons*. Report to Ministry of Labour.
29. Ratchagit T. (2019). Trends in the aging labor market and how to prepare for self- development for the elderly (Online). www.th.hrnote.asia/tips/190613-old-people-prepare-aging-society/. December 9, 2020.
30. Recruitment and Consulting Services Association Australia and New Zealand. (2015). *A survey of attitudes towards older workers* (Online). www.humanrights.gov.au/sites/default/files/Submission%20No%20308%20-%20RCSA%20appendix%202.pdf. March 15, 2021.
31. Soonthornchawakan, N., & Cintakulchai, S. (2009). *Job opportunities creation for the elderly in Thailand*. Report to the Foundation of Thai Gerontology Research and Development institute. December, 2009. Economic Research and Training Center Faculty of Economics Thammasat University.

32. Suksiriserekul, S. (2013). The study of the continuation of elderly working career effect on relieving the lack of experience workforce.
33. Taylor, P. E., & Walker, A. (1994). The ageing workforce: Employers' attitudes towards older people. *Work, Employment and Society, 8*(4), 569–591.
34. United Nation Population Fund and Office of the National Economic and Social Development Board. (2011). *Impact of demographic change in Thailand*. United Nation Population Fund.
35. Vanichyanukoh, D., Anantho, S., & Thongrin, S. (2017). The process of internet acceptance for developing the quality of life of senior citizens. *Dhonburi Rajabhat University Journal, 11*(2), 60–73.
36. Zacher, H. (2015). Successful aging at work. *Work, Aging and Retirement, 1*(1), 4–25.
37. Zagata, L., & Sutherland, L. A. (2015). Deconstructing the young farmer problem in Europe: Towards a research agenda. *Journal of Rural Studies, 38*, 39–51.

Nonlinear Forecasting of Exchange Rate Volatility Using Google Search

Chatchai Khiewngamdee, Napon Hongsakulvasu, and Asama Liammukda

Abstract The purpose of this research is to widen previous studies by looking at the relationship between the Google Search Volume Index: GSVI and foreign currency rates volatility. We also test whether Google search queries can provide effective and efficient ways to predict exchange rates. Firstly, we compare the effective prediction of the traditional model and the new model using GAM method, which is allowed both linear and non-linear relationship in the model. Finally, we employ Granger Causality test to test if GSVI is causal to exchange rate volatility. The results show that the GAM model is more efficient than the classical linear model and it has significantly increased the forecasting potential. However, the intention of adding investor attention variables is insufficient to prove that Google trend factors can improve forecast accuracy.

Keywords Exchange rate · Google trend · Prediction · Nonlinear model

1 Introduction

Traditionally, Macroeconomic data have been used to forecast the movement of many financial market variables, particularly equity, credit, and foreign exchange markets, since they provide useful fundamental information for market analysis in each economic state of an economy. Thus, there are large number of papers have studied how macroeconomic factors affect the financial market variables and their predictability. Nevertheless, since the boom of internet search engines and information, investor attention has recently become a newly emerging concept that can be considered as an important factor in price movements and fluctuations [21]. Moreover, many studied have proved that investor attention helps improve the forecast accuracy of price in

C. Khiewngamdee (✉) · N. Hongsakulvasu
Faculty of Economics, Chiang Mai University, Chiang Mai, Thailand
e-mail: getliecon@gmail.com

A. Liammukda
Faculty of Science, Department of Statistics, Chiang Mai University, Chiang Mai, Thailand

S. Sriboonchitta et al. (eds.), *Credible Asset Allocation, Optimal Transport Methods, and Related Topics*, Studies in Systems, Decision and Control 429,
https://doi.org/10.1007/978-3-030-97273-8_47

many financial markets including the application for company and economic data [2, 3, 7, 17, 19] and the widely used proxy to measure the investor attention in financial markets is internet-based data such as the Google Search Volume Index (GSVI). Since before investors make decision about their trade, they tend to search for relevant information and hence search data are likely a direct and early signal reflecting their position changes. Consequently, it is expected that the GSVI series drives the changes of investors' trading positions.

The risk of exchange rate volatility has become a major source of concern for investors, entrepreneurs, policymakers, and most importantly exporters and importers since the fall of the Breton Woods agreement in 1973. Generally, exchange rate uncertainty has several impacts on economy. For instance, the effect of exchange rate fluctuations on trade earnings and trade volumes for international trade, both in agricultural sector and industrial sector, and foreign investments and hence make international trade and investment decisions more difficult because volatility increases exchange rate risk. More importantly, for agricultural producing countries such as Thailand, exchange rate fluctuations play an important role in the national economy through export and also cause farm income uncertainty. Furthermore, exchange rate volatility influences economic policy and make policymaker faces difficult choices. For example, for inflation targeting regime countries, central banks have to revise the expected inflation target frequently because of changes in the level and volatility in exchange rate [18]. Therefore, many studies have tried to find ways to predict exchange rate change and the factors that affect exchange rate movement in order to support policymakers and investors to manage and moderate risk.

Since the Google Search Volume Index has been widely used to forecast many economic variables such as sale volume, unemployment, commodity prices, and stock prices as a proxy of investor attention, many studies found that there is a strong relationship between Google search queries and change in those variables and it also improves the forecasting performance. However, while most of the GSVI studies focused on forecasting commodity and stock prices, there are few papers who investigate the forecast ability of GSVI on foreign exchange rate volatility which is also a major concern of the economy.

Therefore, this study aims to widen those studies by investigate the relationship between GSVI and foreign exchange rates volatility. We also test whether Google search queries can provide effective and efficient ways to predict exchange rates. Firstly, we test the stationary of our data using the Augmented Dickey-Fuller (ADF) test. Then, we employ Granger Causality test to test if GSVI is causal to exchange rate volatility. Finally, we conduct in- and out-of-sample forecasts between autoregressive model and one including GSVI by applying concept of the autoregressive GARCH model and compare their forecasting performance using root mean square error (RMSE) and mean square error (MSE). We also check the preciseness of our models using a robustness test. The results of our study would be widely useful for investors, entrepreneurs, farmers and agricultural commodity exporters, and policymakers.

2 Literature Review

For the past decade, many studies used search volume index of search engines to forecast economic variables. For example, Bank et al. [3] found that Google search queries causes an increase in trading activity and stock liquidity. They also suggested that increase in search volume is associated with temporarily higher future returns. In 2013, Rao and Srivastava [19] investigate a predictability of Google search volume index and Twitter sentiment on Down Jones stock index, oil price, gold price, and Euro exchange rate. The results presented that the forecasting performance of the Google search volume index outperforms. Moreover, Li et al. [17] used the Google search volume index to predict crude oil prices by employing recursive one-week-ahead forecasts. The empirical results showed that Google search volume index helps improve the forecast accuracy of crude oil price. In addition to Google. other search engines are also used to represent investor attention, for instance, Zhang et al. [29] examined the impact of investor attention, proxy by Baidu index search queries, on stock's abnormal return. The results present that there is a strong relationship between investor attention and abnormal return. They also suggest that open-source information can enhance the market efficient.

In the research field of economic forecasting, Askitas and Zimmermann [2] employed econometric models to forecast the unemployment in Germany using Google search volume index and found that there are strong correlations between keyword searches and unemployment rates. In addition, Guzman [11] employed Granger Causality tests to predict inflation expectation using Google Inflation Search Index compared with other indicators. The out-of-sample forecasts result indicates that the Google Inflation Search Index has the lowest forecast error of all the inflation expectations indicators tested. Furthermore, in asset market, Humphrey [14] investigated whether Google search query improves national and local existing home sales forecasts by employing in-sample and out-of-sample testing. The results show that Google search query data improves the forecasting accuracy of national and local existing home sales. Swallow and Labbé [6] examined the forecasting performance of Google search queries on automobile sales in Chile. The empirical results indicated that Google search volume index outperforms benchmark specifications in both in–sample and out–of–sample forecasts.

In the field of financial analysis, the GSVI was considered as a favorable indicator of investor attention. Da et al. [7] was the first to illustrate the usefulness of the Google search volume index in financial applications by applying the search volume of firms' products to forecast revenue and earnings surprises. They found that search volume index positively affects revenue surprises, and it is a value-relevant leading indicator for a firm's future cashflow. In 2011, Gao et al. [7] employed co-integration and the modified EGARCH model to analyze the impact of market concerns, derived from search query volumes in Google, on oil volatility. They found that there is a long-run relationship between oil prices and market concern for oil prices and oil demand. They also suggest that market concern transmitted through the Internet supports the linkage between oil price changes and external events by influencing the expectation of traders. Goddard et al. [10] analyzed the relationship among investor attention on stocks, stock volatility, and foreign exchange markets. Investor attention was measured by a Google search volume index. The empirical results present that investor attention comoves with contemporaneous FX market volatility and is related to the currency risk premium. Besides, Bank et al. [3] suggested that an increase in Google search queries is associated with a rise in trading activity and stock liquidity.

3 Method and Econometric Model

In order to capture the how Google search affects the exchange rate volatility, we follow Afkhami et al. [1] by using a two-step GARCH approach which proposed by Engle and Sheppard [8]. The benefit of a two-step approach is to avoid the problem of 'Dimensionality curse'. When we add the Google search variable into the model in second step, it can be easily upgraded by using the superior method such as a Generalized Additive Model (GAM) from Wood [27].

For the first step, we use GARCH (1,1) framework which is proposed by Engle [9] and Bollerslev [4].

$$r_t = c + \varepsilon_t \tag{1}$$

$$\varepsilon_t = \xi_t \sqrt{(h_t)}, \tag{2}$$

$$h_t = \omega + \alpha \varepsilon_{t-1}^2 + \beta h_{t-1}, \tag{3}$$

where r_t is the log returns of each exchange rate at time t, $\mathbb{E}\left(\varepsilon_t | \mathcal{F}_{t-1}\right) = 0$, $\left(\varepsilon_t^2 | \mathcal{F}_{t-1}\right) = h_t$, h_t is the conditional variance, $\mathcal{F}_{t-1}$ is the information set available at time $t-1$ and ξ_t is a sequence of independent and identically distributed (i.i.d.) random variables with mean zero and variance 1, $\omega > 0$, $\gamma \geq 0$, $\beta \geq 0$, and $(\alpha + \beta) < 1$.

For the second step, the return innovation, ε_t, and the conditional variance, h_t, are extracted from the GARCH (1,1) in Eqs. (1)–(3). The natural log of return innovation squared, $\ln\left(\varepsilon_t^2\right)$, is used as the dependent variable since the expectation of itself is the volatility of exchange rate. The conditional variance of the previous time lagged, h_{t-1}, and the google search volume of the previous time lagged, G_{t-1}, are used as the independent variables as it's shown in Eq. (4) while z_t is $i.i.d.\ N(0, \sigma_z^2)$ random variables of Eq. (4). In second step, Afkhami et al. [1] use the simple linear regression to estimate the parameters in Eq. (4). The drawback of Afkhami et al. [1] paper is to limit the relationship between google search volume and the volatility in only linear form.

$$\ln\left(\varepsilon_t^2\right) = \beta_0 + \beta_1 h_{t-1} + \beta_2 G_{t-1} + z_t \tag{4}$$

Our study's contribution is to use the superior regression method that can relax the linear assumption between google search volume and the exchange rate volatility. In this paper, we propose to use the generalized additive model (GAM) method from Wood [27] to estimate the equation in the second step. The benefit of GAM is the flexible of functional term which can capture the relationship between the independent variable and dependent variable in both linear and non-linear form. The second step equation which applies for GAM method is shown in Eq. (5). With GAM method, the linear parameters, β_1 and β_2, were replaced by $f_1()$ and $f_2()$.

$$\ln\left(\varepsilon_t^2\right) = \beta_0 + f_1(h_{t-1}) + f_2(G_{t-1}) + z_t \tag{5}$$

To understand how to estimate $f_1()$ and $f_2()$ using GAM method, we set the example by simplified the model to have only one independent variable which is shown in Eq. (6).

$$y_t = f(x_t) + \varepsilon_t, \tag{6}$$

where y_t is any dependent variable, x_t is any independent variable, $f()$ is a smooth function and ε_t $N(0, \sigma_\varepsilon^2)$.

By using regression splines method, the smooth function, $f()$, can be estimated as it's shown in Eq. (7). In this step, we need to choose the type of basis function $B()$ and the number of basis function, k , Then $f()$, can be estimated by using a simple linear regression.

$$f(x) = b_1 B_1(x) + b_2 B_2(x) + b_3 B_3(x) + \cdots + b_K B_K(x) \tag{7}$$

For a simple example, we can use a 4th order polynomial as the basis function in equation, then we will have $B_1(x) = 1$, $B_2(x) = x$, $B_3(x) = x^2$, $B_4(x) = x^3$ and $B_5(x) = x^4$. So, the Eq. (7) becomes

$$f(x) = b_1 + b_2 x + b_3 x^2 + b_4 x^3 + b_5 x^4 \tag{8}$$

GAM is a data driven approach because each basis function can be calculated from the data. The smooth function, $f()$, is very flexible which can be linear or non-linear shape.

To estimate a smooth function $f()$ by using a regression spline alone can create a "wiggliness" problem. According to Wood [27], we can control the smoothness of the model by using the penalty least square as it's shown in Eq. (9). On the right side of Eq. (9) is the penalty term and the parameter, λ, is the smoothing parameter of the model. The shape of a smooth function $f()$ will depend on the smoothing parameter, λ. If λ is high, then the wiggliness penalty will be large and a smooth function $f()$ will be close to linear shape. On the other hand, if λ is low, the wiggliness penalty will be small and a smooth function $f()$ will be a non-linear shape.

$$\sum_{t=1}^{T} [y_t - f(x_t)]^2 + \lambda \int_{-\infty}^{+\infty} [f''(x_t)]^2 dt \tag{9}$$

To find the optimal value of λ, we follow Wahba [24] and Wood [27] to use the generalized cross validation (GCV) score which is shown in Eq. (10). By minimizing

Eq. 10, we will get the optimal value of λ and then we can estimate the smooth function $f()$ which will show how variable x_t affects variable y_t.

$$GCV = \frac{T \sum_{t=1}^{T} \left(y_t - \widehat{f}(x_t)\right)^2}{\left[tr\,(I - A)\right]^2} \tag{10}$$

where, A is the corresponding influence matrix.

After we get the idea about how to estimate the smooth function in Eq. (6), then we can extend the equation to include the additional independent variable as we use in our model in Eq. (5). In Eq. (5), we have 2 independent variables which the first smooth function, $f_1()$, will show how the previous period of exchange rate volatility affect the current exchange rate volatility while the second smooth function, $f_2()$, will show how google search volume in previous period affects the current exchange rate volatility.

Unlike what we show in the simple explanation above that we use a 4th order polynomial as the basis function, in our model, we follow the suggestion of Wood [27] by using a thin-plate regression splines to be a basis function. Wood [28] shows the benefits of a thin-plate regression splines which there is no need to provide the knot locations and it performs very well when we have many independent variables in the model.

For the identifiability problem, we put the constraint on the constant parameter inside basis function of one of the smooth functions, $f_i()$, to be equal to zero.

In the path of variable selection, we also mention the method as the tool for computing Minimum Bayes factor (MBF) with an emphasis on the two-sided p-based Bayes factor. The formula of this method is provided as the following,

$$MBF\,(p) = \begin{cases} -\exp{(1)}p\log p & for\ p < 1/\exp(1) \\ 1 & for\ p \geq 1/\exp(1) \end{cases} \tag{11}$$

when p is p-value [22]. However, The MBF is different from p-value approach, The categories of Bayes factor is provided in Table 1 [12].

Table 1 Minimum Bayes factor detail

MBF	Interpretation
1–1/3	Weak evidence for H_1.
1/3–1/10	Moderate evidence for H_1
1/10–1/30	Substantial evidence for H_1
1/30–1/100	Strong evidence for H_1
1/100–1/300	Very strong evidence for H_1
$< 1/300$	Decisive evidence for H_1

Table 2 Variable detail

Variable	Abbreviations
US dollar index	USDX
Thai Baht	THB
China Yuan	CNY
Japan Yen	JPY
South Korea Won	KRW
Dow Jones index	DJI
Stock exchange of Thailand index	SET
China securities index	CSI300
Nikkei index	NIKKEI
The Korea composite stock price index	KOSPI
Google trend of US dollar	GUSD
Google trend of Thai Baht	GTHB
Google trend of China Yuan	GCNY
Google trend of Japan Yen	GJPY
Google trend of South Korea Won	GKRW
Oil future price	OIL
Gold future price	GOLD

4 Data

The data used in this study are Google Search Volume Index (GSVI) and major foreign exchange rates for Thai agricultural international trading, namely, US dollar (USD), Euro (EUR), Japanese Yen (JPY), and Chinese Yuan (CNY) including Thai Baht (THB). Additionally, we use US dollar index (UXY) to represent the US dollar, while other exchange rates are defined by in terms of dollar, for instance, EUR-USD, USD-JPY, and GBP-USD. We collected the monthly data from Google Trends and Thomson Reuters database covering the period from January 2004 to April 2021 at a monthly frequency. In addition, the value of GSVI at the week t is noted as

$$GSVI_{K_i,t} = \frac{S_{K_i,t}}{\sum_{K_1,t}^{K_m,t} S_{K_i,t}}, \tag{12}$$

where $S_{K_i,t}$. is the search volume of the specific term, K_i. at week t. $\sum_{K_1,t}^{K_m,t} S_{K_i,t}$ represents the search volumes of all terms during time t.in Google and $K_{m,t}$ is the total search terms during time t.Therefore, The GSVI is based on the query ratio and presents the value on a scale from 0 to 100.

In order to extract the GSVI from Google Trends, we start with keywords related to each foreign exchange rate and build our set of exchange rate-related keywords for each exchange rate such as JPY-related keywords, EUR-related keywords. Then, we

Table 3 Descriptive statistics

Variable	Mean	S.D.	Skewness	Kurtosis
USDX	87.005	7.841	0.077	1.800
THB	33.615	3.176	1.025	3.127
CNY	6.914	0.667	0.863	2.555
JPY	104.030	12.477	−0.716	2.574
KRW	1106.743	96.417	0.419	5.074
DJI	16132.967	6224.199	0.838	2.572
SET	1145.435	402.787	−0.092	1.524
CSI300	2676.668	869.021	0.498	4.125
NIKKEI	15539.492	5091.600	0.382	2.296
KOSPI	1826.023	466.514	−0.323	3.241
GUSD	37.144	26.907	0.496	1.792
GTHB	46.808	25.656	0.284	1.590
GCNY	29.837	16.531	1.595	6.458
GJPY	40.947	27.287	0.337	1.746
GKRW	33.789	22.708	0.386	1.849
OIL	68.581	22.827	0.443	2.462
GOLD	1169.171	418.608	−0.363	2.173

add keywords to the initial set based on Google Search's suggestions on the keywords and repeat this step until time series data for all terms are gathered. We also filter out the words for which Google Trends does not have enough data to generate time series. Finally, according to the methodology proposed by Li et al. [17], we use descriptive, correlation and stability analysis to obtain five GSVI series that most represents investor attention on each exchange rate dynamics.

The data employed in this study are monthly data, which were retrieved from The Thomson Reuters data-based and Google trend data-based. After that, the in-sample period starts from January 2004 to July 2016 and out of sample period starts from August 2016 to April 2021 (total 208 observations). These variables were transformed into logarithms for the final econometric analysis. Foremost, the descriptive statistics of the data are given in Table 3. First, the data is all positive. For Skewness, we found that all variables do not equal 0 [15] and the Kurtosis is greater than 3 [16] called leptokurtic. This confirms that all data have a skewed and do not have a normal distribution.

5 Results and Discussion

This research transforms the p-value into maximum bayes factors (MBF) to provide the statistic significant because the p-value is not reasonable to the hypothesis test

Table 4 Model result

Currency	Variable	edf	Ref.df	F	MBF
USD	USDXL1	5.0087	6.0740	64.3416	0.0000
	DJI	8.2808	8.8168	3.4014	0.0155
	OIL	1.8780	2.2982	1.3546	0.9165
	GOLD	1.0000	1.0000	0.0369	1.0000
	GUSD	1.9583	2.4559	1.0138	1.0000
THB	THBL1	3.1335	3.8572	78.8215	0.0000
	SET	1.2500	1.4134	0.6315	1.0000
	OIL	1.3004	1.5140	0.8136	1.0000
	GOLD	5.0634	6.1812	2.3881	0.2891
	GTHB	1.0000	1.0000	0.3311	1.0000
CNY	CNYL1	4.2135	5.1658	89.4658	0.0000
	CSI300	1.0000	1.0000	0.0211	1.0000
	OIL	1.0000	1.0000	4.4930	0.3245
	GOLD	1.6220	2.0355	3.1366	0.3790
	GCNY	6.0083	7.1156	3.6974	0.0194
JPY	JPYL1	2.7833	3.5080	79.1874	0.0000
	NIKKEI	2.5313	3.1486	1.9442	0.7091
	OIL	2.0172	2.4851	0.8778	1.0000
	GOLD	1.0000	1.0000	0.0659	1.0000
	GJPY	5.8571	7.0011	2.7687	0.1267
KRW	KRWL1	2.1215	2.6199	14.6698	0.0000
	KOSPI	2.0194	2.5213	4.7072	0.0622
	OIL	2.6547	3.2241	2.3922	0.5082
	GOLD	1.0001	1.0003	0.4624	1.0000
	GKRW	1.0000	1.0000	11.9450	0.0143

Note Minimum Bayes factor (MBF) details are show in Table 1

and probability wrong [25]. Firstly, we test whether the data is non-linear for using the GAM model. The result show that the conditional variances lag 1 are statistically significant at 1% for all currency and the EDF values of them as more than 1. This provide that a nonlinear effect have in all variables. Observably, all currency has a strongly significant with themselves in the past (time $t-1$). The result does not surprise because most of time series data always determine by themselves [5, 26]. Furthermore, this study considers commodity futures prices, including gold and oil, as a control variable. It generally will represent the volatility in storge of value assets and the cost inflation rate, respectively. It is evident that the gold price does not cause in only USD and KRW volatility, while the CNY and KRW volatility are statistically affected by the oil price (Table 4).

Foremost, the research' aim has included the investor attention proxy in financial markets as a Google Search Volume Index (GSVI) among each moment. This addition will have resulted in improved forecast model accuracy. The result of this estimation is that each google trend has a statistically significant only in 3 currency including CNY, JPY and KRW. While google trend is insignificant in THB and USD. This mean that only google trend in China, Japan and South Korea have affected on each currency volatility. In addition to this, the above table is only a non-linearity and statistical test of the model. Although the model is unable to determine the coefficient exactly, the coefficient results has shown in the following Fig. 1.

Figure 1 displays the relation-coefficient between each currency volatility and their covariable (x) following by Table 2. There are showing in rows and columns, respectively. However, the constraints of the non-linear model do not provide the exact coefficient. Their result of GAM model will identify the effect of the changing level of covariable x on the currency fluctuation. The coefficient results basically provide by slope of the graph. Then, if the slope is increase, the coefficient will have a positive value and on the contrary. In addition, the research only focuses on variables that are statistically significant (previous part) because those covariables (x) have an effect on currency fluctuations.

Observably, The GAM model's coefficient result is split into three simple characteristics: (i) when the variable volatility is low, the correlation is positive. Nevertheless, as the covariates has high volatility, the correlation becomes constant. This feature is often found in their historical data ($t-1$). It has provided a strong effect on present fluctuation by yesterday's data and is hardly found in the stock market and oil price volatility, especially in the Korean Won (KRW). (ii) Negative associations; this case represents a negative correlation between currency fluctuations and covariables that only the influence of the gold price on the Thai Baht and Chinese yuan. It means that the Thai Baht and the Chinese Yuan are both affected by gold future price volatility, and (iii) Mixed relationships; they have both influences the currency volatility. We have almost found in google trend. On the other hand, inadmissible covariates result almost seems close to a linear line that do not change in slop. This result means that they do not have a contribution to currency volatility.

Definitely, this study pays attention to GSVI variables that exactly affect currency volatility. the results of the study, the relationship has been mixed. As a result, the Google trend (GSVI) is used to describe investor activity. Initially, investors began to pay attention to the news. Following that, investors attempt to emotionally adjust as well as manage their portfolios. As a result, the later impact has been the opposite on coefficient [13, 20, 23]. Furthermore, a few GSVI variables has been admitted by this model. The study's findings were rather startling. The structure has resulted in Google trends have no effect on currency volatility. This is mean the investors do not concern on the news in all currency. Therefore, caution is advised while using it.

After we estimate the model parameters and test the relationship between variables and dependent variables (Currency volatility) by mean square error (MSE), this result will prove a currency volatility forecasting model by the GAM model, which is a non-linear model. Firstly, we evaluate the models in three characteristics: (i) with all covariates (Full), (ii) with their historical data (at $t-1$) and google trend (Lag & Google trend) and (iii) with only their historical data (at $t-1$) (Lag) that show in Table 5. Foremost, the result provide that the GAM model is suitable for all exchange rates more than a classical linear regression model. However, we have only

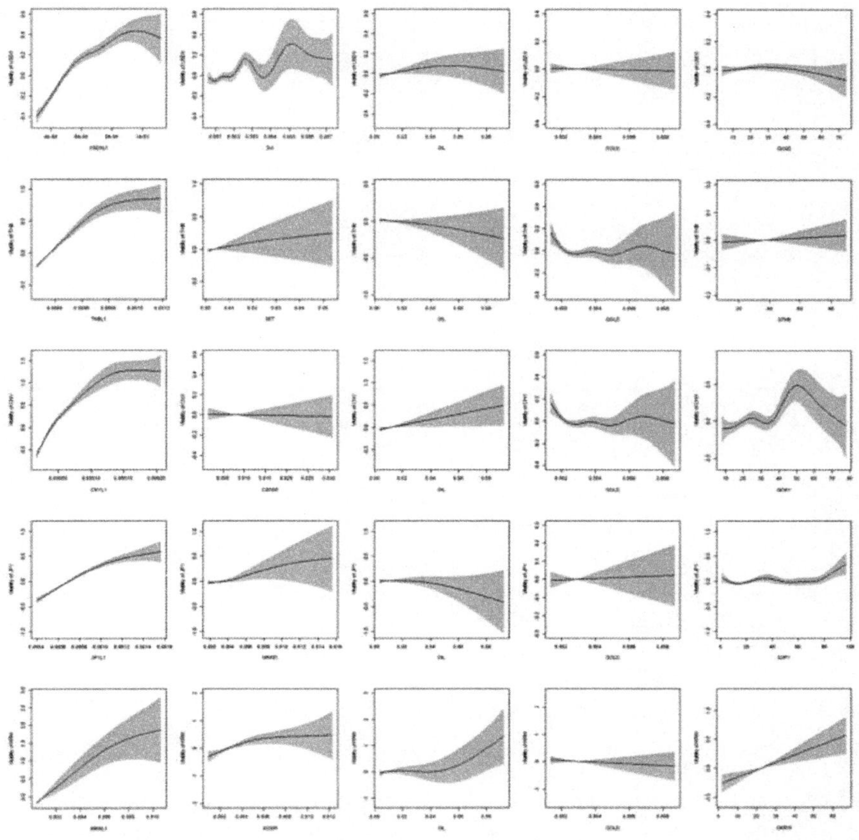

Fig. 1 The coefficient result of GAM model

Table 5 Forecasting model selections

Currency	Full		Lag & Google trend		Lag	
	GAM	Linear	GAM	Linear	GAM	Linear
USDX	0.0119*	0.0206	0.0069*	0.0156	0.0053*	0.0089
THB	0.0578	0.0211*	0.0116	0.0115*	0.0115*	0.0115
CNY	0.4631*	1.8223	0.2769*	1.4534	0.1711*	1.6338
JPY	0.1699	0.0538*	0.0494*	0.0542	0.0261*	0.0433
KRW	5.0256	0.5608*	0.2639	0.1050*	0.0544*	0.0561

Note "*" denote the lowest MSE of the model

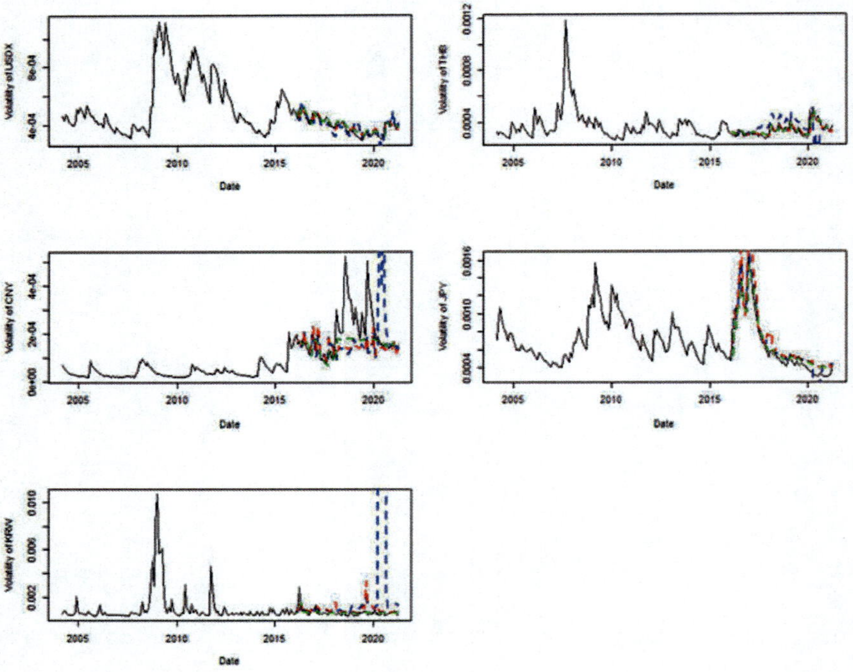

Fig. 2 The prediction result of GAM model

the US dollar (USDX) and Chinese Yuan (CNY) that are reasonable for all factors in the model (Full), and historical data in the GAM model (Lag & Google trend) is suitable for Thai Bath (THB) and South Korea Won (KRW).The result is quite surprising due to GSVI determined almost all dependent variables prediction, but some were not. It could be stated that some currency volatilities are not always based on investor attention. Then, the prediction result provides in Fig. 2. Consequently, adding a GSVI variable has a significant increase the potential of the almost currency volatility forecast. However, it is necessary to be careful when adding this google trend variables.

In particular, the study strongly confirmed that the non-linear model is more efficient than the classical linear model and it has significantly increased the forecasting potential compared to the MSE value. However, the intention of adding investor attention variables as a GSVI. The results are not sufficient to confirm that Google trend variables could increase the accuracy of predictions in all currency volatility.

6 Conclusion

The purpose of this research is to widen previous studies by looking at the relationship between the GSVI and foreign currency rates. We also test whether Google search queries can provide effective and efficient ways to predict exchange rates. Firstly, we test the stationary of our data using the Augmented Dickey-Fuller (ADF) test. Then, we employ Granger Causality test to test if GSVI is causal to exchange rate volatility. Finally, we use the idea of the Markov switching autoregressive GARCH model to conduct in- and out-of-sample predictions between an autoregressive model and one that includes GSVI and compare their forecasting ability using root mean square error (RMSE) and mean square error (MSE) (MSE). The result has provided that the non-linear model was shown to be more efficient than the traditional linear model and has considerably enhanced forecasting potential when compared to the MSE value. The purpose of investor attention factors as a GSVI. The findings are insufficient to prove that Google trend factors can improve forecast accuracy for all types of currency volatility.

References

1. Afkhami, M., Cormack, L., & Ghoddusi, H. (2017). Google search keywords that best predict energy price volatility. *Energy Economics, 67*, 17–27.
2. Askitas, N., & Zimmermann, K. F. (2009). Google econometrics and unemployment forecasting.
3. Bank, M., Larch, M., & Peter, G. (2011). Google search volume and its influence on liquidity and returns of German stocks. *Financial Markets and Portfolio Management, 25*(3), 239.
4. Bollerslev, T. (1986). Generalized autoregressive conditional heteroskedasticity. *Journal of Econometrics, 31*(3), 307–327.
5. Box, G. E., Jenkins, G. M., Reinsel, G. C., & Ljung, G. M. (2015). *Time series analysis: Forecasting and control*. Wiley.
6. Carrière-Swallow, Y., & Labbé, F. (2013). Nowcasting with Google Trends in an emerging market. *Journal of Forecasting, 32*(4), 289–298.
7. Da, Z., Engelberg, J., & Gao, P. (2011). In search of attention. *Journal of Finance, 66*, 1461–1499.
8. Engle, R. F. & Sheppard, K. (2001). *Theoretical and empirical properties of dynamic conditional correlation multivariate GARCH (No. w8554)*. National Bureau of Economic Research.
9. Engle, R. F. (1982). Autoregressive conditional heteroscedasticity with estimates of the variance of United Kingdom inflation. *Econometrica*, 987–1007.
10. Goddard, J., Kita, A., & Wang, Q. (2015). Investor attention and FX market volatility. *Journal of International Financial Markets, Institutions and Money, 38*, 79–96.
11. Guzman, G. (2011). Internet search behavior as an economic forecasting tool: The case of inflation expectations. *Journal of economic and social measurement, 36*(3), 119–167.
12. Held, L., & Ott, M. (2016). How the maximal evidence of p-values against point null hypotheses depends on sample size. *The American Statistician, 70*(4), 335–341.
13. Huang, M. Y., Rojas, R. R., & Convery, P. D. (2020). Forecasting stock market movements using google trend searches. *Empirical Economics, 59*(6), 2821–2839.
14. Humphrey, B. D. (2010). *Forecasting existing home sales using Google search engine queries*. Graduation with Distinction in Economics in Trinity College of Duke University.

15. Hutchinson, M., & Gallagher, L. S. (2006). *kurtosis and convertible arbitrage hedge fund performance.*
16. Kallner, A. (2014). Laboratory statistics, handbook of formulas and terms. *Chemistry International, 36*(4), 23–23.
17. Li, X., Zhang, C., Li, W., Ricard, R., Meng, Q., & Zhang, W. (2015). Assessing street-level urban greenery using Google Street View and a modified green view index. *Urban Forestry & Urban Greening, 14*(3), 675–685.
18. McKenzie, M. D. (1999). The impact of exchange rate volatility on international trade flows. *Journal of economic Surveys, 13*(1), 71–106.
19. Rao, T., & Srivastava, S. (2013, May). Modeling movements in oil, gold, forex and market indices using search volume index and twitter sentiments. In Proceedings of the 5th Annual ACM Web Science Conference (pp. 336–345).
20. Song, Y., Ji, Q., Du, Y. J., & Geng, J. B. (2019). The dynamic dependence of fossil energy, investor sentiment and renewable energy stock markets. *Energy Economics, 84,* 104564.
21. Vlastakis, N., & Markellos, R. N. (2012). Information demand and stock market volatility. *Journal of Banking & Finance, 36*(6), 1808–1821.
22. Vovk, V. G. (1993). A logic of probability, with application to the foundations of statistics. *Journal of the Royal Statistical Society, Series B, 55,* 317–351.
23. Vozlyublennaia, N. (2014). Investor attention, index performance, and return predictability. *Journal of Banking & Finance, 41,* 17–35.
24. Wahba, G. (1980). Spline bases, regularization, and generalized cross validation for solving approximation problems with large quantities of noisy data. In E. Cheney (Ed.), *Approximation Theory III.* Academic Press.
25. Wasserstein, R. L., & Lazar, N. A. (2016). The ASA statement on p-values: Context, process, and purpose.
26. Wei, W. W. (2006). Time series analysis. In *The Oxford handbook of quantitative methods in psychology* (vol. 2).
27. Wood, S. N. (2006). *Generalized additive models: An introduction with R.* Chapman and Hall/CRC.
28. Wood, S. N. (2003). Thin plate regression splines. *Journal of the Royal Statistical Society, Series B, 65,* 95–114.
29. Zhang, W., Shen, D., Zhang, Y., & Xiong, X. (2013). Open source information, investor attention, and asset pricing. *Economic Modelling, 33,* 613–619.

Detection of Buy and Sell Signals Using Technical Indicators with a Prediction Model Based on Neural Networks

Francisco Solano López Rodríguez and José Manuel Zurita López

Abstract In the world of financial and stock markets, predicting price movements has been a topic of great importance investigated by many investors for decades, which goes against the efficient market theory, which considers that successive changes in prices follow a random walk, due to the immediate adjustment of prices to new information. However, research in this field has not stopped and new prediction methods continue to be published in an attempt to solve an extremely complex problem, as a result of the large number of variables that can influence share prices, such as economic, political and social factors or certain current events or news. In recent years, technological advances have led to the use of new ways of prediction based on machine learning techniques. This paper aims to use these techniques to predict trends in stock prices, for which we will make use of neural networks, a very powerful prediction model, which has the ability to learn from data and take advantage of this learning to make decisions. Our model will be trained with numerous technical indicators used in the stock market to detect buy and sell signals and will also be fed back with new information, so that the model is as up-to-date as possible in order to improve its predictive capacity. The performance of the model will be evaluated on some stocks of companies belonging to the NASDAQ-100 and EURO STOXX 50 stock indexes.

Keywords Neural network · Technical indicator · Stock market

1 Introduction

The efficient market hypothesis, introduced by Malkiel [1], states that the market reflects all the current information available, which tells us that the market follows a random walk and that it is therefore impossible to predict price movements. Despite this hypothesis, trying to predict market trends has been an important field of study

F. S. López Rodríguez (✉) · J. M. Zurita López
E.T.S. de Ingenierías Informática y de Telecomunicación,
C/Periodista Daniel Saucedo Aranda S/N, 18071 Granada, Spain
e-mail: fransol0728@gmail.com

S. Sriboonchitta et al. (eds.), *Credible Asset Allocation, Optimal Transport Methods, and Related Topics*, Studies in Systems, Decision and Control 429,
https://doi.org/10.1007/978-3-030-97273-8_48

for many years, in fact there are studies [2–4] in which it has been proven that the movements are not completely random, but that there is some dependence on historical data. The most common prediction methods make use of classical techniques of stock market analysis, where on the one hand we can find strategies based on fundamental analysis focused on the study of the economic structure of a company. On the other hand we find strategies based on technical analysis, which is based on the study of market behavior through chart analysis, where in order to eliminate the subjectivity that can be derived from it, we usually resort to the use of technical indicators such as moving averages or other indicators to generate buy and sell signals. In this paper we will focus our attention on the use of technical indicators, which have been used by many investors to obtain trading rules and elaborate a multitude of buy and sell strategies by combining these rules.

With the advancement of technology, computers have increasing computing capacity, which has led to artificial intelligence being more and more present and has been transferred to innumerable fields, among which is the area of finance. An important area of artificial intelligence is Machine Learning, which aims to give computers the ability to learn without being explicitly programmed, for this purpose it seeks to build algorithms that can learn from data and build models that allow decisions to be made. This area has attracted particular attention in the world of trading, with the aim of detecting patterns in the time series of market prices and thus being able to forecast price movements. One of the best known computational models of Machine Learning is neural networks, a mathematical model inspired by the structure of brain neurons.

There are many papers on neural networks with the aim of predicting the future of market prices, where we can highlight two different approaches. The first approach is to predict the price value that the asset will have in the future and thus determine whether it is a good time to buy or sell depending on whether the predicted value is above or below the current value [5, 6]. The other approach is, instead of determining what the asset will be worth, simply determining the direction of the price, using the network in this case as a classifier, where the neural network prediction will indicate whether the future trend will be bearish, bullish or sideways [7–9].

Our paper aims to contribute to the existing research on the use of Machine Learning techniques for investing in the stock market, specifically resorting to neural networks with the objective in our case of using it as a binary classifier to predict the price trend. To do this, we will use the information provided by the technical indicators and let the neural network obtain patterns in the data, to develop a predictive model that will help us in decision making. In this way, for each market, the neural network performs a training with historical data, to obtain the optimal way to act at each moment with the information available to it. This is particularly interesting, since without the need for human intervention, the algorithm finds the optimal prediction model for each market.

Therefore, our main objective in this research is to obtain a neural network-based model that provides us with buy and sell signals for stocks and thus can help us to make investment decisions. In order to achieve this, our research has some merits that differentiate it from others. A novelty we propose in this research is the fact that after the training of the neural network, during validation, the model will continue to be fed back with new information, and in this way it will be updated to the new changes in the market. Another point to highlight about this paper is the clever way of labeling the training data, depending on which action turns out to be more beneficial to carry out, which will be explained in Sect. 3.3. On the other hand, we propose to add to the training data set, features obtained from technical indicators commonly used in the stock market, namely, we will make use of simple moving averages, exponential moving averages, rate of change, stochastic oscillator, relative strength index and standard deviation.

2 Background

Before presenting the suggested approach, in this section, we will review the neural network model that is the main element of our project. We will also review some technical indicators that will be used as additional information to feed the neural model.

2.1 Multilayer Perceptron

The neural model we will use for this research is the multilayer perceptron, a variant of the perceptron model proposed in [10]. This is a computational model used to solve classification problems, with the ability to approximate non-linear relationships between data and filter the noise between them, which makes it a suitable model to be applied in a large number of complex problems.

This model, inspired by biological neural networks, has an architecture organized in layers of different levels, where each layer is composed of a series of nodes called neurons. In the multilayer perceptron, connections are forward directed, so that neurons in one layer connect to neurons in the next layer (Fig. 1).

Three types of layers can be distinguished: the input layer, the hidden layers and the output layer. The neurons in the input layer are only responsible for receiving signals from the outside and propagating them to all neurons in the next layer. The neurons in the hidden layers are responsible for performing nonlinear processing of the input data, propagating the result to the neurons in the next layer. The last layer is the output layer, which provides the response of the neural network to the outside for each of the patterns it has received as input.

We will call $X = (x_1, \ldots, x_N)$ the vector of input data, where N is the number of features in the input data. The connection between a neuron and any neuron of the next layer has a numerical value called synaptic weight, which we will denote by w^l_{ji} in the case of the connection that connects neuron i of layer l with neuron j of layer $l+1$ and we will denote by w^l_j to the vector of synaptic weights that connects

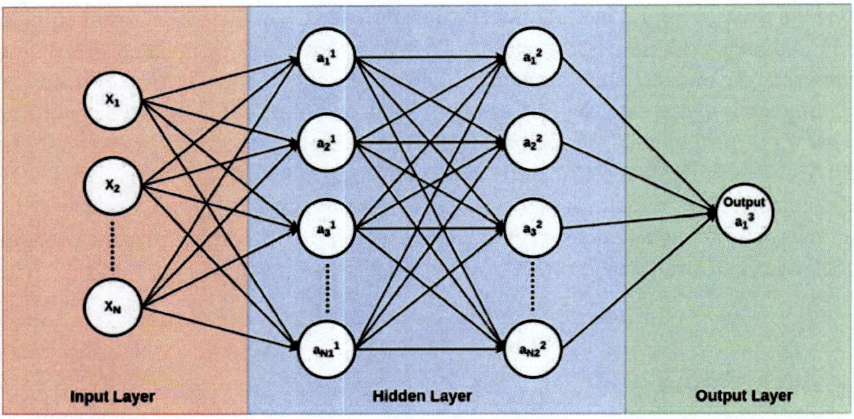

Fig. 1 Example of multilayer perceptron with two hidden layers, *N* nodes in the input layer and one node in the output layer

each neurons of layer l with the j neurons of layer $l+1$. The activation of the i-th neuron of layer l will be denoted as a_i^l and the vector of activations of layer l as A^l. In the case of input nodes the activation of the neurons will match the value of the input data ($A^1 = X$). In the case of the neurons of the hidden layers and of the output layer, to obtain the activation of the j-th neuron of the $l+1$ layer, we will calculate the linear combination between the value of the nodes of the l layer and the value of the synaptic weights linking these nodes with the node to be calculated, to this result we will add a threshold called bias, which we will denote by θ_i^{l+1} and finally we will apply a function f to the result obtained.

$$a_j^{l+1} = f\left(\sum_i w_{ji}^l a_i^l + \theta_j^{l+1}\right) = f(w_j^l \cdot A^l + \theta_j^{l+1}), \quad \text{where } l > 1 \tag{1}$$

In the above formula the operator $\cdot$ represents the scalar product. The function f is called activation function and thanks to it this model can solve nonlinear classification problems. For this function we usually use functions such as sigmoid or hyperbolic, but for efficiency and also because of the good results it has been shown to give, in this work we will use the ReLU function in the case of neurons in the hidden layers. For the output layer we will use the hyperbolic tangent function.

$$ReLU : \mathbb{R} \to [0, +\infty], \quad ReLU(x) = \max(0, x) \tag{2}$$

$$\tanh : \mathbb{R} \to (-1, 1), \tanh(x) = \frac{e^x - e^{-x}}{e^x + e^{-x}} \tag{3}$$

For the perceptron model to be able to classify correctly for a given input data, we must find the synaptic weights and thresholds of the nodes, so that the difference between the output of the network with respect to the expected output is minimal. To this end, we will work with a training data set and minimize an error formula, based on the mean square error, which we will optimize by using the backpropagation algorithm first introduced for training neural networks in [11]. This algorithm is based on the use of the gradient descent, a computational method to optimize functions, in particular in this research we will make use of the stochastic gradient descent method [12].

In order not to fall into the well-known problem of overfitting, we will use a regularization method, whose objective is to reduce the complexity of the model, giving rise to simpler models that tend to generalize better. In the case of neural networks, reducing the complexity of the model is equivalent to reducing the absolute value of the synaptic weights. If a synaptic weight reaches zero it would mean that the link between the corresponding neurons would disappear (and thus reduce the complexity of the neural network). For this purpose we will modify the objective function, so that not only the error function but also the complexity will be minimized. Thus the objective function will be as follows:

$$Cost(W,\theta) = MSE(W,\theta) + \lambda Complexity(W) \tag{4}$$

For simplicity in the above formula we have denoted $W = (w_1, \ldots, w_{N_W})$ to the vector of all synaptic weights and θ to the vector of all thresholds. MSE is the mean square error and λ is a parameter that regulates the importance given to the minimization of the complexity in the cost function. In this research we will make use of the Ridge or L2 regularization, which penalizes high absolute values in the synaptic weights, causing them to tend to decrease. The formula is as follows:

$$Complexity(W) = \frac{1}{2N_W}\sum_{i=1}^{N_W} w_i^2 \tag{5}$$

where N_W is the total number of synaptic weights. Note that the complexity function does not make use of thresholds, since it is only intended to reduce the value of the synaptic weights.

2.2 Technical Indicators

Technical indicators are mathematical formulas that are applied to a series of prices and volumes with the intention of assisting in investment decisions. We can distinguish between trend followers and oscillators. Trend followers allow us to see the evolution of the price of an asset and can help us to determine when a trend is losing strength or when a new one is starting. Oscillators allow us to detect overbought or oversold areas, divergences and convergences that help us to anticipate trend changes. Due to the information that certain technical indicators typically used in trading can offer us, we will select some of them to be used as additional features for the input nodes of our neural network.

Moving averages

These technical indicators are intended to smooth out fluctuations and facilitate trend spotting. There are many types of moving averages, but in this paper we will use only two types. One of the types we will use is the simple moving average, which consists of a proportional average of k moving observations over time. On the other hand, we will also use the exponential moving average, which employs an exponential type of weighting, giving more relevance to the most recent quotes [13].

Standard deviation

The standard deviation is a measure of dispersion, which tells us how far away the values are from the mean. Applied to quoted prices it gives us a measure of market volatility, which can indicate the investment risk at a given time. The calculation of the standard deviation over k observations would be calculated as follows:

$$\sigma = \sqrt{\frac{1}{k}\sum_{i=1}^{k}(x_i - \mu)^2}, \quad \text{where } \mu = \frac{1}{k}\sum_{i=1}^{k} x_i \tag{6}$$

Stochastic oscillator

It is an oscillator type indicator, which measures the relative position of the closing price of a day, with respect to the closing prices of the previous days, in a considered time interval. The calculation of the stochastic oscillator over k observations would be calculated as follows:

$$\%K_n = 100 \times \frac{C_n - \min}{\max - \min}, \quad \%D_n = \frac{1}{k}\sum_{i=0}^{k-1} \%K_{n-i} \tag{7}$$

where n is the current period, C_n is the current closing price, min the minimum value of the analyzed period, max the maximum value of the analyzed period [14].

Relative strength index (RSI)

This indicator calculates the rate of increase or decrease in an asset over time. This information can help us to detect overbought and oversold areas and can also provide us with buy and sell signals. It moves on a scale from 0 to 100 and is calculated as follows:

$$RSI = 100 - \frac{100}{1 - RS}, \quad RS = \frac{AUP}{ADP} \tag{8}$$

where AUP is defined as the average gain of the up periods during the specified time period and ADP as the average gain of the down periods [15].

Rate of Change

This oscillator type indicator measures the price variation between two periods, it calculates the difference between the current closing price C_n and the closing price of k periods ago C_{n-k} and divides it by C_{n-k}. It can be used to identify buy and sell signals and also allows us to detect divergences, where the price trend is losing strength and a change of direction in the price movement is possible.

3 Proposed Approach

3.1 Feature Selection

To feed the prediction model we will make a selection of features with relevant information, so that the data set we obtain to train the network is as rich as possible in order to get a better predictive performance. To do this we will make use on the one hand of data relating to an OHLC chart, consisting of the opening price, the high, the low and the close of each time period considered, in addition we will accompany this information with the trading volume of the period, on the other hand we will add as additional information, the values of certain technical indicators calculated from the closing price, which will provide us with extra information such as the strength of the trend at a given time or the volatility of the market at that time. The technical indicators that we will calculate will be the following:

- Simple moving average of 7, 14 and 21 periods.
- Exponential moving average of 7, 14 and 21 periods.
- Rate of change of 13 and 21 periods.
- Stochastic oscillator of 7, 14 and 21 periods.
- Relative strength index of 9, 14 and 21 periods.
- Standard deviation of 7, 14 and 21 periods.

3.2 Neural Network Architecture

The implemented neural network is composed of 6 layers, the first of which corresponds to the input layer, which has one node for each of the features described in the previous section. The nodes of the inner layers have 128, 64, 16 and 8 nodes respectively. Finally we have the output layer, which has only one node. The activation function to be used for the nodes of the hidden layers of the neural network will be the ReLU function, whose simplicity of calculation will have an impact on the training time, as well as on the time taken to get the predictions. The function used for the output node will be the hyperbolic tangent function.

3.3 Neural Network Output

In order to train the neural network, we must first label each element of the training set with the desired output for that input data. In our case, the objective of our model is to predict the price trend to determine which is the best action to perform at each moment, for this we will label a record with 0 if it is a good time to sell or +1 in case it is more advisable to buy. The labeling method we will describe below is similar to the one used in the paper [16].

To make the decision of which action to assign to each period of the training data, a simulation of d days will be carried out and the action that turns out to be more beneficial will be assigned as the action. In this simulation, g (gain) will be considered as the maximum gain allowed, l (loss) will be the limit of loss that can be tolerated, c (commission $\times$ 2) which is the commission of each operation multiplied by 2 and d the maximum number of days of duration of the simulation.

With the above parameters, the way to allocate an action will be as follows:

- It will be considered a good time to buy if the value of g is exceeded before d days have elapsed since the beginning of the simulation or if after the end of d days the profit obtained is above c.
- It is considered to be a good time to sell if before d days have elapsed the loss limit l is exceeded or if after the end of d days the profit obtained is below c (Figs. 2 and 3).

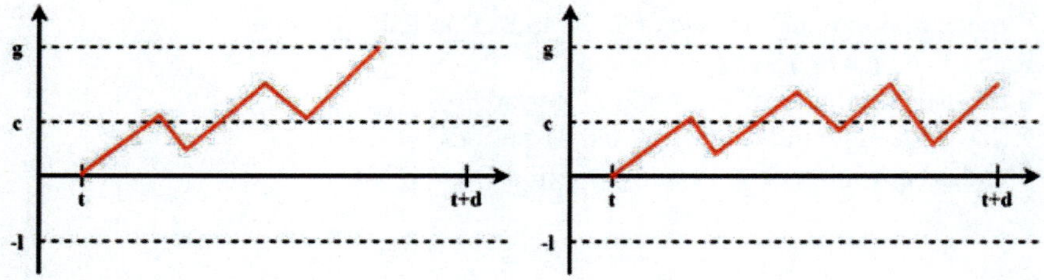

Fig. 2 Period t examples labeled as +1. In the left case because of exceeding the established gain limit and right case because of be above c

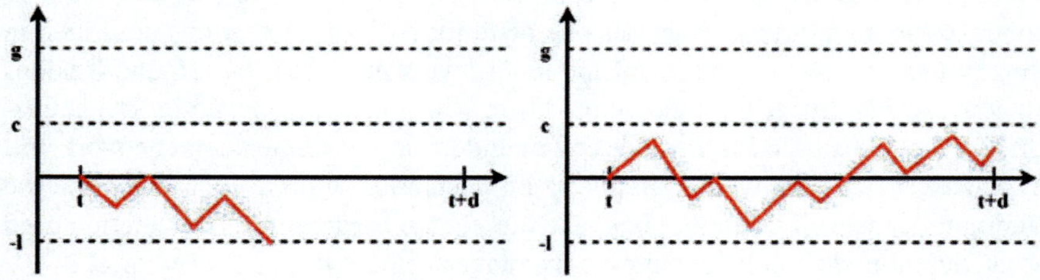

Fig. 3 Period t examples labeled as 0. In the left case because of exceeding the established loss limit and right case because of be below c

3.4 Neural Network Training

Before using the neural network to make predictions, it is necessary to train it. For this purpose, we will take a data set dated before the period that will be used to validate the model. With this data set, the network will be trained for 300 epochs, where the optimization method that we will use in the backpropagation algorithm will be the stochastic gradient descent, with a learning rate of $\eta = 0.001$. The objective function to be minimized with this algorithm will be the mean square error function. In addition, the L2 regularization method will be applied, with which the complexity of the model will be reduced to avoid losing the generalization capacity of the model.

After training, the model will be ready to make predictions during the simulation in the validation period. In addition, in each period t during the simulation, we will take the period between the interval $(t - d - 30, t - d)$ to retrain the model with more current information so that with this feedback the model adapts to the new market situation. As we can observe, the interval with which we retrain comprises a period of 30 days, moreover we take a time window prior to d days of the period t, because for the labeling of the data we performed a simulation of d days and taking later days would be cheating in the validation.

3.5 Strategy Developed Based on the Neural Network

The strategy to follow will be based on taking actions based on the buy and sell signals generated by the neural network. To do this we will enter the feature set based on technical indicators for the period t and the network will return an output value. Since the training data was labeled as 0 or +1 depending on whether it was a good time to sell or buy respectively, then the output of the network will return values around the interval [0, +1]. In the case of executing a buy order, as many shares as possible will be bought with the available capital. In the case of sell orders, all available shares will be sold, i.e. we will go all or nothing.

It will be considered that the output of the neural network is a buy signal when the value of this is higher than 0.6, on the other hand if the output returns a value lower than 0.4 we will consider it as a sell signal. The interval [0.4, 0.6] being very close to 0.5, is considered an interval of indecision, where the response of the network is not clear enough to make a decision, therefore if the output value is in this interval we will not perform any action.

4 Study Cases and Experimental Results

4.1 Metrics Used

In this section we will describe the metrics that we will use to evaluate the performance of the proposed strategy. Our objective is obviously to develop a trading strategy that achieves the maximum possible profit, so one of the metrics we will use will be the percentage of profit obtained (Profit (%)). We will also include the number of total operations carried out (Total oper.), the number of positive operations (Oper.+) and the number of negative operations (Oper.−) (by operation we mean the purchase of a security and the subsequent sale of it). On the other hand, we will calculate the average profit of each operation (A. Oper.), the average of the negative operations (A. Oper.−) and the average of the positive operations (A. Oper.+). We will also obtain the result of dividing the average of the positive ones by the average of the negative ones (+/−), if for example the result is less than 1, it means that to obtain profits it is necessary to make more positive operations than negative ones. Finally we will obtain the maximum DrawDown (Max DD (%)), a risk indicator that measures the reduction of capital after a series of losing trades.

4.2 Experiments Performed

In this paper we will perform the same experiments as those carried out in the research Fuzzy modeling of stock trading with fuzzy candlesticks [17], so that we can compare our results with those of another research focused on stock trading. As in the aforementioned paper, the training periods of the model will comprise the time interval from 22-Dec-2009 to 21-Dec-2011. Once the neural network is trained we will validate the model obtained in the 2 subsequent years, i.e. from 22-Dec-2011 to 22-Dec-2013. The data on which the validations will be carried out correspond to quotation values belonging to the Nasdaq-100 and Eurostoxx. The initial investment portfolio will be 6 thousand dollars and the cost of each operation performed will be 1%.

On the other hand, in addition to the fuzzy logic trading paper, we will compare the results with the buy-and-hold strategy and with a strategy based on the crossover of moving averages.

Nasdaq-100

The 9 stocks considered for the Nasdaq-100 are: Apple (AAPL), Adobe (ADBE), ADP (ADP), Autodesk (ADSK), Akamai Technologies (AKAM), Alexion Pharmaceuticals (ALXN), Applied Materials (AMAT), Amgen (AMGN) and Amazon (AMZN).

One of the problems of the period analyzed for the selected Nasdaq-100 markets is that most of them are in a bullish period, so as we will see, the buy-and-hold strategy will be a winner in most of the simulations. However, in those periods that

Table 1 Nasdaq-100 strategy results

Strategy: buy and hold

Stock	Profit (%)	Max DD (%)	Total oper.	Oper.+	Oper.−	A. Oper.	A. Oper. +	A. Oper. −	+/−
AAPL	35.68	−44.18	0	0	0	0.00	0.00	0.00	Nan
ADBE	110.22	−13.98	0	0	0	0.00	0.00	0.00	Nan
ADP	48.80	−10.54	0	0	0	0.00	0.00	0.00	Nan
ADSK	60.62	−28.51	0	0	0	0.00	0.00	0.00	Nan
AKAM	48.47	−29.08	0	0	0	0.00	0.00	0.00	Nan
ALXN	91.18	−29.12	0	0	0	0.00	0.00	0.00	Nan
AMAT	66.50	−23.93	0	0	0	0.00	0.00	0.00	Nan
AMGN	76.77	−16.48	0	0	0	0.00	0.00	0.00	Nan
AMZN	124.74	−15.31	0	0	0	0.00	0.00	0.00	Nan
Total	73.68	–	0	0	0				

Strategy: moving averages cross

Stock	Profit (%)	Max DD (%)	Total oper.	Oper.+	Oper.−	A. Oper.	A. Oper. +	A. Oper. −	+/−
AAPL	36.41	−23.47	6	3	3	0.06	0.17	−0.06	2.83
ADBE	29.76	−16.81	6	3	3	0.01	0.06	−0.04	1.50
ADP	27.37	−9.07	4	2	2	0.05	0.12	−0.02	6.00
ADSK	20.15	−13.59	5	1	4	−0.01	0.11	−0.05	2.20
AKAM	−4.83	−26.59	6	3	3	−0.00	0.08	−0.09	0.89
ALXN	−15.43	−47.19	7	1	6	−0.03	0.13	−0.06	2.17
AMAT	12.62	−28.29	8	4	4	0.02	0.10	−0.06	1.67
AMGN	24.20	−16.45	6	4	2	0.04	0.07	−0.02	3.50
AMZN	70.29	−16.08	4	4	0	0.07	0.07	0.00	NaN
Total	22.28	–	52	25	27				

Strategy: neural networks

Stock	Profit (%)	Max DD (%)	Total oper.	Oper.+	Oper.−	A. Oper.	A. Oper. +	A. Oper. −	+/−
AAPL	73.34	−18.98	14	8	6	0.05	0.10	−0.03	3.33
ADBE	61.06	−11.41	9	8	1	0.05	0.06	−0.02	3.00
ADP	26.09	−10.21	9	5	4	0.03	0.06	−0.01	6.00
ADSK	56.45	−25.33	15	13	2	0.03	0.05	−0.10	0.50
AKAM	19.50	−17.09	11	7	4	0.02	0.07	−0.06	1.17
ALXN	46.19	−28.24	9	6	3	0.05	0.10	−0.05	2.00
AMAT	25.38	−11.89	7	5	2	0.04	0.07	−0.04	1.75
AMGN	27.98	−16.26	11	5	6	0.02	0.07	−0.02	3.50
AMZN	83.71	−15.60	15	13	2	0.04	0.05	−0.04	1.25
Total	46.63	–	100	70	30				

Strategy: fuzzy candlesticks

Stock	Profit (%)	Max DD (%)	Total oper.	Oper.+	Oper.−	A. Oper.	A. Oper. +	A. Oper. −	+/−
AAPL	−1.01	−8.29	30	16	14	0.00	0.17	−0.20	0.83
ADBE	−3.30	−7.93	22	9	13	−0.13	1.69	−1.38	1.22
ADP	3.63	−3.16	38	22	16	0.07	0.63	−0.68	0.92
ADSK	9.77	−3.50	23	14	9	0.74	2.41	−1.79	1.35
AKAM	−28.98	−2.85	24	16	8	1.73	3.49	−1.36	2.57
ALXN	3.92	−8.23	22	13	9	0.08	1.00	−1.24	0.81
AMAT	−0.05	−7.86	23	13	10	−0.09	3.88	−5.25	0.74
AMGN	18.05	−6.60	26	16	10	0.42	1.12	−0.71	1.58
AMZN	12.55	−3.89	28	16	12	0.10	0.32	−0.20	1.58
Total	7.74	–	264	148	116				

are not bullish, we can see how the strategy based on neural networks obtains the highest profits. The results of these experiments are shown in Table 1.

The following two charts show a comparison of the buy-and-hold strategy, the strategy based on moving averages and the strategy based on neural networks. As we can see, the fuzzy logic investment chart is not shown since we do not have such information. In left chart we can observe a bullish period and how buy and hold get the highest profits. On the other hand, in the chart on the right, we can observe a significant drop at the end of 2012, in this case the neural network strategy shows the highest performance.

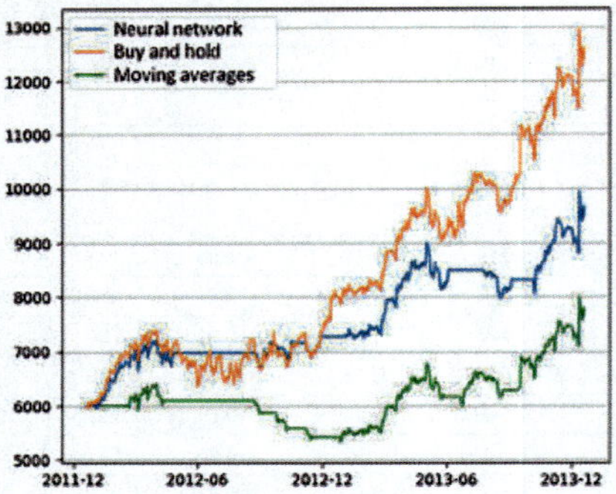

(a) ADBE amount profit

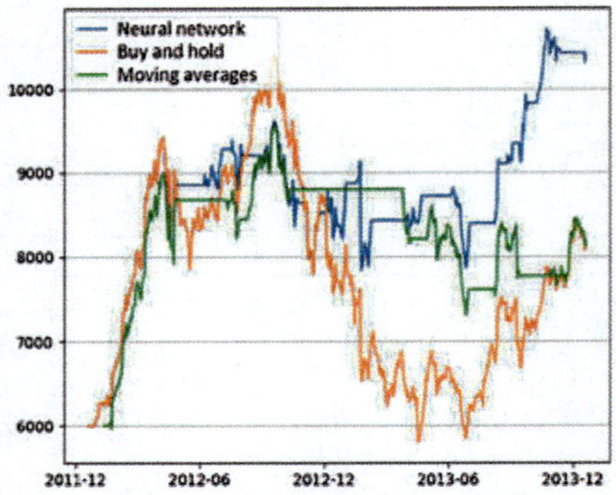

(b) AAPL amount profit

Eurostoxx

The 9 stocks belonging to Eurostoxx are: Aegon (AGN), Allianz (ALV), Deutsche Boerse (DB1), Banco de Sabadell (SAB), Santander (SAN), BME (BME), Credit Agricole (ACA), SAMPO (SAMAS) and British Land Company (BLND).

The results of these experiments are shown in Table 2. The following two graphs show the accumulated gains of the buy and hold strategies, the one based on the moving average crossover and the one based on neural networks. Again we observe that although it seems that in bullish periods buy and hold seems to be the best option, in bearish or sideways periods as in the case of the right graph the neural network outperforms the results obtained by the other two strategies.

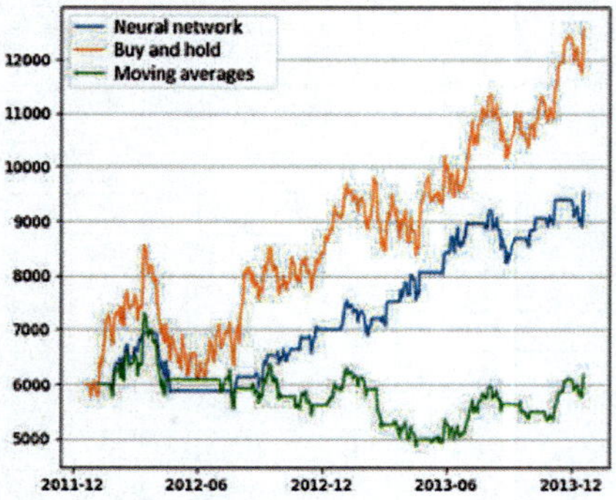

(c) AGN amount profit

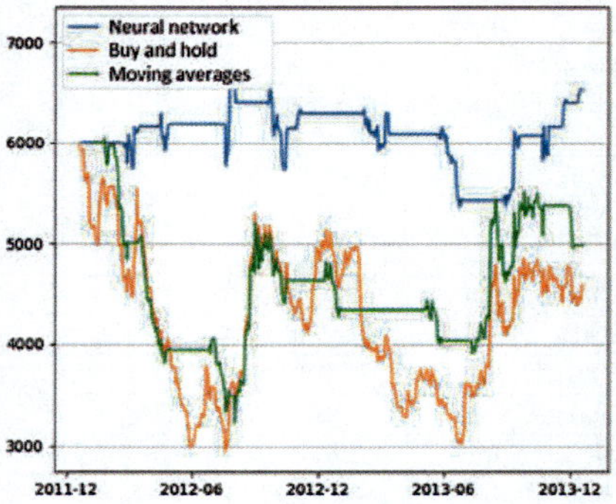

(d) SAB amount profit

Table 2 Eurostoxx strategy results

Strategy: buy and hold

Stock	Profit (%)	Max DD (%)	Total oper.	Oper.+	Oper.−	A. Oper.	A. Oper. +	A. Oper. −	+/−
ACA	109.32	−46.89	0	0	0	0.00	0.00	0.00	NaN
AGN	109.97	−29.16	0	0	0	0.00	0.00	0.00	NaN
ALV	68.93	−26.08	0	0	0	0.00	0.00	0.00	NaN
BLND	36.31	−16.44	0	0	0	0.00	0.00	0.00	NaN
BME	29.72	−36.06	0	0	0	0.00	0.00	0.00	NaN
DB1	37.78	−29.53	0	0	0	0.00	0.00	0.00	NaN
SAB	−23.40	−50.77	0	0	0	0.00	0.00	0.00	NaN
SAMPO	85.66	−18.25	0	0	0	0.00	0.00	0.00	NaN
SAN	15.84	−43.92	0	0	0	0.00	0.00	0.00	NaN
Total	52.21	–	0	0	0				

Strategy: moving averages cross

Stock	Profit (%)	Max DD (%)	Total oper.	Oper.+	Oper.−	A. Oper.	A. Oper. +	A. Oper. −	+/−
ACA	40.51	−41.12	5	2	3	0.04	0.27	−0.11	2.45
AGN	2.99	−33.46	9	3	6	−0.01	0.06	−0.04	1.50
ALV	4.40	−21.88	7	3	4	0.01	0.11	−0.06	1.83
BLND	2.94	−16.54	6	3	3	0.01	0.04	−0.03	1.33
BME	17.39	−16.82	6	2	4	0.04	0.23	−0.06	3.83
DB1	30.54	−14.80	5	3	2	0.06	0.11	−0.02	5.50
SAB	−16.94	−46.57	8	2	6	−0.00	0.33	−0.12	2.75
SAMPO	26.09	−16.28	6	3	3	0.04	0.13	−0.04	3.25
SAN	−13.89	−34.66	5	1	4	−0.03	0.14	−0.07	2.00
Total	10.45	–	57	22	35				

Strategy: neural networks

Stock	Profit (%)	Max DD (%)	Total oper.	Oper.+	Oper.−	A. Oper.	A. Oper. +	A. Oper. −	+/−
ACA	43.32	−23.41	8	4	4	0.04	0.14	−0.05	2.80
AGN	58.88	−19.88	20	18	2	0.02	0.04	−0.09	0.44
ALV	23.64	−13.64	8	5	3	0.02	0.04	−0.03	1.33
BLND	12.59	−15.77	14	9	5	0.01	0.03	−0.03	1.00
BME	16.73	−15.19	12	6	6	0.02	0.07	−0.04	1.75
DB1	11.96	−24.67	11	8	3	0.01	0.04	−0.07	0.57
SAB	8.81	−24.90	13	9	4	0.01	0.03	−0.05	0.60
SAMPO	41.36	−10.91	12	8	4	0.03	0.06	−0.03	2.00
SAN	9.86	−17.86	8	4	4	0.02	0.07	−0.03	2.33
Total	25.24	–	106	71	35				

Strategy: fuzzy candlesticks

Stock	Profit (%)	Max DD (%)	Total oper.	Oper.+	Oper.−	A. Oper.	A. Oper. +	A. Oper. −	+/−
ACA	11.61	−3.59	17	10	7	5.65	18.66	−12.95	1.44
AGN	31.33	−0.81	15	13	2	26.20	33.50	−8.16	4.11
ALV	11.92	−4.25	29	16	13	0.29	0.86	−0.38	2.23
BLND	8.05	−2.73	23	15	8	0.03	0.09	−0.06	1.45
BME	0.79	−4.76	18	8	10	0.08	3.52	−2.67	1.32
DB1	12.05	−5.52	31	18	13	0.40	1.53	−1.07	1.43
SAB	−24.50	−25.24	35	11	24	−24.91	56.11	−66.19	0.85
SAMPO	12.39	−0.62	12	10	2	2.81	3.71	−1.69	2.20
SAN	13.93	−3.49	23	12	11	6.88	20.22	−7.67	2.64
Total	8.62	–	203	113	90				

Table 3 Bearish periods results

Stock	Profit (%)	Max DD (%)	Total oper.	Oper.+	Oper.−	A. Oper.	A. Oper. +	A. Oper. −	+/−
Strategy: buy and hold									
SAN	−58.51	−63.78	0	0	0	0.00	0.00	0.00	NaN
BBVA	−54.01	−57.94	0	0	0	0.00	0.00	0.00	NaN
AMAT	−5.46	−50.15	0	0	0	0.00	0.00	0.00	NaN
Total	−39.33	–	0	0	0				
Strategy: moving averages cross									
Stock	Profit (%)	Max DD (%)	Total oper.	Oper.+	Oper.−	A. Oper.	A. Oper. +	A. Oper. −	+/−
SAN	−40.65	−43.04	7	0	7	−0.07	0.00	−0.07	−0.00
BBVA	−20.17	−23.62	4	0	4	−0.05	0.00	−0.05	−0.00
AMAT	−17.56	−37.62	5	1	4	−0.05	0.09	−0.08	1.12
Total	−26.12	–	16	1	15				
Strategy: neural networks									
Stock	Profit (%)	Max DD (%)	Total oper.	Oper.+	Oper.−	A. Oper.	A. Oper. +	A. Oper. −	+/−
SAN	22.33	−17.08	13	11	2	0.02	0.03	−0.06	0.50
BBVA	16.64	−15.88	10	6	4	0.02	0.05	−0.03	1.67
AMAT	7.67	−32.65	8	5	3	0.01	0.06	−0.09	0.67
Total	15.55	–	31	22	9				

Bearish periods

Since in the experiments carried out in [17] the markets were mostly bullish in the analyzed period, some additional case studies to fuzzy candlesticks research have been included, to be able to evaluate the strategy developed in less favorable scenarios, which is where it really has merit make a profit. These experiments have been carried out in the following scenarios:

- **Santander (SAN)** stock belonging to Eurostoxx stock index, training the model in the time period from July-2012 to July-2014 and validated from July-2014 to July-2016.
- **Banco Bilbao Vizcaya Argentaria (BBVA)** stock belonging to Ibex 35 stock index, training the model in the time period from July-2012 to July-2014 and validated from July-2014 to July-2016.
- **Applied Materials (AMAT)** stock belonging to Nasdaq-100 stock index, training the model in the time period from June-2015 to June-2017 and validated from June-2017 to June-2016.

The results of these experiments are shown in Table 3.

Table 4 Dow Jones results

Stock	NN Total cap. ($)	Total cap. ($) [18]	NN Profit (%)	Profit (%) [18]
MMM	15518.99	15234.16	**55.18**	52.34
AXP	18585.82	14727.15	**85.85**	47.27
AAPL	24816.99	14742.93	**148.16**	47.42
BA	15472.42	17010.05	54.72	**70.10**
CAT	13418.27	10252.42	**34.18**	2.52
CVX	18029.31	17907.21	**80.29**	79.07
CSCO	17751.18	21182.93	77.51	**111.82**
KO	22302.72	17258.98	**123.02**	72.58
DIS	22168.38	28859.03	121.68	**188.59**
DD	14123.92	17750.91	41.23	**77.50**
Average	18218.80	17492.57	**82.182**	74.921

This is the comparison between NN Profit (%) column and Profit (%) [18] column, where the bold letter refers to the highest value in both columns for each row

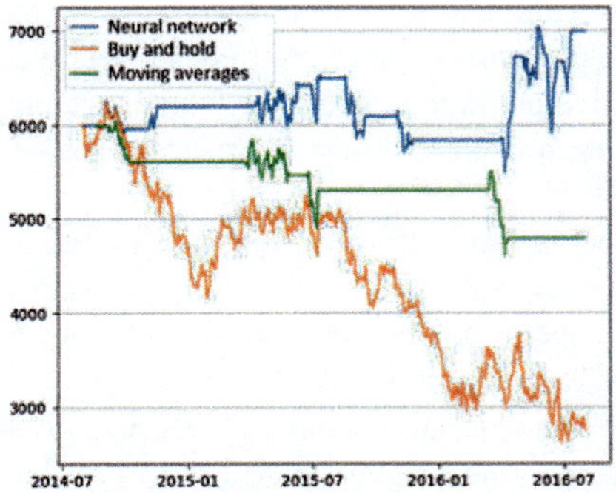

(e) BBVA Amount profit

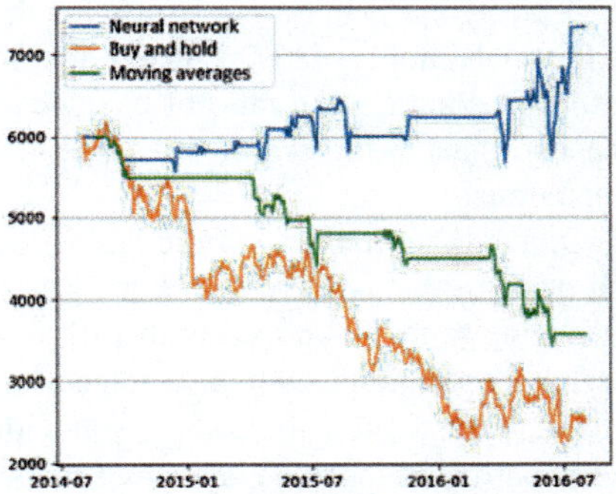

(f) SAN Amount profit

Comparison with previous research on neural networks

Finally, a comparison has been made with previous research on neural networks, specifically paper [18], where a neural network trained with the technical indicators RSI, MACD and Williams %R is presented. In this research, the proposed model has been validated with data obtained from the share prices of Dow Jones companies. The test set covers the time period from January 2007 to January 2017. Table 4 shows the comparison of the results of our research with the results of the paper, where we can see the total capital earned by our neural network (NN Total cap.) and by Sezer et al. [18] (Total cap.) and the percentage of profit earned by our neural network (NN Profit (%)) and by Sezer et al. [18] (Profit (%)).

4.3 Analysis of the Results

In the study cases analyzed in the Nasdaq-100 and Eurostoxx sections, we can see that the best results have generally been obtained by the buy and hold strategy, due to the fact that the markets studied in the period analyzed were in a bullish period. However, in those cases where this was not the case, as for example in the case of Sabadell in the Eurostoxx study where we can appreciate a bearish sideways trend or Apple in Nasdaq-100 where we can appreciate a strong fall at the end of 2012, we see how the neural network overcomes all the other strategies. On the other hand, we can see how in the case of bull markets, after the buy and hold strategy, the neural network is the one that obtains the best results.

If we compare our results with those of the paper with fuzzy logic, we see that in all cases we improve the results with a much higher profit, which could be explained perhaps due to the high number of operations performed by the strategy based on fuzzy logic, which has possibly caused a lower performance due to the commission costs per operation.

A positive aspect for our strategy is the percentage of positive trades with respect to the total number of trades, where about 70% of the trades have resulted in a positive profit. In addition, the ratio of positive/negative operations is higher than 1, which means higher profits in positive operations compared to lower losses in negative operations.

On the other hand, a very remarkable aspect is that in those simulations on bearish periods, the neural network has been the only one capable of obtaining profits, managing to maintain the number of positive operations higher than the number of negative operations, with an average profit of 15% in a period of 2 years.

Finally, regarding the comparison with previous research on neural networks, the results are quite positive, since our model has obtained higher gains in most of the cases and also as can be seen the average of the gains of all validations is higher for the case of our neural network.

5 Conclusions and Future Work

In this work we have seen how to apply neural networks to predict the best times to buy and sell. The results obtained have been quite satisfactory, since we have seen how the defined strategy, taking decisions only by means of the network predictions, has been able to obtain profit, even in bearish periods. From the investors' point of view, it is very interesting, as it shows an automatic way of investing, by means of buy and sell signals different from the classical strategies normally used.

One of the problems that the neural network may present is its inability to explain its decisions, being a black box that returns predictions without interpretability. However, the neural network has been provided with a large amount of historical data and quality characteristics obtained from different technical indicators, from which it has been able to extract hidden relationships in the data, which justifies its response and provides robustness and reliability. Furthermore, the results support its use, as considerable benefits have been obtained even in unfavorable scenarios.

Future work remains, such as the use of other technical indicators or the selection of characteristics with one of the existing techniques. On the other hand, it could be interesting to test different architectures for the neural network by modifying the number of layers and nodes to obtain the optimal structure using for example genetic algorithms.

References

1. Malkiel, B. G. (1973). *A random walk down wall street*. Norton.
2. Poterba, J. M., & Summers, L. H. (1988). Mean reversion in stock prices: Evidence and implications. *Journal of Financial Economics, 22*(1), 27–59.
3. Chaudhuri, K., & Wu, Y. (2003). Mean reversion in stock prices: Evidence from emerging markets. *Managerial Finance, 29*, 22–37.
4. Bali, T. G., Demirtas, K. O., & Levy, H. (2008). Nonlinear mean reversion in stock prices. *Journal of Banking & Finance, 32*(5), 767–782.
5. Zhang, Y., Yan, B., & Aasma, M. (2020). A novel deep learning framework: Prediction and analysis of financial time series using ceemd and lstm. *Expert Systems with Applications, 159*, 113609.
6. Saud, A. S., & Shakya, S. (2020). Analysis of look back period for stock price prediction with rnn variants: A case study on banking sector of nepse. *Procedia Computer Science, International Conference on Computational Intelligence and Data Science, 167*, 788–798.
7. Long, J., Chen, Z., He, W., Wu, T., & Ren, J. (2020). An integrated framework of deep learning and knowledge graph for prediction of stock price trend: An application in chinese stock exchange market. *Applied Soft Computing, 91*, 106205.
8. Long, W., Lu, Z., & Cui, L. (2019). Deep learning-based feature engineering for stock price movement prediction. *Knowledge-Based Systems, 164*, 163–173.
9. Kia, A. N., Haratizadeh, S., & Shouraki, S. B. (2020). Network-based direction of movement prediction in financial markets. *Engineering Applications of Artificial Intelligence, 88*, 103340.
10. Rosenblatt, F. (1958). *The perceptron: A theory of statistical separability in cognitive systems*. United States Department of Commerce.
11. Rumelhart, D. E., Hinton, G. E., & Williams, R. J. (1986). Learning Representations by Back-propagating Errors. *Nature, 323*(6088), 533–536.
12. Amari, S. (1993). Backpropagation and stochastic gradient descent method. *Neurocomputing, 5*(4), 185–196.
13. Raudys, A., Lenčiauskas, V., & Malčius, E. (2013). Moving averages for financial data smoothing. In T. Skersys, R. Butleris, & R. Butkiene (Eds.), *Information and Software Technologies* (pp. 34–45). Springer.
14. Naved, M., & Srivastava, P. (2015). Profitability of oscillators used in technical analysis for financial market. *Advances in Economics and Business Management (AEBM), 2*, 925–931.
15. Bartkus, C. (2018). Using the relative strength index for active investments in the foreign exchange market. *Modern Management Review, 23*, 12.
16. Lauretto, M. D. S., Silva, B. B. C., & Andrade, P. M. (2013). Evaluation of a supervised learning approach for stock market operations. *ArXiv*, vol. abs/1301.4944.
17. Naranjo, R., Arroyo, J., & Santos, M. (2018). Fuzzy modeling of stock trading with fuzzy candlesticks. *Expert Systems with Applications, 93*, 15–27.
18. Sezer, O. B., Ozbayoglu, A. M., & Dogdu, E. (2017). An artificial neural network-based stock trading system using technical analysis and big data framework. In *Proceedings of the SouthEast Conference*, ACM SE '17, New York, NY, USA, pp. 223–226, Association for Computing Machinery.

An Analysis of the Effects of Tourism Demand, Yield Curve, and Stock Returns on Economic Growth of Thailand: A Comparison Between the Bayesian DCC-GARCH and Bayesian Change-Point Methods

Pathairat Pastpipatkul, Pairach Piboonrungroj, and Panicha Subsai

Abstract In this paper, the Bayesian DCC-GARCH model is applied to find the relationship among economic growth, tourism demand, yield curve, and stock return in Thailand from 2013 to 2020, based on 30 quarterly observations. Only the time-varying correlation between real GDP and US tourist arrivals is positive and the volatility significantly increased around quarter 25 to quarter 28. The Bayesian change-point method is used to detect the sudden changes in the dependent variable time-series. There is one change point at quarter 28. Both methodologies provide similar evidence in that there exists a time-varying correlation, and a change-point in quarter 28 (July–October 2020) which coincides with the reopening time of Thailand after the Covid-19 lockdown. Even though two approaches are different, similar results are obtained in the relationship of various explanatory variables with economic growth. return from stocks, yield curve, and tourism demand are found to positively affect economic growth. Moreover, it is advised that the two methods could be used in future study to describe the real-world situation with different characteristics of the data set to provide more evidence.

Keywords Economic growth · Tourism demand · Yield curve · Stock return · DCC-GARCH · Change point · Bayesian · Thailand

P. Pastpipatkul · P. Piboonrungroj · P. Subsai (✉)
Faculty of Economics, Chiang Mai University, Chiang Mai, Thailand
e-mail: panicha.s.y@gmail.com

P. Pastpipatkul
e-mail: pathairat.past@cmu.ac.th

P. Piboonrungroj
e-mail: me@pairach.com

S. Sriboonchitta et al. (eds.), *Credible Asset Allocation, Optimal Transport Methods, and Related Topics*, Studies in Systems, Decision and Control 429,
https://doi.org/10.1007/978-3-030-97273-8_49

1 Introduction

This study aims to find the relationship of three markets (tourism market, yield curve as financial market, and stock return) with economic growth in Thailand using two different Bayesian methods. While yield curve and stock return are the proxies of macroeconomic conditions, tourism that operates like a global trade has recently become important for the livelihoods of many economic segments in Thailand which is one of the top ten destination countries globally considering the number of international tourist arrivals [37].

Tourism is one of the fastest-developing sectors in the global economy and its contribution to the economies of developing countries is extraordinarily significant. The report of the World Travel and Tourism Council shows that in 2019 tourism sector globally generated 8.9 trillion US dollars and 330 million jobs which is a 10% of the global employment [40]. This sector has played a remarkable role, directly and indirectly, in economic activities especially in the tourism dependent economies. Tourism's effect on the economy spreads beyond economic growth through interlinkages to increased employment, added income earnings, being a source of government revenue, foreign exchange earnings and balance of payment support [5, 13, 30]. In the case of Thailand, American tourists has significantly increased over the past two decades, with an increase around 35.83% from 2000 to 2019 [36]. The theoretical justification of such relationship is known as the tourism-led growth hypothesis (TLGH) particularly by Balaguer and Cantavella-Jordá [3]. The literature on the TLGH is supported by Dritsakis [11], Durbarry [13], and Oh [30]. Meanwhile, the competing economic-led tourism growth hypothesis (ELTGH) can be found in Lanza et al. [24] and Narayan [28].

As both current and prospective economic conditions have implication on economic growth, the present researchers rely on various data and methods for use as economic predictors. The yield curve, which plots the yield of Treasury bonds against their maturity, is one of the most watched financial indicators [22]. Dependence between the yield curve and GDP comes from its connection with state monetary policy, as bond yields react to monetary policy while monetary policy responds to the output of the economy [39]. The attention on yield curve has been increasing recently by many researchers. Studies by Hamilton and Kim [21], Stock and Watson [32], Estrella and Mishkin [16] found that term spreads can predict output growth and recessions up to one year in advance and also that the usefulness of yield curve varies across countries and over time. Estrella and Hardouvelis [16] documented that the yield spread between the ten-year Treasury bond rate and the three-month Treasury bill rate is a useful predictor of future growth in output and the probability of a recession. Hvozdenska [23] found ten-year and three-month yield spread has a significant predictive power for the real GDP growth trajectory. When the yield curve (the spread between sovereign 10-year and 3-month bonds) becomes flat (negative), it may signal a GDP decrease (recession).It is widely believed that there exists the relationship between stock return and economic growth (see [2, 26]). Generally speaking, the performance of equity investments is linked to economic growth because the source of value for equity investments is driven by economic activity. For a long period of time, Thailand's growth has relied on agriculture as a main driver. Many industries in Thailand rely on the agricultural sector as a key supplier of some important raw materials [8, 34, 38]. In Thailand, agriculture is one of the foundations for the rapid development of many industries, in the forms of raw materials, intermediate goods, and local food.

The focus in this study is on a specific multivariate GARCH model called Dynamic Conditional Correlation (DCC-GARCH). GARCH models are tools for forecasting and analyzing volatility of time series when the volatility varies over time. The DCC-GARCH model was first introduced by Engle and Sheppard [15] and can be viewed as an extension of the Constant Conditional Correlation (CCC) GARCH model presented by Bollerslev [4]. Eagle earlier proposed a GARCH model whose construction made it possible to analyze inter-dependence between markets by estimating the time-varying conditional correlation. Many previous studies have used the static economic framework, which does not allow for the potential dynamic character. Only a handful of studies have recently concentrated on the time-varying relationship between tourism and economic growth, using such frameworks as the Diebold and Yilmaz spillover index, multivariate GARCH models and rolling-window Granger causality [1, 12, 25, 33]. Moreover, this study explores a Bayesian method for the detection of sudden changes in the generated parameters of a data series and contributes to this limited number of studies by employing the Dynamic Conditional Correlation (DCC) model of Engle [15] to assess the time-varying correlation of tourism demand, yield curve, and stock returns with economic growth which is still an area to which very little attention has been paid in Thailand. This study draws attention to the effects of the three markets on economic growth using the Bayesian DCC-GARCH model, in which parameters show correlations between variables when modelling the conditional variance. After that if there are (any) time varying correlations, then the question is where is the change-point when using the same model and dataset? Furthermore, this study also considers the COVID-19 pandemic which has had a significant impact on people's lives worldwide [9]. The main contribution of this study is the performance comparison between two methodologies or models in providing consistent results to describe the real situation. This paper is organized as follows. In Sect. 2, econometric frameworks are described in detail. Section 3 presents the data description. Results from the two different methods are shown and discussed in Sect. 4. Finally, in Sect. 5, conclusions are presented.

2 Methodology

2.1 Multivariate GARCH Models

GARCH models are tools for forecasting and analyzing volatility of time series when the volatility varies over time and many multivariate GARCH models exist.

Nascimento et al. [29] defined the general form of the multivariate GARCH model can be shown as follows.

$$y_t = y_{t1}, ..., y_{tm}, \tag{1}$$

$$y_t = H_t^{\frac{1}{2}} \epsilon_t, \tag{2}$$

where y_t is a vector at time t for m time series, H_t is the matrix of conditional covariance and $H_t^{\frac{1}{2}}$ is the square root of the positive definite matrix H. The errors vector ϵ_t has the order $m1$ such that $E(\epsilon_t) = 0$, and $cov(\epsilon_t) = I_m$.

This paper is on a specific multivariate GARCH model called Dynamic Conditional Correlation (DCC- GARCH) which is a generalization of the CCC-GARCH model, which analyzes interdependence by estimating the time-varying conditional correlation [18].

$$H_t = D_t R_t D_t, \tag{3}$$

where H_t represents the conditional covariance matrix, D_t is the diagonal matrix from a univariate GARCH (1,1) and R_t is the conditional correlation matrix which is computed using:

$$D_t = diag(h_{d,t}^{\frac{1}{2}}, h_{mm,t}^{\frac{1}{2}}), \tag{4}$$

$$R_t = diag(Q_t)^{-\frac{1}{2}} Q_t diag(Q_t)^{\frac{1}{2}}, \tag{5}$$

$$Q_t = (1 - a - b)R + a\mu_{t-1}\hat{\mu}_{t-1} + bQ_{t-1}, \tag{6}$$

where Q_t is the conditional covariance between series. The parameter constraints are $a > 0$, $b > 0$ and $a + b < 1$.

2.2 Bayesian Theorem

Considering θ as the set of all model parameters and I is the information, the joint distribution as a function of the probability density chosen for the errors ϵ_t. The likelihood, likelihood function, or $Pr(y|\theta)$ contains the available information provided by the sample. The likelihood is

$$p(y_t|I_t) = |H_t|^{-\frac{1}{2}} p_{\epsilon_t}(H_t^{-\frac{1}{2}} y_t); \quad I_t = y_{t-1}, y_{t-2}, ...; \quad t = 1, ..., T, \tag{7}$$

$$L(\theta) = \prod_{i=1}^{n} |H_t|^{-\frac{1}{2}} P_{\epsilon_t}(H_t^{-\frac{1}{2}} y_t) \tag{8}$$

Hence, the DCC-GARCH model is defined as

$$L(\theta) = \prod_{i=1}^{T}[\prod_{i=1}^{m} h_{ii,t}^{-\frac{1}{2}}]|R_t|^{-\frac{1}{2}} P_{\epsilon_t}(D_t R_t D_t^{-\frac{1}{2}} y_t). \tag{9}$$

where P_{ϵ_t} is the joint density function for ϵ_t.The choice of ϵ_t distribution has a crucial role in the modeling. For the estimation of the parameters, the function bayesDcc-Garch in R package bayesDccGarch was used, as proposed by Fioruci et al. [19], and defined in 1000 thousand the number of simulations with 500 thousand being the sample of burn-in.

2.3 Change-Point Model

Change-point models have been applied in different areas including economics [20]. Mohammad-Djafari and Féron [27] noted vector from observed data in equation (10).

$$x_n = [x(t_n), x(t_n + 1), ..., x(t_{n+1})] \quad ; n = 0, ..., N, \tag{10}$$

where x is the vector that containing the data observed from time t, N represent the data samples in each segment. Then, the model in each segment of data by using Gauss-Makov chain is

$$p(x(t_n))N(\mu_n, \sigma_n^2), \tag{11}$$

$$p(x(t_n + l)x(t_n + l - 1)) = N(\rho_n x(t_n + l - 1) + (1 - \rho_n)\mu_n, \sigma_n^2(1 - \rho_n^2)); \tag{12}$$
$$l = 1, ..., l_n - 1 \quad ; l_n = t_{n+1} - t_n + 1 = dim[x_n]$$

Noting by $t = [t_1, ..., t_N]$ represent the vector of the change points and $\theta = \mu_n, \sigma_n, \rho_n, n = 0, ..., N$, and assuming that the samples from any two segments are independent, the model can write below:

$$p(x|t,\theta,N)=\prod_{n=0}^{N}N(\mu_n 1,\Sigma_n)$$
$$=(\prod_{n=0}^{N}\frac{|\Sigma_n|^{-\frac{1}{2}}}{2\pi(l_n/2)})exp-\frac{1}{2}\sum_{n=0}^{N}(x_n-\mu_n 1)'\Sigma_n^{-1}(x_n-\mu_n 1) \quad (13)$$

In Bayesian approach assigns prior probability laws on both t and θ and use the posterior probability law $p(t,\theta|x)$ as a tool for doing any inference as model is the following:

$$t_n=t_{n-1}+\epsilon_n \textbf{ with } \epsilon_n \sim P(\lambda), \quad (14)$$

where ϵ_n are assumed Independent and Identically Distributed Data (iid) and λ is prior mean of time intervals (t_n-t_{n-1}), with N is the number of change point, so $\lambda=\frac{T}{(N+)}$. The prior selection is

$$p(x,t|\theta,N)=p(x,|\theta,N)p(t|\lambda,N) \quad (15)$$

In Bayesian approach assigns prior probability laws to the hyperparameters θ, i.e., $p(\theta)$ and then the joint a posterior notes $\theta=\mu_n,\sigma_n,\rho_n, n=0,...,N$. Thus a posterior is written:

$$p(t,\theta|x,\lambda,N)=p(x,t|\theta,N)p(t|\lambda,N)p(\theta,N). \quad (16)$$

This function generates samplings from the posterior distribution of linear Gaussian model with multiple change-points. This paper uses the Markov Chain Monte Carlo method of Chib [7] and total of 10,000 approximate samples were generated from the MCMC sampling. The first 1000 were discarded as the burn-in and the remaining 9000 were kept for estimation.

3 Data Description

There are four variables to be observed, namely American tourist arrivals (GUS_i) as a proxy of tourism demand, economic growth as real GDP growth $(rgdp_i)$, and yield curve $(yield_i)$ of Thailand at time t, and stock return of the agricultural industry $(agro_i)$. According to our objective, comparison will be made in the performance between two methodologies Bayesian DCC-GARCH and Bayesian change-point models) which describe consequences with the real situation.

For the purpose of this study, this paper uses quarterly data from 2013 to 2020. For tourist arrivals, used is the data of GUS (United States' tourist arrival). Stock return of the Stock Exchange of Thailand (SET) is represented by the data of agro and food industry. As with previous studies, the yield curve is defined as the 10-

year government bonds minus 3-month treasury yields [16, 17]. The formula used to calculate yield curve is as follows:

$$y_t^n = \frac{(y_i + y_1^e + \cdots + y_{n-1}^e)}{} = L_n^t, \quad (17)$$

where y_t^n refers to the bond's yield at n maturity, at t time in country i, which in turn represents the expected yield at time $t + i$ based on available information. L_t^n represents the liquidity premium of n period' bond at time t. All variables are measured in percentage form with the data retrieved from CEIC, Tourism Authority of Thailand, Thaibma, Thomson reuters Eikon, and investing.com, respectively.

4 Results

4.1 Descriptive Summary

Table 1 reports descriptive statistics. On average, $rgdp$ (Real GDP) growth is around 0.61%, only GUS (United States' tourist arrival) faces a negative growth. For $yield$ (yield curve), the smallest volatility is presented. The data show negative skewness. The excess kurtosis statistics are equal to −0.35, −0.62, −0.63 and −0.66 indicating the platykurtic characteristics of the distribution which are higher in fat tail and standard deviation than the normal distribution. All series are stationary based on the ADF test.

4.2 Bayesian Estimation of the DCC- GARCH (1,1) Model

This article deals with the time-varying correlation of GUS (United States' tourist arrival), $agro$ (agricultural return) and $yield$ (yield curve) with economic growth as $rgdp$ (real GDP) by first using the Bayesian DCC-GARCH. After that, the Change-point model is used to find the change point from regression, with the same three independent variables to obviously explain such dependent variable as $rgdp$ (real GDP).

Table 1 Descriptive statistics for the data (2013–2020)

Variable	Mean	SD	Max	Min	Skew	Kurtosis	ADF
rgdp	0.61	6.13	9.87	−15.84	−0.39	−0.35	−3.81***
Yield	1.03	0.42	1.79	0.07	−0.26	−0.62	−3.35***
Agro	0.07	6.52	12.3	−14.05	−0.29	−0.63	−4.07***
GUS	−7.36	39.38	47.61	−100	−0.75	0.66	−3.33***

Note ***, ** and * denote statistical significance at the 1, 5 and 10% level, respectively

Table 2 Estimation of Information criteria for the DCC-GARCH models

Model	EAIC	EBIC	DIC
A	762.69	788.50	733.15
B	781.86	809.10	751.66
C	760.97	788.21	729.97

Note A: model with the expected value of the Akaike information criterion. B: model with the expected value of the Bayesian information criterion. C: model with the deviance information criterion

Table 3 Summary of the MCMC results for the DCC-GARCH model

	Mean	SD	2.50%	25%	50%	75%	97.50%
δ	1.35	0.23	0.96	1.18	1.33	1.49	1.85
γ_1	1.06	0.15	0.78	0.96	1.06	1.16	1.39
ω_1	6.9	3.94	0.64	3.82	6.47	9.55	15.21
α_1	0.35	0.17	0.08	0.22	0.34	0.48	0.7
β_1	0.35	0.16	0.07	0.24	0.35	0.46	0.68
γ_2	0.26	0.11	0.06	0.18	0.27	0.34	0.45
ω_2	1.03	0.49	0.14	0.69	1	1.33	2.13
α_2	0.46	0.22	0.03	0.29	0.46	0.62	0.87
β_2	0.24	0.23	0	0.05	0.15	0.36	0.83
γ_3	0.78	0.19	0.5	0.65	0.75	0.86	1.27
ω_3	12.38	6.58	0.98	7.45	12.06	16.78	26.27
α_3	0.22	0.17	0.01	0.09	0.19	0.32	0.63
β_3	0.55	0.21	0.12	0.41	0.57	0.71	0.89
γ_4	0.34	0.21	0.02	0.17	0.33	0.48	0.78
ω_4	14.34	7.34	1.12	9.14	14.05	19.08	29.7
α_4	0.15	0.05	0.05	0.11	0.15	0.18	0.26
β_4	0.83	0.05	0.72	0.79	0.83	0.86	0.93
a	0.02	0.02	0	0	0.01	0.02	0.06
b	0.38	0.28	0.02	0.14	0.33	0.61	0.93
$a+b$	0.4	0.29	0.02	0.15	0.34	0.63	0.99

Table 2 presents the goodness of fit statistics for the model comparison about how data were generated and which model is best supported by the available evidence. In the model selection was performed upon the expected value of the Akaike information criterion (EAIC), the expected value of the Bayesian information criterion (EBIC) and the deviance information criterion (DIC) of each model that the best performance due to the minimization of the values [14]. As the lowest EAIC, EBIC and DIC values correspond to the best model, the Model C which has the smallest values from all criteria is selected for the estimation. Table 3 contains the posterior means, standard deviation (SD), and quantile for each variable. The conditional variance results are highly significant, indicating the existence of the GARCH effect in the

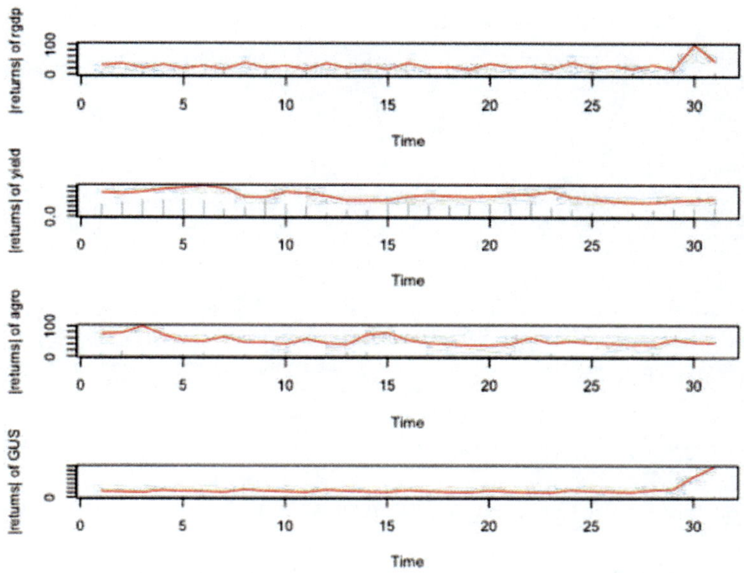

Fig. 1 Volatilities of variables in the models. *Note* Vertical axis (|returns|) presents the volatilities in each variable over time and horizontal axis represents time

series. The estimated coefficients in Table 3 reject the posterior distribution of the CCC condition, from the CCC model hypothesis, the coefficients (a and b) are equal to 0. Moreover, the correlation's estimation will satisfy the necessary condition of a + b <1, which justifies the Bayesian DCC-GARCH model in measuring the time-varying conditional correlations [10].

This study analyzed the time-varying correlation (Bayesian DCC-GARCH). It is possible to see how the conditional correlations evolve through time by considering the MCMC approach instead of the Maximum likelihood. This study also pursued further whether there are co-movements among variables.

Figure 1 shows the volatilities of tourist's arrival, stock return, yield curve, and economic growth over the study period. One can observe that in most recent years, the volatilities of $rgdp$ (real GDP) and GUS (United States' tourist arrival) became significantly higher. Figure 2, meanwhile, shows that the correlations of the pairwise variables are time-varying or dynamic and sometimes changed sharply.

From the GARCH (1,1) model estimates in Table 3 and graphical information in Fig. 2, it shows that the time-varying correlation between $rgdp$ (real GDP) and $yieldcurve$ (black line) as well as the co-movement between $yield$ (yield curve) and GUS (United States' tourist arrival) (light blue line) climbed up and down dramatically around zero to -0.15 and dropped sharply in Thailand reopening after Covid-19. Then, the correlations fluctuated steadily mainly around -0.05 and -0.1 in the pairs between $rgdp$ (real GDP) and $agro$ (agricultural return) shown as red line and $agro$ (agricultural return) and GUS (United States' tourist arrival) shown as

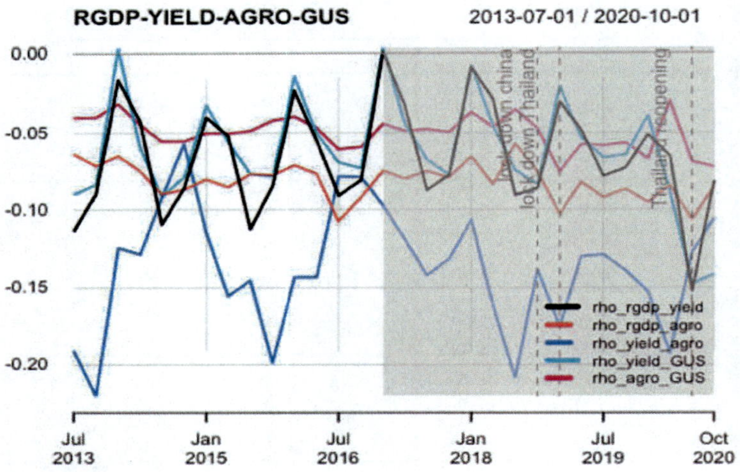

Fig. 2 The dynamic correlations from the DCC-GARCH models

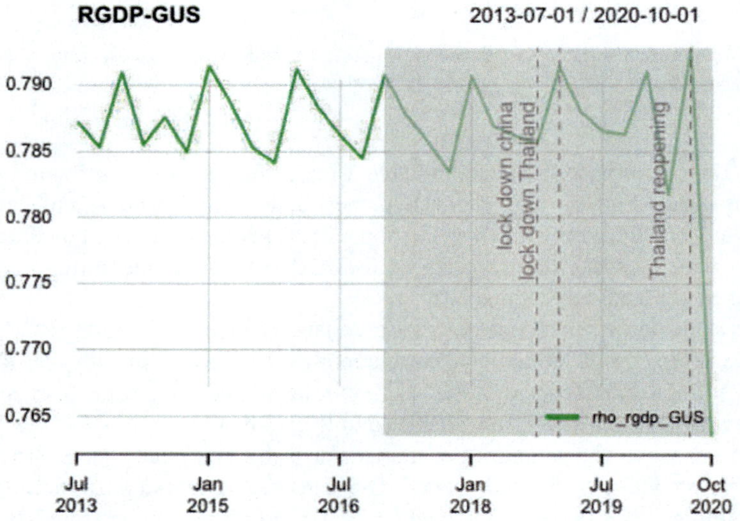

Fig. 3 The dynamic correlations from the DCC-GARCH models. *Note 1* Dot line shows the important events related to the COVID-19 pandemic including lockdown in China, lockdown in Thailand and Thailand reopening. *Note 2* Correlation between real GDP and yield curve (black line). Correlation between real GDP and agricultural return (red line). Correlation between real GDP and American tourist (green line). Correlation between yield curve and agricultural return (blue line). Correlation between yield curve and American tourist (light blue line). Correlation between agricultural return and American tourist (purple line)

Table 4 The matrix of the natural log

	Model 0	Model 1	Model 2	Model 3	Model 4
Model 0	0.00	−5.24	2.23	7.96	13.04
Model 1	5.24	0.00	7.47	13.20	18.28
Model 2	−2.23	−7.47	0.00	5.73	10.82
Model 3	−7.96	−13.20	−5.73	0.00	5.09
Model 4	−13.04	−18.28	−10.82	−5.09	0.00

Note Model 0, Model 1, Model 2, Model 3, and Model 4 represent the model with no change- point, one change-point, two change-points, and three change-points, respectively

purple line with a sharp increase (purple line) and a sharp drop (red line) in Thailand reopening period. The blue line representing the correlation between $yield$ (yield curve) and $agro$ (agricultural return) reveals the continuous and dramatic shift around −0.20 to −0.05 over the sample period.

The correlation of $rgdp$ (real GDP) with GUS (United States' tourist arrival) as plotted in Fig. 3 was positive and fluctuating around 0.790 with the highest value at 0.795 in the period of Thailand reopening, and sharp drop to lower than 0.765. It is surprising to find that only the pair of $rgdp$ (real GDP) and GUS (United States' tourist arrival) has positive correlation. According to official report, in October 2020, Thailand had a low cumulative number of infections and the Thai Ministry of Tourism and Sports proposed special tourist visas to attract tourists from cities not having a strong outbreak who have high purchasing power to visit Thailand for driving the tourism economy.

4.3 *Change-Point Model*

For the change-point analysis, we find the change-point and selecting the best model upon the Bayes factor as presented in Table 4.

In Table 4, it shows the results from the change-point test by Bayes factor. The Bayes factor values give the probabilities that indicate the best model to be model 1 which means there is one change-point in the model.

In Fig. 4, it reveals that there is one state change point in quarter 28 or the period of Thailand reopening, and both provide the results from the Change point method. This finding is also in line with the results of the Bayesian DCC-GARCH model estimation.

The posterior means, standard deviation (SD), and percentile for each variable are presented in Table 5. For regime 1, yield curve and agricultural return have −0.13 and −0.03 as the coefficients, indicating that if the term spread and agricultural return each increases by one percent, real GDP will decrease by 0.13 and 0.03% respectively. Meanwhile, the coefficient of GUS is positive and at every level of percentile indicating the increase in US tourist arrival will link to an increase in

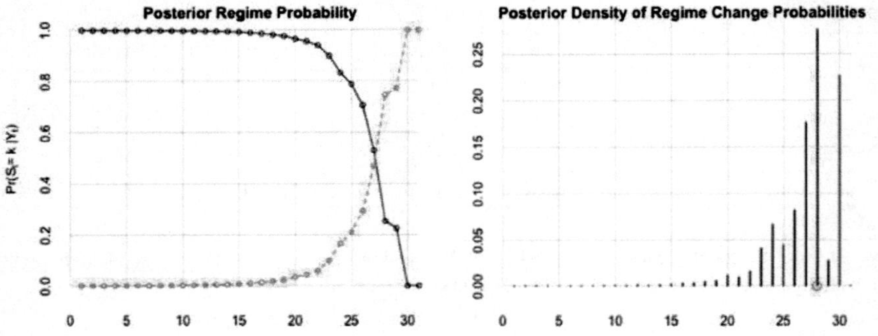

Fig. 4 Plot of the state of posterior regime probability and posterior density of regime change probability

Table 5 Summary of empirical mean, standard deviation, and percentiles for each variable

	Mean	SD	2.50%	25%	50%	75%	97.50%
Regime 1							
(Intercept)	0.25	0.76	−1.27	−0.27	0.27	0.77	1.74
Yield	−0.13	0.65	−1.4	−0.56	−0.13	0.3	1.16
Agro	−0.03	0.09	−0.2	−0.08	−0.03	0.03	0.15
GUS	0.2	0.03	0.14	0.18	0.2	0.22	0.26
Sigma2	6.72	2.71	3.27	4.89	6.15	7.86	13.49
Regime 2							
(Intercept)	0.99	0.86	−0.69	0.41	1	1.57	2.69
Yield	1.01	0.86	−0.68	0.43	1.01	1.59	2.72
Agro	0.05	0.28	−0.5	-0.14	0.04	0.22	0.61
GUS	0.06	0.03	0	0.04	0.06	0.08	0.12
Sigma2	35.08	12.17	18.84	26.66	32.65	40.9	64.67

Note Mean is average value, SD is standard deviation for each covariate. 2.50, 25, 50, 75, and 97.50% represent coefficients in each percentile distribution in 2.50, 25, 50, 75, and 97.50% respectively

real GDP.In context of regime 2, yield curve, agricultural return and GUS have the coefficients equal to 1.01, 0.05 and 0.06, indicating that if these variables increase by one percent, real GDP will increase by 1.01, 0.05 and 0.06%, respectively.

5 Conclusions and Policy Implications

In this paper, the Bayesian DCC-GARCH and the Bayesian change-point models are used to find the relationship among economic growth, tourism demand, yield curve and stock return in Thailand from 2013 to 2020. From the Bayesian DCC-GARCH model, only the correlation between $rgdp$ (real GDP) and GUS (United

States' tourist arrival) was found to be time-varying and positive; and both variables show a significant increase in their volatilities around quarter 25 to quarter 28 when Thailand pursued a reopening. Before the disruption by COVID-19, American tourist arrivals increased due to the 2.4% growth in the US economy that helped elevate both employment and private consumption levels [35]. Another support came from the Tourism Authority of Thailand [36] that ran a tourism promotion campaign during the high-seasoned winter 2018–summer 2019 with the involvement of Thai restaurants as well as airline partners in the United States. Following the disruption, the Thai government have tried to create various campaigns such as free visa on arrival and free double-entry tourist visa for calling confidence back and expanding the tourism market. Furthermore, the Global COVID-19 Index ranked Thailand at the first position in Asia and the sixth from 184 countries around the world for the best health and epidemic management with Covid-19 [6]. The result in this paper provides implications for strategic policy, especially that future National Tourism Plans, government and related agencies should attempt to develop tourism services by giving priority to the most positive relationship with American travelers. According to the Bayesian change-point model, there is one change point at quarter 28. Indeed, from Table 5, for the long run if government holds agricultural return and yield curve relatively stable and draws attention from GUS (United States' tourist arrival), real GDP will increase. There are several explanations for supporting the above statements. Firstly, agricultural return index comes from return in businesses that are related to cultivation, forestry, livestock, processing of produce because many industries in Thailand have to rely on the agricultural sector as a supplier of some important raw materials. However, at one point a change took place with a significant decrease in the overall agricultural output around the first half of 2019 consequential to the unusually hot and dry weather with scant rainfall, crop damage from a Tropical storm and flood that imperiled not only agricultural productivity and export but also the linkage-sectors. Furthermore, the Thai general election was held on 24 March 2019 that was the first election since the 2014 which was drafted under the ruling military. For suggestion, the political situation might affect decisions of stockholders, investors and tourists. When comparing the Bayesian DCC-GARCH model and the Bayesian change-point model in terms of tourism demand and ag-sector stock return relations with economic growth of Thailand, both methodologies provide similar answers that there was correlation and a change point in quarter 28 or the Thailand reopening period. For the recommendation, the government can increase economic growth considering the yield curve, agricultural sector and tourism sectors generally have positive correlation in normal periods. Moreover, future analysis can use these two methods to describe real situation with different characteristics of dataset for identifying more evidence. We also suggest employing the Markov Switching Dynamic Copula to investigate the nonlinear dependence among the variables considered in this study [31].

References

1. Antonakakis, N., Dragouni, M., & Filis, G. (2015). Tourism and growth: The times they area changing. *Annals of Tourism Research, 50*, 165–169.
2. Baker, D., Delong, J. B., & Krugman, P. R. (2005). Asset returns and economic growth. *Brookings Papers on Economic Activity, 1*, 289–315.
3. Balaguer, J., & Cantavella-Jorda, M. (2002). Tourism as a long-run economic growth factor: The Spanish case. *Applied Economics, 34*(7), 877–884.
4. Bollerslev, T. (1990). Modelling the coherence in short-run nominal exchange rates: A multivariate generalized arch model. *The Review of Economics and Statistics, 72*, 498–505.
5. Brida, J. G., Gomez D. M., & Segarra V. (2020). On the empirical relationship between tourism and economic growth. *Tourism Management, 81*.
6. Center for Health Security & Nuclear Threat Initiative. (2019). Global Health Security (GHS) Index. Nuclear Threat Initiative.
7. Chib, S. (1998). Estimation and comparison of multiple change-point models. *Journal of Econometrics, 86*, 221–241.
8. Chinprateep, A. (2013). Agricultural based to manufacturing oriented growth: In the case of Thailand. *NIDA Case Research Journal, 5*(1), 101–125.
9. Chriscaden, K. (2020). Impact of COVID-19 on people's livelihoods, their health, and our food systems. World Health Organization.
10. Chodchuangnirun, B., Yamaka, W., & Khiewngamdee, C. (2018). A regime switching for dynamic conditional correlation and GARCH: Application to agricultural commodity prices and market risks. In *International Symposium on Integrated Uncertainty in Knowledge Modelling and Decision Making* (pp. 289–301). Springer, Cham.
11. Dritsakis, N. (2004). Tourism as a long-run economic growth factor: An empirical investigation for Greece using causality analysis. *Tourism Economics, 10*(3), 305–316.
12. Dragouni, M., Filis, G., Gavriilidis, K., & Santamaria, D. (2016). Sentiment, mood and outbound tourism demand. *Annals of Tourism Research, 60*, 80–96.
13. Durbarry, R. (2002). The economic contribution of tourism in Mauritius. *Annals of Tourism Research, 29*, 862–865.
14. Ebrahim, H. K. M. S., Teymour, M., Mohammad, R. S., & Reza, T. (2019). Bayesian estimation of the relationship between return volatility and the trading volume of pharmaceutical index in Tehran stock exchange using dynamic conditional correlation model. *Journal of Advanced Pharmacy Education & Research, 10*, 112–119.
15. Engle, R., & Sheppard, K. (2001). Theoretical and empirical properties of dynamic conditional correlation multivariate GARCH. NBER Working Papers, No. 8554.
16. Estrella, A., & Hardouvelis, G. A. (1991). The term structure as a predictor of real economic activity. *The Journal of Finance, 46*(2), 555–576.
17. Estrella, A., & Mishkin, F. S. (1998). Predicting U.S. recessions: Financial variables as leading indicators. *Review of Economics and Statistics, 80*, 45–61.
18. Engle, R. F. (2009). *Anticipating correlations a new paradigm for risk management*. Princeton University Press.
19. Fioruci, A. J., Ehlers, R. S., & Andrade Filho, M. G. (2013). Bayesian multivariate GARCH models with dynamic correlations and asymmetric error distributions. *Journal of Applied Statistics, 41*(2), 320–331.
20. Ghilagaber, G. & Munezero, P. (2019). Bayesian change-point modelling of the effects of 3-points-for- a-win rule in football. Journal of Applied Statistics.
21. Hamilton, J. D., & Kim, D. H. (2002). A reexamination of the predictability of economic activity using the yield spread. *Journal of Money, Credit, and Banking, 34*(2), 340–360.
22. Haubrich, J. G. & Dombrosky, A. M. (1966). Predicting real growth using the yield curve (pp. 26–35). The Federal Reserve Bank of Cleveland.
23. Hvozdenska, J. (2015). The yield curve as a predictor of gross domestic product growth in nordic countries. *Procedia Economics and Finance, 26*, 438–445.

24. Lanza, A., Temple, P., & Urga, G. (2003). The implications of tourism specialisation in the long run: An econometric analysis for 13 OCDE economies. *Tourism Management, 24*, 315–321.
25. Lean, H. H., & Tang, C. F. (2010). Is the tourismfiled growth hypothesis stable for Malaysia? *International Journal of Tourism Research, 12*(4), 375–378.
26. Madsen, J. B., Dzhumashev, R., & Yao, H. (2013). Stock returns and economic growth. *Applied Economics, 45*(10), 1257–1271.
27. Mohammad-Djafari, A., & Féron, O. (2006). Bayesian approach to change points detection in time series. *International Journal of Imaging Systems and Technology, 16*.
28. Narayan, B., Rajendran, C., Sai, L. P., & Gopalan, R. (2009). Dimensions of service quality in tourism–An Indian perspective. *Total Quality Management Business Excellence, 20*(1), 61–89.
29. Nascimento, D., Xavier, C., Felipe, I., & Neto, F., L. (2019). Dynamic conditional correlation GARCH: A multivariate time series novel using a Bayesian approach. *Journal of Modern Applied Statistical Methods, 18*(1).
30. Oh, C. O. (2005). The contribution of tourism development to economic growth in the Korea economy. *Tourism Management, 26*(1), 39–44.
31. Pastpipatkul, P., Yamaka, W., & Sriboonchitta, S. (2016). Dependence structure of and co-movement between Thai currency and international currencies after introduction of quantitative easing. *Causal inference in econometrics* (pp. 545–564). Springer.
32. Stock, J. H., & Watson, M. W. (2003). Forecasting output and inflation: The role of asset prices. *Journal of Economic Literature, 41*, 788–829.
33. Tang, C. F., & Tan, E. C. (2013). How stable is the tourism-led growth hypothesis in Malaysia? Evidence from disaggregated tourism markets. *Tourism Management, 37*, 52–57.
34. Tansuchat, R., Maneejuk, P., Wiboonpongse, A., & Sriboonchitta, S. (2016). Price transmission mechanism in the Thai rice market. In Causal Inference in Econometrics (pp. 451–461). Springer.
35. Trade Policy and Strategy Office. (2019). Economic overview of the month of October, 2019. *Journal of Trade Policy and Strategy Office, 9*(88).
36. Tourism Authority of Thailand. (2018). *Summary of the tourism promotion action plan*. Tourism Authority of Thailand (Headquarters).
37. UNWTO (2019). *Tourism highlights 2019 edition*. United Nations World Tourism Organisation.
38. Wattanutchariya, W., Tansuchat, R., & Ruennareenard, J. (2016). Supply chain management of Thai parboiled rice for export. In International Conference on Industrial Engineering and Operations Management ,8-10.
39. Wheelock, D. C., & Wohar, M. E. (2009). Can the term spread predict output growth and recessions? A survey of the literature. *Federal Reserve Bank of St. Louis Review, 91*(5), 419–440.
40. WTTC. (2020). *Travel & Tourism Global Economic Impact & Trends*. World Travel & Tourism Council.

Relationship Among International Trade, Financial Development, and Economic Growth: The Case of ASEAN

Pathairat Pastpipatkul, Kunchon Wattanakul, and Chutikan Khumchui

Abstract This study investigated the existence of relationship among international trade, financial development, and economic growth for ASEAN as a group and individual countries, using annual data covering the period 2010–2019 for the analysis by Bayesian regression and Bayesian Model Averaging (BMA) methods. In this study separated financial development in two group of variables consist of financial development variables and cashless variables. For this study defines definition of cashless that mean not using money in the form of coins and notes or operating with credit and debit cards and electronic systems. Therefore, cashless variables in this study include credit card transaction value (CRC), debit card transaction value (DBC), and e-money transaction value (EM). The study results reveal that, for ASEAN as a group, there are four variables most appropriate to describe real GDP namely domestic credit to private sector (DCP), inflation (I), stock trade (ST), and export (EXP) but the cashless variables do not show effect on real GDP. For individual countries, the results reveal the sets of variables explaining real GDP to be different across countries. Cashless variables in particular affect real GDP of only some countries in the group. Based on the results, this paper offers some imperative policy recommendations.

1 Introduction

The financial sector is one of the important sectors driving economic growth and development in many countries [42]. Financial markets and financial institutions are the important measurement of the status of financial development and financial intermediaries, respectively which are the largest part of the financial system in most countries.

P. Pastpipatkul · K. Wattanakul · C. Khumchui (✉)
Faculty of Economics, Chiang Mai University, Chiang Mai, Thailand
e-mail: chutikan.chtkk@gmail.com

P. Pastpipatkul
e-mail: pathairat.past@cmu.ac.th

K. Wattanakul
e-mail: kunchon.w@cmu.ac.th

S. Sriboonchitta et al. (eds.), *Credible Asset Allocation, Optimal Transport Methods, and Related Topics*, Studies in Systems, Decision and Control 429,
https://doi.org/10.1007/978-3-030-97273-8_50

It is well documented in the literature of both growth and finance that financial development and economic growth are highly related (see, for example [32, 34]) and have reciprocal relationship. The financial market is an important source for mobilizing and generating funds [42]. A well-developed financial market channelizes better information on profitable investments and promotes appropriate capital allocation [12, 17, 18, 38] and also reduces the cost of corporate governance [6]. To sum up, development in the financial system can be defined as and improvement in the depth, access, and efficiency of financial markets, with the increased access to financial markets [42]. Meanwhile, developed financial intermediaries boost the financial technological innovations [28], help reduce the cost of obtaining information and executing contracts, and help permit executing transactions more efficiently [26, 27]. Moreover, expanding financial access provides dynamic efficiency in the system, bringing about a structural change through increased innovation and welfare to the entire economy [18]. However, the nexus between economic growth and financial development generally has been unclear [9].

Nowadays, the advancement of technology and innovation plays an important role in many sectors, especially in the financial sector. Technology and innovation drive and improve the channels of payments to make the whole business more convenient for service users. The cashless payment is the economic transaction that eliminates the usage of cash as a medium of exchange for goods and services by allowing electronic payments or non-electronic payments [10].

World Payment Report [43] revealed cashless transaction increased 12% during 2016–2017 to reach 539 billion US dollars globally. In 2017, developing markets are close to contributing a half of all cashless transactions. Emerging Asia plays a key role in recording the growth in global cashless transactions because the widespread availability of digital wallets and the success of e-commerce platforms, and innovation in mobile payments are driving the growth in the Emerging Asia region.

The integration of countries into the world economy is often seen as an important factor for increasing income and growth, and it has proven to be an effective way for countries to promote economic growth and development. Over the past 20 years, world trade' growth has averaged 6% per year, twice as fast as world output. But trade has been an engine of growth for much longer [23]. Results on the direction of relationship between international trade and economic growth are still inconclusive [3]. Yet, a fair number of recent studies have shown that international trade is crucial for economic growth of many countries [5, 11, 24, 29, 37, 41].

In the existing growth literature, much attention has been given to assessing the impact of trade on economic growth. The role of trade policy in economic growth has been a key matter of debate in the economic growth literature. So, the nexus between economic growth and trade has mainly been examined with respect to three competing hypotheses (the export-led growth hypothesis, the growth-led export hypothesis, and the feedback hypothesis). In the literature, one of the most enduring areas of research is the so-called export-led growth (ELG) hypothesis (see, for example [4, 13, 30, 33, 40]). Export expansion and openness to foreign markets is viewed as a key determinant of economic growth because of the positive externalities it provides [20]. Furthermore, exports can provide foreign exchange that allows for more imports of intermediate goods which in turn increase capital formation and thus stimulate output growth [2]. Several studies have also shown that it is possible to have growth-led export (GLE) which has the reverse causal flow from economic growth to export growth. In the GLE case, export expansion could be stimulated by productivity gains caused by increases in domestic levels of skilled-labor and technology [7, 25].

The Association of Southeast Asia Nations (ASEAN) was established in 1967 and currently includes ten countries: Brunei Darussalam, Cambodia, Indonesia, Laos, Malaysia, Myanmar, the Philippines, Singapore, Thailand, and Vietnam. The formation of ASEAN was primarily driven by political and security motivations, with a view to promoting economic cooperation. ASEAN is a vibrant region of over 600 million people living, which provides an opportunity for the member countries to improve the socio-economic conditions of their people through infrastructure and communications networks, as well as the movement of people, goods and services within the region. These developments lead investors to diversify their investment portfolios by and benefit from investing in ASEAN countries.However, since Brunei Darussalam, Cambodia, and Myanmar do not have financial markets yet, Lao securities exchange was first operated in January 2011 while the Philippines and Vietnam have some information missing, this paper selects for its study the four remaining ASEAN countries consisting of Indonesia, Malaysia, Singapore, and Thailand.

According to a vast literature regarding the relationship between financial development and economic growth, financial development assists economic growth through various channels, including export expansion [35]. A well-functioning financial sector of a country can have positive impact on export in addition to its influence on output growth [21, 36]. Hence, to get a higher world trade' export shares, it is important for the country to have a well-developed financial system.

There is a vast literature that investigates the relationship among GDP growth, export, and financial development; but literature on countries in ASEAN is limited.This study aims to investigate the existence of relationship between international trade, financial development, and economic growth for ASEAN as a group and individual countries.

The article proceeds as follows: Sect. 2 defines data and methodology of the study. Section 3 provides results and discussions, and the paper concludes with Sect. 4.

2 Data and Methodology

2.1 Data Description

Data used in this paper are annual figures covering the period 2010–2019. The variables of interest are real GDP (RGDP), Domestic credit to private sector (DCP), Commercial bank branches (CBB), Automated teller machines (ATM), Interest rate spread (IRS), Stocks traded (ST), Market capitalization of listed domestic companies (MCL), Stock turnover ratio of domestic share (STR), Credit card transaction value (CRC), Debit card transaction value (DBC), E-money transaction value (EM), Labor force (L), Inflation (I). Data were gathered from World Development Indicators and CEIC.

2.2 *Bayes' Theorem*

Bayes' Theorem (Bayes' Rule) was originally developed by Thomas Bayes. The idea of this theorem is the conditional probability. The conditional probability of an event is the probability obtained with additional information that another event has occurred. The basic theorem states simply:

$$P(A\,|B\,) = \frac{P(A\,|B\,)P(A)}{P(A)} \tag{1}$$

The theorem states that the conditional probability for event B given event A is equal to the conditional probability of event A, given event B, multiplied by the incremental probability for event B, and divided by the marginal probability for event A.

Maximum Likelihood Estimation (MLE) is a method of estimating the model's parameters. This is the most widely used method for statistical estimation. The likelihood function, or $p(y_i\,|\theta\,)$, contains the information provided by the sample. The likelihood function is given by

$$p(y\,|\theta\,) = \Pi_{i=1}^{n}(y_i\,|\theta\,) \tag{2}$$

In the Bayesian approach, before observing data, an unknown parameter was modeled as a random variable Θ, with $f_\Theta\,(\theta)$ probability distribution called prior distribution. Posterior distribution demonstrates the knowledge about the parameter θ after having observed the data. This is obtained by multiplying the prior distribution total parameters by the likelihood function, so that the posterior is determined by:

$$f_{\Theta|x}\,(x\,|\theta\,)\,\alpha\; f_{\Theta|x}\,(x\,|\theta\,)\,f_\Theta(\theta) \tag{3}$$

where the symbol α hides the proportional factor $f_x(X) = \int f_{\Theta|x}\,(x\,\left|\theta'\right.)\,f_\Theta(\theta')d\theta'$, which is independent of θ

Bayesian panel data analysis
The analysis of panel data must take into account the specific structure of various observations for each individual. There are two popular methods for determining the special time structures known as the fixed- and the random-effects models [31].

Fixed effects model
The fixed effects model has individual specific dummy variables and assumes that the difference between units can capture the difference in its constant term. The fixed effects model can be known as an OLS model with the matrix of descriptive variables X, being augmented by vectors of dummy variables for each individual.

The fixed effects model is determined by: The fixed effects model has individual specific dummy variables and assumes that the difference between units can capture the difference in its constant term. The fixed effects model can be known as an OLS model with the matrix of descriptive variables X, being augmented by vectors of dummy variables for each individual. The fixed effects model is determined by:

$$y_i = X_i\beta + \alpha_i j_T + \in_{i,} \in_i \perp X_i, \beta, \alpha_i, \in_i \sim n(\tau I_T) \tag{4}$$

where y_i is a vector of length T, X_i is a $T \times k$ matrix and β is a $k \times 1$ vector of coefficients.$\in_i$ is the error term of length $T \times 1$ and is normal, homoscedastic, not autocorrelated and independent of X_i, α_i and β. The term of I_T represents the identity matrix of size T.

Random effects model The random effects model assumes that all constants are similar and located around a common value. Each effect is assumed to vary normally around $\bar{\alpha}$. There are two resulting equations:

$$y_i = X_i\beta + \alpha_i j_T + \in_{i,} \tag{5}$$

$$\alpha_i = \bar{\alpha} + \eta_i \tag{6}$$

where $\in_i$ and η_i are independent and normally distributed.

2.3 Markov Chain Monte-Carlo (MCMC)

The Markov Chain Monte Carlo (MCMC) methods were initially developed by Los Alamos in the 1940's [19]. This class of methods is now often employed to solve complex, integration, and optimization problem in large dimensional spaces [1]. The Markov Chain Monte Carlo is a simulation method that extracts the value of θ from the approximate distribution and then constructing the target distribution $p\,(\theta\,|y)$ [16].

The ideas of aperiodictiy and irreducibility are the main theory behind the MCMC methods that are important properties for the chain to converge to the stationary distribution. Another important property is reversibility where:

$$p\left(\theta^t\right) p\left(\theta^{t-1}\left|\theta^t\right.\right) = p\left(\theta^{t-1}\right) p\left(\theta^t\left|\theta^{t-1}\right.\right). \tag{7}$$

There are many types of MCMC algorithms. The two most basically used are Gibbs sampling and Metropolis-Hastings algorithm. Gibbs sampling is a technique to generate random variables indirectly from a posterior distribution. Assuming that the parameter vector θ is divided into d components or sub-vectors, $\theta = (\theta 1, ..., \theta d)$ Each reiteration of the Gibbs sampler is repeated through the sub-vectors of θ, drawing each subsection conditionally with all other values, creating it a reiteration with d

steps. The order of the d sub-vectors of θ is selected, on the other hand, θ_j^t is sampled from the conditional distribution given all the other components of θ:

$$p\left(\theta_j \left|\theta_{-j}^{t-1}, y\right.\right) \tag{8}$$

where θ_{-j}^{t-1} refers to all the components of θ except for θ_j. In each sub-vector θ_j is conditionally updated on the last values of the other components of θ, which are the reiteration t values for the components already updated and the reiteration $t-1$ values for the others [16]. In general, a conditional distribution of the requisite equilibrium distribution is accepted in Gibbs sampling [8].

Metropolis-Hastings algorithm is a common feature of the Metropolis algorithm. The difference is the $j\ (\theta^* \left|\theta\right.)$ distribution that is not necessary a symmetric one [16]. Metropolis-Hastings relates with the proposal distribution $q\ (\theta^* \left|\theta\right.)$ in sampling value θ^* process with a stationary distribution $p\ (\theta \left|y\right.)$ After that, Markov chain moves to θ^* by using an acceptance probability.

$$\alpha\left(\theta^{t-1}, \theta^*\right) = \min\left\{1, \frac{p\left(\theta^* \left|y\right.\right) q\left(\theta^{(t-1)}\right)}{p\left(\theta^{(t-1)} \left|y\right.\right) q\left(\theta^*\right)}\right\}, \tag{9}$$

otherwise, it remains at θ.

2.4 *Bayesian Model Averaging*

Bayesian Model Averaging (BMA) refers to a canonical regression problem in model uncertainty [44]. This technique is an extension of the usual Bayesian inference methods that one does not only model parameter uncertainty through the prior distribution, but also model uncertainty to obtain posterior distribution using Bayes' theorem. Hence, the direct model consists of combined estimation and prediction [15]. Suppose that y is the dependent variable, α_γ is a constant, β_γ is coefficient, and ε is the error term with variance σ^2 , the general structure of the linear model is determined by:

$$y = \alpha_\gamma + \beta_y X_y + \varepsilon,\ \varepsilon \sim N\left(0, \sigma^2 I\right), \tag{10}$$

The BMA solves the problem by estimating models for all possible combinations of $\{X\}$ and generating a weighted average of all of them. If X contains K possible variables, which means estimating 2^K variable combinations and 2^K models. The model weights of this averaging from posterior model probabilities that occur from Bayes' theorem:

$$p\left(M_y\,|y,X\right)=\frac{p\left(y\,|M_y,X\right)p\left(M_y\right)}{p\left(y\,|X\right)}=\frac{p\left(y\,|M_y,X\right)p\left(M_y\right)}{\sum_{S}^{2^K}p\left(y\,|M_s,X\right)p\left(M_s\right)} \tag{11}$$

Therefore, $p\left(y\,|X\right)$ denotes the integrated likelihood, which is constant over all models, $p\left(M_y\,|y,X\right)$ is the marginal likelihood of the model $p\left(y\,|M_y,X\right)$ times a prior model probability $p\left(M_y\right)$.

$$p\left(\theta\,|y,X\right)=\sum_{y-1}^{2^K}p\left(\theta\,|M_y,y,X\right)\frac{p\left(M_y\,|X,y\right)p\left(M_y\right)}{\sum_{s-1}^{2^K}p\left(M_s\,|y,X\right)p\left(M_s\right)} \tag{12}$$

Then the reconditioning leads to PMPs. A popular option is to uniformly define the prior probability for each type $p\left(M_y\right)\alpha 1$ to denote a lack of prior knowledge.

2.5 *Bayesian Linear Models and Zellner's g prior*

Zellner's g prior is a specific prior structure to use with Bayesian regression linear model [44]. Before looking at the data, the prior coefficients in the normal distribution must be determined with the specified mean and variance. Assuming the prior mean is zero, the variance structure is determined according to

$$\textbf{Zellner's } g\sim\sigma^2\left(\frac{1}{g}X_y'X_y\right)^{-1} \tag{13}$$

$$\beta_\gamma\,\Big|\,g\sim N\left(0,\sigma^2\left(\frac{1}{g}X_\gamma'X_\gamma\right)^{-1}\right) \tag{14}$$

Hence, β_γ is the crucial prior of regression coefficients. g is the hyperparameter to represent how certain is that coefficients are indeed zero: A small g means small prior coefficient variances and implies quite certain that coefficient is indeed zero. On the contrary, a large g means that coefficients are very uncertain.

Table 1 Descriptive statistics of financial development variables, economic growth, and export

Variable	REAL	DCP	CBB	ATM	IRS	ST	MCL	STR	CRC	DBC	EM	EXP	L	I
Mean	1.15	4.51	2.44	4.08	1.21	3.63	4.66	3.57	−3.49	1.92	0.36	4.91	1.91	3.74
Sd	0.64	0.56	0.24	0.46	0.46	0.82	0.6	0.49	9.09	1.77	0.58	12.95	2.35	1.04
Skewness	−1.47	−1.06	0.45	−0.91	−0.43	−0.66	−0.21	0.62	−1.08	1.12	2.23	0.66	1.18	−0.23
Kurtosis	3.13	−0.69	−0.82	2.3	−1.25	−1.04	−1.21	−0.89	−0.8	1.53	5.1	−0.48	3.61	−1.02
ADF	−3.58*	−2.27**	−5.66*	−2.55*	−2.12**	−2.33*	−2.82*	−2.46*	−2.90*	−5.56*	6.49	−4.38*	−1.51***	−3.49*

Note ADF denotes Augmented Dickey-Fuller unit root test with 10, 5 and 1%. ***, ** and * denote the significant 1, 5 and the 10% levels, respectively

3 Results

3.1 Descriptive Summary

Table 1 provides descriptive statistics for the variables of economic growth or real GDP, domestic credit to private sector, commercial bank branches, automated teller machines, interest rate spread, stocks traded, market capitalization of listed domestic companies, stocks turnover ratio of domestic share, credit card transaction value, debit card transaction value, e-money transaction value, labor force, inflation, and export. Starting with mean, export (EXP) has maximum value. In contrast, credit card transaction value (CRC) has the lowest mean value and smaller volatility than the others. The variables except commercial bank branches (CBB), stocks turnover ratio of domestic share (STR), debit card transaction value (DBC), e-money transaction value (EM), export (EXP), and labor (L) that have negative skewness. The values of excess kurtosis statistic of 3.13, 2.3, 1.53, 5.1, and 3.61 for real GDP(REAL), automated teller machines (ATM), debit card transaction value (DBC), e-money transaction value (EM), and labor (L), respectively, point to the Leptokurtic characteristics of the distribution. The platykurtic characteristics of the distribution which are higher in fat tail and standard deviation than the normal distribution illustrates in the rest variables. The ADF test results imply that all variables are stationary.

3.2 Bayesian Model Averaging and Markov Chain Monte Carlo (MCMC)

At the beginning, this paper used the Bayesian Model Averaging method to select the best model and independent variables for describing real GDP. There are five model specifications in this paper for the ASEAN group as a whole and each country (Indonesia, Malaysia, Singapore, and Thailand). Then, these five equations are estimated by Bayesian regression using MCMC package in R program and were generated 50 thousand approximated samples as the number of simulations with 25 thousand being the sample of burn-in.

3.2.1 Bayesian Model Averaging

This study used the Bayesian Model Averaging method to explain which variables best describe the model. This method can solve the small observation problem in model estimation. The results of Bayesian Model Averaging are shown in Tables 2 and 3.

Table 2 presents the results of Bayesian Model Averaging for the group. The first column is the importance of the variable to explain the data, with PIP that shows posterior inclusion probabilities, i.e., domestic credit to private sector (DCP) has

Table 2 Descriptive statistics of financial development variables, economic growth, and export

Variable	PIP	Post mean	Post SD	Cond. Pos. Sign
DCP	0.82851	−1.53995	0.98496	0.00377
ST	0.59552	0.82077	1.40429	0.95978
I	0.58009	−0.39482	0.41965	0.00791
MCL	0.45861	0.47703	1.05091	0.95802
STR	0.41295	0.05870	0.77334	0.55856
EXP	0.29424	0.07500	0.16425	0.99812
IRS	0.29345	−0.12051	0.27881	0.02955
CRC	0.27543	0.04283	0.29662	0.69316
CBB	0.26659	0.08285	0.20590	0.96237
ATM	0.24099	0.04378	0.19349	0.77492
EM	0.18553	0.03137	0.19056	0.67349
DBC	0.17630	−0.01449	0.15303	0.37252
L	0.16451	0.01328	0.07850	0.87039

83% of PIP, meaning stocks traded is significant to explain real GDP. The second column Mean displays the coefficients averaged of all models, and the comparatively large coefficient seems to be most significant, e.g., stocks traded (ST) has a large coefficient meaning that stocks traded is significant to the model. Furthermore, the coefficients' posterior standard deviations (SD) reflect more evidence to describe the direction of variables.

Table 3 illustrates the results on variables explaining real GDP in each country. *PIP* presents the posterior inclusion probability of the corresponding variable to be important to describe real GDP. First, Indonesia has four variables that are significant to explain real GDP consisting of domestic credit to private sector (DCP), automated teller machines (ATM), export (EXP), and stocks traded (ST) where domestic credit to private sector (DCP) has 99.9% of PIP, automated teller machines (ATM) has 74.3%, export (EXP) has 49.2%, and stocks traded (ST) has 44.1%. Second, Malaysia has credit card transaction value (CRC), inflation (I), interest rate spread (IRS), and automated teller machines (ATM) of which *PIP* are 99.7, 96.6, 82.8 and 76.5%, respectively. Third, the variables significant to explain real GDP of Singapore consist of inflation (I), debit card transaction value (DBC), automated teller machines (ATM), and credit card transaction value (CRC), as their *PIP* are 99.99, 93.1, 82.2, and 78.5% respectively. Last, Thailand has credit card transaction value (CRC), and inflation (I) with PIP of 100, and 45.8%, respectively. Next, $Post\ Mean$ shows the coefficients averaged, all variables of Indonesia have three small coefficients meaning that they are less significant to the model. Malaysia has only one positive coefficient variable that is credit card transaction value (CRC). Next, Singapore has four variables with positive coefficient meaning that all variables are significant to the model. Furthermore, Thailand has two negative variables that mean two of them are not significant to the model. Moreover, *Post SD* reflect more evidence, because

Table 3 Results of Bayesian model averaging for individual countries

Variable	PIP	Post mean	Post SD	Cond. Pos. Sign
Indonesia				
DCP	0.99993	−0.72828	0.11953	0.00000
ATM	0.74333	0.11588	0.08608	0.99997
EXP	0.49273	−0.00123	0.00156	0.00002
ST	0.44126	−0.04989	0.12540	0.00214
Malaysia				
CRC	0.99689	0.02997	0.00750	1.00000
I	0.96554	−0.21608	0.08410	0.00000
IRS	0.82773	−0.32366	0.21203	0.00000
ATM	0.76460	−0.18013	0.14466	0.00159
Singapore				
I	0.99985	0.30369	0.04838	1.00000
DBC	0.93151	0.06570	0.02967	1.00000
ATM	0.82231	0.26324	0.17204	0.99989
CRC	0.78506	0.01229	0.00801	0.99576
Thailand				
CRC	1.00000	−0.04628	0.00263	0.00000
I	0.45781	−0.01401	0.02034	0.00867

all variables in each country are positive. The last, *Cond. Pos. Sign* can presuppose the coefficient sign. The coefficient sign is negative when the corresponding number of the variable is close to zero, i.e., domestic credit to private sector (DCP), export (EXP), and stocks traded (ST) of Indonesia, inflation (I), and interest rate spread (IRS) of Malaysia, and credit card transaction value (CRC) of Thailand have the values close to zero suggesting that the coefficient sign of these variables will be negative.

Moreover, the results of Bayesian Model Averaging for the entire group and the individual countries reveal variables to describe the model are different from one case to another. For the entire group, financial development variables and variable of trade can explain real GDP, but cashless variables are insignificant to affect real GDP of the group. In contrast, at the country level, some cashless variables were found to affect real GDP, e.g., CRC has 99.7% of *PIP* to describe real GDP of Malaysia.

Table 4 shows the results of the best models explaining real GDP of each country and the group. Each best model was selected according to the Posterior Model Probability (PMP). The best model for the group with the PMP of 2.7% has four variables that can explain real GDP including domestic credit to private sector (DCP), stocks traded (ST), inflation (I), market capitalization of listed domestic companies (MCL). For individual countries starting from Indonesia, the best model with 2.1% PMP has the variables of domestic credit to private sector (DCP), automated teller

Table 4 The results of the best model

Variable	$PIP_{\text{Group of counties}}$	Country			
		$PIP_{\text{Indinesia}}$	PIP_{Malaysia}	$PIP_{\text{Singapore}}$	PIP_{Thailand}
DCP	0.83(−)	1.00(−)	0.50(+)	0.51(+)	0.29(+)
CBB	0.27(+)	0.22(+)	0.30(+)	0.70(−)	0.23(−)
ATM	0.24(+)	0.74(+)	0.76(−)	0.82(+)	0.18(−)
IRS	0.29(−)	0.22(+)	0.83(−)	0.31(+)	0.38(−)
ST	0.60(+)	0.44(−)	0.48(+)	0.26(+)	0.23(+)
MCL	0.46(+)	0.34(−)	0.26(+)	0.34(+)	0.24(+)
STR	0.41(+)	0.25(−)	0.26(+)	0.26(−)	0.16(+)
CRC	0.28(+)	0.31(−)	1.00(+)	0.79(+)	1.00(−)
DBC	0.18(−)	0.16(−)	0.71(−)	0.93(+)	0.16(+)
EM	0.19(+)	0.17(−)	0.30(+)	0.15(−)	0.21(+)
EXP	0.29(+)	0.49(−)	0.31(+)	0.24(−)	0.14(+)
L	0.16(+)	0.22(−)	0.24(+)	0.22(−)	0.14(+)
I	0.58(−)	0.20(−)	0.97(−)	1.00(+)	0.46(−)
PMP	0.02654	0.02134	0.0472	0.05612	0.03018
PMP	0.02654	0.02134	0.0472	0.05612	0.03018

Note: (−), (+) denote the direction of variables
PIP is posterior inclusion probability
PMP is the posterior model probability

machines (ATM), export (EXP), and stocks traded (ST) that explain the Indonesian real GDP. PMP of Malaysia is 4.7% and the best model has credit card transaction value (CRC), inflation (I), interest rate spread (IRS), and automated teller machines (ATM) as the best explanatory variables. The best model for Singapore with 5.6% PMP consists of inflation (I), debit card transaction value (DBC), automated teller machines (ATM), and credit card transaction value (CRC) to explain its real GDP. PMP of the best model for Thailand is 3% and the explanatory variables of the Thai real GDP are credit card transaction value (CRC), and inflation (I). Figure 1 provides the graphical overview to explain the best models that are scaled by their posterior model probabilities.

Figure 1 illustrates the real GDP models for ASEAN as a group where blue color corresponds to a positive coefficient and red color corresponds to a negative coefficient. The horizontal axis presents the best models that are scaled by PMPs. The best model with most positive coefficients in blue color only includes stocks traded (ST). On the other hand, domestic credit to private (DCP) is in red color meaning that it has a negative coefficient. Figures 2, 3, 4, and 5 show the best models for each country corresponding to Indonesia, Malaysia, Singapore, and Thailand, respectively. Figure 2 for Indonesia shows the first four variables in red color and thus having a negative coefficient to include domestic credit to private sector (DCP), export (EXP), and stock market (ST), with DCP having the most negative effect. Figure 3 presents Malaysia' real GDP models. Only the variable of credit card transaction value (CRC)

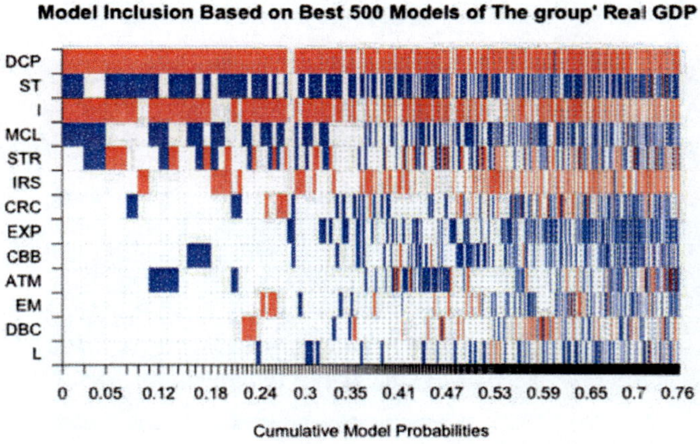

Fig. 1 Model inclusion. *Note* Blue color corresponds to a positive coefficient, red to a negative coefficient, and white to non-inclusion (a zero coefficient). The horizontal axis is scaled by PMP_s

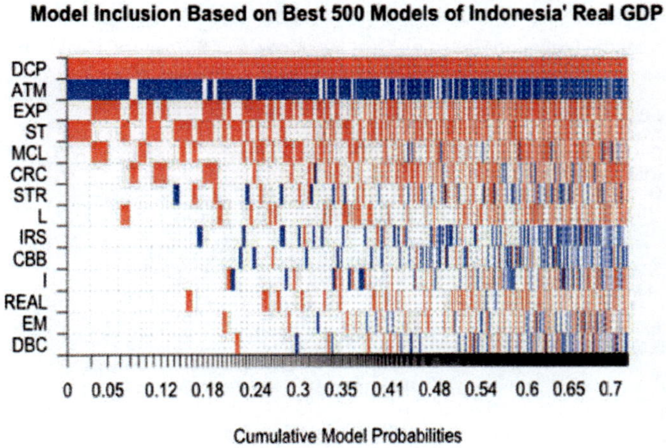

Fig. 2 The best model for Indonesia's real GDP

in blue color has the most positive coefficients while inflation (I) has the most negative coefficients. Figure 4 shows the real GDP models for Singapore where the variables that have the most positive coefficients in blue color include inflation (I), debit card transaction value (DBC), credit card transaction value (CRC), and automated teller machines (ATM), with inflation (I) having the most positive impact. Finally, Fig. 5 presenting the real GDP models for Thailand shows that the best model contains the variables of credit card transaction value (CRC) in red color that has the most negative coefficients.

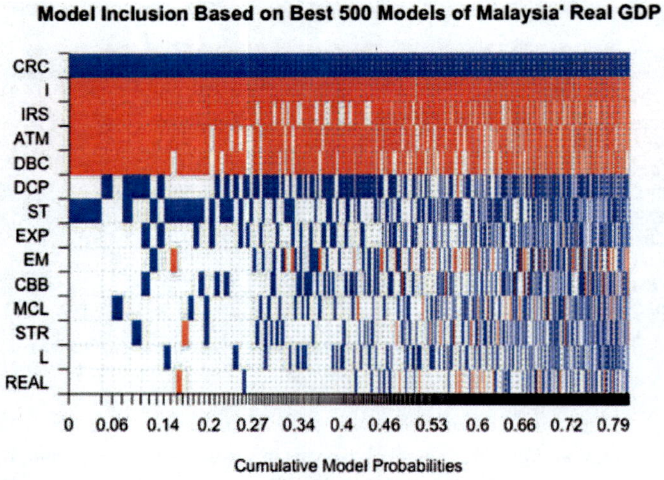

Fig. 3 The best model for Malaysia's real GDP

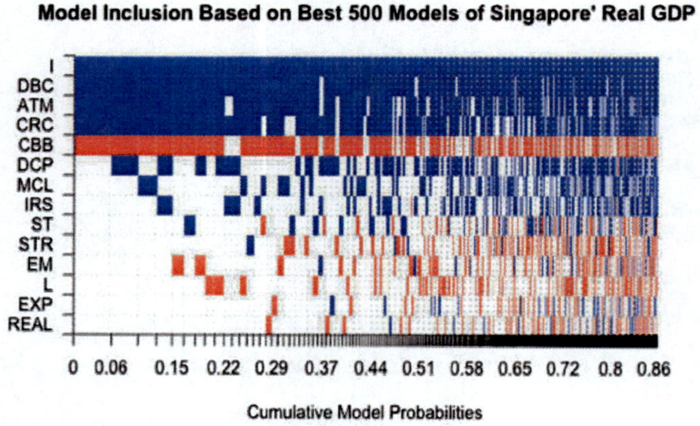

Fig. 4 The best model for Singapore's real GDP

3.2.2 Markov Chain Monte Carlo (MCMC)

Tables 5 and 6 summarize the MCMC simulation results including the posterior means, standard deviation (SD), 2.50% quantile, 25% quantile, 50% quantile, 75% quantile, and 97.50% quantile for each variable. From Table 5, the variable having the highest mean value is stocks traded (ST) and that having the lowest mean value is domestic credit to private sector (DCP). Furthermore, all variables at 2.50 and 25% quantile distribution have an inverse relationship with real GDP but their relationship turn positive at 75% quantile distribution. In particular, the positive effect of DCP and I on real GDP is highly pronounce at 97.50% quantile.

Table 5 MCMC simulation results for the group

Variable	Mean	SD	2.5%	25%	50%	75%	97.5%
(Intercept)	1.31	5.80	−11.09	−2.13	1.45	4.89	13.54
DCP	−1.56	1.17	−3.98	−2.31	−1.57	−0.85	0.72
I	−0.37	0.30	−0.96	−0.55	−0.36	−0.18	0.20
ST	0.77	0.56	−0.34	0.45	0.78	1.08	1.93
MCL	1.20	0.54	0.16	0.87	1.20	1.54	2.30
σ^2	0.31	0.09	0.19	0.25	0.30	0.35	0.51

Table 6 MCMC simulation results for individual countries

Variable	Mean	SD	2.5%	25%	50%	75%	97.5%
Indonesia							
(Intercept)	5.26	5.56	−5.80	1.81	5.26	8.71	16.35
DCP	−0.80	1.49	−3.77	−1.73	−0.8	0.13	2.17
ATM	−0.25	0.58	−1.35	−0.57	−0.25	0.07	0.85
ST	0.05	1.29	−2.34	−0.53	0.06	0.63	2.37
EXP	0.01	0.36	−0.63	−0.13	0.01	0.16	0.67
σ^2	0.36	0.09	0.22	0.3	0.35	0.41	0.58
Malaysia							
(Intercept)	2.35	2.46	−2.25	1.03	2.35	3.68	6.96
CRC	−0.03	0.55	−1.06	−0.26	−0.03	0.19	0.94
I	−0.02	0.54	−0.96	−0.26	−0.02	0.22	0.94
IRS	0.34	1.65	−2.66	−0.45	0.32	1.10	3.37
ATM	−0.29	0.61	−1.42	−0.59	−0.30	0.01	0.84
σ^2	0.45	0.11	0.27	0.37	0.43	0.51	0.72
Singapore							
(Intercept)	2.70	2.83	−2.55	1.13	2.66	4.21	8.20
I	−0.01	0.52	−1.02	−0.25	0.00	0.24	0.91
DBC	−0.08	0.56	−1.18	−0.33	−0.07	0.18	0.99
CRC	−1.26	1.62	−4.48	−1.96	−1.18	−0.5	1.53
ATM	0.11	0.72	−1.05	−0.28	0.06	0.43	1.66
σ^2	0.36	0.09	0.22	0.29	0.34	0.41	0.58
Thailand							
(Intercept)	1.85	1.77	−1.16	1.12	1.85	2.57	4.88
CRC	−0.04	0.35	−0.66	−0.18	−0.04	0.10	0.59
I	−0.15	0.43	−0.94	−0.34	−0.15	0.04	0.62
σ^2	0.43	0.11	0.27	0.35	0.41	0.49	0.68

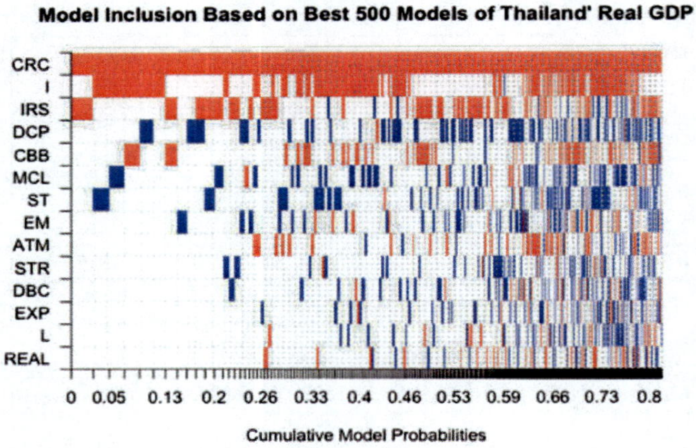

Fig. 5 The best model for Thailand's real GDP

Table 6 presents the results of MCMC simulation for each country. For Indonesia, the variable having the highest mean value is stocks traded (ST) and that having the lowest mean value is domestic credit to private sector (DCP). The variable having the highest mean value for Malaysia is interest rate spread (IRS). Meanwhile, Singapore has automated teller machines (ATM) as the variable with the highest mean value. Furthermore, Thailand has credit card transaction value (CRC) as the variable with the highest mean. Besides, the distribution at 2.50% quantile of all variables in each country decreases real GDP but all variables at 75% quantile distribution will render a positive effect on real GDP.

4 Conclusions and Policy Implications

This paper investigated the relationship among international trade, financial development, and economic growth in ASEAN from 2010 to 2019 using the Bayesian regression and BMA methods. At the beginning, the study separated the investigation into two cases, the first case focuses on the group as a whole and the second focuses on individual countries including Indonesia, Malaysia, Singapore, and Thailand. For ASEAN as a group, four out of the fourteen variables considered were selected by the Bayesian Model Averaging procedures as the most appropriate to describe real GDP namely domestic credit to private sector (DCP), stocks traded (ST), inflation (I), market capitalization of listed domestic companies (MCL) in the best model selection has the most positive coefficients, followed by stocks traded (ST), while domestic credit to private sector (DCP) has the most negative coefficients. For the group, however, the cashless variables do not show effect on real GDP. For individual countries, the results reveal the sets of variables explaining real

GDP to be different across countries. Cashless variables in particular affect real GDP of only some countries in the group. Malaysia, as seen in Table 4, PIP of credit card transaction value (CRC) is 99.7% with positive sign, meaning that its CRC has a positive relationship with its real GDP. This might not be surprising as Malaysian service sector is one of the top three important industries in contribution to real GDP of the country [14]. [22] found that Malaysian people use credit cards to spend on such basic needs as fuel, health services, and other goods and services including hotel, dinning, and groceries shopping. Hence, credit card uses may link to real GDP growth. On the contrary, also can be seen from Table 4, Thailand has PIP of credit card transaction value (CRC) = 100% with a negative sign indicating the negative relationship between CRC and real GDP of Thailand. IPSR [39] shows a piece of statistical evidence that most Thai credit card users are in the Y generation and some activities of this generation do not encourage real GDP growth; for example, 50% of generation Y persons did not have savings, 48% made a late debt repayment, 45% had income less than expense and their lack of financial discipline may negatively affect economic growth. Finally, it is hoped that organizations, government, and financial sector can use the results of this paper to design and implement policy especially in the national financial planning in the future. In particular, related organizations should pay attention to the positive relationship for driving economic growth.

Acknowledgements This work was supported by Faculty of Economics, Chiang Mai University, Chiang Mai, Thailand.

References

1. Andrieu, C., de Freitas, N., Doucet, A., & Jordan, M. I. (2003). An introduction to MCMC for machine learning. *Machine Learning, 50*(1–2), 5–43.
2. Awokuse, T. O. (2007). Causality between exports, imports and economic growth: Evidence from transition economies. *Economics Letters, 94*(3), 389–395.
3. Balaguer, J., & Cantavella, M. (2002). Tourism as a long-run economic growth factor: the Spanish case. *Applied Economics, 34*, 877–84.
4. Balassa, B. (1978). Exports and economic growth: further evidence. *Journal of Development Economics, 5*, 181–189.
5. Bahmani-Oskooee, M., & Alse, J. (1993). Export growth and economic growth: An application of cointegration and error correction modelling. *Journal of Developing Areas, 27*, 535–42.
6. Bencivenga, V. R., & Smith, B. D. (1993). Some consequences of credit rationing in an endogenous growth model. *Journal of Economic Dynamics and Control, 17*(1/2), 97–122.
7. Bhagwati, J. N. (1988). *Protectionism*. MIT Press.
8. Brooks, S., Gelman, A., Jones, G., & Meng, X.-L. (2011). *Handbook of Markov Chain Monte Carlo*. CRC Press.
9. Calderon, C., & Liu, L. (2003). The direction of causality between financial development and economic growth. *Journal of Development Economics, 72*(2003), 321–334.
10. Chen, R., Yamaka, W., & Osathanunkul, R. (2019). Determinants of non-cash payments in Asian countries. *Journal of Physics: Conference Series* (Vol. 1324, No. 1, p. 012103). IOP Publishing.
11. Chow, P. C. Y. (1987). Causality between export growth and industrial development: empirical evidence from the NICs. *Journal of Development Economics, 26*, 55–63.

12. Diamond, D. W. (1984). Financial intermediation and delegated monitoring. *The Review of Economic Studies, 51*(3), 393–414.
13. Faye, E. C. (2001). *Reference to Ethiopian country by: Ethio-Italian cooperation Arsi-Bale rural development project paper prepared for the conference on Annual Meeting of the American.*
14. Federal Government Administrative Centre of Malaysia. (2018). *Eleventh Malaysia Plan.* Economic Planning Unit, Prime Minister's Department, Malaysia, 18.
15. Fragoso, T. M., Bertoli, W., & Louzada, F. (2018). Bayesian model averaging: A systematic review and conceptual classification. *International Statistical Review, 86*(1), 1–28.
16. Gelman, A., Carlin, J. B., Stern, H. S., Dunson, D. B., Vehtari, A., & Rubin, D. B. (2014). *Bayesian data analysis*, 3nd ed. Chapman and Hall/CRC.
17. Greenwood, J., & Jovanovic, B. (1990). Financial development, growth, and the distribution of Income. *The Journal of Political Economy, 98*(5), 1076–1107.
18. Guru, B., & Yadav, I. (2019). Financial development and economic growth: Panel evidence from BRICS. *Journal of Economics, Finance and Administrative Science, 24.* https://doi.org/10.1108/JEFAS-12-2017-0125
19. Haugh, M. (2017). *MCMC and Bayesian modeling. IEOR E4703 Monte-Carlo Simulation.* Columbia University.
20. Helpman, E., & Krugman, P. R. (1985). *Market structure and foreign trade. Increasing returns, imperfect competition, and the international economy.* MIT Press.
21. Hur, R. M., & Riyanto, Y. (2006). Finance and trade: A cross-country empirical analysis on the impact of financial development and assets tangibility on international trade. *World Development, 34*, 1728–1741.
22. Hussin, R. S., Kassim, S., & Jamal, N. (2013). Credit Card Holders in Malaysia: Customer characteristics and credit card usage. *Journal of Economics and Management, 7*(1), 108–122.
23. International Monetary Fund. (2001). *Global trade liberalization and the developing countries.*
24. Jin, J. C. (1995). Export-led Growth and the Four Little Dragons. *The Journal of International Trade and Economic Development, 4*, 203–215.
25. Krugman, P. R. (1984). Import protection as export promotion: International competition in the presence of oligopoly and economies of scale. In H. Kierzkowski (Ed.), *Monopolistic competition and international trade* (pp. 180–193). Clarendon Press.
26. Levine, R. (1997). Financial development and economic growth: Views and agenda. *Journal of Economic Literature, 35*, 688–726.
27. Levine, R. (2005). Finance and growth: Theory and evidence. In P. Aghion & S. Durlauf (Eds.), *Handbook of economic growth.* Elsevier Science.
28. Maneejuk, P., & Yamaka, W. (2020). An analysis of the impacts of telecommunications technology and innovation on economic growth. *Telecommunications Policy, 44*(10), 102038.
29. Martin, A. D., Quinn, K. M., & Park, J. H. (2011). MCMCpack: Markov Chain Monte Carlo in R. *Journal of Statistical Software, 42*(9).
30. Michaely, M. (1977). Exports and Growth: An Empirical Investigation. *Journal of Development Economics, 4*, 49–53.
31. Morawetz, U. (2006). *Bayesian modelling of panel data with individual effects applied to simulated data.* Institut fu ̈r Nachhaltige Wirtschaftsentwicklung.
32. Patrick, H. T. (1966). Financial development and economic growth in underdeveloped countries. *Economic Development and Cultural Change, 14*(2), 174–189.
33. Per-ola, M. (2008). A note on the export-led growth hypothesis: A time series approach. *Cuadernos de Economía, 45*, 293–302.
34. Schumpeter, J. A. (1911). *Theory of economic development.* Harvard University Press.
35. Shahbaz, M., & Rahman, M. M. (2014). Exports, financial development and economic growth in Pakistan. *International Journal of Development Issues, 13*(2), 155–170.
36. Shahbaz, M. (2009). A reassessment of finance-growth nexus for Pakistan: Under the investigation of FMOLS and DOLS techniques. *Journal of Applies Economics, 1*, 65–80.
37. Shan, J., & Sun, F. (1997), On the export-led growth hypothesis for Australia: An empirical re-examination. *Applied Economics Letters* (forthcoming).

38. Stiglitz, J. E., & Weiss, A. (1983). Incentive effects of terminations: Applications to the credit and labor markets. *The American Economic Review, 73*(5), 912–927.
39. The Siam Commercial Bank Public Company Limited. (2018). *Is it true that Gen Y has more debts than other generations?*. Retrieve from: https://www.scb.co.th/en/personal-banking/stories/geny-more-debts.html
40. Wattanutchariya, W., Tansuchat, R., & Ruennareenard, J. (2016). Supply chain management of Thai parboiled rice for export. In *International Conference on Industrial Engineering and Operations Management* (pp. 8–10).
41. Xu, Z. (1996). On the causality between export growth and GDP growth: An empirical reinvestigation. *Review of International Economics, 4*, 172–84.
42. World Bank. (2012). *Financial development*.
43. World Payment Report. (2019). Monetizing quick-win solutions while building a robust ecosystem strategy can help bank stay in the payments game.
44. Zeugner, S., & Feldkircher, M. (2015). Bayesian Model averaging employing fixed and flexible priors: The BMS package for R. *Journal of Statistical Software, 68*(4).